# Handbook of Experimental Pharmacology

Volume 184

AF846064

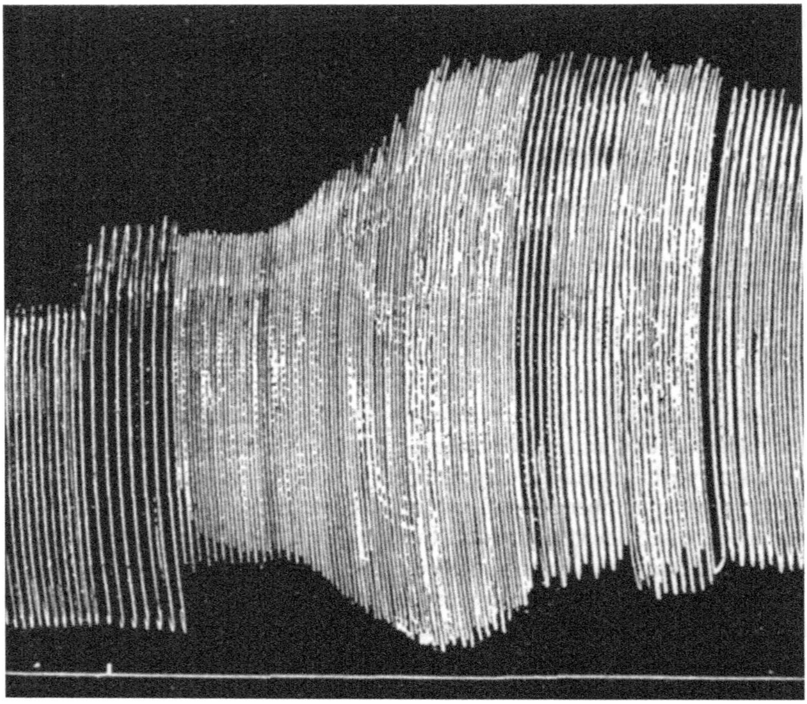

**Fig. 1** Effect of nicotine on a rabbit isolated heart. From Dixon (1924).

nicotine into the isolated perfused heart of a rabbit. Immediately on injection, nicotine slowed the heart rate by stimulating intracardiac vagal ganglion cells. After a few seconds, however, bradycardia was replaced by marked tachycardia and an increase in contraction amplitude. Because the isolated heart does not contain sympathetic ganglion cells (and because an effect on the myocardium can be excluded), nicotine must have acted on the cardiac sympathetic axon terminals, on what we now call presynaptic nicotinic receptors.

Presynaptic nicotinic receptors are ligand-gated ion channels. Many other presynaptic receptors couple to G-proteins. Presynaptic receptors may be targets of bloodborne substances or substances secreted from neighboring cells, including neighboring axon terminals. In 1971 it was noticed with some surprise that many axon terminals even possess receptors for their own transmitter–presynaptic autoreceptors, the $\alpha_2$-autoreceptors for noradrenaline being a prominent example (reviewed in Starke 2001). The various presynaptic ligand-gated ion channels and G-protein-coupled receptors are discussed in the second part of this volume. Questions regarding where the receptors' signal transduction pathways hit the exocytosis cascade and whether the receptors have therapeutic potential will be addressed in all chapters.

We attempt a synthesis of a large amount of information and cannot be expected to be totally successful. Nevertheless, we hope that the various contributions will be useful, and that the book will be of help to scientists in a wide number of fields.

## References

Dixon WE (1924) Nicotin, Coniin, Piperidin, Lupetidin, Cytisin, Lobelin, Spartein, Gelsemin. In: Heffter A (ed) Handbuch der experimentellen Pharmakologie, vol 2 part 2. Springer, Berlin, pp 656–736

Starke K (2001) Presynaptic autoreceptors in the third decade: focus on $\alpha_2$-adrenoceptors. J Neurochem 78:685–93

Südhof TC (2004) The synaptic vesicle cycle. Annu Rev Neurosci 27:509–47

Dallas — *Thomas C. Südhof*
Freiburg i. Br. — *Klaus Starke*

# Contents

# Contributors

S. Boehm
Institute of Pharmacology, Center for Biomolecular Medicine and Pharmacology, Medical University of Vienna, Vienna, Austria, Institute of Experimental and Clinical Pharmacology, Medical University of Graz, Universitätsplatz 4, 8010 Graz, Austria, stefan.boehm@meduni-graz.at

Jean-Luc Boulland
The Biotechnology Centre of Oslo, Centre for Molecular Biology and Neuroscience, University of Oslo, 0317 Oslo, Norway

David A. Brown
Department of Pharmacology, University College, Gower Street, London, WC1E 6BT, United Kingdom, d.a.brown@ucl.ac.uk

May K.L. Bredahl
The Biotechnology Centre of Oslo, University of Oslo, 0317 Oslo, Norway

Farrukh A. Chaudhry
The Biotechnology Centre of Oslo, Centre for Molecular Biology and Neuroscience, University of Oslo, P. O. Box 1125, Blindern, 0317 Oslo, Norway, f.a.chaudhry@biotek.uio.no

M.M. Dorostkar
Institute of Pharmacology, Center for Biomolecular Medicine and Pharmacology, Medical University of Vienna, Waehringer Straße 13a, 1090 Vienna, Austria

Robert H. Edwards
Departments of Neurology and Physiology, University of California, San Francisco School of Medicine, 600 16th Street, GH-N272B San Francisco, California 94143-2140, USA

Robert Feil
Interfakultäres Institut für Biochemie, Universität Tübingen, Hoppe-Seyler-Straße, 4, 72076 Tübingen, Germany

Thomas J. Feuerstein
Neurochirugische Universitätsklinik Breisacherstrasse 64 Freiburg, Germany, thomas.feuerstein@uniklinik-freiburg.de

Ralf Gilsbach
Institute of Experimental and Clinical Pharmacology and Toxicology, University of Freiburg, Albertstrasse 25, 79104 Freiburg, Germany

Jorge Gonalves
Department of Pharmacology, Faculty of Pharmacy, University of Porto, Porto, Portugal, jgonçalves@reit.up.pt

Lutz Hein
Institute of Experimental and Clinical Pharmacology, University of Freiburg, Albertstrasse 25, 79104 Freiburg, Germany, lutz.hein@pharmakol.uni-freiburg.de

Reinhard Jahn
Max Planck Institute for Biophysical Chemistry, Department of Neurobiology, Am Fassberg 11, 37077 Göttingen, Germany, rjahn@gwdg.de

Monica Jenstad
The Biotechnology Centre of Oslo, Centre for Molecular Biology and Neuroscience, University of Oslo, N-0317 Oslo, Norway

M. Kathmann
Institut für Pharmakologie und Toxikologie, Rheinische Friedrich-Wilhelms-Universität, Reuterstrasse 2b, 53113 Bonn, Germany

Ege T. Kavalali
Department of Neuroscience, Department of Physiology, The University of Texas, Southwestern Medical Center, 5323 Harry Hines Blvd. Dallas, Texas 75390-9111, United States

Mikhail Khvotchev
Department of Neuroscience, University of Texas, Southwestern Medical Center, Dallas, Texas 75390-9111, United States

Alexandra E. Kisilevsky
Hotchkiss Brain Institute and Department of Physiology and Biophysics, University of Calgary, Calgary, Canada

Thomas Kleppisch
Institut für Pharmakologie und Toxikologie, Technische Universität München, Biedersteiner Straße 29, D-80802 München, Germany, kleppisch@ipt.med.tu-muenchen.de

Thorsten Lang
Department of Neurobiology, Max Planck Institute for Biophysical Chemistry, Am Fassberg 11, 37077 Göttingen, Germany, tlang@gwdg.de

S Z. Langer
Alpha-2 Pharmaceutica AB, 8 H. Rosenblum Street, Apt. 4650, Tel Aviv 69379, Israel, langer.salomon@gmail.com

David M. Lovinger
Laboratory for Integrative Neuroscience, National Institute on Alcohol Abuse and Alcoholism, National Institutes of Health, 5625 Fishers Lane, Bethesda, MD 20892-9411, USA, lovindav@mail.nih.gov

Cesare Montecucco
Departimento de Scienze Biomediche and Istituto CNR di Neuroscienze, Università di Padova, Viale G. Colombo 3, 35121 Padova, Italy, cesare.montecucco@unipd.it

Glória Queiroz
Department of Pharmacology, Faculty of Pharmacy, University of Porto, Porto, Portugal, gloria@ff.up.pt

M. Raiteri
Department of Experimental Medicine, Pharmacology and Toxicology Section, Center of Excellence for Biomedical Research, University of Genoa, 16132 Genoa, Italy, raiteri@pharmatox.unige.it

Alexis Rohou
Division of Cell and Molecular Biology, Imperial College London, London, SW7 2AY, United Kingdom, alexis.rohou@imperial.ac.uk

Ornella Rossetto
Departimento de Scienze Biomediche and Istituto CNR di Neuroscienze, Università di Padova, Viale G. Colombo 3, 35121 Padova, Italy, ornella.rossetto@unipd.it

E. Schlicker
Institut für Pharmakologie und Toxikologie, Rheinische Friedrich-Wilhelms-Universität, Reuterstrasse 2b, 53113 Bonn, Germany, e.schlicker@uni-bonn.de

Talvinder S. Sihra
Department of Pharmacology, University College London, Gower Street, London, United Kingdom, t.sihra@ucl.ac.uk

Thomas C. Südhof
Departments of Neuroscience and Molecular Genetics, Howard Hughes Medical Institute, The University of Texas, Southwestern Medical Center, Dallas, Texas 75390-9111, USA, thomas.sudhof@utsouthwestern.edu

Shuzo Sugita
Division of Cellular and Molecular Biology, Toronto Western Research Institute, Toronto, Ontario M5T 2S8, Canada, ssugita@uhnres.utoronto.ca

Yuri A. Ushkaryov
Division of Cell and Molecular Biology, Imperial College London, London SW7 2AY, United Kingdom, y.ushkaryov@imperial.ac.uk

Gerald W. Zamponi
Department of Physiology and Biophysics, University of Calgary, 3330 Hospital Drive NW, Calgary T2N 4N1, Canada, Zamponi@ucalgary.ca

# Neurotransmitter Release

**Thomas C. Südhof**



**Abstract** Neurons send out a multitude of chemical signals, called neurotransmitters, to communicate between neurons in brain, and between neurons and target cells in the periphery. The most important of these communication processes is synaptic transmission, which accounts for the ability of the brain to rapidly process information, and which is characterized by the fast and localized transfer of a signal from a presynaptic neuron to a postsynaptic cell. Other communication processes, such as the modulation of the neuronal state in entire brain regions by neuromodulators, provide an essential component of this information processing capacity. A large number of diverse neurotransmitters are used by neurons, ranging from classical fast transmitters such as glycine and glutamate over neuropeptides to lipophilic compounds

---

Thomas C. Südhof
Departments of Neuroscience and Molecular Genetics, and Howard Hughes Medical Institute, The University of Texas Southwestern Medical Center, Dallas, TX 75390-9111, USA
thomas.sudhof@utsouthwestern.edu

T.C. Südhof, K. Starke (eds.), *Pharmacology of Neurotransmitter Release. Handbook of Experimental Pharmacology 184.*

and gases such as endocannabinoids and nitric oxide. Most of these transmitters are released by exocytosis, the i.e. the fusion of secretory vesicles with the plasma membrane, which exhibits distinct properties for different types of neurotransmitters. The present chapter will provide an overview of the process of neurotransmitter release and its historical context, and give a reference point for the other chapters in this book.

## 1 Principles of Neurotransmitter Release

Neurons communicate with each other and their target cells via two principal mechanisms: the secretion and reception of chemical messengers called neurotransmitters, and the direct transfer of intercellular signals via gap junctions. Communication via neurotransmitters occurs in several forms that range from classical synaptic transmission at synapses to diffuse secretion of neuromodulators which mediate volume transmission. Communication via gap junctions occurs at so-called electrical synapses. Almost all of the neuronal communication is mediated by neurotransmitters, and electrical synapses are exceedingly rare in vertebrate brain. Both types of communication are not unique to neurons. Secretion of neuromodulators and neuropeptides is also mediated by endocrine cells and even some highly differentiated cells such as adipocytes, and diffusible neurotransmitters such as nitric oxide are released by many non-neuronal cells. Only the presynaptic secretion of classical neurotransmitters in the context of a synapse is specific to neurons, although the postsynaptic cell can be either a neuron (most of the time) or an effector cell (e.g., a muscle cell). The present book will only deal with communication by neurotransmitters, and only with the release of such transmitters and the pharmacology of this release.

What is a neurotransmitter, and how many different "types" of neurotransmitter release exist? At least five types of neurotransmitter release can be defined.

1. Synaptic neurotransmitter release occurs in a classical, electron microscopically observable synapse, and is mediated by synaptic vesicle exocytosis from nerve terminals (Figure 1; Katz, 1969; Südhof, 2004; note that a "nerve terminal" is not necessarily the end of an axon, but generally is formed by axons en passant as they arborize throughout the brain). Synaptic neurotransmitter release, the first step in synaptic transmission, transfers information extremely rapidly (in milliseconds) in a highly localized manner (restricted to an area of less than a square micrometer; reviewed in Südhof, 2004). Synaptic release secretes "classical" neurotransmitters: GABA, glycine, glutamate, acetylcholine, and ATP. It has been suggested that in addition to neurons, astrocytes also secrete classical neurotransmitters by a similar mechanism (**?**), but this type of secretion has not been directly demonstrated.
2. Monoaminergic neurotransmitters (dopamine, noradrenaline, adrenaline, histamine, and serotonin) are released by exocytosis of small dense-core vesicles from

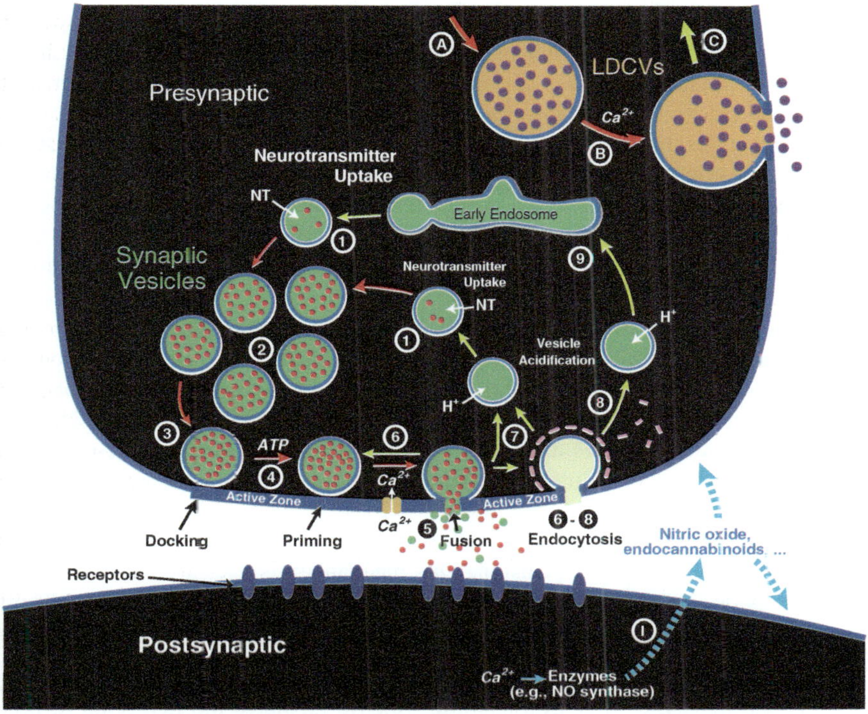


**Fig. 1** Secretory pathways in neurons. The drawing schematically illustrates the three major neurotransmitter release pathways. (**a**) Release of classical neurotransmitters by synaptic vesicle exocytosis (center; steps 1–9). Classical neurotransmitter release depends on an underlying synaptic vesicle cycle that starts when synaptic vesicles are filled with neurotransmitters by active transport (step 1), and form the vesicle cluster (step 2). Filled vesicles dock at the active zone (step 3), where they undergo a priming reaction (step 4) that makes them competent for $Ca^{2+}$-triggered fusion-pore opening (step 5). After fusion-pore opening, synaptic vesicles undergo endocytosis and recycle via three alternative pathways: local reuse (step 6; also called kiss-and-stay), fast recycling without an endosomal intermediate (step 7; also called kiss-and-run), or clathrin-mediated endocytosis (step 8) with recycling via endosomes (step 9). Steps in exocytosis are indicated by red arrows, and steps in endocytosis and recycling by yellow arrows. (**b**) Release of neuropeptides and biogenic amines by LDCV exocytosis. LDCVs are generated in the cell body by budding from the Golgi complex filled with neuropeptides (not shown). LDCVs are then transported from the cell body to the axons or dendrites (step A, as shown for nerve terminals). A $Ca^{2+}$-signal triggers the translocation and fusion of LDCVs with the plasma membrane outside of the active zone (step B). After exocytosis, empty LDCVs recycle and refill by transport to the cell body and recycling via the Golgi complex (step C). (**c**) Release of gaseous or lipidic neurotransmitters, which are synthesized in either the pre- or the postsynaptic neuron (only the postsynaptic synthesis is shown), and secreted by diffusion across the plasma membrane (step I) to act on local extracellular receptors (e.g., CB1 receptors for endocannabinoids) or intracellular targets (e.g., guanylate cyclase for nitric oxide). (Modified from Südhof, 2004).

axonal varicosities that are largely not associated with a specialized postsynaptic structure (i.e., are outside of synapses; Brock and Cunnane, 1987; Stjärne, 2000). However, at least in the case of dopamine, postsynaptic specializations can occur with presynaptic small dense-core vesicles.

3. Neuropeptides are secreted by exocytosis of large dense-core vesicles (LDCVs) outside of synapses (Figure 1; Salio et al., 2006). LDCVs undergo exocytosis in all parts of a neuron, most often in axon terminals and dendrites. Monoamines are often co-stored with neuropeptides in LDCVs and co-secreted with them upon exocytosis. For all intents and purposes, LDCV-mediated secretion resembles hormone secretion in endocrine cells.
4. Classical neurotransmitters and monoamines may rarely be secreted by neurons, not by exocytosis, but by transporter reversal. This mechanism involves the transport of neurotransmitters from the cytosol to the extracellular fluid via transporters that normally remove neurotransmitters from the extracellular fluid. This mechanism appears to account for the burst of dopamine released by amphetamines (Fleckenstein et al., 2007), but its physiological occurrence remains unclear.
5. A fifth pathway, finally, is the well-established secretion of small membrane-permeable mediators by diffusion. This mechanism is used for the secretion of nitric oxide, endocannabinoids, and other important lipidic or gaseous neurotransmitters. The major point of regulation of release here is the synthesis of the respective compounds, not their actual secretion.

Only the first type of neurotransmitter release mediates the fast point-to-point synaptic transmission process at classical synapses (sometimes referred to as wiring transmission). All of the other types of neurotransmitter release effect one or another form of "volume transmission" whereby the neurotransmitter signal acts diffusely over more prolonged time periods (Agnati et al., 1995). Of these volume transmitter pathways, the time constants and volumes involved differ considerably. For example, diffusible neurotransmitters such as nitric oxide act relatively briefly in a localized manner, whereas at least some neuropeptides act on the whole brain, and can additionally act outside of it (i.e., function as hormones). There is an overlap between wiring and volume neurotransmission in that all classical neurotransmitters act as wiring transmitters via ionotropic receptors, and also act as "volume transmitters" via G-protein-coupled receptors. Moreover, neuromodulators in turn feed back onto classical synaptic transmission.

Quantitatively, synaptic transmission is the dominant form of communication between neurons. A single look at an electron micrograph reveals that synapses with their appendant organelles, especially synaptic vesicles, are abundant in brain, whereas LDCVs are only observed occasionally (Figure 2). However, this does not mean that synaptic transmission is more important than the volume transmission pathways. The two principally different signaling pathways play distinct roles in information processing by the brain, and both are essential for brain function.

With the multitude of different types of transmitters, the question arises whether a single neuron can release more than one transmitter. Dale's principle stated that

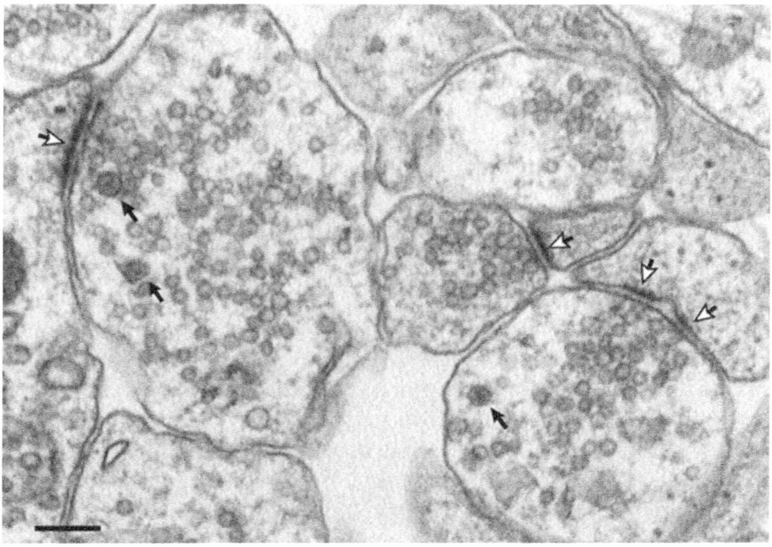

**Fig. 2** Electron micrograph of synapses. The image shows synapses formed by cultured cortical neurons from mouse. Note abundant synaptic vesicles in nerve terminals adjacent to synaptic junctions that are composed of presynaptic active zones and postsynaptic densities (open arrows point to postsynaptic densities of synaptic junctions; synapse on the right contains two junctions). In addition to synaptic vesicles, two of the nerve terminals contain LDCVs (closed arrows). Calibration bar = 500 nm. (Image courtesy of Dr. Xinran Liu, UT Southwestern).

this is not the case, but seems to be incorrect given the fact that virtually all neurons secrete neuropeptides and either classical neurotransmitters or monoamines (Salio et al., 2006). Moreover, many neurons additionally secrete diffusible neurotransmitters. Thus, a neuron usually operates by multiple neurotransmitter pathways simultaneously. To add to the complexity of these parallel signaling pathways, the relatively small number of neurons that secrete monoamines from axonal varicosities may also secrete classical neurotransmitters in separate classical synapses (Trudeau, 2004). Despite this complexity, however, Dale has to be given credit for his principle because the multiple transmitters secreted by a given neuron generally operate in distinct secretory and effector pathways. A given neuron usually releases only one type of classical neurotransmitter (with a few exceptions), suggesting that a modified Dale principle is still correct cotransmission.

## 2 Very Short History of the Analysis of Neurotransmitter Release

Our current concept of synaptic transmission, as mediated by intercellular junctions formed by one neuron with another neuron or target cell, is fairly recent. This concept was proposed in the second half of the 19th century, and proven only in the 20th century. It was embedded in a larger debate of whether neurons form a "reticular" network of connected cells, or a network of cells whose connections are discontinuous (the so-called neuron theory). Like with everything else in neuroscience, Ramón y Cajal is usually credited with the major discoveries in this field, but the actual concept predates him, and the development of the current view of synaptic transmission is due to a team effort. When Ramón y Cajal followed in the footsteps of scientists like Kühne, Koelliker, and His, who had formulated the first concept of synapses, even though the actual term was coined much later, Cajal's elegant prose and the fortunate opposition of Emilio Golgi to the neuron theory enhanced the influence of his writings and somewhat obscured the fact that the actual concepts that Cajal was presenting were already well established in the literature.

The term synapse was coined in 1897 by the physiologist Charles Sherrington in M. Foster's *Textbook of Physiology*, but the idea of the chemical synapse was developed almost half a century earlier in studies on the neuromuscular junction. As always in science, technical advance spawned conceptual breakthroughs. The three technical advances that fueled the progress in neuroscience in the second half of the 19th century were the improvements in light microscopy, chiefly due to Lister's invention of apochromatic lenses, the continuous development of staining methods culminating in Golgi's epynomous stain, and the application of more precise electrical recordings, allowing the emergence of electrophysiology to complement anatomy. Each historical stage in the discovery process is coupled to a particular preparation and technical approach, and major progress was usually achieved when a new technique was applied to a new preparation. This pattern also applies to the discovery of the synapse which was first described, without naming it, at the neuromuscular junction.

In the middle of the 19th century, it was known from the work of Volta, Galvani, and others that the nerve stimulates muscle contractions at the neuromuscular junction, and that electrical signals were somehow involved. Using the tools of cellular neuroanatomists, Kühne (1862) and Krause (1863) first demonstrated that the neuromuscular junction is not composed of a direct cellular connection between nerve and muscle as had been believed, but is discontinuous. Fifteen years later, the electrophysiologist Emil du Bois-Reymond (1877) proposed that the transmission of a synaptic signal is chemical. Subsequent work by Koelliker, Cajal, and Sherrington generalized this concept of a discontinuous synaptic connection that mediates intercellular signaling to the interneuronal synapses. Although the concept of the synapse continued to be disputed until well into the 20th century (e.g., see Golgi's Nobel lecture), the very existence of these disputes should not prevent us from recognizing that the actual description of synaptic transmission, and at least its proof for one particular synapse, the neuromuscular junction, had been established 50 years earlier.

The next major step forward in deciphering the mechanisms of synaptic transmission occurred in the neuropharmacological studies of Henry Dale, Otto Loewi, Wilhelm Feldberg, and their colleagues. Although, as in the discovery of the synapse as an intercellular noncontinuous junction, many individuals contributed, Loewi is generally credited with the single decisive experiment. This is probably fair, since

Loewi demonstrated directly that a chemical mediator (acetylcholine) is responsible for the transmission of the signal from the vagus nerve to the heart (Loewi, 1921). Despite Loewi's, Dale's, and Feldberg's advances, however, doubts lingered as to whether a chemical signal could be fast enough to account for the speed of synaptic transmission. Many scientists, with John Eccles (one of Sherrington's last pupils) as the most vocal protagonist, continued to espouse the view that fast synaptic transmission is essentially electrical, whereas chemical signaling serves only as a slow modulatory event. In other words, these views proposed a clean division of transmission into fast synaptic wiring transmission that is electrical, and slow volume transmission that is chemical. The doubts about the speed of chemical neurotransmission, and its general validity, were only definitively laid to rest by Bernhard Katz's seminal experiments on the frog neuromuscular junction, demonstrating that synaptic transmission operates as a quantal chemical event (Katz, 1969). It is remarkable that from Kühne's to Katz's studies, the major contributions to establishing synaptic transmission as the major mechanism by which neurons communicate came from the neuromuscular junction. The concept of the synapse was first postulated at the neuromuscular junction, the first genuine neurotransmitter was identified with acetylcholine as the neuromuscular junction neurotransmitter, and the chemical quantal nature of synaptic transmission was revealed at the neuromuscular junction.

The findings of Katz and colleagues raised two major questions: what are the mechanisms that allow the fast secretion of neurotransmitters from presynaptic terminals in response to an action potential? What molecules mediate the fast recognition of these neurotransmitters by the postsynaptic cell? The elucidation of the basic mechanisms of release again started with the cholinergic system in the description and isolation of synaptic vesicles as the central organelle, chiefly by Victor Whittaker (Whittaker and Sheridan, 1965). The progress in the field, however, then shifted to central synapses, with the identification of the major molecules involved in release of neurotransmitters, and the description of the mechanism by which $Ca^{2+}$-influx into nerve terminals achieves the fast triggering of release via binding to synaptotagmins (reviewed in Südhof, 2004). The discovery of neurotransmitter receptors and their properties was initiated by classical pharmacological approaches dating back to the British school founded by Langley (1921), but the definitive description of these receptors was enabled by the simultaneous development of patch clamping by Neher and Sakmann (1976) and of molecular cloning of these receptors by S. Numa (Noda et al., 1982).

## 3 Basic Mechanisms of Release by Exocytosis

Most neurotransmitter release occurs by exocytosis of secretory vesicles, which involves the fusion of the secretory vesicles (synaptic vesicles and LDCVs) with the plasma membrane. All intracellular membrane fusion (except for mitochondrial fusion) is thought to operate by the same fundamental mechanism that involves a core machinery composed of four classes of proteins: SNARE-proteins, SM-proteins (for

Sec1/Munc18-like proteins), Rab-proteins, and Rab-effectors (Jahn et al., 2003). The specific isoforms of these proteins that are being used vary tremendously between fusion reactions, but the general principle by which these proteins act seems to be always similar: Rab and Rab-effector proteins appear to proofread the docking and fusion reaction between the two target membranes and may even mediate the docking at least in part, whereas SNARE- and SM-proteins catalyze the actual fusion reaction.

## 3.1 Rab-Proteins and Rab-Effectors

Rab-proteins are GTP-binding proteins that interact with effectors in a GTP-dependent manner. Rab3A, 3B, 3C, and 3D represent a family of Rab-proteins that are highly enriched on synaptic vesicles and other secretory organelles throughout the body. In addition, Rab27A and 27B are also generally found on secretory vesicles, although it is unclear whether they are present on synaptic vesicles (Südhof, 2004). Rab3/27 proteins together function in exocytosis, and mediate vesicle docking at least in part. Two classes of Rab3/27 effectors were described: rabphilins and RIMs. Both effector classes include multiple members encoded by distinct genes. Rabphilins are cytosolic proteins that are recruited to secretory vesicles by Rab3/27, but their function has remained largely obscure. RIMs are components of the detergent-insoluble protein complex that makes up the active zone, the part of the presynaptic plasma membrane where synaptic vesicles dock and fuse (Figure 3). The active zone is composed of the RIM-containing protein complex that includes several other large proteins, in particular Munc13s, piccolo/bassoon, ELKS, and α-liprins, all of which are crucial for normal synaptic vesicle exocytosis. It is noticeable that in most intracellular fusion reactions, Rab-effectors are composed of large complexes that do more than just bind the Rab-protein, but perform several functions in the fusion process, with the Rab-protein often being involved in the docking of the membranes for fusion and in the regulation of the other activities of the complex during the fusion reaction. The same appears to be true for Rab3/27 binding to the RIM-containing active zone protein complex. The whole active zone complex could be considered as a single large Rab-effector complex (Figure 3), and is likely involved not only in the docking of synaptic vesicles, but also in organizing the actual fusion reaction and in synaptic plasticity (see below).

## 3.2 SNARE Proteins

Membrane fusion consists of merging two negatively charged phospholipid bilayers, and thus requires overcoming a major energy barrier (Jahn et al., 2003). SNARE proteins represent a family of membrane proteins that are present on opposing membranes destined to fuse. As first proposed by Jahn, Heuser, Rothman and colleagues

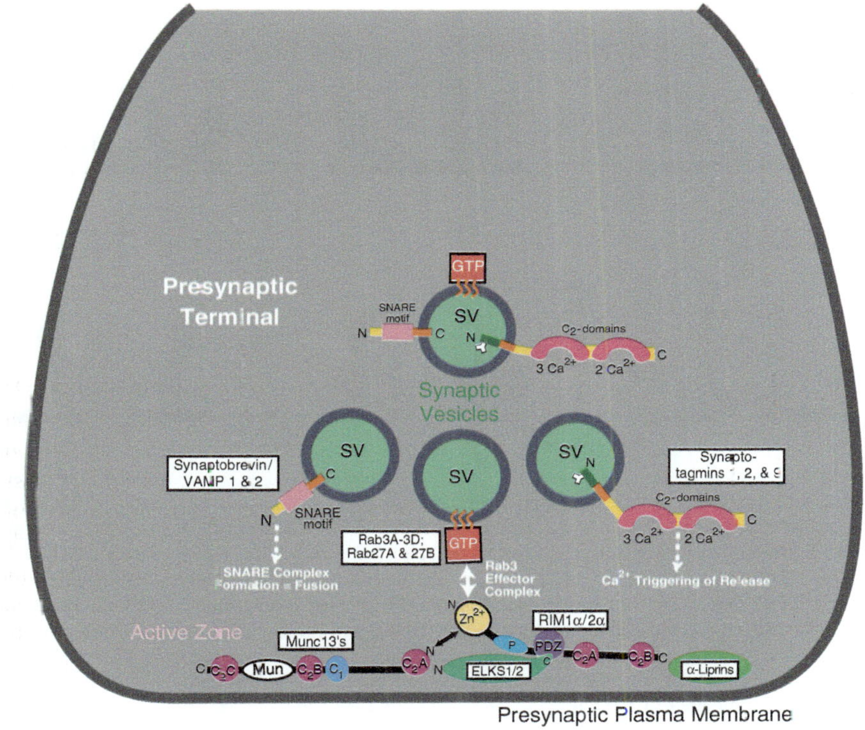


**Fig. 3** Interaction of Rab3 and Rab27 on synaptic vesicles with the active zone protein complex containing Munc13s, RIMs, ELKS, and liprins. The schematic drawing depicts a nerve terminal with a few synaptic vesicles containing the three vesicle proteins that mediate exocytosis: the SNARE protein synaptobrevin/VAMP that participates in fusion (see Figure 4), the Rab-proteins Rab3 and Rab27 that attach synaptic vesicles to the active zone protein complex as shown, and the $Ca^{2+}$-sensor protein synaptotagmin that translates the $Ca^{2+}$-signal into release (Figure 5). The active zone protein complex is composed of Munc13, RIM, ELKS, and liprins, so that RIM binds to all of the three other active zone proteins, and additionally interacts with Rab3/27 via its N-terminal domain. The active zone protein complex likely contains other protein components that are not shown, in particular piccolo/bassoon. (Modified from Südhof, 2004).

(Hanson et al., 1997; Weber et al., 1998), formation of a "trans-complex" by SNARE proteins on opposing membranes forces these membranes together, thereby overcoming the energy barrier (Figure 4). SNARE proteins contain a characteristic 60-residue sequence, the so-called SNARE motif. SNARE complexes are assembled from four types of SNARE motifs (called R, Qa, Qb, and Qc, classified based on sequence homologies and the central residue) that fold into a tight four-helical bundle which always contains one copy for each type of SNARE motif. The close approximation of two membranes by SNARE-complex assembly destabilizes their negatively charged surfaces, thereby initiating the intermixing of their hydrophobic lipid interiors. This is thought to provide the energy for membrane fusion.

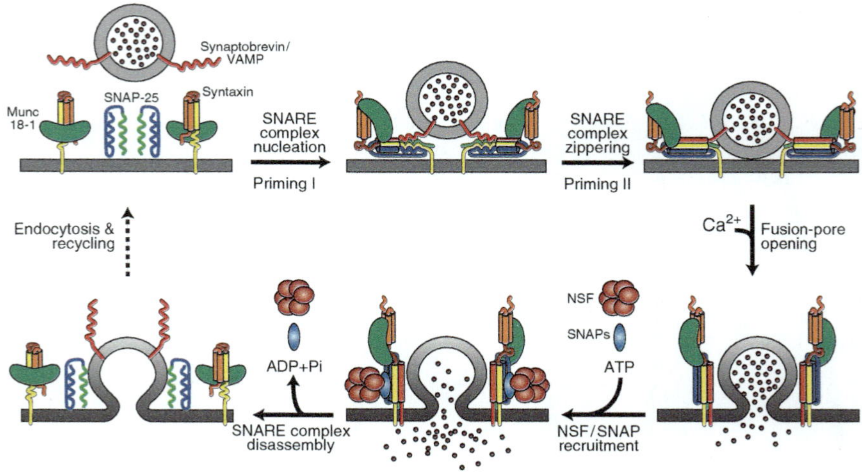


**Fig. 4** Schematic diagram of the SNARE protein/Munc18 cycle. Docked synaptic vesicles (top left) may be attached to the active zone via the Rab/RIM interaction (see Figure 3) but contain SNARE proteins that have not yet formed a complex with each other (synaptobrevin/VAMP on synaptic vesicles and SNAP-25 and syntaxin-1 on the plasma membrane; note that syntaxin-1 is thought to be complexed to the SM-protein Munc18-1). Priming is envisioned to occur in two steps that involve the successive assembly of SNARE-complexes (priming I and II). During priming, Munc18-1 is thought to be continuously associated with syntaxin-1, shifting from a heterodimeric binding mode in which it was attached to syntaxin-1 alone to a heteromultimeric binding mode in which it is attached to the entire SNARE complex (top right). After priming, $Ca^{2+}$ triggers fusion-pore opening to release the neurotransmitters by binding to synaptotagmin (see Figure 5). After fusion-pore opening, SNAPs (no relation to SNAP-25) and NSF (an ATPase) bind to the assembled SNARE complexes, disassemble them with ATP-hydrolysis, thereby allowing synaptic vesicles to undergo re-endocytosis and to recycle with synaptobrevin on the vesicle, while leaving SNAP-25 and syntaxin-1/Munc18-1 on the plasma membrane. Note that the overall effect is that SNARE/Munc18-proteins undergo a cycle of association/dissociation that fuels the membrane fusion reaction which underlies release. (Modified from Rizo and Südhof, 2002).

Synaptic exocytosis involves three SNARE proteins: the R-SNARE synaptobrevin/VAMP (isoforms 1 and 2) on the vesicle, and the Q-SNAREs syntaxin (isoforms 1 and 2) and SNAP-25 on the plasma membrane (Figure 4). Since SNAP-25 has two SNARE-motifs, synaptobrevin, syntaxin, and SNAP-25 together have four SNARE-motifs. Synaptobrevins and SNAP-25 are relatively simple SNARE proteins that are composed of little else besides SNARE motifs and membrane-attachment sequences (a transmembrane region for synaptobrevin, and a cysteine-rich palmitoylated sequence for SNAP-25). Syntaxins, in contrast, are complex proteins. The N-terminal two-thirds of syntaxins include a separate, autonomously folded domain (the so-called $H_{abc}$-domain), while the C-terminal third is composed of a SNARE motif and transmembrane region just like synaptobrevin.

### 3.3 SM Proteins

Genes for SM-proteins were discovered in genetic screens in *C. elegans* (unc18) and yeast (sec1), and their connection to membrane fusion was identified when the SM-protein Munc18-1 was found to directly bind to syntaxin-1 (Brenner, 1974; Novick et al., 1980; Hata et al., 1993). SM-proteins are composed of a conserved ~600 amino acid sequence that folds into an arch-shaped structure. With seven members in mammals and four in yeast, SM-proteins constitute a small family of highly homologous proteins. SM proteins have essential roles in all fusion reactions tested. Three SM proteins (Munc18-1, −2, and −3) are involved in exocytosis, where they are at least as essential as SNARE proteins. For example, deletion of Munc18-1 in mice has more severe consequences for synaptic vesicle exocytosis than deletion of synaptobrevin or SNAP-25 (Verhage et al., 2000).

Initially, Munc18-1 was found to bind only to monomeric syntaxin-1 in a manner that is incompatible with SNARE-complex formation. Puzzlingly, however, other SM proteins were subsequently found to bind to assembled SNARE complexes. This puzzle was resolved with the discovery that Munc18-1 (and presumably −2) participates in two distinct modes of SNARE interactions: the originally defined binding to monomeric syntaxins, and a novel mode of direct binding to assembled SNARE complexes (Dulubova et al., 2007; Shen et al., 2007). These results suggested that all SM-proteins directly or indirectly interact with assembled SNARE complexes in fusion. The additional binding of Munc18-1 to the closed conformation of syntaxin prior to SNARE complex formation renders $Ca^{2+}$-triggered exocytosis unique among fusion reactions, possibly in order to achieve a tighter control of the fusion reaction.

### 3.4 Mechanism of SNARE and SM Protein Catalyzed Fusion

Both SNARE and SM proteins are required as components of the minimal fusion machinery. At the synapse, for example, deletion of Munc18-1 leads to a loss of all synaptic vesicle fusion, revealing Munc18-1 as an essential component of the fusion machine (Verhage et al., 2000). It is likely that SNARE proteins first force membranes together by forming trans-complexes, thereby creating a fusion intermediate that at least for synaptic vesicles appears to consist of a hemifusion stalk (Figure 4). Since the unifying property of SM proteins is to bind to assembled SNARE complexes, they likely act after such a fusion intermediate has formed, but their exact role remains unknown.

Each intracellular fusion reaction exhibits characteristic properties, and involves a different combination of SM and SNARE proteins. The specificity of fusion reactions appears to be independent of SNARE proteins because SNARE complex formation is nonspecific as long as the Q/R-rule is not violated (i.e., the fact that SNARE complexes need to be formed by SNARE proteins containing R-, Qa-, Qb-, and Qc-SNARE motifs), and of SM proteins because SM proteins often function in multiple fusion reactions. Fusion specificity must be determined by other mechanisms, possibly GTP-binding proteins of the rab family.

# 4 Mechanism of $Ca^{2+}$-Triggering: $Ca^{2+}$-Channels, $Ca^{2+}$-Buffering, and Synaptotagmin

Neurotransmitter release is triggered by $Ca^{2+}$ when an action potential invades the nerve terminal and gates the opening of voltage-sensitive $Ca^{2+}$-channels. Thus, there are two determinants of neurotransmitter release: (1) The $Ca^{2+}$-dynamics in the nerve terminal that are dictated by the properties and location of the $Ca^{2+}$-channels; the concentration, affinities, and kinetics of local $Ca^{2+}$-buffers; and the $Ca^{2+}$-extrusion mechanisms and (2) the action of the $Ca^{2+}$-receptors that translate the $Ca^{2+}$-signal into release, with most release being mediated by $Ca^{2+}$-binding to synaptotagmins (see below).

## *4.1* $Ca^{2+}$*-Dynamics*

The $Ca^{2+}$-concentration in a nerve terminal depends on the number and temporal pattern of action potentials, the effectiveness of these action potentials to open $Ca^{2+}$-channels, and the properties and concentrations of $Ca^{2+}$-buffers. Not only the time course of changes in $Ca^{2+}$-concentrations, but also the spatial distribution of $Ca^{2+}$, is important because $Ca^{2+}$ is not uniformly distributed in a nerve terminal. Moreover, the $Ca^{2+}$-dynamics of a nerve terminal differ between nerve terminals, and play a central role in synaptic plasticity (e.g., see Rozov et al., 2001; Zucker and Regehr, 2002).

$Ca^{2+}$-channels are well investigated, have proven to be great drug targets, and will be discussed at length in the chapter by Kisilevsky and Zamponi. Two types of $Ca^{2+}$-channels, the so-called P/Q- and N-type channels (referred to as Cav2.1 and 2.2) account for the vast majority of releases. These $Ca^{2+}$-channels are located in the active zone of the presynaptic terminal (Llinas et al., 1992), although their precise location is unknown. $Ca^{2+}$-channels are – not surprisingly – tightly regulated by several signaling systems. As a result of their non-uniform localization and their stringent regulation, the $Ca^{2+}$-signal produced by the opening of $Ca^{2+}$-channels by a given action potential cannot be predicted, but varies greatly between synapses in amplitude, space and time (Rozov et al., 2001). This variation is increased by differences in $Ca^{2+}$-buffering between synapses. $Ca^{2+}$-buffers are much less understood than $Ca^{2+}$-channels because of the large number of different types of buffers, the difficulty in manipulating them pharmacologically or genetically, and the problems in measuring them. The most important nerve terminal $Ca^{2+}$-buffer likely is ATP, which has a relatively low $Ca^{2+}$-affinity but a high concentration and is highly mobile, rendering it an effective buffer at peak $Ca^{2+}$-concentrations (Meinrenken et al.,

2003). In addition, a number of $Ca^{2+}$-binding proteins of the EF-hand class probably function as nerve terminal $Ca^{2+}$-buffers, including parvalbumin and calbindin. Interestingly, these proteins exhibit a highly selective distribution in the brain, with inhibitory interneurons generally expressing much higher concentrations of selected $Ca^{2+}$-buffer proteins than excitatory neurons, a difference that may account for the distinct short-term plasticity properties of GABA release (Silberberg et al., 2005). $Ca^{2+}$-extrusion, finally, is mediated by multiple mechanisms. Although uptake of nerve terminal $Ca^{2+}$ into mitochondria and endoplasmic reticulum has been widely discussed, the most important step is the transport of $Ca^{2+}$-ions from the cytosol into the extracellular space against a vast concentration gradient. Several mechanisms for this transport exist, of which the $Na^{+}/Ca^{2+}$-exchanger and the plasma membrane $Ca^{2+}$-ATPase appear to be the most important (Meinrenken et al., 2003).

## *4.2 Synaptotagmins as* $Ca^{2+}$*-Sensors for Fast Neurotransmitter Release*

At resting $Ca^{2+}$-concentrations, release occurs spontaneously at a low rate; $Ca^{2+}$ increases this rate $>10,000$-fold in less than a millisecond, faster than most diffusion-controlled chemical reactions. $Ca^{2+}$ thus increases the probability of a rare event that normally occurs all the time. Studies in the Calyx of Held synapse – a large vertebrate synapse that allows simultaneous recordings from pre- and postsynaptic neurons – showed that $Ca^{2+}$ triggers release at micromolar concentrations with an apparent cooperativity of 5 (Meinrenken et al., 2003). For each active zone, the $Ca^{2+}$-signal produced by an action potential has a finite probability of triggering exocytosis (usually between 0.05 and 0.30) that changes as a function of the previous use of a synapse (i.e., is subject to plasticity), and additionally depends on signaling mediated by neuromodulators. In addition to this fast synchronous type of neurotransmitter release, synapses exhibit a second slower, more asynchronous type of release that is also triggered by $Ca^{2+}$. Both types are likely important, but under most physiological conditions the fast synchronous type predominates.

The speed of $Ca^{2+}$-triggered exocytosis suggests that $Ca^{2+}$ acts at a stage at which part of the fusion reaction has already been completed, most likely on vesicles in a hemifusion state that is presumably created by the combined action of SNARE proteins and Munc18-1 (i.e., after priming II in Figure 4). Much of synaptic vesicle exocytosis likely operates in the "kiss-and-run" mode (Harata et al., 2006). Kiss-and-run exocytosis predominates during low-frequency stimulation, while full exocytosis is more important during high-frequency stimulus trains. The kiss-and-run mode allows fast recycling of synaptic vesicles for reuse after exocytosis, but probably has only a minimal effect on the kinetics or amount of release because the total size of synaptic vesicles (~20 nm radius) is only 5- to 10-fold larger than that of the fusion pore.

The definition of the primary structure of synaptotagmin-1, composed of an N-terminal transmembrane region and two C-terminal $C_2$-domains (Figure 3), led to the hypothesis that synaptotagmin-1 functions as the $Ca^{2+}$-sensor for fast neurotransmitter release (Perin et al., 1990). Subsequent studies revealed that synaptotagmin-1 and two other synaptotagmin isoforms, synaptotagmin-2 and -9, act as $Ca^{2+}$-sensors in exocytosis by a strikingly simple mechanism: $Ca^{2+}$ flowing into the nerve terminal when $Ca^{2+}$-channels open during an action potential binds to the two $C_2$-domains of synaptotagmins (Xu et al., 2007). $Ca^{2+}$-binding to the $C_2$-domains induces the interaction of synaptotagmins with phospholipids and with SNARE proteins with a micromolar apparent $Ca^{2+}$-affinity, consistent with the affinity of release, which in turn opens the fusion pore mechanically by pulling the SNARE complexes apart (Südhof, 2004). This model was supported by a host of biochemical and genetic experiments, most importantly the finding that point mutations in the endogenous synaptotagmin gene in mice that alter the apparent $Ca^{2+}$-affinity of synaptotagmin also alter the apparent $Ca^{2+}$-affinity of release in an identical manner (Fernandez-Chacon et al., 2001). However, these findings also raised a crucial question that is only now beginning to be answered: Why doesn't the fusion pore just pop open after priming II (Figure 4)? Recent results suggest that another nerve-terminal protein, a soluble protein called complexin, plays a central role here (Giraudo et al., 2006; Tang et al., 2006). Complexins bind in a $Ca^{2+}$-independent manner only to the C-terminal part of assembled SNARE complexes where the SNARE complexes are anchored in the membrane (Figure 5). Deletion of complexins in mice revealed a selective impairment of fast synchronous $Ca^{2+}$-triggered release, whereas other forms of fusion were unchanged (Reim et al., 2001). This phenotype resembles the synaptotagmin-1 knockout phenotype in its selectivity, but was milder because the complexin knockout phenotype can be rescued by boosting $Ca^{2+}$-influx into the nerve terminal, whereas the synaptotagmin-1 knockout phenotype cannot. Thus complexins function as activators of SNARE complexes for subsequent synaptotagmin-1 action. The mechanism by which the functions of synaptotagmin-1 and complexins interact was revealed in biochemical experiments demonstrating that $Ca^{2+}$-induced binding of synaptotagmin-1 to SNARE complexes displaces complexin from the SNARE complexes (Tang et al., 2006), indicating that synaptotagmin-1 serves as the trigger of a complexin-cocked gun.

These observations suggest a model for how synaptotagmin-1 triggers exocytosis (Figure 5): By binding to SNARE complexes during assembly, complexins force completion of SNARE-complex assembly that creates an activated, frozen fusion intermediate, likely a hemifusion state. When $Ca^{2+}$ binds to synaptotagmin, this induces binding of synaptotagmin to both fusing membranes and to the SNARE complexes. We envision that this binding pulls open the fusion pore by two independent mechanisms: a mechanical force on the membrane by coupling phospholipids to SNARE complexes via the simultaneous binding, and a disinhibition of fusion-pore opening by displacing from the SNARE complexes the complexins that at the same activate and stabilize the complexes. Note that the complexin/synaptotagmin-dependent fusion reaction mediates fast synchronous release, and is distinct from the second type of release, slower asynchronous release which also depends on SNARE-proteins and Munc18-1 but is independent of complexins and synaptotagmin.

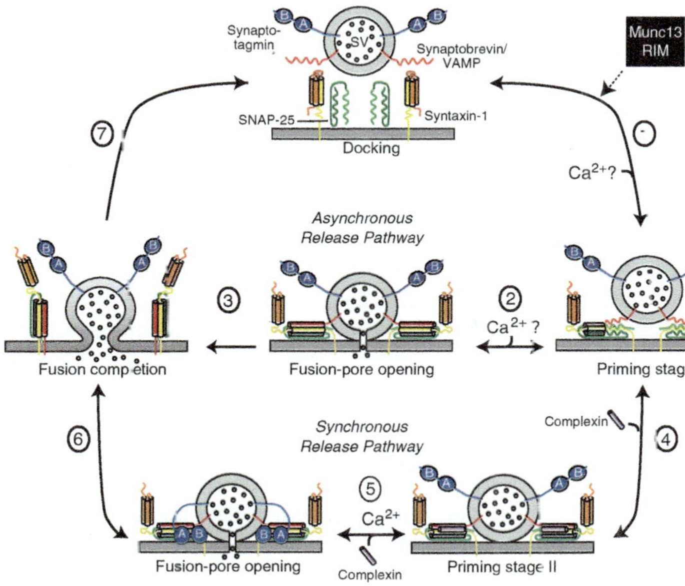


**Fig. 5** Model for the interplay between SNARE, complexin, and synaptotagmin function in $Ca^{2+}$-triggered neurotransmitter release. Docked vesicles containing unassembled SNARE complexes (top) are attached to the active zone via the Rab3/27 interaction with RIMs (Figure 3), and are primed in a two-stage reaction by SNARE-complex assembly (step 1; see also Figure 4). The resulting primed vesicles form the substrate for two release pathways: asynchronous release that is synaptotagmin- and complexin-independent (steps 2 and 3), and synchronous release that depends on synaptotagmin and complexins (steps 4–6). Note that quantitatively, the asynchronous pathway is less important for release than the synchronous pathway which mediates >90% of all release under most conditions. Asynchronous release is triggered by $Ca^{2+}$ by a poorly understood mechanism that, like synchronous release, requires assembly of SNARE complexes, but differs from synchronous release in that it does not appear to stop fusion-pore opening by a "clamp" after SNARE complexes are fully assembled. Asynchronous release becomes important when the $Ca^{2+}$-concentration rises for prolonged time periods to intermediate levels (0.1–1.0 μM) that are too low to trigger synchronous release, but sufficient to trigger asynchronous release. This occurs, for example, during the accumulation of residual $Ca^{2+}$ during high-frequency stimulus trains (Zucker and Regehr, 2002). Synchronous release involves "superpriming" of synaptic vesicles by binding of complexins to assembled SNARE complexes (step 4). Complexin binding activates and freezes SNARE complexes in a metastable state (priming stage II). Superprimed vesicles in which the SNARE complexes have been clamped by complexin are then substrate for fast $Ca^{2+}$-triggering of release when $Ca^{2+}$-binding to synaptotagmin triggers the simultaneous interaction of synaptotagmin with phospholipids and SNARE complexes, with the latter reaction displacing complexin and resulting in fusion-pore opening (step 5). Opened fusion pores can then dilate to complete fusion (step 6), although both steps 2 and 5 are potentially reversible, i.e., lack of dilation of the fusion pore could lead to "kiss-and-stay" or "kiss-and-run" exocytosis in these pathways. Note that steps 1 and 4 are also probably reversible, with a much faster forward than backward speed. It is likely that step 1 is $Ca^{2+}$-dependent, but it is unclear whether or not step 2 is $Ca^{2+}$-dependent, since it is possible that asynchronous release is $Ca^{2+}$-dependent solely because $Ca^{2+}$ accelerates step 1, and step 2 has a finite probability. Thus the nature of $Ca^{2+}$-triggering of asynchronous release could operate either at the priming or at the actual fusion step. Note that the function of Munc18-1 is not included in this diagram, but is thought to be central to the actual fusion reaction (see Figure 4). (Modified from Tang et al., 2006).

In addition to functioning as $Ca^{2+}$-sensors for vesicle exocytosis, synaptotagmins may be involved in vesicle endocytosis, particularly the decision between kiss-and-run versus full exocytosis. Such a role would be economical in linking fusion-pore opening (which is triggered by $Ca^{2+}$-binding to synaptotagmin) to fusion-pore expansion or contraction, but the precise mechanisms involved have not yet been explored.

As mentioned above, three different synaptotagmin isoforms mediate fast synchronous release: synaptotagmin-1, -2, and -9. Why does the vertebrate brain express three different proteins to do the same job? As it turns out, these three different synaptotagmins mediate the $Ca^{2+}$-triggering of release with distinct properties, and are differentially distributed (Xu et al., 2007). Of the three synaptotagmins, synaptotagmin-2 is the fastest, and synaptotagmin-9 the slowest, corresponding with the selective presence of synaptotagmin-2 in the calyx synapse, one of the fastest synapses in brain involved in sound localization, and synaptotagmin-9 in the limbic system involved in emotional reactions. Thus, the properties of the synapses formed by a neuron depend among others on the isoform of synaptotagmin which that neuron expresses, adding additional complexity to the neural circuits formed by synaptic networks.

## 5 Regulation of Release Beyond $Ca^{2+}$-Triggering

We already alluded to the fact that release is not constant, but is highly plastic, i.e., regulated by extrinsic signals and by the intrinsic previous activity of a nerve terminal. Both forms of regulation constitute a type of synaptic memory: the history of the activity of other surrounding neurons and of the neuron to which a terminal belongs strongly influence to what extent this terminal translates an action potential into a neurotransmitter release signal (Zucker and Regehr, 2002). Many different forms of synaptic plasticity exist. Although traditionally postsynaptic forms of plasticity have been appreciated more than presynaptic forms, recent studies have revealed the existence of purely presynaptic forms of synaptic plasticity, and have moreover shown that many forms of plasticity are mediated by the combined action of pre- and postsynaptic mechanisms.

A complete discussion of presynaptic modulation and plasticity of release is beyond the scope of this chapter. Instead, three exemplary forms of presynaptic plasticity will be discussed because of their importance for the pharmacology of neurotransmitter release.

### *5.1 Acetylcholine-Receptor-Mediated* $Ca^{2+}$*-Influx into Presynaptic Nerve Terminals*

Nicotine is an addictive drug that activates a diverse subset of ionotropic acetylcholine receptors. Most cholinergic actions in brain, both by ionotropic and metabotropic receptors, are modulatory, and very few fast synapses exist that utilize acetylcholine as a transmitter. Interestingly, although some ionotropic acetylcholine receptors are postsynaptic, a relatively large proportion, much larger than observed for other neurotransmitters, appears to be presynaptic. A possible pathway accounting for the addictive actions of nicotine was discovered in the observation that presynaptic nicotinic acetylcholine receptors, when activated, mediate the influx of $Ca^{2+}$ into presynaptic terminals, and thereby stimulate the release of neurotransmitters (Wannacott, 1997). Interestingly, this effect appears to operate via a class of nicotinic acetylcholine receptors containing α6 subunits that are relatively enriched in the nigrostriatal pathway, suggesting that nicotine may be addictive by increasing dopamine release (Quik and McIntosh, 2006). Thus the presynaptic facilitation of neurotransmitter release by cholinergic nicotinic receptors is one of the best and physiologically most important systems illustrating how presynaptically acting receptors can regulate release.

## 5.2 $Ca^{2+}$-*Channel Modulation by Presynaptic Receptors*

Presynaptic G-protein coupled receptors for a large number of neurotransmitters, both autoreceptors and receptors for extrinsic signals, suppress $Ca^{2+}$-channel gating in response to an action potential. This mechanism of action appears to be the dominant mechanism involved in short-term plasticity mediated by presynaptic receptors. A typical example is depolarization-induced suppression of inhibition (DSI), which is the short-term suppression of presynaptic GABA-release induced by the depolarization of the postsynaptic cell (Diana and Marty, 2004). DSI is caused when the postsynaptic depolarization causes the release of endocannabinoids from the postsynaptic cell, and the endocannabinoids then bind to presynaptic CB1 receptors whose activation suppresses presynaptic $Ca^{2+}$-channels. Like many other forms of presynaptic suppression mediated by activation of presynaptic receptors, this effect is short-lasting (in the millisecond range). The precise mechanisms by which $Ca^{2+}$-channels are suppressed appear to vary between receptors, but the outcome is always a very effective short-term decrease in synaptic signaling.

## 5.3 *Presynaptic Long-Term Plasticity Mediated by cAMP-Dependent Protein Kinase A (PKA)*

PKA-dependent long-term potentiation and depression was initially discovered in the mossy-fiber synapses of the hippocampus, and later demonstrated in parallel fiber synapses of the cerebellum and corticothalamic synapses of the forebrain (Malenka and Siegelbaum, 2001). This widespread form of plasticity does not involve changes in $Ca^{2+}$-influx, but operates via a direct increase or decrease, respectively, of the amount of vesicle exocytosis that can be triggered by a given $Ca^{2+}$-signal. Interestingly, this form of plasticity appears to depend on the interaction of the synaptic vesicle GTP-binding protein Rab3 with its effectors, RIM-proteins (Figure 3), since deletion of either Rab3A or of RIM1α abolishes LTP in the hippocampal mossy fibers (Castillo et al., 2002). Moreover, recently it was revealed that a special form of endocannabinoid-dependent long-term depression of GABAergic synapses in the hippocampus also operates as a presynaptic process that requires PKA activation and RIM1α. This form of LTD, referred to as i-LTD for inhibitory LTD, is distinct from the DSI discussed above; indeed, DSI is normal in mice lacking RIM1α. Thus, endocannabinoids can trigger two different forms of plasticity depending on the precise conditions of the postsynaptic depolarization. Overall, these results highlight the role of the core release machinery in presynaptic long-term plasticity.

## 6 $Ca^{2+}$-Induced Exocytosis of Small Dense-Core Vesicles and LDCVs

Biogenic amines are generally secreted from small dense-core vesicles (SDCVs) in varicosities of axons that are not associated with classical synapses, and additionally from LDCVs, whereas neuropeptides are secreted – usually together with biogenic amines – only from LDCVs. Different from synaptic vesicle exocytosis where $Ca^{2+}$ is thought to trigger fusion of predocked and primed vesicles, a significant part of SDCV and LDCV exocytosis involves the $Ca^{2+}$-dependent mobilization of vesicles before $Ca^{2+}$-triggered fusion. As a result, dense-core vesicle exocytosis requires more sustained $Ca^{2+}$-transients, but is also slower and longer lasting than synaptic vesicle exocytosis (Stjarne, 2000).

$Ca^{2+}$-triggered exocytosis of SDCVs, LDCVs, and their endocrine equivalents has been examined in several systems (e.g., chromaffin cells, cultured melanotrophic neurons, pancreatic β-cells, and PC12 cells), but the best-studied system is that of chromaffin cells because they allow the highest measurement resolution, and are the only system that has been investigated extensively by a loss-of-function approach. $Ca^{2+}$-triggered chromaffin granule exocytosis is ~10-fold slower than synaptic vesicle exocytosis, but otherwise the two systems are similar. Chromaffin granule exocytosis, like synaptic vesicle exocytosis, involves SNARE and Munc18-proteins. Knockout experiments revealed, however, that $Ca^{2+}$-triggering of chromaffin granule and synaptic vesicle exocytosis are mechanistically distinct. Specifically, chromaffin cell exocytosis is mediated by two different synaptotagmins: synaptotagmin-1, which is shared with synaptic vesicle exocytosis, and synaptotagmin-7, which does not function in synaptic vesicle exocytosis. The synaptotagmin-7 selectively functions as a major $Ca^{2+}$-sensor for chromaffin granule but not for synaptic vesicle exocytosis, indicating a fundamental difference between the two types of exocytosis. However, beyond this difference, little is known about the other molecular components that direct SDCV and LDCV exocytosis and make it different from synaptic vesicle exocytosis, a question that is of obvious importance in understanding the release of biogenic amines and neuropeptides in brain.

## 7 Presynaptic Drug Targets

This introductory chapter discussed the fundamental mechanisms of release because understanding these fundamental mechanisms is crucial for insight into the actions of drugs that act on release. Overall, the neurotransmitter release machinery is a poor drug target, as judged by the number of successful drugs that act on it, and in contrast to the neurotransmitter reception machinery (i.e., neurotransmitter receptors) that is the target of many different drugs. In fact, the quantitatively largest number of drugs that influence neurotransmitter release act on presynaptic neurotransmitter receptors. Nevertheless, several well-established drugs exist that act on the release machinery: reserpine on the vesicular catecholamine transporter, leviracetam on the vesicle protein SV2, and cocaine on the presynaptic re-uptake of dopamine.

It is clear that the presynaptic machinery is an underutilized target for drugs. The actual membrane-trafficking components of that machinery, such as the SNARE, SM, or Rab-proteins or synaptotagmins, are probably not good drug targets because they operate via protein-protein interactions that are difficult to influence with small molecules. However, several synaptic proteins with receptor, transport, or enzyme functions are likely drug targets, but have not yet been explored. These proteins include synapsins, which are ATP-binding proteins (Hosaka and Südhof, 1998), CSPα as a synaptic vesicle co-chaperone (Tobaben et al., 2001), and vesicular GABA- and glutamate transporter (Gasnier, 2000; Takamori, 2006). We hope that the material presented in this book will not only help understand the actions of current drugs, but also stimulate the development of new ones.

## References


Agnati LF, Zoli M, Stromberg I, Fuxe K (1995) Intercellular communication in the brain: wiring versus volume transmission. Neuroscience 69:711–26

Brock JA, Cunnane TC (1987) Relationship between the nerve action potential and transmitter release from sympathetic postganglionic nerve terminals. Nature 326:605–7

Brenner S. (1974) The genetics of Caenorhabditis elegans. Genetics 77:71–94

Castillo PE, Schoch S, Schmitz F, Südhof TC, Malenka RC (2002) RIM1α is required for presynaptic long-term potentiation. Nature 415:327–30

Diana MA, Marty A (2004) Endocannabinoid-mediated short-term synaptic plasticity: depolarization-induced suppression of inhibition (DSI) and depolarization-induced suppression of excitation (DSE).Br J Pharmacol 42:9–19

Du Bois-Reymond E (1877) Gesammelte Abhandlungen zur Allgemeinen Muskel- und Nervenphysik. 2 vols. Leipzig: von Veit Verlag.

Dulubova I, Khvotchev M, Liu S, Huryeva I, Südhof TC, Rizo J (2007) Munc18-1 binds directly to the neuronal SNARE complex. Proc Natl Acad Sci USA 104:2697–2702

Fernandez-Chacon R, Konigstorfer A, Gerber SH, Garcia J, Matos MF, Stevens CF, Brose N, Rizo J, Rosenmund C, Südhof TC (2001) Synaptotagmin I functions as a calcium regulator of release probability. Nature 410:41–9

Fleckenstein AE, Volz TJ, Riddle EL, Gibb JW, Hanson GR (2007) New insights into the mechanism of action of amphetamines. Annu Rev Pharmacol Toxicol 47:681–98

Foster M (1897) A textbook of physiology, 7th ed., Part III. London: Macmillan

Gasnier B (2000) The loading of neurotransmitters into synaptic vesicle. Biochimie 82:327–37

Giraudo CG, Eng WS, Melia TJ, Rothman JE (2006) A clamping mechanism involved in SNARE-dependent exocytosis. Science 313:676–80

Hata Y, Slaughter CA, Südhof TC (1993) Synaptic vesicle fusion complex contains unc-18 homologue bound to syntaxin. Nature 366:347–351

Hanson PI, Roth R, Morisaki H, Jahn R, Heuser JE (1997) Structure and conformational changes in NSF and its membrane receptor complexes visualized by quick-freeze/deep-etch electron microscopy. Cell 90:523–35

Harata NC, Aravanis AM, Tsien,R (2006) Kiss-and-run and full-collapse fusion as modes of exo-endocytosis in neurosecretion. J Neurochem 97:1546–70

Hosaka M, Südhof TC (1998) Synapsins I and II are ATP-binding proteins with differential $Ca^{2+}$ regulation. J Biol Chem 273:1425–9

Jahn R, Lang T, Südhof TC (2003) Membrane fusion. Cell 112:519–33

Katz B (1969) The release of neural transmitter substances. Liverpool: Liverpool University Press

Krause W (1863) Über die Endigung der Muskelnerven. Z Rat Med 18:136–60

Kühne W (1862) Über die peripherischen Endorgane der motorischen Nerven. Leipzig: Engelmann

Llinas R, Sugimori M, Silver RB (1992) Microdomains of high calcium concentration in a presynaptic terminal. Science 256:677–9

Loewi O (1921) Über humorale Übertragbarkeit der Herznervenwirkung. Pflügers Arch. 189: 239–42

Malenka RC, Siegelbaum SA (2001) Synaptic plasticity. In Synapses (Cowan MW, Südhof TC, Stevens CF, eds), The Johns Hopkins University Press, Baltimore, pp 393–453

Meinrenken CJ, Borst JG, Sakmann B (2003) Local routes revisited: the space and time dependence of the $Ca^{2+}$ signal for phasic transmitter release at the rat calyx of Held. J Physiol 547:665–89

Neher E, Sakmann B (1976) Single-channel currents recorded from membrane of denervated frog muscle fibres. Nature 260:799–802

Noda M, Takahashi H, Tanabe T, Toyosato M, Furutani Y, Hirose T, Asai M, Inayama S, Miyata T, Numa S (1982) Primary structure of alpha-subunit precursor of Torpedo californica acetylcholine receptor deduced from cDNA sequence. Nature 299:793–7

Novick P, Field C, Schekman R (1980) Identification of 23 complementation groups required for post-translational events in the yeast secretory pathway. Cell 21:205–15

Perin MS , Fried VA, Mignery GA, Jahn R, Südhof TC (1990) Phospholipid binding by a synaptic vesicle protein homologous to the regulatory region of protein kinase C. Nature 345:260–3

Quik M, McIntosh JM (2006) Striatal alpha6* nicotinic acetylcholine receptors: potential targets for Parkinson's disease therapy. J Pharmacol Exp Ther 316:481–9

Reim K, Mansour M, Varoqueaux F, McMahon HT, Südhof TC, Brose N, Rosenmund C (2001) Complexins regulate a late step in $Ca^{2+}$-dependent neurotransmitter release. Cell 104:71–81

Rizo J, Südhof TC (2002) Snares and Munc18 in synaptic vesicle fusion. Nat Rev Neurosci 3: 641–53

Rozov A, Burnashev N, Sakmann B, Neher E (2001) Transmitter release modulation by intracellular Ca2+ buffers in facilitating and depressing nerve terminals of pyramidal cells in layer 2/3 of the rat neocortex indicates a target cell-specific difference in presynaptic calcium dynamics. J Physiol 531:807–26

Salio C, Lossi L, Ferrini F, Merighi A. (2006) Neuropeptides as synaptic transmitters. Cell Tissue Res 326:583–98

Shen J, Tareste DC, Paumet F, Rothman JE, Melia TJ. (2007) Selective activation of cognate SNAREpins by Sec1/Munc18 proteins. Cell 128:183–95

Silberberg G, Grillner S, LeBeau FE, Maex R, Markram H (2005) Synaptic pathways in neural microcircuits. Trends Neurosci 28:541–51

Stjarne L (2000) Do sympathetic nerves release noradrenaline in "quanta"? J Auton Nerv Syst 81:236–43

Südhof TC (2004) The synaptic vesicle cycle. Annu Rev Neurosci 27, 509–47

Tang J, Maximov A, Shin O-H, Dai H, Rizo J, Südhof TC (2006) A complexin/synaptotagmin-1 switch controls fast synaptic vesicle exocytosis. Cell 126:1175–87

Takamori S (2006) VGLUTs: 'exciting' times for glutamatergic research? Neurosci Res 55:343–51

Tobaben S, Thakur P, Fernandez-Chacon R, Südhof TC, Rettig J, Stahl B (2001) A trimeric protein complex functions as a synaptic chaperone machine. Neuron 31:987–99

Trudeau LE (2004) Glutamate co-transmission as an emerging concept in monoamine neuron function. J Psychiatry Neurosci 29:296–310

Verhage M, Maia AS, Plomp JJ, Brussaard AB, Heeroma JH, et al. (2000) Synaptic assembly of the brain in the absence of neurotransmitter secretion. Science 287:864–9

Wonnacott S (1997) Presynaptic nicotinic ACh receptors. Trends Neurosci 20:92–8

Weber T, Zemelman BV, McNew JA, Westermann B, Gmachl M, Parlati F, Sollner TH, Rothman JE (1998) SNAREpins: minimal machinery for membrane fusion. Cell 92:759–72

Whittaker VP, Sheridan MN (1965) The morphology and acetylcholine content of isolated cerebral cortical synaptic vesicles. J Neurochem 12:363–72

Xu J, Mashimo T, Südhof TC (2007) Synaptotagmin-1, -2, and -9: $Ca^{2+}$ sensors for fast release that specify distinct presynaptic properties in subsets of neurons. Neuron 54:567–81

Zucker RS, Regehr WG (2002) Short-term synaptic plasticity. Annu Rev Physiol 64:355–405

# Pharmacology of Neurotransmitter Release: Measuring Exocytosis

**Mikhail Khvotchev and Ege T. Kavalali(✉)**



**Abstract** Neurotransmission in the nervous system is initiated at presynaptic terminals by fusion of synaptic vesicles with the plasma membrane and subsequent exocytic release of chemical transmitters. Currently, there are multiple methods to detect neurotransmitter release from nerve terminals, each with their own particular advantages and disadvantages. For instance, most commonly employed methods monitor actions of released chemical substances on postsynaptic receptors or artificial substrates such as carbon fibers. These methods are closest to the physiological setting because they have a rapid time resolution and they measure the action of the endogenous neurotransmitters rather than the signals emitted by exogenous probes. However, postsynaptic receptors only indirectly report neurotransmitter release in a form modified by the properties of receptors themselves, which are often nonlinear detectors of released substances. Alternatively, released chemical substances

Ege T. Kavalali
Department of Neuroscience, U.T. Southwestern Medical Center, 5323 Harry Hines Blvd., Dallas, TX 75390-9111, USA
Ege.Kavalali@UTSouthwestern.edu

T.C. Süchof, K. Starke (eds.), *Pharmacology of Neurotransmitter Release. Handbook of Experimental Pharmacology 184.*

can be detected biochemically, albeit on a time scale slower than electrophysiological methods. In addition, in certain preparations, where presynaptic terminals are accessible to whole cell recording electrodes, fusion of vesicles with the plasma membrane can be monitored using capacitance measurements. In the last decade, in addition to electrophysiological and biochemical methods, several fluorescence imaging modalities have been introduced which report synaptic vesicle fusion, endocytosis, and recycling. These methods either take advantage of styryl dyes that can be loaded into recycling vesicles or exogenous expression of synaptic vesicle proteins tagged with a pH-sensitive GFP variant at regions facing the vesicle lumen. In this chapter, we will provide an overview of these methods with particular emphasis on their relative strengths and weaknesses and discuss the types of information one can obtain from them.

# 1 Electrophysiological Detection of Secretion

## *1.1 Electrical Detection of Neurotransmitter Release Using Postsynaptic Ionotropic Receptors*

During fast chemical neurotransmission, released neurotransmitter substances cross the synaptic cleft and activate a set of neurotransmitter receptors on the postsynaptic side (Figure 1). These neurotransmitter receptors are ligand-gated ion channels that are activated rapidly within milliseconds upon neurotransmitter binding. Ion fluxes through neurotransmitter-gated channels in turn can be measured electrically via whole cell recording from the somata or dendrites of postsynaptic neurons (Hamill et al., 1981; Stuart et al., 1993). This electrophysiological method is commonly used to detect synaptic transmission and estimate the properties of neurotransmitter exocytosis. The principal advantage of this method is its direct physiological relevance. The main impact of fast neurotransmitter release is to cause electrical signaling in the postsynaptic neuron and in the case of $Ca^{2+}$ permeable channels such as NMDA receptors to activate second messenger signaling cascades. Therefore, measuring neurotransmitter release by postsynaptic recordings provides a direct readout of its impact on postsynaptic signaling.

The second advantage of this method is its rapidity; electrical signals can be detected reliably with a millisecond time resolution. Furthermore, these whole cell recordings can be performed on any cell type or neuronal process in culture, in brain slices as well as in vivo (Margrie et al., 2002).

Dissociated neuronal cultures provide a versatile system for analysis of mechanisms underlying neurotransmitter release. These cultures can be prepared from fetal or postnatal brain tissue. This preparation has been particularly instrumental in analysis of synapses deficient in key components of the release machinery. For instance, genetic deletion of synaptic SNARE (soluble N-ethylmaleimide-sensitive factor attachment protein receptors) proteins such as synaptobrevin-2 and SNAP-25

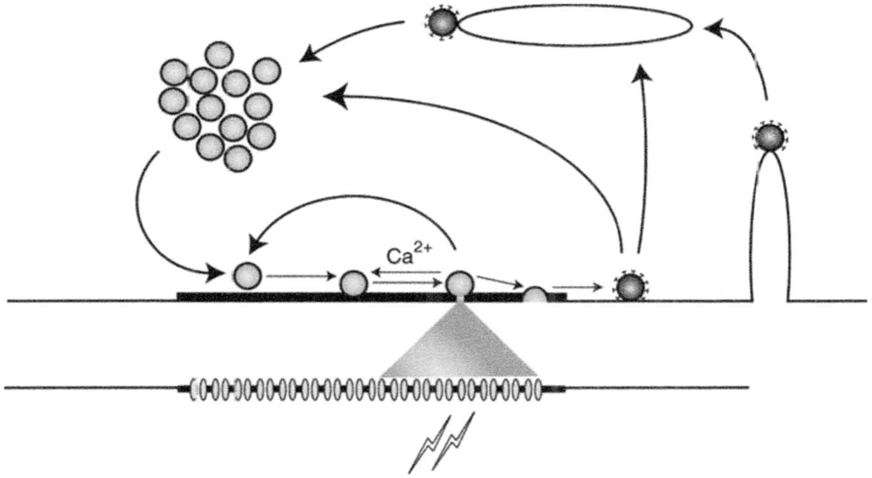


**Fig. 1** The synaptic vesicle cycle. During neurotransmission a rise in $Ca^{2+}$ level leads to vesicle fusion and neurotransmitter release at the active zone, after which vesicles completely collapse onto the plasma membrane. During this process vesicle membrane components are thought to intermix with their plasma membrane counterparts. Subsequently clathrin, through its adaptor proteins, is recruited to the membrane (typically at the periphery of the active zone) and forms coated vesicles by reclustering vesicle membrane components, which eventually bud off the plasma membrane with the help of the GTPase dynamin. Besides the plasma membrane, endocytosis of clathrin-coated vesicles may also occur from membrane infoldings or endosomal cisternae. This classical pathway is thought to operate within 40 and 90 seconds time scale. During an alternative fast (within seconds) pathway of vesicle recycling, synaptic vesicles retain their identity and do not intermix with the plasma membrane or endosomal compartments. This pathway is also referred to as "kiss and run" pathway in which the fusion pore opens transiently without complete collapse of the vesicle and without intermixing of synaptic vesicle membrane components with the plasma membrane. Most commonly used methods to study neurotransmitter release either measure the action of released neurotransmitters or they use exogenous fluorescent probes to monitor membrane and protein turnover in the presynaptic terminal.

lead to lethality at birth, therefore analysis was performed in neuronal cultures obtained from fetal brains (Schoch et al., 2001; Tafoya et al., 2006). In the absence of systemic effects arising from the loss of these proteins essential to neurotransmitter secretion, mutant synapses formed in these cultures readily attain morphological maturity. A critical advantage of neurotransmitter release measurements in dissociated neuronal cultures is their high sensitivity for detecting very low probability fusion events. Other major advantages of synaptic preparations include triggering fusion using non-$Ca^{2+}$-dependent means of release, such as hypertonicity or alpha-latrotoxin. Such approaches are instrumental for identifying selective effects of mutants on $Ca^{2+}$ sensitivity of fusion. A form of dissociated neuronal cultures can be prepared by low-density plating of neurons on isolated glial islands that gives rise single neurons forming synapses onto themselves that are called "autapses." Use of autaptic cultures substantially simplifies the whole cell recording configuration by enabling a single electrode to perform stimulation and recording at the same time. However, local stimulation methods applied to nonautaptic neuronal cultures also bring in several advantages (Maximov et al., 2007).

A key advantage of dissociated neuronal cultures is their amenability to molecular manipulations either through high-efficiency transfections using the calcium phosphate technique (Virmani et al., 2003) or via viral gene delivery, especially by lentiviral infections (Deak et al., 2006). Low-efficiency transfection methods can be useful to assess alterations in postsynaptic parameters after gene delivery to a target neuron. However, when presynaptic properties need to be assessed, then a high-efficiency transfection setting provides a significant advantage, as in this setting nearly all neurons in a culture are transformed and thus all nerve terminals synapsing onto a given target neuron can carry the exogenous protein of interest. Comparison of high-efficiency versus low-efficiency transfections, therefore, can help distinguish cell-autonomous versus non-cell-autonomous effects of a particular gene product (Burrone et al., 2002). Furthermore, this setting may help to sort out pre-versus postsynaptic effects of a particular protein, which are typically hard to deduce from autaptic cultures, as the protein of interest can be present in both pre- and postsynaptic compartments of a given cell (Nelson et al., 2006).

## *1.2 Detection of Probability of Neurotransmitter Release*

During a whole-cell voltage clamp recording experiment an alteration in synaptic efficacy can be detected as an increase or a decrease in the postsynaptic current triggered in response to a presynaptic action potential. However, this change in postsynaptic response may arise from multiple sources. There can be an increase in the number of active synapses or conversion of silent synapses to active ones. There may also be an alteration in the number or neurotransmitter sensitivities of postsynaptic receptors. All-or-none changes in synaptic strength as well as postsynaptic mechanisms typically scale synaptic responses without substantial alterations in their frequency-dependent characteristics. In contrast, an alteration in the probability of neurotransmitter release may result in the same outcome (i.e., an increase or a decrease in synaptic strength) but in turn can fundamentally alter synaptic responses to physiological trains of action potentials. A change in the probability of neurotransmitter release is the most common way neurotransmitter release kinetics are altered during synaptic plasticity. Such a change may arise from an alteration in the number of vesicles available for release or a modification in their fusion propensity in response to an action potential. These alterations can alter the rate of short-term synaptic plasticity. For instance, an increase in the probability of neurotransmitter release, in the absence of rapid replenishment of fused vesicles, may result in faster depletion of readily releasable vesicles and cause enhanced depression. A decrease in the release probability may make synapses less prone to vesicle depletion and tend to decrease the extent of depression. Therefore, systematic measurement of the extent of synaptic depression induced by a train of action potentials at moderate to high frequency stimulation may yield critical information in regard to neurotransmitter release kinetics (Zucker and Regehr, 2002).

In addition to measuring the kinetics of short-term synaptic depression or facilitation from a population of synapses, one can also obtain a more direct measure of neurotransmitter release probability if vesicle fusion from a single release site can be detected. Such a setting can be achieved during paired recordings between sparsely connected neurons (albeit in a labor-intensive way), or using the minimal stimulation method via application of low-intensity stimulation to activate a single release site. Under these conditions one can typically observe a substantial amount of neurotransmission failures arising from the probabilistic nature of vesicle fusion in single synapses. Over a large number of independent trials, a decrease in the number of failures would suggest an increase in probability of release or vice versa.

Use-dependent block of NMDA receptors by MK-801 provides an additional method to estimate the probability neurotransmitter release. MK-801 is a well-characterized blocker of NMDA receptors that penetrates open NMDA channel pores and impairs the ion flux through NMDA receptors (Huettner and Bean, 1988). This highly selective open channel block renders MK-801-dependent inhibition of NMDA currents extremely use-dependent. Therefore, the time course of MK-801 block of synaptically evoked NMDA currents is proportional to the activation of NMDA receptors by vesicle fusion events, providing a measure of the probability of neurotransmitter release (Hessler et al., 1993; Rosenmund et al., 1993). However, recent studies suggest that NMDA receptors in given synapse may not be saturated by the glutamate released in response to a single action potential (Mainen et al., 1999; Oertner et al., 2002). This observation may complicate direct estimation of the absolute release probability. Nevertheless, this method may still allow a relative measure by comparing different experimental conditions as long as there are no alterations in the number of NMDA receptors and in their sensitivities to glutamate between the conditions in question.

As indicated above, detection of evoked quantal responses (either through minimal stimulation or paired recordings) provides a suitable setting to determine neurotransmitter release probability and alterations in rate of vesicle fusion. However, in synapses with multiple release sites, such as the calyx of Held, isolation of evoked quantal responses is nearly impossible and truly quantal release is hard to detect except in the case of spontaneous neurotransmission. Therefore, under these conditions, the rate of synaptic vesicle fusion can be determined by deconvolution of synaptic currents with the quantal unitary current. This approach is valid only when the synaptic current can be assumed to result from the convolution between a quantal current and quantal release rates. This assumption is not valid in cases where postsynaptic mechanisms, such as receptor saturation and desensitization, alter quantal events and thus shape synaptic responses during repetitive stimulation (Neher and Sakaba, 2001).

Despite its wide use and versatility, electrophysiological approaches also possess several caveats. First, this form of detection is limited to fast neurotransmitters that are released rapidly and activate closely juxtaposed ligand-gated channels.

Using this approach it is not possible to examine release kinetics of neuromodulatory transmitters such as dopamine or cathecholamines, as they typically activate G-protein-coupled receptors and only generate slow electrical responses that are not necessarily linearly proportional to release kinetics. Second, in the case of ionotropic receptors the reliance of this physiological setting on receptor properties such as receptor desensitization and saturation brings in a major caveat that hinders direct estimation of presynaptic release properties. Third, synaptic events, which occur at distal sites as well as on thin dendritic branches, are hard to detect due to dendritic filtering. Fourth, in most cases information is difficult to obtain from single release sites; therefore, experimental readout typically provides an analysis of a population of synapses. Lastly, electrical detection of neurotransmitter release is not inherently sensitive to use history of vesicles (i.e., synaptic vesicle recycling). However in cases when endocytosis is impaired (Delgado et al., 2000) or vesicle re-acidification is inhibited using blockers of vacuolar ATPase, electrophysiological readout of neurotransmitter release can provide information in regard to synaptic vesicle recycling (Ertunc et al., 2007).

### *1.3 Studying Synaptic Vesicle Recycling Using Electrophysiological Techniques*

In order to determine the reliance of neurotransmission on synaptic vesicle endocytosis and recycling, one needs to estimate the time point when exocytosed vesicles become re-available for release. This parameter can be estimated from the kinetic difference between the rate of FM dye destaining (see section 2.2) and the time course of neurotransmitter release from a set of synapses. The rationale behind these experiments stems from previous observations that during stimulation, FM dyes, especially the least hydrophobic FM dye FM2-10, can be cleared out of a fused vesicle within a second by departitioning into solution (Ryan, 1996; Klingauf et al., 1998; Kavalali et al., 1999; Pyle et al., 2000), or within milliseconds by lateral diffusion in the neuronal membrane (Zenisek et al., 2002). These time frames are typically faster than the rate of fusion pore closure and endocytic retrieval. Therefore, under most circumstances recycled vesicles would not contain significant amounts of FM dye that could be detected as further destaining, whereas the same vesicles would be refilled with neurotransmitter following endocytosis that could give rise to further synaptic responses. This difference between the two reporters of vesicle mobilization should result in a deviation between the kinetics of FM dye destaining and neurotransmitter release at the time when recycled vesicles start to be reused. The method has been successfully applied to estimate the kinetics of synaptic vesicle recycling in several preparations, including the frog neuromuscular junction and dissociated hippocampal cultures (Betz and Bewick, 1992, 1993; Sara et al., 2002). However, applicability of this technique relies on two basic assumptions. First, loss of FM dye from vesicles upon fusion should be unrestricted leading to negligible dye retention. There are several lines of evidence that this may not be the case especially for the more lipophillic dye FM1-43. Second, to estimate the time course of synaptic vesicle recycling from the kinetic comparison of dye release and postsynaptic electrical signals, we have to assume a linear relationship between the electrical and optical measures. This assumption may also be a problem under circumstances where relative occupancy of postsynaptic receptors changes during trains of action potentials (Ertunc et al., 2007). In the case of glutamatergic neurotransmission, this assumption can be tested by examining the extent of block in the presence of rapidly dissociating AMPA receptor antagonists kynurenate or γ-DGG. These compounds compete with the endogenous levels of glutamate (Diamond and Jahr, 1997), and therefore their effectiveness decreases as the amount of glutamate concentration in the synaptic cleft increases. Furthermore, FM dye release kinetics typically has a low time resolution ($\sim$1 s), which may lead to underestimation of rapid alterations in neurotransmitter release kinetics.

An alternative method to estimate the time course of synaptic vesicle recycling takes advantage of the naturally occurring macrolide antibiotics such as bafilomycin A1 and folimycin (also called concanamycin A). These compounds can reduce synaptic output by selectively decreasing the neurotransmitter refilling (Roseth et al., 1995; Fonnum et al., 1998). This effect is attributed to their blockade of v-ATPase, which acidifies the lumen of recently endocytosed vesicles and generates an electrical and pH gradient (Moriyama and Futai, 1990). The neurotransmitter refilling process is maintained by transporters coupled to this proton pump to create the necessary driving force (Drose and Altendorf, 1997). After inhibition of this vesicular proton pump, endocytosed vesicles cannot be reloaded with neurotransmitter and therefore, when reused, these vesicles cannot evoke synaptic responses for a second time. In this way, postsynaptic recordings from synapses treated with proton pump inhibitors only exhibit responses originating from previously unused vesicles within the recycling pool. Although bafilomycin has recently been used in studies that investigate the synaptic vesicle endocytosis with optical pH sensitive probes (Sankaranarayanan and Ryan, 2001), folimycin's affinity (Ki = 20 pM) for v-ATPase is 25-fold higher than that of bafilomycin (Ki = 500 pM). This property of folimycin requires lower concentrations of vehicle (DMSO), which may have effects of its own at higher concentrations. This approach recently enabled examination of the kinetics of vesicle recycling in synapses onto CA1 pyramidal neurons using electrophysiological measures, which allows measurements at a milliseconds time scale (Ertunc et al., 2007).

## *1.4 Detection of Neurotransmitter Secretion Through Amperometric Recordings*

As indicated in the previous sections, whole-cell recordings are widely used to detect neurotransmission in multiple settings; however, they possess some shortcomings, which hinder their applicability to all forms of neurotransmission. Whole-cell recordings can only be effective to detect fast neurotransmitters that are released rapidly and activate closely juxtaposed ligand-gated channels. Using this approach it is not possible to examine release kinetics of neuromodulatory transmitters such as dopamine, serotonin, or cathecholamines as they typically activate G-protein coupled receptors and only generate slow electrical responses that are not linearly proportional to release kinetics. Furthermore, in the case of ionotropic receptors the reliance of this physiological setting on receptor properties such as receptor desensitization and saturation hinders direct estimation of presynaptic release properties.

The chemical structures of catecholamine or monoamine transmitters make them susceptible to rapid oxidation. These oxidation current signals can be detected using carbon fiber electrodes held at DC voltage above the redox potential of these neurotransmitters. This method proves an accurate depiction of vesicular neurotransmitter content free of nonlinear properties of postsynaptic ionotropic receptors (Bruns et al., 2000). Furthermore, detailed analysis of individual fusion event profiles can provide information in regard to diffusion kinetics of neurotransmitters out of fused vesicles (Chow et al., 1992). For instance, initial application of this method to catecholamine secretion of adrenal chromaffin cells revealed prominent slowly rising "foot" signals preceding the actual rapid rise of neurotransmitter concentration after fusion (Chow et al., 1992). The characteristics of this foot signal have been routinely analyzed in multiple settings to examine mechanisms underlying fusion pore expansion (Jackson and Chapman, 2006). In addition to amperometry, fast cyclic voltametric methods can be used to identify the released substances. In fast cyclic voltametry, a periodic voltage pattern is applied to the carbon electrode to extract the characteristic redox signatures of oxidizable substances in a solution. This method can be useful in distinguishing individual components within a mixture of neurotransmitters in cases where there is co-release of multiple oxidizable neurotransmitters (Zhou et al., 2005).

## *1.5 Detection of Membrane Fusion via Presynaptic Capacitance Measurements*

Synaptic vesicles are membranous organelles, which release their neurotransmitter content after fusion with the plasma membrane. This process can be directly monitored via capacitance measurements. Alteration in total membrane capacitance can be used as a reporter of membrane fusion or endocytosis as long as the magnitude of the capacitance change is above the background noise. A critical requirement for employment of this method is direct electrical access to the secretory cellular component. This requirement significantly impairs the applicability of this method to small synapses of the central nervous system, which typically cover a 1 $\mu m^2$ surface area. Therefore, most capacitance measurements have been limited to secretion from cell bodies of mast cells, adrenal chromaffin cells, or large pituitary nerve terminals. Recent studies on a large presynaptic terminal, called the calyx of Held, located within the medial nucleus of the trapezoid body region of the brain stem, have employed capacitance measurements in examining the coupling between synaptic vesicle exocytosis and endocytosis (Sun and Wu, 2001). In the hippocampus, large mossy fiber terminals originating from dentate granule cells have also been studied using this method, revealing a large pool of recycling vesicles (Hallermann et al., 2003).

Despite its limited applicability to synaptic vesicle fusion, the capacitance measurement method, when available, provides several advantages due to its ability to report the kinetics of exocytosis and endocytosis as well as in some cases properties of fusion pore openings. Nevertheless, capacitance measurements also possess several shortcomings. First, measurement of fusion during repetitive stimulation is problematic, as rapid changes in membrane potential induce massive charge movements across the membrane and render capacitance measurements difficult to interpret. Therefore, the best setting to examine evoked release using this method takes advantage of light-induced $Ca^{2+}$ uncaging within the nerve terminal bypassing voltage-gated $Ca^{2+}$ influx rather than direct depolarization of the nerve terminal. Second, like every method, capacitance measurements are prone to artifacts. These artifacts may originate from multiple sources, including inaccurate estimation of the electrotonic properties of the nerve terminal, which is needed to extract capacitance values from phase changes in the output signal. In some cases fusion events not relevant to neurotransmitter release may complicate direct interpretation of the signals in terms of synaptic vesicle fusion. Some of these artifacts can be alleviated by appropriate control experiments such as testing capacitance changes in the absence of secretion using application of clostridial toxins that block membrane fusion. Recently, capacitance measurements using the cell-attached configuration of the patch clamp technique have been achieved on or near the active zones of the calyx of Held nerve terminals. This configuration has several advantages over traditional whole terminal recoding configuration due to electrical isolation of a small membrane area, thus limiting background noise (He et al., 2006). This setting allowed resolution of single-vesicle fusion events and their properties of endocytosis. However, establishing a tight giga-ohm seal on a release site may also modify the properties of vesicle fusion events by altering membrane tension or the presynaptic scaffolding matrix present in active release sites.

## 2 Fluorescent Visualization of Synaptic Vesicle Fusion and Recycling

### *2.1 Optical Detection of Neurotransmitter Release*

Despite their versatility, electrophysiological methods typically report neurotransmitter release from a population of synapses or release sites. In most cases, unitary quantal transmission can only be detected using recordings of spontaneous neurotransmission or under settings that activate a very small number of synapses. Moreover, these methods usually provide little or no spatial information about the origin of release. Optical experiments using fluorescent probes, on the other hand, report release at the level of individual synapses. These methods either monitor $Ca^{2+}$ influx through postsynaptic NMDA receptors or can directly monitor presynaptic properties using styryl dye imaging and GFP-tagged synaptic vesicle proteins.

NMDA receptor mediated $Ca^{2+}$ influx provides a valuable reporter of postsynaptic receptor activation in response to presynaptic glutamate release. NMDA receptors have a very high affinity for glutamate ($\sim 1\,\mu M$ range), thus they can detect even subtle fusion events that may not release sufficient glutamate to activate AMPA receptors. In addition, in response to a single quantal fusion event postsynaptic NMDA receptors are not saturated, which enables measurements of release with a large dynamic range despite their high affinity. Therefore, this approach has recently been quite informative for providing evidence for multivesicular release in central synapses (Oertner et al., 2002). $Ca^{2+}$ indicator dyes with adequate linear response range and affinity can be used to detect $Ca^{2+}$ signal in a single dendritic spine evoked by quantal release events. This method, by its very nature, is limited to glutamatergic synapses or synapses with reliable postsynaptic $Ca^{2+}$ selective channels (e.g., some subtypes of nicotinic acetylcholine receptors or purinergic receptors such as P2X receptors). Moreover, the properties of neurotransmitter release can only be inferred postsynaptically; therefore, the same caveats that apply to electrophysiological recordings in principle apply to this method as well. However, the ability to monitor neurotransmission at the level of individual synapses and availability of a wide variety of $Ca^{2+}$ indicators makes this setting a method of choice for investigation of synaptic activation in intact tissues such as brain slices using two-photon imaging.

Recently, in addition to postsynaptic $Ca^{2+}$ imaging, several fluorescence imaging modalities have been introduced which primarily report synaptic vesicle fusion and endocytosis. These methods either take advantage of styryl dyes that can be loaded into recycling vesicles or they utilize exogenous expression of synaptic vesicle proteins tagged with a pH-sensitive variant of GFP at regions facing the vesicle lumen. Styryl dyes, such as FM1-43 and FM2-10, are amphipathic compounds that label recycling synaptic vesicles in an activity-dependent manner. This optical approach is also useful for monitoring the kinetics of replenishment for the release-ready pool of vesicles in a synapse. Furthermore, uptake and release of these dyes provide a faithful readout for synaptic vesicle reuse during activity. GFP-tagged synaptic vesicle proteins such as synaptobrevin and synaptophysin, on the other hand, provide a molecularly specific way to examine surface exposure and re-internalization of these proteins during synaptic vesicle exo- and endocytosis.

## *2.2 Functional Analysis of Exocytosis and Vesicle Recycling Using Styryl Dyes*

In the last decade significant advances in our understanding of the dynamics of the synaptic vesicle cycle were made possible by the availability of new optical probes. The first set of probes was the styryl dye FM1-43 and its analogs (Betz et al., 1996).

FM1-43 is a partially lipophilic dye, which increases fluorescence 100-fold when it partitions into membranes; thus it is virtually invisible in aqueous solution. The chemical structure of FM1-43 makes it difficult for it to cross lipid membranes, thus limiting the fluorescence labeling to outer membranes and recycling vesicles. Using this methodology, one can detect activity-dependent vesicular dye uptake and release in synaptic connections in multiple preparations, including central neuronal cultures. Studies using FM dyes revealed that activity-induced dye uptake and release operates with high fidelity. Specifically, the extent of fluorescent dye endocytosed during a particular stimulation matches the magnitude of fluorescence loss due to exocytosis in response to the same stimulation. This observation suggests that synaptic vesicles recycle without intermixing with endosomal compartments and thus the fluorescence probes do not get diluted (Murthy and Stevens, 1998).

The properties of styryl dyes make them useful tools for revealing quantitative information on vesicular recycling within a presynaptic terminal (Figure 2). These dyes can be rapidly inserted into membranes within milliseconds; however, with the increasing lengths of their hydrocarbon chains, they have decreasing rates of dissociation from membranes (Ryan, 1996; Klingauf et al., 1998; Kavalali et al., 1999). Previous measurements have shown that FM1-43 departitions from lipid to aqueous environment with a time constant of 2.5s, whereas its analogs, FM2-10 and FM1-84, departition with time constants of 0.6s and 4.8s, respectively. These differential departitioning properties of FM dyes can be exploited to estimate the kinetics of endocytosis in hippocampal synapses (Ryan, 1996; Klingauf et al., 1998; Kavalali et al., 1999). Because of FM2-10's membrane dissociation rate, destaining experiments with the dye provide the closest estimates of vesicle mobilization and exocytosis. In contrast, the same experiments performed with FM1-43 may be vulnerable to dye trapping due to fast endocytosis. A significant proportion of endocytosis takes place following the cessation of stimulation, albeit on a slower time scale. This slow endocytosis may require formation of clathrin coats, followed by a recycling pathway where vesicles go through an endosomal intermediate structure and recover all of the necessary molecular components. Alternatively, recycling vesicles can be formed from membrane invaginations called "cisternae" (Richards et al., 2000) that develop adjacent to the active zone. Reformation of synaptic vesicles from these endosomal structures requires clathrin and clathrin-adaptor proteins that bind to specific synaptic vesicle membrane proteins. Recent studies in the frog neuromuscular junction have shown that a slow endocytosis pathway, characterized by large membrane invaginations formed in response to prior activity, operates at rest for up to at least 10 minutes or more after cessation of exocytosis and selectively traps FM1-43 but not FM2-10. The slowly endocytosed membrane can form synaptic vesicles, which can in turn mix with the reserve pool and become re-available for release with a delay. FM2-10 uptake, in contrast, occurs in rapidly endocytosing vesicles, which preferentially populate the readily releasable pool and give rise to the fast component of destaining.

In a typical experiment to estimate slow endocytosis, one can stimulate synaptic terminals with 10- to 30-Hz field stimulation for 30 to 60 seconds and rapidly perfuse FM1-43 onto the preparation (for 60 seconds and up to 10 minutes) at gradually

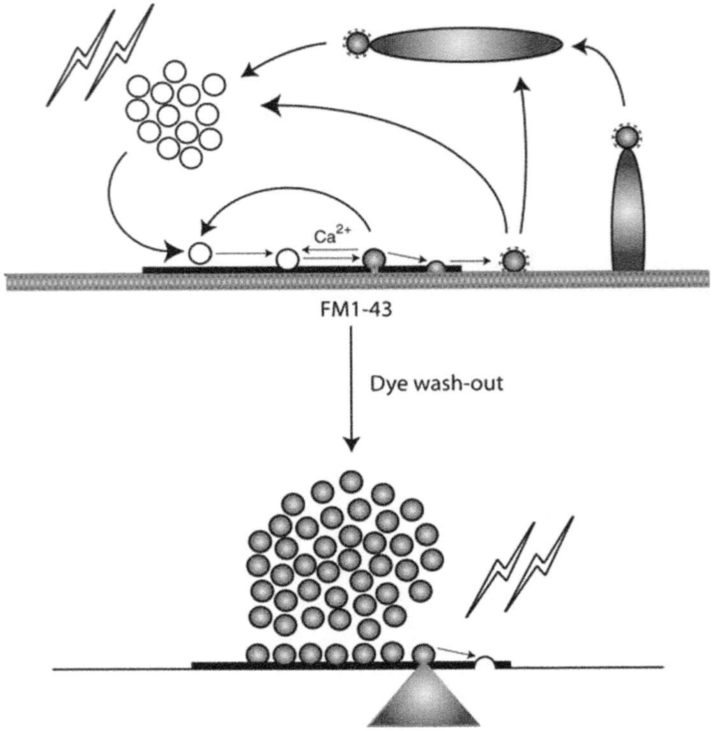


**Fig. 2** Optical monitoring of vesicle recycling using FM dyes. FM1-43 and its analogs are partially lipophilic dyes which fluorescence increases almost 100-fold when they partition into membranes. Therefore, they are virtually invisible in aqueous solution. The chemical structure of FM dyes makes it difficult for them to cross lipid membranes, thus limiting the fluorescence labeling to outer membranes and recycling vesicles. During a typical experiment, stimulation in the presence of extracellular dye leads to uptake of dye molecules into vesicles that have undergone exocytosis followed by endocytosis. Subsequent washout of extracellular dye uncovers fluorescent staining specific to actively recycling vesicles during the stimulation paradigm. Consequent stimulation results in fusion of dye-loaded vesicles and loss of fluorescence trapped in synapses. This fluorescence loss is due to departitioning of the dye into aqueous solution or lateral diffusion of dye in neuronal membrane.

increasing delays following the cessation of stimulation. The time course of decrease in the amount of fluorescence trapped within synapses provides an estimate of the time constant of slow endocytosis. This time constant has previously been shown to slow with increasing extent of depletion of vesicles in the neuromuscular junction (Wu and Betz, 1996). However, in these experiments, parallel control experiments are essential to account for the potential effect of changes in the level of exocytosis. Some evidence indicates that the number of fused vesicles can regulate the rate of endocytosis. Under the same setting more rapid endocytosis can be quantified when dye is applied concomitantly during stimulation and washed out rapidly afterward with no delay. Fast endocytosis is typically tightly coupled to exocytosis, and it is thought to be critical for maintenance of neurotransmission during sustained neurotransmission (Sara et al., 2002).

The measurement of the size of functionally distinct vesicle pools is a critical step in linking the structure and function of presynaptic terminals. Styryl dyes are also quite useful in determination of vesicle pool sizes at the level of individual synapses (Mozhayeva et al., 2002). In these measurements, it is important to distinguish changes in the $Ca^{2+}$ dependence of the release machinery from changes in numbers and maturational (release competent) states of synaptic vesicles. This distinction can be achieved by using saturating levels of $Ca^{2+}$-dependent stimulation or $Ca^{2+}$-independent stimulations, such as hypertonic sucrose application (typically +500 mOsm), to mobilize vesicles. The recycling kinetics of readily releasable pool (RRP) vesicles can be determined using uptake and release of the styryl dye FM2-10 in response to hypertonic stimulation. To determine the total recycling pool, uptake and release of styryl dyes can be induced by sustained depolarization achieved via application of elevated potassium solution. This strong stimulation can induce mobilization of virtually all vesicles capable of recycling within individual synapses. The comparison of the two fluorescence measurements (RRP versus the total recycling pool) can reveal critical information in regard to functional allocation of vesicles to a particular pool in given condition.

Detecting release of FM dyes rapidly after their uptake into the synapse provide a commonly used setting to quantify synaptic vesicle reuse in synapses (Ryan and Smith, 1995). In these experiments, FM dye is applied during and/or immediately after stimulation to label endocytosing vesicles. The dye-containing solution is then swiftly washed out to prevent further dye uptake. This dye application stains a number of vesicles, depending on the stimulation strength as well as the duration of dye presence. Application of a second stimulation at gradually delayed time points (typically 0 to 120 s) causes dye loss due to the re-availability of labeled vesicles. However, the proximity of stimulations inducing dye uptake and release makes it difficult to image actual dye loss during the second stimulation because of the high fluorescence background originating from residual dye. This shortcoming is usually circumvented by monitoring the remaining detectable staining at a later time point (typically ~5 minutes) after complete dye washout. The releasable fluorescence trapped in the synapse gradually decreases following stimulations applied at longer intervals or for longer durations after dye washout. The time course of this decrease in fluorescence provides an estimate of the time constant of vesicular reuse, which is the time it takes for an endocytosed vesicle to become release-ready again. This is a very sensitive method to detect vesicle reuse. As stated above, the major caveat is the background fluorescence, while usually not sufficient to restain vesicles during stimulation, it makes fast imaging nearly impossible. Because of this problem, the vesicle reuse kinetics is typically reconstructed from a population of synapses rather than from repeated measurements at individual presynaptic terminals.

### *2.3 Detection of Synaptic Vesicle Exocytosis and Endocytosis Using pHluorin-Tagged Synaptic Vesicle Proteins*

SynaptopHluorin (spH) is another fluorescent probe that is commonly used to monitor exocytosis and endocytosis of synaptic vesicles (Miesenbock et al., 1998; Sankaranarayanan and Ryan, 2000). Synaptophluorin is a fusion construct of the synaptic vesicle protein synaptobrevin (also called vesicle associated membrane protein, or "VAMP") with a pH-sensitive EGFP at its C-terminal (located in the synaptic vesicle lumen). Synaptic vesicle lumen normally has an acidic pH of approximately 5.5, at which spH fluorescence is quenched (Figure 3). When vesicles fuse, lumenal EGFP is exposed to the extracellular pH, which results in a marked increase in its fluorescence. Studies using spH have provided valuable information on the tight coupling between exocytosis and endocytosis. One advantage of the spH probe is the molecular specificity of the fluorescence signal, since the signal originates from a synaptic vesicle protein as opposed to nonspecific lipid-based fluorescence of FM dyes. Proton diffusion from a synaptic vesicle is not expected to be limited, and the time course of vesicle re-acidification is thought to be fast (see below), without imposing a significant delay in the fluorescence signal during spH endocytosis. Although most studies so far used synaptobrevin as a carrier for the pHluorin tag, recent work took advantage of other synaptic vesicle proteins as pHluorin attachments (Granseth et al., 2006; Voglmaier et al., 2006). Surprisingly, in these experiments, different synaptic vesicle proteins tagged with pHluorin gave different results with respect to their background fluorescence (i.e., surface exposure) and retrieval. For instance, both synaptophysin (a four-transmembrane domain synaptic vesicle protein) and the vesicular glutamate transporter reveal

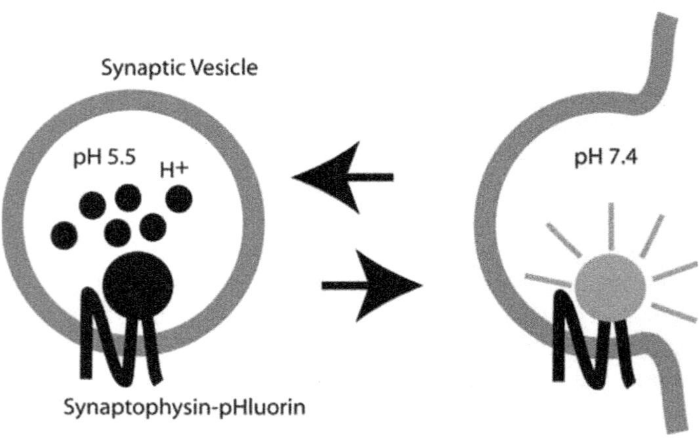


**Fig. 3** Synaptophysin-pHluorin is a fusion construct of the synaptic vesicle protein synaptophysin (a four-transmembrane domain synaptic vesicle protein) with a pH-sensitive EGFP at its C-terminal (located in the synaptic vesicle lumen). Synaptic vesicle lumen normally has an acidic pH of approximately 5.5 at which spH fluorescence is quenched. When vesicles fuse, lumenal EGFP is exposed to the extracellular pH, which results in a marked increase in its fluorescence. During endocytosis, pHluorin fluorescence is re-quenched as vesicle lumen becomes acidic.

very little surface exposure (thus fluorescence) in the absence of stimulation. Once stimulated they show a swift increase in fluorescence without significant lateral diffusion upon fusion, which was a salient feature of synaptopHluorin-based measurements. The dependence of the properties of pHluorin-based measurements on the identity of the tagged protein complicates a straightforward interpretation of these experiments in terms of synaptic vesicle recycling. Despite this caveat, these measurements reveal critical information in regard to copy numbers and specific cell biological attributes of synaptic vesicle proteins.

The vacuolar ATPase blockers discussed above are also useful in dissecting the kinetics of exocytosis and endocytosis in pHluorin-based measurements. In response to brief field stimulation, synaptophysin-pHluorin (or synaptopHluorin) fluorescence shows a swift increase and a gradual slow return to baseline (Granseth et al., 2006), presumably due to slow compensatory endocytosis. However, when the same experiment is repeated in the presence of a vacuolar ATPase blocker (such as folimycin) synaptophysin-pHluorin fluorescence shows a larger increase detectable within the first 500 ms (Ertunc et al., 2007). The difference between these two signals can be taken as a readout of endocytosis with the assumption that re-acidification of vesicles upon endocytosis proceeds rapidly and requenches internalized pHluorin. However, current estimates of synaptic vesicle re-acidification widely differ. According to some estimates re-acidification may take place rapidly within 500 milliseconds or less upon endocytosis (Gandhi and Stevens, 2003; Ertunc et al., 2007). In contrast, recent findings in hippocampal cultures suggest a relatively slow time course (~5 seconds) (Atluri and Ryan, 2006). These slower estimates have originated from measurements of synaptic vesicle re-acidification during delayed endocytosis following stimulation and, therefore, they may have overlooked a fast component of vesicle re-acidification. The vesicles that are endocytosed with a delay after stimulation may end up being slowly re-acidified due to slower reassembly of their protein complements. Whereas vesicles that are endocytosed rapidly during stimulation may also be re-acidified rapidly, as their protein components (including the vacuolar ATPase) are more likely to be intact.

## *2.4 Comparison of Styryl Dye Imaging and pHluorin-Based Visualization of Exo-Endocytosis*

While FM dyes and spH are commonly used to measure endocytosis and exocytosis, there are advantages and disadvantages to both. FM dye measurements have been instrumental in tracing vesicles during their cycle in the synapse. Using uptake and release of FM dyes one could reach conclusions on vesicle mixing with endosomal compartments (Murthy and Stevens, 1998) or on repeated use of the same vesicles during activity (Aravanis et al., 2003), which would be difficult to attain using probes like spH. FM dyes report lipid cycling; spH reports the fate of synaptic vesicle protein synaptobrevin (or others) upon fusion and endocytosis. SpH is a genetically expressed probe, which facilitates more reproducible repeated measurements on the same set of synapses (as it does not require dye uptake). However, measurement of endocytosis with spH is confounded by two factors. First, lateral diffusion of vesicular spH following fusion is commonly observed (Sankaranarayanan and Ryan, 2000; Li and Murthy, 2001). It is unclear whether reuptake of spH truly reflects endocytosis of vesicles or reclustering and re-internalization of synaptobrevin. It is also unknown whether endocytosis of lipid and protein components of synaptic vesicles are distinct processes or whether they are coupled. Second, the dependence of endocytosis measurements on vesicle re-acidification may delay the fluorescence signal after spH internalization. There are also concerns with using FM dyes. Typically, FM dye measurements suffer from nonspecificity. FM dyes stain the surface membrane indiscriminately, and thus they are less specific to synaptic vesicles. Moreover, changes in FM dye fluorescence typically occur slowly, in seconds, making it difficult to detect events in the millisecond time scale, which more closely match the time course of neurotransmission. In addition, after FM dye uptake, preparations need to be washed with dye-free solution for a significant period to reduce background fluorescence, which limits the rapid detection of synaptic vesicle re-availability after endocytosis. In summary, spH provides a molecularly specific probe to monitor the coupling between exocytosis and endocytosis, whereas FM dyes are better at tagging vesicles during their cycling within the synapses. Thus these two optical probes complement each other in monitoring distinct aspects of the synaptic vesicle cycle.

The experiments using these optical tools consistently supported the premise that synaptic vesicle exocytosis and endocytosis are tightly coupled processes. This largely kinetic coupling is also backed up by recent molecular evidence that proteins critical for exocytosis such as synaptotagmin and synaptobrevin, are also essential for triggering endocytosis (Poskanzer et al., 2003; Deak et al., 2004; Nicholson-Tomishima and Ryan, 2004).

## 3 Biochemical Approaches to Measuring Neurotransmitter Release

Synaptic neurotransmission in brain occurs mostly by exocytic release of vesicles filled with chemical substances (neurotransmitters) at presynaptic terminals. Thus, neurotransmitter release can be detected and studied by measuring efflux of neurotransmitters from synapses by biochemical methods. Various methods have been successfully employed to achieve that, including direct measurements of glutamate release by high-performance liquid chromatography of fluorescent derivatives or by enzyme-based continuous fluorescence assay, measurements of radioactive efflux from nerve terminals preloaded with radioactive neurotransmitters, or detection of neuropeptides by RIA or ELISA. Biochemical detection, however, lacks the sensitivity and temporal resolution afforded by electrophysiological and electrochemical approaches. As a result, it is not possible to measure individual synaptic events and apply quantal analysis to verify the vesicular nature of neurotransmitter release.

Carefully controlled experiments are required to ensure that most of the measured neurotransmitter efflux is vesicular and is not significantly contaminated by release from nonvesicular sources (e.g., leakage from the cytoplasmic pool, reversal of plasma membrane neurotransmitter transporters). Use of biochemical methods to measure neurotransmitter release also has a number of significant advantages: it is instrumentally simpler and less expensive; release is averaged over large number of terminals, thus eliminating synapse to synapse variability; detection does not depend on postsynaptic terminal function.

In principle, high-density primary neuronal cultures and acute or chronic brain slice preparations can be used to measure neurotransmitter release biochemically. However, by far the most commonly used preparation is synaptosomes, resealed broken-off nerve terminals obtained by homogenization of brain tissue in isotonic solutions of low ionic strength and differential centrifugation. Synaptosomes retain most of the cytoplasm, synaptic vesicles, mitochondria, and other organelles found in presynaptic terminals. When incubated in oxygenated balanced salt solutions containing glucose, synaptosomes remain metabolically active for several hours ex vivo, generating ATP and maintaining membrane potential and normal channel function (Nicholls, 2003). Depending on the method used, synaptosomes can be prepared quickly (under 30 min) in crude form from a large portion of the brain (e.g., neocortex) or in highly pure form from specific brain region (e.g., CA3 region of hippocampus) after several step gradient centrifugations. Unlike dissociated neuronal cultures and brain slices, synaptosomes are electrically inaccessible. To stimulate neurotransmitter release, synaptosomes are treated with substances modifying cation-selective channels, e.g., by applying high potassium solutions or potassium channel blockers (4-aminopyridine) to open voltage clamped calcium channels or by substances stimulating neurotransmitter release independent of channel function, e.g., hypertonic solutions (0.5 M sucrose), calcium ionophores (ionomycin), or presynaptic neurotoxins ($\alpha$-latrotoxin).

Glutamate release from synaptosomes can be measured by enzyme-based continuous fluorescence assay (Nicholls and Sihra, 1986; Nicholls et al., 1987). This experimental setup allows direct measurement of glutamate efflux from synaptosomes but cannot be easily adapted to measure release of other neurotransmitters. In addition, synaptosomes are subjected to stimulation for prolonged period of time, which results in nonvesicular glutamate release, thus complicating interpretation of data obtained by this method. A different approach is used in superfusion release assays developed to measure neurotransmitter release from synaptosomes preloaded with radioactive neurotransmitters (Lonart et al., 1993; Lonart and Sudhof, 1998). Synaptosomes are captured on glass fiber filters, thus avoiding mechanical damage caused by stirring and sedimentation (Figure 4). This method allows measurements of all major neurotransmitters, and in some cases can be used to monitor release of several neurotransmitters simultaneously (Khvotchev et al., 2000). Superfusion release setup enables rapid and transient stimulation of synaptosomes, which can be applied multiple times. Preloading of synaptosomes with radioactive neurotransmitter and assaying neurotransmitter release should be carried out relatively quickly to avoid metabolization of radioactive compounds.

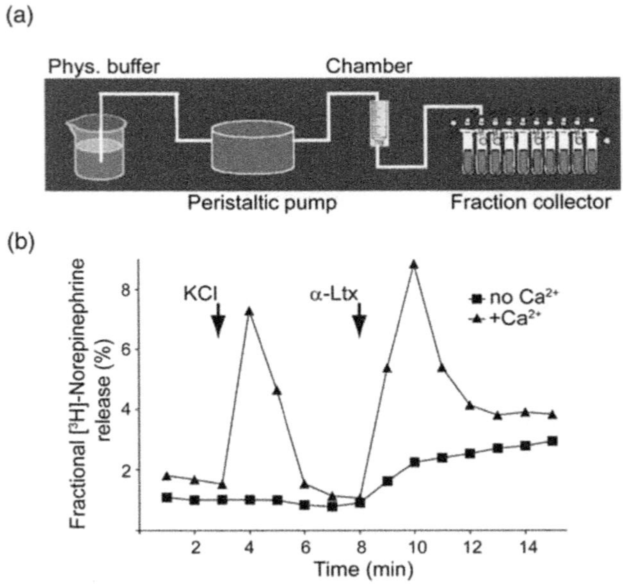


**Fig. 4** Superfusion neurotransmitter release assay in synaptosomes. (**a**) Schematic drawing of a superfusion setup. Synaptosomes are preloaded with radioactive neurotransmitter and captured on fiberglass filters in superfusion chambers under continuous superfusion with gassed physiological salt solution (e.g., Krebs bicarbonate buffer) using a peristaltic pump. After a 10- to 15-minute wash, neurotransmitter release is triggered by rapid switching of superfusion lines to a stimulating buffer (e.g., high-potassium solution). Superfusate is collected on a fraction collector, and radioactivity is measured by liquid scintillation. (**b**) Typical trace recording of tritium-labeled norepinephrine fractional release in rat cortical synaptosomes stimulated by high potassium and α-latrotoxin in the presence or absence of external calcium.

Significant advances in studying secretion were obtained with help of permeabilized cell preparations, e.g., cracked PC12 cells that allows direct biochemical access to intracellular release machinery (Klenchin et al., 1998; Martin and Grishanin, 2003). Similar approaches were relatively unsuccessful in neuronal preparations, possibly due to the complexity and small size of synaptic terminals. However, better-controlled permeabilization (using bacterial toxins) may be a promising approach to develop reliable and reproducible assays to measure neurotransmitter release in these preparations.

## References

Aravanis AM, Pyle JL, Tsien RW (2003) Single synaptic vesicles fusing transiently and successively without loss of identity. Nature 423: 643–7

Atluri PP, Ryan TA (2006) The kinetics of synaptic vesicle reacidification at hippocampal nerve terminals. J Neurosci 26: 2313–20

Betz WJ, Bewick GS (1992) Optical analysis of synaptic vesicle recycling at the frog neuromuscular junction. Science 255: 200–3
Betz WJ, Bewick GS (1993) Optical monitoring of transmitter release and synaptic vesicle recycling at the frog neuromuscular junction. J Physiol 460: 287–309
Betz WJ, Mao F, Smith CB (1996) Imaging exocytosis and endocytosis. Curr Opin Neurobiol 6: 365–71
Bruns D, Riedel D, Klingauf J, Jahn R (2000) Quantal release of serotonin. Neuron 28:205–20
Burrone J, O'Byrne M, Murthy VN (2002) Multiple forms of synaptic plasticity triggered by selective suppression of activity in individual neurons. Nature 420: 414–18
Chow RH, von Ruden L, Neher E (1992) Delay in vesicle fusion revealed by electrochemical monitoring of single secretory events in adrenal chromaffin cells. Nature 356: 60–3
Deak F, Shin OH, Kavalali ET, Südhof TC (2006) Structural determinants of synaptobrevin 2 function in synaptic vesicle fusion. J Neurosci 26: 6668–76
Deak F, Schoch S, Liu X, Sudhof TC, Kavalali ET (2004) Synaptobrevin is essential for fast synaptic-vesicle endocytosis. Nat Cell Biol 6:1102–8
Delgado R, Maureira C, Oliva C, Kidokoro Y, Labarca P (2000) Size of vesicle pools, rates of mobilization, and recycling at neuromuscular synapses of a Drosophila mutant, shibire. Neuron 28: 941–53
Diamond JS, Jahr CE (1997) Transporters buffer synaptically released glutamate on a submillisecond time scale. J Neurosci 17: 4672–87
Drose S, Altendorf K (1997) Bafilomycins and concanamycins as inhibitors of V-ATPases and P-ATPases. J Exp Biol 200 ( Pt 1): 1–8
Ertunc M, Sara Y, Chung C, Atasoy D, Virmani T, Kavalali ET (2007) Fast synaptic vesicle reuse slows the rate of synaptic depression in the CA1 region of hippocampus. J Neurosci 27: 341–54
Fonnum F, Fykse EM, Roseth S (1998) Uptake of glutamate into synaptic vesicles. Prog Brain Res 116: 87–101
Gandhi SP, Stevens CF (2003) Three modes of synaptic vesicular recycling revealed by single-vesicle imaging. Nature 423: 607–13
Granseth B, Odermatt B, Royle SJ, Lagnado L (2006) Clathrin-mediated endocytosis is the dominant mechanism of vesicle retrieval at hippocampal synapses. Neuron 51: 773–86
Hallermann S, Pawlu C, Jonas P, Heckmann M (2003) A large pool of releasable vesicles in a cortical glutamatergic synapse. Proc Natl Acad Sci U S A 100: 8975–80
Hamill OP, Marty A, Neher E, Sakmann B, Sigworth FJ (1981) Improved patch-clamp techniques for high-resolution current recording from cells and cell-free membrane patches. Pflugers Arch 391: 85–100
He L, Wu XS, Mohan R, Wu LG (2006) Two modes of fusion pore opening revealed by cell-attached recordings at a synapse. Nature 444: 102–5
Hessler NA, Shirke AM, Malinow R (1993) The probability of transmitter release at a mammalian central synapse. Nature 366: 569–72
Huettner JE, Bean BP (1988) Block of N-methyl-D-aspartate-activated current by the anticonvulsant MK-801: selective binding to open channels. Proc Natl Acad Sci U S A 85: 1307–11
Jackson MB, Chapman ER (2006) Fusion pores and fusion machines in Ca2+-triggered exocytosis. Annu Rev Biophys Biomol Struct 35: 135–60
Kavalali ET, Klingauf J, Tsien RW (1999) Properties of fast endocytosis at hippocampal synapses. Philos Trans R Soc Lond B Biol Sci 354: 337–46
Khvotchev M, Lonart G, Sudhof TC (2000) Role of calcium in neurotransmitter release evoked by alpha-latrotoxin or hypertonic sucrose. Neuroscience 101: 793–802
Klenchin VA, Kowalchyk JA, Martin TF (1998) Large dense-core vesicle exocytosis in PC12 cells. Methods 16: 204–8
Klingauf J, Kavalali ET, Tsien RW (1998) Kinetics and regulation of fast endocytosis at hippocampal synapses. Nature 394: 581–5
Li Z, Murthy VN (2001) Visualizing postendocytic traffic of synaptic vesicles at hippocampal synapses. Neuron 31: 593–605

Lonart G, Sudhof TC (1998) Region-specific phosphorylation of rabphilin in mossy fiber nerve terminals of the hippocampus. J Neurosci 18: 634–40
Lonart G, Cassels KL, Johnson KM (1993) Nitric oxide induces calcium-dependent [3H]dopamine release from striatal slices. J Neurosci Res 35: 192–8
Mainen ZF, Malinow R, Svoboda K (1999) Synaptic calcium transients in single spines indicate that NMDA receptors are not saturated. Nature 399: 151–5
Margrie TW, Brecht M, Sakmann B (2002) In vivo, low-resistance, whole-cell recordings from neurons in the anaesthetized and awake mammalian brain. Pflugers Arch 444: 491–8
Martin TF, Grishanin RN (2003) PC12 cells as a model for studies of regulated secretion in neuronal and endocrine cells. Methods Cell Biol 71: 267–86
Maximov A, Pang ZP, Tervo DG, Sudhof TC (2007) Monitoring synaptic transmission in primary neuronal cultures using local extracellular stimulation. J Neurosci Methods 161: 75–87
Miesenbock G, De Angelis DA, Rothman JE (1998) Visualizing secretion and synaptic transmission with pH-sensitive green fluorescent proteins. Nature 394: 192–5
Moriyama Y, Futai M (1990) H(+)-ATPase, a primary pump for accumulation of neurotransmitters, is a major constituent of brain synaptic vesicles. Biochem Biophys Res Commun 173: 443–8
Mozhayeva MG, Sara Y, Liu X, Kavalali ET (2002) Development of vesicle pools during maturation of hippocampal synapses. J Neurosci 22: 654–65
Murthy VN, Stevens CF (1998) Synaptic vesicles retain their identity through the endocytic cycle. Nature 392: 497–501
Neher E, Sakaba T (2001) Combining deconvolution and noise analysis for the estimation of transmitter release rates at the calyx of held. J Neurosci 21: 444–61
Nelson ED, Kavalali ET, Monteggia LM (2006) MeCP2-dependent transcriptional repression regulates excitatory neurotransmission. Curr Biol 16: 710–16
Nicholls DG (2003) Bioenergetics and transmitter release in the isolated nerve terminal. Neurochem Res 28: 1433–41
Nicholls DG, Sihra TS (1986) Synaptosomes possess an exocytotic pool of glutamate. Nature 321: 772–3
Nicholls DG, Sihra TS, Sanchez-Prieto J (1987) Calcium-dependent and -independent release of glutamate from synaptosomes monitored by continuous fluorometry. J Neurochem 49: 50–7
Nicholson-Tomishima K, Ryan TA (2004) Kinetic efficiency of endocytosis at mammalian CNS synapses requires synaptotagmin I. Proc Natl Acad Sci U S A 101: 16648–52
Oertner TG, Sabatini BL, Nimchinsky EA, Svoboda K (2002) Facilitation at single synapses probed with optical quantal analysis. Nat Neurosci 5: 657–64
Poskanzer KE, Marek KW, Sweeney ST, Davis GW (2003) Synaptotagmin I is necessary for compensatory synaptic vesicle endocytosis in vivo. Nature 426: 559–63
Pyle JL, Kavalali ET, Piedras-Renteria ES, Tsien RW (2000) Rapid reuse of readily releasable pool vesicles at hippocampal synapses. Neuron 28: 221–31
Richards DA, Guatimosim C, Betz WJ (2000) Two endocytic recycling routes selectively fill two vesicle pools in frog motor nerve terminals. Neuron 27: 551–9
Rosenmund C, Clements JD, Westbrook GL (1993) Nonuniform probability of glutamate release at a hippocampal synapse. Science 262: 754–7
Roseth S, Fykse EM, Fonnum F (1995) Uptake of L-glutamate into rat brain synaptic vesicles: effect of inhibitors that bind specifically to the glutamate transporter. J Neurochem 65: 96–103
Ryan TA (1996) Endocytosis at nerve terminals: timing is everything. Neuron 17: 1035–7
Ryan TA, Smith SJ (1995) Vesicle pool mobilization during action potential firing at hippocampal synapses. Neuron 14: 983–9
Sankaranarayanan S, Ryan TA (2000) Real-time measurements of vesicle-SNARE recycling in synapses of the central nervous system. Nat Cell Biol 2: 197–204
Sankaranarayanan S, Ryan TA (2001) Calcium accelerates endocytosis of vSNAREs at hippocampal synapses. Nat Neurosci 4: 129–36
Sara Y, Mozhayeva MG, Liu X, Kavalali ET (2002) Fast vesicle recycling supports neurotransmission during sustained stimulation at hippocampal synapses. J Neurosci 22: 1608–17

Schoch S, Deak F, Konigstorfer A, Mozhayeva M, Sara Y, Sudhof TC, Kavalali ET (2001) SNARE function analyzed in synaptobrevin/VAMP knockout mice. Science 294: 1117–22

Stuart GJ, Dodt HU, Sakmann B (1993) Patch-clamp recordings from the soma and dendrites of neurons in brain slices using infrared video microscopy. Pflugers Arch 423: 511–18

Sun JY, Wu LG (2001) Fast kinetics of exocytosis revealed by simultaneous measurements of presynaptic capacitance and postsynaptic currents at a central synapse. Neuron 30: 171–82

Tafoya LC, Mameli M, Miyashita T, Guzowski JF, Valenzuela CF, Wilson MC (2006) Expression and function of SNAP-25 as a universal SNARE component in GABAergic neurons. J Neurosci 26: 7826–38

Virmani T, Han W, Liu X, Sudhof TC, Kavalali ET (2003) Synaptotagmin 7 splice variants differentially regulate synaptic vesicle recycling. EMBO J 22: 5347–57

Voglmaier SM, Kam K, Yang H, Fortin DL, Hua Z, Nicoll RA, Edwards RH (2006) Distinct endocytic pathways control the rate and extent of synaptic vesicle protein recycling. Neuron 51: 71–84

Wu LG, Betz WJ (1996) Nerve activity but not intracellular calcium determines the time course of endocytosis at the frog neuromuscular junction. Neuron 17: 769–79

Zenisek D, Steyer JA, Feldman ME, Almers W (2002) A membrane marker leaves synaptic vesicles in milliseconds after exocytosis in retinal bipolar cells. Neuron 35: 1085–97

Zhou FM, Liang Y, Salas R, Zhang L, De Biasi M, Dani JA (2005) Corelease of dopamine and serotonin from striatal dopamine terminals. Neuron 46: 65–74

Zucker RS, Regehr WG (2002) Short-term synaptic plasticity. Annu Rev Physiol 64: 355–405

# Presynaptic Calcium Channels: Structure, Regulators, and Blockers

**Alexandra E. Kisilevsky and Gerald W. Zamponi(✉)**



**Abstract** The central and peripheral nervous systems express multiple types of ligand and voltage-gated calcium channels (VGCCs), each with specific physiological roles and pharmacological and electrophysiological properties. The members of the $Ca_v2$ calcium channel family are located predominantly at presynaptic nerve terminals, where they are responsible for controlling evoked neurotransmitter release. The activity of these channels is subject to modulation by a number of different means, including alternate splicing, ancillary subunit associations, peptide and small organic blockers, G-protein-coupled receptors (GPCRs), protein kinases, synaptic proteins, and calcium-binding proteins. These multiple and complex modes of calcium channel regulation allow neurons to maintain the specific, physiological window of cytoplasmic calcium concentrations which is required for optimal neurotransmission and proper synaptic function. Moreover, these varying means of channel regulation provide insight into potential therapeutic targets for the treatment of

Gerald W. Zamponi
Hotchkiss Brain Department of Physiology and Biophysics, University of Calgary, 3330 Hospital Dr. NW, Calgary T2N 4N1, Canada
Zamponi@ucalgary.ca

T.C. Südhof, K. Starke (eds.), *Pharmacology of Neurotransmitter Release. Handbook of Experimental Pharmacology 184.*

pathological conditions that arise from disturbances in calcium channel signaling. Indeed, considerable efforts are presently underway to identify and develop specific presynaptic calcium channel blockers that can be used as analgesics.

## 1 Subtypes and Physiological Functions of Calcium Channels

The entry of calcium ions into the cytosol mediates a wide range of cellular responses, including the activation of calcium dependent enzymes, gene transcription, and the initiation of calcium triggered membrane fusion – a key aspect of vesicular neurotransmitter release (Dolmetsch et al. 2001; Fields et al. 2005; Martin-Moutot et al. 1996; Reid et al. 2003; Sutton et al. 1999; Wheeler et al. 1994). Calcium may enter cells by a number of different mechanisms, including ligand and voltage-gated calcium channels (VGCCs). The nervous system expresses a number of different types of VGCCs, each with specific physiological functions and electrophysiological and pharmacological properties (Table 1). These include the L-type ($Ca_V1.1$-1.4), P/Q-type ($Ca_V2.1$), N-type ($Ca_V2.2$), R-type ($Ca_V2.3$), and T-type ($Ca_V3.1$-3.3) channels, which are broadly classified into low- and high-voltage activated channels (LVA and HVA channels, respectively) (Catterall 2000). LVA (a.k.a. T-type) calcium channels activate in response to small membrane depolarizations, display rapid gating kinetics, exhibit a small unitary conductance, and play a major role in neuronal pacemaker activity (Mangoni et al. 2006; Nelson et al. 2006). In contrast, HVA channels require larger membrane depolarizations in order to open and can be further categorized, based on their functional characteristics, into N-, P-, Q-, R-, and L-types. These HVA calcium channel subtypes are distinguished by their pharmacological profiles: L-type channels are sensitive to dihydropyridines (Fox et al. 1987), N-type channels are selectively inhibited by ω-conotoxins GVIA, MVIIA, and CVID (Adams et al. 1993; Feng et al. 2003; Olivera et al. 1984; Reynolds et al. 1986), and P- and Q-type channels are differentially sensitive to ω-agatoxin IVA (Adams et al. 1993). R-type channels were originally identified based on their resistance to these pharmacological tools (Randall and Tsien 1995) but have since been shown to be potently inhibited by the spider toxin SNX-482 (Bourinet et al. 2001; Newcomb et al. 1998). These calcium channel subtypes also support distinct physiological functions and exhibit specific subcellular distributions (Table 1). For example, L-type calcium channels are expressed on cell bodies and support calcium-dependent gene transcription (Bading et al. 1993; Dolmetsch et al. 2001; Weick et al. 2003), while both N-type and P/Q-type channels are expressed at presynaptic nerve terminals, where they control evoked neurotransmitter release (Ishikawa et al. 2005; Westenbroek et al. 1992; Westenbroek et al. 1995). Recent evidence also implicates R-type calcium channels in the release of neurotransmitters at certain synapses (Kamp et al. 2005). For the purposes of this chapter, we will focus predominantly on these presynaptic VGCC subtypes.

**Table 1** Comparison of the physiological, pharmacological, and electrophysiological characteristics of voltage-gated calcium channel subunit subtypes

| | Channel Family | Calcium Current | Subunit | (Sub)-Cellular Distribution | Physiological Function | Pharmacological Sensitivity | Knockout Phenotype |
|---|---|---|---|---|---|---|---|
| | $Ca_v1.1$ | L-type | $\alpha_{1S}$ | skeletal muscle | excitation-contraction coupling, gene transcription | dihydropyridines (DHPs) | lethal |
| | $Ca_v1.2$ | L-type | $\alpha_{1C}$ | cardiac muscle, endocrine cells | excitation-contraction coupling, hormone secretion | DHPs | lethal |
| | $Ca_v1.3$ | L-type | $\alpha_{1D}$ | cochlear neurons, endocrine cells | tonic neurotransmitter release, hormone secretion | DHPs | cardiac arrythmias, deaf |
| HVA | $Ca_v1.4$ | L-type | $\alpha_{1F}$ | retinal neurons, immune cells | tonic neurotransmitter release, T lymphocyte activation | DHPs | blind |
| | $Ca_v2.1$ | P/Q-type | $\alpha_{1A}$ | presynaptic nerve terminals | evoked neurotransmitter release | ω-agatoxin IVA, IIIA ω-grammotoxin SIVA | ataxic, absence seizures |
| | $Ca_v2.2$ | N-type | $\alpha_{1B}$ | presynaptic nerve terminals | evoked neurotransmitter release | ω-conotoxin GVIA, MVIIA and CVID, ω-grammotoxin SIVA, farnesol, peptidylamines | hyponociceptive, reduced anxiety, reduced withdrawal symptoms |
| | $Ca_v2.3$ | R-type | $\alpha_{1E}$ | dendrites, presynaptic nerve terminals | evoked neurotransmitter release, neuronal excitability | SNX-482 | seizure resistant, hyponociceptive |

(*Continued*)

**Table 1** Continued

| | Channel Family | Calcium Current | Subunit | (Sub)-Cellular Distribution | Physiological Function | Pharmacological Sensitivity | Knockout Phenotype |
|---|---|---|---|---|---|---|---|
| LVA | $Ca_v3.1$ | T-type | $\alpha_{1G}$ | neurons, cardiac tissue | pacemaker activity, burst firing, oscillatory behaviour, hormone secretion, smooth muscle contraction | Kurtoxin, ethosuximide, nickel (low sensitivity) | seizure resistant, increased visceral pain perception |
| | $Ca_v3.2$ | T-type | $\alpha_{1H}$ | neurons, cardiac tissue, endocrine cells, smooth muscle | | Kurtoxin, ethosuximide, nickel (high sensitivity) | compromised vascular function, reduced pain sensitivity |
| | $Ca_v3.3$ | T-type | $\alpha_{1I}$ | neurons | | Kurtoxin, nickel (low sensitivity) | – |
| ancillary subunits | common to all $\alpha_1$ subunits | | $\beta$ | common to $\alpha_1$ subunit distribution | membrane targeting, alteration of channel kinetics and current densities | – | lethal ($\beta_{1,2a}$), reduced neuronal function ($\beta_3$), ataxic and absence seizures ($\beta_4$) |
| | | | $\alpha_2$-$\delta$ | | | gabapentin, pregabalin | ataxic, absence seizures |

## 2 Molecular Structure of Voltage-Gated Calcium Channels

HVA calcium channels are formed by association of a multimeric protein complex that is comprised of $\alpha_1$, $\beta$, and $\alpha_2 - \delta$ subunits, with, as far as we know, a 1:1:1 stoichiometry (Figure 1) (Catterall 2000). In addition, although its presence remains controversial, some types of HVA calcium channels appear to contain an additional $\gamma$ subunit (Kang and Campbell 2003; Moss et al. 2002). The $\alpha_1$ subunit is the major pore-forming subunit and is by itself sufficient to form a functional channel. However, the ancillary subunits are important for proper membrane targeting, and alteration of the functional characteristics of $\alpha_1$ (Bichet et al. 2000; Stea et al. 1993; Walker and De Waard 1998; Zamponi et al. 1996). All $\alpha_1$ subunits share a common architecture of four major transmembrane domains (I through IV), connected by cytosolic linker regions and cytoplasmic N- and C-termini. Each major domain contains six membrane spanning helices (S1 through S6) and a re-entrant p-loop that is thought to contain the amino acids that line the pore and form the ion

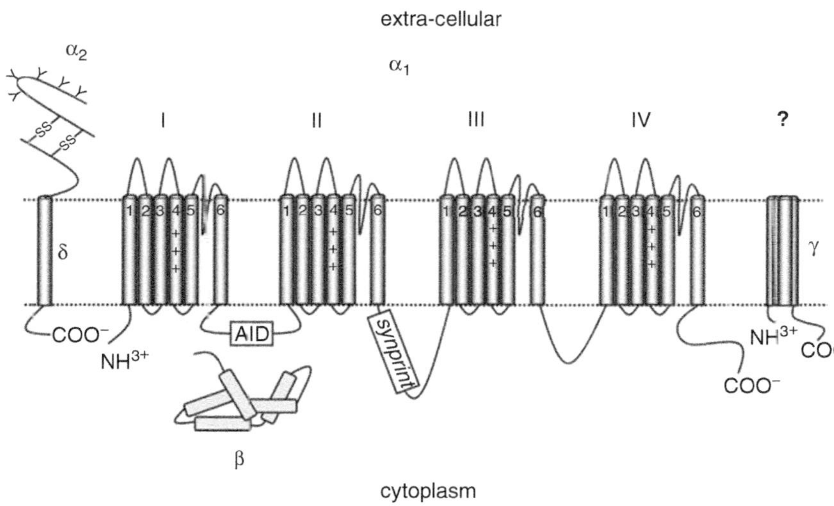


**Fig. 1** Topology of the voltage-gated calcium channel protein complex. The main pore-forming subunit, $\alpha_1$, is composed of four homologous domains (I through IV), connected by cytosolic linkers and flanked by cytosolic N- and C-termini. Each domain consists of six membrane-spanning helices (S1 through S6) and a p-loop between the S5 and S6 segments. The S4 segment contains several positively charged amino acids which act as the voltage sensor. The II-III linker contains a *synprint* region which is important for interactions with synaptic proteins essential for neurotransmitter release. The calcium channel $\beta$ subunit is a cytoplasmic protein that associates with the $\alpha_1$ subunit on the I–II linker alpha interaction domain (AID). The calcium channel $\alpha_2$-$\delta$ subunit is composed of the membrane-spanning $\delta$-subunit, connected via di-sulfide linkages to the heavily glycosylated $\alpha_2$ subunit. Finally, the calcium channel $\gamma$ subunit is comprised of four transmembrane helices with cytoplasmic N- and C-termini. The presence of this subunit in synaptic calcium channels is controversial, hence it is depicted with a question mark (?) overhead. The auxiliary calcium channel subunits are important for proper membrane targeting of the channel as well as for alteration of channel function.

selectivity filter (Figure 1) (Catterall 2000). Moreover, it is the calcium channel $\alpha_1$ subunit that defines the calcium channel subtype. To date, 10 different $\alpha_1$ subunits have been identified and shown to correspond to the native calcium channel isoforms when expressed functionally. They fall into three gene families, $Ca_v1$, $Ca_v2$, and $Ca_v3$ (Table 1). The $Ca_v1$ family encodes four different L-type channel isoforms (Koschak et al. 2003; Mikami et al. 1989; Tomlinson et al. 1993; Williams et al. 1992), the $Ca_v3$ family encodes three different T-type channel isoforms (Cribbs et al. 1998; Klockner et al. 1999; Lee et al. 1999b; McRory et al. 2001; Monteil et al. 2000; Perez-Reyes et al. 1998), and most relevant to this chapter, the $Ca_v2$ family includes the synaptic calcium channel subtypes. $Ca_v2.2$ and $Ca_v2.3$, respectively, correspond to N-type and R-type channels (Dubel et al. 1992; Soong et al. 1993; Williams et al. 1994), and different splice isoforms of $Ca_v2.1$ give rise to P- and Q-type channels (Bourinet et al. 1999). Each of the other types of calcium channel $\alpha_1$ subunits undergoes alternate splicing, often producing calcium channel isoforms with very distinct electrophysiological properties (Lipscombe et al. 2002), and in the case of N-type calcium channels, differential cellular expression patterns (Bell et al. 2004).

The mammalian brain expresses four different types of β subunits ($\beta_1$ through $\beta_4$), which can undergo alternate splicing (Dolphin 2003a; Richards et al. 2004). They share a similar structural arrangement with two highly conserved regions (C1 and C2) of high overall sequence homology (75% and 85%, respectively), separated and flanked by a total of three variable regions (V1 through V3) of much lower homology (35%-55%) (Walker and De Waard 1998). These subunits are cytoplasmic proteins, with the exception of $\beta_{2a}$, whose N-terminus contains a pair of cysteine residues that can be palmitoylated, thus leading to membrane insertion of some splice isoforms of this subunit (Qin et al. 1998). Each of the four β subunits physically binds to a region within the $\alpha_1$ subunit domain I-II linker, which is highly conserved among all HVA calcium channels and is termed the alpha interaction domain (AID) (Figure 1) (Pragnell et al. 1994). In addition, there have been reports of a second calcium channel β subunit interaction site within the C-terminus of $\alpha_1$ (Qin et al. 1997), however, this role of this site remains unclear. The functional effects of β subunit co-expression include changes in channel kinetics and, depending on the channel subtype, an increase in current densities (Bichet et al. 2000; Chien et al. 1995; Yasuda et al. 2004). The latter observation has been linked to the masking of an endoplasmic reticulum (ER) retention signal on the calcium channel $\alpha_1$ subunit (Bichet et al. 2000); however, our recently published work indicates that, at least in mammalian expression systems, the calcium channel $\alpha_1$ subunit can give rise to robust current activity even when expressed alone (Yasuda et al. 2004). Perhaps the most obvious effect of the different β subunits is their regulation of channel inactivation rates, with $\beta_{2a}$ dramatically slowing inactivation, and $\beta_3$ and $\beta_{1b}$ accelerating inactivation (Arikkath and Campbell 2003; Yasuda et al. 2004).

The calcium channel β subunit is the only calcium channel subunit for which there is crystal structure information. The core of this subunit is homologous to membrane-associated guanylate kinases (MAGUKs), with conserved, interacting SH3 and guanylate kinase (GK) domains (Takahashi et al. 2004). Residues in the

GK domain form a hydrophobic groove for high-affinity binding of HVA calcium channels. Binding of the $\beta$ subunit to the AID region is critically dependent on functional association of the SH3 and GK regions (Opatowsky et al. 2003; Van Petegem et al. 2004). The functional significance of the intramolecular SH3-GK interaction is supported by electrophysiological data showing that coexpression of separate cDNA constructs that encode the SH3 and GK domains, respectively, results in normal $\beta$ subunit function, whereas either one of the regions alone is incapable of regulating channel activity (Takahashi et al. 2005).

There are also four different types of calcium channel $\alpha_2$-$\delta$ subunits, each of which is encoded by a single gene that is postranslationally cleaved into $\alpha_2$ and $\delta$ peptides which are then reconnected via a disulfide bond (De Jongh et al. 1990; Klugbauer et al. 2003). The $\alpha_2$ subunit is extracellular and can be heavily glycosylated, whereas the $\delta$ subunit spans the plasma membrane with a single helix (Figure 1).

It has been known for a long time that the skeletal muscle L-type calcium channel complex also contains a $\gamma$ subunit – a four transmembrane helix with cytoplasmic N- and C-termini (Arikkath and Campbell 2003). Seven additional potential candidates for neuronal calcium channel $\gamma$ subunits have been identified; however, it is not clear if these are bona fide calcium channel subunits. Indeed, the first identified neuronal calcium channel $\gamma$ subunit "stargazin" has also been linked to AMPA receptor trafficking and pharmacology (Chen et al. 2000; Tomita et al. 2005). Hence, it remains unclear to what extent these subunits associate with the calcium channel complex in neurons.

Calcium channel $\alpha_1$ and $\beta$ subunits are subject to further heterogeneity due to alternate splicing, thus giving rise to a potentially vast number of different types of calcium channel molecules whose expression and distribution may be dynamically regulated to suit a particular physiological function. It is important to bear this in mind when comparing native calcium channel currents to data obtained in transient expression systems.

## 3 Consequences of Calcium Channel Gene Knockout

The differential physiological roles of individual calcium channel subunits are perhaps best highlighted by knockout (KO) mouse experiments (Table 1). KO animals lacking each of the major calcium channel $\alpha_1$ and $\beta$ subunits have been characterized in detail. KO of the $Ca_v1.1$ channel is lethal due to inability of mice to contract their diaphragm (Strube et al. 1996). KO of the $Ca_v1.2$ channel is embryonic lethal due to compromised cardiac function (Seisenberger et al. 2000). Mice lacking $Ca_v1.3$ exhibit cardiac arrhythmias and are deaf (Platzer et al. 2000), whereas mice lacking $Ca_v1.4$ are blind (Mansergh et al. 2005). These two calcium channel subtypes, respectively, control tonic glutamate release at ribbon synapses in cochlear hair cells and photoreceptors – this type of neurotransmitter release thus differs from the N- and P/Q-type channel mediated action potential evoked release, which is transient. Mice lacking $Ca_V3.1$ channels are resistant to certain types of pharmacologically induced seizures and may have an altered perception of visceral pain (Kim et al. 2001b), whereas mice lacking $Ca_V3.2$ channels are subject to compromised vascular function (Chen et al. 2003). $Ca_V2.3$ KO mice behave normally but show reduced pain responses, as well as resistance to certain types of epileptic seizures (Saegusa et al. 2000; Weiergraber et al. 2006). Although both N-type and P/Q-type channels control neurotransmitter release, they are clearly not created equal. Mice lacking the N-type ($Ca_V2.2$) channel are hyposensitive to pain and show reduced ethanol reward behavior (Hatakeyama et al. 2001; Kim et al. 2001a; Newton et al. 2004). Other than a slightly compromised control of blood pressure, these mice are behaviorally normal and viable. In contrast, mice lacking $Ca_V2.1$ channels are severely ataxic and show absence seizures (Jun et al. 1999). These mice usually die before reaching maturity, which is in stark contrast to the findings with $Ca_V2.2$ KO mice. The ataxic and epileptic phenotype of the $Ca_V2.1$ KO mice is consistent with data from several mouse lines with missense and frame shift mutations (Doyle et al. 1997; Wakamori et al. 1998), or polyglutamine expansions (Zhuchenko et al. 1997), in $Ca_V2.1$, as well as with several $Ca_V2.1$ human mutations found in patients with episodic ataxia type 2 (Denier et al. 2001; Friend et al. 1999; Guida et al. 2001; Ophoff et al. 1996) (see Section 4).

KO of the calcium channel $\beta_1$ subunit is lethal due to skeletal muscle paralysis that results from reduced membrane targeting of the $Ca_V1.1$ channel (Gregg et al. 1996; Strube et al. 1996). KO of the $\beta_{2a}$ subunit is also lethal due to compromised development of the cardiac vasculature (Ball et al. 2002). Mice lacking the $\beta_3$ subunit are viable but show reduced neuronal L-type and N-type channel activity (Namkung et al. 1998). In the context of synaptic physiology, KO of the calcium channel $\beta_4$ subunit is of interest. *Lethargic* is a mouse line with a mutation in the calcium channel $\beta_4$ subunit that leads to a premature stop codon and *de facto* elimination of this subunit. The consequences of this KO are strikingly similar to those observed with KO or mutated $Ca_V2.1$ animals (Burgess et al. 1997), hinting at a potential common effect of these subunits on synaptic function. Similar considerations apply to $\alpha_2$-$\delta$ subunits where a frameshift mutation leads to the elimination of this subunit in the *ducky*$^{2j}$ mouse strain – again, as with $Ca_V2.1$ KO animals, these mice are ataxic and have absence seizures (Brodbeck et al. 2002). These observations suggest that KO of either $\beta_4$ or $\alpha_2$-$\delta$ may lead to altered P/Q-type channel targeting and/or function.

The notion that N-type and P/Q-type channel knockout mice are phenotypically distinct suggests that these two calcium channel subtypes contribute to different aspects of synaptic function. Indeed, is has been reported that P/Q-type channels are more frequently linked to excitatory transmission, whereas N-type channels may contribute more often to inhibitory transmission (Potier et al. 1993). This is an important point to consider in the context of channel modulation as outlined below.

## 4 Calcium Channelopathies Involving P/Q-Type Calcium Channel $\alpha_1$ Subunits

Mutations in the $\alpha_1$ subunits of voltage gated calcium channels have been linked to a number of disorders. These include hypokalemic periodic paralysis ($Ca_V1.1$) (Lapie et al. 1997), Timothy syndrome ($Ca_V1.2$ and $Ca_V3.2$) (Splawski et al. 2005; Splawski et al. 2004; Splawski et al. 2006), congenital stationary night blindness ($Ca_V1.4$) (Bech-Hansen et al. 1998; Hoda et al. 2005), and various forms of absence epilepsy ($Ca_V3.2$) (Khosravani and Zamponi 2006), none of which we will elaborate on further. Mutations in P/Q-type calcium channels of both mice and humans are also associated with a number of pathological conditions. As mentioned earlier, point mutations in $Ca_V2.1$ have been associated with episodic ataxia type 2, as well as with familial hemiplegic migraine (FHM) in humans (Ophoff et al. 1996). When introduced into recombinant channels and expressed in either transient expression systems or neurons, FHM mutations tend to result in decreased channel function (Pietrobon and Striessnig 2003). In contrast, data from knock-in mice carrying certain FHM mutations are consistent with a gain of function mediated by these mutations. To date, this issue has not been resolved. Similarly, patients with spinocerebellar ataxia type 6 show an ataxic phenotype due to a polyglutamine expansion in the $Ca_V2.1$ C-terminus (Zhuchenko et al. 1997) but, as in the case of FHM mutations, it is unclear whether this is due to increased or decreased P/Q-type channel activity. However, in general terms, any alteration in P/Q-type channel function is expected to produce pronounced effects on neurotransmitter release and synaptic function. Thus, increases and decreases in P/Q-type channel activity may well produce similar alterations in neuronal network function depending on whether inhibitory or excitatory inputs are affected.

There are also a number of murine models of P/Q-type channelopathies. *Leaner, Tottering, rolling Nagoya*, and *Rocker* mice all carry mutations in the $Ca_V2.1$ calcium channel $\alpha_1$ subunit that result in either the substitution of individual amino acids or premature truncations of the channel protein. These mice tend to show an ataxic phenotype and, in some cases, absence seizures (Lorenzon et al. 1998; Mori et al. 2000; Noebels and Sidman 1979; Zwingman et al. 2001). In addition, several of these mutations cause cerebellar atrophy (Herrup and Wilczynski 1982). When introduced into recombinant channels, these mutations tend to reduce channel activity which is consistent with reduced P/Q-type currents in neurons isolated from these mouse lines.

To our knowledge, there are no identified N-type or R-type calcium channelopathies, which is consistent with the mild phenotype observed upon KO of the $Ca_V2.2$ or $Ca_V2.3$ genes.

# 5 Calcium Channel Pharmacology

## *5.1 Peptide Blockers*

As mentioned earlier, the pharmacological profile of a given calcium channel is a key means of identifying a particular calcium channel subtype. While small organic molecules such as dihydropyridines have been used to definitively characterize L-type calcium channels (Bean 1984), peptide toxins isolated from fish-hunting marine snails, spiders, scorpions, and snakes have been an invaluable source of selective inhibitors of synaptic calcium channels (Doering and Zamponi 2005). The archetypal N-type calcium channel inhibitor is ω-conotoxin GVIA, a peptide isolated from the fish-hunting cone snail *Conus geographus*. This toxin docks at the outer vestibule of the channel and physically occludes the channel pore, thus preventing calcium entry (Ellinor et al. 1994). Block occurs with high affinity and selectivity over other types of calcium channels and is poorly reversible upon washout (Feng et al. 2001b). ω-Conotoxins MVIIA and CVID are isolated, respectively, from the venoms of *Conus magus* and *Conus catus*, and they too selectively inhibit N-type calcium channels via a pore blocking mechanism (Feng et al. 2001b; Fox 1995; Monje et al. 1993; Olivera et al. 1984). Compared with MVIIA, CVID has even higher selectivity for N-type over P/Q-type calcium channels. Both MVIIA and CVID have been used clinically to treat pain, via intrathecal injection, underscoring the importance of N-type calcium channels in synaptic transmission in the dorsal horn of the spinal cord. Although multiple structural domains likely contribute to conotoxin block, efforts to map the contoxin interaction site on the N-type calcium channel indicate that the domain III S5-S6 region is critical (Ellinor et al. 1994; Feng et al. 2001b). In addition, the ancillary $\alpha_2$-δ subunits have been shown to reduce the affinity of N-type calcium channels for MVIIA and CVID (Mould et al. 2004).

Perhaps the best known peptide blocker of P/Q-type calcium channels is ω-agatoxin IVA, a peptide isolated from the venom of the North American funnel web spider *Agelenopsis aperta*. Unlike the conotoxins which are pore blockers, aga IVA functions as a gating inhibitor (McDonough et al. 1997). Application of aga IVA at concentrations as low as several nanomolars causes rapid inhibition of P-type channel activity and typically does not reverse upon washout (Adams et al. 1993; Turner et al. 1992). However, application of strong membrane depolarizations removes aga IVA block so that the channel effectively undergoes a large depolarizing shift in half activation potential – in essence, the toxin prevents activation gating but can be dislodged from its binding site by membrane depolarization (Mintz et al. 1992). The site of action of aga IVA appears to involve extracellular regions within domain IV of the $Ca_V2.1$ subunit (Winterfield and Swartz 2000). Indeed, alternate splicing of an asparagine-proline motif within the domain IV S3-S4 linker accounts for a 50-fold difference in aga IVA affinity between P-type and Q-type calcium channels (Bourinet et al. 1999). Other examples of peptide gating inhibitors include ω-grammotoxin SIVA (an N- and P/Q-type channel blocker) and SNX-482 (an R-type channel blocker), both isolated from different species of tarantula. Interestingly, the effects of aga IVA and grammotoxin on P/Q-type channel activity are additive, indicating that they bind to separate sites on the channel (McDonough et al. 2002).

The 78 amino acid ω-agatoxin IIIA blocks P/Q-type channels with nanomolar affinity, but also potently inhibits L-type and N-type channels (Mintz 1994; Mintz et al. 1991). This toxin is much larger than the ω-conotoxins (which are typically 25–30 amino acids in length), but nonetheless acts by a pore blocking mechanism and is thus distinct in action from aga IVA (Mintz 1994). Yet, pore block by this peptide is incomplete suggesting that toxin-bound channels can still conduct a small amount of calcium. Due to its nonselective nature, and its failure to elicit complete channel block, this toxin is not very useful experimentally.

Aga IVA and ω-conotoxin GVIA are standard tools in elucidating the roles of P/Q-type and N-type calcium channels in synaptic transmission. In many types of synapses, application of either toxin may mediate moderate inhibition of neurotransmitter release, whereas co-application of both blockers may almost abolish synaptic transmission due to the nonlinear dependence of synaptic release on intracellular calcium concentration. On a final note, we should add that there are many other species of cone snails and spiders that produce active toxins which selectivity inhibit specific calcium channel subtypes (for example, ω-conotoxins GVIB, GVIC, GVIIA, SVIA, SVIB), and it is likely that many more remain to be discovered (Olivera et al. 1994).

## 5.2 Small Organic Compounds Blocking N-type Channels

Besides peptide toxins, presynaptic calcium channels can also be inhibited by divalent metal ions such as cadmium, and by a number of different classes of small organic molecules (Doering and Zamponi 2005). While there are, to our knowledge, no selective small organic inhibitors of P/Q-type calcium channels, substantial efforts are being made to develop small organic N-type channel blockers for use as analgesics which can be administered orally. Such compounds include 4-benzyloxyaniline piperidine derivatives, fatty acid like compounds such as farnesol, and peptidylamines (Hu et al. 1999a; Hu et al. 1999b; Hu et al. 1999c; Roullet et al. 1999) . High throughput screening continues to identify novel compounds which block N-type channel activity and which show efficacy as analgesics. However, in many cases the degree of selectivity of these compounds for N-type channels over other calcium channel subtypes is much less than that observed for peptide toxins.

The anticonvulsant gabapentin (1-(aminomethyl) cyclohexaneacetic acid) and its second-generation cousin pregabalin are highly effective as analgesics in rats and humans (Cheng and Chiou 2006; Zareba 2005). These compounds are unique in that they bind to the calcium channel $\alpha_2$-$\delta$ subunit (Gee et al. 1996). Transgenic animals in which the putative gabapentin binding site on this subunit has been eliminated are resistant to the effects of gabapentin. It remains controversial as to whether the gabapentin-$\alpha_2$-$\delta$ interaction does indeed reduce voltage gated calcium currents, or whether the analgesic effects of gabapentin are mediated by a different mechanism.

# 6 Regulation by G Protein–Coupled Receptors

It is well established that activation of G protein–coupled receptors (GPCRs), in particular those that couple to the $G\alpha_{i/o}$ subunit, can result in potent inhibition of both N-type and P/Q-type calcium channels (Figure 2) (Beech et al. 1992; Bernheim et al. 1991; Caulfield et al. 1994; Dunlap and Fischbach 1981; Golard and Siegelbaum 1993; Ikeda 1992; Ikeda and Schofield 1989; Lipscombe et al. 1989; Mintz and Bean 1993; Shapiro and Hille 1993; Zhu and Ikeda 1993). The physio-

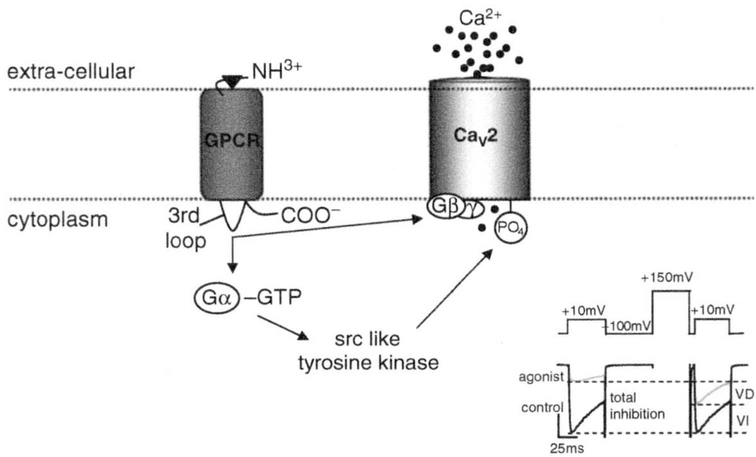


**Fig. 2** Voltage-dependent (VD) and voltage-independent (VI) G-protein mediated inhibition of synaptic calcium channels. In the absence of agonist, the heterotrimeric G-proteins, $G\alpha\beta\gamma$, bind GDP and associate with the third intracellular loop of seven trans-membrane helix spanning G-protein coupled receptors (GPCRs). Agonist (▼) activation of the receptor results in an exchange of GDP for GTP on the G protein $\alpha$-subunit. This exchange leads to dissociation of $G\alpha$-GTP and $G\beta\gamma$ subunits, which are subsequently free to act on downstream effectors such as enzymes or ion channels. $G\beta\gamma$, which remains associated with the plasma membrane, binds to N- and P/Q-type calcium channels on the $\alpha_1$ subunit, stabilizing the closed conformation of the channel and resulting in loss of current activity which is voltage-dependent (VD). Liberated $G\alpha$-GTP is cytoplasmic and activates enzymes such as adenylyl cyclase. Subsequent phosphorylation of synaptic calcium channels on the $\alpha_1$ subunit leads to voltage-independent (VI) inhibition of channel activity. **Inset:** In the presence of GPCR agonist (grey trace), a step to +10 mV results in inhibition of control current (black trace). VD and VI inhibition can be separated by applying a strong, depolarizing pre-pulse (i.e. +150 mV) immediately prior to the test pulse. This pre-pulse relieves $G\beta\gamma$-mediated VD inhibition, by physical removal of $G\beta\gamma$ from the channel, and remaining current inhibition can thus be attributed to VI inhibition.

logical significance of this modulation is underscored by the action of opiates in the pain pathway, where activation of μ-opioid receptors by morphine mediates analgesia in part by inhibition of N-type calcium channel activity (Altier and Zamponi 2004). GPCR-mediated inhibition may have both voltage-dependent (VD) and voltage-independent (VI) components (Figure 2). VD inhibition involves the binding of G protein βγ subunits to a cytoplasmic site on the calcium channel $\alpha_1$ subunit, thereby stabilizing the closed conformation of the channel (Figure 2) (Herlitze et al. 1996; Ikeda 1996). This inhibition is referred to as VD because it is favored at hyperpolarized potentials and can be relieved by strong membrane depolarizations or rapid trains of action potentials, thus allowing for activity-dependent dis-inhibition (Bean 1989; Hille 1994; Zamponi and Snutch 1998a, 1998b). The degree of VD inhibition depends on the calcium channel subtype, with N-type channels typically undergoing a larger degree of inhibition than P/Q-types (Currie and Fox 1997), and on the G protein β subunit isoform (Arnot et al. 2000; Garcia et al. 1998; Ruiz-Velasco and Ikeda 2000; Zhou et al. 2000). There may also be a subtle G protein γ subunit dependent component to N-type channel regulation (Ruiz-Velasco and Ikeda 2000). Furthermore, distinct Gβ subunit structural determinants appear to underlie modulation of these two channel types (Doering et al. 2004; Ford et al. 1998; Mirshahi et al. 2002a; Mirshahi et al. 2002b; Tedford et al. 2006). With different types of GPCRs coupling to specific subsets of G protein βγ subunits, differing degrees of VD inhibition of both calcium channel subtypes can thus be elicited. In addition, the extent of VD inhibition of both N- and P/Q-type channels appears to be dependent on the type of β subunit that is present in the calcium channel complex (Barrett and Rittenhouse 2000; Bourinet et al. 1996; Canti et al. 2000; Feng et al. 2001a). While many of the molecular details governing this regulation have been described over the past decade and have been reviewed in recent literature (Dolphin 2003b) new and important details concerning the structural underpinnings of VD inhibition are still emerging. For example, although both the calcium channel domain I-II linker and N-terminal regions have long been implicated in G protein modulation, it is now known that an interaction between these two regions is necessary for VD modulation (Agler et al. 2005). It is not completely resolved as to whether G protein βγ subunits facilitate the interaction between the N-terminus and the I-II linker region, or whether this interaction is needed to allow Gβγ binding to the channel. It has also been recently demonstrated that the C-terminal region of N-type calcium channels serves to enhance the affinity of the channel for Gβγ (Li et al. 2004), suggesting that a complex interplay among multiple cytoplasmic regions of the channel underlies VD inhibition. The aforementioned calcium channel β subunit dependence of VD G protein modulation may be due to the close proximity of the calcium channel β subunit interaction site and putative G protein binding sites within the domain I-II linker region (De Waard et al. 1997; Zamponi et al. 1997). However, the underlying mechanism remains subject to investigation. Indeed, there are differing views as to whether G protein βγ subunits compete with the calcium channel β subunit for binding, and whether or not G protein interactions with the channel cause dissociation of the calcium channel β subunit from the channel complex. However, irrespective of these molecular details, it is clear that G protein βγ subunits interact with cytoplasmic regions of the calcium channel $\alpha_1$ subunit resulting in a reduction of N- and P/Q-type channel activity that culminates in reduced calcium entry into the cell and thus reduced neurotransmitter release.

VD G protein modulation can be fine tuned by activation of other intracellular signaling cascades. For example, the activation of protein kinase C (PKC), which regulates channel activity directly, also interferes with the ability of N-type channels to undergo VD inhibition (Barrett and Rittenhouse 2000; Swartz 1993; Swartz et al. 1993). Intriguingly, this effect is dependent on the G protein $\beta\gamma$ subunit subtype, so that only G$\beta_1$-mediated responses are antagonized (Cooper et al. 2000). This is due to a unique pair of asparagine residues localized to the N-terminal helix of the G$\beta_1$ subunit (Doering et al. 2004). The structural basis for this effect has been resolved, as it has been shown that a single threonine residue located in the putative G protein $\beta\gamma$ subunit interaction site is the key locus for PKC-mediated effects (Hamid et al. 1999). In essence, this cross talk allows the channel to integrate two separate second messenger pathways in order to modulate current activity in a kinase- and G$\beta$ subunit-specific manner. The complexity of VD G protein modulation may allow synapses to precisely regulate the amount of calcium entering cells, and thus to adjust synaptic output.

In addition to VD G$\beta\gamma$ mediated inhibition of N-type channel activity, GPCR-induced VI modulation has been described for N-type calcium channels (Figure 2) (Beech et al. 1992; Beedle et al. 2004; Bernheim et al. 1991; Delmas et al. 1998a; Delmas et al. 1998b; Kammermeier et al. 2000; Schiff et al. 2000; Shapiro et al. 1999; Surmeier et al. 1995). VI inhibition is insensitive to membrane depolarizations and occurs over a slower time course than VD inhibition, implicating intracellular signaling cascades such as tyrosine kinase activation in this process. Furthermore, for P/Q-type channels, a form of VI modulation has been shown to involve an association of the G protein $\alpha_o$ subunit with the $Ca_V2.1$ C-terminus (Kinoshita et al. 2001). More recently, two studies have revealed a novel form of VI inhibition of N-type calcium channels that involves physical removal of calcium channels from the plasma membrane. N-type calcium channels and nociceptin (or ORL1) receptors have been shown to form stable, physical signaling complexes (Beedle et al. 2004). Prolonged agonist application results in internalization of the entire receptor-channel complex and to its subsequent degradation in lysosomes (Altier et al. 2006). The resultant loss of channel numbers in the plasma membrane and reduced current densities are, in effect, a form of VI inhibition. Channel internalization has also been shown to occur in response to $GABA_B$ receptor activation, although by an entirely different molecular mechanism that involves tyrosine kinase activity and cytoskeletal elements (Tombler et al. 2006). Moreover, this effect spontaneously reverses with receptor desensitization. That said, as in the case of nociceptin receptor activation, numbers of channels in the plasma membrane are reduced. It is likely that other types of GPCRs show similar receptor-mediated channel trafficking, and that this mechanism may emerge as a novel means of regulating N-type channels and thus synaptic activity.

## 7 Regulation of Calcium Channels by Synaptic Proteins

The activity of N-type and P/Q-type calcium channels is also regulated by proteins that form part of the synaptic vesicle release machinery (Figure 3). These channel subtypes contain a specific synaptic protein interaction site (termed *synprint*) in the

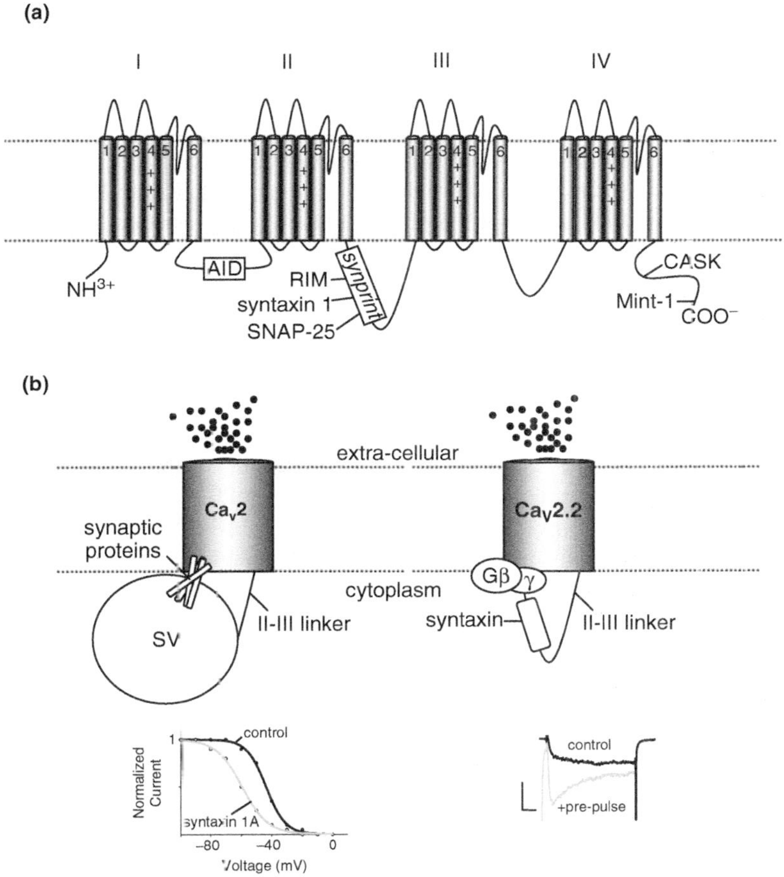


**Fig. 3** Regulation of calcium channels by synaptic proteins. (**a**) Putative interactions sites of synaptic and adaptor proteins with $Ca_v2$ channels are indicated on the calcium channel $\alpha_1$ subunit. RIM, syntaxin 1, and SNAP-25 are believed to interact with synaptic calcium channels on the II-III linker *synprint* region, while CASK and Mint-1 are thought to interact with the $\alpha_1$ C-terminus. (**b**) The physical associations between calcium channels and synaptic proteins are believed to colocalize synaptic vesicles (SVs) to sources of cellular calcium entry (left). In addition, several of these $Ca_v2$-synaptic protein interactions are believed to reduce channel activity directly by shifting half-inactivation channel kinetics to more hyperpolarized potentials (bottom, left). In the absence of GPCR activation, N-type calcium channel interactions with syntaxin 1A are speculated to result in tonic, Gβγ-mediated channel inhibition. This tonic, Gβγ-mediated inhibition is evident following its removal by a strong depolarizing pre-pulse (bottom, right).

intracellular loop linking domains II and III (Sheng et al. 1994). However, it is important to note that there are channel splice variants that lack this region and that, although invertebrates are capable of neurotransmission, no equivalent *synprint* region is found in invertebrate synaptic calcium channel homologs (Spafford and Zamponi 2003; Zamponi 2003). In the case of the N-type calcium channel, the *synprint* region is known to bind syntaxin 1, SNAP-25, and Rim (Coppola et al. 2001; Jarvis and Zamponi 2001b; Sheng et al. 1994; Sheng et al. 1997). The P/Q-type channel *synprint* region also interacts with syntaxin 1 and SNAP-25, however, unlike the N-type channel, it also binds synaptotagmin 1 (Charvin et al. 1997). The *synprint* motif in these channels appears to serve multiple functions. It has been suggested that the interactions between the *synprint* region and the synaptic proteins help to co-localize synaptic vesicles to sources of calcium entry (Figure 3ii) (Zamponi 2003). There is also evidence that the *synprint* region is involved in targeting calcium channels to presynaptic sites, as substitutions or deletions to this region lead to a loss of presynaptic localization of both N- and P/Q-type channels (Harkins et al. 2004; Mochida et al. 2003). However, it should be noted that other regions of the channels, such as the C-terminus by virtue of its interactions with the adaptor proteins Mint-1 and CASK, appear equally critical for synaptic targeting (Maximov and Bezprozvanny 2002; Spafford et al. 2003). Finally, the *synprint* region appears to serve as an important modulatory site that allows synaptic proteins to regulate channel activity *per se*. In both transient expression systems and neurons, binding of syntaxin 1 and SNAP-25 results in a hyperpolarizing shift in the half-inactivation potential of both P/Q-type and N-type calcium channels, thus reducing channel availability (Figure 3ii) (Bezprozvanny et al. 1995; Jarvis and Zamponi 2001a; Stanley 2003). This negative shift is ablated in the concomitant presence of both proteins (and in the case of P/Q-type channels, in the presence of synaptotagmin) (Zhong et al. 1999), is abolished in the presence of munc-18, and does not occur when syntaxin is in its "open" conformation. Because the association between syntaxin 1 and munc-18, as well as the conformational state of syntaxin, vary during the vesicle release cycle, channel availability can be dynamically regulated at various stages in the neurotransmitter release process. This in turn may help to ensure the appropriate amount of calcium entering the nerve terminal during neurotransmission.

Our group also identified a second action of syntaxin 1A (but not 1B) on N-type calcium channels. Co-expression of this protein with N-type calcium channels results in a tonic G protein–mediated inhibition of channel activity that does not involve receptor activation (Jarvis et al. 2000). This appears to be due to a syntaxin-mediated co-localization of the channel and Gβγ that ultimately culminates in tonic channel inhibition (Figure 3ii). Indeed, syntaxin 1A binds to Gβγ *in vitro* on a region that is distinct from the Gβγ site involved in interactions with the *synprint* motif. These data are consistent with the idea of a sandwich formed by the channel, syntaxin and Gβγ (Jarvis et al. 2002). Intriguingly, although syntaxin 1B can also bind to Gβγ *in vitro*, no tonic G protein inhibition of channel activity ensues (Lu et al. 2001). This suggests that, compared with syntaxin 1A, syntaxin 1B adopts a slightly different channel-bound conformation. The syntaxin-mediated effect on G protein inhibition appears to be much more robust than the effect on channel availability as it occurs when syntaxin is in the open conformation and is maintained upon coexpression of SNAP-25. Furthermore, although binding of syntaxin 1A to the *synprint* region is destabilized upon activation of PKC (Yokoyama et al. 2005; Yokoyama et al. 1997), the functional effects of PKC activation are limited to elimination of syntaxin-mediated channel availability and not to G protein inhibition (Jarvis and Zamponi 2001a, 2005). For completeness, we note that a similar SNARE protein dis-inhibition of P/Q-type calcium channels is observed for cyclin-dependent kinase 5, which has been shown to increase P/Q-type channel activity by blocking interactions of the channel with SNAP-25 (Tomizawa et al. 2002).

It is worth noting that the binding interaction between synaptic proteins and Gβγ may allow G proteins to regulate synaptic activity independently of N-type channel activity. Taken together, the interactions between calcium channels, G proteins, kinases, and synaptic proteins such as syntaxin provide for complex mechanisms that regulate N-type calcium channel activity both in expression systems and in neurons.

## 8 Regulation of Presynaptic Calcium Channel Activity by Protein Kinases

Many types of kinases have been shown to regulate calcium channel activity, and this topic in itself would be sufficient for a free-standing chapter. Here, we will give only a few select examples of the major pathways of this type of presynaptic calcium channel regulation.

PKC is an important regulator of presynaptic calcium channel activity. PKC activity has been shown to upregulate R-type calcium channels (Kamatchi et al. 2000; Shekter et al. 1997; Stea et al. 1995), as well as to increase N-type calcium channel activity (Hamid et al. 1999; Stea et al. 1995). This, in addition to the aforementioned antagonistic effects of PKC on syntaxin and VD G protein inhibition of N-type channels, produces an increase in N-type current activity in response to PKC activation. It has recently been shown that the PKC binding protein, enigma homolog (ENH) serves to co-localize PKC-ε and N-type channels to facilitate PKC-dependent phosphorylation of the $Ca_v2.2$ subunit (Maeno-Hikichi et al. 2003), indicating the formation of a macromolecular signaling complex. In contrast, there is little direct PKC regulation of P/Q-type calcium channels (Wu et al. 2002).

Protein kinase A (PKA) has been reported to increase P/Q-type channel activity in both neurons (Huang et al. 1998) and in transient expression systems (Fournier et al. 1993; Fukuda et al. 1996; Kaneko et al. 1998). In contrast, N-type calcium channels appear to be relatively insensitive to PKA. PKA has been shown to regulate P/Q-type calcium channel activity indirectly by interfering with phosphatidylinositol 4,5 biphosphate ($PIP_2$) mediated regulation (Wu et al. 2002). Under normal circumstances, $PIP_2$ slows rundown of P/Q-type channel activity and produces VD current inhibition. Activation of PKA selectively antagonizes the VD inhibition mediated by $PIP_2$, perhaps by preventing $PIP_2$ binding to the inhibitory site on the channel while sparing $PIP_2$ action on the site linked to current rundown. Once again, this cross talk underscores the ability of voltage-gated calcium channels to integrate multiple signaling pathways.

Calcium channel β subunits are also targets for phosphorylation events. It was recently reported that a phosphatidylinositol 3-kinase dependent elevation of phosphatidyl-inositol-4,5 triphosphate ($PIP_3$) is critical for the ability of the calcium channel $\beta_{2a}$ subunit to promote N-type calcium channel trafficking to the plasma membrane (Viard et al. 2004). Increased levels of $PIP_3$ result in activation of protein kinase B, which phosphorylates $\beta_{2a}$ and allows for trafficking of N-type channels to the plasma membrane. Along similar lines, it has been shown that Ras/extracellular signal–regulated kinase (ERK) phosphorylation of both the $\alpha_1$ and the $\beta_{1b}$ subunit mediates an upregulation of $Ca_v2.2$ current activity (Martin et al. 2006). Collectively, these findings clearly underscore the importance of ancillary calcium channel subunits as targets for second messenger regulation.

## 9 Feedback Regulation by Calcium Binding Proteins

Another important regulatory mechanism of calcium channel activity is via calcium binding proteins such as calmodulin (Figure 4) (Halling et al. 2005). It is well established that L-type calcium channels undergo robust calcium-dependent inactivation (CDI) – a reduction of current activity in response to elevated intracellular calcium levels which serves to protect cells form calcium overload (Imredy and Yue 1994). In the absence of calcium ions, calmodulin is pre-associated with the C-terminus of the channel at a highly conserved region containing pre-IQ and IQ motifs. Upon calcium entry into the cell, calmodulin binds $Ca^{2+}$, accelerating the inactivation kinetics of the channel. CDI in L-type channels requires the high-affinity calcium binding sites of calmodulin, thus being relatively resistant to standard calcium buffers (Soldatov 2003). Although CDI was once believed to be a unique property of L-type channels, it was subsequently shown that all types of HVA channels are subject to this process (Lee et al. 1999a; Liang et al. 2003). However, unlike L-type channels, R-, N-, and P/Q-type channels undergo CDI only when calcium is weakly buffered (Figure 4) (Liang et al. 2003). Indeed, the CDI process in these channels involves the low-affinity calcium binding sites of calmodulin. This key difference means that calcium entry through an individual L-type channel is sufficient to induce CDI, whereas CDI in non-L-type HVA channels requires global rises in intracellular calcium, increases generated by calcium entry though numerous channels.

P/Q-type channels undergo a second calcium-dependent process whereby current activity increases following a strong membrane depolarization or a train of action potentials. This increase is only observed in the presence of external calcium, is insensitive to calcium buffers, and requires the high-affinity calcium binding sites on calmodulin (Chaudhuri et al. 2004; DeMaria et al. 2001). This process, termed calcium-dependent facilitation (CDF), is not observed with N-type or R-type channels and is $Ca_v2.1$ channel splice isoform dependent (Chaudhuri et al. 2004). The

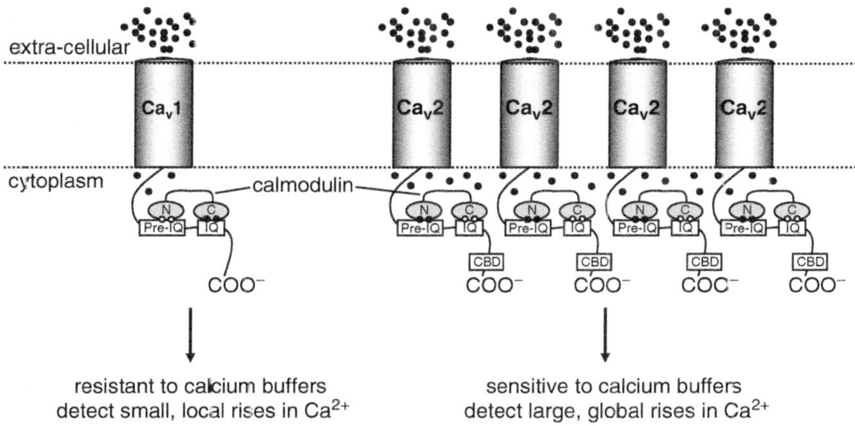


**Fig. 4** Modulation of calcium channels by calcium binding proteins. Calcium-dependent inactivation (CDI) in L-type channels (Ca$_V$1) involves binding of calcium ions (•) to the high-affinity binding sites located on the C-lobe of calmodulin. In contrast, CDI in synaptic calcium channels involves binding of calcium ions to the low-affinity binding sites located on the N-lobe of calmodulin. Both types of CDI result in a loss of current activity, in response to calcium, as the inactivation kinetics of the channel are accelerated. However, because of the difference in occupation of low-versus high-affinity calcium binding sites, between channel types, different forms of CDI exhibit different properties. In the case of L-type channels, CDI is resistant to calcium buffers and detects small, local rises in calcium generated by a single ion channel (left), whereas, in the case of synaptic channels, CDI is sensitive to standard calcium buffers and detects large, global rises in calcium which require calcium entry through numerous channels (right). P/Q-type channels are also subject to calcium-dependent facilitation (CDF), and to modulation by the calcium binding proteins, calcium binding protein 1 (CaBP1) and visin-like protein 2 (VILIP-2). Both of these proteins bind to the calcium binding domain (CBD) on the channel's C-terminus. In contrast to calmodulin, CaBP1 antagonizes calcium channel opening in a calcium-independent manner, and VILIP-2 slows the inactivation kinetics of the channel.

distinct roles of the high- and low-affinity sites of calmodulin in CDF and CDI may thus manifest themselves in an initial upregulation of P/Q-type channel activity, followed by negative feedback inhibition as intracellular calcium concentration is increased.

We now know of other types of calcium binding proteins that mediate similar feedback regulation of P/Q-type calcium channels, including calcium binding protein 1 (CaBP1) and visin-like protein 2 (VILIP-2). Both of these proteins bind to the calcium binding domain (CBD) within the channel C-terminus and, compared with calmodulin, differentially regulate channel activity. CaBP1 antagonizes channel opening in a calcium-independent manner (Lee et al. 2002), whereas VILIP-2 slows the inactivation kinetics of the channel (Lautermilch et al. 2005). Both proteins must undergo myristoylation in order to mediate their effects (Few et al. 2005). To our knowledge, this type of regulation has only been described for P/Q-type calcium channels.

Overall, the regulation of presynaptic calcium channels by different types of calcium binding proteins may provide for mechanisms by which neurons can fine tune the amount of calcium entering presynaptic nerve terminals, shift the relative contributions of N-type and P/Q-type channels to calcium entry, and thus regulate the amount of neurotransmitter that can be released from the synapse.

## 10 Summary

The entry of calcium into neurons via presynaptic calcium channels is a key step in evoked neurotransmitter release. Compromised calcium channel function can lead to severe neurological consequences, and yet the pharmacological inhibition of specific calcium channel subtypes can be beneficial in the treatment of conditions such as neuropathic pain. Because of the importance of these channels, neurons have evolved complex means for regulating calcium channel activity, including activation of second messenger pathways by G protein coupled receptors and feedback inhibition by calcium binding proteins. By these means, neurons are able to maintain the fine balance of cytoplasmic calcium levels that is required for optimal neurotransmitter release.

## References

Adams ME, Myers RA, Imperial JS, Olivera BM (1993) Toxityping rat brain calcium channels with omega-toxins from spider and cone snail venoms. Biochemistry 32: 12566–70

Agler HL, Evans J, Tay LH, Anderson MJ, Colecraft HM, Yue DT (2005) G protein-gated inhibitory module of N-type (ca(v)2.2) ca2+ channels. Neuron 46: 891–904

Altier C, Khosravani H, Evans RM, Hameed S, Peloquin JB, Vartian BA, Chen L, Beedle AM, Ferguson SS, Mezghrani A, Dubel SJ, Bourinet E, McRory JE, Zamponi GW (2006) ORL1 receptor-mediated internalization of N-type calcium channels. Nat Neurosci 9: 31–40

Altier C, Zamponi GW (2004) Targeting Ca2+ channels to treat pain: T-type versus N-type. Trends Pharmacol Sci 25: 465–70

Arikkath J, Campbell KP (2003) Auxiliary subunits: essential components of the voltage-gated calcium channel complex. Curr Opin Neurobiol 13: 298–307

Arnot MI, Stotz SC, Jarvis SE, Zamponi GW (2000) Differential modulation of N-type 1B and P/Q-type 1A calcium channels by different G protein subunit isoforms. J Physiol 527 Pt 2: 203–12

Bading H, Ginty DD, Greenberg ME (1993) Regulation of gene expression in hippocampal neurons by distinct calcium signaling pathways. Science 260: 181–6

Ball SL, Powers PA, Shin HS, Morgans CW, Peachey NS, Gregg RG (2002) Role of the beta(2) subunit of voltage-dependent calcium channels in the retinal outer plexiform layer. Invest Ophthalmol Vis Sci 43: 1595–603

Barrett CF, Rittenhouse AR (2000) Modulation of N-type calcium channel activity by G-proteins and protein kinase C. J Gen Physiol 115: 277–86

Bean BP (1984) Nitrendipine block of cardiac calcium channels: high-affinity binding to the inactivated state. Proc Natl Acad Sci U S A 81: 6388–92

Bean BP (1989) Neurotransmitter inhibition of neuronal calcium currents by changes in channel voltage dependence. Nature 340: 153–6

Bech-Hansen NT, Naylor MJ, Maybaum TA, Pearce WG, Koop B, Fishman GA, Mets M, Musarella MA, Boycott KM (1998) Loss-of-function mutations in a calcium-channel alpha1-subunit gene in Xp11.23 cause incomplete X-linked congenital stationary night blindness. Nat Genet 19: 264–7

Beech DJ, Bernheim L, Hille B (1992) Pertussis toxin and voltage dependence distinguish multiple pathways modulating calcium channels of rat sympathetic neurons. Neuron 8: 97–106

Beedle AM, McRory JE, Poirot O, Doering CJ, Altier C, Barrere C, Hamid J, Nargeot J, Bourinet E, Zamponi GW (2004) Agonist-independent modulation of N-type calcium channels by ORL1 receptors. Nat Neurosci 7: 118–25

Bell TJ, Thaler C, Castiglioni AJ, Helton TD, Lipscombe D (2004) Cell-specific alternative splicing increases calcium channel current density in the pain pathway. Neuron 41: 127–38

Bernheim L, Beech DJ, Hille B (1991) A diffusible second messenger mediates one of the pathways coupling receptors to calcium channels in rat sympathetic neurons. Neuron 6: 859–67

Bezprozvanny I, Scheller RH, Tsien RW (1995) Functional impact of syntaxin on gating of N-type and Q-type calcium channels. Nature 378: 623–6

Bichet D, Cornet V, Geib S, Carlier E, Volsen S, Hoshi T, Mori Y, De Waard M (2000) The I-II loop of the Ca2+ channel alpha1 subunit contains an endoplasmic reticulum retention signal antagonized by the beta subunit. Neuron 25: 177–90

Bourinet E, Soong TW, Stea A, Snutch TP (1996) Determinants of the G protein-dependent opioid modulation of neuronal calcium channels. Proc Natl Acad Sci U S A 93: 1486–91

Bourinet E, Soong TW, Sutton K, Slaymaker S, Mathews E, Monteil A, Zamponi GW, Nargeot J, Snutch TP (1999) Splicing of alpha 1A subunit gene generates phenotypic variants of P- and Q-type calcium channels. Nat Neurosci 2: 407–15

Bourinet E, Stotz SC, Spaetgens RL, Dayanithi G, Lemos J, Nargeot J, Zamponi GW (2001) Interaction of SNX482 with domains III and IV inhibits activation gating of alpha(1E) (Ca(V)2.3) calcium channels. Biophys J 81: 79–88

Brodbeck J, Davies A, Courtney JM, Meir A, Balaguero N, Canti C, Moss FJ, Page KM, Pratt WS, Hunt SP, Barclay J, Rees M, Dolphin AC (2002) The ducky mutation in Cacna2d2 results in altered Purkinje cell morphology and is associated with the expression of a truncated alpha 2 delta-2 protein with abnormal function. J Biol Chem 277: 7684–93

Burgess DL, Jones JM, Meisler MH, Noebels JL (1997) Mutation of the Ca2+ channel beta subunit gene Cchb4 is associated with ataxia and seizures in the lethargic (lh) mouse. Cell 88: 385–92

Canti C, Bogdanov Y, Dolphin AC (2000) Interaction between G proteins and accessory subunits in the regulation of 1E calcium channels in Xenopus oocytes. J Physiol 527 Pt 3: 419–32

Catterall WA (2000) Structure and regulation of voltage-gated Ca2+ channels. Annu Rev Cell Dev Biol 16: 521–55

Caulfield MP, Jones S, Vallis Y, Buckley NJ, Kim GD, Milligan G, Brown DA (1994) Muscarinic M-current inhibition via G alpha q/11 and alpha-adrenoceptor inhibition of Ca2+ current via G alpha o in rat sympathetic neurones. J Physiol 477 Pt 3: 415–22

Charvin N, L'Eveque C, Walker D, Berton F, Raymond C, Kataoka M, Shoji-Kasai Y, Takahashi M, De Waard M, Seagar MJ (1997) Direct interaction of the calcium sensor protein synaptotagmin I with a cytoplasmic domain of the alpha1A subunit of the P/Q-type calcium channel. Embo J 16: 4591–6

Chaudhuri D, Chang SY, DeMaria CD, Alvania RS, Soong TW, Yue DT (2004) Alternative splicing as a molecular switch for Ca2+/calmodulin-dependent facilitation of P/Q-type Ca2+ channels. J Neurosci 24: 6334–42

Chen CC, Lamping KG, Nuno DW, Barresi R, Prouty SJ, Lavoie JL, Cribbs LL, England SK, Sigmund CD, Weiss RM, Williamson RA, Hill JA, Campbell KP (2003) Abnormal coronary function in mice deficient in alpha1H T-type Ca2+ channels. Science 302: 1416–8

Chen L, Chetkovich DM, Petralia RS, Sweeney NT, Kawasaki Y, Wenthold RJ, Bredt DS, Nicoll RA (2000) Stargazin regulates synaptic targeting of AMPA receptors by two distinct mechanisms. Nature 408: 936–43

Cheng JK, Chiou LC (2006) Mechanisms of the antinociceptive action of gabapentin. J Pharmacol Sci 100: 471–86

Chien AJ, Zhao X, Shirokov RE, Puri TS, Chang CF, Sun D, Rios E, Hosey MM (1995) Roles of a membrane-localized beta subunit in the formation and targeting of functional L-type Ca2+ channels. J Biol Chem 270: 30036–44

Cooper CB, Arnot MI, Feng ZP, Jarvis SE, Hamid J, Zamponi GW (2000) Cross-talk between G-protein and protein kinase C modulation of N-type calcium channels is dependent on the G-protein beta subunit isoform. J Biol Chem 275: 40777–81

Coppola T, Magnin-Luthi S, Perret-Menoud V, Gattesco S, Schiavo G, Regazzi R (2001) Direct interaction of the Rab3 effector RIM with Ca2+ channels, SNAP-25, and synaptotagmin. J Biol Chem 276: 32756–62

Cribbs LL, Lee JH, Yang J, Satin J, Zhang Y, Daud A, Barclay J, Williamson MP, Fox M, Rees M, Perez-Reyes E (1998) Cloning and characterization of alpha1H from human heart, a member of the T-type Ca2+ channel gene family. Circ Res 83: 103–9

Currie KP, Fox AP (1997) Comparison of N- and P/Q-type voltage-gated calcium channel current inhibition. J Neurosci 17: 4570–9

De Jongh KS, Warner C, Catterall WA (1990) Subunits of purified calcium channels. Alpha 2 and delta are encoded by the same gene. J Biol Chem 265: 14738–41

De Waard M, Liu H, Walker D, Scott VE, Gurnett CA, Campbell KP (1997) Direct binding of G-protein betagamma complex to voltage-dependent calcium channels. Nature 385: 446–50

Delmas P, Abogadie FC, Dayrell M, Haley JE, Milligan G, Caulfield MP, Brown DA, Buckley NJ (1998a) G-proteins and G-protein subunits mediating cholinergic inhibition of N-type calcium currents in sympathetic neurons. Eur J Neurosci 10: 1654–66

Delmas P, Brown DA, Dayrell M, Abogadie FC, Caulfield MP, Buckley NJ (1998b) On the role of endogenous G-protein beta gamma subunits in N-type Ca2+ current inhibition by neurotransmitters in rat sympathetic neurones. J Physiol 506 Pt 2: 319–29

DeMaria CD, Soong TW, Alseikhan BA, Alvania RS, Yue DT (2001) Calmodulin bifurcates the local Ca2+ signal that modulates P/Q-type Ca2+ channels. Nature 411: 484–9

Denier C, Ducros A, Durr A, Eymard B, Chassande B, Tournier-Lasserve E (2001) Missense CACNA1A mutation causing episodic ataxia type 2. Arch Neurol 58: 292–5

Doering CJ, Kisilevsky AE, Feng ZP, Arnot MI, Peloquin J, Hamid J, Barr W, Nirdosh A, Simms B, Winkfein RJ, Zamponi GW (2004) A single Gbeta subunit locus controls cross-talk between protein kinase C and G protein regulation of N-type calcium channels. J Biol Chem 279: 29709–17

Doering CJ, Zamponi GW (2005) Molecular pharmacology of non-L-type calcium channels. Curr Pharm Des 11: 1887–98

Dolmetsch RE, Pajvani U, Fife K, Spotts JM, Greenberg ME (2001) Signaling to the nucleus by an L-type calcium channel-calmodulin complex through the MAP kinase pathway. Science 294: 333–9

Dolphin AC (2003a) Beta subunits of voltage-gated calcium channels. J Bioenerg Biomembr 35: 599–620

Dolphin AC (2003b) G protein modulation of voltage-gated calcium channels. Pharmacol Rev 55: 607–27

Doyle J, Ren X, Lennon G, Stubbs L (1997) Mutations in the Cacnl1a4 calcium channel gene are associated with seizures, cerebellar degeneration, and ataxia in tottering and leaner mutant mice. Mamm Genome 8: 113–20

Dubel SJ, Starr TV, Hell J, Ahlijanian MK, Enyeart JJ, Catterall WA, Snutch TP (1992) Molecular cloning of the alpha-1 subunit of an omega-conotoxin-sensitive calcium channel. Proc Natl Acad Sci U S A 89: 5058–62

Dunlap K, Fischbach GD (1981) Neurotransmitters decrease the calcium conductance activated by depolarization of embryonic chick sensory neurones. J Physiol 317: 519–35

Ellinor PT, Zhang JF, Horne WA, Tsien RW (1994) Structural determinants of the blockade of N-type calcium channels by a peptide neurotoxin. Nature 372: 272–5

Feng ZP, Arnot MI, Doering CJ, Zamponi GW (2001a) Calcium channel beta subunits differentially regulate the inhibition of N-type channels by individual Gbeta isoforms. J Biol Chem 276: 45051–8

Feng ZP, Doering CJ, Winkfein RJ, Beedle AM, Spafford JD, Zamponi GW (2003) Determinants of inhibition of transiently expressed voltage-gated calcium channels by omega-conotoxins GVIA and MVIIA. J Biol Chem 278: 20171–8

Feng ZP, Hamid J, Doering C, Bosey GM, Snutch TP, Zamponi GW (2001b) Residue Gly1326 of the N-type calcium channel alpha 1B subunit controls reversibility of omega-conotoxin GVIA and MVIIA block. J Biol Chem 276: 15728–35

Few AP, Lautermilch NJ, Westenbroek RE, Scheuer T, Catterall WA (2005) Differential regulation of CaV2.1 channels by calcium-binding protein 1 and visinin-like protein-2 requires N-terminal myristoylation. J Neurosci 25: 7071–80

Fields RD, Lee PR, Cohen JE (2005) Temporal integration of intracellular Ca2+ signaling networks in regulating gene expression by action potentials. Cell Calcium 37: 433–42

Ford CE, Skiba NP, Bae H, Daaka Y, Reuveny E, Shekter LR, Rosal R, Weng G, Yang CS, Iyengar R, Miller RJ, Jan LY, Lefkowitz RJ, Hamm HE (1998) Molecular basis for interactions of G protein betagamma subunits with effectors. Science 280: 1271–4

Fournier F, Bourinet E, Nargeot J, Charnet P (1993) Cyclic AMP-dependent regulation of P-type calcium channels expressed in Xenopus oocytes. Pflugers Arch 423: 173–80

Fox AP, Nowycky MC, Tsien RW (1987) Kinetic and pharmacological properties distinguishing three types of calcium currents in chick sensory neurones. J Physiol 394: 149–72

Fox JA (1995) Irreversible and reversible blockade of IMR32 calcium channel currents by synthetic MVIIA and iodinated MVIIC omega-conopeptides. Pflugers Arch 429: 873–5

Friend KL, Crimmins D, Phan TG, Sue CM, Colley A, Fung VS, Morris JG, Sutherland GR, Richards RI (1999) Detection of a novel missense mutation and second recurrent mutation in the CACNA1A gene in individuals with EA-2 and FHM. Hum Genet 105: 261–5

Fukuda K, Kaneko S, Yada N, Kikuwaka M, Akaike A, Satoh M (1996) Cyclic AMP-dependent modulation of N- and Q-type Ca2+ channels expressed in Xenopus oocytes. Neurosci Lett 217: 13–6

Garcia DE, Li B, Garcia-Ferreiro RE, Hernandez-Ochoa EO, Yan K, Gautam N, Catterall WA, Mackie K, Hille B (1998) G-protein beta-subunit specificity in the fast membrane-delimited inhibition of Ca2+ channels. J Neurosci 18: 9163–70

Gee NS, Brown JP, Dissanayake VU, Offord J, Thurlow R, Woodruff GN (1996) The novel anticonvulsant drug, gabapentin (Neurontin), binds to the alpha2delta subunit of a calcium channel. J Biol Chem 271: 5768–76

Golard A, Siegelbaum SA (1993) Kinetic basis for the voltage-dependent inhibition of N-type calcium current by somatostatin and norepinephrine in chick sympathetic neurons. J Neurosci 13: 3884–94

Gregg RG, Messing A, Strube C, Beurg M, Moss R, Behan M, Sukhareva M, Haynes S, Powell JA, Coronado R, Powers PA (1996) Absence of the beta subunit (cchb1) of the skeletal muscle dihydropyridine receptor alters expression of the alpha 1 subunit and eliminates excitation-contraction coupling. Proc Natl Acad Sci U S A 93: 13961–6

Guida S, Trettel F, Pagnutti S, Mantuano E, Tottene A, Veneziano L, Fellin T, Spadaro M, Stauderman K, Williams M, Volsen S, Ophoff R, Frants R, Jodice C, Frontali M, Pietrobon D (2001) Complete loss of P/Q calcium channel activity caused by a CACNA1A missense mutation carried by patients with episodic ataxia type 2. Am J Hum Genet 68: 759–64

Halling DB, Aracena-Parks P, Hamilton SL (2005) Regulation of voltage-gated Ca2+ channels by calmodulin. Sci STKE 2005: re15

Hamid J, Nelson D, Spaetgens R, Dubel SJ, Snutch TP, Zamponi GW (1999) Identification of an integration center for cross-talk between protein kinase C and G protein modulation of N-type calcium channels. J Biol Chem 274: 6195–202

Harkins AB, Cahill AL, Powers JF, Tischler AS, Fox AP (2004) Deletion of the synaptic protein interaction site of the N-type (CaV2.2) calcium channel inhibits secretion in mouse pheochromocytoma cells. Proc Natl Acad Sci U S A 101: 15219–24

Hatakeyama S, Wakamori M, Ino M, Miyamoto N, Takahashi E, Yoshinaga T, Sawada K, Imoto K, Tanaka I, Yoshizawa T, Nishizawa Y, Mori Y, Niidome T, Shoji S (2001) Differential nociceptive responses in mice lacking the alpha(1B) subunit of N-type Ca(2+) channels. Neuroreport 12: 2423–7

Herlitze S, Garcia DE, Mackie K, Hille B, Scheuer T, Catterall WA (1996) Modulation of Ca2+ channels by G-protein beta gamma subunits. Nature 380: 258–62

Herrup K, Wilczynski SL (1982) Cerebellar cell degeneration in the leaner mutant mouse. Neuroscience 7: 2185–96

Hille B (1994) Modulation of ion-channel function by G-protein-coupled receptors. Trends Neurosci 17: 531–6

Hoda JC, Zaghetto F, Koschak A, Striessnig J (2005) Congenital stationary night blindness type 2 mutations S229P, G369D, L1068P, and W1440X alter channel gating or functional expression of Ca(v)1.4 L-type Ca2+ channels. J Neurosci 25: 252–9

Hu LY, Ryder TR, Rafferty MF, Cody WL, Lotarski SM, Miljanich GP, Millerman E, Rock DM, Song Y, Stoehr SJ, Taylor CP, Weber ML, Szoke BG, Vartanian MG (1999a) N,N-dialkyl-dipeptidylamines as novel N-type calcium channel blockers. Bioorg Med Chem Lett 9: 907–12

Hu LY, Ryder TR, Rafferty MF, Dooley DJ, Geer JJ, Lotarski SM, Miljanich GP, Millerman E, Rock DM, Stoehr SJ, Szoke BG, Taylor CP, Vartanian MG (1999b) Structure-activity relationship of N-methyl-N-aralkyl-peptidylamines as novel N-type calcium channel blockers. Bioorg Med Chem Lett 9: 2151–6

Hu LY, Ryder TR, Rafferty MF, Feng MR, Lotarski SM, Rock DM, Sinz M, Stoehr SJ, Taylor CP, Weber ML, Bowersox SS, Miljanich GP, Millerman E, Wang YX, Szoke BG (1999c) Synthesis of a series of 4-benzyloxyaniline analogues as neuronal N-type calcium channel blockers with improved anticonvulsant and analgesic properties. J Med Chem 42: 4239–49

Huang CC, Wang SJ, Gean PW (1998) Selective enhancement of P-type calcium currents by isoproterenol in the rat amygdala. J Neurosci 18: 2276–82

Ikeda SR (1992) Prostaglandin modulation of Ca2+ channels in rat sympathetic neurones is mediated by guanine nucleotide binding proteins. J Physiol 458: 339–59

Ikeda SR (1996) Voltage-dependent modulation of N-type calcium channels by G-protein beta gamma subunits. Nature 380: 255–8

Ikeda SR, Schofield GG (1989) Somatostatin blocks a calcium current in rat sympathetic ganglion neurones. J Physiol 409: 221–40

Imredy JP, Yue DT (1994) Mechanism of Ca(2+)-sensitive inactivation of L-type Ca2+ channels. Neuron 12: 1301–18

Ishikawa T, Kaneko M, Shin HS, Takahashi T (2005) Presynaptic N-type and P/Q-type Ca2+ channels mediating synaptic transmission at the calyx of Held of mice. J Physiol 568: 199–209

Jarvis SE, Barr W, Feng ZP, Hamid J, Zamponi GW (2002) Molecular determinants of syntaxin 1 modulation of N-type calcium channels. J Biol Chem 277: 44399–407

Jarvis SE, Magga JM, Beedle AM, Braun JE, Zamponi GW (2000) G protein modulation of N-type calcium channels is facilitated by physical interactions between syntaxin 1A and Gbetagamma. J Biol Chem 275: 6388–94

Jarvis SE, Zamponi GW (2001a) Distinct molecular determinants govern syntaxin 1A-mediated inactivation and G-protein inhibition of N-type calcium channels. J Neurosci 21: 2939–48

Jarvis SE, Zamponi GW (2001b) Interactions between presynaptic Ca2+ channels, cytoplasmic messengers and proteins of the synaptic vesicle release complex. Trends Pharmacol Sci 22: 519–25

Jarvis SE, Zamponi GW (2005) Masters or slaves? Vesicle release machinery and the regulation of presynaptic calcium channels. Cell Calcium 37: 483–8

Jun K, Piedras-Renteria ES, Smith SM, Wheeler DB, Lee SB, Lee TG, Chin H, Adams ME, Scheller RH, Tsien RW, Shin HS (1999) Ablation of P/Q-type Ca(2+) channel currents, altered synaptic transmission, and progressive ataxia in mice lacking the alpha(1A)-subunit. Proc Natl Acad Sci U S A 96: 15245–50

Kamatchi GL, Tiwari SN, Durieux ME, Lynch C, 3rd (2000) Effects of volatile anesthetics on the direct and indirect protein kinase C-mediated enhancement of alpha1E-type Ca(2+) current in Xenopus oocytes. J Pharmacol Exp Ther 293: 360–9

Kammermeier PJ, Ruiz-Velasco V, Ikeda SR (2000) A voltage-independent calcium current inhibitory pathway activated by muscarinic agonists in rat sympathetic neurons requires both Galpha q/11 and Gbeta gamma. J Neurosci 20: 5623–9

Kamp MA, Krieger A, Henry M, Hescheler J, Weiergraber M, Schneider T (2005) Presynaptic 'Ca2.3-containing' E-type Ca channels share dual roles during neurotransmitter release. Eur J Neurosci 21: 1617–25

Kaneko S, Akaike A, Satoh M (1998) Differential regulation of N- and Q-type Ca2+ channels by cyclic nucleotides and G-proteins. Life Sci 62: 1543–7

Kang MG, Campbell KP (2003) Gamma subunit of voltage-activated calcium channels. J Biol Chem 278: 21315–8

Khosravani H, Zamponi GW (2006) Voltage-gated calcium channels and idiopathic generalized epilepsies. Physiol Rev 86: 941–66

Kim C, Jun K, Lee T, Kim SS, McEnery MW, Chin H, Kim HL, Park JM, Kim DK, Jung SJ, Kim J, Shin HS (2001a) Altered nociceptive response in mice deficient in the alpha(1B) subunit of the voltage-dependent calcium channel. Mol Cell Neurosci 18: 235–45

Kim D, Song I, Keum S, Lee T, Jeong MJ, Kim SS, McEnery MW, Shin HS (2001b) Lack of the burst firing of thalamocortical relay neurons and resistance to absence seizures in mice lacking alpha(1G) T-type Ca(2+) channels. Neuron 31: 35–45

Kinoshita M, Nukada T, Asano T, Mori Y, Akaike A, Satoh M, Kaneko S (2001) Binding of G alpha(o) N terminus is responsible for the voltage-resistant inhibition of alpha(1A) (P/Q-type, Ca(v)2.1) Ca(2+) channels. J Biol Chem 276: 28731–8

Klockner U, Lee JH, Cribbs LL, Daud A, Hescheler J, Pereverzev A, Perez-Reyes E, Schneider T (1999) Comparison of the Ca2+ currents induced by expression of three cloned alpha1 subunits, alpha1G, alpha1H and alpha1I, of low-voltage-activated T-type Ca2+ channels. Eur J Neurosci 11: 4171–8

Klugbauer N, Marais E, Hofmann F (2003) Calcium channel alpha2delta subunits: differential expression, function, and drug binding. J Bioenerg Biomembr 35: 639–47

Koschak A, Reimer D, Walter D, Hoda JC, Heinzle T, Grabner M, Striessnig J (2003) Cav1.4alpha1 subunits can form slowly inactivating dihydropyridine-sensitive L-type Ca2+ channels lacking Ca2+-dependent inactivation. J Neurosci 23: 6041–9

Lapie P, Lory P, Fontaine B (1997) Hypokalemic periodic paralysis: an autosomal dominant muscle disorder caused by mutations in a voltage-gated calcium channel. Neuromuscul Disord 7: 234–40

Lautermilch NJ, Few AP, Scheuer T, Catterall WA (2005) Modulation of CaV2.1 channels by the neuronal calcium-binding protein visinin-like protein-2. J Neurosci 25: 7062–70

Lee A, Westenbroek RE, Haeseleer F, Palczewski K, Scheuer T, Catterall WA (2002) Differential modulation of Ca(v)2.1 channels by calmodulin and Ca2+-binding protein 1. Nat Neurosci 5: 210–7

Lee A, Wong ST, Gallagher D, Li B, Storm DR, Scheuer T, Catterall WA (1999a) Ca2+/calmodulin binds to and modulates P/Q-type calcium channels. Nature 399: 155–9

Lee JH, Daud AN, Cribbs LL, Lacerda AE, Pereverzev A, Klockner U, Schneider T, Perez-Reyes E (1999b) Cloning and expression of a novel member of the low voltage-activated T-type calcium channel family. J Neurosci 19: 1912–21

Li B, Zhong H, Scheuer T, Catterall WA (2004) Functional role of a C-terminal Gbetagamma-binding domain of Ca(v)2.2 channels. Mol Pharmacol 66: 761–9

Liang H, DeMaria CD, Erickson MG, Mori MX, Alseikhan BA, Yue DT (2003) Unified mechanisms of Ca2+ regulation across the Ca2+ channel family. Neuron 39: 951–60

Lipscombe D, Kongsamut S, Tsien RW (1989) Alpha-adrenergic inhibition of sympathetic neurotransmitter release mediated by modulation of N-type calcium-channel gating. Nature 340: 639–42

Lipscombe D, Pan JQ, Gray AC (2002) Functional diversity in neuronal voltage-gated calcium channels by alternative splicing of Ca(v)alpha1. Mol Neurobiol 26: 21–44

Lorenzon NM, Lutz CM, Frankel WN, Beam KG (1998) Altered calcium channel currents in Purkinje cells of the neurological mutant mouse leaner. J Neurosci 18: 4482–9

Lu Q, AtKisson MS, Jarvis SE, Feng ZP, Zamponi GW, Dunlap K (2001) Syntaxin 1A supports voltage-dependent inhibition of alpha1B Ca2+ channels by Gbetagamma in chick sensory neurons. J Neurosci 21: 2949–57

Maeno-Hikichi Y, Chang S, Matsumura K, Lai M, Lin H, Nakagawa N, Kuroda S, Zhang JF (2003) A PKC epsilon-ENH-channel complex specifically modulates N-type Ca2+ channels. Nat Neurosci 6: 468–75

Mangoni ME, Couette B, Marger L, Bourinet E, Striessnig J, Nargeot J (2006) Voltage-dependent calcium channels and cardiac pacemaker activity: from ionic currents to genes. Prog Biophys Mol Biol 90: 38–63

Mansergh F, Orton NC, Vessey JP, Lalonde MR, Stell WK, Tremblay F, Barnes S, Rancourt DE, Bech-Hansen NT (2005) Mutation of the calcium channel gene Cacna1f disrupts calcium signaling, synaptic transmission and cellular organization in mouse retina. Hum Mol Genet 14: 3035–46

Martin-Moutot N, Charvin N, Leveque C, Sato K, Nishiki T, Kozaki S, Takahashi M, Seagar M (1996) Interaction of SNARE complexes with P/Q-type calcium channels in rat cerebellar synaptosomes. J Biol Chem 271: 6567–70

Martin SW, Butcher AJ, Berrow NS, Richards MW, Paddon RE, Turner DJ, Dolphin AC, Sihra TS, Fitzgerald EM (2006) Phosphorylation sites on calcium channel alpha1 and beta subunits regulate ERK-dependent modulation of neuronal N-type calcium channels. Cell Calcium 39: 275–92

Maximov A, Bezprozvanny I (2002) Synaptic targeting of N-type calcium channels in hippocampal neurons. J Neurosci 22: 6939–52

McDonough SI, Boland LM, Mintz IM, Bean BP (2002) Interactions among toxins that inhibit N-type and P-type calcium channels. J Gen Physiol 119: 313–28

McDonough SI, Mintz IM, Bean BP (1997) Alteration of P-type calcium channel gating by the spider toxin omega-Aga-IVA. Biophys J 72: 2117–28

McRory JE, Santi CM, Hamming KS, Mezeyova J, Sutton KG, Baillie DL, Stea A, Snutch TP (2001) Molecular and functional characterization of a family of rat brain T-type calcium channels. J Biol Chem 276: 3999–4011

Mikami A, Imoto K, Tanabe T, Niidome T, Mori Y, Takeshima H, Narumiya S, Numa S (1989) Primary structure and functional expression of the cardiac dihydropyridine-sensitive calcium channel. Nature 340: 230–3

Mintz IM (1994) Block of Ca channels in rat central neurons by the spider toxin omega-Aga-IIIA. J Neurosci 14: 2844–53

Mintz IM, Bean BP (1993) GABAB receptor inhibition of P-type Ca2+ channels in central neurons. Neuron 10: 889–98

Mintz IM, Venema VJ, Adams ME, Bean BP (1991) Inhibition of N- and L-type Ca2+ channels by the spider venom toxin omega-Aga-IIIA. Proc Natl Acad Sci U S A 88: 6628–31

Mintz IM, Venema VJ, Swiderek KM, Lee TD, Bean BP, Adams ME (1992) P-type calcium channels blocked by the spider toxin omega-Aga-IVA. Nature 355: 827–9

Mirshahi T, Mittal V, Zhang H, Linder ME, Logothetis DE (2002a) Distinct sites on G protein beta gamma subunits regulate different effector functions. J Biol Chem 277: 36345–50

Mirshahi T, Robillard L, Zhang H, Hebert TE, Logothetis DE (2002b) Gbeta residues that do not interact with Galpha underlie agonist-independent activity of K+ channels. J Biol Chem 277: 7348–55

Mochida S, Westenbroek RE, Yokoyama CT, Zhong H, Myers SJ, Scheuer T, Itoh K, Catterall WA (2003) Requirement for the synaptic protein interaction site for reconstitution of synaptic transmission by P/Q-type calcium channels. Proc Natl Acad Sci U S A 100: 2819–24

Monje VD, Haack JA, Naisbitt SR, Miljanich G, Ramachandran J, Nasdasdi L, Olivera BM, Hillyard DR, Gray WR (1993) A new Conus peptide ligand for Ca channel subtypes. Neuropharmacology 32: 1141–9

Monteil A, Chemin J, Bourinet E, Mennessier G, Lory P, Nargeot J (2000) Molecular and functional properties of the human alpha(1G) subunit that forms T-type calcium channels. J Biol Chem 275: 6090–100

Mori Y, Wakamori M, Oda S, Fletcher CF, Sekiguchi N, Mori E, Copeland NG, Jenkins NA, Matsushita K, Matsuyama Z, Imoto K (2000) Reduced voltage sensitivity of activation of P/Q-type Ca2+ channels is associated with the ataxic mouse mutation rolling Nagoya (tg(rol)). J Neurosci 20: 5654–62

Moss FJ, Viard P, Davies A, Bertaso F, Page KM, Graham A, Canti C, Plumpton M, Plumpton C, Clare JJ, Dolphin AC (2002) The novel product of a five-exon stargazin-related gene abolishes Ca(V)2.2 calcium channel expression. Embo J 21: 1514–23

Mould J, Yasuda T, Schroeder CI, Beedle AM, Doering CJ, Zamponi GW, Adams DJ, Lewis RJ (2004) The alpha2delta auxiliary subunit reduces affinity of omega-conotoxins for recombinant N-type (Cav2.2) calcium channels. J Biol Chem 279: 34705–14

Namkung Y, Smith SM, Lee SB, Skrypnyk NV, Kim HL, Chin H, Scheller RH, Tsien RW, Shin HS (1998) Targeted disruption of the Ca2+ channel beta3 subunit reduces N- and L-type Ca2+ channel activity and alters the voltage-dependent activation of P/Q-type Ca2+ channels in neurons. Proc Natl Acad Sci U S A 95: 12010–15

Nelson MT, Todorovic SM, Perez-Reyes E (2006) The role of T-type calcium channels in epilepsy and pain. Curr Pharm Des 12: 2189–97

Newcomb R, Szoke B, Palma A, Wang G, Chen X, Hopkins W, Cong R, Miller J, Urge L, Tarczy-Hornoch K, Loo JA, Dooley DJ, Nadasdi L, Tsien RW, Lemos J, Miljanich G (1998) Selective peptide antagonist of the class E calcium channel from the venom of the tarantula Hysterocrates gigas. Biochemistry 37: 15353–62

Newton PM, Orr CJ, Wallace MJ, Kim C, Shin HS, Messing RO (2004) Deletion of N-type calcium channels alters ethanol reward and reduces ethanol consumption in mice. J Neurosci 24: 9862–9

Noebels JL, Sidman RL (1979) Inherited epilepsy: spike-wave and focal motor seizures in the mutant mouse tottering. Science 204: 1334–6

Olivera BM, McIntosh JM, Cruz LJ, Luque FA, Gray WR (1984) Purification and sequence of a presynaptic peptide toxin from Conus geographus venom. Biochemistry 23: 5087–90

Olivera BM, Miljanich GP, Ramachandran J, Adams ME (1994) Calcium channel diversity and neurotransmitter release: the omega-conotoxins and omega-agatoxins. Annu Rev Biochem 63: 823–67

Opatowsky Y, Chomsky-Hecht O, Kang MG, Campbell KP, Hirsch JA (2003) The voltage-dependent calcium channel beta subunit contains two stable interacting domains. J Biol Chem 278: 52323–32

Ophoff RA, Terwindt GM, Vergouwe MN, van Eijk R, Oefner PJ, Hoffman SM, Lamerdin JE, Mohrenweiser HW, Bulman DE, Ferrari M, Haan J, Lindhout D, van Ommen GJ, Hofker MH, Ferrari MD, Frants RR (1996) Familial hemiplegic migraine and episodic ataxia type-2 are caused by mutations in the Ca2+ channel gene CACNL1A4. Cell 87: 543–52

Perez-Reyes E, Cribbs LL, Daud A, Lacerda AE, Barclay J, Williamson MP, Fox M, Rees M, Lee JH (1998) Molecular characterization of a neuronal low-voltage-activated T-type calcium channel. Nature 391: 896–900

Pietrobon D, Striessnig J (2003) Neurobiology of migraine. Nat Rev Neurosci 4: 386–98

Platzer J, Engel J, Schrott-Fischer A, Stephan K, Bova S, Chen H, Zheng H, Striessnig J (2000) Congenital deafness and sincatrial node dysfunction in mice lacking class D L-type Ca2+ channels. Cell 102: 89–97

Potier B, Dutar P, Lamour Y (1993) Different effects of omega-conotoxin GVIA at excitatory and inhibitory synapses in rat CA1 hippocampal neurons. Brain Res 616: 236–41

Pragnell M, De Waard M, Mori Y, Tanabe T, Snutch TP, Campbell KP (1994) Calcium channel beta-subunit binds to a conserved motif in the I-II cytoplasmic linker of the alpha 1-subunit. Nature 368: 67–70

Qin N, Platano D, Olcese R, Costantin JL, Stefani E, Birnbaumer L (1998) Unique regulatory properties of the type 2a Ca2+ channel beta subunit caused by palmitoylation. Proc Natl Acad Sci U S A 95: 4690–5

Qin N, Platano D, Olcese R, Stefani E, Birnbaumer L (1997) Direct interaction of gbetagamma with a C-terminal gbetagamma-binding domain of the Ca2+ channel alpha1 subunit is responsible for channel inhibition by G protein-coupled receptors. Proc Natl Acad Sci U S A 94: 8866–71

Randall A, Tsien RW (1995) Pharmacological dissection of multiple types of Ca2+ channel currents in rat cerebellar granule neurons. J Neurosci 15: 2995–3012

Reid CA, Bekkers JM, Clements JD (2003) Presynaptic Ca2+ channels: a functional patchwork. Trends Neurosci 26: 683–7

Reynolds IJ, Wagner JA, Snyder SH, Thayer SA, Olivera BM, Miller RJ (1986) Brain voltage-sensitive calcium channel subtypes differentiated by omega-conotoxin fraction GVIA. Proc Natl Acad Sci U S A 83: 8804–7

Richards MW, Butcher AJ, Dolphin AC (2004) Ca2+ channel beta-subunits: structural insights AID our understanding. Trends Pharmacol Sci 25: 626–32

Roullet JB, Spaetgens RL, Burlingame T, Feng ZP, Zamponi GW (1999) Modulation of neuronal voltage-gated calcium channels by farnesol. J Biol Chem 274: 25439–46

Ruiz-Velasco V, Ikeda SR (2000) Multiple G-protein betagamma combinations produce voltage-dependent inhibition of N-type calcium channels in rat superior cervical ganglion neurons. J Neurosci 20: 2183–91

Saegusa H, Kurihara T, Zong S, Minowa O, Kazuno A, Han W, Matsuda Y, Yamanaka H, Osanai M, Noda T, Tanabe T (2000) Altered pain responses in mice lacking alpha 1E subunit of the voltage-dependent Ca2+ channel. Proc Natl Acad Sci U S A 97: 6132–7

Schiff ML, Siderovski DP, Jordan JD, Brothers G, Snow B, De Vries L, Ortiz DF, Diverse-Pierluissi M (2000) Tyrosine-kinase-dependent recruitment of RGS12 to the N-type calcium channel. Nature 408: 723–7

Seisenberger C, Specht V, Welling A, Platzer J, Pfeifer A, Kuhbandner S, Striessnig J, Klugbauer N, Feil R, Hofmann F (2000) Functional embryonic cardiomyocytes after disruption of the L-type alpha1C (Cav1.2) calcium channel gene in the mouse. J Biol Chem 275: 39193–9

Shapiro MS, Hille B (1993) Substance P and somatostatin inhibit calcium channels in rat sympathetic neurons via different G protein pathways. Neuron 10: 11–20

Shapiro MS, Loose MD, Hamilton SE, Nathanson NM, Gomeza J, Wess J, Hille B (1999) Assignment of muscarinic receptor subtypes mediating G-protein modulation of Ca(2+) channels by using knockout mice. Proc Natl Acad Sci U S A 96: 10899–904

Shekter LR, Taussig R, Gillard SE, Miller RJ (1997) Regulation of human neuronal calcium channels by G protein betagamma subunits expressed in human embryonic kidney 293 cells. Mol Pharmacol 52: 282–91

Sheng ZH, Rettig J, Takahashi M, Catterall WA (1994) Identification of a syntaxin-binding site on N-type calcium channels. Neuron 13: 1303–13

Sheng ZH, Yokoyama CT, Catterall WA (1997) Interaction of the synprint site of N-type Ca2+ channels with the C2B domain of synaptotagmin I. Proc Natl Acad Sci U S A 94: 5405–10

Soldatov NM (2003) Ca2+ channel moving tail: link between Ca2+-induced inactivation and Ca2+ signal transduction. Trends Pharmacol Sci 24: 167–71

Soong TW, Stea A, Hodson CD, Dubel SJ, Vincent SR, Snutch TP (1993) Structure and functional expression of a member of the low voltage-activated calcium channel family. Science 260: 1133–6

Spafford JD, Munno DW, Van Nierop P, Feng ZP, Jarvis SE, Gallin WJ, Smit AB, Zamponi GW, Syed NI (2003) Calcium channel structural determinants of synaptic transmission between identified invertebrate neurons. J Biol Chem 278: 4258–67

Spafford JD, Zamponi GW (2003) Functional interactions between presynaptic calcium channels and the neurotransmitter release machinery. Curr Opin Neurobiol 13: 308–14

Splawski I, Timothy KW, Decher N, Kumar P, Sachse FB, Beggs AH, Sanguinetti MC, Keating MT (2005) Severe arrhythmia disorder caused by cardiac L-type calcium channel mutations. Proc Natl Acad Sci U S A 102: 8089–96; discussion 8086–8

Splawski I, Timothy KW, Sharpe LM, Decher N, Kumar P, Bloise R, Napolitano C, Schwartz PJ, Joseph RM, Condouris K, Tager-Flusberg H, Priori SG, Sanguinetti MC, Keating MT (2004) Ca(V)1.2 calcium channel dysfunction causes a multisystem disorder including arrhythmia and autism. Cell 119: 19–31

Splawski I, Yoo DS, Stotz SC, Cherry A, Clapham DE, Keating MT (2006) CACNA1H mutations in autism spectrum disorders. J Biol Chem 281: 22085–91

Stanley EF (2003) Syntaxin I modulation of presynaptic calcium channel inactivation revealed by botulinum toxin C1. Eur J Neurosci 17: 1303–5

Stea A, Dubel SJ, Pragnell M, Leonard JP, Campbell KP, Snutch TP (1993) A beta-subunit normalizes the electrophysiological properties of a cloned N-type Ca2+ channel alpha 1-subunit. Neuropharmacology 32: 1103–16

Stea A, Soong TW, Snutch TP (1995) Determinants of PKC-dependent modulation of a family of neuronal calcium channels. Neuron 15: 929–40

Strube C, Beurg M, Powers PA, Gregg RG, Coronado R (1996) Reduced Ca2+ current, charge movement, and absence of Ca2+ transients in skeletal muscle deficient in dihydropyridine receptor beta 1 subunit. Biophys J 71: 2531–43

Surmeier DJ, Bargas J, Hemmings HC, Jr., Nairn AC, Greengard P (1995) Modulation of calcium currents by a D1 dopaminergic protein kinase/phosphatase cascade in rat neostriatal neurons. Neuron 14: 385–97

Sutton KG, McRory JE, Guthrie H, Murphy TH, Snutch TP (1999) P/Q-type calcium channels mediate the activity-dependent feedback of syntaxin-1A. Nature 401: 800–4

Swartz KJ (1993) Modulation of Ca2+ channels by protein kinase C in rat central and peripheral neurons: disruption of G protein-mediated inhibition. Neuron 11: 305–20

Swartz KJ, Merritt A, Bean BP, Lovinger DM (1993) Protein kinase C modulates glutamate receptor inhibition of Ca2+ channels and synaptic transmission. Nature 361: 165–8

Takahashi SX, Miriyala J, Colecraft HM (2004) Membrane-associated guanylate kinase-like properties of beta-subunits required for modulation of voltage-dependent Ca2+ channels. Proc Natl Acad Sci U S A 101: 7193–8

Takahashi SX, Miriyala J, Tay LH, Yue DT, Colecraft HM (2005) A CaVbeta SH3/guanylate kinase domain interaction regulates multiple properties of voltage-gated Ca2+ channels. J Gen Physiol 126: 365–77

Tedford HW, Kisilevsky AE, Peloquin JB, Zamponi GW (2006) Scanning mutagenesis reveals a role for serine 189 of the heterotrimeric G-protein beta 1 subunit in the inhibition of N-type calcium channels. J Neurophysiol 96: 465–70

Tombler E, Cabanilla NJ, Carman P, Permaul N, Hall JJ, Richman RW, Lee J, Rodriguez J, Felsenfeld DP, Hennigan RF, Diverse-Pierluissi MA (2006) G protein-induced trafficking of voltage-dependent calcium channels. J Biol Chem 281: 1827–39

Tomita S, Adesnik H, Sekiguchi M, Zhang W, Wada K, Howe JR, Nicoll RA, Bredt DS (2005) Stargazin modulates AMPA receptor gating and trafficking by distinct domains. Nature 435: 1052–8

Tomizawa K, Ohta J, Matsushita M, Moriwaki A, Li ST, Takei K, Matsui H (2002) Cdk5/p35 regulates neurotransmitter release through phosphorylation and downregulation of P/Q-type voltage-dependent calcium channel activity. J Neurosci 22: 2590–7

Tomlinson WJ, Stea A, Bourinet E, Charnet P, Nargeot J, Snutch TP (1993) Functional properties of a neuronal class C L-type calcium channel. Neuropharmacology 32: 1117–26

Turner TJ, Adams ME, Dunlap K (1992) Calcium channels coupled to glutamate release identified by omega-Aga-IVA. Science 258: 310–3

Van Petegem F, Clark KA, Chatelain FC, Minor DL, Jr. (2004) Structure of a complex between a voltage-gated calcium channel beta-subunit and an alpha-subunit domain. Nature 429: 671–5

Viard P, Butcher AJ, Halet G, Davies A, Nurnberg B, Heblich F, Dolphin AC (2004) PI3K promotes voltage-dependent calcium channel trafficking to the plasma membrane. Nat Neurosci 7: 939–46

Wakamori M, Yamazaki K, Matsunodaira H, Teramoto T, Tanaka I, Niidome T, Sawada K, Nishizawa Y, Sekiguchi N, Mori E, Mori Y, Imoto K (1998) Single tottering mutations responsible for the neuropathic phenotype of the P-type calcium channel. J Biol Chem 273: 34857–67

Walker D, De Waard M (1998) Subunit interaction sites in voltage-dependent Ca2+ channels: role in channel function. Trends Neurosci 21: 148–54

Weick JP, Groth RD, Isaksen AL, Mermelstein PG (2003) Interactions with PDZ proteins are required for L-type calcium channels to activate cAMP response element-binding protein-dependent gene expression. J Neurosci 23: 3446–56

Weiergraber M, Kamp MA, Radhakrishnan K, Hescheler J, Schneider T (2006) The Ca(v)2.3 voltage-gated calcium channel in epileptogenesis-Shedding new light on an enigmatic channel. Neurosci Biobehav Rev

Westenbroek RE, Hell JW, Warner C, Dubel SJ, Snutch TP, Catterall WA (1992) Biochemical properties and subcellular distribution of an N-type calcium channel alpha 1 subunit. Neuron 9: 1099–115

Westenbroek RE, Sakurai T, Elliott EM, Hell JW, Starr TV, Snutch TP, Catterall WA (1995) Immunochemical identification and subcellular distribution of the alpha 1A subunits of brain calcium channels. J Neurosci 15: 6403–18

Wheeler DB, Randall A, Tsien RW (1994) Roles of N-type and Q-type Ca2+ channels in supporting hippocampal synaptic transmission. Science 264: 107–11

Williams ME, Feldman DH, McCue AF, Brenner R, Velicelebi G, Ellis SB, Harpold MM (1992) Structure and functional expression of alpha 1, alpha 2, and beta subunits of a novel human neuronal calcium channel subtype. Neuron 8: 71–84

Williams ME, Marubio LM, Deal CR, Hans M, Brust PF, Philipson LH, Miller RJ, Johnson EC, Harpold MM, Ellis SB (1994) Structure and functional characterization of neuronal alpha 1E calcium channel subtypes. J Biol Chem 269: 22347–57

Winterfield JR, Swartz KJ (2000) A hot spot for the interaction of gating modifier toxins with voltage-dependent ion channels. J Gen Physiol 116: 637–44

Wu L, Bauer CS, Zhen XG, Xie C, Yang J (2002) Dual regulation of voltage-gated calcium channels by PtdIns(4,5)P2. Nature 419: 947–52

Yasuda T, Chen L, Barr W, McRory JE, Lewis RJ, Adams DJ, Zamponi GW (2004) Auxiliary subunit regulation of high-voltage activated calcium channels expressed in mammalian cells. Eur J Neurosci 20: 1–13

Yokoyama CT, Myers SJ, Fu J, Mockus SM, Scheuer T, Catterall WA (2005) Mechanism of SNARE protein binding and regulation of Cav2 channels by phosphorylation of the synaptic protein interaction site. Mol Cell Neurosci 28: 1–17

Yokoyama CT, Sheng ZH, Catterall WA (1997) Phosphorylation of the synaptic protein interaction site on N-type calcium channels inhibits interactions with SNARE proteins. J Neurosci 17: 6929–38

Zamponi GW (2003) Regulation of presynaptic calcium channels by synaptic proteins. J Pharmacol Sci 92: 79–83

Zamponi GW, Bourinet E, Nelson D, Nargeot J, Snutch TP (1997) Crosstalk between G proteins and protein kinase C mediated by the calcium channel alpha1 subunit. Nature 385: 442–6

Zamponi GW, Snutch TP (1998a) Decay of prepulse facilitation of N type calcium channels during G protein inhibition is consistent with binding of a single Gbeta subunit. Proc Natl Acad Sci U S A 95: 4035–9

Zamponi GW, Snutch TP (1998b) Modulation of voltage-dependent calcium channels by G proteins. Curr Opin Neurobiol 8: 351–6

Zamponi GW, Soong TW, Bourinet E, Snutch TP (1996) Beta subunit coexpression and the alpha1 subunit domain I-II linker affect piperidine block of neuronal calcium channels. J Neurosci 16: 2430–43

Zareba G (2005) Pregabalin: a new agent for the treatment of neuropathic pain. Drugs Today (Barc) 41: 509–16

Zhong H, Yokoyama CT, Scheuer T, Catterall WA (1999) Reciprocal regulation of P/Q-type Ca2+ channels by SNAP-25, syntaxin and synaptotagmin. Nat Neurosci 2: 939–41

Zhou JY, Siderovski DP, Miller RJ (2000) Selective regulation of N-type Ca channels by different combinations of G-protein beta/gamma subunits and RGS proteins. J Neurosci 20: 7143–8

Zhu Y, Ikeda SR (1993) Adenosine modulates voltage-gated Ca2+ channels in adult rat sympathetic neurons. J Neurophysiol 70: 610–20

Zhuchenko O, Bailey J, Bonnen P, Ashizawa T, Stockton DW, Amos C, Dobyns WB, Subramony SH, Zoghbi HY, Lee CC (1997) Autosomal dominant cerebellar ataxia (SCA6) associated with small polyglutamine expansions in the alpha 1A-voltage-dependent calcium channel. Nat Genet 15: 62–9

Zwingman TA, Neumann PE, Noebels JL, Herrup K (2001) Rocker is a new variant of the voltage-dependent calcium channel gene Cacna1a. J Neurosci 21: 1169–78

# Pharmacology of Neurotransmitter Transport into Secretory Vesicles

**Farrukh A. Chaudhry(✉), Jean-Luc Boulland, Monica Jenstad, May K. L. Bredahl, and Robert H. Edwards**



**Abstract** Many neuropsychiatric disorders appear to involve a disturbance of chemical neurotransmission and the mechanism of available therapeutic agents supports this impression. Postsynaptic receptors have received considerable attention as drug targets, but some of the most successful agents influence presynaptic processes, in particular neurotransmitter reuptake. The pharmacological potential of many other presynaptic elements, and in particular the machinery responsible for loading transmitter into vesicles, has received only limited attention. The similarity of vesicular transporters to bacterial drug resistance proteins and the increasing evidence for regulation of vesicle filling and recycling suggest that the pharmacological potential of vesicular transporters has been underestimated. In this review, we discuss the pharmacological effects of psychostimulants and therapeutic agents on transmitter release.

Farrukh A. Chaudhry
The Biotechnology Centre of Oslo & The Centre for Molecular Biology and Neuroscience, University of Oslo, P.O.Box 1125, Blindern, 0317 OSLO, Norway
f.a.chaudhry@biotek.uio.no

T.C. Südhof, K. Starke (eds.), *Pharmacology of Neurotransmitter Release. Handbook of Experimental Pharmacology 184.*

## 1 Introduction: Neurotransmitter Recycling

Synaptic transmission involves the exocytotic release of neurotransmitter from a presynaptic site and the activation of specific receptors on the membrane of postsynaptic or surrounding cells. Given the high rates of firing observed for many neuronal populations, sustained signaling also requires replenishment of the released transmitter. Indeed, the nerve terminal expresses a series of transport activities designed to recycle transmitter. Among these activities, one class confers the reuptake of dopamine and other classical transmitters across the presynaptic plasma membrane, regulating their levels in the synaptic cleft, and recycling them for subsequent release (Figure 1) (Raiteri et al., 2002; Torres et al., 2003). Although the amino acid transmitters glutamate and particularly GABA recycle in part by reuptake into the nerve terminal, most of the released glutamate and a substantial proportion of the released GABA undergo transport into surrounding glial cells, where they are converted into glutamine (Danbolt, 2001; Bak et al., 2006). Transported back to the nerve terminal, glutamine serves to regenerate both glutamate and GABA (through conversion from glutamate by glutamic acid decarboxylase) (Figures 1C, 1D) (Chaudhry et al., 2002). Essentially all of the plasma membrane neurotransmitter transporters rely on a $Na^+$ gradient produced by the $Na^+$ pump, cotransporting $Na^+$ with transmitter, but vary in the stoichiometry of ionic coupling and in coupling to other ions such as chloride.

The exocytotic release of classical transmitters requires their transport into synaptic vesicles. In contrast to the plasma membrane transporters, the vesicular transport activities depend on a $H^+$ gradient created by the vacuolar-type $H^+$-ATPase rather than a $Na^+$ gradient, and function as $H^+$ exchangers. Biochemical studies have also revealed multiple, distinct vesicular transport activities. The monoamines dopamine, norepinephrine, and serotonin enter secretory vesicles through a common carrier. Distinct activities have been identified for the vesicular transport of acetylcholine (ACh), GABA, and glutamate (Schuldiner et al., 1995; Liu and Edwards, 1997). Well-characterized drugs acting on plasma membrane transporters do not affect the vesicular activities, but a variety of other drugs inhibiting vesicular transport have dramatic effects on brain function, and psychostimulants also interact with vesicular transport mechanisms.

Molecular identification of the proteins mediating vesicular neurotransmitter transport has revealed similarity to bacterial proteins, rather than to plasma membrane transporters (Figure 2). In contrast to concentrative uptake by plasma membrane transporters, the direction of vesicular transport resembles the efflux of many drugs from bacteria. Indeed, the first vesicular transmitter transporter was identified by expression cloning using selection in the neurotoxin N-methyl-4-phenylpyridinium ($MPP^+$) (Liu et al., 1992). The vesicular monoamine transporters protect against $MPP^+$ by sequestering the toxin inside vesicles, away from its primary site of action in mitochondria. The vesicular transporters thus represent important drug targets with a diversity of potential applications.

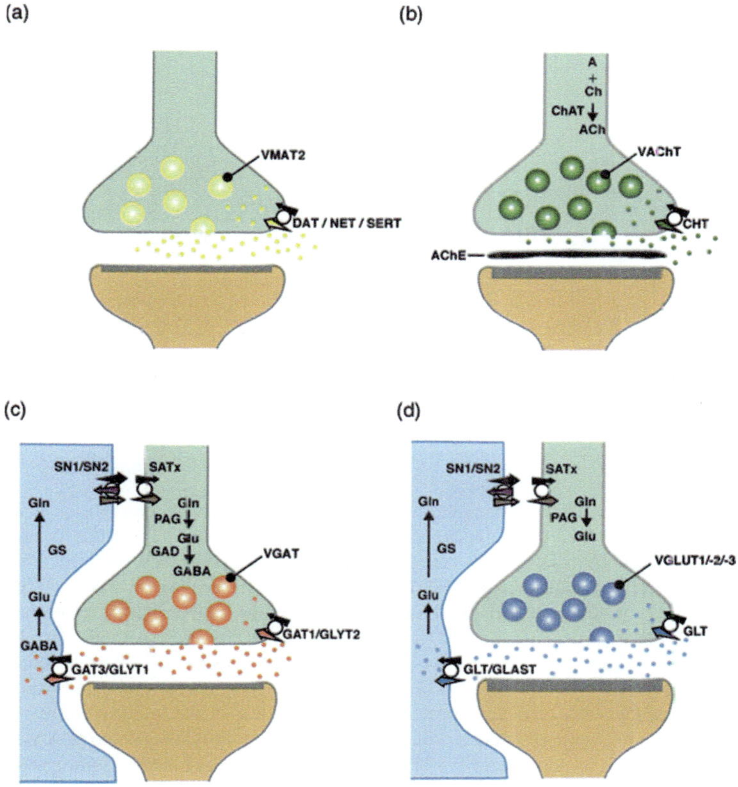


**Fig. 1** Neurotransmitter recycling. Classical neurotransmitters are released by exocytosis from the nerve terminal, and act on specific postsynaptic receptors. The signal is terminated and transmitter recycled by specific transporter proteins located at the nerve terminal and/or in surrounding glial cells. Transport by these proteins is generally energized by the electrochemical gradient of $Na^+$. (**a**) Three distinct transporters mediate transport of monoamines across the plasma membrane. The dopamine transporter DAT, the norepinephrine and epinephrine transporter NET and the serotonin transporter SERT, are responsible for the reuptake of released monoamine at dopaminergic, noradrenergic and serotonergic nerve terminals (yellow arrow), respectively. The vesicular monoamine transporter VMAT2 then recognizes all of these monoamines and transports them inside synaptic vesicles for subsequent exocytotic release. (**b**) Cholinergic signaling is terminated by metabolism of acetylcholine (ACh) to the inactive choline and acetate by acetylcholinesterase (AChE) located in the synaptic cleft. Choline (Ch) is transported back into the nerve terminal (green arrow) by the choline transporter (CHT) where choline acetyltransferase (ChAT) subsequently catalyzes acetylation (**a**) of choline to reform ACh. (**c**) At GABAergic and glycinergic nerve terminals, the GABA transporter (GAT1) and the glycine transporter (GLYT2) recycle GABA and glycine (red arrow), respectively. In addition, GABA may be taken up by surrounding glial cells (e.g. by GAT3), and converted to glutamine which is transported back to the nerve terminal by the concerted action of the system N (SNs) and system A transporters (SATs) (brown arrows). The glial transporter GLYT1 also contributes to the clearance of glycine. (**d**) Reuptake of glutamate at glutamatergic terminals has recently been demonstrated for a GLT isoform (blue arrow). However, the majority of glutamate is accumulated by surrounding glial cells (e.g., by GLT and GLAST) for conversion to glutamine which is subsequently transported back to the nerve terminals. System N transporters and a system A transporter (SATx) enriched in glutamatergic neurons mediate glutamine transfer from glia to excitatory neurons (brown arrows).

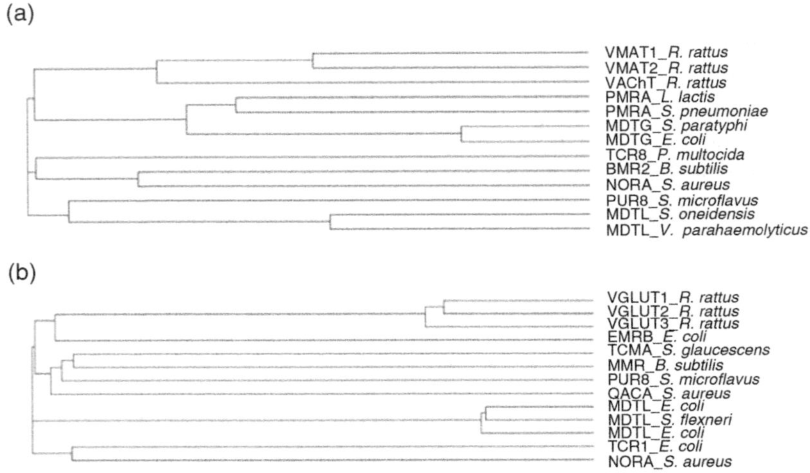


**Fig. 2** Cladogram showing sequence relationship of the vesicular transporters to a selected number of bacterial drug resistance proteins. (**a**) VMAT1, VMAT2 and VAChT are highly homologous. They are more distantly related to a variety of bacterial drug resistance proteins. PMRA, multidrug resistance efflux pump pmrA; MDTG, multidrug resistance protein mdtG; TCR8, a tetracycline resistance protein, class H (TETA(H)); BMR2, multidrug resistance protein 2; NORA, quinolone resistance protein norA; MDTL, multidrug resistance protein mdtL. (**b**) The rat homologs VGLUT 1, 2 and 3 show sequence similarity to a set of bacterial drug resistence proteins: EMRB, multidrug resistance protein B; TCMA, tetracenomycin C resistance and export protein; MMR, methylenomycin A resistance protein (MMR peptide); PUR8, puromycin resistance protein pur8; QACA, antiseptic resistance protein; MDTL, multidrug resistance protein mdtL; TCR1, tetracycline resistance protein, class A (TETA(A)); NORA, Quinolone resistance protein norA. *L. lactis*: *Lactococcus lactis*; *S. pneumoniae*: *Streptococcus pneumoniae*; *S. paratyphi*: *Salmonella paratyphi*; *E. coli*: *Escherichia coli*; *P. multocida*: *Pasteurella multocida*; *B. subtilis*: *Bacillus subtilis*; *S. aureus*: *Staphylococcus aureus*; *S. microflavus*: *Streptomyces microflavus*; *S. oneidensis*: *Shewanella oneidensis*; *V. parahaemolyticus*: *Vibrio parahaemolyticus*; *S. glaucescens*: *Streptomyces glaucescens*; *S. flexneri*: *Shigella flexneri*..

# 2 $H^+$ Electrochemical Gradient

## *2.1 Vacuolar* $H^+$*-ATPase*

An $H^+$ electrochemical gradient ($\Delta\mu_{H+}$) provides the energy required for active transport of all classical neurotransmitters into synaptic vesicles. The $Mg^{2+}$-dependent vacuolar-type $H^+$-ATPase (V-ATPase) that produces this gradient resides on internal membranes of the secretory pathway, in particular endosomes and lysosomes (vacuole in yeast) as well as secretory vesicles (Figure 3). In terms of both structure and function, this pump resembles the F-type ATPases of bacteria, mitochondria and chloroplasts, and differs from the P-type ATPases expressed at the plasma membrane of mammalian cells (e.g., the $Na^+/K^+$-, gastric $H^+/K^+$- and muscle $Ca^{2+}$-ATPases) (Forgac, 1989; Nelson, 1992). The vacuolar and F0F1

ATPases are multisubunit complexes comprising a cytoplasmic complex ($V_1$ or $F_1$) that hydrolyzes or synthesizes ATP and a membrane-bound complex ($V_0$ or $F_0$) which translocates $H^+$ (Wilkens, 2005; Beyenbach and Wieczorek, 2006). Subunit c of $V_0$ consists of six proteolipid subunits that form an $H^+$-binding ring, and the ring rotates during ATP hydrolysis and $H^+$ pumping. The macrolide antibiotics bafilomycin and concanamycin inhibit the V-ATPase by intercalating between helices of the c subunit and preventing their rotation (Figure 3) (Bowman et al., 2004).

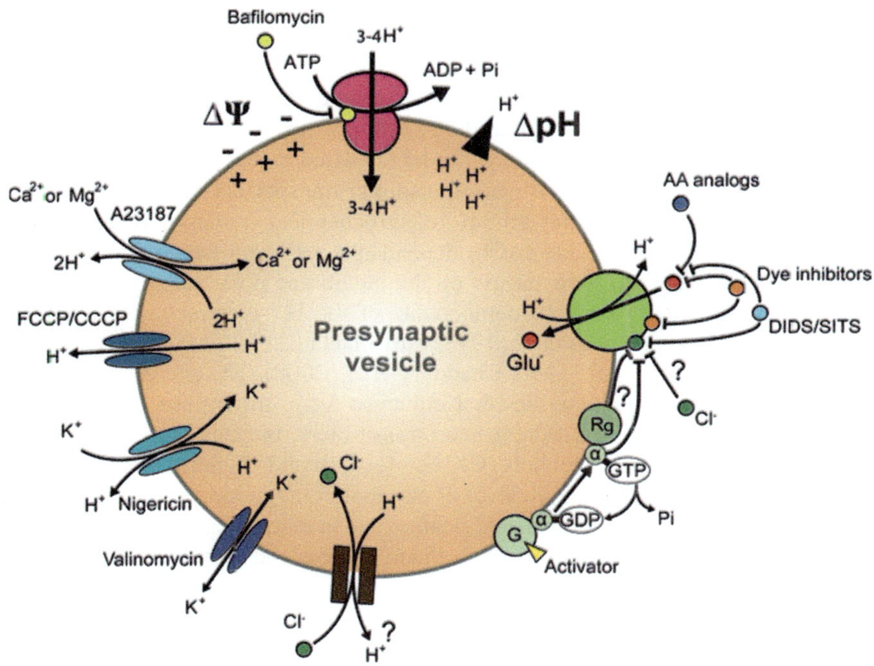


**Fig. 3** Pharmacological targeting of VGLUT containing vesicles. Schematic representation of a VGLUT-expressing vesicle and the main activators/inhibitors of vesicular glutamate uptake. The $Mg^{2+}$-dependent v-ATPase creates an electrochemical gradient of protons ($\Delta\mu_{H+} = \Delta\Psi + \Delta pH$) by coupling the energy produced by hydrolysis of an ATP to proton translocation into synaptic vesicles. Bafilomycin inhibit the V-ATPase by intercalating between helices of the c subunit of $V_0$ and preventing their rotation. The vesicular glutamate transporter couples the transport of $H^+$ down its electrochemical gradient to glutamate flux into synaptic vesicles. Several glutamate analogues, dyes and other compounds compete at the glutamate binding sites, or noncompetitively at distinct sites on the transport proteins. VGLUTs have an allosteric chloride binding site and may mediate chloride flux. Several inhibitors, such as DIDS/SITS, block the chloride binding site. Heterotrimeric G proteins can interact directly or through an as yet unidentified regulator (Rg) protein of the chloride conductance. Chloride channels, such as ClC-3, have been demonstrated on the vesicle membrane. Recent studies suggest that these proteins may be $H^+$ exchangers rather than channels, and yet still mediate net vesicle acidification. Finally, different ionophores such as valinomycin, nigericin, FCCP/CCCP and A23187 can act on the vesicle membrane to perturb the electrochemical gradient of protons and thus the vesicular uptake of glutamate.

These drugs have thus been used extensively to generate models for the function of V-ATPases. Hydrolysis of ATP by the vesicular $H^+$-ATPase is coupled to movement of $H^+$ into the vesicle lumen, creating a $H^+$ electrochemical gradient. Measurements from cholinergic and aminergic vesicles show a $\Delta pH \sim 1.4$ units and an electrical potential $\sim +39\,mV(\Delta\Psi)$ (Whittaker, 1987; Schuldiner et al., 1998).

## 2.2 Chloride Channels

The vesicular $H^+$-ATPase is potently regulated by $Cl^-$. Influx of $Cl^-$ neutralizes the positive charge on luminal $H^+$ and hence dissipates the electrical gradient, allowing the $H^+$ pump to transport more $H^+$ and thereby increase $\Delta pH$. Other mechanisms may serve to increase $\Delta\Psi$ at the expense of $\Delta pH$. The two components of the $H^+$ electrochemical gradient may thus be regulated independently. Interestingly, the activity of different vesicular neurotransmitter transporters depends to differing extents on the two components of the $H^+$ electrochemical gradient. The transport of monoamines and acetylcholine (ACh) depends primarily on the chemical component $\Delta pH$, glutamate transport mostly on the membrane potential $\Delta\Psi$, and GABA transport more equally on both components of the $H^+$ electrochemical gradient (Johnson, Jr., 1988b; Parsons et al., 1993; Carlson et al., 1989; Maycox et al., 1988; Kish et al., 1989; Hell et al., 1990; Chaudhry et al., 2008). The expression of vesicular chloride channels or activities that influence $\Delta\mu_{H+}$ might thus be expected to vary in vesicles that store different neurotransmitters, but this remains unknown. The intracellular chloride channels (ClC-3, -4, -5, -6 and -7) are widely expressed, but ClC-3 has a particularly high level of expression in the brain, and disruption of ClC-3 leads to impaired vesicular acidification and to neural degeneration (Stobrawa et al., 2001) through mechanisms that remain unclear and may involve expression on the postsynaptic plasma membrane (Wang et al., 2006). Indeed, it has recently become clear that many of the intracellular ClCs may function as $Cl^-/H^+$ exchangers rather than $Cl^-$ channels (Figure 3) (Accardi et al., 2004; Scheel et al., 2005; Picollo and Pusch, 2005). Although it seems counterintuitive that influx into the vesicles of $Cl^-$ through a $Cl^-/H^+$ exchanger would mediate vesicle acidification, the shown stoichiometry of 2 $Cl^-$: 1 $H^+$ would move relatively large amounts of charge and still enable net acidification by secondary activation of the $H^+$ pump. Why then couple the entry of chloride to proton efflux? Given the activity of the $H^+$-ATPase, perhaps it serves to concentrate chloride inside vesicles, for a function that is not yet understood (Jentsch, 2006).

## 2.3 Ionophores

In addition to drugs acting on the V-ATPase, several agents are capable of dissipating the electrochemical gradient for $H^+$ (Figure 3). The ionophore nigericin dissipates the chemical gradient for $H^+$ ($\Delta pH$) by exchanging $H^+$ for $K^+$. Similarly,

A23187 exchanges $H^+$ for $Ca^{2+}$ or $Mg^{2+}$. FCCP/CCCP allows $H^+$ to run down their electrochemical gradient, but will be limited by the development of membrane potential. In contrast, valinomycin is a $K^+$ ionophore, influencing specifically $\Delta\Psi$. These compounds dissipate the electrochemical gradient of $H^+$ and would therefore be expected to reduce all vesicular neurotransmitter transport activities. However, they will have different effects, depending on the proportion of $\Delta pH$ and $\Delta\Psi$ expressed in a given vesicle population, and on the stoichiometry of ionic coupling by the neurotransmitter transporters.

# 3 Vesicular Transport Proteins

## *3.1 Vesicular Monoamine Transporters (VMAT1 and VMAT2)*

Studies on monoamine uptake by chromaffin granules from the adrenal medulla show that the activity relies more on $\Delta pH$ than $\Delta\Psi$. In particular, the uptake of one cytoplasmic monoamine is coupled to the movement of two $H^+$ in the opposite direction (Knoth et al., 1981). However, the activity recognizes the protonated and hence charged form of the substrates, resulting in a net efflux of only one charge despite the two protons. Due to the $H^+$ electrochemical gradient and this coupling stoichiometry, chromaffin granules accumulate monoamines at molar concentrations. Indeed, monoamines interact with other vesicular components to form insoluble aggregates within chromaffin granules (Johnson, 1988a).

Vesicular monoamine transport can also protect against exogenous and endogenous toxins by sequestering them into synaptic vesicles. Systemic administration of 1-methyl-4-phenyl-1,2,3,6-tetragydropyridine (MPTP) in rats has been used as an animal model for idiopathic Parkinson's disease (Langston et al., 1983). The lipophilic MPTP penetrates the blood-brain barrier relatively easily and is metabolized by glial monoamine oxidase B (MAO-B) to the active metabolite N-methyl-4-phenylpyridinium ($MPP^+$) (Heikkila et al., 1984; Langston et al., 1984). $MPP^+$ resembles protonated monoamines, and is recognized by plasma membrane monoamine transporters that drive its accumulation in monoamine cells (Javitch et al., 1985), where it enters mitochondria, inhibits oxidative phosphorylation, and triggers apoptosis (Ramsay et al., 1986; Nicklas et al., 1985; Yang et al., 1998). However, the vesicular monoamine transporters have a high affinity for $MPP^+$, sequester the toxin inside vesicles, and thus protect against its mitochondrial toxicity. Indeed, selection in $MPP^+$ was used to isolate the first cDNA encoding a vesicular neurotransmitter transporter (Liu et al., 1992). Since dopamine is itself toxic due to the generation of hydrogen peroxide and production of highly reactive quinones (Michel and Hefti, 1990), the vesicular monoamine transporters (VMATs) may also protect against intrinsic toxicity of the endogenous transmitter, but the role of cytosolic dopamine in the pathogenesis of Parkinson's disease remains unknown.

Molecular cloning has identified two closely related but distinct vesicular monoamine transporters, VMAT1 and VMAT2 (Liu et al., 1992; Erickson et al., 1992; Liu and Edwards, 1997). Sequence analysis predicts 12 transmembrane domains with N- and C-termini in the cytoplasm, and the proteins show no sequence similarity to plasma membrane monoamine transporters (Nguyen et al., 1983; Neal and Chater, 1987; Neyfakh et al., 1991). Rather, they show similarity to bacterial proteins involved in detoxification (Figure 2). These bacterial proteins all function as $H^+$ exchangers, and several are even inhibited by the same drugs that inhibit VMATs (e.g., reserpine).

VMAT1 is expressed by endocrine tissues (e.g., chromaffin cells in the adrenal medulla) and paracrine cells (e.g., enterochromaffin cells in the stomach and intestines, and small intensely fluorescent (SIF) cells in sympathetic ganglia) (Erickson et al., 1996). In contrast, VMAT2 is expressed by monoamine neurons in the central nervous system (e.g., substantia nigra, locus coeruleus, and the raphe nucleus), and in selected peripheral endocrine populations (Table 1) (Weihe et al., 1994; Peter et al., 1995; Nirenberg et al., 1996). In general, VMAT2 has a higher affinity for most monoamines (by ~three-fold) than VMAT1 (Peter et al., 1994; Erickson et al., 1996), but both VMAT1 and 2 can protect against $MPP^+$ toxicity (Table 1). However, only VMAT2 appears to recognize histamine, which may bind to a site different from that of other monoamines (Merickel et al., 1995).

The VMATs are also among the very few vesicular neurotransmitter transporters whose turnover number is known. At 29° C, they transport ~5 molecules of serotonin per second and up to 20 molecules of dopamine (Peter et al., 1994). Since synaptic vesicles contain 5 to 20,000 molecules of transmitter and can recycle within at least 20 seconds (Ryan and Smith, 1995; Rizzoli et al., 2003), this rate has important implications for quantal size. At 5 molecules/second, the vesicle would contain only 100 molecules of transmitter after 20 seconds—if there were only one transporter per vesicle. Recent estimates suggest several transporters per vesicle (Takamori et al., 2006), but these might still not suffice to fill a rapidly cycling vesicle with monoamine unless the turnover was substantially higher at 37° C, where it is more difficult to measure transport accurately due to increased membrane leakiness. Indeed, the ability to determine the turnover of VMATs has been enabled by the availability of ligands to quantify the transporter and hence provide a denominator for measurements of transport.

### 3.1.1 Reserpine

The alkaloid reserpine is very effective in treating hypertension, but frequently produces the side effect of depression. Both actions are due to the inhibition of VMAT, the former in the peripheral sympathetic nervous system, and the latter in the brain. Indeed, the inhibition of VMAT by reserpine gave rise to the monoamine hypothesis of affective disorders (Carlsson et al., 1965; Beers and Passman, 1990). In terms of mechanism, reserpine competes with monoamines for binding to the VMATs and has three-fold higher affinity for VMAT2 than for VMAT1 (Ki 12 nM

**Table 1** A summary of the vesicular transporters, their differential cellular localization, properties and inhibitors. For details and references, see text

| | Main cellular localization | Substrates | Specific inhibitors |
|---|---|---|---|
| VMAT1 | Endocrine tissue<br>Sympathetic ganglia<br>Enterochromaffin cells | Monoamines (Km ~ low μM range)<br>Monoamine analogues<br>Toxins (e.g. MPP+) | Reserpine (Ki ~ 34 nM)<br>Psychostimulantia (e.g. Amphetamine, cocaine) |
| VMAT2 | Locus ceruleus<br>Substantia nigra<br>Arcuate nucleus<br>Groups of cells in the midbrain<br>Raphe nuclei<br>Postganglionic sympathetic neurons | Monoamines (Km ~ low μM range)<br>Monoamine analogues<br>Histamine<br>Toxins (e.g. MPP+) | Reserpine (Ki ~ 12 nM)<br>Tetrabenazine<br>Psychostimulantia (e.g. Amphetamine, cocaine) |
| VAChT | Medial septum - diagonal band of Broca<br>Magnocellular forebrain nuclei<br>Striatum<br>Preganglionic neurons<br>Postganglionic parasympathic neurons<br>Motoneurons | Acethylcholine (Km ~ high mM range)<br>Variety of cationic compounds | vesamicol (Kd ~ 5 nM)<br>variety of ACh analogues |
| VGAT | Majority of interneurons of the CNS<br>A- and B-cells of islets of Langerhans | GABA (Km ~ 5 mM)<br>Glycine (Km > 10 mM) | Nipecotic acid<br>γ-vinyl-GABA<br>β-alanine |

(*Continued*)

**Table 1** Continued

| | Main cellular localization | Substrates | Specific inhibitors |
|---|---|---|---|
| VGLUT1 | Cerebral cortex<br>Hippocampus<br>Cerebellar cortex<br>Glial cells<br>A – cells of islets of Langerhans | Glutamate (Km∼1 – 3 mM) | ***Amino acid analogs (Ki for glutamate)***<br>Erythro-4 Methyl-L-glutamate (Ki ∼ 0.7 mM)<br>Trans aminocyclopentane-1,3-dicarboxylic acid (t-ACDP) (Ki ∼ 0.7 mM)<br>4-methylene-L-glutamate (Ki ∼ 3 mM)<br>Kynurenic acid (Ki ∼ 1.3 mM)<br>Xanthurenic acid (Ki ∼ 0.19 mM)<br>7-chloro-kynurenic acid (Ki ∼ 0.59 mM)<br>6-biphenyl-4-yl-quinoline-2,4-dicarboxylic acid (Ki ∼ 41 uM)<br>6-(4′-phenylstyryl)-quinoline-2,4-dicarboxylic acid (Ki ∼ 64 uM)<br>***Biphenyl derivative of naphthalene disulfonic acid***<br>Evans blue (Ki ∼ 40 nM)<br>Chicago sky blue (Ki ∼ 190 nM)<br>Trypan blue (IC50 ∼ 20 nM)<br>Naphtol blue black, Reactive blue 2, Benzopurpurin 4B, Ponceau SS, Direct blue 71, Acid red 114: IC50 ∼ 330 to 1670 nM<br>***Stilbene derivatives***<br>4-acetamido-4′-isothiocyanostilbene-2,2′- disulfonic acid (SITS)<br>4,4′-diisothiocyanostilbene-2,2′-disulfonic acid (DIDS)<br>5-nitro-2-(3-phenylpropylamino)benzoic acid (N144)<br>***Polyhalogenated fluorescein derivative***<br>Rose bengal (IC50 ∼19 nM) |
| VGLUT2 | Thalamus<br>Brainstem<br>Cerebellar nuclei<br>Glial cells<br>A – cells of islets of Langerhans | Glutamate (Km∼1 – 3 mM) | |
| VGLUT3 | Subset of GABAergic, cholinergic, aminergic and glutamatergic neurons<br>Motoneurons<br>Glial cells<br>B – cells of islets of Langerhans<br>Muscle | Glutamate (Km∼1 – 3 mM) | |

and 34 nM, respectively), consistent with differences in the recognition of substrates by the two transporters (Peter et al., 1994; Erickson et al., 1996). Interestingly, the imposition of $\Delta\mu_{H+}$ accelerates reserpine binding to the VMATs and monoamines inhibit reserpine binding with potencies similar to their apparent affinity as substrates (Weaver and Deupree, 1982; Rudnick et al., 1990; Schuldiner et al., 1993a). Moreover, the membrane-impermeable analog, reserpic acid, cannot inhibit transport from the vesicular lumen (Chaplin et al., 1985). Thus, the binding of reserpine and monoamines occurs when the substrate recognition site is oriented to the cytoplasmic face of the vesicle membrane. With monoamine bound, the transporter reorients to the inside of the vesicle, releasing monoamine (possibly due to a reduction in affinity for the substrate) and rebinding $H^+$, which reorients the substrate recognition site back to the cytoplasm (Liu and Edwards, 1997). At high concentrations, reserpine can act like a detergent, and recent studies have revealed other targets for reserpine in neurons, including sites on other proteins. Indeed, reserpine inhibits the plasma membrane dopamine transporter function in a noncompetitive manner at high concentrations (Metzger et al., 2002). In vivo, reserpine administration does not alter the level of VMATs, but single injections can block the binding site for another major VMAT inhibitor, tetrabenazine (Naudon et al., 1996).

### 3.1.2 Tetrabenazine

In contrast to reserpine, the binding of tetrabenazine to VMAT is affected only by very high concentrations of monoamines, indicating a distinct site of interaction (Darchen et al., 1989). However, there is an interaction between the binding of reserpine and tetrabenazine, suggesting either that the sites are overlapping, or that the drugs interact with conformationally distinct stages of the transport cycle (Darchen et al., 1989; Liu and Edwards, 1997). In addition, tetrabenazine shows much higher affinity for VMAT2 than VMAT1, both in human and particularly in rat, where VMAT1 is essentially resistant to inhibition by tetrabenazine (Schuldiner et al., 1993a; Erickson et al., 1996; Adamson et al., 1997). Ketanserin, a serotonin receptor antagonist, binds to the same site as tetrabenazine, and a photoactivatable analog (7-Azido-8-iodoketanserin) has been used to demonstrate that the binding site is located at the N-terminus of bovine VMAT2 (Sievert and Ruoho, 1997; Sagne et al., 1997). Lobeline, an alkaloid resembling nicotinic receptor agonists (Dwoskin and Crooks, 2002), also inhibits VMAT by interacting with the tetrabenazine binding site (Teng et al., 1998). Thus, VMAT inhibitors appear to fall into two classes, the ones that bind to the reserpine site and those that bind to the tetrabenazine site.

### 3.1.3 Psychostimulants

Psychostimulants generally act by increasing the level of extracellular monoamine. The mechanisms responsible include inhibition of plasma membrane monoamine transport and, in particular, stimulation of flux reversal, which may occur by exchange or by efflux promoted in the case of the dopamine transporter by phosphorylation of the N-terminus (Fischer and Cho, 1979; Rudnick and Wall, 1992; Schuldiner et al., 1993b; Sulzer and Galli, 2003; Wall et al., 1995; Khoshbouei et al., 2004; Fog et al., 2006). However, psychostimulants also influence the flux of monoamine across the vesicle membrane. Amphetamine is a lipophilic weak base that can deplete monoamine stores by dissipating the $\Delta\mu_{H+}$ which drives monoamine uptake. Second, amphetamines may inhibit VMAT2 directly (Peter et al., 1994; Gonzalez et al., 1994; Sulzer and Rayport, 1990; Sulzer et al., 1993; Sulzer et al., 1995). The analysis of VMAT2 knockout mice illustrates the importance of interaction with VMAT for the effect of amphetamine. In the absence of VMAT2, amphetamine still induces dopamine release in culture, and administration in vivo can increase survival of the knockout mice, which usually die within a few days after birth (Fon et al., 1997). Presumably, dopamine efflux promoted by amphetamine bypasses the defect in exocytotic release in the VMAT2 knockout. However, dopamine release in response to amphetamine is greatly reduced in the knockout, indicating that amphetamine acts at least in part by releasing monoamine stores. In addition, mice heterozygous for VMAT2, which appear grossly normal, exhibit an increased locomotor response to amphetamine that occludes the usual increase seen with the sensitization to repeated administration (Wang et al., 1997).

Fleckenstein and co-workers have also demonstrated that amphetamine and its analogs (methamphetamine [METH], 3,4-methylenedioxymethamphetamine [MDMA]) alter the subcellular location of VMAT2-containing vesicles (Riddle et al., 2005). The "reuptake blockers" cocaine and methylphenidate (MPD), which inhibit the plasma membrane dopamine transporter (DAT), mobilize VMAT2-expressing vesicles. In vivo, cocaine exposure increases the Vmax of VMAT2 for dopamine and the number of binding sites for dihydrotetrabenazine (Brown et al., 2001). In addition, it mobilizes a synapsin-dependent reserve pool of dopamine-containing synaptic vesicles, thereby increasing the release of dopamine (Venton et al., 2006). MPD increases VMAT2 immunoreactivity and dihydrotetrabenazine (DHTBZ) binding, apparently through the redistribution of VMAT2-containing vesicles. The vesicle-mobilizing action of both the "releasers" (such as amphetamines) and the "reuptake blockers" (such as cocaine) has been shown to involve dopamine receptors since antagonists abolish the effects (Sandoval et al., 2002; Riddle et al., 2005).

### 3.1.4 Toxins

The VMATs were identified through their ability to protect against toxicity of $MPP^+$, presumably by sequestering the compound inside secretory vesicles, and away from its primary site of action in mitochondria. Indeed, the sequence of the VMATs shows homology to many bacterial drug resistance genes, including the toxin-extruding antiporters (TEXANs) (Figure 2). These transporters recognize many of the same drugs as the mammalian multidrug resistance (MDR) transporters, including P-glycoprotein, but use an $H^+$ exchange mechanism rather than ATP hydrolysis to provide the energy for efflux. Several of the transporters, including the tetracycline resistance protein from Tn10, even show inhibition by reserpine (Ruiz et al., 2005). Considerable functional as well as structural analysis of the multimeric mini-TEXAN EmrE has begun to elucidate the mechanism of active transport by this family of proteins (Schuldiner et al., 2001; Gottschalk et al., 2004). In addition, cytotoxic compounds such as doxorubicin, rhodamine, ethidium, isometamidium, and tacrine are recognized by the VMATs as well as the TEXANs (Yelin and Schuldiner, 1995). The VMATs also recognize several environmental toxins, such as brominated flame retardants (BFRs), which inhibit vesicular dopamine transport (Mariussen and Fonnum, 2003). Finally, meta-iodobenzylguanidine (MIBG), a noradrenaline analog, is used for scintigraphy and radiation therapy of neuroendocrine (NE) tumors, and accumulation of this compound is inhibited by reserpine, indicating its accumulation in secretory vesicles in vivo (Kolby et al., 2003).

### 3.1.5 Nonspecific Effects

Pharmacological agents acting primarily on other proteins may also influence the VMATs. Phencyclidine, a noncompetitive NMDA receptor antagonist and DAT inhibitor, at high doses increases vesicular dopamine transport and binding of tetrabenazine (Crosby et al., 2002). However, the effect of phencyclidine is not mimicked by dizocilpine (another NMDA antagonist), but is antagonized by eticlopride, a D2 receptor blocker. Clozapine, an atypical antipsychotic, is very effective at reducing negative symptoms of schizophrenia as well as depression (Buchanan et al., 1998; Buckley, 1999), and leads to increased tetrabenazine binding, whereas haloperidol has no effect on VMAT binding capacity (Rehavi et al., 2002).

The antiarrythmic drug amiodarone inhibits reserpine binding in a concentration-dependent manner and has a sympatholytic, anti-adrenergic effect in the heart, inhibiting the uptake of norepinephrine (Haikerwal et al., 1999). Blockers of VMAT and L-type calcium channels have also been reported to exert reciprocal inhibitory actions (Mahata et al., 1995).

## *3.2 Vesicular Acetylcholine Transporter (VAChT)*

A distinct synaptic vesicle transport activity for acetylcholine (ACh) has biochemical properties that resemble the VMATs. Vesicular ACh transport depends more on $\Delta pH$ than $\Delta\Psi$, but is nonetheless electrogenic (Anderson et al., 1981). Since ACh is permanently positively charged, the transport activity is thus believed to move more than one proton, similar to vesicular monoamine transport. However, vesicular ACh transport appears much less efficient than transport by VMAT in generating high concentration gradients of transmitter. Since the coupling would predict extremely high gradients, it has been speculated that the protein itself leaks protons and hence dissipates the driving force for transport (Parsons et al., 1993).

Organophosphorus toxins inhibit the enzyme acetylcholinesterase (AChE) which degrades ACh in the synaptic cleft (Figure 1). In *Caenorhabditis elegans*, resistance to such a toxin reflects a defect in ACh release and was thus used to screen for genes required for the release of ACh. Mutations in a sequence similar to VMAT (*unc-17*) led to identification of the vesicular acetylcholine transporter (VAChT) (Alfonso et al., 1993). The rat ortholog (rVAChT) shows 40% identity and 65% similarity to the rat VMATs (Roghani et al., 1994; Erickson et al., 1994), and the predicted topology also suggests 12 transmembrane domains. However, the Km for VAChT is several orders of magnitude higher (mM) than that for the VMATs (low μM) (Varoqui and Erickson, 1996). As anticipated for a protein with these properties, VAChT associates with synaptic vesicles in cholinergic neurons of the central nervous system, and the mRNA co-localizes with mRNA for the biosynthetic enzyme choline acetyltransferase (ChAT) (Schafer et al., 1994; Roghani et al., 1994; Roghani et al., 1996). The turnover of VAChT has also been determined, and it appears to be even slower than the VMATs, with a rate ∼1/second (Varoqui and Erickson, 1996).

### 3.2.1 Pharmacology of VAChT

Although closely related to the VMATs in sequence (Figure 2), the pharmacological profile of VAChT differs markedly (Table 1). The substrate selectivity is poor, and many cationic compounds are recognized by the transporter (Clarkson et al., 1992; Rogers and Parsons, 1989; Rogers et al., 1993). On the other hand, it does not recognize the precursor choline, and is not inhibited by reserpine (Bravo et al., 2005; Scheunemann et al., 2004). Rather, (−)-trans-2-(4-phenylpiperidino)-cyclohexanol (vesamicol) inhibits ACh uptake in synaptic vesicles from the electric organ of *Torpedo californica* (Anderson et al., 1983), and is the best characterized inhibitor of VAChT in higher organisms. Vesamicol does not inhibit the VMATs. Vesamicol is stereospecific with the (−)-enantiomer 25 times more potent than the (+) isomer (Rogers and Parsons, 1989), and it has a dissociation constant of ∼5 nM (Bahr et al., 1992). The inhibition of transport is noncompetitive (Bahr and Parsons, 1986), although substrates compete with vesamicol for binding at equilibrium (Bahr et al., 1992), perhaps due to a distinct ACh binding site that regulates the access of vesamicol (Van der Kloot, 2003). During transport, the binding site for ACh translocates inward, releasing ACh into the vesicle lumen. This allows vesamicol to bind in two distinct states, to either the inwardly facing substrate recognition site or, after $H^+$ exchange, to the outwardly facing recognition site (Bravo et al., 2004). This model predicts more binding of vesamicol during steady-state transport than at rest (Bravo et al., 2004). In addition, the inhibition of VAChT with vesamicol appears to generate an efflux of the most recently accumulated ACh (Anderson et al., 1986)—presumably due to blocking of uptake of ACh that has leaked out of the vesicle nonspecifically. Indeed, cationic transmitters such as monoamines and ACh appear to leak nonspecifically out of synaptic vesicles, presumably across the lipid bilayer, and this may account for the regulation of quantal size by the number of vesicular transporters. Overexpression of the VMATs and VAChT increases quantal size (Pothos et al., 2000; Song et al., 1997), presumably by off-setting the leak across the vesicle membrane.

A variety of other ACh analogs have been identified (Kaufman et al., 1989; Rogers and Parsons, 1989). Some of these, such as cetiedil (Israel et al., 1987; Rogers and Parsons, 1989), inhibit VAChT competitively, presumably by interfering with the binding of ACh to VAChT (Van der Kloot, 2003). Drugs acting on proteins involved in the ACh metabolism may also influence VAChT. Rivastigmine, an AChE inhibitor used clinically for memory disorders, increases VAChT immunoreactivity (Tayebati et al., 2004).

## *3.3 Vesicular GABA and Glycine Transporter (VGAT)*

A distinct vesicular transport activity for GABA relies on both $\Delta pH$ and $\Delta\Psi$ (Kish et al., 1989; Takamori et al., 2000; McIntire et al., 1997), presumably because as a zwitterion with no net charge, the number of protons exchanged for GABA equals the amount of charge moved. In terms of substrate recognition, vesicular GABA uptake was not inhibited by glycine in several studies, suggesting distinct transport proteins for the two inhibitory neurotransmitters (Kish et al., 1989). However, other studies have suggested that the same activity recognizes both GABA and glycine (Christensen et al., 1990; Christensen and Fonnum, 1991; Burger et al., 1991; Roseth and Fonnum, 1995; McIntire et al., 1997; Sagné et al., 1997). Further, physiological studies have documented the corelease of glycine and GABA from the same synaptic vesicles (Jonas et al., 1998).

Consistent with the differences in ionic coupling, vesicular GABA transport is encoded by a protein with no discernible sequence similarity to the VMATs or VAChT. Genetic analysis in *C. elegans* identified a protein with multiple transmembrane domains (UNC-47) required for GABAergic transmission. Expressed in a heterologous expression system, the mammalian ortholog VGAT conferred GABA uptake dependent on both $\Delta pH$ and $\Delta\Psi$ (McIntire et al., 1997). The expressed transporter exhibits low-affinity uptake of GABA (Km $\sim$5 mM), consistent with results from native synaptic vesicles, and glycine inhibits GABA transport (IC50 $\sim$25 mM) (Sagne et al., 1997; McIntire et al., 1997). Furthermore, immunogold electron microscopy demonstrated VGAT expression in at least a subpopulation of GABAergic, glycinergic as well as mixed terminals (Chaudhry et al., 1998). Indeed, disruption of VGAT reduces release of GABA as well as glycine (Wojcik et al., 2006), consistent with the evidence for co-release in vivo (Jonas et al., 1998). However, it is important to note that the release of glycine depends heavily on the high-affinity neuronal glycine transporter GLYT2. GLYT2 presumably generates the high cytoplasmic concentrations of glycine necessary to drive the low-affinity vesicle transporter (Figure 1) (Gomeza et al., 2003).

### 3.3.1 Pharmacology of VGAT

No drugs acting specifically on VGAT have thus far been identified. However, several structural analogues of GABA have been tested for inhibition of VGAT (Table 1). Nipecotic acid and N-butyric acid weakly inhibit VGAT activity (McIntire et al., 1997). The potent anticonvulsant $\gamma$-vinyl-GABA (vigabatrin), known to inhibit GABA transaminase, was also found to inhibit VGAT with low affinity (McIntire et al., 1997).

Interestingly, the fate of GABA differs from that of monoamines after genetic disruption of the cognate vesicular transporter. In mice, the loss of VMAT2 results in the depletion of all monoamines (to 1%-5% wild-type levels), apparently due to metabolism of the cytosolic transmitter by monoamine oxidase, as well as other factors (Fon et al., 1997). In contrast, GABA accumulates to levels above wild type in the *C. elegans* mutant *unc-47* (McIntire et al., 1993). In addition, a vesicle membrane-bound form of the biosynthetic enzyme glutamic acid decarboxylase (GAD65) can apparently be inhibited by disrupting vesicular $\Delta\mu_{H+}$ with the proton pump inhibitor bafilomycin A1, the protonophore carbonyl cyanide m-chorophenylhydrazone or the ionophore gramicidin. Direct coupling between GAD and VGAT has thus been suggested, although the mechanism remains unclear (Jin et al., 2003).

## *3.4 Vesicular Glutamate Transporters (VGLUT1-3)*

Characterization of the glutamate transport activity in native synaptic vesicles has demonstrated that it recognizes specifically glutamate (not aspartate) and depends predominantly on $\Delta\Psi$ rather than $\Delta$pH (Naito and Ueda, 1985; Maycox et al., 1988; Carlson et al., 1989; Carlson and Ueda, 1990; Roseth, 1995). In addition, a low concentration of $Cl^-$ (2–10 mM) is required for optimal vesicular glutamate transport, with no or high $Cl^-$ reducing but not entirely eliminating transport activity. Although chloride entry might be expected to dissipate the membrane potential that drives glutamate uptake, it is thus also required to activate transport and apparently interacts with an allosteric regulatory site (Figure 3) (Naito and Ueda, 1983; Hartinger and Jahn, 1993; Wolosker et al., 1996).

Molecular cloning has identified three vesicular proteins that exhibit the biochemical properties described for glutamate uptake by native synaptic vesicles from brain (Reimer and Edwards, 2004; Erickson et al., 2006; Takamori, 2006). They all show stereoselectivity for L-glutamate and a Km $\sim$1–3 mM. None of them recognizes aspartate or the inhibitory neurotransmitters GABA or glycine. Dissipation of $\Delta\Psi$ with the $K^+$ ionophore valinomycin substantially reduces transport, but addition of the $H^+$ ionophore nigericin further reduces glutamate uptake, suggesting a reliance on $\Delta$pH as well. Interestingly, dissipation of $\Delta$pH alone can sometimes increase vesicular glutamate transport, presumably by increasing $\Delta\Psi$, but the loss of $\Delta$pH in addition to $\Delta\Psi$ clearly reduces transport further, supporting the $H^+$

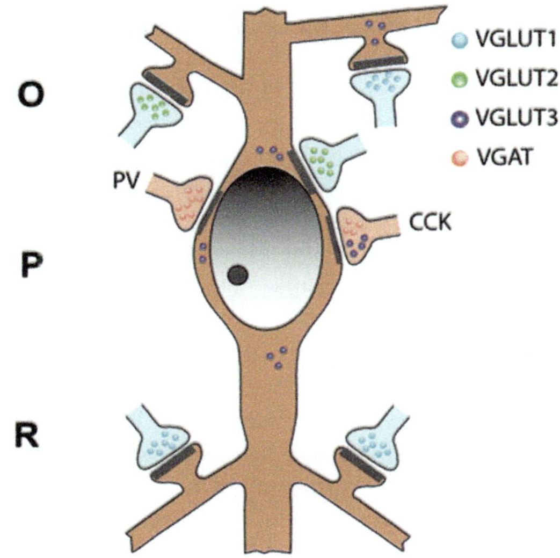


**Fig. 4** Differential subcellular localization of the three VGLUTs. Nerve terminals containing VGLUT1, −2 or −3, all target the pyramidal cells of hippocampal region CA2. However, the inputs differ in origin and they target distinct regions of the pyramidal cell membrane. VGLUT1$^{+}$ nerve terminals derive from Schaffer collaterals and form synapses on the spines in stratum radiatum and oriens. VGLUT2$^{+}$ terminals originate in subcortical areas and form synapses onto spines, proximal denrites or cell bodies. VGLUT3 localizes to the subset of interneurons that contain cholecystokinin (CCK), and form synapses onto the cell body and proximal dendrites. VGLUT3 also mediates glutamate release from the cell bodies and dendrites of selected neurons in the cortex and striatum.

exchange mechanism (Bellocchio et al., 2000). Moreover, all three isoforms show a biphasic dependence on [$Cl^-$], with the optimal activity ~2–4 mM. However, the isoforms differ in expression at cellular and subcellular levels (Figure 4; Table 1), as well as through development (Boulland et al., 2004). VGLUT1 and VGLUT2 show a complementary expression pattern, occurring in different subpopulations of glutamatergic nerve terminals (Fremeau, Jr. et al., 2001; Schafer et al., 2002). In particular, VGLUT1 is expressed by glutamatergic cells in cerebral cortex, hippocampus, and cerebellar cortex, whereas VGLUT2 predominates in glutamatergic cells of the thalamus, brainstem, and deep cerebellar nuclei. The two homologs are also involved in glial glutamate release (Bezzi et al., 2004; Haydon and Carmignoto, 2006). In contrast, VGLUT3 is expressed throughout the brain but restricted to a subpopulation of glial cells and neurons, including serotonergic, cholinergic, GABAergic and glutamatergic cells (Gras et al., 2002; Fremeau, Jr. et al., 2002). In contrast to VGLUT1 and 2, which localize only to axon terminals, VGLUT3 localizes as well to cell bodies and dendrites, where it participates in retrograde signaling (Harkany et al., 2004). VGLUT2 and VGLUT3 are also transiently expressed at high levels in specific subpopulations of cells early in life, suggesting additional developmental roles for glutamatergic neurotransmission (Boulland et al., 2004).

### 3.4.1 Pharmacology of Vesicular Glutamate Transport

The drugs that inhibit vesicular glutamate transport (Table 1) differ from those acting on plasma membrane glutamate transporters. However, several of the drugs were initially identified as glutamate receptor blockers, or at least derived from these agents. Evans blue, a biphenyl derivative of naphthalene disulfonic acid, blocks specifically the excitatory AMPA/kainate receptor responses mediated by glutamate receptor subunits GluR1, GluR1,2, GluR1,3, and GluR2,3, and inhibition appears to be noncompetitive (Keller et al., 1993). The group of F. Fonnum has characterized the effects of Evans blue and related dyes on vesicular transport of glutamate (Table 1; Figure 3). At low concentrations, Evans blue inhibits glutamate uptake competitively (Ki $\sim$ 40 nM), while at higher concentrations it affects membrane potential and the pH gradient as well (Fykse et al., 1992; Roseth et al., 1995). The structural analog Chicago sky blue also blocks vesicular glutamate uptake competitively, but is much less effective (Roseth et al., 1995). Neither Evans blue nor Chicago sky blue has any effect on $Na^+$-dependent glutamate uptake across the cell membrane (Roseth et al., 1995). The authors of this study suggest that the distance of $\sim$6 Å between sulfonic acid groups structurally fixed by aromatic rings and the electron-donating groups in close vicinity of the negative charge are critical for the inhibition of the glutamate uptake (Roseth et al., 1995). Although we still lack the structural information required to interpret the observed structure-activity relationships, the distance between negatively charged groups is smaller in aspartate than glutamate, possibly accounting for the lack of recognition. Further, dyes structurally similar to Evans blue with biphenyl and amino- and sulphonic-acid-substituted naphthyl groups potently inhibit glutamate uptake. The $IC_{50}$ value for Trypan blue is 49 nM, whereas Naphtol blue black, Reactive blue 2, Benzopurpurin 4B, Ponceau SS, Direct blue 71, and Acid Red 114 have $IC_{50}$ values ranging from 330 to 1670 nM (Roseth et al., 1998). However, all of these dyes are cell impermeant and thus of limited use in vivo or even in vitro with intact cells. Some of the dyes (e.g., Evans blue and Chicago sky blue) also inhibit VGAT, albeit less potently.

Another dye derived from polyhalogenated fluorescein, Rose Bengal, acts as a potent noncompetitive inhibitor of vesicular glutamate uptake with Ki $\sim$19 nM (Ogita et al., 2001). The site of Rose Bengal action remains to be determined, but there are indications that it involves an allosteric site (Ogita et al., 2001). Structure-activity analysis has also shown that the phenyl group, multiple iodine groups, and the bridging oxygen of xanthine are required for potent inhibition of vesicular glutamate uptake (Bole and Ueda, 2005). Since Rose Bengal is membrane permeable, it is much more useful in vivo and in brain slice experiments where it has been shown to impair glutamate release (Montana et al., 2004).

Glutamate analogs have also been tested for their ability to interact with the VGLUTs (Table 1). Although several act on plasma membrane transporters, they have either no or only partial effects on vesicular glutamate uptake (Naito and Ueda, 1985; Tabb and Ueda, 1991). Interestingly, several amino acid analogs used to block the glutamate-binding domain of glutamate receptors and/or selectively inhibit plasma membrane glutamate transport also inhibit the VGLUTs (Ortwine et al., 1992; Bridges et al., 1999). Quinoline-2,4-dicarboxylic acids (QDCs) were tested as inhibitors of vesicular glutamate transport due to their structural similarity to quinoline-2-carboxylic acids (QCAs), which have an embedded glutamate structure (Carrigan et al., 1999). The Ki values of optimized QDCs are as low as 41 μM, and QDCs resembling Evans blue and Chicago sky blue have been reported as particularly promising (Carrigan et al., 2002). Table 1 summarizes the Ki values of these glutamate analogs.

### 3.4.2 Other Targets in Vesicular Glutamate Transport

Although it is well known that optimal glutamate uptake requires low concentrations of $Cl^-$, the mechanisms involved are not completely known. The group of F. Fonnum has demonstrated that the stilbene derivatives and plasma membrane anion channel blockers 4,4'-diisothiocyanatostilbene-2,2'-disulfonic acid (DIDS), 4-acetamido-4'-isothiocyanatostilbene-2,2'-disulfonic acid (SITS), and 5-nitro-2-(3-phenylpropylamino)-benzoic acid (N144) inhibit glutamate uptake through competition directly with glutamate (Figure 3) (Roseth et al., 1995). These organic acids have two centers of negative charge and resemble the dyes mentioned above. The loop diuretics furosemide and bumetanide, acting on the $Na^+/K^+/2Cl^-$ cotransporter, also inhibit glutamate uptake. Like the stilbene derivatives, these drugs block glutamate uptake competitively and the $Cl^-$ conductance noncompetitively (Roseth et al., 1995). However, there is compelling evidence that an allosteric effect not involving the glutamate binding site also contributes. The group of R. Jahn has shown inhibition of vesicular glutamate transport by DIDS which could be overcome by chloride, but not glutamate, suggesting a distinct, DIDS-sensitive chloride binding site on the cytoplasmic face of the vesicular glutamate transporter (Figure 3) (Hartinger and Jahn, 1993). Consistent with this, Wolosker and co-workers have shown that although chloride flux would influence the driving force for glutamate transport, the regulation by chloride does not depend on $\Delta pH$ or $\Delta\Psi$, but rather represents an allosteric form of regulation (Wolosker et al., 1996). However, the regulation of glutamate transport may not involve the VGLUTs directly, but an associated intracellular chloride channel (Jentsch et al., 1999; Stobrawa et al., 2001). Genetic inactivation of ClC-3 has been shown to produce severe retinal and hippocampal degeneration (Stobrawa et al., 2001). However, the mechanism for this degeneration and the role of ClC-3 and other intracellular chloride channels in the allosteric regulation of VGLUTs and their observed chloride channel activity remains unknown. Recent reconstitution of purified VGLUT2 from a heterologous expression system has demonstrated the same biphasic dependence on chloride observed for the native protein (Juge et al., 2006), suggesting that the chloride binding site is not encoded by another polypeptide. However, the authors also suggest that VGLUT2 encodes a $Na^+$-dependent phosphate cotransport activity (Juge et al., 2006). Originally described as phosphate cotransporters (Ni et al., 1994), vesicular glutamate transport catalyzed by the VGLUTs is not affected by phosphate or $Na^+$ (Bellocchio et al., 2000; Fremeau, Jr. et al., 2001), so it is surprising that they encode a mechanistically and pharmacologically distinct activity. Indeed, we have observed that heterologous expression of the VGLUTs confers a highly variable phosphate uptake activity that can be $Na^+$-independent (R.H.E., unpublished observations). Previous work on the related transporter NaPi-1 also demonstrates a poor correlation between expression of the protein and phosphate uptake activity (Busch et al., 1996; Broer et al., 1998).

Heterotrimeric G proteins also appear to regulate vesicular glutamate transport by influencing the allosteric chloride site. Activation of the $G\alpha_{o2}$-subunit shifts the sensitivity of vesicular glutamate transport to lower chloride concentrations (Winter et al., 2005). In the absence of $G\alpha_{o2}$, vesicular glutamate transport exhibits no clear dependence on chloride, a result quite remarkable in light of the reconstitution studies demonstrating preservation of the chloride dependence with the purified protein (Juge et al., 2006). However, it remains unclear whether the effect of G proteins involves a direct interaction with the VGLUTs or with an unidentified "regulator" protein (Figure 3). An Inhibitory Protein Factor (IPF) that blocks vesicular glutamate transport has been purified from bovine brain cytosol, and apparently corresponds to a proteolytic fragment of fodrin (Ozkan et al., 1997).

A variety of other drugs also regulate the activity and/or expression of VGLUTs. Antidepressant agents, including fluoxetine, desipramine, the atypical antipsychotic clozapine, and lithium increase VGLUT1 mRNA and protein in neurons of the cerebral cortex and hippocampus (Moutsimilli et al., 2005; Tordera et al., 2005). Antidepressants and their analogs may thus influence the release of glutamate. Bilirubin, L-homocysteate, L-a-aminoadipate, the alkaloid bromocriptine, several fatty acids, and some other compounds also inhibit vesicular glutamate transport, and their analogs suggest additional strategies for drug development (Roseth et al., 1995; Roseth et al., 1998; Thompson et al., 2005). Nonetheless, despite the wide range of compounds that have been reported to affect VGLUT function, there are few if any specific antagonists, and certainly none that really distinguish between the three isoforms.

## 3.5 Other Transporters on Vesicles

Since many of the proteins found on synaptic vesicles still lack a clearly identified function, a number of these may contribute to filling the vesicles with transmitter. Indeed, the synaptic vesicle protein SV2 was identified many years ago through the production of monoclonal antibodies to purified synaptic vesicles (Buckley and Kelly, 1985). The sequence derived from molecular cloning (Bajjalieh et al., 1992; Feany et al., 1992) predicted 12 transmembrane domains, and multiple isoforms have since been identified. SV2A appears to be the most ubiquitous isoform, but SV2B and C exhibit an overlapping distribution (Janz et al., 1999; Bajjalieh et al., 1993; Bajjalieh et al., 1994). A related sequence, SVOP, has also been identified in invertebrates as well as mammals (Janz et al., 1998). However, the function of SV2 remains unknown. The knockout of SV2A alone exhibits seizures and early lethality (Crowder et al., 1999), while physiological analysis of the double SV2A and B knockout shows a calcium-sensitive increase in transmitter release, suggesting a role in $Ca^{2+}$ homeostasis (Janz et al., 1999). On the other hand, the analysis of release from endocrine cells as well as neurons shows a relatively specific role for SV2 in promoting release from the readily releasable pool of vesicles (Xu and Bajjalieh, 2001; Custer et al., 2006). The N-terminal, cytoplasmic domain of SV2 has also been shown to interact with selected isoforms of the synaptotagmin family (Schivell et al., 1996; Pyle et al., 2000), but the polytopic structure of SV2 and its clear relationship to sugar transporters strongly suggests a role as channel, or more likely, transporter. Interestingly, the widely use anticonvulsant levetiracetam (Keppra) has recently been found to bind specifically to SV2A (and not the other isoforms) (Lynch et al., 2004). The affinity of binding by related compounds correlates with their potency as anticonvulsants, further implicating SV2 as the site of action for these drugs. However, it seems paradoxical that the SV2A knockout develops seizures and a drug that binds to the same site can prevent seizures, suggesting that levetiracetam might activate rather than inhibit SV2A.

Remarkably, SV2 also appears to be the cellular receptor for botulinum neurotoxin A (Dong et al., 2006). In particular, a segment of the large lumenal loop binds to the toxin, and all of the SV2 isoforms appear capable of serving as receptors. Thus, levetiracetam, even if it is bound to the same site as the toxin, seems unlikely to protect against poisoning since it interacts only with SV2A.

A number of transporters that function on the plasma membrane also reside on synaptic vesicles, providing a mechanism for their regulated delivery to the cell surface. The $Na^{+}$- and $Cl^{-}$-dependent proline transporter resides primarily on synaptic vesicles (Renick et al., 1999; Crump et al., 1999), but the physiological role of this transporter and hence the biological context for its regulatory exocytosis remains unclear. More recently, it has been found that the $Na^{+}$-dependent choline transporter involved in the recycling of choline after cleavage from ACh by extracellular acetylcholinesterase, also localizes predominantly to synaptic vesicles (Ferguson et al., 2003). In this case, it was demonstrated that synaptic vesicle exocytosis produces a concomitant increase in choline uptake. Since loss of the choline transporter dramatically impairs cholinergic transmission in mammals as well as *C. elegans* (Ferguson and Blakely, 2004; Matthies et al., 2006), the biological role of the transporter is clear. Its localization to synaptic vesicles presumably serves to coordinate expression of the transporter with activity, but it remains unclear why this transporter normally resides in vesicles, whereas other proteins involved in reuptake remain primarily on the cell surface. Nonetheless, membrane trafficking of the plasma membrane neurotransmitter transporters remains a major mechanism for their regulation.

## 4 Conclusions

The vesicular transporters are capable of concentrating neurotransmitters from 10- to 100,000-fold, with important implications for $Ca^{2+}$-dependent quantal release. These transporters also protect neurons from the toxicity of exogenous compounds such as the parkinsonian neurotoxin MPTP and against endogenous molecules such as the normal neurotransmitter dopamine. The molecular analysis of vesicular neurotransmitter transport also provides an important new perspective to understand the physiological regulation of transmitter release and its role in disease. The proteins responsible and even orphan transporters of the synaptic vesicle also represent potent targets for the development of therapeutic agents.

## References


Accardi A, Kolmakova-Partensky L, Williams C, Miller C (2004) Ionic currents mediated by a prokaryotic homologue of CLC Cl-channels. J Gen Physiol 123:109–19.

Adamson CR, Emley TE, Herbig LJ, Near JA (1997) Effects of nerve growth factor on dihydrotetrabenazine binding to PC12 cells. Neurochem Int 30:411–15.

Alfonso A, Grundahl K, Duerr JS, Han HP, Rand JB (1993) The Caenorhabditis elegans unc-17 gene: a putative vesicular acetylcholine transporter. Science 261:617–19.

Anderson DC, King SC, Parsons SM (1981) Uncoupling of acetylcholine uptake from the Torpedo cholinergic synaptic vesicle ATPase. Biochem Biophys Res Commun 103:422–8.

Anderson DC, King SC, Parsons SM (1983) Pharmacological characterization of the acetylcholine transport system in purified Torpedo electric organ synaptic vesicles. Mol Pharmacol 24:48–54.

Anderson KJ, Monaghan DT, Cangro CB, Namboodiri MAA, Neale JH, Cotman CW (1986) Localization of N-acetylaspartylglutamate-like immunoreactivity in selected areas of the rat brain. Neurosci Lett 72:14–20.

Bahr BA, Clarkson ED, Rogers GA, Noremberg K, Parsons SM (1992) A kinetic and allosteric model for the acetylcholine transporter-vesamicol receptor in synaptic vesicles. Biochemistry 31:5752–5762.

Bahr BA, Parsons SM (1986) Demonstration of a receptor in Torpedo synaptic vesicles for the acetylcholine storage blocker L-trans-2-(4-phenyl[3,4-3H]-piperidino)cyclohexanol. Proc Natl Acad Sci USA 83:2267–2270.

Bajjalieh SM, Frantz GD, Weimann JM, McConnell SK, Scheller RH (1994) Differential expression of synaptic vesicle protein 2 (SV2) isoforms. J Neurosci 14:5223–5235.

Bajjalieh SM, Peterson K, Linial M, Scheller RH (1993) Brain contains two forms of synaptic vesicle protein 2. Proc Natl Acad Sci USA 90:2150–2154.

Bajjalieh SM, Peterson K, Shinghal R, Scheller RH (1992) SV2, a brain synaptic vesicle protein homologous to bacterial transporters. Science 257:1271–1273.

Bak LK, Schousboe A, Waagepetersen HS (2006) The glutamate/GABA-glutamine cycle: aspects of transport, neurotransmitter homeostasis and ammonia transfer. J Neurochem 98:641–653.

Beers MH, Passman LJ (1990) Antihypertensive medications and depression. Drugs 40:792–799.

Bellocchio EE, Reimer RJ, Fremeau RT, Jr., Edwards RH (2000) Uptake of glutamate into synaptic vesicles by an inorganic phosphate transporter [see comments]. Science 289:957–960.

Beyenbach KW, Wieczorek H (2006) The V-type H+ ATPase: molecular structure and function, physiological roles and regulation. J Exp Biol 209:577–589.

Bezzi P, Gundersen V, Galbete JL, Seifert G, Steinhauser C, Pilati E, Volterra A (2004) Astrocytes contain a vesicular compartment that is competent for regulated exocytosis of glutamate. Nat Neurosci 7:613–620.

Bole DG, Ueda T (2005) Inhibition of vesicular glutamate uptake by Rose Bengal-related compounds: structure-activity relationship. Neurochem Res 30:363–369.

Boulland JL, Qureshi T, Seal RP, Rafiki A, Gundersen V, Bergersen LH, Fremeau RT, Jr., Edwards RH, Storm-Mathisen J, Chaudhry FA (2004) Expression of the vesicular glutamate transporters during development indicates the widespread corelease of multiple neurotransmitters. J Comp Neurol 480:264–280.

Bowman EJ, Graham LA, Stevens TH, Bowman BJ (2004) The bafilomycin/concanamycin binding site in subunit c of the V-ATPases from Neurospora crassa and Saccharomyces cerevisiae. J Biol Chem 279:33131–33138.

Bravo DT, Kolmakova NG, Parsons SM (2005) New transport assay demonstrates vesicular acetylcholine transporter has many alternative substrates. Neurochem Int 47:243–247.

Bravo DT, Kolmakova NG, Parsons SM (2004) Transmembrane reorientation of the substrate-binding site in vesicular acetylcholine transporter. Biochemistry 43:8787–8793.

Bridges RJ, Kavanaugh MP, Chamberlin AR (1999) A pharmacologicalal review of competitive inhibitors and substrates of high-affinity, sodium-dependent glutamate transport in the central nervous system. Curr Pharm Des 5:363–379.

Broer S, Schuster A, Wagner CA, Broer A, Forster I, Biber J, Murer H, Werner A, Lang F, Busch AE (1998) Chloride conductance and Pi transport are separate functions induced by the expression of NaPi-1 in Xenopus oocytes. J Membr Biol 164:71–77.

Brown JM, Hanson GR, Fleckenstein AE (2001) Regulation of the vesicular monoamine transporter-2: a novel mechanism for cocaine and other psychostimulants. J Pharmacol Exp Ther 296:762–767.

Buchanan RW, Breier A, Kirkpatrick B, Ball P, Carpenter WT, Jr. (1998) Positive and negative symptom response to clozapine in schizophrenic patients with and without the deficit syndrome. Am J Psychiatry 155:751–760.

Buckley K, Kelly RB (1985) Identification of a transmembrane glycoprotein specific for secretory vesicles of neural and endocrine cells. J Cell Biol 100:1284–1294.

Buckley PF (1999) New antipsychotic agents: emerging clinical profiles. J Clin Psychiatry 60 Suppl 1 12–17.

Burger PM, Hell J, Mehl E, Krasel C, Lottspeich F, Jahn R (1991) GABA and glycine in synaptic vesicles: storage and transport characteristics. Neuron 7:287–293.

Busch AE, Schuster A, Waldegger S, Wagner CA, Zempel G, Broer S, Biber J, Murer H, Lang F (1996) Expression of a renal type I sodium/phosphate transporter (NaPi-1) induces a conductance in Xenopus oocytes permeable for organic and inorganic anions. Proc Natl Acad Sci USA 93:5347–5351.

Carlson MD, Kish PE, Ueda T (1989) Characterization of the solubilized and reconstituted ATP-dependent vesicular glutamate uptake system. J Biol Chem 264:7369–7376.

Carlson MD, Ueda T (1990) Accumulated glutamate levels in the synaptic vesicle are not maintained in the absence of active transport. Neurosci Lett 110:325–330.

Carlsson A, Dahlstroem A, Fuxe K, Hillarp NA (1965) Failure of reserpine to deplete noradrenaline neurons of alpha-methylnoradrenaline formed from alpha-methyl dopa. Acta Pharmacol Toxicol (Copenh) 22:270–276.

Carrigan CN, Bartlett RD, Esslinger CS, Cybulski KA, Tongcharoensirikul P, Bridges RJ, Thompson CM (2002) Synthesis and in vitro pharmacology of substituted quinoline-2,4-dicarboxylic acids as inhibitors of vesicular glutamate transport. J Med Chem 45:2260–2276.

Carrigan CN, Esslinger CS, Bartlett RD, Bridges RJ, Thompson CM (1999) Quinoline-2,4-dicarboxylic acids: synthesis and evaluation as inhibitors of the glutamate vesicular transport system. Bioorg Med Chem Lett 9:2607–2612.

Chaplin L, Cohen AH, Huettl P, Kennedy M, Njus D, Temperley SJ (1985) Reserpic acid as an inhibitor of norepinephrine transport into chromaffin vesicle ghosts. J Biol Chem 260:10981–10985.

Chaudhry FA, Reimer RJ, Bellocchio EE, Danbolt NC, Osen KK, Edwards RH, Storm-Mathisen J (1998) The vesicular GABA transporter, VGAT, localizes to synaptic vesicles in sets of glycingergic as well as GABAergic neurons. J Neurosci 18:9733–9750.

Chaudhry FA, Reimer RJ, Edwards RH (2002) The glutamine commute: take the N line and transfer to the A. J Cell Biol 157:349–355.

Chaudhry FA, Edwards RH, Fonnum F (2008) Vesicular neurotransmitter transporters as targets for endogenous and exogenous toxic substances. Annu Rev Pharmacol Toxicol, 48 [sep 20 Epub ahead of print]

Christensen H, Fonnum F (1991) Uptake of glycine, GABA and glutamate by synaptic vesicles isolated from different regions of rat CNS. Neurosci Lett.

Christensen H, Fykse EM, Fonnum F (1990) Uptake of glycine into synaptic vesicles isolated from rat spinal cord. J Neurochem 54:1142–1147.
Clarkson ED, Rogers GA, Parsons SM (1992) Binding and active transport of large analogues of acetylcholine by cholinergic synaptic vesicles in vitro. J Neurochem 59:695–700.
Crosby MJ, Hanson JE, Fleckenstein AE, Hanson GR (2002) Phencyclidine increases vesicular dopamine uptake. Eur J Pharmacol 438:75–78.
Crowder KM, Gunther JM, Jones TA, Hale BD, Zhang HZ, Peterson MR, Scheller RH, Chavkin C, Bajjalieh SM (1999) Abnormal neurotransmission in mice lacking synaptic vesicle protein 2A (SV2A). Proc Natl Acad Sci USA 96:15268–15273.
Crump FT, Fremeau RT, Craig AM (1999) Localization of the brain-specific high-affinity l-proline transporter in cultured hippocampal neurons: molecular heterogeneity of synaptic terminals. Mol Cell Neurosci 13:25–39.
Custer KL, Austin NS, Sullivan JM, Bajjalieh SM (2006) Synaptic vesicle protein 2 enhances release probability at quiescent synapses. J Neurosci 26:1303–1313.
Danbolt NC (2001) Glutamate uptake. Prog Neurobiol 65:1–105.
Darchen F, Scherman D, Henry JP (1989) Reserpine binding to chromaffin granules suggests the existence of two conformations of the monoamine transporter. Biochemistry 28:1692–1697.
Dong M, Yeh F, Tepp WH, Dean C, Johnson EA, Janz R, Chapman ER (2006) SV2 is the protein receptor for botulinum neurotoxin A. Science 312:592–596.
Dwoskin LP, Crooks PA (2002) A novel mechanism of action and potential use for lobeline as a treatment for psychostimulant abuse. Biochem Pharmacol 63:89–98.
Erickson JD, De GS, Varoqui H, Schafer MK, Weihe E (2006) Activity-dependent regulation of vesicular glutamate and GABA transporters: a means to scale quantal size. Neurochem Int 48:643–649.
Erickson JD, Eiden LE, Hoffman BJ (1992) Expression cloning of a reserpine-sensitive vesicular monoamine transporter. Proc Natl Acad Sci USA 89:10993–10997.
Erickson JD, Schafer MK, Bonner TI, Eiden LE, Weihe E (1996) Distinct pharmacologicalal properties and distribution in neurons and endocrine cells of two isoforms of the human vesicular monoamine transporter. Proc Natl Acad Sci USA 93:5166–5171.
Erickson JD, Varoqui H, Schafer MK, Modi W, Diebler MF, Weihe E, Rand J, Eiden LE, Bonner TI, Usdin TB (1994) Functional identification of a vesicular acetylcholine transporter and its expression from a "cholinergic" gene locus. J Biol Chem 269:21929–21932.
Feany MB, Lee S, Edwards RH, Buckley KM (1992) The synaptic vesicle protein SV2 is a novel type of transmembrane transporter. Cell 70:861–867.
Ferguson SM, Blakely RD (2004) The choline transporter resurfaces: new roles for synaptic vesicles? Mol Interv 4:22–37.
Ferguson SM, Savchenko V, Apparsundaram S, Zwick M, Wright J, Heilman CJ, Yi H, Levey AI, Blakely RD (2003) Vesicular localization and activity-dependent trafficking of presynaptic choline transporters. J Neurosci 23:9697–9709.
Fischer JF, Cho AK (1979) Chemical release of dopamine from striatal homogenates: evidence for an exchange diffusion model. J Pharmacol Exp Ther 208:203–209.
Fog JU, Khoshbouei H, Holy M, Owens WA, Vaegter CB, Sen N, Nikandrova Y, Bowton E, McMahon DG, Colbran RJ, Daws LC, Sitte HH, Javitch JA, Galli A, Gether U (2006) Calmodulin kinase II interacts with the dopamine transporter C terminus to regulate amphetamine-induced reverse transport. Neuron 51:417–429.
Fon EA, Pothos EN, Sun BC, Killeen N, Sulzer D, Edwards RH (1997) Vesicular transport regulates monoamine storage and release but is not essential for amphetamine action. Neuron 19:1271–1283.
Forgac M (1989) Structure and function of vacuolar class of ATP-driven proton pumps. Physiol Rev 69:765–796.
Fremeau RT, Jr., Burman J, Qureshi T, Tran CH, Proctor J, Johnson J, Zhang H, Sulzer D, Copenhagen DR, Storm-Mathisen J, Reimer RJ, Chaudhry FA, Edwards RH (2002) The identification of vesicular glutamate transporter 3 suggests novel modes of signaling by glutamate. Proc Natl Acad Sci USA 99:14488–14493.

Fremeau RT Jr., Troyer MD, Pahner I, Nygaard GO, Tran CH, Reimer RJ, Bellocchio EE, Fortin D, Storm-Mathisen J, Edwards RH (2001) The expression of vesicular glutamate transporters defines two classes of excitatory synapse. Neuron 31:247–260.

Fykse EM, Iversen EG, Fonnum F (1992) Inhibition of L-glutamate uptake into synaptic vesicles. Neurosci Lett 135:125–128.

Gomeza J, Ohno K, Hulsmann S, Armsen W, Eulenburg V, Richter DW, Laube B, Betz H (2003) Deletion of the mouse glycine transporter 2 results in a hyperekplexia phenotype and postnatal lethality. Neuron 40:797–806.

Gonzalez AM, Walther D, Pazos A, Uhl GR (1994) Synaptic vesicular monoamine transporter expression: distribution and pharmacological profile. Brain Res Mol Brain Res 22:219–226.

Gottschalk KE, Soskine M, Schuldiner S, Kessler H (2004) A structural model of EmrE, a multidrug transporter from Escherichia coli. Biophys J 86:3335–3348.

Gras C, Herzog E, Bellenchi GC, Bernard V, Ravassard P, Pohl M, Gasnier B, Giros B, El Mestikawy S (2002) A third vesicular glutamate transporter expressed by cholinergic and serotoninergic neurons. J Neurosci 22:5442–5451.

Haikerwal D, Dart AM, Little PJ, Kaye DM (1999) Identification of a novel, inhibitory action of amiodarone on vesicular monoamine transport. J Pharmacol Exp Ther 288:834–837.

Harkany T, Holmgren C, Hartig W, Qureshi T, Chaudhry FA, Storm-Mathisen J, Dobszay MB, Berghuis P, Schulte G, Sousa KM, Fremeau RT, Jr., Edwards RH, Mackie K, Ernfors P, Zilberter Y (2004) Endocannabinoid-Independent Retrograde Signaling at Inhibitory Synapses in Layer 2/3 of Neocortex: Involvement of Vesicular Glutamate Transporter 3. J Neurosci 24:4978–4988.

Hartinger J, Jahn R (1993) An anion binding site that regulates the glutamate transporter of synaptic vesicles. J Biol Chem 268:23122–23127.

Haydon PG, Carmignoto G (2006) Astrocyte control of synaptic transmission and neurovascular coupling. Physiol Rev 86:1009–1031.

Heikkila RE, Manzino L, Cabbat FS, Duvoisin RC (1984) Protection against the dopaminergic neurotoxicity of 1-methyl-4-phenyl-1,2,5,6-tetrahydropyridine by monoamine oxidase inhibitors. Nature 311:467–469.

Hell JW, Maycox PR, Jahn R (1990) Energy dependence and functional reconstitution of the gamma-aminobutyric acid carrier from synaptic vesicles. J Biol Chem 265:2111–2117.

Israel M, Manaranche R, Morot Gaudry-Talarmain Y, Lesbats B, Gulik-Krzywicki T, Dedieu JC (1987) Effect of cetiedil on acetylcholine release and intramembrane particles in cholinergic synaptosomes. Biol Cell 61:59–63.

Janz R, Goda Y, Geppert M, Missler M, Sudhof TC (1999) SV2A and SV2B function as redundant Ca2+ regulators in neurotransmitter release. Neuron 24:1003–1016.

Janz R, Hofmann K, Sudhof TC (1998) SVOP, an evolutionarily conserved synaptic vesicle protein, suggests novel transport functions of synaptic vesicles. J Neurosci 18:9269–9281.

Javitch JA, D'Amato RJ, Strittmatter SM, Snyder SH (1985) Parkinsonism-inducing neurotoxin, N-methyl-4-phenyl-1,2,3,6 – tetrahydropyridine: uptake of the metabolite N-methyl-4-phenylpyridine by dopamine neurons explains selective toxicity. Proc Natl Acad Sci USA 82:2173–2177.

Jentsch TJ (2006) Chloride and the endosomal/lysosomal pathway-emerging roles of CLC chloride transporters. J Physiol.

Jentsch TJ, Friedrich T, Schriever A, Yamada H (1999) The CLC chloride channel family. Pflugers Arch 437:783–795.

Jin H, Wu H, Osterhaus G, Wei J, Davis K, Sha D, Floor E, Hsu CC, Kopke RD, Wu JY (2003) Demonstration of functional coupling between gamma -aminobutyric acid (GABA) synthesis and vesicular GABA transport into synaptic vesicles. Proc Natl Acad Sci USA 100:4293–4298.

Johnson RG, Jr. (1988b) Accumulation of biological amines into chromaffin granules: a model for hormone and neurotransmitter transport. Physiol Rev 68:232–307.

Johnson RG, Jr. (1988a) Accumulation of biological amines into chromaffin granules: a model for hormone and neurotransmitter transport. Physiol Rev 68:232–307.

Jonas P, Bischofberger J, Sandkuhler J (1998) Corelease of two fast neurotransmitters at a central synapse [see comments]. Science 281:419–424.

Juge N, Yoshida Y, Yatsushiro S, Omote H, Moriyama Y (2006) Vesicular glutamate transporter contains two independent transport machineries. J Biol Chem.

Kaufman R, Rogers GA, Fehlmann C, Parsons SM (1989) Fractional vesamicol receptor occupancy and acetylcholine active transport inhibition in synaptic vesicles. Mol Pharmacol 36:452–458.

Keller BU, Blaschke M, Rivosecchi R, Hollmann M, Heinemann SF, Konnerth A (1993) Identification of a subunit-specific antagonist of alpha-amino-3-hydroxy-5-methyl-4-isoxazolepropionate/kainate receptor channels. Proc Natl Acad Sci USA 90:605–609.

Khoshbouei H, Sen N, Guptaroy B, Johnson L, Lund D, Gnegy ME, Galli A, Javitch JA (2004) N-terminal phosphorylation of the dopamine transporter is required for amphetamine-induced efflux. PLoS Biol 2:E78.

Kish PE, Fischer-Bovenkerk C, Ueda T (1989) Active transport of gamma-aminobutyric acid and glycine into synaptic vesicles. Proc Natl Acad Sci USA 86:3877–3881.

Knoth J, Zallakian M, Njus D (1981) Stoichiometry of H+-linked dopamine transport in chromaffin granule ghosts. Biochemistry 20:6625–6629.

Kolby L, Bernhardt P, Levin-Jakobsen AM, Johanson V, Wangberg B, Ahlman H, Forssell-Aronsson E, Nilsson O (2003) Uptake of meta-iodobenzylguanidine in neuroendocrine tumours is mediated by vesicular monoamine transporters. Br J Cancer 89:1383–1388.

Langston JW, Ballard P, Tetrud JW, Irwin I (1983) Chronic Parkinsonism in humans due to a product of meperidine-analog synthesis. Science 219:979–980.

Langston JW, Forno LS, Rebert CS, Irwin I (1984) Selective nigral toxicity after systemic administration of 1-methyl-4-phenyl-1,2,5,6-tetrahydropyrine (MPTP) in the squirrel monkey. Brain Res 292:390–394.

Liu Y, Edwards RH (1997) The role of vesicular transport proteins in synaptic transmission and neural degeneration. Annu Rev Neurosci 20:125–156.

Liu Y, Peter D, Roghani A, Schuldiner S, Prive GG, Eisenberg D, Brecha N, Edwards RH (1992) A cDNA that suppresses MPP+ toxicity encodes a vesicular amine transporter. Cell 70:539–551.

Lynch BA, Lambeng N, Nocka K, Kensel-Hammes P, Bajjalieh SM, Matagne A, Fuks B (2004) The synaptic vesicle protein SV2A is the binding site for the antiepileptic drug levetiracetam. Proc Natl Acad Sci USA 101:9861–9866.

Mahata M, Mahata SK, Parmer RJ, O'Connor DT (1996) Vesicular monoamine transport inhibitors. Novel action at calcium channels to prevent catecholamine secretion. Hypertension 28:414–420.

Mariussen E, Fonnum F (2003) The effect of brominated flame retardants on neurotransmitter uptake into rat brain synaptosomes and vesicles. Neurochem Int 43:533–542.

Matthies DS, Fleming PA, Wilkes DM, Blakely RD (2006) The Caenorhabditis elegans choline transporter CHO-1 sustains acetylcholine synthesis and motor function in an activity-dependent manner. J Neurosci 26:6200–6212.

Maycox PR, Deckwerth T, Hell JW, Jahn R (1988) Glutamate uptake by brain synaptic vesicles. J Biol Chem 263:15423–15428.

McIntire SL, Jorgensen E, Horvitz HR (1993) Genes required for GABA function in Caenorhabditis elegans. Nature 364:334–337.

McIntire SL, Reimer RJ, Schuske K, Edwards RH, Jorgensen EM (1997) Identification and chacterization of the vesicular GABA transporter. Nature 389:870–876.

Merickel A, Rosandich P, Peter D, Edwards RH (1995) Identification of residues involved in substrate recognition by a vesicular monoamine transporter. J Biol Chem 270:25798–25804.

Metzger RR, Brown JM, Sandoval V, Rau KS, Elwan MA, Miller GW, Hanson GR, Fleckenstein AE (2002) Inhibitory effect of reserpine on dopamine transporter function. Eur J Pharmacol 456:39–43.

Michel PP, Hefti F (1990) Toxicity of 6-hydroxydopamine and dopamine for dopaminergic neurons in culture. J Neurosci Res 26:428–435.

Montana V, Ni Y, Sunjara V, Hua X, Parpura V (2004) Vesicular glutamate transporter-dependent glutamate release from astrocytes. J Neurosci 24:2633–2642.
Moutsimilli L, Farley S, Dumas S, El MS, Giros B, Tzavara ET (2005) Selective cortical VGLUT1 increase as a marker for antidepressant activity. Neuropharmacology 49:890–900.
Naito S, Ueda T (1985) Characterization of glutamate uptake into synaptic vesicles. J Neurochem 44:99–109.
Naito S, Ueda T (1983) Adenosine triphosphate-dependent uptake of glutamate into protein I-associated synaptic vesicles. J Biol Chem 258:696–699.
Naudon L, Raisman-Vozari R, Edwards RH, Leroux-Nicollet I, Peter D, Liu Y, Costentin J (1996) Reserpine affects differentially the density of the vesicular monoamine transporter and dihydrotetrabenazine binding sites. Eur J Neurosci 8:842–846.
Neal RJ, Chater KF (1987) Nucleotide sequence analysis reveals similarities between proteins determining methylenomycin A resistance in Streptomyces and tetracycline resistance in eubacteria. Gene 58:229–241.
Nelson N (1992) The vacuolar H(−)-ATPase–one of the most fundamental ion pumps in nature. J Exp Biol 172:19–27.
Neyfakh AA, Bidnenko VE, Chen LB (1991) Efflux-mediated multidrug resistance in Bacillus subtilis: similarities and dissimilarities with the mammalian system. Proc Natl Acad Sci USA 88:4781–4785.
Nguyen TT, Postle K, Bertrand KP (1983) Sequence homology between the tetracycline-resistance determinants of Tn10 and pBR322. Gene 25:83–92.
Ni B, Rosteck PR, Jr., Nadi NS, Paul SM (1994) Cloning and expression of a cDNA encoding a brain-specific Na(+)-dependent inorganic phosphate cotransporter. Proc Natl Acad Sci USA 91:5607–5611.
Nicklas WJ, Vyas I, Heikkila RE (1985) Inhibition of NADH-linked oxidation in brain mitochondria by 1-methyl-4-phenyl-pyridine, a metabolite of the neurotoxin, 1-methyl-4-phenyl-1,2,5,6-tetrahydropyridine. Life Sci 36:2503–2508.
Nirenberg MJ, Chan J, Liu Y, Edwards RH, Pickel VM (1996) Ultrastructural localization of the vesicular monoamine transporter-2 in midbrain dopaminergic neurons: potential sites for somatodendritic storage and release of dopamine. J Neurosci 16:4135–4145.
Ogita K, Hirata K, Bole DG, Yoshida S, Tamura Y, Leckenby AM, Ueda T (2001) Inhibition of vesicular glutamate storage and exocytotic release by Rose Bengal. J Neurochem 77:34–42.
Ortwine DF, Malone TC, Bigge CF, Drummond JT, Humblet C, Johnson G, Pinter GW (1992) Generation of N-methyl-D-aspartate agonist and competitive antagonist pharmacophore models. Design and synthesis of phosphonoalkyl-substituted tetrahydroisoquinolines as novel antagonists. J Med Chem 35:1345–1370.
Ozkan ED, Lee FS, Ueda T (1997) A protein factor that inhibits ATP-dependent glutamate and gamma-aminobutyric acid accumulation into synaptic vesicles: purification and initial characterization. Proc Natl Acad Sci USA 94:4137–4142.
Parsons SM, Prior C, Marshall IG (1993) Acetylcholine transport, storage, and release. Int Rev Neurobiol 35:279–390.
Peter D, Jimenez J, Liu Y, Kim J, Edwards RH (1994) The chromaffin granule and synaptic vesicle amine transporters differ in substrate recognition and sensitivity to inhibitors. J Biol Chem 269:7231–7237.
Peter D, Liu Y, Sternini C, de Giorgio R, Brecha N, Edwards RH (1995) Differential expression of two vesicular monoamine transporters. J Neurosci 15:6179–6188.
Picollo A, Pusch M (2005) Chloride/proton antiporter activity of mammalian CLC proteins ClC-4 and ClC-5. Nature 436:420–423.
Pothos EN, Larsen KE, Krantz DE, Liu Y, Haycock JW, Setlik W, Gershon MD, Edwards RH, Sulzer D (2000) Synaptic vesicle transporter expression regulates vesicle phenotype and quantal size. J Neurosci 20:7297–7306.
Pyle RA, Schivell AE, Hidaka H, Bajjalieh SM (2000) Phosphorylation of synaptic vesicle protein 2 modulates binding to synaptotagmin. J Biol Chem 275:17195–17200.

Raiteri L, Raiteri M, Bonanno G (2002) Coexistence and function of different neurotransmitter transporters in the plasma membrane of CNS neurons. Prog Neurobiol 68:287–309.

Ramsay RR, Salach JI, Singer TP (1986) Uptake of the neurotoxin 1-methyl-4-phenylpyridine (MPP+) by mitochondria and its relation to the inhibition of the mitochondrial oxidation of NAD+-linked substrates by MPP+. Biochem Biophys Res Commun 134:743–748.

Rehavi M, Roz N, Weizman A (2002) Chronic clozapine, but not haloperidol, treatment affects rat brain vesicular monoamine transporter 2. Eur Neuropsychopharmacol 12:261–268.

Reimer RJ, Edwards RH (2004) Organic anion transport is the primary function of the SLC17/type I phosphate transporter family. Pflugers Arch 447:629–635.

Renick SE, Kleven DT, Chan J, Stenius K, Milner TA, Pickel VM, Fremeau RT, Jr. (1999) The mammalian brain high-affinity L-proline transporter is enriched preferentially in synaptic vesicles in a subpopulation of excitatory nerve terminals in rat forebrain. J Neurosci 19:21–33.

Riddle EL, Fleckenstein AE, Hanson GR (2005) Role of monoamine transporters in mediating psychostimulant effects. AAPS J 7:E847–E851.

Rizzoli SO, Richards DA, Betz WJ (2003) Monitoring synaptic vesicle recycling in frog motor nerve terminals with FM dyes. J Neurocytol 32:539–549.

Rogers GA, Kornreich WD, Hand K, Parsons SM (1993) Kinetic and equilibrium characterization of vesamicol receptor-ligand complexes with picomolar dissociation constants. Mol Pharmacol 44:633–641.

Rogers GA, Parsons SM (1989) Inhibition of acetylcholine storage by acetylcholine analogs in vitro. Mol Pharmacol 36:333–341.

Roghani A, Feldman J, Kohan SA, Shirzadi A, Gundersen CB, Brecha N, Edwards RH (1994) Molecular cloning of a putative vesicular transporter for acetylcholine. Proc Natl Acad Sci USA 91:10620–10624.

Roghani A, Shirzadi A, Kohan SA, Edwards RH, Butcher LL (1996) Differential distribution of the putative vesicular transporter for acetylcholine in the rat central nervous system. Brain Res Mol Brain Res 43:65–76.

Roseth S, Fonnum F (1995) A study of the uptake of glutamate, gamma-aminobutyric acid (GABA), glycine and beta-alanine in synaptic brain vesicles from fish and avians. Neurosci Lett 183:62–66.

Roseth S, Fykse EM, Fonnum F (1995) Uptake of L-glutamate into rat brain synaptic vesicles: effect of inhibitors that bind specifically to the glutamate transporter. J Neurochem 65:96–103.

Roseth S, Fykse EM, Fonnum F (1998) The effect of arachidonic acid and free fatty acids on vesicular uptake of glutamate and gamma-aminobutyric acid. Eur J Pharmacol 341:281–288.

Rudnick G, Steiner-Mordoch SS, Fishkes H, Stern-Bach Y, Schuldiner S (1990) Energetics of reserpine binding and occlusion by the chromaffin granule biogenic amine transporter. Biochemistry 29:603–608.

Rudnick G, Wall SC (1992) The molecular mechanism of "ecstasy" [3,4-methylenedioxy-methamphetamine (MDMA)]: serotonin transporters are targets for MDMA-induced serotonin release. Proc Natl Acad Sci USA 89:1817–1821.

Ruiz J, Ribera A, Jurado A, Marco F, Vila J (2005) Evidence for a reserpine-affected mechanism of resistance to tetracycline in Neisseria gonorrhoeae. APMIS 113:670–674.

Ryan TA, Smith SJ (1995) Vesicle pool mobilization during action potential firing at hippocampal synapses. Neuron 14:983–989.

Sagne C, El Mestikawy S., Isambert MF, Hamon M, Henry JP, Giros B, Gasnier B (1997) Cloning of a functional vesicular GABA and glycine transporter by screening of genome databases. FEBS Lett 417:177–183.

Sagne C, Isambert MF, Vandekerckhove J, Henry JP, Gasnier B (1997) The photoactivatable inhibitor 7-azido-8-iodoketanserin labels the N terminus of the vesicular monoamine transporter from bovine chromaffin granules. Biochemistry 36:3345–3352.

Sandoval V, Riddle EL, Hanson GR, Fleckenstein AE (2002) Methylphenidate redistributes vesicular monoamine transporter-2: role of dopamine receptors. J Neurosci 22:8705–8710.

Schafer MK, Varoqui H, Defamie N, Weihe E, Erickson JD (2002) Molecular cloning and functional identification of mouse vesicular glutamate transporter 3 and its expression in subsets of novel excitatory neurons. J Biol Chem 277:50734–50748.

Schafer MK, Weihe E, Varoqui H, Eiden LE, Erickson JD (1994) Distribution of the vesicular acetylcholine transporter (VAChT) in the central and peripheral nervous systems of the rat. J Mol Neurosci 5:1–26.

Scheel O, Zdebik AA, Lourdel S, Jentsch TJ (2005) Voltage-dependent electrogenic chloride/proton exchange by endosomal CLC proteins. Nature 436:424–427.

Scheunemann M, Sorger D, Wenzel B, Heinitz K, Schliebs R, Klingner M, Sabri O, Steinbach J (2004) Synthesis of novel 4- and 5-substituted benzyl ether derivatives of vesamicol and in vitro evaluation of their binding properties to the vesicular acetylcholine transporter site. Bioorg Med Chem 12:1459–1465.

Schivell AE, Batchelor RH, Bajjalieh SM (1996) Isoform-specific, calcium-regulated interaction of the synaptic vesicle proteins SV2 and synaptotagmin. J Biol Chem 271:27770–27775.

Schuldiner S, Granot D, Mordoch SS, Ninio S, Rotem D, Soskin M, Tate CG, Yerushalmi H (2001) Small is mighty: EmrE, a multidrug transporter as an experimental paradigm. News Physiol Sci 16:130–134.

Schuldiner S, Liu Y, Edwards RH (1993a) Reserpine binding to a vesicular amine transporter expressed in Chinese hamster ovary fibroblasts. J Biol Chem 268:29–34.

Schuldiner S, Shirvan A, Linial M (1995) Vesicular neurotransmitter transporters: from bacteria to humans. Physiol Rev 75:369–392.

Schuldiner S, Steiner-Mordoch S, Yelin R (1998) Molecular and biochemical studies of rat vesicular monoamine transporter. Adv Pharmacol 42:223–227.

Schuldiner S, Steiner-Mordoch S, Yelin R, Wall SC, Rudnick G (1993b) Amphetamine derivatives interact with both plasma membrane and secretory vesicle biogenic amine transporters. Mol Pharmacol 44:1227–1231.

Sievert MK, Ruoho AE (1997) Peptide mapping of the [125I]Iodoazidoketanserin and [125I]2-N-[(3′-iodo-4′-azidophenyl)propionyl]tetrabenazine binding sites for the synaptic vesicle monoamine transporter. J Biol Chem 272:26049–26055.

Song H, Ming G, Fon E, Bellocchio E, Edwards RH, Poo M (1997) Expression of a putative vesicular acetylcholine transporter facilitates quantal transmitter packaging. Neuron 18:815–826.

Stobrawa SM, Breiderhoff T, Takamori S, Engel D, Schweizer M, Zdebik AA, Bosl MR, Ruether K, Jahn H, Draguhn A, Jahn R, Jentsch TJ (2001) Disruption of ClC-3, a chloride channel expressed on synaptic vesicles, leads to a loss of the hippocampus. Neuron 29:185–196.

Sulzer D, Chen TK, Lau YY, Kristensen H, Rayport S, Ewing A (1995) Amphetamine redistributes dopamine from synaptic vesicles to the cytosol and promotes reverse transport. J Neurosci 15:4102–4108.

Sulzer D, Galli A (2003) Dopamine transport currents are promoted from curiosity to physiology. Trends Neurosci 26:173–176.

Sulzer D, Maidment NT, Rayport S (1993) Amphetamine and other weak bases act to promote reverse transport of dopamine in ventral midbrain neurons. J Neurochem 60:527–535.

Sulzer D, Rayport S (1990) Amphetamine and other psychostimulants reduce pH gradients in midbrain dopaminergic neurons and chromaffin granules: a mechanism of action. Neuron 5:797–808.

Tabb JS, Ueda T (1991) Phylogenetic studies of the synaptic vesicle glutamate transport system. J Neurosci 11:1822–1828.

Takamori S (2006) VGLUTs: 'exciting' times for glutamatergic research? Neurosci Res 55:343–351.

Takamori S, et al. (2006) Molecular anatomy of a trafficking organelle. Cell 127:831–846.

Takamori S, Rhee JS, Rosenmund C, Jahn R (2000) Identification of a vesicular glutamate transporter that defines a glutamatergic phenotype in neurons [see comments]. Nature 407:189–194.

Tayebati SK, Di Tullio MA, Amenta F (2004) Effect of treatment with the cholinesterase inhibitor rivastigmine on vesicular acetylcholine transporter and choline acetyltransferase in rat brain. Clin Exp Hypertens 26:363–373.

Teng L, Crooks PA, Dwoskin LP (1998) Lobeline displaces [3H]dihydrotetrabenazine binding and releases [3H]dopamine from rat striatal synaptic vesicles: comparison with d-amphetamine. J Neurochem 71:258–265.
Thompson CM, Davis E, Carrigan CN, Cox HD, Bridges RJ, Gerdes JM (2005) Inhibitor of the glutamate vesicular transporter (VGLUT). Curr Med Chem 12:2041–2056.
Tordera RM, Pei Q, Sharp T (2005) Evidence for increased expression of the vesicular glutamate transporter, VGLUT1, by a course of antidepressant treatment. J Neurochem 94:875–883.
Torres GE, Gainetdinov RR, Caron MG (2003) Plasma membrane monoamine transporters: structure, regulation and function. Nat Rev Neurosci 4:13–25.
Van der Kloot W (2003) Loading and recycling of synaptic vesicles in the Torpedo electric organ and the vertebrate neuromuscular junction. Prog Neurobiol 71:269–303.
Varoqui H, Erickson JD (1996) Active transport of acetylcholine by the human vesicular acetylcholine transporter. J Biol Chem 271:27229–27232.
Venton BJ, Seipel AT, Phillips PE, Wetsel WC, Gitler D, Greengard P, Augustine GJ, Wightman RM (2006) Cocaine increases dopamine release by mobilization of a synapsin-dependent reserve pool. J Neurosci 26:3206–3209.
Wall SC, Gu H, Rudnick G (1995) Biogenic amine flux mediated by cloned transporters stably expressed in cultured cell lines: amphetamine specificity for inhibition and efflux. Mol Pharmacol 47:544–550.
Wang XQ, Deriy LV, Foss S, Huang P, Lamb FS, Kaetzel MA, Bindokas V, Marks JD, Nelson DJ (2006) CLC-3 channels modulate excitatory synaptic transmission in hippocampal neurons. Neuron 52:321–333.
Wang YM, Gainetdinov RR, Fumagalli F, Xu F, Jones SR, Bock CB, Miller GW, Wightman RM, Caron MG (1997) Knockout of the vesicular monoamine transporter 2 gene results in neonatal death and supersensitivity to cocaine and amphetamine. Neuron 19:1285–1296.
Weaver JA, Deupree JD (1982) Conditions required for reserpine binding to the catecholamine transporter on chromaffin granule ghosts. Eur J Pharmacol 80:437–438.
Weihe E, Schafer MK, Erickson JD, Eiden LE (1994) Localization of vesicular monoamine transporter isoforms (VMAT1 and VMAT2) to endocrine cells and neurons in rat. J Mol Neurosci 5:149–164.
Whittaker VP (1987) Cholinergic synaptic vesicles from the electromotor nerve terminals of Torpedo. Composition and life cycle. Ann N Y Acad Sci 493:77–91.
Wilkens S (2005) Rotary molecular motors. Adv Protein Chem 71:345–382.
Winter S, Brunk I, Walther DJ, Holtje M, Jiang M, Peter JU, Takamori S, Jahn R, Birnbaumer L, hnert-Hilger G (2005) Galphao2 regulates vesicular glutamate transporter activity by changing its chloride dependence. J Neurosci 25:4672–4680.
Wojcik SM, Katsurabayashi S, Guillemin I, Friauf E, Rosenmund C, Brose N, Rhee JS (2006) A shared vesicular carrier allows synaptic corelease of GABA and glycine. Neuron 50:575–587.
Wolosker H, de Souza DO, de ML (1996) Regulation of glutamate transport into synaptic vesicles by chloride and proton gradient. J Biol Chem 271:11726–11731.
Xu T, Bajjalieh SM (2001) SV2 modulates the size of the readily releasable pool of secretory vesicles. Nat Cell Biol 3:691–698.
Yang L, Matthews RT, Schulz JB, Klockgether T, Liao AW, Martinou JC, Penney JB, Jr., Hyman BT, Beal MF (1998) 1-Methyl-4-phenyl-1,2,3,6-tetrahydropyride neurotoxicity is attenuated in mice overexpressing Bcl-2. J Neurosci 18:8145–8152.
Yelin R, Schuldiner S (1995) The pharmacologicalal profile of the vesicular monoamine transporter resembles that of multidrug transporters. FEBS Lett 377:201–207.

# Core Proteins of the Secretory Machinery

**Thorsten Lang and Reinhard Jahn(✉)**



**Abstract** Members of the Rab, SM- and SNARE-protein families play key roles in all intracellular membrane trafficking steps. While SM- and SNARE-proteins become directly involved in the fusion reaction at a late stage, Rabs and their effectors mediate upstream steps such as vesicle budding, delivery, tethering, and transport. Exocytosis of synaptic vesicles and regulated secretory granules are among the best-studied fusion events and involve the Rab3 isoforms Rab3A-D, the SM protein munc18-1, and the SNAREs syntaxin 1A, SNAP-25, and synaptobrevin 2. According to the current view, syntaxin 1A and SNAP-25 at the presynaptic membrane form a complex with synaptic vesicle-associated synaptobrevin 2. As complex formation proceeds, the opposed membranes are pulled tightly together, enforcing the

Reinhard Jahn
Max Planck Institute for Biophysical Chemistry, Dep. Neurobiology, Am Fassberg 11, 37077 Göttingen, Germany
rjahn@gwdg.de

T.C. Südhof, K. Starke (eds.), *Pharmacology of Neurotransmitter Release. Handbook of Experimental Pharmacology 184.*

fusion reaction. Munc18-1 is essential for regulated exocytosis and interacts with syntaxin 1A alone or with SNARE complexes, suggesting a role for munc18-1 in controlling the SNARE-assembly reaction. Compared to other intracellular fusion steps, special adaptations evolved in the synapse to allow for the tight regulation and high membrane turnover rates required for synaptic transmission. Synaptic vesicle fusion is triggered by the intracellular second messenger calcium, with members of the synaptotagmin protein family being prime candidates for linking calcium influx to fusion in the fast phase of exocytosis. To compensate for the massive incorporation of synaptic vesicles into the plasma membrane during exocytosis, special adaptations to endocytic mechanisms have evolved at the synapse to allow for efficient vesicle recycling.

# 1 SNAREs

SNAREs comprise a superfamily of proteins that function in all membrane fusion steps of the secretory pathway within eukaryotic cells. They are small proteins that vary in structure and size (see Section 1.1), but share an evolutionary conserved stretch of 60–70 amino acids containing eight heptad repeats, which is termed SNARE motif (Brunger 2005). The number of different SNAREs varies between different organisms, ranging from 25 in yeast, 36 in mammals, to over 50 in plants. Each fusion step requires a specific set of four different SNARE motifs that is contributed by three or four different SNAREs, and each of the membranes destined to fuse contains at least one SNARE with a membrane anchor.

Membrane traffic usually consists of a sequence of steps involving the generation of a transport vesicle by budding from a precursor compartment, the transport of the vesicle to its destination, and finally the docking and fusion of the vesicle with the target compartment. SNAREs operate in the very last step of this sequence (Jahn and Scheller 2006). SNAREs on opposed membranes form a complex in "trans" that is mediated by the SNARE motifs and that progressively assembles from the N-terminal tips toward the C-terminal membrane anchors, thus clamping the two membranes together. The energy released during assembly is probably used for overcoming the fusion barrier. During fusion, the complex reorients from "trans" to "cis." Cis-SNARE complexes are unusually stable, and disassembly requires the action of an AAA-ATPase and ATP. Hence, SNAREs undergo a conformational cycle that is crucial for fusion. The cycle is controlled by an array of regulatory factors that are only partially understood (see Section 1.2).

While each fusion step appears to be mediated by a specific set of SNAREs, some SNAREs operate in multiple transport steps where they each interact with different SNARE partners. Conversely, SNAREs of the same subfamily (see Section 1.1) can substitute for each other, at least to a certain degree, in a given transport step. In vitro, there is less specificity in SNARE assembly, suggesting that additional control mechanisms are involved. According to current concepts, specificity in membrane traffic is achieved by successive layers of regulation that operate upstream of SNARE assembly and that involve members of conserved protein families. Rab-proteins are thought to orchestrate the initial contact between membranes destined to fuse (see Section 4) and to assure that only appropriate organelles are tethered. SM proteins are involved in preparing and proofreading SNAREs for trans-complex formation (see Section 2). The combination of these and possibly additional still unknown mechanisms would guarantee the required specificity in intracellular membrane trafficking.

The SNAREs involved in the fusion of synaptic vesicles and of secretory granules in neuroendocrine cells, referred to as neuronal SNAREs, have been intensely studied and serve as a paradigm for all SNAREs. They include syntaxin 1A and SNAP-25 at the presynaptic membrane and synaptobrevin 2 (also referred to as VAMP 2) at the vesicle membrane. Their importance for synaptic neurotransmission is documented by the fact that the block in neurotransmitter release caused by botulinum and tetanus neurotoxins is due to proteolysis of the neuronal SNAREs (Schiavo et al. 2000). Genetic deletion of these SNAREs confirmed their essential role in the last steps of neurotransmitter release. Intriguingly, analysis of chromaffin cells from KO mice lacking synaptobrevin or SNAP-25 showed that these proteins can be at least partially substituted by SNAP-23 and cellubrevin, respectively (Sorensen et al. 2003; Borisovska et al. 2005), i.e., the corresponding SNAREs involved in constitutive exocytosis.

In the following sections, we limit our discussion to the neuronal SNARE complex that, however, is paradigmatic for most SNARE complexes studied so far.

## *1.1 Structure of the Neuronal SNAREs*

### 1.1.1 Syntaxin 1A

Syntaxin 1A is a protein composed of 288 amino acids (all numbers refer to rat proteins). Its structure is typical for most SNAREs: the SNARE motif is flanked by an independently folded N-terminal domain and a single transmembrane domain at the C-terminus (Figure 1). The N-terminal domain is composed of an antiparallel three-helix bundle with a small N-terminal extension and is linked to the SNARE motif via a long flexible linker region. Syntaxin is able to interact intramolecularly by folding back its N-terminal domain onto the SNARE motif, resulting in the so-called closed conformation of syntaxin in which the linker is structured and part of the SNARE motif assumes a helical conformation. Syntaxin 1 is highly abundant in neurons and neuroendocrine cells (approximately 1% of total brain protein) but lacking in non-neuronal cells.

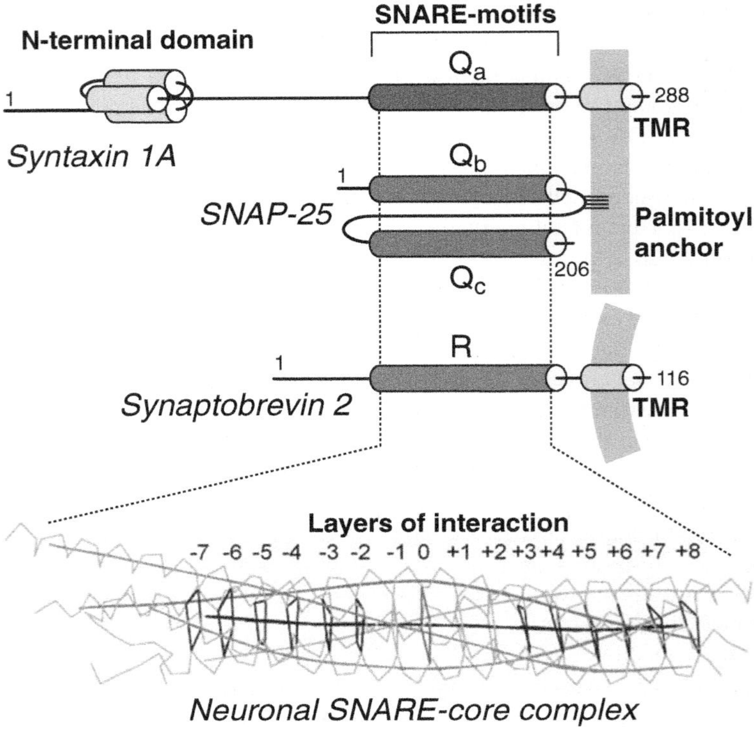


**Fig. 1** Structure of the neuronal SNAREs. Upper panel: domain structure of the three neuronal SNARE proteins involved in synaptic vesicle fusion. Syntaxin 1A and SNAP-25 (contains two SNARE motifs) are associated with the presynaptic membrane, whereas synaptobrevin 2 is synaptic vesicle associated. The SNARE motifs form a stable complex (core complex) whose crystal structure has been analyzed (lower panel). In the complex, each of the SNARE motifs adopts an alpha-helical structure, and the four alpha-helices are aligned in parallel forming a twisted bundle (modified from Sutton et al. 1998). Stability of the complex is mediated by layers of interaction ($-7$ to $+8$) in which amino acids from each of the four alpha-helices participate (see text).

### 1.1.2 Synaptobrevin 2/VAMP 2

Synaptobrevin 2 is a small protein composed of 118 amino acids. It contains a SNARE motif with a short N-terminal proline-rich extension but lacks an independently folded N-terminal domain. Like syntaxin 1, the protein possesses a C-terminal transmembrane domain that is connected to the SNARE motif by a short linker (Figure 1). Synaptobrevin is palmitoylated at cysteine residues close to its transmembrane domain. Synaptobrevin 2 is highly expressed in neurons and neuroendocrine cells, but unlike syntaxin 1 it is also present in many non-neuronal tissues albeit at low levels.

### 1.1.3 SNAP-25

SNAP-25, a protein of 208 amino acids, deviates from the typical SNARE structure in that it has two SNARE motifs, joined by a flexible linker region, but lacks a transmembrane domain (Figure 1). The linker contains a cluster of four palmitoylated cysteine residues by which the protein is anchored at the plasma membrane. SNAP-25 can be phosphorylated at positions Thr138 and Ser187 by cAMP-dependent protein kinase (PKA) and protein kinase C (PKC), respectively. SNAP-25 represents a small subgroup of SNAREs with a similar structure, including SNAP-23, SNAP-29, and SNAP-47. In contrast to the neuron-specific SNAP-25 these SNAREs are ubiquitously expressed.

### 1.1.4 The Neuronal SNARE Complex

Syntaxin 1A, SNAP-25, and synaptobrevin 2 undergo structural changes when they assemble during membrane fusion. The crystal structure of the core-region of the neuronal SNARE complex (the assembled SNARE motifs) has been analyzed (Figure 1) and turned out to be paradigmatic for all SNARE-core complexes. In the complex, all SNARE motifs adopt an alpha-helical structure and are aligned in parallel, forming a twisted coiled-coil (Sutton et al. 1998). Along the longitudinal axis in the center of the bundle 16 stacked layers of interacting side chains have been identified (Figure 1). Each layer is formed by four amino acids, each contributed by a different SNARE motif. The layers are largely hydrophobic, with the exception of one ionic central layer that contains three glutamines and one arginine, all highly conserved (Fasshauer et al. 1998). The central layer is used as a center of reference for the remaining layers and termed "0"-layer. The layers from the N-terminal region of the SNARE motif to the 0-layer are termed $-7$ to $-1$, those upward from the 0-layer $+1$ to $+8$. An attractive working hypothesis is that SNARE motifs resemble a molecular zipper as they assemble from their N-terminal toward their C-terminal regions. According to this model, layers of interaction form sequentially one after another, pulling the opposed membranes stepwise together. During formation of the last layers, membranes would be forced so closely together that they fuse.

### 1.1.5 Q/R Classification of SNAREs

Initially, SNAREs were classified functionally into t-SNAREs (target-membrane SNAREs, e.g., syntaxin 1A and SNAP-25) or v-SNAREs (vesicle-membrane SNAREs, e.g., synaptobrevin 2). However, this concept cannot be applied to homotypic fusion events and is misleading because SNAREs are grouped together that belong to different subfamilies. A complete and unambiguous grouping is accomplished by the Q-/R-SNARE classification referring to the conserved amino acids present in the "0"-layer. According to the position of the SNARE motif in the structurally conserved SNARE complex, SNARE motifs are classified into $Q_a$-,

$Q_b$-, $Q_c$- and R-SNAREs (Bock and Scheller 2001). Following this classification, syntaxin 1A, SNAP-25. and synaptobrevin 2 represent the $Q_a$-, $Q_b$- and $Q_c$-, and R-SNAREs, respectively (Fasshauer et al. 1998). It turned out later that actually all functional SNARE complexes assigned to trafficking steps in yeast and mammals have a $Q_aQ_bQ_cR$-composition (Hong 2005; Jahn and Scheller 2006).

## *1.2 Assembly and Disassembly of SNAREs: Mechanistic Considerations*

As discussed above, fusion is driven by the assembly of SNAREs mediated by their SNARE motifs. Assembly is associated with a major release of energy, and consequently the SNAREs need to be refueled with energy by the generation of free SNAREs before they are reusable for another round of fusion. Thus, SNAREs undergo cyclic assembly and disassembly, and together the individual reactions involved are referred to as the conformational cycle of SNAREs (Figure 2).

### 1.2.1 Assembly and Fusion

Free SNAREs are presumably short-lived, as they can form complexes among themselves, including homophilic oligomerization into clusters or with SNARE interacting proteins. It is becoming apparent that initial trans-contact between SNAREs

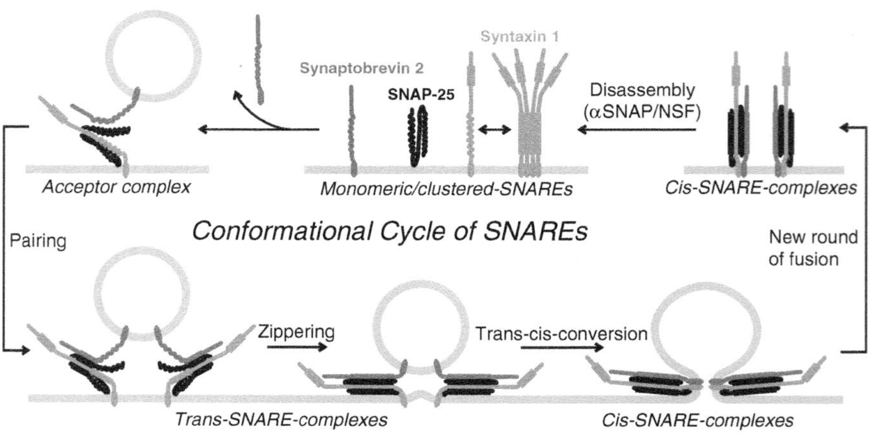


**Fig. 2** The conformational cycle of SNAREs. SNAREs cycle between two extreme conformations, the unstructured monomeric SNAREs and the fully assembled cis-SNARE complexes. Initially, SNAREs on the membranes destined to fuse establish trans-SNARE complexes between the opposed membranes. Proceeding SNARE complex assembly forces the membranes tightly together enforcing membrane fusion. The resulting cis-SNARE complexes are disassembled into free SNAREs by the ATPase NSF and its co-factor, a process that consumes ATP and fuels the SNAREs with energy for undergoing a new SNARE cycle (for details see text).

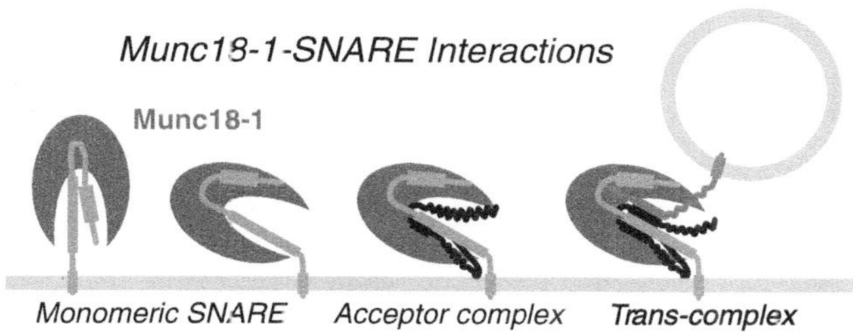


**Fig. 3** munc18-1 SNARE-binding modes. The following munc18-1 interactions with monomeric/assembled SNAREs have been proposed. From left, binding of munc18-1 to a closed conformation of syntaxin 1A (Misura et al. 2000), to a half-open conformation of syntaxin or to an acceptor complex formed by syntaxin and SNAP-25 (Zilly et al. 2006), and to an assembled SNARE complex (Dulubova et al. 2007). It is possible that each of the proposed complexes represents an intermediate on a munc18-1 controlled molecular pathway of specific SNARE complex assembly.

(also referred to as "nucleation") is tightly controlled by cellular factors. A key role in preparing SNAREs, and perhaps in controlling nucleation, is assigned to the SM proteins, including its neuronal variant munc18-1, which are discussed more fully below. (see Section 3 and Figure 3).

It is not clear at present whether assembly is regulated after nucleation of trans-complexes or whether nucleation invariantly proceeds toward fusion. In vitro, assembly is completed rapidly after nucleation (Pobbati et al. 2006). However, mutagenesis of side chains in the central layers of the SNARE complex resulted in phenotypes that are best explained by energy minima during SNARE zippering, representing partially zippered and metastable intermediates in the molecular pathway of SNARE assembly. A partially assembled SNARE complex thus may account for the primed state of secretory granules in neuroendocrine cells (Sorensen et al. 2006). A protein thought to act on partially assembled trans-complexes is complexin, a small, soluble protein of 15 kDa (Marz and Hanson 2002). In vitro, complexin binds to SNARE complexes with high affinity, with the central part of complexin forming an α-helix that binds in an antiparallel orientation in the groove between synaptobrevin and syntaxin. While this interaction is thought to stabilize partially assembled SNARE complexes, deletion of complexin in mice lowers the sensitivity of neuronal exocytosis to calcium, resulting in a phenotype resembling that of synaptotagmin knockout mice (see section 3.2). Recently it has been proposed that complexin operates by stabilizing an otherwise metastable trans-SNARE intermediate that requires calcium-dependent displacement of complexin by synaptotagmin to be activated for fusion (Figure 4) (Tang et al. 2006).

The final steps in fusion are only incompletely understood. For instance, it is controversial whether the non-bilayer transition states in fusion are initiated primarily by force, transmitted from the "pull" of the assembling SNARE motifs via the linkers onto the membrane, as suggested by mutagenesis of the linker domain. Alternatively, the function of the SNARE motifs may be confined to close apposition of the membranes, with the final steps being catalyzed by other factors such as a

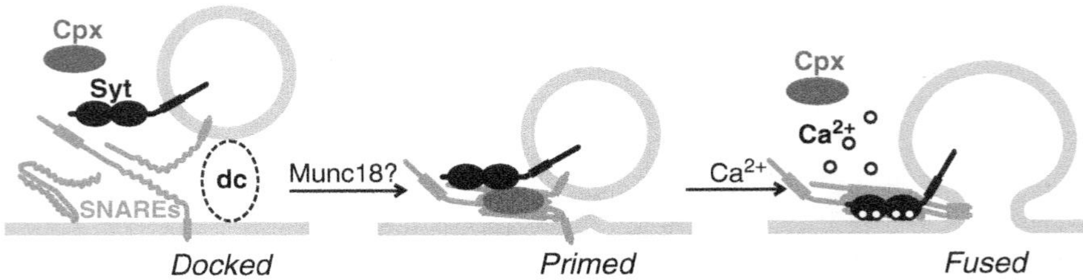


**Fig. 4** Stages in synaptic vesicle exocytosis. Putative intermediate steps on the molecular pathway to synaptic vesicle fusion. Vesicle delivery and tethering to the presynaptic membrane most likely involves Rab-proteins and their effectors. So far, the nature of a speculative docking complex (dc) is unclear, but docking appears to be independent from SNARE proteins. In the primed state, SNAREs have assembled into a complex probably stabilized by complexin (Cpx). The fusion reaction is arrested until the intracellular calcium concentration increases. The putative calcium sensor for fast neurotransmitter release, synaptotagmin 1 (Syt), binds to intracellular calcium and in turn triggers fusion by associating with the presynaptic membrane and interacting with the SNARE complex, thereby displacing complexin (Tang et al. 2006).

perturbing "fusogenic" function of the transmembrane domains. For instance, it has been suggested that hetero-dimerization of the transmembrane domains of syntaxin 1A and synaptobrevin 2 facilitate the conversion from a hypothetical hemifusion intermediate state to full fusion (Ungermann and Langosch 2005). Recent evidence supports the involvement of hemifusion intermediates in the SNARE fusion pathway (see, e.g., Yoon et al. 2006). Furthermore, it is still controversial whether the fusion pore, i.e., the first aqueous connection between the vesicles and the extracellular space, is primarily lipidic or whether proteins (e.g., the transmembrane domains of the SNAREs) are part of the transition state.

### 1.2.2 Other Proteins Involved in the Regulation of SNAREs

In addition to the proteins discussed above, neuronal SNAREs were reported to interact with numerous other proteins in a specific manner, but in most cases both the structural basis and the biological function of these interactions need to be defined. For instance, synaptophysin, a membrane protein of synaptic vesicles, forms a complex with synaptobrevin in which synaptobrevin is not available for interactions with its partner SNAREs syntaxin 1A and SNAP-25, suggesting that this complex represents a reserve pool of recruitable synaptobrevin (Becher et al. 1999) or regulates interactions between the vesicle-associated synaptobrevin and the plasmalemmal SNAREs. Alternatively, it has been suggested that this complex is involved in synaptobrevin sorting to synaptic vesicles.

Munc13 is a 200 kDa protein essential for synaptic vesicle priming. As double knockouts of munc13 and syntaxin in *Caenorhabditis elegans* are rescued by constitutively open syntaxin, it has been suggested that munc13 mediates the transition from closed to open syntaxin (Brunger 2005). Recently, a ternary complex composed of Rab3, RIM, and munc13 was described, suggesting a function in targeting synaptic vesicles to the priming machinery.

Tomosyn is a soluble protein of 130 kDa with a C-terminal R-SNARE motif that is capable of replacing synaptobrevin in the neuronal SNARE complex. Most available data indicate that tomosyn negatively regulates exocytosis by competing with synaptobrevin in the formation of SNARE complexes (Brunger 2005), thereby leading to the inhibition of synaptic vesicle priming (McEwen et al. 2006).

### 1.2.3 Disassembly of SNARE Complexes

After membrane fusion, all neuronal SNAREs reside in the plasma membrane. Their assembled SNARE motifs are aligned in parallel and the TMRs of syntaxin 1A and synaptobrevin 2 are close to each other in the same membrane. These cis-SNARE complexes are of remarkable stability and do not disassemble spontaneously into free SNAREs. Reactivation of the SNAREs is mediated by the ATPase NSF (N-ethylmaleimide-sensitive factor). NSF is required for all intracellular trafficking steps, and its function is to disassemble cis-SNARE complexes into free SNAREs. NSF is a member of the AAA-protein family (ATPases associated with other activities) that generally appear to be involved in disentangling protein complexes and protein aggregates (Hanson and Whiteheart 2005). Unlike many other AAA-ATPases, NSF cannot act alone–it needs co-factors termed SNAPs (soluble NSF attachment protein), represented by three isoforms termed $\alpha$-, $\beta$-, and $\gamma$-SNAP. SNAPs bind to the cis-SNARE complex first and in turn recruit NSF, followed by stimulation of its ATPase activity. The hydrolysis of ATP induces major conformational changes resulting in the disassembly of the entire complex into its free constituents. In this uncomplexed state, the neuronal SNAREs are probably most susceptible to cleavage by clostridial neurotoxins.

## 2 Sec1/Munc18 (SM) Proteins

SM proteins were initially discovered during genetic screens in yeast and *C. elegans* for mutants showing defects in membrane traffic and secretion (Toonen and Verhage 2003). They comprise a small family of cytosolic proteins of 650–700 amino acids with seven members in mammals and four members in yeast. Although not yet documented unequivocally in each case, it appears that each trafficking step catalyzed by SNAREs is dependent on one of the SM proteins. Due to the small number of SM proteins, it is evident that some of them operate in more than one fusion reaction. In contrast to SNAREs, no functional redundancy has been observed so far, but SM proteins from distant species are capable of replacing each other provided they participate in the same trafficking step. Wherever investigated, genetic deletion of an SM protein leads to a block of the corresponding fusion reaction, indicating that their role in membrane fusion is essential. SM proteins most likely exert their function upon regulating SNARE assembly, although also a role in docking has been suggested (Voets et al. 2001). Although significant progress has been made in recent years, it has turned out to be surprisingly tricky to unravel how they work, with major questions still being open.

## 2.1 SM Protein Interactions with SNAREs

All SM proteins interact with SNAREs, either directly or indirectly in complex with other proteins. Furthermore, strong genetic interactions have been documented between SM proteins and SNAREs. In some cases deletion of an SM protein is associated with a reduced expression level of its respective SNARE binding partner (Gallwitz and Jahn 2003).

The crystal structure of mammalian and squid munc18 and of yeast Sly1p shows a remarkable degree of structural conservation. SM proteins are composed of three domains that form an arch-shaped molecule with a central cleft (Misura et al. 2000; Bracher and Weissenhorn, 2002). Surprisingly, however, a confusing variety of binding modes between SM proteins and SNAREs has been observed (see, e.g., Figure 3). Both munc18 and Sly1p directly bind to the corresponding Qa-SNAREs syntaxin 1 and Sed5p (syntaxin 5), respectively, but in a completely different manner. In the crystal structure of the munc18-1-syntaxin 1A complex, syntaxin is arrested in the closed conformation, being inserted in the central cleft of munc18-1. In stark contrast, Sed5p binds only with a short N-terminal peptide that precedes its helical N-terminal domain, and the binding site on Sly1p is represented by a small groove on the surface of the SM protein, with no involvement of the central cleft. The latter binding mode has also been described for yeast and mammalian SM proteins involved in trafficking steps of the ER, the Golgi, the trans-Golgi network, and early endosomes (Toonen and Verhage 2003; Rizo and Südhof 2002), and it thus appears that the binding mode between munc18 and syntaxin 1 is unique among the family.

## 2.2 Munc18-1–an Oddity among the SM Proteins?

Munc18-1 (and its ortholog unc-18 in *C. elegans*) are essential for exocytosis. Knockout of munc18-1 in mice results in a nervous system that initially develops normally but in which synapses are totally silent – one of the most dramatic phenotypes of synaptic proteins that highlights the essential role and the lack of redundancy of the protein. Despite this dramatic phenotype, it has been remarkably difficult to reconcile the physiological findings obtained from synapses containing deleted, overexpressed, or otherwise manipulated munc18 with its biochemical properties studied in vitro. As discussed above, munc18-1 binds to syntaxin 1A with high affinity in a manner that clamps the N-terminal domain of syntaxin onto its

SNARE motif, effectively preventing syntaxin from binding to its partner SNAREs SNAP-25 and synaptobrevin 2. However, an inactivation of syntaxins' SNARE function is exactly the opposite of what one would expect from an essential protein, and (except for a debated study in *Drosophila*) even massive overexpression of munc18 does not appear to affect exocytosis (Gallwitz and Jahn 2003). Thus, it has been debated whether the "closed" conformation of syntaxin 1 in the munc18 represents a nonphysiological extreme situation that does not occur in intact cells, particularly since no other SM protein interferes with the formation of SNARE complexes (see below). Hence, the search has been on for munc18 SNARE complexes with different properties that may be closer to the physiological situation. Recently, evidence was provided showing that in native membranes munc18-1 stabilizes a half-closed conformation of syntaxin that still is capable of engaging in SNARE assembly (Zilly et al. 2006). Furthermore, it has been shown that munc18-1 activates SNARE-mediated membrane fusion in a reconstituted liposome system (Shen et al. 2007), and can bind directly to the assembled SNARE complex (Dulubova et al. 2007). Thus it is becoming apparent that munc18, in addition to its ability to form a complex with closed syntaxin 1, is capable of interacting with the neuronal SNAREs in other binding modes that are not inhibitory but rather may promote assembly of SNARE complexes (Figure 3). Such binding modes are more compatible with all other SM proteins that bind to partially or even fully assembled SNARE complexes. Although much more work needs to be done, a picture is emerging according to which SM proteins may assist in the formation of SNARE acceptor complexes needed for trans-SNARE interaction, and in doing so they may also be involved in proofreading of SNAREs, thus differentiating between cognate and noncognate SNAREs (Peng and Gallwitz 2002).

With the focus clearly being on the SNAREs, it needs to be borne in mind that SM proteins interact with a diverse array of additional proteins, in some cases even forming stable complexes (e.g., the SM protein Vps33p is part of the HOPS/VpsC complex needed for vacuole fusion in yeast). For instance, munc18 binds to the cytoplasmic protein Mint and it has been suggested that munc18-1 binding to Mint could regulate exocytosis by syntaxin-independent interactions (Schütz et al. 2005; Ciufo et al. 2005).

## 3 Synaptotagmins

### *3.1 Synaptotagmin Family*

Synaptotagmins comprise a small family of single-membrane spanning proteins that are expressed in neurons and neuroendocrine cells. So far 16 members have been identified in vertebrates (Craxton 2004). They contain an N-terminal transmembrane domain followed by a variable linker region and two C2 domains which are connected by a short linker (Südhof 2002). Some synaptotagmins have additional short

N-terminal domains that in some cases are glycosylated (Syt 1 and 2) or carry a disulfide bond (Syt 3, 5, 6, and 10). The two C2 domains (C2 stands for second constant sequence, as defined when the first C2 domains were identified in protein kinase C isoforms) are termed C2A and C2B and generally bind three and two calcium ions, respectively, although some synaptotagmins do not bind calcium. Synaptotagmins are found both on synaptic and secretory vesicles (Syt 1, 2, and 9) and on the plasma membrane (e.g., Syt 3 and 7). More recently, additional proteins have been discovered that are similar to synaptotagmins in that they possess C2 domains and membrane anchors. Best characterized are the ferlins which comprise a membrane protein family with four to seven C2 domains and a single TMR at their C-terminus (see, e.g., Washington and Ward 2006). As for the synaptotagmins, recent evidence suggests that ferlins also play a role in $Ca^{2+}$-dependent exocytosis such as that involved in membrane repair in muscle cells (requiring dysferlin) and in vesicle release in the hair cells of the inner ear (otoferlin).

Synaptotagmin 1 is the founding member of the synaptotagmin family, and it is also the most intensely studied isoform. Together with Syt2, Syt1 functions as synaptic $Ca^{2+}$-sensor that couples $Ca^{2+}$-influx with fast transmitter release (Chapman 2002). Intriguingly, some of the plasma membrane associated synaptotagmins have a 10-fold higher binding affinity for $Ca^{2+}$. Together with the finding that Syt7 can function as $Ca^{2+}$-sensor for exocytosis in chromaffin cells but not in neurons, it is conceivable that low-calcium-affinity synaptotagmins (Syt1 and 2) trigger fast neurotransmitter release in neurons (see Section 3.2), whereas plasma-membrane-associated Syt7 may function as calcium sensor for exocytosis in neuroendocrine cells. In addition, having synaptotagmins on both the plasma and the vesicle membrane may result in a system of complementary $Ca^{2+}$-sensors regarding the sensitivity to intracellular calcium signals.

## 3.2 *Synaptotagmin 1 as* $Ca^{2+}$*-Sensor for Fast Neurotransmitter Release*

Genetic deletion of Syt1 in *Drosophila* and mice leads to the loss of the fast, $Ca^{2+}$-dependent phase of transmitter release that follows the arrival of an action potential in the presynaptic nerve terminal (Chapman 2002). Disruption of $Ca^{2+}$-binding to either of the C2 domains severely inhibits the function of synaptotagmin in mediating fast synchronous transmitter release, with the disruption of the C2B-domain being more severe than that of the C2A domain. Furthermore, when mutant synaptotagmins exhibiting either reduced or increased $Ca^{2+}$-affinity are expressed in mice lacking synaptotagmin I a close correlation was observed between the $Ca^{2+}$-affinity and the $Ca^{2+}$ -dependence of neurotransmitter release, all confirming that Syt1 is an essential link between $Ca^{2+}$-influx and the synaptic fusion machinery. Interestingly, lack of synaptotagmin does not abolish exocytosis, as a "normal" SNARE-dependent exocytotic response is attainable when exocytosis is triggered by $\alpha$-latrotoxin (the active ingredient of black widow spider venom).

### 3.3 Molecular Mechanism of Synaptotagmin 1

C2 domains are represented by stable, mostly β-stranded folds, with the $Ca^{2+}$-binding site at one end of the elongated domain. The $Ca^{2+}$-binding site only incompletely coordinates the $Ca^{2+}$-ions. Consequently, the affinity of the free C2 domains for $Ca^{2+}$-is rather low but dramatically increases in the presence of acidic phospholipids. Thus, the C2 domains mediate $Ca^{2+}$-binding to membranes that in the case of Syt1 is further enhanced in the presence of phosphatidylinositol (4, 5) bisphosphate ($PIP_2$). Furthermore, synaptotagmin binds to both isolated syntaxin 1 and SNAP-25 as well as to partially (containing SNAP-25 and syntaxin) and fully assembled SNARE complexes in vitro. For SNARE binding, both $Ca^{2+}$-dependent and $Ca^{2+}$-independent interactions have been described. Mutagenesis studies involving partial or full disruption of $Ca^{2+}$-dependent binding to phospholipids and of binding to SNAREs have suggested that each of these interactions is required for the function of Syt1.

As in the case of munc18-1, it has been tricky to delineate the molecular mechanism of Syt1 action, and many details are still unclear (Rizo et al. 2006). The problems in understanding the mechanism of synaptotagmin, despite major efforts, highlight the fact that we do not yet have a good understanding of the status of the fusion machinery including SNAREs and associated proteins in the docked and primed state that synaptotagmin is acting upon. $Ca^{2+}$-dependent binding of the C2-domains to membranes may clamp the membranes tightly together (as recently suggested). Furthermore, membrane binding may destabilize bilayers at the site of fusion, facilitating the formation of transition states (such as hemifusion) or destabilizing the transition states themselves. Furthermore, as discussed above, Syt1 is capable of displacing complexin from the surface of the SNARE complex in a $Ca^{2+}$-dependent manner. It was proposed that these proteins act in sequence, with complexin stabilizing a labile SNARE trans-intermediate that then is driven toward fusion by Syt1 upon displacement of complexin.

## 4 Rab Proteins

Rabs are small (20–29 kDa) ubiquitously expressed proteins. They represent monomeric GTPases which belong to the Ras GTPase superfamily. So far 11 members are known in yeast and more than 60 in mammalian cells (Schultz et al. 2000), whereas the numbers reflect the complexity of membrane trafficking pathways in these different organisms. Rabs cycle between the cytosol and the membrane of the trafficking organelle. This cycle is controlled by conformational changes that are regulated by guanine-nucleotides, thus providing a molecular switch, with the membrane-bound GTP-form being "on" and the GDP-form being "off." GTP-Rabs bind to proteins termed effectors that only recognize Rabs in their GTP-bound state. A large diversity of Rab effectors is known, many being specific for a single or for small subsets of Rabs.

Rabs function as master regulators of membrane docking and fusion. Fusion can only occur if membrane contact is established, and numerous lines of evidence suggest that contact is orchestrated by Rabs. In fact, selective activation of Rabs at a given site ensures directionality and specificity of membrane docking and fusion (Grosshans et al. 2006), similar to the GTPases Sar1 and Arfs that are involved in coat recruitment during vesicle budding. In some cases, distinct Rab domains are present on the same organelle where they are involved in different transport steps.

As Rabs share a common structure and are highly homologous, the structural diversity of their effectors probably reflects the versatile functions of these GTPases as molecular switches. This diversity is highlighted by the functional diversity of Rab effectors. For instance Rab27a regulates transport of melanosomes to the cell periphery by binding to its effector melanophilin. Melanophilin associates with the actin motor myosin-Va. Rab-mediated tethering of membranes in preparation for fusion involves multimeric complexes. One of the best-studied examples is the exocyst, an octameric protein complex that tethers secretory vesicles to the plasma membrane in yeast. For homotypic membrane fusion of early endosomes (or of yeast vacuoles), Rab effectors interact with SNAREs such as the Rab5 effector EEA1 that binds to syntaxin-13.

## *4.1 Rab3*

Despite major efforts, the precise role of Rab proteins in synaptic exocytosis is still not clear. One of the most abundant synaptic Rab proteins is Rab3, which is selectively localized to synaptic vesicles and that is represented by four homologous isoforms (Rab3A, Rab3B, Rab3C, and Rab3D). Of these, Rab3A is the most abundant and best studied (Südhof 2004). Rab3A undergoes a synaptic vesicle association and dissociation cycle coupled to calcium-stimulated exocytosis and recovery after stimulation.

Like all Rabs, the GTP-bound form of Rab3A is anchored in the synaptic vesicle membrane via a covalently bound geranylgeranyl moiety. When exocytosis is triggered, Rab3A-bound GTP is hydrolyzed to GDP, and the resulting Rab3A-GDP forms a complex with GDI (guanine dissociation inhibitor) in which the geranylgeranyl anchors are enveloped, leading to the dissociation of the GDI-Rab3A-GDP complex from the synaptic vesicle membrane. Rab3A is then recruited again to the synaptic vesicle membrane by a poorly understood mechanism involving binding of a specific GEF (guanine nucleotide exchange factor) to Rab3A and GDP exchange by GTP.

The tight coupling of exocytosis to the Rab3A association and dissociation cycle suggested a key role for Rab3A and its effectors in mediating directionality of synaptic vesicle traffic. However, mice lacking Rab3A are viable and have an only moderate synaptic dysfunction. In neurons derived from the hippocampal CA1 region, an alteration of the short-term plasticity was observed. In contrast, in neurons derived from the hippocampal CA3 region short-term plasticity was unaltered, but

Rab3A was essential for mossy-fiber long-term potentiation. However, a complete genetic analysis in mice showed that Rab3 is essential for survival in mice and that the Rab3 isoforms are functionally redundant (Schlüter et al. 2004).

Presently, it cannot be excluded that despite the apparently highly specific function of certain Rabs in intracellular trafficking pathways there is redundancy with respect to vesicle docking in the synapse. The surprising diversity of Rabs on highly purified synaptic vesicles (more than 30 different Rabs) supports the view that multiple Rabs are required for synaptic vesicle recycling, which may have overlapping functions.

### *4.2 Rab3 Effectors*

Rabphilin and RIM1$\alpha$/2$\alpha$ represent two different classes of effectors that bind to Rab3-GTP (Südhof 2004). They have structural similarities, as both contain two $C_2$ domains, an N-terminally located zinc-finger domain that mediates binding to GTP-Rab3, and sites for phosphorylation by PKA that are located in the center of the proteins.

Rabphilin 3A is a soluble protein that is recruited to the membrane of synaptic vesicles by Rab3A and C in a GTP-dependent manner, closely coupling it to the Rab3 cycle. Like synaptotagmin, it has two functional C2 domains in the C-terminal region, but unlike synaptotagmins, it does not contain a transmembrane domain.). Rabphilin is phosphorylated by various kinases in a stimulation-dependent manner. Rabphilin knockout mice are viable and do not show a major synaptic phenotype, thus providing no clue for the function of this protein.

In contrast, genetic deletion of RIM1$\alpha$ in mice revealed that the protein is required for long-term potentiation both in the hippocampus and in the cerebellum. Biochemical experiments revealed that RIM1$\alpha$ is part of a presynaptic protein scaffold containing "active zone proteins" that is required for normal release of neurotransmitters. No change in the number of docked vesicles was observed, however, suggesting that other proteins are needed for the docking of synaptic vesicles at active zones.

## 5 Endocytic Proteins

Exocytosis of synaptic vesicles or of regulated secretory granules results in incorporation of membrane into the plasma membrane. For maintaining the cell surface area constant, homeostatic mechanisms are required that assure a rapid and efficient re-internalization of the incorporated vesicle membranes. Different types of vesicle recycling pathways are discussed for synapses (Figure 5), including fast retrieval of the vesicle at the site of exocytosis, called "kiss-and-run," a slower clathrin-dependent pathway, and other clathrin-independent retrieval pathways (Royle and

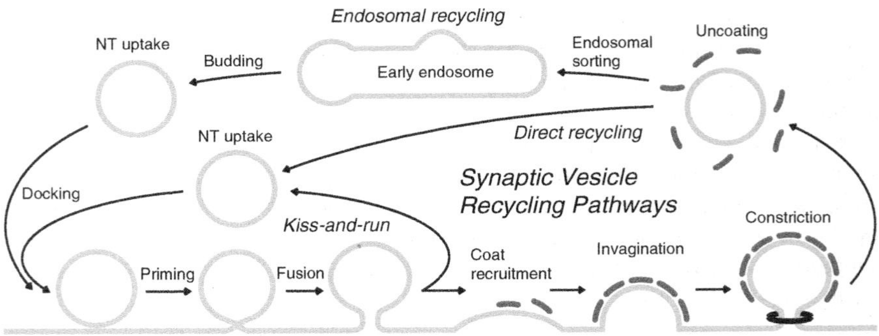


**Fig. 5** Synaptic vesicle recycling in the synapse. For synaptic vesicle recycling, several endocytic mechanisms appear to co-exist in synaptic nerve terminals. In the case of fast "kiss-and-run" exocytosis/endocytosis, the fused vesicle does not collapse into the membrane but is retrieved directly by a fast process. The molecular machinery underlying this pathway is unknown. Vesicles that have fully collapsed into the membrane are recycled by clathrin-mediated endocytosis. Clathrin, along with other proteins, is involved in membrane invagination (see figure and text) and leads finally to the formation of a constricted pit. The GTPase dynamin (black ring) mediates membrane scission of the constricted pit. After removal of the clathrin coat, two pathways are possible (direct recycling and recycling via the early endosome). In all cases, before fusion the recycled vesicles have to be loaded with neurotransmitters (NT).

Lagnado 2003). Which of the pathways dominates seems to depend on the type of neuron and the duration and intensity of the exocytotic stimulus. Apparently, the fast "kiss-and-run" pathway is preferred at low stimulation frequency for retrieving vesicles of the readily releasable pool. At higher stimulation frequencies the slow clathrin-dependent, endosomal recycling pathway is activated (Rizzoli and Betz 2005). Whereas no synapse is functional without an intact clathrin-dependent recycling pathway, similarly firm evidence for the need of the "kiss-and-run" pathway is not available. In fact, most of the evidence for "kiss-and-run" is either at the detection limit of the respective techniques, or of indirect nature. Furthermore, no protein machinery that is specific for this pathway has been identified. For these reasons, doubts persist whether such a pathway does exist at all in synapses. Another debated issue relates to the function of endosomal intermediates. Although synapses possess functional early endosomes it is unclear whether endocytosed vesicles must pass through an endosomal intermediate during each recycling or whether endosomes can be bypassed, with synaptic vesicles reforming directly after clathrin uncoating (Südhof 2004).

## *5.1 Kiss-and-Run Exocytosis/Endocytosis*

According to the classical view, synaptic vesicles completely flatten during exocytosis, which is followed by retrieval of the membrane components by clathrin-dependent endocytosis. Evidence for direct retrieval (kiss-and-run) was provided more than 30 years ago by Bruno Ceccarelli. At the neuromuscular endplate, Ceccarelli and co-workers were unable to observe reduction of vesicle numbers and the appearance of coated vesicles during low-stimulation frequencies, although synaptic vesicles acquired an extracellular fluid-phase marker in a stimulus-dependent fashion (Ceccarelli et al. 1973). When exocytotic events were captured by electron microscopy using rapid shock-freezing of stimulated synapses, exocytotic events were visible, with vesicles opening to the extracellular space by means of a pore. It was proposed that synaptic vesicles do not necessarily flatten into the membrane but instead can undergo rapid and transient fusion with the presynaptic membrane, a mechanism that was termed "kiss-and-run" (Fesce et al. 1994). This mechanism would allow for neurotransmitter release without the loss of synaptic vesicle identity, and therefore without any need of membrane recycling. Indeed, evidence for "reversible" fusion has since been provided from secretory cells with much larger vesicles. For instance, capacitance measurements on a variety of secretory cells showed that exocytosis, observed as a stepwise increase in capacitance, is frequently followed by a decrease of similar size (Fernandez et al. 1984; Breckenridge and Almers 1987). The molecular mechanisms that would mediate the scission of the fused, but incompletely flattened synaptic vesicles are still unclear. In chromaffin cells, in which the regulated organelles are large secretory granules, available data indicate that dynamin-1 may be responsible for the direct rapid retrieval pathway while dynamin 2 is involved in slow clathrin-mediated endocytosis (Artalejo et al. 2002). However, it is unclear how these findings are exactly related to the many types of nonclassical recycling pathways in nerve terminals that may differ with respect to retention of shape, protein, or lipid of the fused vesicle before its reinternalization. However, as a generally accepted terminology for the different fast nonclassical modes is not available, they are usually termed "kiss-and-run" exocytosis/endocytosis.

## *5.2 Clathrin-Mediated Endocytosis (CME)*

As has been shown by genetic approaches and by in vitro reconstitution, the molecular machinery for clathrin-mediated endocytosis (CME) in nerve terminals is generally similar to that involved in CME of non-neuronal cells. However, at the synapse internalization and recycling of vesicle membranes occurs in seconds, much faster than, e.g., ligand-induced, receptor-mediated endocytosis that occurs at a slower time scale. In fact the synapse possesses several adaptations of the CME pathway that may be responsible for these differences. First, endocytic proteins are highly enriched in synapses, including clathrin, AP2, epsin, eps15 (epidermal growth factor pathway subunit 15), amphiphysin, and synaptojanin. In addition, several of the major CME proteins express neuron-specific isoforms, such as dynamin, AP180, syndapin 1, clathrin light chain, and intersectin. Second, within the nerve terminal the endocytic machinery is localized close to sites of vesicle exocytosis (Roos and Kelly 1999). Hence, there is no need for long-distance diffusion from exocytotic to endocytic sites when the clathrin machinery is activated. Third, recent evidence suggests that the components of synaptic vesicles may remain clustered prior to internalization instead of dispersing into the membrane (see, e.g., Willig et al. 2006), thereby avoiding the need for elaborate sorting to reconcentrate vesicle components in the plasma membrane.

CME involves sequential and morphologically distinguishable steps, including coat recruitment and assembly on the membrane, invagination, formation of a constricted pit, fission, and uncoating. Coat recruitment and invagination are initiated by endocytic adaptors like AP-2 and stonin that function in the selection of cargo molecules and the initiation of the assembly of the clathrin coat and other factors which are required for the shaping of the vesicle. The endocytic adaptors bind to a type of lipid enriched in the cytoplasmic leaflet of the presynaptic membrane (phosphatidylinositol (4,5)-bisphosphate), to clathrin and other accessory proteins, and to cargo-sorting signals of, e.g., the synaptic vesicle transmembrane protein synaptotagmin (Maldonado-Báez and Wendland 2006; Di Paolo and De Camilli 2006). Although it appears that adaptor-lipid binding initiates the process, stable recruitment of adaptors to the plasma membrane requires cooperation between theses three types of adaptor-interactions (coincidence detection). Adaptor binding to clathrin and the neural specific AP180 leads to the formation of lattice-like structures with basic units of a trimer of clathrin (named triskelion). These assemble into larger baskets, and the interaction of epsin and AP-180 with the polymerizing clathrin lattice is supposed to promote membrane curvature, finally leading to the invagination of the membrane. Such invaginations are morphologically visible in electron micrographs as clathrin-coated pits. As has been revealed more recently by using fluorescent proteins for the study of endocytosis, clathrin-coated pits are not static but dynamic structures exchanging components with the soluble pool of endocytic proteins (Edeling et al. 2006).

The next step is the formation of a constricted pit, followed by fission of the membrane. Membrane fission is mediated by the mechanochemical GTPase dynamin that appears to wrap around the neck of the clathrin-coated pit (Takei and Haucke 2001). Recruitment of dynamin to its site of action is facilitated by amphiphysin, a protein that not only binds to dynamin but also to AP2 and clathrin. Furthermore, dynamin binds to phosphatidylinositol (4,5)-bisphosphate by its pleckstrin homology domain. The binding of GTP to dynamin in complex with amphiphysin has been suggested to redistribute dynamin close to the neck region, resulting in the formation of a constricted pit. GTP-hydrolysis leads to a conformational change of dynamin accompanied by a constriction of the dynamin ring around the neck. This mechanism probably generates the membrane fission event, resulting in a clathrin-coated vesicle ready for transport into the cytosol. Dynamin may thus be regarded as the counterpart of the SNAREs, with the SNAREs mediating fusion and dynamin mediating fission.

Uncoating requires an interaction with the uncoating ATPase Hsc70. Apparently, however, hydrolysis of phosphatidylinositol (4,5)-bisphosphate is required, which is carried out by the protein synaptojanin. Synaptojanin has two phosphatase domains, and in its absence clathrin-coated vesicles accumulate. Furthermore, the coated vesicle-associated protein auxilin is needed for uncoating. This protein recruits Hsc70 and stimulates its ATPase activity.

It is still unclear whether uncoated recycling vesicles must pass through an endosomal intermediate before synaptic vesicles are re-formed. Synaptic vesicles contain high concentrations of the protein machinery required for endosome fusion (Takamori et al. 2006), and synaptic organelles that have just undergone endocytosis are capable of homotypic fusion. However, it is possible that the endosomal intermediate serves as a backup rather than a mandatory intermediate, with vesicles being directly regenerated after uncoating without the involvement of an additional fusion and budding step.

## *5.3 Coupling Exocytosis to Endocytosis*

In the "kiss-and-run" mode exocytosis and endocytosis are directly coupled to each other, while in the case of classical complete vesicle fusion, exocytosis and slow clathrin-mediated endocytosis are timely and spatially separated. However, it appears that also in the latter case exocytosis and endocytosis occur coordinated, as both are stimulated by an increase of the cytoplasmic calcium concentration. It has been shown that after calcium entry the enzyme phospho-inositol-5 kinase I$\gamma$, which is enriched in the synapse, catalyzes the synthesis of phosphatidylinositol (4,5)-bisphosphate and that this mechanism is important for synaptic vesicle trafficking (Di Paolo et al. 2004). As many proteins involved in clathrin-mediated endocytosis are recruited to the plasma membrane by binding to phosphatidylinositol (4,5)-bisphosphate (e.g., amphiphysin, dynamin, epsin, AP-180, and AP-2) it is attractive to speculate that elevated levels of calcium mediate the recruitment of endocytic proteins to the plasma membrane by this mechanism. The increased level of phosphatidylinositol (4,5)-bisphosphate could be in part degraded by synaptojanin that thereby initiates the disassembly of the clathrin coat. Hence, calcium-induced transient increases in the level of phosphatidylinositol (4,5)-bisphosphate appear to play a central role for coupling exocytosis to clathrin-mediated endocytosis. In addition, it has been demonstrated that calcium also leads to the dephosphorylation of endocytic proteins as amphiphysin, dynamin, and synaptojanin, which in vitro is important for efficient coat assembly (Cousin and Robinson 2001).

## References

Artalejo CR, Elhamdani A, Palfrey HC (2002) Sustained stimulation shifts the mechanism of endocytosis from dynamin-1-dependent rapid endocytosis to clathrin- and dynamin-2-mediated slow endocytosis in chromaffin cells. Proc Natl Acad Sci USA 99:6358–63

Becher A, Drenckhahn A, Pahner I, Margittai M, Jahn R, Ahnert-Hilger G (1999) The synaptophysin-synaptobrevin complex: a hallmark of synaptic vesicle maturation. J Neurosci 19:1922–31

Borisovska M, Zhao Y, Tsytsyura Y, Glyvuk N, Takamori S, Matti U, Rettig J, Südhof T, Bruns D (2005) v-SNAREs control exocytosis of vesicles from priming to fusion. EMBO J. 24:2114–26
Bracher A, Weissenhorn W (2002) Structural basis for the Golgi membrane recruitment of Sly1p by Sed5p. EMBO J 21:6114–24
Breckenridge LJ, Almers W (1987) Final steps in exocytosis observed in a cell with giant secretory granules. Proc Natl Acad Sci USA 84:1945–49
Brunger AT (2005) Structure and function of SNARE and SNARE-interacting proteins. Q Rev Biophys 38:1–47
Ceccarelli B, Hurlbut WP, Mauro A (1973) Turnover of transmitter and synaptic vesicles at the frog neuromuscular junction. J Cell Biol 57:499–524
Ciufo LF, Barclay JW, Burgoyne RD, Morgan A (2005) Munc18-1 regulates early and late stages of exocytosis via syntaxin-independent protein interactions. Mol Biol Cell. 16:470–82
Chapman ER (2002) Synaptotagmin: a Ca(2+) sensor that triggers exocytosis? Nat Rev Mol Cell Biol 3:498–508
Cousin MA, Robinson PJ (2001) The dephosphins: dephosphorylation by calcineurin triggers synaptic vesicle endocytosis. Trends Neurosci 24:659–65
Craxton M (2004) Synaptotagmin gene content of the sequenced genomes. BMC Genomics 5:43
Di Paolo G, De Camilli P (2006) Phosphoinositides in cell regulation and membrane dynamics. Nature 443:651–7
Di Paolo G, Moskowitz HS, Gipson K, Wenk MR, Voronov S, Obayashi M, Flavell R, Fitzsimonds RM, Ryan TA, De Camilli P (2004) Impaired PtdIns(4,5)P2 synthesis in nerve terminals produces defects in synaptic vesicle trafficking. Nature 431:415–22
Dulubova I, Khvotchev M, Liu S, Huryeva I, Südhof TC, Rizo J (2007) Munc18-1 binds directly to the neuronal SNARE complex. Proc Natl Acad Sci USA. 2007 Feb 14; 104, 2697–702
Edeling MA, Smith C, Owen D (2006) Life of a clathrin coat: insights from clathrin and AP structures. Nat Rev Mol Cell Biol 7:32–44
Fasshauer D, Sutton RB, Brunger AT, Jahn R. (1998) Conserved structural features of the synaptic fusion complex: SNARE proteins reclassified as Q- and R-SNAREs. Proc Natl Acad Sci USA. 95:15781–6
Fernandez JM, Neher E, Gomperts BD (1984) Capacitance measurements reveal stepwise fusion events in degranulating mast cells. Nature 312:453–5
Fesce R, Grohovaz F, Valtorta F, Meldolesi J (1994) Neurotransmitter release: fusion or 'kiss-and-run'? Trends Cell Biol 4:1–4
Gallwitz D, Jahn R. (2003) The riddle of the Sec1/Munc-18 proteins – new twists added to their interactions with SNAREs. Trends Biochem Sci 28:113–16
Grosshans BL, Ortiz D, Novick P (2006) Rabs and their effectors: achieving specificity in membrane traffic. Proc Natl Acad Sci U S A 103:11821–7
Hanson PI, Whiteheart SW (2005) AAA+ proteins: have engine, will work. Nat Rev Mol Cell Biol 6:519–29
Hong W (2005) SNAREs and traffic. Biochim Biophys Acta 1744:493–517
Jahn R, Scheller RH. (2006) SNAREs–engines for membrane fusion. Nat Rev Mol Cell Biol 7:631–43
Maldonado-Baez L, Wendland B (2006) Endocytic adaptors: recruiters, coordinators and regulators. Trends Cell Biol 16:505–13
Marz KE, Hanson PI (2002) Sealed with a twist: complexin and the synaptic SNARE complex. Trends Neurosci 25:381–3
McEwen JM, Madison JM, Dybbs M, Kaplan JM (2006) Antagonistic regulation of synaptic vesicle priming by Tomosyn and UNC-13. Neuron 51:303–15
Misura KM, Scheller RH, Weis WI (2000) Three-dimensional structure of the neuronal-Sec1-syntaxin 1a complex. Nature 404:355–62
Peng R, Gallwitz D (2002) Sly1 protein bound to Golgi syntaxin Sed5p allows assembly and contributes to specificity of SNARE fusion complexes. J Cell Biol 157:645–55
Pobbati AV, Stein A, Fasshauer D (2006) N- to C-terminal SNARE complex assembly promotes rapid membrane fusion. Science 313:673–6

Rizo J, Chen X, Arac D (2006) Unraveling the mechanisms of synaptotagmin and SNARE function in neurotransmitter release. Trends Cell Biol 16:339–50
Rizo J, Südhof TC (2002) Snares and Munc18 in synaptic vesicle fusion. Nat Rev Neurosci 3:641–53
Rizzoli SO, Betz WJ (2005) Synaptic vesicle pools. Nat Rev Neurosci 6:57–69
Roos J, Kelly RB (1999) The endocytic machinery in nerve terminals surrounds sites of exocytosis. Curr Biol 9:1411–14
Royle SJ, Lagnado L (2003) Endocytosis at the synaptic terminal. J Physiol 553:345–55
Schiavo G, Matteoli M, Montecucco C (2000) Neurotoxins affecting neuroexocytosis. Physiol Rev 80:717–66
Schlüter OM, Schmitz F, Jahn R, Rosenmund C, Sudhof TC (2004) A complete genetic analysis of neuronal Rab3 function. J Neurosci 24:6629–37
Schultz J, Doerks T, Ponting CP, Copley RR, Bork P (2000) More than 1,000 putative new human signalling proteins revealed by EST data mining. Nat Genet 25:201–04
Schütz D, Zilly F, Lang T, Jahn R, Bruns D (2005) A dual function for Munc-18 in exocytosis of PC12 cells. Eur J Neurosci 21:2419–32
Shen J, Tareste DC, Paumet F, Rothman JE, Melia TJ (2007) Selective activation of cognate SNAREpins by Sec1/Munc18 proteins. Cell 128:183–95
Sorensen JB, Wiederhold K, Müller EM, Milosevic I, Nagy G, de Groot BL, Grubmüller H, Fasshauer D (2006) Sequential N- to C-terminal SNARE complex assembly drives priming and fusion of secretory vesicles. EMBO J 25:955–66
Sorensen JB, Nagy G, Varoqueaux F, Nehring RB, Brose N, Wilson MC, Neher E. (2003) Differential control of the releasable vesicle pools by SNAP-25 splice variants and SNAP-23. Cell 114:75–86
Südhof TC (2004) The synaptic vesicle cycle. Annu Rev Neurosci 27:509–47
Südhof TC (2002) Synaptotagmins: why so many? J Biol Chem 277:7629–32
Sutton RB, Fasshauer D, Jahn R, Brunger AT (1998) Crystal structure of a SNARE complex involved in synaptic exocytosis at 2.4 A resolution. Nature 395:347–53
Takamori S, Holt M, Stenius K, Lemke EA, Gronborg M, Riedel D, Urlaub H, Schenck S, Brugger B, Ringler P, Muller SA, Rammner B, Grater F, Hub JS, De Groot BL, Mieskes G, Moriyama Y, Klingauf J, Grubmüller H, Heuser J, Wieland F, Jahn R (2006) Molecular anatomy of a trafficking organelle. Cell 127:831–46
Takei K, Haucke V (2001) Clathrin-mediated endocytosis: membrane factors pull the trigger. Trends Cell Biol 11:385–91
Tang J, Maximov A, Shin OH, Dai H, Rizo J, Südhof TC (2006) A complexin/synaptotagmin 1 switch controls fast synaptic vesicle exocytosis. Cell 126:1175–87
Toonen RF, Verhage M (2003) Vesicle trafficking: pleasure and pain from SM genes. Trends Cell Biol. 13:177–86
Ungermann C, Langosch D (2005) Functions of SNAREs in intracellular membrane fusion and lipid bilayer mixing. J Cell Sci 118:3819–28
Voets T, Toonen RF, Brian EC, de Wit H, Moser T, Rettig J, Südhof TC, Neher E, Verhage M (2001) Munc18-1 promotes large dense-core vesicle docking. Neuron 31:581–91
Washington NL, Ward S (2006) FER-1 regulates Ca2+ – mediated membrane fusion during C. elegans spermatogenesis. J Cell Sci 119:2552–62
Willig KI, Rizzoli SO, Westphal V, Jahn R, Hell SW (2006) STED microscopy reveals that synaptotagmin remains clustered after synaptic vesicle exocytosis. Nature 440:935–9
Yoon TY, Okumus B, Zhang F, Shin YK, Ha T (2006) Multiple intermediates in SNARE-induced membrane fusion. Proc Natl Acad Sci USA 103:19731–6
Zilly FE, Sorensen JB, Jahn R, Lang T (2006) Munc18-bound syntaxin readily forms SNARE complexes with synaptobrevin in native plasma membranes. PLoS Biol 4:e330

# Presynaptic Neurotoxins with Enzymatic Activities

**Ornella Rossetto and Cesare Montecucco(✉)**



**Abstract** Toxins that alter neurotransmitter release from nerve terminals are of considerable scientific and clinical importance. Many advances were recently made in the understanding of their molecular mechanisms of action and use in human therapy. Here, we focus on presynaptic neurotoxins, which are very potent inhibitors of

Cesare Montecucco
Dipartimento di Scienze Biomediche and Istituto CNR di Neuroscienze, Universit di Padova, Viale G. Colombo 3, 35121 Padova, Italy
cesare.montecucco@unipd.it

T.C. Südhof, K. Starke (eds.), *Pharmacology of Neurotransmitter Release. Handbook of Experimental Pharmacology 184.*

the neurotransmitter release because they are endowed with specific enzymatic activities: (1) clostridial neurotoxins with a metallo-proteolytic activity and (2) snake presynaptic neurotoxins with a phospholipase A2 activity.

Tetanus and botulinum neurotoxins produced by anaerobic bacteria of the genus *Clostridium* enter peripheral cholinergic nerve terminals and cleave proteins of the neuroexocytosis apparatus, causing a persistent, but reversible, inhibition of neurotransmitter release. Botulinum neurotoxins are used in the therapy of many human syndromes caused by hyperactive cholinergic nerve terminals. At variance, some snake presynaptic neurotoxins block nerve terminals by binding to the plasma membrane, where they hydrolyze phospholipids with production of lysophospholipids and fatty acids. These compounds change the membrane conformation causing enhanced neurotransmitter release and, at the same time, inhibition of synaptic vesicle recycling. It is possible to envisage pharmacological applications of the lysophospholipid/fatty acid mixture to inhibit hyperactive superficial nerve terminals.

## 1 Introduction

Many toxins act selectively on the nervous tissue and the most poisonous toxins are neurotoxins. Given the essential role of the nervous system in animal physiology, even a minor biochemical modification of a few neurons may result in a profound behavioral alteration. In general, neurotoxins block in one way or another the transmission of the nerve impulse. The majority of neurotoxins act by binding specifically to ion channels and the ensuing strong alteration of ion permeability of the plasma membrane results in the blockade of the transmission of nerve signals. In contrast, clostridial neurotoxins with metalloproteolytic activity and snake neurotoxins with phospholipase $A_2$ activity interfere directly with neurotransmitter release by exploiting their highly specific enzymatic activity, which makes them the most poisonous substances.

Tetanus neurotoxin (TeNT) and botulinum neurotoxins (BoNTs) are the sole cause of tetanus and botulism. This permitted their early detection in culture filtrates of anaerobic and spore-forming bacteria of the genus *Clostridium* (Faber 1890; Tizzoni and Cattani 1890; Van Ermengem 1897). However, it took over a century to achieve a molecular understanding of the pathogenesis of the diseases they cause, and still, some steps of the intoxication process are ill-known. TeNT inhibits neurotransmitter release of synapses of the central nervous system (CNS) causing the spastic paralysis of tetanus; BoNT inhibits the release of acetylcholine (ACh) at peripheral cholinergic nerve terminals causing the flaccid paralysis of botulism. Hence, these neurotoxins cause two diseases (tetanus and botulism) with opposite clinical symptoms, yet they impair the same very essential neuronal function: *neuroexocytosis* (Brooks et al. 1955; Burgen et al. 1949; Rossetto et al. 2006; Schiavo et al. 2000). To date, one TeNT and seven (A–G) serologically distinct BoNTs have been characterized. The latter can be further divided into subserotypes; for example, there are four subserotypes of BoNT/A (termed A1-A4) (Arndt et al. 2006;

Smith et al. 2005), which are approximately 90% identical in their amino acid sequences.

Presynaptic snake neurotoxins endowed with PLA2 activity (SPANs) are major components of the venom of four families of venomous snakes (*Crotalidae, Elapidae, Hydrophiidae, and Viperidae*). These neurotoxins play a crucial role in envenomation of the prey (Harris 1997) by causing a persistent blockade of neurotransmitter release from nerve terminals with a peripheral paralysis very similar to that of botulism (Connolly et al. 1995; Gutiérrez et al. 2006; Kularatne 2002; Prasarnpun et al. 2005; Theakston et al. 1990; Trevett et al. 1995; Warrell et al. 1983).

The botulinum neurotoxins and the snake presynaptic PLA2 neurotoxins share three levels of interest: (1) they are pathogenic to humans and animals, (2) they contribute to the understanding of the molecular steps of neurotransmission, and (3) their present and future clinical applications. In this chapter, these neurotoxins are considered in terms of mode of action and in relation to their potential use in cell biology and neuroscience research as well as therapeutics in some human neurodisorders.

## 2 Toxicity of BoNTs and of SPANs

Tetanus and botulinum neurotoxins are the most potent toxins known, as few nanograms/Kg are sufficient to kill most mammals. When injected peripherally the mouse $LD_{50}$ s of TeNT and BoNTs are between 0.4 ng and 1 ng of toxin per Kg of body weight (Gill 1982). When BoNTA and BoNT/B are injected into the brain the $LD_{50}$ s are comparable with those determined by intraperitoneal injection in mice (Luvisetto et al. 2003).

The SPANs are comparatively less potent and show a much wider range of toxicity, with mouse $LD_{50}$ values comprised between the 1 μg/kg of textilotoxin (present in the venom of the Australian snake *Pseudonaja textilis*) to the 1300 μg/kg of pseudexin A (present in the venom of *Pseudechis porphyriacus*) (Kini 1997; Montecucco and Rossetto 2000; Pearson et al. 1993), when injected intraperitoneally. In addition, they display largely different toxicities versus different animal classes, probably reflecting different predator-prey relationships in different venomous snakes. Even though SPANs do not cross the blood-brain barrier, they are active on the central nervous system and β-bungarotoxin is three orders of magnitude more potent when administered into rat brain (Hanley and Emson 1979). These high toxicities derive from: (1) the specificity of these toxins for the nervous tissue, whose complete functionality is essential for survival, particularly in the wilderness, and (2) their enzymatic activities localized on nerve terminals.

# 3 The Diseases

## 3.1 Tetanus

It is now well established that TeNT is the sole cause of tetanus, though it took many centuries to establish a direct mechanistic link between the neurotoxin and the disease it causes. In fact, tetanus was first described in medical terms by Hippocrates, who defined this paralysis as "tetanus" (τετανοσ in Greek means contraction). Tetanus may develop in different forms, and the most common one is the *generalized tetanus* caused by contamination of even minor wounds or skin scratches with spores of toxigenic *Clostridium tetani*. Between the time of injury and the first symptoms there is a lag phase varying from few days up to four weeks, which corresponds to the time necessary for (1) germination of spores, (2) toxin production and release, and (3) toxin diffusion, binding, transport to its target cells within the spinal cord. Tetanus usually begins with a characteristic lockjaw (*risus sardonicus*), with difficulty in swallowing and neck stiffness (Bleck 1989). With time, the muscle paralysis extends downwards to the muscles of the trunk, abdomen, and legs. The typical tetanic seizure is characterized by a sudden burst of tonic contraction of skeletal muscles with extension to the lower extremities. Seizures are very painful and can be triggered by minor stimuli such as a small light or noise, with the patient remaining completely conscious during such episodes. Later on, autonomic symptoms develop with alterations of blood pressure and of the cardiac rhythm and sweating. Dysphagia and glottal and laryngeal spasm may cause cyanosis and asphyxia, which can be relieved by tracheotomy and mechanically assisted respiration.

A milder form of the disease is *local tetanus*, with rigidity of the muscles close to the site of injury and release of the toxin. This local tetanus may persist for a considerable period of time without further developments or it may proceed to generalized tetanus. It is due to dysfunction of the spinal cord inhibitory interneurons which inhibit the alpha motor neurons of the affected muscles, with little, if any, further spread through the central nervous system. Tetanus is often fatal, and death follows body exhaustion and usually intervenes by respiratory failure or heart failure (Bleck 1989). The mortality has decreased owing to modern intensive care techniques, but it is still high because of the usually advanced age of patients and because their respiration has to be mechanically assisted for long periods of time with the risk of developing pulmonary infections. Following vaccination with tetanus toxoid (formaldehyde-treated tetanus toxin), tetanus has almost disappeared from the more developed countries, but it still takes hundreds of thousand of lives in those regions of the world where vaccination is not performed (Galazka and Gasse 1995). Here, the major form of tetanus is *tetanus neonatorum*, which develops following the nonsterile cut of the umbilical cord of babies born from nonimmunized mothers. This condition is prevalent in communities that employ traditional midwifery practices such as cutting the cord with dirty scissors or rubbing manure on the umbilical stump.

## 3.2 Botulism

Botulism was described much later than tetanus (Kerner 1817; Midura and Arnon 1976; Pickett et al. 1976), and this delayed recognition is due to its much less evident symptoms, which include a generalized muscular weakness with diplopia, ptosis, dysphagia, facial paralysis, and reduced salivation and lacrimation. The paralysis then progressively descends to affect the muscles of the trunk, including respiratory and visceral muscles. All the symptoms of botulism can be ascribed to the blockade of skeletal and autonomic peripheral cholinergic nerve terminals (Tacket and Rogawski 1989).

The gravity of the illness depends on the amount and type of BoNT, which is usually acquired via the oral route. Death follows the blockade of respiratory muscles, but if ventilated mechanically, the patient will eventually recover completely. In general terms, botulism is much less dangerous than tetanus also because, in most cases, the amount of toxin that reaches the general circulation is not sufficient to block respiration. BoNT/A, /B and /E account for most cases of human botulism, and the disease caused by BoNT/A is more dangerous with symptoms persisting much longer.

The incidence and the types of botulism depend on the occurrence of toxigenic strains of *Clostridium botulinum, C. barati, and C. butyrricum* in the environment and subsequently in foods, and on the cooking practices. The geographical distribution of the different types of botulism corresponds to the distribution of the different toxinotypes of *Clostridium* in the environment. Consequently, type E botulism is mainly found in the colder regions of the northern hemisphere, and aquatic animals are usually involved (Hauschild 1993), while types A and B botulism generally occur in the temperate countries. Most of botulism outbreaks are caused by home-prepared or – stored/fermented food contaminated by spores of BoNT producing strains of *Clostridium* and kept under anaerobic conditions which permit bacterial germination and production of the poison; contamination of foods prepared by food industries is very rare (Hauschild 1993).

A less common form of the disease is that of *infant botulism*, which follows the ingestion of spores of neurotoxigenic *Clostridium* that germinate and multiply within the intestinal tract. Rather than a primary intoxication, this disease is an intoxication following a previous infection of the infant's intestinal tract, which lacks the protective bacterial flora of the adult, allowing colonization by *Clostridium*. It has been estimated that fewer than 100 spores are sufficient to cause such disease (Arnon 1980). The affected infants present constipation, weak sucking, hypotonia, and ptosis. In the more severe cases, the patient becomes lethargic and loses head posture control. The disease progresses to a flaccid paralysis which may extend to respiratory muscles with arrest. Type A toxin is usually associated with more severe diseases and a higher mortality rate than type B or E toxin, and the recovery time is accordingly longer (Arnon 1980). "Wound botulism" following spore contamination of wounds is very rare. On the contrary, animal botulism is rather common, and it can also be caused by BoNT/C and /D.

As a consequence of the fact that a single protein is responsible for all the clinical symptoms of tetanus and botulism, these diseases can be completely prevented by anti-toxin specific antibodies (Galazka and Gasse 1995; Middlebrook and Brown 1995). Toxin neutralizing antibodies can be acquired passively by injection of immunoglobulins isolated from immunized donors or, actively, as a result of vaccination with toxoids, obtained by treating TeNT or BoNT with formaldehyde (Ramon and Descombey 1925; Galazka and Gasse 1995). Tetanus toxoid is very immunogenic, and it is used as a standard immunogen in a variety of immunological studies (Corradin and Watts 1995). Antitetanus and antibotulism vaccines based on the recombinant binding domain (Hc) of TeNT and BoNT have been recently developed based on the observation that the majority of protecting antibodies are directed against this part of the molecule (Byrne and Smith 2000).

## *3.3 Features of Envenoming from Bites of Snakes Producing Venoms Containing Large Amounts of PLA2 Presynaptic Neurotoxins*

At variance from most neurotoxigenic clostridia, which release only BoNT or TeNT, snake venoms contain a multitude of pharmacologically active molecular components. In general, they can be divided into three classes based on their physiological targets: (1) the neuromuscular system, similarly to BoNTs, (2) the blood coagulation system, and (3) the control of blood pressure (Gutiérrez et al. 2006). In the case of many highly poisonous snakes of Asiatic and Australian origin, the major role in envenomation is played by presynaptic neurotoxins endowed with PLA2 activity (SPAN) (Connolly et al. 1995; Prasarnpun et al. 2005). SPANs cause a persistent blockade of neurotransmitter release from peripheral nerve terminals with no evidence of central activities, though they bind to and act on neurons isolated from brain and spinal cord (Kini 1997; Lambeau and Lazdunski 1999; Othman et al. 1982; Rehm and Betz 1982; Rigoni et al. 2004, 2005; Rossetto et al. 2006; Rugolo et al. 1986). Venoms from *Elapidae* and *Hydrophiidae* snakes also contain paralytic α-toxins, which act postsynaptically by binding and inhibiting the acetylcholine receptors. In fact, independently on the anatomical site of biting, patients develops symptoms comparable to those of botulism. They are reported to have ptosis, diplopia, ophthalmoplegia, difficulty in swallowing, respiratory paralysis, abdominal pain, and autonomic symptoms. Mechanical ventilation is often required, but most patients recover within days/weeks (Connolly et al. 1995; Kularatne 2002; Prasarnpun et al. 2005; Theakston et al. 1990; Trevett et al. 1995; Warrell et al. 1983).

There is no accurate estimation of the incidence and number of deaths caused by snake bites in the world. It is possible that tens of thousands of individuals die each year following snake envenomation in Asia, which is the most affected area (Gutiérrez et al. 2006). The death rate has strongly decreased in Australia following the introduction of first-aid procedures and of anti-venoms animal antisera (Hodgson and Wickramaratna 2002, 2006).

## 4 Structural Organization of Tetanus and Botulinum Neurotoxins

The similar effect of the eight clostridial neurotoxins (CNTs) on nerve terminals is the result of a closely related protein structure. They are synthesized in the bacterial cytosol as a 150 kDa polypeptide chain without a leader sequence, and are released in the culture medium only after bacterial autolysis. No protein is associated with TeNT, whereas the BoNTs are released as multimeric complexes with a variable set of nontoxic proteins encoded by genes present in the neurotoxin locus (Inoue et al. 1996; Minton 1995; Sakaguchi 1982). Some BoNT-associated proteins have hemagglutinating activity: HA of 17 kDa (HA17), HA of 34 kDa (HA34) and HA of 70 kDa (HA70) (Fujita et al. 1995; Henderson et al. 1997). In addition, a large nontoxic nonhemagglutinating protein (NTNH) of similar size to BoNT is always present, and it has been suggested that it nests the formation of the BoNT-NTNH complex, which then may, or may not, progress to the formation of larger complexes (Figure 1). In fact, three forms of progenitor toxins have been characterized: extra-large size (LL sediments at 19S, approximately 900 kDa); large size (L sediments at 16S, 500 kDa), and medium size (M, sediments at 12S, 300 kDa). It is noteworthy that BoNT interacts with the nontoxic portion in the complex via the $H_C$ domain (Chen et al. 1997), raising the possibility that NTNH prevents its interaction with the presynaptic terminal receptors. Complexed BoNTs are more stable than isolated BoNTs to proteolysis and denaturation induced by temperature, solvent removal, or acid pH (Chen et al. 1998a; Sakaguchi 1982), and therefore they have an increased resistance to the environment in which they are released and to their passage through the alimentary tract (Ohishi et al. 1977). There is evidence that accessory proteins do protect BoNT within the gut, but it is not clear if they also promote their uptake from the intestinal lumen into the general body fluid circulations (Fujinaga et al. 2000; Maksymowych et al. 1999). The BoNTs transcytose from the apical to the basolateral side of intestinal epithelial cells (Ahsan et al. 2005; Maksymowych and Simpson 1998; Simpson 2000), and the toxin form appearing in the lymph is the 150 kDa "naked" protein (Heckly et al. 1960). Once in the tissue fluids, they then reach their specific targets which are the peripheral cholinergic nerve terminals (Montecucco et al. 2004; Rossetto et al. 2006; Simpson 2000). This gastrointestinal passage is bypassed by TeNT, which enters the general circulation directly form the infected wounded site. The target specificity of TeNT and BoNTs is more evident if one considers that they are lethal at pico-femtomolar concentrations and yet they are able to reach their target presynaptic terminals, which represent an infinitesimal part of the body in terms of exposed cell surfaces (Montecucco et al. 2004).

The single chain 150 kDa CNTs are inactive and are activated by specific proteolysis within a surface exposed loop subtended by a highly conserved disulfide bridge.

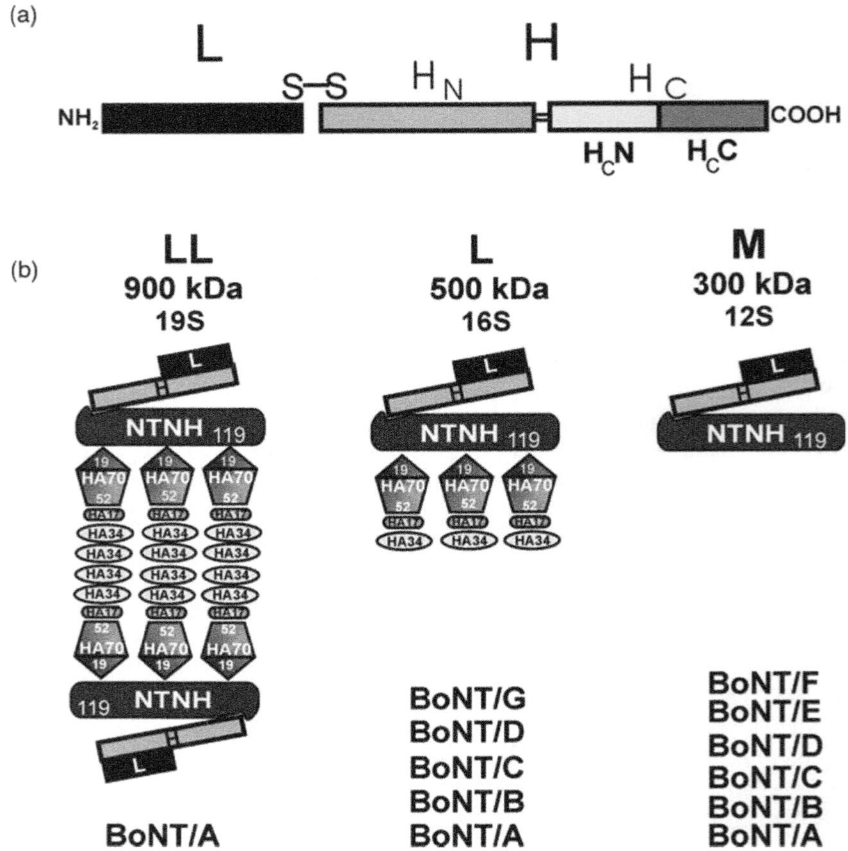


**Fig. 1** (**a**) Schematic structure of activated di-chain clostridial neurotoxins. The neurotoxins are composed of a two-polypeptide chain held together by a single disulfide bridge. The C-terminal portion of the heavy chain (H, 100 kDa) is responsible for neurospecific binding (domain $H_C$), while the N-terminus ($H_N$) is implicated in the translocation of the light chain in cytosol and pore formation. Structurally $H_C$ can be further subdivided into two portions of 25 kDa, $H_CN$, and $H_CC$. The light chain (L, 50 kDa) is a zinc-endopeptidase responsible for the intracellular activity of CNTs. (**b**) Botulinum neurotoxins are produced as multimeric progenitor toxins. The neurotoxin moiety interacts with a large, nontoxic, nonhemagglutinin protein (NTNH) via its C-terminal domain. This dimeric complex (Mw 300 kDa, 12S) serves as a scaffold for the assembly of larger complexes formed by the incorporation of three types of additional proteins with hemagglutinin activity and with Mw of 17 kDa (HA17), 35 kDa (HA35), and 70 kDa (HA70, present in a cleaved state of 19 and 52 kDa). These additional forms are the large L (16S, Mw 500 kDa) and the extra-large LL (19S, Mw 900 kDa) progenitor toxins.

Several bacterial and tissue proteinases are able to generate the active di-chain neurotoxin (Dasgupta 1994; Krieglstein et al. 1991). The heavy chain (H, 100 kDa) and the light chain (L, 50 kDa) remain associated via noncovalent interactions and via the conserved interchain S-S bond, whose integrity is essential for neurotoxicity (Figure 1) (de Paiva et al. 1993; Schiavo et al. 1990; Simpson et al. 2004).

The length of the polypeptide chains of CNTs varies from the 1251 amino acid residues of *C. butyricum* BoNT/E to the 1297 residues of BoNT/G and the 1315 residues of TeNT (Minton 1995; Niemann 1991). The exact length of the L and H chains depends on the site of proteolytic cleavage within the exposed loop. The H chains vary in size from the 829 amino acid residues of BoNT/E to the 857 residues of TeNT. The L chains range in size from the 419 amino acid residues of BoNT/E to the 449 residues of TeNT.

Since the first molecular glimpse of BoNT/A in 1998 (Lacy et al. 1998) there has been a profusion of X-ray crystal structures. There are now 33 BoNT holotoxin or domain structures available, including BoNT/A and BoNT/B holotoxins, TeNT and BoNT/B binding domain, and LC structures of BoNT/A–/G (structures are available at the protein data bank: http://www.rcsb.org/pdb/). CNTs are organized in three distinct 50 kDa domains (Lacy et al. 1998; Swaminathan and Eswaramoorthy 2000): (1) an N-terminal metalloprotease domain (LC), (2) a membrane translocation domain ($H_N$), and (3) a binding domain ($H_C$) (see Figure 2) (Lacy et al. 1998; Swaminathan and Eswaramoorthy 2000; Umland et al. 1997).

## 4.1 The Binding Domain

The $H_C$ binding domain of these CNTs consist of two distinct subdomains: the N-terminal half ($H_CN$) and the C-terminal half ($H_CC$), with little protein-protein contacts among them. $H_CN$ is enriched in β-strands arranged in a jelly-roll motif closely similar to some carbohydrate binding proteins (legume lectins). The amino acid sequence of $H_CN$ is highly conserved among CNTs. The $H_CC$ subdomain contains a modified β-trefoil folding motif present in several proteins involved in recognition and binding functions such as IL-1, fibroblast growth factor, and the Kunitz-type trypsin inhibitors. Its sequence is poorly conserved among CNTs. It harbors one binding site for the oligosaccharide portion of polysialogangliosides in BoNT/A and /B (Figure 2), while the $H_CC$ of TeNT has two such sites (Rummel et al. 2003, 2004b). $H_CC$ of BoNT/B harbors a cleft which fits in segment 44–60 of synaptotagmin belonging to the luminal domain of this SV protein (Chai et al. 2006; Jin et al. 2006; Rummel et al. 2007).

## 4.2 The Translocation Domain

The $H_N$ portions of BoNT/A and /B are very similar, and their sequences are highly homologous among the various CNTs. $H_N$ has an elongated shape determined by a pair of unusually long and twisted α–helices, corresponding to segment 685-827 of BoNT/A (Figure 2) (Lacy et al. 1998; Swaminathan and Eswaramoorthy 2000). At both ends of the pair, there is a shorter α-helix which lies parallel to the main helices and, in addition, several strands pack along the two core helices. The overall

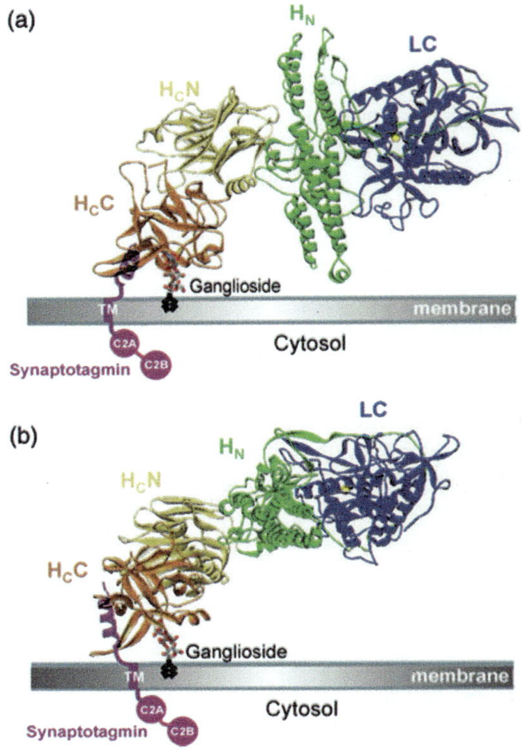


**Fig. 2** Binding of botulinum neurotoxin type B to the presynaptic membrane. Two different models have been proposed (Chai et al. 2006; Jin et al. 2006). The C-terminal part of the binding domain ($H_CC$ in orange) interacts specifically both with polysialogangliosides and with the luminal domain of synaptotagmin (magenta), which adopts an helical conformation upon BoNT/B binding. The remaining part of the synaptotamin molecule is drawn as a transmembrane domain (TM) and two C2 cytoplasmic domains (magenta balls). The helices of the toxin translocation domain ($H_N$) may sit orthogonal to the plane of the membrane surface (panel **a**) (Chai et al. 2006) or may be oriented parallely (panel **a**) (Jin et al. 2006). The light chain (LC), the N-terminal part of the heavy chain ($H_N$), and the two C-terminal subdomains of the heavy chain ($H_C$) are shown in blue, green, yellow and orange, respectively. The yellow sphere represents the atom of zinc of the active site of the L chain metalloprotease.

structure of $H_N$ resembles that of some viral proteins undergoing an acid-driven conformational change (Bullough et al. 1994; Kielian and Rey 2006; Weissenhorn et al. 1997). Two models of BoNT/B bound to both Syt-II and ganglioside have been generated with a different orientation of the translocation domain with respect to the cell membrane (Baldwin et al. 2007; Chai et al. 2006; Jin et al. 2006; Rossetto and Montecucco 2007; Schiavo 2006) (Figure 2). Jin et al. (2006) suggest that the helices of the translocation domain sit orthogonal to the plane of the membrane (Figure 2a). On the contrary, Chai et al. (2006) propose a parallel orientation, which would place hydrophobic areas of the helices close to the inner surface of the vesicle, facilitating their insertion into the membrane (Figure 2b).

### 4.3 The Catalytic Domain

The crystal structures of LC/A (Segelke et al. 2004), LC/B (Hanson and Stevens 2000), LC/E (Agarwal et al. 2004), LC/D (Arndt et al. 2006), LC/F (Agarwal et al. 2005), LC/G (Arndt et al. 2005), and LC/TeNT (Breidenbach and Brunger 2005a) have been determined. They have a unique structure among metalloproteases. The active site zinc atom is coordinated by the imidazole rings of two histidines, a water molecule bound to the glutamic acid of the motif and a glutamic acid which is conserved among CNTs. The Glu residue of the motif is particularly important because it coordinates the water molecule directly implicated in the hydrolytic reaction of proteolysis. Its mutation leads to complete inactivation of these neurotoxins (Li et al. 2000). The active site is similar to that of thermolysin and identifies a primary sphere of residues essential for catalysis, which coincides with the zinc coordinating residues. In addition, there is a secondary layer of residues, less close to the zinc centre, including Arg362 and Tyr365 (numbering of BoNT/A), which appears to be directly involved in the hydrolysis of the substrate (Binz et al. 2002; Rigoni et al. 2001; Rossetto et al. 2001a; Yamasaki et al. 1994). The active site of BoNTs is a long cleft which becomes accessible to the substrate upon reduction of the interchain disulfide bridge.

## 5 Structure of the Snake Presynaptic PLA2 Neurotoxins

At variance with the high structural similarity of BoNTs, the SPANs vary greatly, both in terms of structure and of PLA2 activity. The crystallographic structure of few of the simpler toxins has been determined (Arni and Ward 1996; Carredano et al. 1998; Kwong et al. 1995; Scott 1997; Singh et al. 2001; Westerlund et al. 1992) (see structures at Protein Data Bank: http://www.rcsb.org/pdb/). They include a phospholipase A2 domain with a flattened ellipsoid shape of approximate dimensions 45 Å × 30 Å × 20 Å. The structure is stabilized by six intrachain disulfide bonds and by a catalytic $Ca^{2+}$ atom. Two long antiparallel and disulfide linked α-helices characterize their structure and include the four key residues of a catalytic site shielded from the water solvent (His48, Asp49, Tyr52 and Asp99) (Figure 3). The histidine residue hydrogen binds the water molecule used for hydrolysis, while Asp49 coordinates and positions the $Ca^{2+}$ ion which binds both the phosphate and the sn-2 carbonyl groups of the phospholipid molecule during hydrolysis. In addition, there is an interfacial binding site, which mediates the absorption of the enzyme onto the lipid-water interface, which is strongly promoted by anionic amphipatic molecules such as fatty acids (Scott 1997). Another characteristic feature of PLA2s is the presence inside the molecule of an hydrophobic channel where a phospholipid molecules diffuses from the membrane into the catalytic site.

Chemical modification and structural comparison studies implicate the extended loop and β region between residues 59 and 100 as the main responsible for neurotoxicity (Dufton and Hider 1983; Kini and Evans 1989; Yang 1997). However,

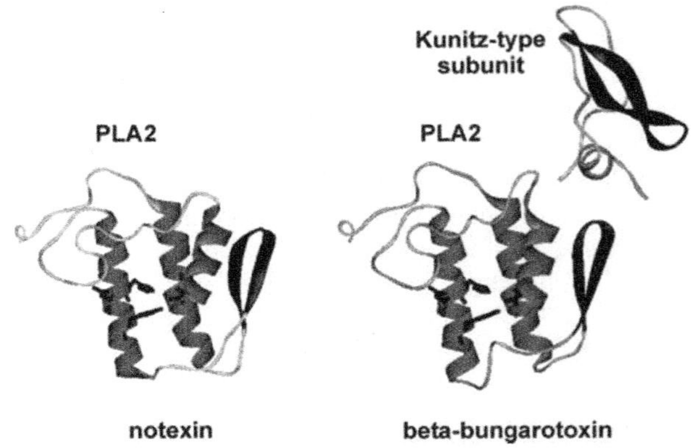


**Fig. 3** Ribbon diagram of the single subunit notexin (PDB number 1AE7, Westerlund et al. 1992) and the β-bungarotoxin (PDB number 1BUN, Kwong et al. 1995). The structure of the PLA2 reveals the characteristic three α-helices (dark grey) with two β-strands (black). In β-bungarotoxin the PLA2 subunit is disulfide linked to a 7 kDa protein with a sequence and a structure very similar to that of the Kunitz-type trypsin inhibitor and to the three-foil subdomain $H_CN$ of clostridial neurotoxins. The residues of the catalytic site His48, Asp49, Tyr52, and Asp99 (Asp94 in β-bungarotoxin) are shown.

this leaves open the possibility that additional regions and/or subunits may be involved, particularly in the case of multi-subunit toxins. In fact, SPANs vary greatly in terms of their quaternary structure, from the single subunit notexin (from the Australian tiger snake *Notechis scutatus*), to β-bungarotoxin (from *Bungarus multicinctus*), which consists of two covalently linked subunits, the larger of which is a PLA2 (Figure 3). Similarly, only one of the three subunits of taipoxin (from *Oxyuranus scutellatus*) has PLA2 activity. In contrast, textilotoxin (from the Australian elapid *Pseudonaja textilis*) consists of 5 subunits (A, B, C and two D subunits), all endowed with PLA2 activity (Kini 1997).

## 6 The Mode of Action of Clostridial Neurotoxins

The structural organization of CNTs is functionally related to the fact that they intoxicate neurons via a four-step mechanism consisting of (1) binding, (2) internalization, (3) membrane translocation, and (4) enzymatic target modification (Figure 4) (Montecucco et al. 1994; Montecucco and Schiavo 1995; Rossetto et al. 2006). The LC is responsible for the intracellular catalytic activity, the amino-terminal 50 kDa domain of the H chain ($H_N$) is implicated in membrane translocation, and the carboxy-terminal part ($H_C$) is mainly responsible for the neurospecific binding.

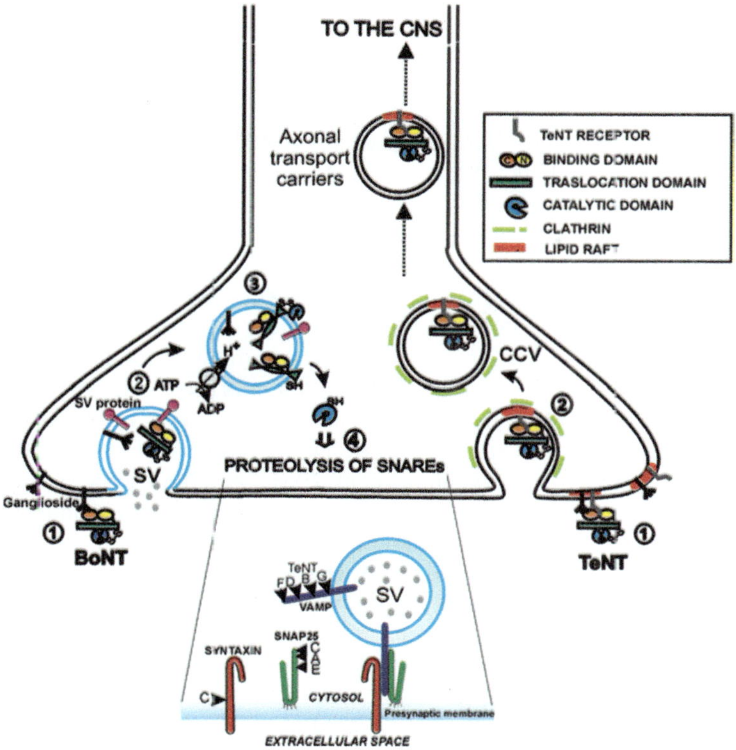


**Fig. 4** Binding and entry of BoNTs and TeNT at peripheral nerve terminal. (**1**) The BoNT binding domain associates with the presynaptic membrane of α-motoneurons through interaction with ganglioside and with the exposed luminal domain of a synaptic vesicle protein upon SV membrane fusion. (**2**) BoNTs are endocytosed within synaptic vesicles via their retrieval to be refilled with neurotransmitter. TeNT exploits a pathway requiring lipid rafts and the clathrin machinery by binding to a lipid–protein receptor complex containing the ganglioside GD1b. Once internalized in clathrin-coated vesicles (CCV), TeNT is sorted into vesicle carriers of the axonal retrograde transport pathway. (**3**) At the low pH generated by the v-ATPase in the vesicle lumen, BoNTs change conformation, insert into the lipid bilayer of the vesicle membrane and translocate the L chain into the cytosol. (**4**) Inside the cytosol the L chain catalyses the proteolysis of one of the three SNARE proteins VAMP, SNAP-25 and syntaxin, depicted in the lower panel. The same four-step pathway of entry of BoNTs into peripheral nerve terminals is believed to be followed by TeNT in the inhibitory interneurons of the spinal cord, which are reached after retroaxonal transport and release from the peripheral motoneurons.

## *6.1 Binding*

From the site of adsorption, BoNTs diffuse in the body fluids and eventually bind very specifically to the presynaptic membrane of cholinergic terminals. At variance, the pathway of TeNT is more complex, as it involves an additional retroaxonal transport to the spinal cord and a specific binding to inhibitory interneurons

(Lalli et al. 2003). The $H_C$ domain plays the major role in the neurospecific binding (Chai et al. 2006; Jin et al. 2006). Polysialogangliosides are involved in CNT binding (reviewed in Montecucco et al. 2004; Rossetto et al. 2006; Schiavo et al. 2000; Verderio et al. 2006; Yowler and Schengrund 2004). In fact, the $H_CC$ subdomain of BoNT bind polysialogangliosides, particularly GD1a, GT1b, and GQ1b, via the conserved peptide motif H...SXWY...G (Rummel et al. 2004b). Pre-incubation of BoNTs with polysialogangliosides partly prevents cell intoxication and the pre-treatment of cultured cells with polysialogangliosides increases their sensitivity to BoNT/A; conversely, depletion of gangliosides in neuroblastoma or PC12 cells impairs BoNT/A and BoNT/B internalization. Neuraminidase removes sialic-acid residues and decreases BoNT cell binding. Knockout mice defective in the production of polysialogangliosides show reduced sensitivity to BoNT/A and BoNT/B (Bullens et al. 2002; Chai et al. 2006; Kitamura et al. 2005; Schiavo et al. 2000; Yowler et al. 2002). More recently, polysialogangliosides GD1b and GT1b and phosphatidylethanolamine were reported to be the functional receptors of BoNT/C and BoNT/D, respectively (Tsukamoto et al. 2005). At variance, two sugar-binding sites have been identified within the TeNT $H_CC$ fragment and this double binding, possibly to a polysialoganglioside molecule and to a glycoprotein, is essential for the toxicity of TeNT (Rummel et al. 2003; Herreros et al. 2000).

It is well established that both glycolipid and protein receptors are crucial for high-affinity binding to nerve cell membranes (Rummel et al. 2007). As BoNTs and TeNT are toxic at concentrations estimated to be $<$ picomolar (Montecucco et al. 2004), binding has to be interpreted as the simultaneous occupation of the ganglioside binding pocket(s) and of the protein binding site. Indeed, TeNT was shown to interact with a GPI-anchored glycoprotein and to bind to lipid rafts (Herreros and Schiavo 2002; Herreros et al. 2001; Munro et al. 2001). Synaptic vesicle (SV) proteins participate with their luminal domains to the neurospecific binding of BoNT/A, /B, and /G, and this is probably true for all the other BoNTs (for a recent review, see Verderio et al. 2006). BoNT/B and BoNT/G interact with the luminal domain of the SV proteins synaptotagmin I (Syt-I) and Syt-II. BoNT/B has a higher affinity for Syt-II (Dong et al. 2003; Nishiki et al. 1996), whereas BoNT/G interacts preferentially with Syt-I (Rummel et al. 2004a). The crystal structures of BoNT/B in complex with the luminal domain of Syt-II (Chai et al. 2006; Jin et al. 2006) shows that $H_CC$ accommodates the $\alpha$-helical segment 45–59 of Syt-II in a cleft adjacent to the polysialogangliosides-binding site of the toxin (Figure 2). Mutations in the synaptotagmin-binding cleft and in the polysialoganglioside-binding pocket greatly reduce the toxicity of BoNT/B and /G, and double mutants at the two binding sites abolish neurotoxicity (Rummel et al. 2007).

At variance, BoNT/A binds specifically to the luminal domain of SV2 (Dong et al. 2006; Mahrhold et al. 2006). SV2 is a highly glycosylated transmembrane protein of SV implicated in vesicle recruitment to the plasma membrane in endocrine cells (Iezzi et al. 2005). Vertebrates express three distinct, but homologous, SV2 proteins termed SV2A, SV2B, and SV2C (Bajjalieh et al. 1992; Bajjalieh et al. 1994). Interestingly, motor neuron terminals express the receptor isoforms that bind BoNTs more strongly, such as Syt-II (Juzans et al. 1996; Li et al. 1994; Marquèze et al.

1995) and SV2C (Janz and Südhof 1999). Conversely, high-affinity isoforms are poorly expressed in CNS neurons and terminals, in which Syt-I, SV2A, and SV2B are the prevalent isoforms (reviewed in Verderio et al. 2006). Recently Baldwin and Barbieri (2007) demonstrated that BoNT/A and /B binding domains associate with synaptic vesicle membrane proteins and suggested that BoNT protein receptor may be a component of a larger protein complex.

Both TeNT and BoNTs bind the presynaptic membrane of α-motoneurons, but then TeNT follows a different intracellular trafficking route and this must be determined by yet unidentified specific receptor(s).

## 6.2 Internalization

Available evidence indicates that CNTs do not enter the cell directly from the plasma membrane, but are endocytosed inside the lumen of vesicular structures in a temperature- and energy-dependent process (Black and Dolly 1986; Critchley et al. 1985; Dolly et al. 1984; reviewed in Schiavo et al. 2000). The finding of SV proteins as receptors of BoNTs support the proposal (Montecucco and Schiavo 1995) that, after binding, BoNTs are endocytosed within synaptic vesicles (SV) via their retrieval to be refilled with neurotransmitter, an hypothesis originally advanced to account for the increased rate of poisoning with NMJ activity (Figure 4) (Hughes and Whaler 1962).

The protein receptor of TeNT would be responsible for its inclusion in an endocytic vesicle that moves in a retrograde direction all along and inside the axon (Figure 4). TeNT-$H_C$ internalization at the motor nerve terminal occurs via a specialized clathrin-dependent pathway, which is distinct from SV endocytosis and is preceded by a lateral sorting from its lipid raft–associated ligand GD1b. (Deinhardt et al. 2006; Roux et al. 2005). The TeNT-carrying vesicles reach the cell body of the motoneurons which reside within the spinal cord and then move to dendritic terminals to release the toxin in the intersynaptic space. TeNT then binds and enters the inhibitory interneurons of the spinal cord via SV endocytosis (Matteoli et al. 1996). Consequently, it blocks the release of inhibitory neurotransmitters and impairs the system which controls the balanced movement of the skeleton (Kandel et al. 2000), giving rise to a spastic paralysis. It is not known if, and if so, to what an extent, BoNTs migrate retroaxonally. Direct measurements of retroaxonal transport indicated that $^{125}$I-BoNT/A does not reach the central nervous system (Habermann and Dreyer 1986).

## 6.3 Membrane Translocation

The L chains of CNTs block neuroexocytosis by acting in the cytosol, and therefore at least this toxin domain must reach the cell cytosol. In order to do so, the L chain must cross the hydrophobic barrier of the vesicle membrane, and compelling evidence indicates that CNTs have to be exposed to low pH for this step to occur (Matteoli et al. 1996; Simpson et al. 1994; Williamson and Neale 1994). In fact, acidic pH causes a conformational change from a water-soluble "neutral" structure to an "acid" structure with exposure of hydrophobic patches on the surface of both the H and L chains which then enter into the hydrocarbon core of the lipid bilayer (Montecucco et al. 1989; Puhar et al. 2004; Schiavo et al. 2000; Shone et al. 1987). Following this low pH-induced membrane insertion, BoNTs and TeNT form transmembrane ion channels (reviewed in Montecucco and Schiavo 1995) and in PC12 membranes (Sheridan 1998). The cleavages of SNAP-25 by BoNT/A and BoNT/E is differentially affected by inhibitors of the vacuolar ATPase proton pump (Keller et al. 2004), but this is unlikely to be due to a difference in the toxin acid-triggered conformational change, as the low pH-driven toxin conformational change takes place in a very narrow range of pH: 4.4–4.6 for TeNT and BoNT/A, /B, /C, /E, and /F (Puhar et al. 2004). This similarity indicates that the key residues for the conformational change are conserved among CNTs and that they behave very similarly with respect to lowering pH. Currently, membrane translocation is the least understood step of the mechanism of cell intoxication of CNTs, and further investigation is required to uncover the extent and implication of the structural changes that occur at low pH (for a recent discussion see Rossetto and Montecucco 2004). Koriazova and Montal (2003) have proposed that the H chain channel acts as a trans-membrane chaperone for the L chain, preventing its hydrophobicity-driven aggregation and maintaining its unfolded conformation during translocation. The L chain is then released into the neutral cytosol where it refolds into its native enzymatically active conformation. In this vision, additional chaperones are not implicated. However, a complex of cytosolic chaperones and thioredoxin present on the cytosolic face of endosomes was recently found to assist the translocation of the enzymatic chain of other bacterial toxins acting in the cytosol (Haug et al. 2003; Ratts et al. 2003). Four different chaperone systems have been found so far within neuron terminals and are essential for normal function of the synaptic vesicle cycle during neurotransmitter release (Evans et al. 2003; Zinsmaier and Bronk 2001). In light of the proposed role of SVs as vesicular carriers of BoNTs inside peripheral neurons, it is noteworthy that two out of four synaptic terminals chaperone complexes includes the cystein string protein of SVs. It is therefore possible that the membrane translocation and refolding of the L chain is assisted by SV chaperones in addition to the H chain, but this remains to be proven.

## *6.4 SNARE Proteins' Specific Metalloproteolytic Activity*

The L chains of BoNTs and TeNT are highly specific proteases that recognize and cleave only three proteins, the so-called SNARE proteins, which form the core of the neuroexocytosis apparatus (Figure 4) (Schiavo et al. 2000). TeNT, BoNT/B, /D, /F and /G cleave VAMP, a protein of the SV membrane, at different single peptide bonds; BoNT/C cleaves both syntaxin and SNAP-25, two proteins of the presynaptic membrane; BoNT/A and /E cleave SNAP-25 at different sites within the COOH-terminus (reviewed in Humeau et al. 2000; Schiavo et al. 2000). BoNT-poisoned nerve terminals have a normal size and appearance with normal content and shape of synaptic vesicles and mitochondria (Duchen 1971; Thesleff 1960). Possibly the only morphological sign of BoNT/A poisoning is noticeable at the frog NMJ with the disappearance of the small outward curvatures of the membrane close to the active zones (Harlow et al. 2001; Pumplin and Reese 1977).

VAMP, SNAP-25, and syntaxin form an heterotrimeric coiled-coil complex, termed the SNARE complex, which induces the juxtaposition of vesicle to the target membrane (Jahn et al. 2003; Südhof 2004). Several SNARE complexes assemble into a rosette in order to bring the SV membrane close enough to the cytosolic face of the presynaptic membrane to permit their fusion with subsequent release of the vesicle neurotransmitter content into the synaptic cleft (Montecucco et al. 2005). Proteolysis of one SNARE protein prevents the formation of a functional SNARE complex and, consequently, the release of neurotransmitter. This is not true for BoNT/A and /C, which cleave SNAP-25 within few residues from the C terminus giving rise to a truncated SNAP-25, which can still form a SNARE complex but not a rosette of SNARE complexes. Given the fact that a rosette of SNARE complexes is necessary for the exocytosis of one synaptic vesicle, the presence of a single defective SNARE complex will have a dominant negative effect on neurotransmitter release. This rationale explains the experimental finding that incomplete proteolysis of SNAP-25 at nerve terminals by BoNT/A is sufficient to cause full inhibition of neurotransmitter release (Bruns et al. 1997; Foran et al. 1996; Jurasinski et al. 2001; Keller et al. 2004; Osen-Sand et al. 1996; Raciborska et al. 1998; Williamson et al. 1996). This also explains the fact that the paralysis induced by BoNT/A and /C is long lasting, as the inhibitory activity of C-terminal truncated SNAP-25 on the assembly of the rosette of SNARE complexes overlaps and functionally extends the lifetime of the metalloproteolytic activity of the L chain of these two BoNTs (for a detailed discussion see Rossetto et al. 2006). These explanations, however, may not be sufficient to account for the astonishing fact that the duration of the BoNT/A inhibition of autonomic nerve terminals is more than one year (Naumann and Jost 2004).

The molecular basis of the CNTs specificity for the three SNAREs resides on protein-protein interactions which extend well beyond the proteolysed SNARE regions, and a major role is played by a nine-residue-long motif present within the SNARE proteins. This motif is characterized by three carboxylate residues alternated with hydrophobic and hydrophilic residues (Breidenbach and Brunger 2004; Evans et al. 2005; Pellizzari et al. 1996; Rossetto et al. 1994; Vaidyanathan et al. 1999; Washbourne et al. 1997). This motif is present in two copies in VAMP and syntaxin and four copies in SNAP-25. The various CNTs differ with respect to the specific interaction with the recognition motif (Rossetto et al. 2001b). Only protein segments including at least one copy of the motif are cleaved by the toxins, and mutations within the motif inhibit the proteolysis (Fang et al. 2006; Schiavo et al. 2000). Moreover antibodies against the SNARE motif inhibit the proteolytic activity of the neurotoxins (Pellizzari et al. 1997). A recent co-crystal structure of LC/A and SNAP25-(146–204) defined additional regions of interaction external to the cleavage site and to the motif (Breidenbach and Brunger 2004; Breidenbach and Brunger 2005b). More recently, the fitting of an extended region of the substrate (residues 189–203) within the long active site cleft was defined following extensive mutagenesis of LC/A and SNAP-25 (Chen and Barbieri 2006; Chen et al. 2007).

There are no quantitative data on the number of L chains required to intoxicate a nerve terminal. In *Aplyisa californica* cholinergic neurons, few molecules of toxin appear to be sufficient to block neuroexocytosis within an hour at room temperature (Poulain, personal communication). It is even more likely that few copies of L chain are sufficient in warm-blooded animals. It is evident that as long as the toxin is present in an active form, the nerve signal cannot be transmitted.

The half-time of action of the different BoNTs has been estimated in mice cerebellar granular neurons in culture by Foran et al. (2003b): BoNT/A (31 days) > BoNT/C (25) > BoNT/B (10) > BoNT/F (2) > BoNT/E (0.8). Comparable data were found in the mice NMJ *in vivo* (Meunier et al. 2003; Morbiato et al., 2007). These data parallel those observed in human patients injected with different BoNTs within skeletal muscles: BoNT/A (2–4 months) > BoNT/C (12–16 weeks) > BoNT/B (5–10 w) > BoNT/F (5–8 w) > BoNT/E (4–6 w). In general, humans recover more slowly than do rodents. The lifetime of LC inside the nerve terminal is clearly involved in determining the duration of paralysis. The finding that LC/A overexpressed in cultured cells binds to the cytosolic face of cell membrane, while LC/E is prevalently cytosolic (Fernández-Salas et al. 2004), has provided a possible explanation, within the limits of the significance of protein overexpression with respect to their intracellular localization. However, it is hard to conceive that this is the sole explanation, particularly if one considers that the duration of the effect of BoNT/A in autonomic disfunction in humans is 12–15 weeks (Naumann and Jost 2004). As discussed above, experimental evidence supports the proposal that an important role is also played by the dominant negative effect on neurotransmission exerted by the BoNT/A- or /C-truncated SNAP-25 (Montecucco et al. 2005; Rossetto et al. 2006). Nevertheless, even the addition of the lifetimes of LC and of truncated SNAP-25 may not be sufficient to provide a satisfactory explanation for the very long time of recovery of intoxicated autonomic terminals of exocrine glands.

## 7 The Mode of Action of PLA2 Snake Presynaptic Neurotoxins

Contrary to TeNT and BoNTs, which act inside nerve, there is a consensus that, at least in the initial phases of intoxication, SPANs act on the presynaptic plasma membrane. Many investigations performed with different SPANs using different NMJ preparations have documented that the progression of muscle paralysis consists of three phases:

(1) A brief initial phase of weak inhibition of Ach release, lasting few minutes, which is absent with some toxins such as notexin and paradaxin, and which has a different relevance in mice and rats NMJ preparations.
(2) A more prolonged second phase (10–30 min) which consists in a facilitated neurotransmitter release with generation of sufficient intersynaptic Ach to stimulate muscle twitch. The increased ACh release, induced by SPANs, is due to an increased frequency of m.e.p.p.s with no change in m.e.p.p.s amplitude, indicating an unaltered response of endplates; also e.p.p.s are larger in this phase of SPAN action and, in some cases, giant m.e.p.p.s are induced.
(3) A third phase of progressive decline of neurotransmission with strong decrease of e.p.p.s and m.e.p.p.s and no alteration of muscle contractile properties (reviewed in Harris 1997; Howard and Gundersen 1980; Montecucco and Rossetto 2000; Rowan 2001).

Moreover, the time course of the SPAN-induced paralysis is greatly shortened by nerve stimulation, which increases the tightly coupled events of SV exocytosis and endocytosis (Sudhof 2004); on the contrary depressing nerve terminal activity decreases the rate of intoxication (Chang et al. 1973). The action of SPANs is also characterized by the fact that the toxic effects can be neutralized by antitoxin antibodies, or $Sr^{2+}$ or washing with fresh medium only if performed within a short time period after toxin addition (Kamenskaya and Thesleff 1974; Simpson et al. 1993; Wilson et al. 1995).

Phase 1 is promoted by $Ca^{2+}$ and is largely unaffected by temperature and by treatments that depress the PLA2 activity of SPANs. In addition this phase is saturated at low toxin concentrations (Caratsch et al. 1981, 1985; Simpson et al. 1993), and the binding of radioactively labeled SPANs to the central nervous tissue indicates high affinity and limited number of binding sites, mainly located at presynaptic nerve terminals (Lambeau and Lazdunski 1999; Lambeau et al. 1989; Othman et al. 1982; Rehm and Betz 1982; Strong et al. 1977). The different SPANs do not compete one with another for binding to the presynaptic membrane. β-bungarotoxin is rather specific for motor nerve terminals, and sensory discharges are not affected for several hours of incubation with the toxin. At variance in the central nervous system this neurotoxin appear to bind more specifically to cholinergic and GABAergic terminals (Abe et al. 1976; Rehm and Betz 1982). These data suggest that phase 1 represents SPAN binding to the nerve terminal, a biological activity not overlapping the enzymatic one. Why the mere SPAN binding on the extracellular surface of presynaptic terminals should inhibit neuroexocytosis is not clear. This is even more puzzling if one considers that lysophosphatidylserine, lysophosphatidylglycerol, and lysophosphatidic acid also induce a phase 1 in the mouse hemidiaphragm (Caccin et al. 2006). These authors have suggested that SPANs and some lysophospholipids bind to selected portions of the presynaptic membrane corresponding to the active zones of neuroexocytosis (AZ), which face the Ach receptor rich in foldings of the muscle membrane (Couteaux and Pécot-Dechavassine 1970; Harlow et al. 2001). SPAN binding causes by itself, or following a limited and local phospholipid hydrolysis, a structural alteration of these sites of neurotransmitter release resulting in a lowered probability of membrane fusion; it is conceivable that certain lysophospholipids can as well alter the AZ, making them less prone to membrane fusion with synaptic vesicles. Toxin binding to the AZ also accounts for the lack of binding competition among the different SPANs, and, at the same time, for the presynaptic neurospecificity of SPANs. Additionally, it is possible that SPANs display their hydrolytic activity within AZ boundaries where some lipid mismatching and a reduced amount of cholesterol may exist. This suggestion follows the finding that β-bungarotoxin is most active at the liquid crystalline-gel phase transition of phosphatidylcholines and virtually inactive in the presence of cholesterol (Kelly et al. 1979; Strong and Kelly 1977). Different pharmacokinetics and/or different affinities of binding of SPANs to the AZ also account for the lack of correlation between their PLA2 activity and toxicity (Montecucco and Rossetto 2000) and the very large difference in toxicity between SPANs and the pancreatic PLA2. In fact, in this view, binding with consequent strict localization of the PLA2 activity is a major event in the entire intoxication procession.

The progressive increase of lysophospholipids within the AZ due either to SPAN PLA2 mediated hydrolysis of phospholipids, or to partition of lysophospholipids present in the bathing medium (Rigoni et al. 2005), is suggested to cause a transition from inhibition (phase 1) to promotion of neuroexocytosis (phase 2) with the mechanisms discussed here below.

Most studies indicate that the PLA2 activity is implicated in phase 3, though it is not clear how and to what an extent (Howard and Gundersen 1980; Kini 1997; Montecucco and Rossetto 2000). SPANs were indeed shown to hydrolyse phospholipids with production of lysophosphatidylcholine (lysoPC) and fatty acids (FA) with no preference for a particular FA chain (Rigoni et al. 2005). Moreover, it was found that an equimolar mixture of these two lipids (lysoPC + FA) inhibit nerve terminals in a manner identical to that of SPANs, conclusively demonstrating that phase 3 is caused by the PLA2 activity of nerve-bound toxins. These findings clearly show that, at least in the initial phase of their action, SPANs acts on the outside of neurons. This does not exclude that later on SPANs may enter the paralyzed nerve terminal and contribute to further damage. Indeed. ammodytoxin (from the venom of *Vipera ammodytes ammodytes*) binds to cytosolic and mitochondrial proteins and has been detected inside hippocampal neurons (Petrovic et al. 2004; Sribar et al. 2001, 2003a, 2003b). In addition, taipoxin was found to bind to a 49-kDa calcium-binding protein located on the endoplasmic reticulum (Dodds et al. 1997; Kirkpatrick et al. 2000) and has been immunodetected inside chromaffin cells (Neco et al. 2003).

The involvement of the PLA2 activity in phase 2 is more controversial, as it was implicated in the case of the frog NMJ (Abe and Miledi 1978; Abe et al. 1977; Caratsch et al. 1981; Kondo et al. 1978; Oberg and Kelly 1976) and in some studies employing rat and mice NMJ preparations (Strong et al. 1976, 1977) but not in others (Chang 1979; Chang and Su 1982; Chang et al. 1977). However, an enhanced ACh release can be also induced by lysophosphatidylserine and lysophosphatidylglycerol with no toxin present, providing compelling evidence for the involvement of PLA2 in phase 2 (Caccin et al. 2006).

Electron microscopic pictures taken during phase 2 show an increased number of clathrin-coated Ω-shaped plasma membrane invaginations, which were initially attributed to synaptic vesicle caught in the process of fusion with the presynaptic membrane (Abe et al. 1976; Chen and Lee 1970), but later studies interpreted them as incomplete endocytosis of SV (Dixon and Harris 1999; Harris et al. 2000). We have suggested that SPAN halts endocytosis at its very last step, the fission of the vesicle from the plasma membrane which requires the neck closure, i.e., exactly the reverse process of membrane fusion (Montecucco and Rossetto 2000; Rossetto et al. 2006).

Frog, mice, and rat preparations fixed in phase 3 show enlarged axon terminals with large depletion of synaptic vesicles, including those close to the membrane (ready to release pool) and those occupying a more internal position (reserve pool). There are several clathrin-coated Ω-shaped plasma membrane invaginations, also in areas not facing the muscle membrane foldings. Mitochondria are swollen with altered cristae, so much as to appear in some cases as large vacuoles. Smaller vacuoles are also present (Abe et al. 1976; Chen and Lee 1970; Cull-Candy et al. 1976; Dixon and Harris 1999; Gopalakrishnakone and Hawgood 1984; Hamilton et al. 1980; Harris et al. 2000; Lee et al. 1984; Prasarnpun et al. 2004). It should be emphasized that even at this late stage of SPAN intoxication, no damage to muscle, fibroblasts, and Schwann cells is noticeable. Observations made at later stages (several hours after treatment) both in frog and rat muscles show that the nerve terminal of the motor neuron axon virtually disappears from the endplate. This is accompanied by protrusions of Schwann cells establishing extensive contacts with the muscle fiber, similar to what happens after several days of denervation achieved by transecting the motor axon or by crushing the nerve terminal (Abe et al. 1976; Dixon and Harris 1999; Harris et al. 2000; Kang et al. 2003; Son and Thompson 1995). Prasarnpun et al. 2005 also documented a substantial axonal degeneration with loss of mitochondria, neurofilaments, and synaptophysin staining. The robust extracellular matrix structure of the NMJ, which binds ACh esterase, remains in place.

Loss of SVs and alterations of mitochondria are also caused by SPANs on neurons in culture or by incubation with the LysoPC + OA lipid mixture (Bonanomi et al. 2005; Rigoni et al. 2004, 2005, 2007). A stricking and typical morphological alteration of cultured neurons and of uncovered *Drosophila* NMJ is the bulging of their nerve terminals (Megighian et al. 2007; Rigoni et al. 2004). The kinetics and the morphological changes induced by SPANs at NMJs and these *in vitro* findings led us to propose that SPANs promote fusion of SV with the presynaptic membrane, and at the same time, inhibit their retrieval (Figure 5). These alterations are the consequence of the same membrane lesion which consists in the hydrolysis of the cylindrically shaped phospholipids into the cone-shaped fatty acids and the inverted cone-shaped lysophospholipids. At least in the initial stages of intoxication, this hydrolysis takes place on the external leaflet of the presynaptic membrane, close to the SPAN binding sites, as indicated by the finding that LysoPC is the major product of SPAN hydrolysis with very little hydrolysis of phospholipids present on the cytosolic membrane leaflet (Rigoni et al. 2005). FAs rapidly equilibrate by

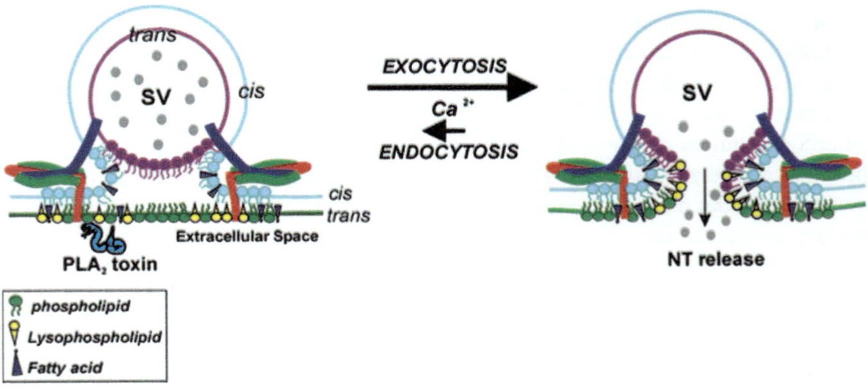


**Fig. 5** The snake PLA2 neurotoxin is depicted here as a snake, which binds to an active zone, i.e., a synaptic vesicle (SV) release site, and hydrolyses the phospholipids of the external layer of the presynaptic membrane (green) with formation of the inverted-cone shaped lysophospholipid (yellow) and the cone-shaped fatty acid (dark blue). Fatty acids rapidly equilibrate by trans-bilayer movement among the two layers of the presynaptic membrane. In such a way lysophospholipids, which induce a positive curvature of the membrane, are present in trans and fatty acid, which induce a negative curvature, are present also in cis, with respect to the fusion site. This membrane conformation facilitates the transition from a hemifusion intermediate to a pore. Thus, the action of the toxin promotes exocytosis of neurotransmitter (NT) (from the left to the right panel) and, for the same membrane topological reason, it inhibits the opposite process, i.e., the fission of the synaptic vesicle.

trans-bilayer movement among the two layers of the presynaptic membrane (Kamp et al. 1995), while lysoPC remains largely confined to the outer membrane leaflet. LysoPC, which induces a positive curvature of the membrane, is present in trans, and FAs, which induce a negative curvature, are also present in cis with respect to the synaptic vesicle-membrane fusion site. This membrane conformation is fusogenic because it facilitates the transition from an hemifusion lipid intermediate to a pore, which develops by intermixing the cis monolayers of the two membranes that have come together following the formation of the rosette of SNARE complexes (Chernomordik and Kozlov 2003; Chernomordik et al. 2006) (Figure 5).The existence of a lipid hemifusion intermediate has been theoretically predicted long ago (Kozlov and Markin 1983) and experimentally documented in the case of viral membrane fusion (Chernomordik et al. 1997), and, more recently, in the case of SNARE-mediated fusion (Giraudo et al. 2005; Reese and Mayer 2005; Xu et al. 2005, Yoon et al. 2006). It is possible that the ready-to-release SVs are indeed already hemifused with the presynaptic membrane. and this would account for the ultra-fast event of neuroexocytosis that takes place within few hundreds of microseconds from the entry of calcium (Zimmerberg and Chernomordik 2005). For the same reason, the membrane configuration induced by the SPAN hydrolytic action inhibits membrane fission, which is the reverse of the membrane fusion, and is essential for the recycling of the empty SV (Figure 5). This scenario accounts for the bulging of nerve terminals and the Ω-shaped membrane indentation induced by SPANs. It also explains the acceleration of SPAN intoxication induced by electrical stimulation of the NMJ, mentioned above, as there is a tight coupling between SV exocytosis and retrieval and increasing nerve stimulation increases the rate of SV exo-endocytosis (Cremona and De Camilli 1997; Sudhof 2004). In addition, it explains the finding that SPAN injected into the cytosol inhibit exocytosis (Wei et al. 2003) because this generates an anti-fusogenic membrane configuration with LysoPC on the cytosolic leaflet and FA on both sides.

The SPANs PLA2 activity is not fatty acid specific. Eicosopentanoic acid and arachidonic acid were recently found to promote exocytosis as deduced from increased membrane capacitance in PC-12 cells (Ong et al. 2006). Moreover, arachidonic acid, and some detergents, were found to stimulate the release of munc18 from syntaxin, thus allowing formation of SNARE complexes, a passage which is expected to increase exocytosis (Rickman and Davletov 2005). Moreover, omega-3 and omega-6 polyunsaturated fatty acids were found to stimulate membrane expansion at the nerve growth cone by a mechanism requiring syntaxin 3 (Darios and Davletov 2006). Accordingly, it is possible that these fatty acids contribute to the increased neuroexocytosis triggered in phase 2, but their role may be limited *in vivo* as the lysophospholipid was found to be much more effective than the fatty acid in the induction of paralysis (Caccin et al. 2006; Rigoni et al. 2005).

This rationale explains the induced neuroexocytosis of the ready-to-release SV pool, but does not well account for the massive vesicle depletion observed in the later phases of intoxication. However, one should consider that, with time, more and more lysophospholipids and FAs are generated and their accumulation may cause additional phenomena. Among them, increased plasma membrane permeability to ions caused by lysophospholipids (Leung et al. 1998; Wilson-Ashworth et al. 2004; Woodley et al. 1991) and partition of FAs inside mitochondria where they act as uncouplers of respiration with inhibition of ATP generation (Wojtczak and Schönfeld 1993). A specific, site-directed permeabilization of the membrane was proposed to be implicated in the intoxication by β-bungarotoxin (Rugolo et al. 1986), but how this would generate the observed alterations of morphology was not discussed. Clearly the FA uncoupling of mitochondria proposed by Wernicke et al. 1975 plays little role, if any, in SPAN intoxication because FA alone does not inhibit the NMJ (Rigoni et al. 2005). The products of SPAN phospholipid hydrolysis are likely to act at the plasma membrane level, as the SPAN-induced NMJ paralysis is rapidly reversed by albumin washing (Caccin et al. 2006). It was recently found that the bulged nerve terminals accumulate calcium independently on the agent responsible for bulging, whether it is SPAN or the LysoPC + FA mixture (Rigoni et al. 2007). $Ca^{2+}$ ions entered from the medium and an increased cytosolic $Ca^{2+}$ concentration within the nerve terminal of a synapse is by itself sufficient to trigger the release of all vesicles, including the SV reserve pool (Ceccarelli et al. 1972; Rizzoli and Betz 2005). Such high $Ca^{2+}$ has an additional pathological consequence owing to the accumulation of this ion into the mitochondrial matrix with precipitation of calcium phosphate (Carafoli 1986).

The last event of this SPAN nerve terminal intoxication scenario may be the entry of the SPAN inside the nerve terminal, as suggested by two studies performed with cells in culture (Neco et al. 2003; Petrovic et al. 2004). A SPAN inside a nerve terminals in the presence of $Ca^{2+}$ could damage profoundly mitochondria and other organelles via its PLA2 activity as shown with isolated mitochondria (Wagner et al. 1974; Wernicke et al. 1975). In fact, β-bungarotoxin is much more active on low cholesterol membranes than on the plasma membrane (Lee et al. 1972; Strong and Kelly 1977; Kelly et al. 1979).

## 8 Regeneration of the Skeletal Neuromuscular Junction and the Innervated Muscle Fibers after Poisoning by Botulinum or Snake Neurotoxins

None of the clostridial or snake neurotoxins, when used at concentrations corresponding to the mouse lethal doses, kill intoxicated neurons *in vivo*, although they are extremely toxic to the animal because of the unique role of synaptic transmission in animal physiology and behavior. A small amount of BoNT, dissolved in a minimal volume of carrier solution and injected into a muscle, does not spread around significantly. NMJs present at and around the injection site become paralyzed and lose their functionality, but the motoneuron and the innervated muscle fiber remain alive. Anatomical contacts between nerve and muscle are maintained, and there is no apparent loss of motor axons (Eleopra et al. 2002). However, the muscle undergoes a transient atrophy with loss of acetylcholine esterase staining and dispersion of acetylcholine receptors from the end plate (Angaut-Petit et al. 1990; Bambrick and Gordon 1994; Comella et al. 1993). Thin, amyelinated neuritic processes are emitted from their original parent nerve ending (terminal sprouting) or from the most proximal node of Ranvier (nodal sprouting), ultimately growing into the muscle (de Paiva et al. 1999; Son and Thompson 1995). Sprouting from the poisoned nerve terminal is preceded and guided by the nerve terminal Schwann cells, which are, at least in part, activated to migrate by the lack of supply of ACh consequent to the persistent inhibition of ACh release caused by the toxin (Kang et al. 2003; Son and Thompson 1995). Nerve sprouting is a remarkable example of synaptic plasticity. The number of motor end plates on single muscle fibers increases as well as the number of fibers innervated by a single motor axon. Moreover, single muscle fibers may become innervated by more than one motor axon. Following recovery of the normal function by the original NMJ, the sprouts degenerate and ACh esterase and ACh receptors reconcentrate exclusively at the junctions; the muscle eventually regains its normal size (Borodic et al. 1994; de Paiva et al. 1999; Meunier et al. 2003; reviewed in Foran et al. 2003a; Meunier et al. 2002).

The time necessary to recover the function after BoNT paralysis mainly depends on the type of nerve terminal and the type of BoNT. The entire process of recovery from BoNT/A intoxication typically requires two to four months at the mammalian NMJ, but much longer ($>1$ year) at glands (Naumann and Jost 2004). BoNT/A has been the first type of BoNT used for therapy in humans (Schantz and Johnson 1997; Scott et al. 1989), and luckily it is the one that gives the longest lasting effects.

In fact, the other BoNT serotypes are very effective paralyzing agents, but their action is much shorter, except in the case of BoNT/C, which is comparable to BoNT/A (Eleopra et al. 1997, 1998a, 2002, 2004; Ludlow et al. 1992; Morbiato et al. 2007). This parallels the well-known fact that botulism caused by BoNT/A in humans is more dangerous than that caused by BoNT/B and /E (Smith and Sugiyama 1988) and observations performed at the rat NMJ (Kauffman et al. 1985; Molgó et al. 1989; Sellin et al. 1983).

The biochemical/cellular events at the basis of the different duration of action of the different toxins are unknown, but several factors may contribute: (1) the lifetime of the L chain in the cytosol, (2) the turnover of the truncated SNARE proteins, (3) secondary biochemical events triggered by the production of truncated SNARE proteins and/or the released peptides, and (4) other effects. The lifetime of intracellular acting toxins is usually quite long, and it is estimated to be on the order of weeks (Bartels et al. 1994; Habig et al. 1986; Keller et al. 1999; Marxen and Bigalke 1991; O'Sullivan et al. 1999). Moreover, it was documented that in chromaffin cells and at the frog NMJ, the truncated SNAP-25 generated by BoNT/A or /C persists for a long time and does interfere with exocytosis (Criado et al. 1999; Huang et al. 1998; O'Sullivan et al. 1999; Raciborska and Charlton 1999).

The cotreatment of human endplates with BoNT/A and E results in a rapid recovery of neuromuscular function, equivalent to that of BoNT/E alone (Eleopra et al. 1998b). This suggests that the lifetime of the protease activities does not account for the duration of paralysis at the motor nerve endings in situ. Meunier et al. (2003) have recently shown that the prolonged paralysis induced by BoNT/A arises from the slow replacement of BoNT/A truncated SNAP-25 fragment 1–197 (SNAP-25A), which is known to be inhibitory when overexpressed (O'Sullivan et al. 1999). Moreover, type E was shown to hasten the removal of inhibitory SNAP-25A from BoNT/A-treated mouse neuromuscular synapses by converting it to SNAP-25E (fragment 1–180). SNAP-25E can be replaced rapidly by newly synthesized SNAP-25, thereby accelerating resumption of synaptic transmission. Co-injection of BoNT/A and BoNT/E in rats and mice does not produce the same effect as in humans (Keller et al. 1999), raising the question of how comparable are data obtained in different animal species. Even larger are the differences found with cultured neurons (Foran et al. 2003b; O'Sullivan et al. 1999).

Fewer data are available on the regeneration of the NMJ intoxicated by SPANs (Dixon and Harris 1999; Gopalakrishnakone and Hawgood 1984; Grubb et al. 1991; Harris et al. 2000). As mentioned above, SPANs impair the nerve terminal function more rapidly than BoNTs, and morphological alterations are much more extensive. Within few hours nerve terminals degenerate with loss of neurofilaments to such an extent that some terminals are barely discernible. Schwann cells are unaffected and extend their processes into the synaptic cleft. Also, the muscle fibers may degenerate and be invaded by phagocytic cells. Re-innervation of the muscle fibers is achieved by pre- and postterminal sprouts within three to five days from intoxication. This is accompanied by muscle regeneration with the appearance of smaller fibers with a typical central nucleus, which persists in this location for months. By 7 to 10 days more than 90% of the muscle fibers are capable of generating an indirect action potential, and by three to four weeks the process of regeneration is complete and few examples of multiple innervation originated by the sprouting process persist. The short duration of the effects of SPAN with the relatively rapid recovery with respect to BoNT/A or /C can be rationalized by considering the different nature of the SPAN hydrolysis products, which are highly diffusible. Moreover, phagocytic cells are recruited, possibly following the release of inflammatory mediators, either directly by the SPAN via release of arachidonic acid or indirectly by other chemotactic substances, or both.

## 9 Clostridial Neurotoxins in Cell Biology

It is presently very difficult to isolate and culture peripheral motoneurons, and this has hampered research on the morphological and functional consequences of CNT intoxication. However, dissociated spinal cord neurons in culture, including primary motorneurons, are very sensitive to CNTs. In addition, brain neurons derived from the cortex or the hippocampus or the cerebellum are sensitive to these neurotoxins. In these cells, BoNT/A was found to cause no detectable morphological changes. In contrast BoNT/C, which cleaves both SNAP-25 and syntaxin, brings about rapid swelling of synaptic terminals followed by degeneration of axons and dendrites followed by cell death (Williamson and Neale 1998) and the collapse of growth cones in chick dorsal root ganglia (Igarashi et al. 1996). BoNT/C also kills cortical and hippocampal neurons (Osen-Sand et al. 1996) and cerebellar granular cells (Berliocchi et al. 2005; Foran et al. 2003b). In contrast, when injected in different human muscles, BoNT/C does not cause loss of motoneurons (Eleopra et al. 2002). Taken together, these results demonstrate the central role played by syntaxin in the control of the integrity of synaptic contacts, in addition to its essential function in exocytosis.

Given the general role of SNAREs in vesicular trafficking, which is paralleled by a strong structural similarity, the use of CNTs is not limited to neuronal cells possessing receptors (for comprehensive reviews see Rossetto et al. 2001b; Schiavo et al. 2000). If the simple incubation with CNTs is sufficient to cause inhibition of neurotransmitter release in nerve cells or synaptosomes following SNARE cleavage, exocytosis can be inhibited in many non-neuronal cells as well, provided that they are permeabilized or microinjected. Incubation with very high doses of CNTs may be also sufficient to elicit effects with cells characterized by a large fluid phase endocytosis. Alternatively, cells can be transfected with the gene encoding for the light chain. The fact that the SNARE protein isoforms are involved in a variety of intracellular vesicle fusion events (Jahn and Scheller 2006; Takamori et al. 2006), in addition to exocytosis, has extended their potential range of use, but a word of caution is called for because more than one SNARE isoform can be inactivated by a given toxin within the same cell. It is not straightforward to predict the cleavability of a SNARE by a CNT from its sequence. Moreover, low sensitivity of GABA release to BoNT/A and BoNT/E intoxication, relative to the release of ACh, noradrenaline or glutamate has been reported by different studies (Ashton and

Dolly 1988; Bigalke et al. 1981). This could result from the different expression of the proteolytic substrate in different neuronal populations. Indeed, the virtual absence of SNAP-25, as detected by immunofluorescence at inhibitory terminals of CNS neurons (Frassoni et al. 2005; Verderio et al. 2004, 2007) parallels the low sensitivity of GABAergic exocytosis to BoNT/A and BoNT/E. On the same line, the higher sensitivity to these neurotoxins of synaptic vesicle recycling in immature GABAergic cells parallels the higher degree of SNAP-25 expression in developing inhibitory neurons as compared to the mature neurons (Frassoni et al. 2005; Verderio et al. 2007). Different amounts of the expressed SNAREs or their different intracellular distributions relative to the toxin location, as well as the expression of toxin-insensitive SNARE isoforms, could therefore explain the distinct properties or the different efficacy of intoxication in specific neuronal subpopulation. The finding that BoNT/A and BoNT/E, used at suitable concentrations, only partially affect GABAergic transmission, while impairing glutamatergic signaling, opened the possibility of using these toxins to treat CNS diseases characterized by imbalances between excitatory and inhibitory transmission such as epilepsy (Costantin et al. 2005; Verderio et al. 2007).

The absolute neurospecificity of clostridial neurotoxins, and the ability of TeNT to undergo axonal retrograde transport, make them ideal tools to study endocytosis and sorting at the synapse, and of retro-axonal transport both *in vitro* and *in vivo* (reviewed in Deinhardt and Schiavo 2005). These processes, which are still poorly understood at the molecular level, represent an exciting area of application for clostridial neurotoxins and their binding fragments.

## 10 Therapeutic Uses

The demonstration that the inhibition of the nerve-muscle impulse is followed by a functional recovery of the NMJ provides the scientific basis of the rapidly growing use of BoNTs in the therapy of a variety of human diseases caused by hyperfunction of cholinergic terminals and other neurodiseases (Jankovic 2006; Montecucco et al. 1996; Scott et al. 1989; Truong and Jost 2006). Injections of minute amounts of BoNT into the muscle(s) to be paralyzed lead to a depression of the symptoms lasting months. Owing to the long lasting duration of its effect, BoNT/A has almost invariably been used. Since the NMJ paralysis is reversible, the injection has to be repeated, and the possibility of an immune response with production of BoNT/A-neutralizing antibodies can occur (Brin 1997; Hanna et al. 1999; Jankovic and Schwartz 1995; Sankhla et al. 1998; Zuber et al. 1993). The injection of a different BoNT serotype could overcome this drawback. BoNT/E and BoNT/F were tested in humans and were found to have beneficial effects of short duration (Aoki 2001; Billante et al. 2002; Chen et al. 1998b; Eleopra et al. 1998b, 2006; Mezaki et al. 1995). Also BoNT/B has a rather short duration of action, and longer paralysis can only be achieved with very high doses, thus increasing the possibility of an immune response (Brin et al. 1999; Dressler and Bigalke 2005; Dressler and Eleopra

2006; Lew et al. 1997; Sloop et al. 1997). Studies performed in humans (reviewed in Eleopra et al. 2006) and in mice (Morbiato et al. 2007), show that BoNT/C has a general profile of action similar to that of BoNT/A.

Injection of BoNT is currently recognized as the best available treatment for dystonias and for certain types of strabismus, and new uses are continuously found (Bhidayasiri and Truong 2005; Montecucco and Molgó 2005). In addition, BoNT/A inhibit ACh release at autonomic nerve terminals which innervate the glands and smooth muscle, and it is currently used to treat diseases such as hypersalivation and hypersudoration (Brisinda et al. 2004; Naumann and Jost 2004).

The use of BoNT/A has been increasingly reported in many conditions of pathological pain, including migraine and other headache disorders (Aoki 2003; Binder and Blitzer 2003), musculoskeletal pain, such as myofascial pain, low back pain, and other chronic pain syndromes (Luvisetto et al. 2007; Reilich et al. 2004; Sycha et al. 2004).

BoNTs are not known to be retrotransported to the CNS, but they are highly toxic when injected directly in the brain (Luvisetto et al. 2003). Both the intraventricular and the hind paw injection of minute amounts of BoNT/A in mice strongly depress the perception of inflammatory pain induced by formalin (Luvisetto et al. 2006). On the basis of the differential effect of BoNT on glutamatergic and GABAergic nerve terminals in hippocampal neurons (Verderio et al. 2004), BoNT/E was tested as a potential therapeutic of epilepsy, and it was found to effectively lower the symptoms induced in mice by the epileptogenic kaininic acid (Costantin et al. 2005).

TeNT has not yet been used as a therapeutic but it has a series of interesting biological properties. When injected in the hippocampus of rats it induces an epileptic-like syndrome (Bagetta and Nisticò 1994). The carboxy-terminal third of TeNT, which retains most, but not all, of the neuronal binding and uptake properties of the entire toxin, has been used as a carrier of lysosomal hydrolase (Dobrenis et al. 1992; Jiang et al. 2005), superoxide dismutase inside cells in culture (Figueiredo et al. 1997; Francis et al. 1995), or β-galactosidase in mouse embryos (Coen et al. 1997). These studies open the possibility of using TeNT as carrier of various biologicals from peripheral sites of injections to selected areas of the CNS.

**Acknowledgements** This work is supported by by research grants from Telethon-Italia GP0272Y01, the University of Padova, and FIRBRBNE01RHZM.

## References

Abe T, Miledi R (1978) Inhibition of beta-bungarotoxin action by bee venom phospholipase A2. Proc R Soc Lond B Biol Sci 200:225–30

Abe T, Limbrick AR, Miledi R (1976) Acute muscle denervation induced by beta-bungarotoxin. Proc R Soc Lond B Biol Sci 194:545–53

Abe T, Alemá S, Miledi R (1977) Isolation and characterization of presynaptically acting neurotoxins from the venom of bungarus snakes. Eur J Biochem 80:1–12

Agarwal R, Eswaramoorthy S, Kumaran D, Binz T, Swaminathan S (2004) Structural analysis of botulinum neurotoxin type e catalytic domain and its mutant Glu212–>Gln reveals the pivotal role of the Glu212 carboxylate in the catalytic pathway. Biochemistry 43:6637–44

Agarwal R, Binz T, Swaminathan S (2005) Structural analysis of botulinum neurotoxin serotype F light chain: implications on substrate binding and inhibitor design. Biochemistry 44:11758–65

Ahsan CR, Hajnóczky G, Maksymowych AB, Simpson LL (2005) Visualization of binding and transcytosis of botulinum toxin by human intestinal epithelial cells. J Pharmacol Exp Ther 315:1028–35

Angaut-Petit D, Molgó J, Comella JX, Faille L, Tabti N (1990) Terminal sprouting in mouse neuromuscular junctions poisoned with botulinum type A toxin: morphological and electrophysiological features. Neuroscience 37:799–808

Aoki KR (2001) A comparison of the safety margins of botulinum neurotoxin serotypes A, B, and F in mice. Toxicon 39:1815–20

Aoki KR (2003) Evidence for antinociceptive activity of botulinum toxin type A in pain management. Headache 43 Suppl 1:S9–15

Arndt JW, Yu W, Bi F, Stevens RC (2005) Crystal structure of botulinum neurotoxin type g light chain: serotype divergence in substrate recognition. Biochemistry 44:9574–80

Arndt JW, Chai Q, Christian T, Stevens RC (2006a) Structure of botulinum neurotoxin type D light chain at 1.65 a resolution: repercussions for vamp-2 substrate specificity. Biochemistry 45:3255–62

Arni RK, Ward RJ (1996) Phospholipase A2–a structural review. Toxicon 34:827–41

Arnon SS (1980) Infant botulism. Annu Rev Med 31:541–60

Ashton AC, Dolly JO (1988) Characterization of the inhibitory action of botulinum neurotoxin type A on the release of several transmitters from rat cerebrocortical synaptosomes. J Neurochem 50:1808–16

Bagetta G, Nistico G (1994) Tetanus toxin as a neurobiological tool to study mechanisms of neuronal cell death in the mammalian brain. Pharmacol Ther 62:29–39

Bajjalieh SM, Peterson K, Shinghal R, Scheller RH (1992) SV2, a brain synaptic vesicle protein homologous to bacterial transporters. Science 257:1271–3

Bajjalieh SM, Frantz GD, Weimann JM, McConnell SK, Scheller RH (1994) Differential expression of synaptic vesicle protein 2 (SV2) isoforms. J Neurosci 14:5223–35

Baldwin MR, Barbieri JT (2007) Association of botulinum neurotoxin serotypes a and b with synaptic vesicle protein complexes. Biochemistry 46:3200–10

Baldwin MR, Kim JP, Barbieri JT (2007) Botulinum neurotoxin b-host receptor recognition: it takes two receptors to tango. Nat Struct Mol Biol 14:9–10

Bambrick L, Gordon T (1994) Neurotoxins in the study of neural regulation of membrane proteins in skeletal muscle. J Pharmacol Toxicol Methods 32:129–38

Bartels F, Bergel H, Bigalke H, Frevert J, Halpern J et al. (1994) Specific antibodies against the Zn(2+)-binding domain of clostridial neurotoxins restore exocytosis in chromaffin cells treated with tetanus or botulinum a neurotoxin. J Biol Chem 269:8122–7

Berliocchi L, Fava E, Leist M, Horvat V, Dinsdale D et al. (2005) Botulinum neurotoxin C initiates two different programs for neurite degeneration and neuronal apoptosis. J Cell Biol 168:607–18

Bhidayasiri R, Truong DD (2005) Expanding use of botulinum toxin. J Neurol Sci 235:1–9

Bigalke H, Heller I, Bizzini B, Habermann E (1981) Tetanus toxin and botulinum A toxin inhibit release and uptake of various transmitters, as studied with particulate preparations from rat brain and spinal cord. Naunyn Schmiedebergs Arch Pharmacol 316:244–51

Billante CR, Zealear DL, Billante M, Reyes JH, Sant'Anna G et al. (2002) Comparison of neuromuscular blockade and recovery with botulinum toxins A and F. Muscle Nerve 26:395–403

Binder WJ, Blitzer A (2003) Treatment of migraine headache with botulinum toxin type A. Facial Plast Surg Clin North Am 11:465–75

Binz T, Bade S, Rummel A, Kollewe A, Alves J (2002) Arg(362) and tyr(365) of the botulinum neurotoxin type A light chain are involved in transition state stabilization. Biochemistry 41:1717–23

Black JD, Dolly JO (1986) Interaction of 125I-labeled botulinum neurotoxins with nerve terminals. II. autoradiographic evidence for its uptake into motor nerves by acceptor-mediated endocytosis. J Cell Biol 103:535–44

Bleck (1989) Clinical aspects of tetanus. In: Simpson LL (ed) Botulinum neurotoxin and tetanus toxin. Academic Press, San Diego, CA, pp 379–98

Bonanomi D, Pennuto M, Rigoni M, Rossetto O, Montecucco C et al. (2005) Taipoxin induces synaptic vesicle exocytosis and disrupts the interaction of synaptophysin I with VAMP2. Mol Pharmacol 67:1901–8

Borodic GE, Ferrante R, Pearce LB, Smith K (1994) Histologic assessment of dose-related diffusion and muscle fiber response after therapeutic botulinum A toxin injections. Mov Disord 9:31–9

Breidenbach MA, Brunger AT (2004) Substrate recognition strategy for botulinum neurotoxin serotype A. Nature 432:925–9

Breidenbach MA, Brunger AT (2005a) 2.3 A crystal structure of tetanus neurotoxin light chain. Biochemistry 44:7450–7

Breidenbach MA, Brunger AT (2005b) New insights into clostridial neurotoxin-SNARE interactions. Trends Mol Med 11:377–81

Brin MF (1997) Botulinum toxin: chemistry, pharmacology, toxicity, and immunology. Muscle Nerve Suppl 6:S146–68

Brin MF, Lew MF, Adler CH, Comella CL, Factor SA et al. (1999) Safety and efficacy of neurobloc (botulinum toxin type B) in type A-resistant cervical dystonia. Neurology 53:1431–38

Brisinda G, Bentivoglio AR, Maria G, Albanese A (2004) Treatment with botulinum neurotoxin of gastrointestinal smooth muscles and sphincters spasms. Mov Disord 19 Suppl 8:S146–56

Brooks VB, Curtis DR, Eccles JC (1955) Mode of action of tetanus toxin. Nature 175:120–1

Bruns D, Engers S, Yang C, Ossig R, Jeromin A et al. (1997) Inhibition of transmitter release correlates with the proteolytic activity of tetanus toxin and botulinus toxin A in individual cultured synapses of hirudo medicinalis. J Neurosci 17:1898–1910

Bullens RWM, O'Hanlon GM, Wagner E, Molenaar PC, Furukawa K et al. (2002) Complex gangliosides at the neuromuscular junction are membrane receptors for autoantibodies and botulinum neurotoxin but redundant for normal synaptic function. J Neurosci 22:6876–84

Bullough PA, Hughson FM, Skehel JJ, Wiley DC (1994) Structure of influenza haemagglutinin at the ph of membrane fusion. Nature 371:37–43

Burgen ASV, Dickens F, Zatman LJ (1949) The action of botulinum toxin on the neuro-muscular junction. J Physiol 109:10–24

Byrne MP, Smith LA (2000) Development of vaccines for prevention of botulism. Biochimie 82:955–66

Caccin P, Rigoni M, Bisceglie A, Rossetto O, Montecucco C (2006) Reversible skeletal neuromuscular paralysis induced by different lysophospholipids. FEBS Lett 580:6317–21

Carafoli E (1986) Mitochondrial pathology: an overview. Ann N Y Acad Sci 488:1–18

Caratsch CG, Maranda B, Miledi R, Strong PN (1981) A further study of the phospholipase-independent action of beta-bungarotoxin at frog end-plates. J Physiol 319:179–91

Caratsch CG, Miledi R, Strong PN (1985) Influence of divalent cations on the phospholipase-independent action of beta-bungarotoxin at frog neuromuscular junctions. J Physiol 363:169–79

Carredano E, Westerlund B, Persson B, Saarinen M, Ramaswamy S et al. (1998) The three-dimensional structures of two toxins from snake venom throw light on the anticoagulant and neurotoxic sites of phospholipase A2. Toxicon 36:75–92

Ceccarelli B, Hurlbut WP, Mauro A (1972) Depletion of vesicles from frog neuromuscular junctions by prolonged tetanic stimulation. J Cell Biol 54:30–8

Chai Q, Arndt JW, Dong M, Tepp WH, Johnson EA et al. (2006) Structural basis of cell surface receptor recognition by botulinum neurotoxin B. Nature 444:1096–1100

Chang CC (1979) The action of the snake venoms and on nerve and muscle. In: Lee C (ed) Snake venoms. Springer, Berlin/Heidelberg/New York, pp 309–76

Chang CC, Su MJ (1982) Presynaptic toxicity of the histidine-modified, phospholipase A2-inactive, beta-bungarotoxin, crotoxin and notexin. Toxicon 20:895–905
Chang CC, Huang MC, Lee CY (1973) Mutual antagonism between botulinum toxin and -bungarotoxin. Nature 243:166–7
Chang CC, Lee JD, Eaker D, Fohlman J (1977) Short communications the presynaptic neuromuscular blocking action of taipoxin. A comparison with beta-bungarotoxin and crotoxin. Toxicon 15:571–6
Chen IL, Lee CY (1970) Ultrastructural changes in the motor nerve terminals caused by beta-bungarotoxin. Virchows Arch B Cell Pathol 6:318–25
Chen F, Kuziemko GM, Amersdorfer P, Wong C, Marks JD et al. (1997) Antibody mapping to domains of botulinum neurotoxin serotype A in the complexed and uncomplexed forms. Infect Immun 65:1626–30
Chen F, Kuziemko GM, Stevens RC (1998a) Biophysical characterization of the stability of the 150-kilodalton botulinum toxin, the nontoxic component, and the 900-kilodalton botulinum toxin complex species. Infect Immun 66:2420–5
Chen R, Karp BI, Hallett M (1998b) Botulinum toxin type F for treatment of dystonia: long-term experience. Neurology 51:1494–6
Chen S, Barbieri JT (2006) Unique substrate recognition by botulinum neurotoxins serotypes A and E. J Biol Chem 281:10906–11
Chen S, Kim JP, Barbieri JT (2007) Mechanism of substrate recognition by botulinum neurotoxin serotype A. J Biol Chem 282:9621–27
Chernomordik LV, Kozlov MM (2003) Protein-lipid interplay in fusion and fission of biological membranes. Annu Rev Biochem 72:175–207
Chernomordik LV, Leikina E, Frolov V, Bronk P, Zimmerberg J (1997) An early stage of membrane fusion mediated by the low pH conformation of influenza hemagglutinin depends upon membrane lipids. J Cell Biol 136:81–93
Chernomordik LV, Zimmerberg J, Kozlov MM (2006) Membranes of the world unite! J Cell Biol 175:201–7
Coen L, Osta R, Maury M, Brûlet P (1997) Construction of hybrid proteins that migrate retrogradely and transynaptically into the central nervous system. Proc Natl Acad Sci U S A 94:9400–5
Comella JX, Molgo J, Faille L (1993) Sprouting of mammalian motor nerve terminals induced by in vivo injection of botulinum type-D toxin and the functional recovery of paralysed neuromuscular junctions. Neurosci Lett 153:61–4
Connolly S, Trevett AJ, Nwokolo NC, Lalloo DG, Naraqi S et al. (1995) Neuromuscular effects of papuan taipan snake venom. Ann Neurol 38:916–20
Corradin G, Watts C (1995) Cellular immunology of tetanus toxoid. Curr Top Microbiol Immunol 195:77–87
Costantin L, Bozzi Y, Richichi C, Viegi A, Antonucci F et al. (2005) Antiepileptic effects of botulinum neurotoxin E. J Neurosci 25:1943–51
Couteaux R, Pécot-Dechavassine M (1970) Synaptic vesicles and pouches at the level of "active zones" of the neuromuscular junction C R Acad Sci Hebd Seances Acad Sci D 271:2346–9
Cremona O, De Camilli P (1997) Synaptic vesicle endocytosis. Curr Opin Neurobiol 7:323–30
Criado M, Gil A, Viniegra S, Gutiérrez LM (1999) A single amino acid near the C terminus of the synaptosome associated protein of 25 kKa (SNAP-25) is essential for exocytosis in chromaffin cells. Proc Natl Acad Sci U S A 96:7256–61
Critchley DR, Nelson PG, Habig WH, Fishman PH (1985) Fate of tetanus toxin bound to the surface of primary neurons in culture: evidence for rapid internalization. J Cell Biol 100:1499–1507
Cull-Candy SG, Fohlman J, Gustavsson D, Lüllmann-Rauch R, Thesleff S (1976) The effects of taipoxin and notexin on the function and fine structure of the murine neuromuscular junction. Neuroscience 1:175–80
Darios F, Davletov B (2006) Omega-3 and omega-6 fatty acids stimulate cell membrane expansion by acting on syntaxin 3. Nature 440:813–17

Dasgupta B (1994) Structures of botulinum neurotoxin, its functional domains, and perspectives on the cristalline type a toxin. In: Jankovic J, Hallett M (eds) Therapy with botulinum toxin. Marcel Dekker, New York, pp 15–39
de Paiva A, Poulain B, Lawrence GW, Shone CC, Tauc L et al. (1993) A role for the interchain disulfide or its participating thiols in the internalization of botulinum neurotoxin A revealed by a toxin derivative that binds to ecto-acceptors and inhibits transmitter release intracellularly. J Biol Chem 268:20838–44
de Paiva A, Meunier FA, Molgó J, Aoki KR, Dolly JO (1999) Functional repair of motor endplates after botulinum neurotoxin type A poisoning: biphasic switch of synaptic activity between nerve sprouts and their parent terminals. Proc Natl Acad Sci U S A 96:3200–5
Deinhardt K, Schiavo G (2005) Endocytosis and retrograde axonal traffic in motor neurons. Biochem Soc Symp :139–50
Deinhardt K, Berninghausen O, Willison HJ, Hopkins CR, Schiavo G (2006) Tetanus toxin is internalized by a sequential clathrin-dependent mechanism initiated within lipid microdomains and independent of epsin1. J Cell Biol 174:459–71
Dixon RW, Harris JB (1999) Nerve terminal damage by beta-bungarotoxin: its clinical significance. Am J Pathol 154:447–55
Dobrenis K, Joseph A, Rattazzi MC (1992) Neuronal lysosomal enzyme replacement using fragment C of tetanus toxin. Proc Natl Acad Sci U S A 89:2297–2301
Dodds DC, Omeis IA, Cushman SJ, Helms JA, Perin MS (1997) Neuronal pentraxin receptor, a novel putative integral membrane pentraxin that interacts with neuronal pentraxin 1 and 2 and taipoxin-associated calcium-binding protein 49. J Biol Chem 272:21488–94
Dolly JO, Black J, Williams RS, Melling J (1984) Acceptors for botulinum neurotoxin reside on motor nerve terminals and mediate its internalization. Nature 307:457–60
Dong M, Richards DA, Goodnough MC, Tepp WH, Johnson EA et al. (2003) Synaptotagmins i and ii mediate entry of botulinum neurotoxin b into cells. J Cell Biol 162:1293–1303
Dong M, Yeh F, Tepp WH, Dean C, Johnson EA et al. (2006) SV2 is the protein receptor for botulinum neurotoxin A. Science 312:592–6
Dressler D, Bigalke H (2005) Botulinum toxin type B de novo therapy of cervical dystonia: frequency of antibody induced therapy failure. J Neurol 252:904–7
Dressler D, Eleopra R (2006) Clinical use of non-A botulinum toxins: botulinum toxin type B. Neurotox Res 9:121–5
Duchen LW (1971) An electron microscopic study of the changes induced by botulinum toxin in the motor end-plates of slow and fast skeletal muscle fibres of the mouse. J Neurol Sci 14:47–60
Dufton MJ, Hider RC (1983) Classification of phospholipases A2 according to sequence. evolutionary and pharmacological implications. Eur J Biochem 137:545–51
Eleopra R, Tugnoli V, Rossetto O, Montecucco C, De Grandis D (1997) Botulinum neurotoxin serotype C: a novel effective botulinum toxin therapy in human. Neurosci Lett 224:91–4
Eleopra R, Tugnoli V, De Grandis D, Montecucco C (1998a) Botulinum toxin serotype C treatment in subjects affected by focal dystonia and resistant to botulinum toxin serotype A Neurology 50 (Suppl. 4):A72
Eleopra R, Tugnoli V, Rossetto O, De Grandis D, Montecucco C (1998b) Different time courses of recovery after poisoning with botulinum neurotoxin serotypes A and E in humans. Neurosci Lett 256:135–8
Eleopra R, Tugnoli V, Quatrale R, Gastaldo E, Rossetto O et al. (2002) Botulinum neurotoxin serotypes A and C do not affect motor units survival in humans: an electrophysiological study by motor units counting. Clin Neurophysiol 113:1258–64
Eleopra R, Tugnoli V, Quatrale R, Rossetto O, Montecucco C (2004) Different types of botulinum toxin in humans. Mov Disord 19 Suppl 8:S53–9
Eleopra R, Tugnoli V, Quatrale R, Rossetto O, Montecucco C et al. (2006) Clinical use of non-A botulinum toxins: botulinum toxin type C and botulinum toxin type F. Neurotox Res 9:127–31
Evans ER, Sutton JM, Gravett A, Shone CC (2005) Analysis of the substrate recognition domain determinants of botulinum type B toxin using phage display. Toxicon 46:446–53

Evans GJO, Morgan A, Burgoyne RD (2003) Tying everything together: the multiple roles of cysteine string protein (csp) in regulated exocytosis. Traffic 4:653–9

Faber K (1890) Die pathogenie des tetanus. Berl klin Wochenschr 27:717–20

Fang H, Luo W, Henkel J, Barbieri J, Green N (2006) A yeast assay probes the interaction between botulinum neurotoxin serotype B and its SNARE substrate. Proc Natl Acad Sci U S A 103:6958–63

Fernández-Salas E, Ho H, Garay P, Steward LE, Aoki KR (2004) Is the light chain subcellular localization an important factor in botulinum toxin duration of action? Mov Disord 19 Suppl 8:S23–34

Figueiredo DM, Hallewell RA, Chen LL, Fairweather NF, Dougan G et al. (1997) Delivery of recombinant tetanus-superoxide dismutase proteins to central nervous system neurons by retrograde axonal transport. Exp Neurol 145:546–54

Foran P, Lawrence GW, Shone CC, Foster KA, Dolly JO (1996) Botulinum neurotoxin C1 cleaves both syntaxin and SNAP-25 in intact and permeabilized chromaffin cells: correlation with its blockade of catecholamine release. Biochemistry 35:2630–6

Foran PG, Davletov B, Meunier FA (2003a) Getting muscles moving again after botulinum toxin: novel therapeutic challenges. Trends Mol Med 9:291–9

Foran PG, Mohammed N, Lisk GO, Nagwaney S, Lawrence GW et al. (2003b) Evaluation of the therapeutic usefulness of botulinum neurotoxin B, C1, E, and F compared with the long lasting type A. basis for distinct durations of inhibition of exocytosis in central neurons. J Biol Chem 278:1363–71

Francis JW, Hosler BA, Brown RHJ, Fishman PS (1995) CuZn superoxide dismutase (SOD-1):tetanus toxin fragment C hybrid protein for targeted delivery of SOD-1 to neuronal cells. J Biol Chem 270:15434–42

Frassoni C, Inverardi F, Coco S, Ortino B, Grumelli C et al. (2005) Analysis of SNAP-25 immunoreactivity in hippocampal inhibitory neurons during development in culture and in situ. Neuroscience 131:813–23

Fujinaga Y, Inoue K, Nomura T, Sasaki J, Marvaud JC et al. (2000) Identification and characterization of functional subunits of clostridium botulinum type a progenitor toxin involved in binding to intestinal microvilli and erythrocytes. FEBS Lett 467:179–83

Fujita R, Fujinaga Y, Inoue K, Nakajima H, Kumon H et al. (1995) Molecular characterization of two forms of nontoxic-nonhemagglutinin components of clostridium botulinum type a progenitor toxins. FEBS Lett 376:41–4

Galazka A, Gasse F (1995) The present status of tetanus and tetanus vaccination. Curr Top Microbiol Immunol 195:31–53

Gill DM (1982) Bacterial toxins: a table of lethal amounts Microbiol. Rev. 46:86–94

Giraudo CG, Hu C, You D, Slovic AM, Mosharov EV et al. (2005) SNAREs can promote complete fusion and hemifusion as alternative outcomes. J Cell Biol 170:249–60

Gopalakrishnakone P, Hawgood BJ (1984) Morphological changes induced by crotoxin in murine nerve and neuromuscular junction. Toxicon 22:791–804

Grubb BD, Harris JB, Schofield IS (1991) Neuromuscular transmission at newly formed neuromuscular junctions in the regenerating soleus muscle of the rat. J Physiol 441:405–21

Gutiérrez JM, Theakston RDG, Warrell DA (2006) Confronting the neglected problem of snake bite envenoming: the need for a global partnership. PLoS Med 3:e150

Habermann E, Dreyer F (1986) Clostridial neurotoxins: handling and action at the cellular and molecular level. Curr Top Microbiol Immunol 129:93–179

Habig WH, Bigalke H, Bergey GK, Neale EA, Hardegree MC et al. (1986) Tetanus toxin in dissociated spinal cord cultures: long-term characterization of form and action. J Neurochem 47:930–7

Hamilton RC, Broad AJ, Sutherland SK (1980) Effects of australian eastern brown snake (pseudonaja textilis) venom on the ultrastructure of nerve terminals on the rat diaphragm. Neurosci Lett 19:45–50

Hanley MR, Emson PC (1979) Neuronal degeneration induced by stereotaxic injection of beta-bungarotoxin into rat brain. Neurosci Lett 11:143–8

Hanna PA, Jankovic J, Vincent A (1999) Comparison of mouse bioassay and immunoprecipitation assay for botulinum toxin antibodies. J Neurol Neurosurg Psychiatry 66:612–16

Hanson MA, Stevens RC (2000) Cocrystal structure of synaptobrevin-II bound to botulinum neurotoxin type B at 2.0 A resolution. Nat Struct Biol 7:687–92

Harlow ML, Ress D, Stoschek A, Marshall RM, McMahan UJ (2001) The architecture of active zone material at the frog's neuromuscular junction. Nature 409:479–84

Harris JB (1997) Toxic phospholipases in snake venom: an introductory review. Symp. zool. Soc. Lond. 70:235–50

Harris JB, Grubb BD, Maltin CA, Dixon R (2000) The neurotoxicity of the venom phospholipases A(2), notexin and taipoxin. Exp Neurol 161:517–26

Haug G, Wilde C, Leemhuis J, Meyer DK, Aktories K et al. (2003) Cellular uptake of clostridium botulinum C2 toxin: membrane translocation of a fusion toxin requires unfolding of its dihydrofolate reductase domain. Biochemistry 42:15284–91

Hauschild A (1993) Epidemiology of human foodborne botulism. In: Hauschild A, Dodds KL (eds) Clostridium botulinum: ecology and control in foods. Marcel Dekker, Inc. New York, pp 69–104

Heckly RJ, Hildebrand GJ, Lamanna C (1960) On the size of the toxic particle passing the intestinal barrier in botulism. J Exp Med 111:745–59

Henderson I, Davis T, Elmore M, Minton N (1997) The genetic basis of toxin production in clostridium botulinum and clostridium tetani. In: Rood I (ed) The clostridia: molecular biology and pathogenesis. Academic Press, New York, pp 261–94

Herreros J, Schiavo G (2002) Lipid microdomains are involved in neurospecific binding and internalisation of clostridial neurotoxins. Int J Med Microbiol 291:447–53

Herreros J, Lalli G, Montecucco C, Schiavo G (2000) Tetanus toxin fragment C binds to a protein present in neuronal cell lines and motoneurons. J Neurochem 74:1941–50

Herreros J, Ng T, Schiavo G (2001) Lipid rafts act as specialized domains for tetanus toxin binding and internalization into neurons. Mol Biol Cell 12:2947–60

Hodgson WC, Wickramaratna JC (2002) In vitro neuromuscular activity of snake venoms. Clin Exp Pharmacol Physiol 29:807–14

Hodgson WC, Wickramaratna JC (2006) Snake venoms and their toxins: an australian perspective. Toxicon 48:931–40

Howard BD, Gundersen CBJ (1980) Effects and mechanisms of polypeptide neurotoxins that act presynaptically. Annu Rev Pharmacol Toxicol 20:307–36

Huang X, Wheeler MB, Kang YH, Sheu L, Lukacs GL et al. (1998) Truncated SNAP-25 (1–197), like botulinum neurotoxin A, can inhibit insulin secretion from HIT-T15 insulinoma cells. Mol Endocrinol 12:1060–70

Hughes R, Whaler BC (1962) Influence of nerve-ending activity and of drugs on the rate of paralysis of rat diaphragm preparations by cl. botulinum type a toxin. J Physiol 160:221–33

Humeau Y, Doussau F, Grant NJ, Poulain B (2000) How botulinum and tetanus neurotoxins block neurotransmitter release. Biochimie 82:427–46

Iezzi M, Theander S, Janz R, Loze C, Wollheim CB (2005) SV2A and SV2C are not vesicular $Ca2+$ transporters but control glucose-evoked granule recruitment. J Cell Sci 118:5647–60

Igarashi M, Kozaki S, Terakawa S, Kawano S, Ide C et al. (1996) Growth cone collapse and inhibition of neurite growth by botulinum neurotoxin C1: a t-SNARE is involved in axonal growth. J Cell Biol 134:205–15

Inoue K, Fujinaga Y, Watanabe T, Ohyama T, Takeshi K et al. (1996) Molecular composition of clostridium botulinum type a progenitor toxins. Infect Immun 64:1589–94

Jahn R, Scheller RH (2006) SNAREs–engines for membrane fusion. Nat Rev Mol Cell Biol 7:631–43

Jahn R, Lang T, Südhof TC (2003) Membrane fusion. Cell 112:519–33

Jankovic J (2006) Botulinum toxin therapy for cervical dystonia. Neurotox Res 9:145–8

Jankovic J, Schwartz K (1995) Response and immunoresistance to botulinum toxin injections. Neurology 45:1743–6

Janz R, Südhof TC (1999) SV2C is a synaptic vesicle protein with an unusually restricted localization: anatomy of a synaptic vesicle protein family. Neuroscience 94:1279–90

Jiang K, Watson DJ, Wolfe JH (2005) A genetic fusion construct between the tetanus toxin C fragment and the lysosomal acid hydrolase beta-glucuronidase expresses a bifunctional protein with enhanced secretion and neuronal uptake. J Neurochem 93:1334–44

Jin R, Rummel A, Binz T, Brunger AT (2006) Botulinum neurotoxin b recognizes its protein receptor with high affinity and specificity. Nature 444:1092–5

Jurasinski CV, Lieth E, Dang Do AN, Schengrund CL (2001) Correlation of cleavage of SNAP-25 with muscle function in a rat model of botulinum neurotoxin type A induced paralysis. Toxicon 39:1309–15

Juzans P, Molgo J, Faille L, Angaut-Petit D (1996) Synaptotagmin II immunoreactivity in normal and botulinum type A treated mouse motor nerve terminals. Pflugers Arch 431:R283–4

Kamenskaya MA, Thesleff S (1974) The neuromuscular blocking action of an isolated toxin from the elapid (oxyuranus scutellactus). Acta Physiol Scand 90:716–24

Kamp F, Zakim D, Zhang F, Noy N, Hamilton JA (1995) Fatty acid flip-flop in phospholipid bilayers is extremely fast. Biochemistry 34:11928–37

Kandel E, Schwartz J, Jessel T (2000) Principles of neural science, 4th edn. McGrawHill, USA

Kang H, Tian L, Thompson W (2003) Terminal schwann cells guide the reinnervation of muscle after nerve injury. J Neurocytol 32:975–85

Kauffman JA, Way JFJ, Siegel LS, Sellin LC (1985) Comparison of the action of types A and F botulinum toxin at the rat neuromuscular junction. Toxicol Appl Pharmacol 79:211–17

Keller JE, Neale EA, Oyler G, Adler M (1999) Persistence of botulinum neurotoxin action in cultured spinal cord cells. FEBS Lett 456:137–42

Keller JE, Cai F, Neale EA (2004) Uptake of botulinum neurotoxin into cultured neurons. Biochemistry 43:526–32

Kelly RB, von Wedel RJ, Strong PN (1979) Phospholipase-dependent and phospholipase-independent inhibition of transmitter relase by beta-bungarotoxin. Adv Cytopharmacol 3:77–85

Kerner (1817) Vergiftung durch verborbene Würste. Tübinger Blatter 3:1–25

Kielian M, Rey FA (2006) Virus membrane-fusion proteins: more than one way to make a hairpin. Nat Rev Microbiol 4:67–76

Kini RM (1997) Venom phospholipase a2 enzymes. John Wiley & Sons, Chichester

Kini RM, Evans HJ (1989) A model to explain the pharmacological effects of snake venom phospholipases a2. Toxicon 27:613–35

Kirkpatrick LL, Matzuk MM, Dodds DC, Perin MS (2000) Biochemical interactions of the neuronal pentraxins. Neuronal pentraxin (NP) receptor binds to taipoxin and taipoxin-associated calcium-binding protein 49 via NP1 and NP2. J Biol Chem 275:17786–92

Kitamura M, Igimi S, Furukawa K, Furukawa K (2005) Different response of the knockout mice lacking b-series gangliosides against botulinum and tetanus toxins. Biochim Biophys Acta 1741:1–3

Kondo K, Toda H, Narita K (1978) Characterization of phospholipase A activity of beta1-bungarotoxin from bungarus multicinctus venom. II. identification of the histidine residue of beta1-bungarotoxin modified by p-bromophenacyl bromide. J Biochem (Tokyo) 84:1301–8

Koriazova LK, Montal M (2003) Translocation of botulinum neurotoxin light chain protease through the heavy chain channel. Nat Struct Biol 10:13–18

Kozlov MM, Markin VS (1983) Possible mechanism of membrane fusion Biofizika 28:242–7

Kriegstein KG, Henschen AH, Weller U, Habermann E (1991) Limited proteolysis of tetanus toxin. relation to activity and identification of cleavage sites. Eur J Biochem 202:41–51

Kularatne SAM (2002) Common krait (bungarus caeruleus) bite in anuradhapura, sri lanka: a prospective clinical study, 1996–98. Postgrad Med J 78:276–80

Kwong PD, McDonald NQ, Sigler PB, Hendrickson WA (1995) Structure of beta 2-bungarotoxin: potassium channel binding by kunitz modules and targeted phospholipase action. Structure 3:1109–19

Lacy DB, Tepp W, Cohen AC, DasGupta BR, Stevens RC (1998) Crystal structure of botulinum neurotoxin type a and implications for toxicity. Nat Struct Biol 5:898–902

Lalli G, Bohnert S, Deinhardt K, Verastegui C, Schiavo G (2003) The journey of tetanus and botulinum neurotoxins in neurons. Trends Microbiol 11:431–7

Lambeau G, Lazdunski M (1999) Receptors for a growing family of secreted phospholipases A2. Trends Pharmacol Sci 20:162–70

Lambeau G, Barhanin J, Schweitz H, Qar J, Lazdunski M (1989) Identification and properties of very high affinity brain membrane-binding sites for a neurotoxic phospholipase from the taipan venom. J Biol Chem 264:11503–510

Lee CY, Chang CC, Chen YM (1972) Reversibility of neuromuscular blockade by neurotoxins from elapid and sea snake venoms. Taiwan Yi Xue Hui Za Zhi 71:344–9

Lee CY, Tsai MC, Chen YM, Ritonja A, Gubensek F (1984) Mode of neuromuscular blocking action of toxic phospholipases A2 from vipera ammodytes venom. Arch Int Pharmacodyn Ther 268:313–24

Leung YM, Xion Y, Ou YJ, Kwan CY (1998) Perturbation by lysophosphatidylcholine of membrane permeability in cultured vascular smooth muscle and endothelial cells. Life Sci 63:965–73

Lew MF, Adornato BT, Duane DD, Dykstra DD, Factor SA et al. (1997) Botulinum toxin type B: a double-blind, placebo-controlled, safety and efficacy study in cervical dystonia. Neurology 49:701–7

Li JY, Jahn R, Dahlström A (1994) Synaptotagmin I is present mainly in autonomic and sensory neurons of the rat peripheral nervous system. Neuroscience 63:837–50

Li L, Binz T, Niemann H, Singh BR (2000) Probing the mechanistic role of glutamate residue in the zinc-binding motif of type A botulinum neurotoxin light chain. Biochemistry 39:2399–2405

Ludlow CL, Hallett M, Rhew K, Cole R, Shimizu T et al. (1992) Therapeutic use of type F botulinum toxin. N Engl J Med 326:349–50

Luvisetto S, Rossetto O, Montecucco C, Pavone F (2003) Toxicity of botulinum neurotoxins in central nervous system of mice. Toxicon 41:475–81

Luvisetto S, Marinelli S, Lucchetti F, Marchi F, Cobianchi S et al. (2006) Botulinum neurotoxins and formalin-induced pain: central vs. peripheral effects in mice. Brain Res 1082:124–31

Luvisetto S, Marinelli S, Cobianchi S, Pavone F (2007) Anti-allodynic efficacy of botulinum neurotoxin A in a model of neuropathic pain. Neuroscience 145:1–4

Mahrhold S, Rummel A, Bigalke H, Davletov B, Binz T (2006) The synaptic vesicle protein 2C mediates the uptake of botulinum neurotoxin A into phrenic nerves. FEBS Lett 580:2011–14

Maksymowych AB, Reinhard M, Malizio CJ, Goodnough MC, Johnson EA et al. (1999) Pure botulinum neurotoxin is absorbed from the stomach and small intestine and produces peripheral neuromuscular blockade. Infect Immun 67:4708–12

Maksymowych AB, Simpson LL (1998) Binding and transcytosis of botulinum neurotoxin by polarized human colon carcinoma cells. J Biol Chem 273:21950–7

Marquèze B, Boudier JA, Mizuta M, Inagaki N, Seino S et al. (1995) Cellular localization of synaptotagmin i, ii, and iii mrnas in the central nervous system and pituitary and adrenal glands of the rat. J Neurosci 15:4906–17

Marxen P, Bigalke H (1991) Tetanus and botulinum A toxins inhibit stimulated F-actin rearrangement in chromaffin cells. Neuroreport 2:33–6

Matteoli M, Verderio C, Rossetto O, Iezzi N, Coco S et al. (1996) Synaptic vesicle endocytosis mediates the entry of tetanus neurotoxin into hippocampal neurons. Proc Natl Acad Sci U S A 93:13310–15

Megighian A, Rigoni M, Caccin P, Zordan MA, Montecucco C (2007) A lysolecithin/fatty acid mixture promotes and then blocks neurotransmitter release at the drosophila melanogaster larval neuromuscular junction. Neurosci Lett 416:6–11

Meunier FA, Schiavo G, Molgó J (2002) Botulinum neurotoxins: from paralysis to recovery of functional neuromuscular transmission. J Physiol Paris 96:105–13

Meunier FA, Lisk G, Sesardic D, Dolly JO (2003) Dynamics of motor nerve terminal remodeling unveiled using SNARE-cleaving botulinum toxins: the extent and duration are dictated by the sites of SNAP-25 truncation. Mol Cell Neurosci 22:454–66

Mezaki T, Kaji R, Kohara N, Fujii H, Katayama M et al. (1995) Comparison of therapeutic efficacies of type A and F botulinum toxins for blepharospasm: a double-blind, controlled study. Neurology 45:506–8

Middlebrook JL, Brown JE (1995) Immunodiagnosis and immunotherapy of tetanus and botulinum neurotoxins. Curr Top Microbiol Immunol 195:89–122

Midura TF, Arnon SS (1976) Infant botulism. identification of clostridium botulinum and its toxins in faeces. Lancet 2:934–6

Minton NP (1995) Molecular genetics of clostridial neurotoxins. Curr Top Microbiol Immunol 195:161–94

Molgó J, Siegel LS, Tabti N, Thesleff S (1989) A study of synchronization of quantal transmitter release from mammalian motor endings by the use of botulinal toxins type A and D. J Physiol 411:195–205

Montecucco C, Molgó J (2005) Botulinal neurotoxins: revival of an old killer. Curr Opin Pharmacol 5:274–9

Montecucco C, Rossetto O (2000) How do presynaptic PLA2 neurotoxins block nerve terminals? Trends Biochem Sci 25:266–70

Montecucco C, Schiavo G (1995) Structure and function of tetanus and botulinum neurotoxins. Q Rev Biophys 28:423–72

Montecucco C, Schiavo G, Dasgupta BR (1989) Effect of pH on the interaction of botulinum neurotoxins A, B and E with liposomes. Biochem J 259:47–53

Montecucco C, Papini E, Schiavo G (1994) Bacterial protein toxins penetrate cells via a four-step mechanism. FEBS Lett 346:92–8

Montecucco C, Schiavo G, Tugnoli V, de Grandis D (1996) Botulinum neurotoxins: mechanism of action and therapeutic applications. Mol Med Today 2:418–24

Montecucco C, Rossetto O, Schiavo G (2004) Presynaptic receptor arrays for clostridial neurotoxins. Trends Microbiol 12:442–6

Montecucco C, Schiavo G, Pantano S (2005) SNARE complexes and neuroexocytosis: how many, how close? Trends Biochem Sci 30:367–72

Morbiato L, Carli L, Johnson E, Montecucco C, Molgo J et al. (2007) Neuromuscular paralysis and recovery in mice injected with botulinum toxin A and C Eur J Neurosci 25:2697–704

Munro P, Kojima H, Dupont JL, Bossu JL, Poulain B et al. (2001) High sensitivity of mouse neuronal cells to tetanus toxin requires a GPI-anchored protein. Biochem Biophys Res Commun 289:623–9

Naumann M, Jost W (2004) Botulinum toxin treatment of secretory disorders. Mov Disord 19 Suppl 8:S137–41

Neco P, Rossetto O, Gil A, Montecucco C, Gutiérrez LM (2003) Taipoxin induces F-actin fragmentation and enhances release of catecholamines in bovine chromaffin cells. J Neurochem 85:329–37

Niemann H (1991) Molecular biology of clostridial neurotoxins. In: Alouf J, Freer J (eds) A sourcebook of bacterial protein toxins. Academic Press, London, pp 303–48

Nishiki T, Tokuyama Y, Kamata Y, Nemoto Y, Yoshida A et al. (1996) The high-affinity binding of clostridium botulinum type b neurotoxin to synaptotagmin ii associated with gangliosides gt1b/gd1a. FEBS Lett 378:253–7

O'Sullivan G, Mohammed N, Foran P, Lawrence G, Dolly O (1999) Rescue of exocytosis in botulinum toxin A-poisoned chromaffin cells by expression of cleavage-resistant SNAP-25. Identification of the minimal essential C-terminal residues. J Biol Chem 274:36897–904

Oberg SG, Kelly RB (1976) The mechanism of beta-bungarotoxin action. I. modification of transmitter release at the neuromuscular junction. J Neurobiol 7:129–41

Ohishi I, Sugii S, Sakaguchi G (1977) Oral toxicities of clostridium botulinum toxins in response to molecular size. Infect Immun 16:107–9

Ong W, Jiang B, Tang N, Ling S, Yeo J et al. (2006) Differential effects of polyunsaturated fatty acids on membrane capacitance and exocytosis in rat pheochromocytoma-12 cells. Neurochem Res 31:41–8

Osen-Sand A, Staple JK, Naldi E, Schiavo G, Rossetto O et al. (1996) Common and distinct fusion proteins in axonal growth and transmitter release. J Comp Neurol 367:222–34

Othman IB, Spokes JW, Dolly JO (1982) Preparation of neurotoxic 3H-beta-bungarotoxin: demonstration of saturable binding to brain synapses and its inhibition by toxin I. Eur J Biochem 128:267–76

Pearson JA, Tyler MI, Retson KV, Howden ME (1993) Studies on the subunit structure of textilotoxin, a potent presynaptic neurotoxin from the venom of the australian common brown snake (pseudonaja textilis). 3. the complete amino-acid sequences of all the subunits. Biochem Biophys Acta 1161:223–9

Pellizzari R, Rossetto O, Lozzi L, Giovedi' S, Johnson E et al. (1996) Structural determinants of the specificity for synaptic vesicle-associated membrane protein/synaptobrevin of tetanus and botulinum type B and G neurotoxins. J Biol Chem 271:20353–8

Pellizzari R, Mason S, Shone CC, Montecucco C (1997) The interaction of synaptic vesicle-associated membrane protein/synaptobrevin with botulinum neurotoxins D and F. FEBS Lett 409:339–42

Petrovic U, Sribar J, Paris A, Rupnik M, Krzan M et al. (2004) Ammodytoxin, a neurotoxic secreted phospholipase A(2), can act in the cytosol of the nerve cell. Biochem Biophys Res Commun 324:981–5

Pickett J, Berg B, Chaplin E, Brunstetter-Shafer MA (1976) Syndrome of botulism in infancy: clinical and electrophysiologic study. N Engl J Med 295:770–2

Prasarnpun S, Walsh J, Harris JB (2004) Beta-bungarotoxin-induced depletion of synaptic vesicles at the mammalian neuromuscular junction. Neuropharmacology 47:304–14

Prasarnpun S, Walsh J, Awad SS, Harris JB (2005) Envenoming bites by kraits: the biological basis of treatment-resistant neuromuscular paralysis. Brain 128:2987–96

Puhar A, Johnson EA, Rossetto O, Montecucco C (2004) Comparison of the pH-induced conformational change of different clostridial neurotoxins. Biochem Biophys Res Commun 319:66–71

Pumplin DW, Reese TS (1977) Action of brown widow spider venom and botulinum toxin on the frog neuromuscular junction examined with the freeze-fracture technique. J Physiol 273:443–57

Raciborska DA, Charlton MP (1999) Retention of cleaved synaptosome-associated protein of 25 kDa (SNAP-25) in neuromuscular junctions: a new hypothesis to explain persistence of botulinum A poisoning. Can J Physiol Pharmacol 77:679–88

Raciborska DA, Trimble WS, Charlton MP (1998) Presynaptic protein interactions in vivo: evidence from botulinum A, C, D and E action at frog neuromuscular junction. Eur J Neurosci 10:2617–28

Ramon G, Descombey P (1925) Sur l'immunization antitetanique et sur la production de l'antitoxine tetanique Compt Rend Soc Biol 93:508–98

Ratts R, Zeng H, Berg EA, Blue C, McComb ME et al. (2003) The cytosolic entry of diphtheria toxin catalytic domain requires a host cell cytosolic translocation factor complex. J Cell Biol 160:1139–50

Reese C, Mayer A (2005) Transition from hemifusion to pore opening is rate limiting for vacuole membrane fusion. J Cell Biol 171:981–90

Rehm H, Betz H (1982) Binding of beta-bungarotoxin to synaptic membrane fractions of chick brain. J Biol Chem 257:10015–22

Reilich P, Fheodoroff K, Kern U, Mense S, Seddigh S et al. (2004) Consensus statement: botulinum toxin in myofascial [corrected] pain. J Neurol 251 Suppl 1:I36–8

Rickman C, Davletov B (2005) Arachidonic acid allows SNARE complex formation in the presence of Munc18. Chem Biol 12:545–53

Rigoni M, Caccin P, Johnson EA, Montecucco C, Rossetto O (2001) Site-directed mutagenesis identifies active-site residues of the light chain of botulinum neurotoxin type A. Biochem Biophys Res Commun 288:1231–7

Rigoni M, Schiavo G, Weston AE, Caccin P, Allegrini F et al. (2004) Snake presynaptic neurotoxins with phospholipase a2 activity induce punctate swellings of neurites and exocytosis of synaptic vesicles. J Cell Sci 117:3561–70

Rigoni M, Caccin P, Gschmeissner S, Koster G, Postle AD et al. (2005) Equivalent effects of snake pla2 neurotoxins and lysophospholipid-fatty acid mixtures. Science 310:1678–80
Rigoni M, Pizzo P, Schiavo G, Weston AE, Zatti G et al. (2007) Calcium influx and mitochondrial alterations at synapses exposed to snake neurotoxins or their phospholipid hydrolysis products. J Biol Chem 282:11238–45
Rizzoli SO, Betz WJ (2005) Synaptic vesicle pools. Nat Rev Neurosci 6:57–69
Rossetto O, Montecucco C (2004) Clostridial neurotoxins. In: Proft C (ed) Microbial toxins. molecular and cellular biology. Horizon Scientific Press, Norfolk, UK, pp 149–78
Rossetto O, Montecucco C (2007) Peculiar binding of botulinum neurotoxins. ACS Chem Biol 2:96–8
Rossetto O, Schiavo G, Montecucco C, Poulain B, Deloye F et al. (1994) Snare motif and neurotoxins. Nature 372:415–16
Rossetto O, Caccin P, Rigoni M, Tonello F, Bortoletto N et al. (2001a) Active-site mutagenesis of tetanus neurotoxin implicates tyr-375 and glu-271 in metalloproteolytic activity. Toxicon 39:1151–9
Rossetto O, Seveso M, Caccin P, Schiavo G, Montecucco C (2001b) Tetanus and botulinum neurotoxins: turning bad guys into good by research. Toxicon 39:27–41
Rossetto O, Morbiato L, Rossetto et al. 2006 Caccin P, Rigoni M, Montecucco C (2005) Presynaptic enzymatic neurotoxins. J Neurochem 97:1534–45
Roux S, Colasante C, Saint Cloment C, Barbier J, Curie T et al. (2005) Internalization of a GFP-tetanus toxin C-terminal fragment fusion protein at mature mouse neuromuscular junctions. Mol Cell Neurosci 30:572–82
Rowan EG (2001) What does beta-bungarotoxin do at the neuromuscular junction? Toxicon 39:107–18
Rugolo M, Dolly JO, Nicholls DG (1986) The mechanism of action of beta-bungarotoxin at the presynaptic plasma membrane. Biochem J 233:519–23
Rummel A, Bade S, Alves J, Bigalke H, Binz T (2003) Two carbohydrate binding sites in the H(CC)-domain of tetanus neurotoxin are required for toxicity. J Mol Biol 326:835–47
Rummel A, Karnath T, Henke T, Bigalke H, Binz T (2004a) Synaptotagmins I and II act as nerve cell receptors for botulinum neurotoxin G. J Biol Chem 279:30865–70
Rummel A, Mahrhold S, Bigalke H, Binz T (2004b) The $H_CC$ -domain of botulinum neurotoxins A and B exhibits a singular ganglioside binding site displaying serotype specific carbohydrate interaction. Mol Microbiol 51:631–43
Rummel A, Eichner T, Weil T, Karnath T, Gutcaits A et al. (2007) Identification of the protein receptor binding site of botulinum neurotoxins B and G proves the double-receptor concept. Proc Natl Acad Sci U S A 104:359–64
Sakaguchi G (1982) Clostridium botulinum toxins. Pharmacol Ther 19:165–94
Sankhla C, Jankovic J, Duane D (1998) Variability of the immunologic and clinical response in dystonic patients immunoresistant to botulinum toxin injections. Mov Disord 13:150–54
Schantz EJ, Johnson EA (1997) Botulinum toxin: the story of its development for the treatment of human disease. Perspect Biol Med 40:317–27
Schiavo G (2006) Structural biology: dangerous liaisons on neurons. Nature 444:1019–20
Schiavo G, Matteoli M, Montecucco C (2000) Neurotoxins affecting neuroexocytosis. Physiol Rev 80:717–66
Schiavo G, Papini E, Genna G, Montecucco C (1990) An intact interchain disulfide bond is required for the neurotoxicity of tetanus toxin. Infect Immun 58:4136–41
Scott AB, Magoon EH, McNeer KW, Stager DR (1989) Botulinum treatment of strabismus in children. Trans Am Ophthalmol Soc 87:174–180; discussion 180–4
Scott D (1997) Phospholipase A2: structure and catalytic properties. In: Kini R (ed) Venom phospholipase A2 enzymes: structure, function and mechanism. John Wiley & Sons, Chichester, p 97–128.
Segelke B, Knapp M, Kadkhodayan S, Balhorn R, Rupp B (2004) Crystal structure of clostridium botulinum neurotoxin protease in a product-bound state: evidence for noncanonical zinc protease activity. Proc Natl Acad Sci U S A 101:6888–93

Sellin LC, Kauffman JA, Dasgupta BR (1983) Comparison of the effects of botulinum neurotoxin types A and E at the rat neuromuscular junction. Med Biol 61:120–5
Sheridan RE (1998) Gating and permeability of ion channels produced by botulinum toxin types A and E in PC12 cell membranes. Toxicon 36:703–17
Shone CC, Hambleton P, Melling J (1987) A 50-kDa fragment from the NH2-terminus of the heavy subunit of clostridium botulinum type A neurotoxin forms channels in lipid vesicles. Eur J Biochem 167:175–80
Simpson LL (2000) Identification of the characteristics that underlie botulinum toxin potency: implications for designing novel drugs. Biochimie 82:943–53
Simpson LL, Lautenslager GT, Kaiser II, Middlebrook JL (1993) Identification of the site at which phospholipase A2 neurotoxins localize to produce their neuromuscular blocking effects. Toxicon 31:13–26
Simpson LL, Coffield JA, Bakry N (1994) Inhibition of vacuolar adenosine triphosphatase antagonizes the effects of clostridial neurotoxins but not phospholipase A2 neurotoxins. J Pharmacol Exp Ther 269:256–62
Simpson LL, Maksymowych AB, Park J, Bora RS (2004) The role of the interchain disulfide bond in governing the pharmacological actions of botulinum toxin. J Pharmacol Exp Ther 308:857–64
Singh G, Gourinath S, Sharma S, Paramasivam M, Srinivasan A et al. (2001) Sequence and crystal structure determination of a basic phospholipase A2 from common krait (bungarus caeruleus) at 2.4 A resolution: identification and characterization of its pharmacological sites. J Mol Biol 307:1049–59
Sloop RR, Cole BA, Escutin RO (1997) Human response to botulinum toxin injection: type B compared with type A. Neurology 49:189–94
Smith L, Sugiyama D (1988) Botulism: the organism, its toxins, the disease. Charles C Thomas Publisher, Springfield, IL
Smith TJ, Lou J, Geren IN, Forsyth CM, Tsai R et al. (2005) Sequence variation within botulinum neurotoxin serotypes impacts antibody binding and neutralization. Infect Immun 73:5450–7
Son YJ, Thompson WJ (1995) Schwann cell processes guide regeneration of peripheral axons. Neuron 14:125–32
Sribar J, Copic A, Paris A, Sherman NE, Gubensek F et al. (2001) A high affinity acceptor for phospholipase A2 with neurotoxic activity is a calmodulin. J Biol Chem 276:12493–6
Sribar J, Copic A, Poljsak-Prijatelj M, Kuret J, Logonder U et al. (2003a) R25 is an intracellular membrane receptor for a snake venom secretory phospholipase A(2). FEBS Lett 553:309–14
Sribar J, Sherman NE, Prijatelj P, Faure G, Gubensek F et al. (2003b) The neurotoxic phospholipase A2 associates, through a non-phosphorylated binding motif, with 14-3-3 protein gamma and epsilon isoforms. Biochem Biophys Res Commun 302:691–6
Strong PN, Goerke J, Oberg SG, Kelly RB (1976) Beta-bungarotoxin, a pre-synaptic toxin with enzymatic activity. Proc Natl Acad Sci U S A 73:178–82
Strong PN, Heuser JE, Kelly RB (1977) Selective enzymatic hydrolysis of nerve terminal phospholipids by beta-bungarotoxin: biochemical and morphological studies. Prog Clin Biol Res 15:227–49
Strong PN, Kelly RB (1977) Membranes undergoing phase transitions are preferentially hydrolyzed by beta-bungarotoxin. Biochim Biophys Acta 469:231–35
Südhof TC (2004) The synaptic vesicle cycle. Annu Rev Neurosci 27:509–47
Swaminathan S, Eswaramoorthy S (2000) Structural analysis of the catalytic and binding sites of clostridium botulinum neurotoxin b. Nat Struct Biol 7:693–9
Sycha T, Kranz G, Auff E, Schnider P (2004) Botulinum toxin in the treatment of rare head and neck pain syndromes: a systematic review of the literature. J Neurol 251 Suppl 1:119–30
Tacket C, Rogawski M (1989) Botulism. In: Simpson L. L (ed) Botulinum neurotoxins and tetanus toxin. Academic Press, San Diego, pp 351–78
Takamori S, Holt M, Stenius K, Lemke EA, Grønborg M et al. (2006) Molecular anatomy of a trafficking organelle. Cell 127:831–46

Theakston RD, Phillips RE, Warrell DA, Galagedera Y, Abeysekera DT et al. (1990) Envenoming by the common krait (bungarus caeruleus) and sri lankan cobra (naja naja naja): efficacy and complications of therapy with haffkine antivenom. Trans R Soc Trop Med Hyg 84:301–8

Thesleff S (1960) Supersensitivity of skeletal muscle produced by botulinum toxin. J Physiol 151 598–607

Tizzoni G, Cattani G (1890) Uber das tetanusgift Zentralbl Bakt 8:69–73

Trevett AJ, Lalloo DG, Nwokolo NC, Naraqi S, Kevau IH et al. (1995) Electrophysiological findings in patients envenomed following the bite of a papuan taipan (oxyuranus scutellatus canni). Trans R Soc Trop Med Hyg 89:415–17

Truong DD, Jost WH (2006) Botulinum toxin: clinical use. Parkinsonism Relat Disord 12:331–55

Tsukamoto K, Kohda T, Mukamoto M, Takeuchi K, Ihara H et al. (2005) Binding of clostridium botulinum type C and D neurotoxins to ganglioside and phospholipid. novel insights into the receptor for clostridial neurotoxins. J Biol Chem 280:35164–71

Umland TC, Wingert LM, Swaminathan S, Furey WF, Schmidt JJ et al. (1997) Structure of the receptor binding fragment hc of tetanus neurotoxin. Nat Struct Biol 4:788–92

Vaidyanathan VV, Yoshino K, Jahnz M, Dörries C, Bade S et al. (1999) Proteolysis of SNAP-25 isoforms by botulinum neurotoxin types A, C, and E: domains and amino acid residues controlling the formation of enzyme-substrate complexes and cleavage. J Neurochem 72:327–37

Van Ermengem E (1897) Über ein neuenanaeroben bacillus und seine beziehungen zum botulismus. Z Hyg Infektkr 26:1–56

Verderio C, Pozzi D, Pravettoni E, Inverardi F, Schenk U et al. (2004) SNAP-25 modulation of calcium dynamics underlies differences in gabaergic and glutamatergic responsiveness to depolarization. Neuron 41:599–610

Verderio C, Rossetto O, Grumelli C, Frassoni C, Montecucco C et al. (2006) Entering neurons: botulinum toxins and synaptic vesicle recycling. EMBO Rep 7:995–9

Verderio C, Grumelli C, Raiteri L, Coco S, Paluzzi S et al. (2007) Traffic of botulinum toxins A and E in excitatory and inhibitory neurons. Traffic 8:142–53

Wagner GM, Mart PE, Kelly RB (1974) Beta-bungarotoxin inhibition of calcium accumulation by rat brain mitochondria Biochem Biophys Res Commun 58:475–81

Warrell DA, Looareesuwan S, White NJ, Theakston RD, Warrell MJ et al. (1983) Severe neurotoxic envenoming by the malayan krait bungarus candidus (linnaeus): response to antivenom and anticholinesterase. Br Med J (Clin Res Ed) 286:678–80

Washbourne P, Pellizzari R, Baldini G, Wilson MC, Montecucco C (1997) Botulinum neurotoxin types A and E require the SNARE motif in SNAP-25 for proteolysis. FEBS Lett 418:1–5

Wei S, Ong WY, Thwin MM, Fong CW, Farooqui AA et al. (2003) Group IIa secretory phospholipase A2 stimulates exocytosis and neurotransmitter release in pheochromocytoma-12 cells and cultured rat hippocampal neurons. Neuroscience 121:891–8

Weissenhorn W, Dessen A, Harrison SC, Skehel JJ, Wiley DC (1997) Atomic structure of the ectodomain from hiv-1 gp41. Nature 387:426–30

Wernicke JF, Vanker AD, Howard BD (1975) The mechanism of action of beta-bungarotoxin. J Neurochem 25:483–96

Westerlund B, Nordlund P, Uhlin U, Eaker D, Eklund H (1992) The three-dimensional structure of notexin, a presynaptic neurotoxic phospholipase A2 at 2.0 A resolution. FEBS Lett 301:159–64

Williamson LC, Neale EA (1994) Bafilomycin A1 inhibits the action of tetanus toxin in spinal cord neurons in cell culture. J Neurochem 63:2342–5

Williamson LC, Neale EA (1998) Syntaxin and 25-kDa synaptosomal-associated protein: differential effects of botulinum neurotoxins C1 and A on neuronal survival. J Neurosci Res 52:569–83

Williamson LC, Halpern JL, Montecucco C, Brown JE, Neale EA (1996) Clostridial neurotoxins and substrate proteolysis in intact neurons: botulinum neurotoxin C acts on synaptosomal-associated protein of 25 kDa. J Biol Chem 271:7694–9

Wilson HI, Nicholson GM, Tyler MI, Howden ME (1995) Induction of giant miniature end-plate potentials during blockade of neuromuscular transmission by textilotoxin. Naunyn Schmiedebergs Arch Pharmacol 352:79–87

Wilson-Ashworth HA, Judd AM, Law RM, Freestone BD, Taylor S et al. (2004) Formation of transient non-protein calcium pores by lysophospholipids in S49 lymphoma cells. J Membr Biol 200:25–33

Wojtczak L, Schönfeld P (1993) Effect of fatty acids on energy coupling processes in mitochondria. Biochim Biophys Acta 1183:41–57

Woodley SL, Ikenouchi H, Barry WH (1991) Lysophosphatidylcholine increases cytosolic calcium in ventricular myocytes by direct action on the sarcolemma. J Mol Cell Cardiol 23:671–80

Xu Y, Zhang F, Su Z, McNew JA, Shin Y (2005) Hemifusion in SNARE-mediated membrane fusion. Nat Struct Mol Biol 12:417–22

Yamasaki S, Hu Y, Binz T, Kalkuhl A, Kurazono H et al. (1994) Synaptobrevin/vesicle-associated membrane protein (VAMP) of aplysia californica: structure and proteolysis by tetanus toxin and botulinal neurotoxins type D and F. Proc Natl Acad Sci U S A 91:4688–92

Yang C (1997) Chemical modification and functional sites of phospholipases A2. In: Kini R (ed) Venom phospholipase A2 enzymes: structure, function and mechanism. Wiley & Sons, Chichester, pp 185–204

Yoon TY, Okumus B, Zhang F, Shin YK, Ha T (2006) Multiple intermediates in SNARE-induced membrane fusion. Proc Natl Acad Sci U S A103:19731–6.

Yowler BC, Schengrund C (2004) Glycosphingolipids-sweets for botulinum neurotoxin. Glycoconj J 21:287–93

Yowler BC, Kensinger RD, Schengrund C (2002) Botulinum neurotoxin a activity is dependent upon the presence of specific gangliosides in neuroblastoma cells expressing synaptotagmin I. J Biol Chem 277:32815–19

Zimmerberg J, Chernomordik LV (2005) Neuroscience. Synaptic membranes bend to the will of a neurotoxin. Science 310:1626–7

Zinsmaier KE, Bronk P (2001) Molecular chaperones and the regulation of neurotransmitter exocytosis. Biochem Pharmacol 62:1–11

Zuber M, Sebald M, Bathien N, de Recondo J, Rondot P (1993) Botulinum antibodies in dystonic patients treated with type A botulinum toxin: frequency and significance. Neurology 43:1715–18

# $\alpha$-Latrotoxin and Its Receptors

**Yuri A. Ushkaryov(✉), Alexis Rohou, and Shuzo Sugita**



**Abstract** $\alpha$-Latrotoxin ($\alpha$-LTX) from black widow spider venom induces exhaustive release of neurotransmitters from vertebrate nerve terminals and endocrine cells. This 130-kDa protein has been employed for many years as a molecular tool to study exocytosis. However, its action is complex: in neurons, $\alpha$-LTX induces massive secretion both in the presence of extracellular $Ca^{2+}$ ($Ca^{2+}{}_e$) and in its absence; in endocrine cells, it usually requires $Ca^{2+}{}_e$. To use this toxin for further dissection of secretory mechanisms, one needs an in-depth understanding of its

Yuri A. Ushkaryov
Division of Cell and Molecular Biology, Imperial College London, London, SW7 2AY, UK
y.ushkaryov@imperial.ac.uk

T.C. Südhof, K. Starke (eds.), *Pharmacology of Neurotransmitter Release.*
*Handbook of Experimental Pharmacology 184.*

functions. One such function that explains some α-LTX effects is its ability to form cation-permeable channels in artificial lipid bilayers. The mechanism of α-LTX pore formation, revealed by cryo-electron microscopy, involves toxin assembly into homotetrameric complexes which harbor a central channel and can insert into lipid membranes. However, in biological membranes, α-LTX cannot exert its actions without binding to specific receptors of the plasma membrane. Three proteins with distinct structures have been found to bind α-LTX: neurexin Iα, latrophilin 1, and receptor-like protein tyrosine phosphatase σ. Upon binding a receptor, α-LTX forms channels permeable to cations and small molecules; the toxin may also activate the receptor. To distinguish between the pore- and receptor-mediated effects, and to study structure-function relationships in the toxin, α-LTX mutants have been used. At least one non-pore-forming α-LTX mutant can activate latrophilin, a G protein-coupled receptor, causing release of $Ca^{2+}$ from intracellular stores. Latrophilin action still requires $Ca^{2+}{}_e$ and may trigger transmitter secretion either by itself or by activating $Ca^{2+}$-channels and/or inducing $Ca^{2+}$ waves. These results reveal two $Ca^{2+}{}_e$-dependent mechanisms of α-LTX action (membrane pore formation and signalling via latrophilin), but how α-LTX triggers $Ca^{2+}{}_e$-independent neurotransmitter release still remains unexplained. Hypotheses for this action include direct interaction with intracellular components involved in exocytosis or the effects of α-LTX pores.

## 1 α-LTX and Release of Neurotransmitters

The notorious black widow spider (*Latrodectus* genus) and its venom, first studied by ancient Greeks (Aristotle 350 B.C.), entered the era of modern science in the 1930's with the discovery of its proteinacious active principles (D'Amour et al. 1936) and, following the demonstration that it affects synaptic neurotransmitter release (Longenecker et al. 1970) and the isolation of toxic components (Frontali et al. 1976), gripped the attention of neurobiologists. It played a role in the debate leading to the $Ca^{2+}$ (Augustine et al. 1987) and vesicular quantum (Ceccarelli and Hurlbut 1980) paradigms of neurotransmission. However, some details of α-LTX actions still remain mysterious, despite its widespread use as a stimulant of neurosecretion.

The venom contains at least 86 unique proteins (Duan et al. 2006), including several homologous LTXs which play a role in its toxicity to insects and crustaceans (Grishin 1998), with only one, α-LTX, targeting vertebrates specifically; reviewed by Rosenthal and Meldolesi (1989). α-LTX is usually isolated from spider venom by conventional chromatography (Frontali et al. 1976; Tzeng et al. 1978), but to achieve homogeneity and remove contaminants (Volkova et al. 1995; Pescatori et al. 1995) that may endow the preparation with uncharacteristic properties (Umbach et al. 1998), preparative native electrophoresis should ideally be used (Ashton et al. 2000).

## 1.1 Target Cells

The toxin induces strong and sustained release of neurotransmitters and hormones from secretory cells capable of regulated exocytosis (Rosenthal et al. 1990). The effects of α-LTX on neurosecretion were first described on a cellular level in the 1970s at frog neuromuscular junctions (NMJ) (Longenecker et al. 1970; Frontali et al. 1976); and later in mouse brain slices (Tzeng et al. 1978), in synaptosomes (isolated brain nerve terminals) from rat (Grasso et al. 1978), dog (Tzeng and Siekevitz 1979), and guinea pig (Nicholls et al. 1982); primary cerebellar granule cell cultures (Grasso and Mercanti-Ciotti 1993); and in hippocampal organotypic cultures (Capogna et al. 1996). In addition, catecholamine-secreting chromaffin cells (Picotti et al. 1982) and PC12 cell line (Robello et al. 1987), insulin-secreting pancreatic β-cells in primary or derived cell cultures (Lang et al. 1998), oxytocin- and vasopressin-secreting neurohypophysis cells (Hlubek et al. 2003) and luteinizing hormone-secreting rat gonadotropes (Tse and Tse 1999) were also used to characterize the toxin's effects on secretion from non-neuronal excitable cells. A secretory cell not sensitive to α-LTX still remains to be found.

## 1.2 Site and Mode of Action

From the earliest description of the toxin's actions on neuronal systems, it emerged that α-LTX affects specifically the presynaptic element, from which it causes massive neurotransmitter release (e.g., Longenecker et al. 1970). The toxin has no major enzymatic activities (Frontali et al. 1976). Crucially, α-LTX has been discovered to create $Ca^{2+}$-permeable channels in lipid bilayers (Finkelstein et al. 1976), and a large body of evidence shows that $Ca^{2+}$ influx through membrane channels induced by α-LTX in the presynaptic membrane accounts for a major part of its effect. Pore formation occurs in all the biological systems mentioned above, but the features of α-LTX-triggered release cannot be fully explained by the toxin pore.

The effect of α-LTX at NMJs is usually delayed and develops fully after ~10 min, although synaptosomes react much faster. It is detected electrophysiologically as an increase in the frequency of spontaneous miniature postsynaptic potentials at NMJs (Longenecker et al. 1970; Misler and Hurlbut 1979; Ceccarelli and Hurlbut 1980; Tsang et al. 2000) and excitatory or inhibitory postsynaptic currents at central synapses (Capogna et al. 1996). In addition, α-LTX affects action potential-evoked, synchronous release, and does this in a time-dependent manner: initially, it enhances evoked potentials, but eventually inhibits (Capogna et al. 1996) or blocks them (Longenecker et al. 1970; Hurlbut and Ceccarelli 1979; Liu and Misler 1998). Finally, when used in higher concentrations, α-LTX can cause morphological deformation and cell death, as reviewed by Südhof (2001).

However, the most surprising feature of the α-LTX-evoked secretion is that it can occur both in the presence and absence of $Ca^{2+}{}_e$.

### 1.3 $Ca^{2+}$-*Independent Release*

The ability of α-LTX to trigger neurotransmitter exocytosis in the absence of extracellular $Ca^{2+}$ remains particularly interesting and inexplicable to the field (Longenecker et al. 1970; Ceccarelli et al. 1979; see also Südhof (2001) and Ushkaryov et al. (2004) for review). This is clearly different from depolarization-induced exocytosis, which is $Ca^{2+}$-dependent, but not unlike the effect of hypertonic sucrose. The possibility that α-LTX-induced release involves an unknown, $Ca^{2+}$-independent mechanism which may also occur during normal synaptic activity has provided the *casus belli* for many a quest for α-LTX structure and receptors that could trigger neurotransmission *via* intracellular mechanisms.

In neurons, the $Ca^{2+}$-independent secretion is restricted to small synaptic vesicles, as demonstrated by synaptosomal and NMJ experiments, where glutamate, GABA, and acetylcholine are released in the absence of $Ca^{2+}$, while catecholamines or peptides are not (Matteoli et al. 1988; Davletov et al. 1998; Khvotchev et al. 2000). $Ca^{2+}$-independent release does not normally occur in endocrine cells (Grasso et al. 1980; Michelena et al. 1997; Silva et al. 2005), although in some cultured cells it does (Meldolesi et al. 1983; Lang et al. 1998; Tse and Tse 1999).

The characteristics of $Ca^{2+}$-independent release are peculiar: it requires the presence of divalent cations, such as $Mg^{2+}$, which can be added or removed in succession, causing respective bouts of secretion or its cessation (Misler and Hurlbut 1979). In the absence of $Mg^{2+}$, this release can be supported by slightly hypertonic sucrose, by itself insufficient to cause secretion (Misler and Hurlbut 1979). The $Ca^{2+}$-independent release can be blocked by millimolar $La^{3+}$ (Rosenthal et al. 1990) or concanavalin A (Grasso et al. 1978; Boehm and Huck 1998). It may involve release of $Ca^{2+}$ from mitochondria, as observed in peripheral (Tsang et al. 2000) and not central synapses (Adam-Vizi et al. 1993), but it is unclear if stored $Ca^{2+}{}_i$ itself can trigger release.

As $Ca^{2+}$ is important for endocytosis, the sustained secretory activity in its absence eventually depletes NMJ terminals (Longenecker et al. 1970; Ceccarelli and Hurlbut 1980), but not synaptosomes (Watanabe and Meldolesi 1983), of all vesicles. Electrophysiological recordings at NMJs show that in the absence of $Ca^{2+}$α-LTX causes a large but slow rise in the frequency of spontaneous exocytotic events, which then slow down and cease altogether; bursts of miniatures are never observed.

### 1.4 $Ca^{2+}$-*Dependent Release*

In the presence of $Ca^{2+}{}_e$, α-LTX also causes a slow and large increase in the frequency of miniature events, overlaid by bursts of release (Ceccarelli et al. 1979). As in the absence of $Ca^{2+}$, depletion of vesicles and block of exocytosis also occurs, but because vesicles are able to recycle for some time, the total number of quanta released is twice higher in $Ca^{2+}$ than in its absence (Fesce et al. 1986; Auger and

Marty 1997). It appears that the gradual rise and fall of the frequency of vesicle fusion events is a common feature of α-LTX-induced release both with and without $Ca^{2+}{}_e$, and that $Ca^{2+}$ adds an extra component, increasing the total release.

Influx of $Ca^{2+}$ through α-LTX pores might explain $Ca^{2+}$-dependent secretion, but not the bursts of miniatures. These may be due to $Ca^{2+}$ waves caused by activation of phospholipase C (PLC) (Vicentini and Meldolesi 1984; Davletov et al. 1998) and release of intracellular $Ca^{2+}$ ($Ca^{2+}{}_i$). In fact, U73122 (blocks PLC activation by G proteins), thapsigargin (depletes $Ca^{2+}$ stores), and 2-APB (inhibits activation of $Ca^{2+}$ stores) block α-LTX action (Davletov et al. 1998; Ashton et al. 2001; Capogna et al. 2003), implicating a G protein cascade in $Ca^{2+}$-dependent toxin-induced release.

## 2 The Structure of α-LTX

In order to understand and explain the complex activities of α-LTX, its primary and three-dimensional (3D) structure has been thoroughly studied.

### 2.1 *Sequence Analysis*

α-LTX is synthesized as a 157 kDa polypeptide (Kiyatkin et al. 1990) (Figure 1a) by free ribosomes in the cytosol of secretory epithelial cells of the venom glands (Cavalieri et al. 1990). These cells disintegrate (Smith and Russell 1966) and expel toxin into the gland lumen together with various proteases (Duan et al. 2006). Here, it is cleaved at both termini by a furin-like protease (Volynski et al. 1999), producing an active α-LTX of ~131 kDa (Kiyatkin et al. 1990; Ichtchenko et al. 1998).

The most striking feature of the α-LTX primary structure is a series of 22 ankyrin repeats that account for about two thirds of the sequence (Figure 1b) (Kiyatkin et al. 1995). Ankyrin repeats are found in a wide variety of proteins and generally mediate protein-protein contacts but, unlike other protein-binding motifs, they take part in a wide range of interactions and do not have a specific target (Sedgwick and Smerdon 1999). The N-terminal third of α-LTX shows no significant homology to other proteins. This region contains three conserved cysteines important for structural stability and activity of all LTXs (Kiyatkin et al. 1995; Ichtchenko et al. 1998).

α-LTX is a mostly hydrophilic protein that does not have a classical hydrophobic signal peptide or clear transmembrane regions (TMRs). The hydrophilic nature of α-LTX seemed for a long time at odds with its membrane insertion capabilities, but a study of its 3D structure (Orlova et al. 2000) shed light on this issue (Section 2.2).

### 2.2 *Higher-Order Structures*

So far, crystallization of α-LTX has not been reported. However, the toxin has been successfully studied by cryo-electron microscopy (cryo-EM) (Orlova et al. 2000),

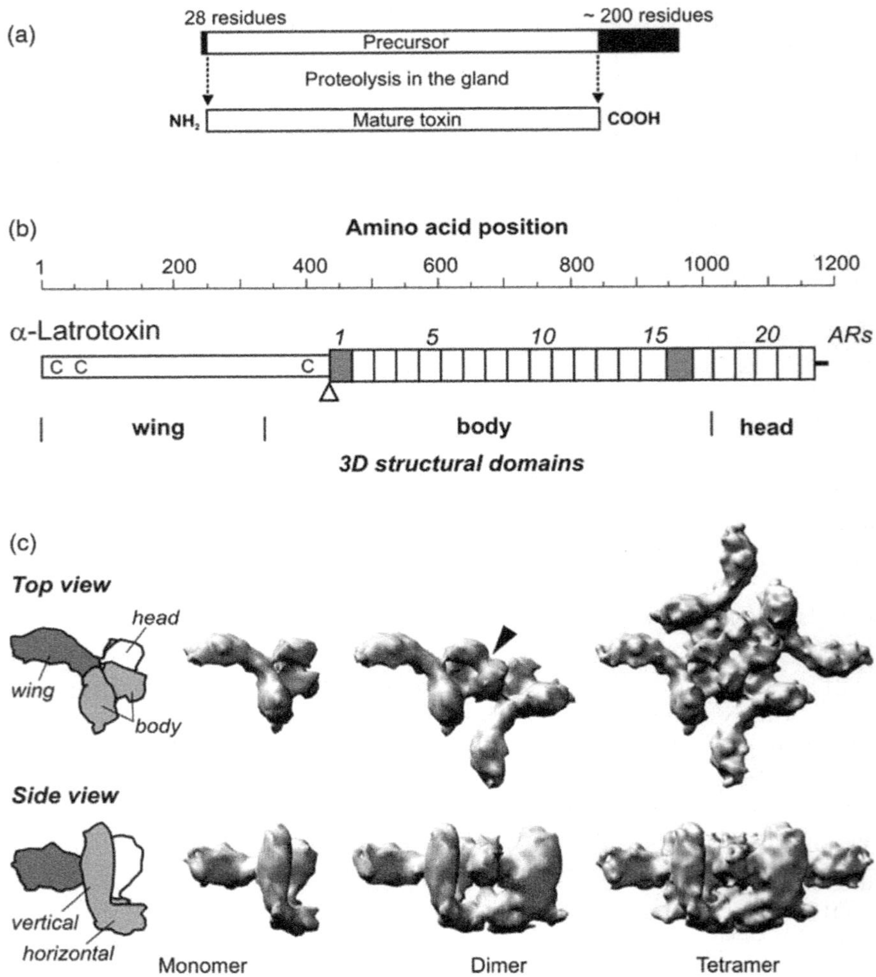


**Fig. 1** The structure of α-LTX. (**a**) Schematic of α-LTX processing in the venom gland. (**b**) Primary and domain structure. The numbered boxes, ankyrin repeats (ARs). Grey, imperfect repeats; C, conserved cysteines residues in the N-terminal domain; open arrowhead, insert in the mutant α-LTX$^{N4C}$. Protein domains identified from the 3D structure (Orlova et al. 2000) are delimited below. (**c**) 3D reconstructions of the α-LTX monomer, dimer and tetramer, viewed from the top and side. The monomer has been computationally extracted from the experimentally determined tetramer structure. Left-most image, a scheme of the monomer, with the domains designated by different shades of grey. Filled arrowhead, strong association of the head domains in the dimer.

a method that allows visualization of proteins instantly frozen in their native conformations. The size of the toxin tetramer (∼520 kDa) proved ideal for this approach to reveal medium-resolution (14 Å) information, and even the dimer (∼260 kDa) was reconstructed to below 20 Å resolution, revealing detailed information on the domain organisation of the toxin in its two oligomeric states (Orlova et al. 2000).

### 2.2.1 Domain Structure of the Monomer

The 3D cryo-EM reconstructions of the α-LTX tetramer and dimer (Orlova et al. 2000) show that the monomer consists of three domains (Figure 1c): (1) The wing, at the N-terminus, has an estimated molecular mass of 36 kDa. Thought to comprise mainly α-helices, it has no close homologues among proteins of known 3D structure. The wing links to the rest of the toxin molecule through a narrow rigid connection. (2) The body comprises the first 17 ankyrin repeats of the α-LTX sequence. Its estimated molecular mass is 76 kDa. The ankyrin repeats forming the "vertical" part of the body are tightly packed into a characteristic structure, reminiscent of ankyrin. However, the lower part of the body is bent and has a more complex arrangement. (3) At the C-terminus is the head, the smallest domain, with an estimated 18.5-kDa mass. It consists of 4.5 ankyrin repeats and demonstrates high spatial homology to ankyrin-repeat-containing proteins with known 3D structures (Orlova et al. 2000).

Despite having a large number of ankyrin repeats (superseded only by the 24 repeats in ankyrin itself), α-LTX does not assume the monotonous arch-like shape so characteristic of proteins with multiple ankyrin repeats, such as the 12-repeat ankyrin fragment (Michaely et al. 2002), the human protein phosphatase HEAT (15 repeats), or the porcine ribonuclease inhibitor LRR (16 repeats) (Andrade et al. 2001). Instead, the ankyrin repeat-containing part of α-LTX is broken in several places and forms clearly delimited domains and subdomains, probably reflecting sequence divergence of some repeats and the functional specialization of the subdomains.

### 2.2.2 Quaternary Structures

Although some monomers have been observed by cryo-EM in EDTA-treated α-LTX, the toxin almost always exists as a stable dimer (Orlova et al. 2000; Ashton et al. 2001). In the asymmetric dimer, the two monomers are associated "head to tail" by tight multipoint contacts mediated by the "horizontal" parts of the bodies and by the sides of the head domains (arrowhead in Figure 1c).

Association of dimers, strongly catalyzed by $Mg^{2+}$, produces a cyclical structure (Figure 1c) that can contain four monomers only. The tetramer has C4 rotational symmetry and resembles a bowl, in which the bottom is formed by the "horizontal" parts of the bodies. This part is important for penetration into lipid bilayers, and it is likely that structural rearrangements required for tetramerization expose the surface regions favorable to interaction with lipid bilayers. In addition, this part represents the intracellular mouth of the channel, with a large (30 Å) central hole in its center.

Above this, in the centre of the "bowl," the four heads form a cylindrical assembly surrounding the channel (Figure 1c), which is restricted at one point to 10 Å (Figure 2; see also Section 3.4.7). This constriction most probably corresponds to the cation binding site (selectivity filter) of the α-LTX channel (Section 3.4.1).

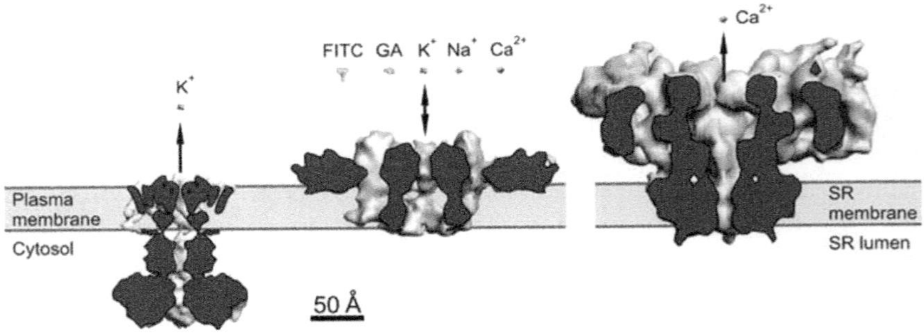


**Fig. 2** Membrane topography of the α-LTX pore. Cross-section of the α-LTX tetramer embedded in a membrane (as observed in liposomes) (Orlova et al. 2000) is shown alongside the cut-open voltage-dependent $K^+$ channel (Kv1.2) (Long et al. 2005) and $Ca^{2+}$ release channel (ryanodine receptor) (Serysheva et al. 2005). Fully hydrated cations and molecules known to permeate through the respective channels are shown next to each reconstruction (FITC, fluoresceine isothiocyanate; NE, norepinephrine). The narrowest part of the α-LTX channel is 10 Å. Molecular images were produced using the UCSF Chimera package (Pettersen et al. 2004).

The wings extend sideways from the body domains perpendicular to the central symmetry axis of the tetramer and could participate in the binding to some receptors (see Sections 2.3 and 4). They also seem to mediate homotypic interactions, causing tetramers to assemble into flat 2D crystals, often containing large numbers of tetramers (Lunev et al. 1991). These lattices could underlie the frequently described phenomenon of α-LTX channel clusterization (Robello et al. 1987; Krasilnikov and Sabirov 1992; Filippov et al. 1994; Van Renterghem et al. 2000).

## 2.3 Recombinant α-LTX

Detailed structure-function analysis of α-LTX is impossible without generating mutants. The difficulty inherent to this approach is to ensure proper folding of this large protein. Two groups relied on baculovirus expression and successfully purified active recombinant toxins. Ichtchenko et al. (1998) used two 8-histidine tags for purification, whereas Volynski et al. (1999, 2003) utilized a monoclonal antibody.

Recently, active recombinant α-LTX has been generated using bacteria in which both thioredoxin reductase and glutathione reductase are inactivated to improve the formation of disulphide bonds in expressed proteins (Li et al. 2005). The toxin is expressed as a fusion with glutathione-S-transferase (GST), which is used for affinity purification of the recombinant toxin and can be subsequently removed by selective proteolysis. Considering the relative ease of generating recombinant proteins in bacteria, this approach will facilitate structure-function studies of α-LTX.

Using these methods, several useful observations regarding the functions of the α-LTX domains have been made by analyzing various mutants of recombinant toxin. These results are discussed in relevant sections below.

## 3 Membrane Pore

How the hydrophilic α-LTX inserts into lipid membranes and makes cation-permeable pores is not fully known, but an in-depth insight into the mechanisms of channel formation has been gained by combining cryo-EM, biochemical and biophysical studies with toxin mutagenesis. α-LTX pore formation consists of at least three steps: toxin tetramerisation, interaction with a specific cell-surface receptor and, finally, membrane insertion. Many experimental procedures can affect some of these steps and thereby prevent or assist channel formation.

### *3.1 Tetramerization*

The tetrameric state of α-LTX is divalent cation-dependent. Accordingly, treatment with EDTA renders purified α-LTX mostly dimeric, while subsequent addition of millimolar $Ca^{2+}$ or $Mg^{2+}$ promptly restores tetramerization (Ashton et al. 2000). This transition, requiring conformation changes (Orlova et al. 2000) and catalyzed by divalent cations, may explain the dependence of α-LTX pore-mediated actions on $Mg^{2+}$ (e.g. Misler and Hurlbut 1979). Tetramerization can also be triggered by amphipathic molecules (Ashton et al. 2000), possibly membrane lipids.

Interestingly, $La^{3+}$ at a concentration higher than 100 μM also greatly reduces the number of tetramers (Ashton et al. 2001). In addition, this trivalent cation is able to block the channels of membrane-inserted tetramers (see Section 3.4.6), making $La^{3+}$ useful in studying pore-independent toxin effects.

The role of tetramerization in toxin pore formation has been vividly illustrated by mutagenesis of α-LTX. In particular, Ichtchenko et al. (1998) have generated an interesting mutant (α-LTX$^{N4C}$) that contains a four-amino-acid insert between the N-terminal domain and the ankyrin repeats (Figure 1b). This insert apparently causes a conformational change enabling α-LTX$^{N4C}$ to form dimers but not cyclical tetramers (Volynski et al. 2003). Accordingly, although α-LTX$^{N4C}$ still binds its receptors (Ichtchenko et al. 1998), it fails to incorporate into the membrane and form pores (Ashton et al. 2001; Volynski et al. 2003), providing a powerful argument that the tetramer is the molecular species that inserts into membranes.

### *3.2 Receptor Interaction*

Although α-LTX is able to insert into pure lipid membranes (Finkelstein et al. 1976), reconstituted receptors greatly enhance the rate of insertion (Scheer et al. 1986). Biological membranes seem even more refractive to the toxin: when cells do not possess α-LTX receptors, no pore formation can be detected (Hlubek et al. 2000; Van Renterghem et al. 2000; Volynski et al. 2000), whereas expression of exogenous receptors allows abundant α-LTX insertion and concomitant channel formation. Receptors, thus, confer specificity to the pore-mediated effects of α-LTX. It is not clear whether receptors are directly involved in membrane insertion, simply concentrate toxin near the membrane, or organize membrane lipid domains to make them accessible to α-LTX. Importantly for some aspects of toxin-evoked secretion, interaction with one of the receptors, neurexin (NRX), is $Ca^{2+}$-dependent, while interaction with the other two, latrophilin 1 (LPH1) and protein tyrosine phosphatase σ (PTPσ), is completely $Ca^{2+}$-independent (see Section 4 for more details).

The role of the structural domains of α-LTX in receptor interaction has been studied by mutagenesis. Ichtchenko et al. (1998) have generated three point-mutants by replacing conserved cysteines in the N-terminal domain (Figure 1b) with serines. These mutants (C14S, C71S, C393S) are inactive in release, probably because they do not bind α-LTX receptors. Accordingly, chemical reduction of disulphide bonds in α-LTX abolishes receptor binding (Ichtchenko et al. 1998). Thus, proper folding of the N-terminal domain, aided by disulphide bonds, is essential for toxin-receptor interactions.

In another work, Li et al. (2005) have deleted the C-terminal ankyrin repeats 15 to 22, comprising the head and part of the body (Figure 1b), and found that this mutant fails to bind LPH1 but still interacts with NRX Iα. This suggests that α-LTX complex with LPH1 requires an intact C-terminal quarter of the toxin molecule. Whether this sequence binds LPH1 directly or affects the conformation of other α-LTX regions that actually interact with LPH remains to be addressed. Clearly, the C-terminal region is not necessary for binding of NRX Iα.

### 3.3 Membrane Insertion

The insertion of the α-LTX channel into black lipid membranes is unidirectional (Robello et al. 1984), suggesting that the toxin has evolved specialized domains to help it enter lipid bilayers. The direct observation of tetramers inserted into artificial lipid membrane (Orlova et al. 2000) indicates that the "bowl" formed by the body domains is immersed deeply in the lipid, while the wing domains remain splayed on, and slightly buried into, the extracellular surface of the bilayer, while the heads are exposed to the extracellular space (Figure 2). In this model, some regions of the body domains are exposed to the cytosol, consistent with the observation that protease treatment from that side of the lipid bilayer modifies the inserted channel (Robello et al. 1984; Chanturia and Lishko 1992) and supportive of the hypotheses that α-LTX could interact with intracellular release machinery (Khvotchev et al. 2000) or act as a fusogen catalyzing vesicle fusion (Sokolov et al. 1987; Lishko et al. 1990).

In addition to the presence of receptors, the ionic composition of buffer (Finkelstein et al. 1976) and the makeup of lipid bilayer (Robello et al. 1984) can influence the rate of α-LTX insertion into lipid membranes. It is possible that the membrane microenvironment affects the insertion of a receptor-bound tetramer also in a physiological context. For example, at 0° C, α-LTX binds to receptor-containing membranes, but it is not inserted (Khvotchev et al. 2000; Volynski et al. 2000).

Another feature of α-LTX is that it co-purifies with a small protein, called LMWP (Volkova et al. 1995) or latrodectin (Pescatori et al. 1995). It does not affect the toxicity of α-LTX nor has secretory activity of its own (Volkova et al. 1995). However, when LMWP is removed (Ashton et al. 2000), the toxin becomes unable to form channels in artificial lipid bilayers, although its receptor-mediated permeation of biological membranes is unhindered (Volynski et al. 2000). It is tempting to speculate that LMWP helps wild-type α-LTX to penetrate lipid bilayers (see also Grishin et al. 1993). To test this, recombinant toxins lacking LMWP need to be studied in pure lipid membranes.

The receptor-aided insertion of α-LTX into biological membranes has been recently investigated using recombinant toxin N-terminally fused to GST and expressed in bacteria (Li et al. 2005). When, after purification, the GST moiety is removed, the resulting wild-type recombinant α-LTX inserts into the plasma membrane and induces strong cation currents in tsA cells (HEK-293-derived) transfected with NRX Iα, whereas the GST-fused toxin is much less effective. As both the GST-free and GST-fused toxins bind well to the receptors (Li et al. 2005), it is clear that GST attached to the α-LTX N-terminus decreases the efficiency of toxin incorporation into membranes. However, despite its large size and hydrophilicity, GST does not block channel formation completely, indicating that when α-LTX inserts into the membrane, the wing does not need to penetrate the bilayer.

This idea is further supported by the analysis of a series of α-LTX truncation mutants, lacking 1, 2, 3, or 8 C-terminal ankyrin repeats (Li et al. 2005). When tested in NRX Iα-transfected tsA cells, some of these constructs induce cation conductance; others completely lack this ability. As all these mutants possess unaltered N-terminal wing domains, the wing apparently does not participate in pore formation, while the structure of the C-terminal head is crucial for membrane insertion. This is consistent with the model of the membrane-inserted α-LTX tetramer in Figure 2.

## *3.4 Features of the Pore*

### 3.4.1 Ion Selectivity

Once formed, the α-LTX channel only mediates cationic currents, probably because negatively charged acidic side chains line the channel (Finkelstein et al. 1976). Most significantly, the α-LTX channel is permeable to $Ca^{2+}$ (Finkelstein et al. 1976; Krasil'nikov et al. 1982; Mironov et al. 1986), and it is this aspect of the α-LTX channel that is most often considered. However, this channel is not very selective, and $Ba^{2+}$, $Sr^{2+}$, $Mg^{2+}$ as well as $Li^{+}$, $Cs^{+}$ and, importantly, $Na^{+}$ and $K^{+}$ currents are also carried by the channel (Krasil'nikov et al. 1982; Mironov et al. 1986).

A detailed analysis of the permeability of α-LTX channels to cations and non-electrolytes (Mironov et al. 1986; Robello 1989; Krasilnikov and Sabirov 1992) postulates that the negatively charged selectivity filter of the channel is located close to the extracellular side of the membrane, fitting in with the hypothesis that the C-terminal head domains form the selectivity filter (see Section 2.2.2).

This hypothesis finds additional confirmation in the features of the α-LTX truncation mutants described above (Section 3.3) (Li et al. 2005). These mutants have 1 to 8 ankyrin repeats removed from their C-termini and form cation channels of dramatically different conductivities. For example, α-LTxΔ1 mutant mediates an enormous conductance, by far exceeding that of wild-type toxin. It is possible that the removal of the last ankyrin repeat lifts an obstruction for cation movements in the extracellular mouth of the channel. In contrast, α-LTxΔ2 and α-LTxΔ3 make very inefficient channels, probably due to perturbations of the channel lining. Finally, α-LTxΔ8 does not induce any cation currents, as it seemingly cannot form tetramers.

### 3.4.2 $Ca^{2+}$ Currents

Although $Ca^{2+}$ only carries a small proportion of currents through cell membrane-inserted α-LTX channels (Hurlbut et al. 1994; Tse and Tse 1999), the influx of $Ca^{2+}$ through presynaptically-targeted α-LTX channels is most often referred to, because of the well-established link between presynaptic $[Ca^{2+}]$ and neurotransmitter release. There is a wealth of evidence indicating that in conditions favorable to channel formation (e.g., in the presence of divalent cations), influx of extracellular $Ca^{2+}$ through α-LTX channels is an important aspect of α-LTX action.

Increases in $[Ca^{2+}]_i$ upon α-LTX addition in the presence of $Ca^{2+}_e$ have been detected in synaptosomes (Nicholls et al. 1982; Meldolesi et al. 1984), native PC12 (Grasso et al. 1980; Meldolesi et al. 1984), and chromaffin (Barnett et al. 1996) cells, as well as in receptor-transfected COS (Volynski et al. 2000; Ashton et al. 2001), BHK (Krasnoperov et al. 1997) and neuroblastoma cells (Volynski et al. 2004).

While the bulk of this effect is due to calcium influx through the toxin channel, it is possible that the rise in $[Ca^{2+}]_i$ is contributed to by $Ca^{2+}$ release from intracellular stores (Tse and Tse 1999; Tsang et al. 2000; Liu et al. 2005) or influx through endogenous $Ca^{2+}$ channels (Nicholls et al. 1982; Lajus et al. 2006). In LPH-transfected neuroblastoma cells, $Ca^{2+}_i$ waves caused by LPH signalling are observed in response to non-pore-forming α-LTX (Volynski et al. 2004). Nevertheless, influx of $Ca^{2+}$ through the channel formed by wild-type toxin appears more dramatic.

### 3.4.3 $Na^+$ Currents

α-LTX channels inserted in artificial membranes are permeable to $Na^+$ (Krasil'nikov et al. 1982; Robello et al. 1984), which is modulated by $Ca^{2+}$ added to the topologically extracellular side of the bilayer (Mironov et al. 1986). Inward $Na^+$ currents are observed electrophysiologically in *Xenopus* oocytes, PC12 and neuroblastoma cells (Wanke et al. 1986; Filippov et al. 1994; Hurlbut et al. 1994), and α-LTX elevates $[Na_i^+]$ in synaptosomes detectable with a fluorescent dye (Deri and Adam-Vizi 1993), although not with radioactive $Na^+$ (Storchak et al. 1994).

Such currents probably account for at least one of α-LTX actions: tetrodotoxin-insensitive depolarization of artificial and biological membranes (Grasso and Senni 1979; Nicholls et al. 1982; Scheer et al. 1986). Although depolarization could cause neurotransmitter release by activating voltage-gated $Ca^{2+}$ channels, $Ca^{2+}$ flow through the α-LTX channel probably overwhelms this effect (Nicholls et al. 1982), so the role of depolarization in α-LTX action is unclear.

In addition, $Na^+$ currents through α-LTX channels can underlie some of the α-LTX-induced nonvesicular neurotransmitter release by causing, for example, the collapse of the cross-membrane $Na^+$ gradient and the reversal of some (McMahon et al. 1990) but not all (Deri and Adam-Vizi 1993) transmitter uptake pumps.

α-LTX-mediated $Na^+$ influx can also induce mitochondria of frog NMJ synapses (Tsang et al. 2000) and neurohypophysis terminals (Hlubek et al. 2003) to release large amounts of $Ca^{2+}$ into the cytosol because of the reversal of their $Na^+/Ca^{2+}$ exchangers, although this seems to be inconsequential for secretion in the absence of $Ca^{2+}_e$. In contrast, increases in $[Ca^{2+}]_i$, partially dependent on $Na^+_e$, correlate with acetylcholine release in synaptosomes (Deri and Adam-Vizi 1993). Generally, a rise $[Na^+]_i$ in nerve terminals correlates with enhanced secretion (Meiri et al. 1981). Thus, $Na^+$ currents represent yet another facet of the α-LTX channel actions at the synapse and could account for some of the toxin-evoked $Ca^{2+}_e$-independent vesicular release. However, α-LTX may be able to induced release even in $Ca^{2+}$-free and $Na^+$-free medium in synaptosomes (Storchak et al. 1994) and frog NMJs (Tsang et al. 2000).

### 3.4.4 Factors Affecting Conductance

A perplexing feature of the literature that explores α-LTX channels is the extremely wide range of single-channel conductances, ranging from 15 pS (Wanke et al. 1986; Krasilnikov and Sabirov 1992) to 100-300 pS (Finkelstein et al. 1976; Van Renterghem et al. 2000) or even 1100 pS (Krasilnikov and Sabirov 1992). To put these figures in perspective, the voltage-gated $K^+$ channel Kv1.2 and the ryanodine-sensitive $Ca^{2+}$-release channels have conductances of 14-18 pS and ~700 pS, respectively (Lindsay et al. 1994; Gutman et al. 2005) (Figure 2).

Several plausible hypotheses can explain this variability. Firstly, the venom contains a range of pore-forming toxins with varying conductances. Secondly, there is evidence that the make-up of the permeated membrane (Robello et al. 1984; Scheer et al. 1986; Krasilnikov and Sabirov 1992) and variations in lipid packing and order (Chanturia and Lishko 1992), but not the type of receptor present (Hlubek et al. 2000; Van Renterghem et al. 2000), can affect the properties of α-LTX channels.

Interestingly, $Ca^{2+}$ inhibits the conductance of α-LTX channels to monovalent cations, causing a "flickery block," in artificial membranes (Mironov et al. 1986; Krasil'nikov et al. 1988), neuroblastoma cells (Hurlbut et al. 1994), and embryonic kidney cells (Hlubek et al. 2000), although this is not apparent in receptor-expressing oocytes (Filippov et al. 1994). $Mg^{2+}$ positively modulates the conductivity of toxin channel for $Ca^{2+}$ (Davletov et al. 1998; Van Renterghem et al. 2000).

Finally, the frequent observation of bursts of conductance, likely corresponding to synchronized opening and closing events, has led many authors to suggest that α-LTX channels form clusters in the membrane (Robello et al. 1987; Krasilnikov and Sabirov 1992; Filippov et al. 1994; Van Renterghem et al. 2000). Such clustering is plausible from a structural point of view, as crystalline arrays of α-LTX tetramers are observed by EM (Lunev et al. 1991) (see Section 2.2.2), and it may account for some of the variability in conductances reported in the literature.

### 3.4.5 Gating

The channel is open most of the time, with an open probability of 0.8 (Filippov et al. 1994). Closure events, albeit rare, have been observed in many systems (Finkelstein et al. 1976; Wanke et al. 1986; Hurlbut et al. 1994), and several channels can open and close in concert (Krasilnikov and Sabirov 1992; Filippov et al. 1994; Tse and Tse 1999). Membrane depolarization may increase the likelihood of closure events (Filippov et al. 1994; Van Renterghem et al. 2000), and $Ca^{2+}$ and $Mg^{2+}$ modulate gating (Filippov et al. 1994; Hurlbut et al. 1994). The molecular basis of gating is currently unknown, but is likely to involve head domains.

### 3.4.6 Divalent and Trivalent Cations

The α-LTX pore is permeable to alkaline earth cations, whose affinities for the channel decrease in the following sequence: $Mg^{2+} > Ca^{2+} > Sr^{2+} \geq Ba^{2+}$ (Mironov et al. 1986). Transition metal cations ($Cd^{2+} > Co^{2+} \geq Ni^{2+} \geq Zn^{2+} > Mn^{2+}$) strongly block $Ca^{2+}$ and $K^{+}$ currents through α-LTX channels in artificial membranes (Mironov et al. 1986). This block is only effective when the cation is applied from the *cis*-side (equivalent to the extracellular side) of the membrane.

Millimolar concentrations of trivalent cations, such as $Yb^{3+}$, $Gd^{3+}$, $Y^{3+}$, $La^{3+}$ and $Al^{3+}$, block the pore-mediated effects of α-LTX in synaptosomes (Scheer 1989). The nature of this effect is complex. For example, $Al^{3+}$ blocks the binding of α-LTX to synaptosomes (Scheer 1989), while $La^{3+}$ prevents tetramerization (Ashton et al. 2000) and thereby channel insertion and formation.

Most trivalent cations, at 50–100 μM, block previously inserted channels (Scheer 1989; Hurlbut et al. 1994; Van Renterghem et al. 2000) and inhibit α-LTX-mediated $Ca^{2+}$-uptake, while $La^{3+}$ blocks depolarization as well (Scheer 1989). Channel inhibition by trivalent cations is very important because $La^{3+}$ is essentially the only reagent that blocks the $Ca^{2+}$-independent α-LTX-evoked neurotransmitter secretion in neurons (Scheer 1989; Capogna et al. 2003).

The use of trivalent cations for the purpose of dissecting the mechanisms of α-LTX action, however, is complicated by their well-described ability to trigger neurotransmitter release in the absence of other stimuli (Scheer 1989). Furthermore, at low concentrations, both $La^{3+}$ (at 15 μM) and $Cd^{2+}$ (at 50 μM) support, rather than block, the secretagogue activity of the toxin (Misler and Falke 1987).

### 3.4.7 Permeation of Molecules/Leakage

It is possible that the water-filled α-LTX channel, which is relatively wide (~10 Å at its narrowest (Krasilnikov and Sabirov 1992; Orlova et al. 2000), can pass small molecules. Indeed, α-LTX channels inserted in the membranes of synaptosomes, NMJ nerve terminals, and receptor-transfected COS7 cells appear to pass fluorescein (Stokes-Einstein radius, $R_e = 4.5$ Å) and norepinephrine ($R_e \leq 4$ Å) (Davletov et al. 1998; Rahman et al. 1999; Volynski et al. 2000), shown in Figure 2 for comparison with 8-hydrated calcium ion ($R_e = 4.2$ Å) and the toxin channel. Analysis of "impermeant" cations commonly used in channel studies reveals that α-LTX channels are poorly permeable (Hurlbut et al. 1994) to glucosamine $H^+$ ($R_e = 4.6$ Å) and not significantly permeable (Tse and Tse 1999) to N-methyl-D-glucamine ($R_e = 5.2$ Å), thus limiting the pore diameter by ~10 Å.

These observations are particularly relevant to the experimental use of α-LTX in neurotransmission studies, since α-LTX has been shown in several systems to cause nonvesicular release by allowing leakage of cytoplasmic neurotransmitters (McMahon et al. 1990; Deri et al. 1993; Davletov et al. 1998). This flux could be mediated by the α-LTX channel itself, by local disruptions of cellular membranes, or by reversal of transmitter uptake pumps driven by $Na^+$ gradient (see Section 3.4.3). Synaptosomes seem to be particularly sensitive to an increase in hydrostatic pressure, which may occur when influx of $Na^+$ or $Ca^{2+}$ leads to a concomitant influx of water.

## 3.5 *Can the Pore Explain Everything?*

It is now universally accepted that α-LTX can insert into, and permeabilize, artificial and biological membranes. Cation currents can explain some, but not all, of the toxin's effects. For example, it is not clear how the α-LTX channel could mediate $Ca_e^{2+}$-independent exocytosis in neurons. Although it would be tempting to assign some of α-LTX actions in the absence of $Ca_e^{2+}$ to $Na^+$ currents, lack of $Na^+$ does not prevent $Ca^{2+}$-independent secretion (Tsang et al. 2000). Cation flux-associated incursion of terminals by water could be involved, but α-LTX effect on intracellular osmotic pressure has not been characterized yet.

Based on the results of α-LTX mutagenesis, strong correlation exists between pore formation and stimulation of $Ca^{2+}$-dependent exocytosis from neuroendocrine cells. However, in some experiments with chromaffin cells, α-LTX action does not involve $Ca^{2+}$ entry (Michelena et al. 1997). In addition, α-LTX sensitizes chromaffin cells to $Ca^{2+}$ even when the cells are permeabilized and toxin pores should have no effect; this involves protein kinase C (PKC) activation (Bittner and Holz 2000). Furthermore, the ability of α-$LTX^{N4C}$ to induce $Ca^{2+}$-dependent exocytosis without forming pores implicates a stimulating mechanism other than pore formation.

Therefore, α-LTX receptors, which are so crucial for the toxin's action in all biological systems, have been comprehensively studied.

All NRXs have a single transmembrane region (TMR) and a short cytoplasmic tail. In α-NRXs, the large ectodomains are composed of six LNS domains (Figure 3a) characteristic of many adhesion molecules. Pairs of LNS(A) and LNS(B) domains, with EGF-like sequences in the middle, form three "major repeats" (Figure 3a). In contrast, β-NRXs have one LNS(B) domain and a unique N-terminal sequence. All NRXs contain an extensively O-glycosylated Thr- and Ser-rich region upstream of the TMR (Ushkaryov et al. 1992; Ushkaryov et al. 1994).

### 4.1.2 Distribution

NRXs are present mostly in brain (Ushkaryov et al. 1992) but also in pancreas and lung (Occhi et al. 2002). They are differentially expressed in brain regions, showing partially overlapping patterns of distribution of their six main forms and region-specific splice variants (Ullrich et al. 1995). It is difficult to localize NRX proteins at the cellular level, partly because the respective α- and β-forms are almost identical immunologically. As demonstrated by light microscopy, NRXs are expressed at synapses (Ushkaryov et al. 1992; Dean et al. 2003), but in the absence of EM data, their distribution between the pre- and postsynaptic membranes is not clear. Based on the presynaptic action of α-LTX, NRX Iα is implicitly presynaptic; β-NRXs are assigned to presynaptic membranes due to their interaction with neuroligins, indeed localized in the postsynaptic membrane by EM (Song et al. 1999). It remains to be seen whether any NRXs may be expressed postsynaptically.

### 4.1.3 Binding of α-LTX

Analysis of NRX isolated from brain by α-LTX-affinity chromatography (Petrenko et al. 1990; Petrenko et al. 1993) shows that out of the six main isoforms only NRX Iα binds toxin with sufficient affinity (Davletov et al. 1995), and this interaction strictly requires $Ca^{2+}$ (Petrenko et al. 1990; Davletov et al. 1995).

The α-LTX binding site has not been mapped in a systematic effort, but incidental evidence implicates the third major repeat, LNS(A3/B3). In particular, α-LTX binds neither a NRX Iα mutant lacking the LNS-domains B2, A3, and B3, nor the full-size NRX Iα with an insert in LNS(B3) (Davletov et al. 1995). Although LNS(B3) domains of all three β-NRXs are able to interact with α-LTX in a splicing-dependent manner under mild conditions (Sugita et al. 1999), they are not sufficient for strong toxin binding: NRX Iβ does not pull down α-LTX (Davletov et al. 1995), nor are β-NRXs isolated from brain by α-LTX chromatography (Petrenko et al. 1990). Finally, a NRX Iα mutant, containing only LNS(A3/B3), binds the toxin well (Li et al. 2005). Thus, the strong interaction of α-LTX with NRX Iα is mediated by both LNS(A3) and LNS(B3), while splicing in LNS(B3) regulates this binding.

### 4.1.4 Proposed Functions

The structure of NRXs is consistent with a function in cell adhesion and/or ligand binding. Indeed, the extracellular domains of α-NRXs interact specifically and strongly with neurexophilins (Petrenko et al. 1993; Petrenko et al. 1996; Missler et al. 1998b), a family of glycosylated, proteolytically processed hormone-like 30-kDa proteins with species-specific differential expression in mammalian tissues (Petrenko et al. 1996; Missler and Südhof 1998). The biological role of neurexophilins is unknown (Missler et al. 1998b), but their tight binding to the LNS(B1) domain of α-NRXs implies long-term/structural, rather than short-term/signaling, interactions.

In addition, both α- and β-NRXs can interact with neuroligins, a family of neuronal proteins that contain a region of homology to esterases in their extracellular domains and a single TMR (Ichtchenko et al. 1995; Ichtchenko et al. 1996) (Nguyen and Südhof 1997; Boucard et al. 2005). Neuroligin-1 has been identified owing to its ability to bind all three β-NRXs – only if they lack an insert at splice site 4 (Ichtchenko et al. 1995) (Figure 3). Neuroligins are also differentially spliced, and the absence of an insert at splice site B enables neuroligins to bind also α-NRXs (Boucard et al. 2005). These proteins are localized on the postsynaptic membrane of excitatory (Song et al. 1999) and inhibitory (Varoqueaux et al. 2004) synapses. Their $Ca^{2+}$-dependent, splicing-controlled interaction with β-NRXs (Ichtchenko et al. 1996; Nguyen and Südhof 1997; Comoletti et al. 2006) is important for synapse formation (Scheiffele et al. 2000; Dean et al. 2003; Chih et al. 2006), differentiation (Levinson et al. 2005), and maturation (Dresbach et al. 2004; Varoqueaux et al. 2006).

In the cytoplasmic tail of NRXs, there is a short C-terminal site that binds to PDZ domain-containing scaffold proteins, CASK (Hata et al. 1996) and Mint (Biederer and Südhof 2000). These proteins interact with each other and with $Ca^{2+}$ channels. Mint also makes complexes munc18, a protein involved in vesicle exocytosis (Verhage et al. 2000). Therefore, NRXs may participate in arranging transmitter release mechanisms. Indeed, knockout of all three α-NRXs has demonstrated that they play a role in organizing N- and P/Q-type $Ca^{2+}$ channels (Zhang et al. 2005) in nerve terminals (Missler et al. 2003) and endocrine cells (Dudanova et al. 2006). Even though β-NRXs have the same C-terminal sequences as the respective α-NRXs, the two isoforms participate differently in synaptogenesis and transmitter secretion. In particular, α-NRXs are apparently not involved in synapse formation (Missler et al. 2003) but may influence the expression of postsynaptic NMDA (but not AMPA) glutamate receptors (Kattenstroth et al. 2004).

## *4.2 Latrophilin*

Because NRX Iα requires $Ca^{2+}$ to bind α-LTX, it cannot mediate the toxin's effects in the absence of $Ca^{2+}$. The quest for a $Ca^{2+}$-independent receptor continued, and a major protein was eventually isolated by α-LTX-chromatography

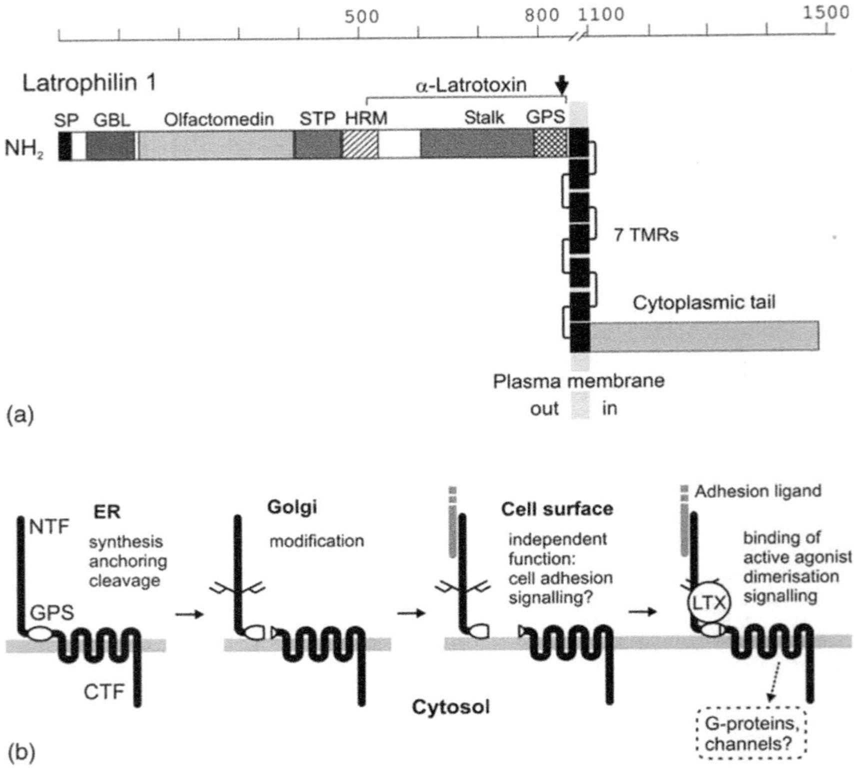


**Fig. 4** LPH structure and processing. (**a**) Generalized domain structure of latrophilins. 7TMRs, seven transmembrane regions and corresponding loops; arrow, cleavage point; bracket, the site of α-LTX binding in LPH1; GBL, galactose-binding lectin; GPS, GPCR-proteolysis site domain; HRM, hormone receptor motif; STP, Ser-, Thr- and Pro-rich domain. (**b**) Scheme of LPH processing and functioning (Volynski et al. 2004). See text for details. ER, endoplasmic reticulum.

of solubilized brain proteins and termed latrophilin (LPH) (Figure 4) (Davletov et al. 1996; Lelianova et al. 1997), or $Ca^{2+}$-independent receptor of α-LTX (CIRL) (Krasnoperov et al. 1996; Krasnoperov et al. 1997). LPH binds toxin in the presence or absence of divalent cations (Davletov et al. 1996). It was not discovered earlier because its 120-kDa toxin-binding fragment was confused on SDS gels with α-LTX, while the second (65 kDa) fragment was lost due to its quantitative precipitation upon boiling with SDS.

### 4.2.1 Structure

LPH is a large (185 kDa) heptahelical receptor. It comprises three major domains (Figure 4a): (1) a long, glycosylated N-terminal extracellular domain; (2) seven hydrophobic TMRs; and (3) a long cytoplasmic tail. The ectodomain contains regions of homology to: galactose-binding lectin; the surface-attached extracellular matrix protein olfactomedin (Snyder et al. 1991; Loria et al. 2004); a hormone receptor motif (HRM) found also in other G-protein-coupled receptors (GPCRs) and probably involved in ligand binding; a "Stalk" domain important for proteolytic cleavage (Chang et al., 2003); and the G-protein-coupled receptor proteolysis site (GPS). The seven TMRs are similar to the corresponding regions of the secretin/calcitonin GPCRs that bind peptide hormones and induce release of various substances (Harmar 2001).

Based on this homology and on the length of ectodomains, LPH and similar receptors have been initially classified as "Long N-terminus group B" GPCRs (Hayflick 2000; Stacey et al. 2000). However, it is recognized now (Fredriksson and Schioth 2005) that they form a separate group, termed "adhesion GPCRs." This nomenclature stems from the fact that these receptors contain various cell adhesion modules in their ectodomains. The group includes 33 proteins identified by their function/activity or through genome searches (Fredriksson and Schioth 2005).

LPH has two very similar homologues, LPH2 and LPH3 (Matsushita et al. 1999). The family is also called CIRL1-3 (Ichtchenko et al. 1999) and CL1-3 (an abbreviation combining the two other names of the protein, CIRL and LPH) (Sugita et al. 1998). The LPHs have seven sites of alternative splicing (Sugita et al. 1998; Matsushita et al. 1999). Some splice variants encode receptors with an altered third cytoplasmic loop or a truncated cytoplasmic tail (Matsushita et al. 1999), modifications that are likely to affect coupling to G proteins.

### 4.2.2 Unusual Architecture

"Adhesion GPCRs" have an unusual architecture (Gray et al. 1996; Krasnoperov et al. 1997; Volynski et al. 2004). They contain two functional parts: an N-terminal cell adhesion receptor-like extracellular domain and a C-terminal domain with seven TMRs and a signaling potential (Hayflick 2000; Stacey et al. 2000). Intriguingly, all these receptors are post-translationally cleaved at the GPS domain (Krasnoperov et al. 1997) (Figure 4a, b) into N- and C-terminal fragments (NTF and CTF), which correspond to the two functional parts. This constitutive proteolysis, occurring in the endoplasmic reticulum, is prerequisite for LPH cell surface delivery (Krasnoperov et al. 2002a; Volynski et al. 2004). The GPS is localized upstream of the first TMR, yet the ectodomain is not released into the medium, apparently because it is attached to the membrane by a hydrophobic anchor (Volynski et al. 2004).

The peculiar biology of LPH has been studied in neuroblastoma cells stably expressing recombinant LPH1 (Volynski et al. 2004). Following delivery to the plasma membrane, the two receptor fragments can dissociate and behave as independent cell-surface proteins. However, under certain conditions, the fragments reassociate. This reassembly, mediated only by seven most N-terminal amino acids of the CTF, is induced by detergents and also by α-LTX binding to the NTF (Volynski et al. 2004). Ternary α-LTX-NTF-CTF complexes form large patches in the plasma membrane, triggering formation of molecular dimers of the CTF. This process temporally and spatially correlates with signal transduction, which can be visualized using the non-pore-forming mutant α-LTX$^{N4C}$ (see Section 2.3). This signaling requires the full-size CTF and involves activation of PLC and release of $Ca^{2+}$ from intracellular stores (Volynski et al. 2004). It is interesting that both in this model system and in central synapses (Ashton et al. 2001; Capogna et al. 2003), α-LTX$^{N4C}$ induces $Ca^{2+}$ signals only in the presence of $Ca_e^{2+}$. Calcium ions may serve here as a cofactor for LPH activation or they may enter the cytosol through an LPH-activated $Ca^{2+}$ channel(s).

### 4.2.3 Distribution

LPHs are differentially distributed in mammalian tissues: while LPH1 and LPH3 are enriched in brain, LPH2 features more prominently in other tissues (Sugita et al. 1998; Ichtchenko et al. 1999; Matsushita et al. 1999). Similar to NRX Iα (Ushkaryov et al. 1992; Occhi et al. 2002), very small levels of LPH1 mRNA can be detected outside brain, especially in kidneys and pancreas, and both receptors reside in endocrine cells, such as pancreatic β-cells (Lang et al. 1998). Based on α-LTX binding and action, which are dramatically inhibited by the LPH gene knockout, this receptor is localized in presynaptic terminals. In a yeast two-hybrid system, the C-terminal cytoplasmic tail of LPH can bind Shank (Tobaben et al. 2000; Kreienkamp et al. 2000), an ankyrin repeat-containing protein of postsynaptic density, although the specificity and significance of this interaction is unclear.

### 4.2.4 Binding of α-LTX

LPH1 has a very high affinity for α-LTX (Kd ~0.1–1.5 nM) (Volynski et al. 2000; Ichtchenko et al. 1999), while the interaction of LPH 2 with toxin is more than 10-fold weaker (Ichtchenko et al. 1999). α-LTX binds the Stalk/GPS domains of the NTF (Krasnoperov et al. 1999), while the lectin and olfactomedin regions upstream could participate in adhesion interactions with the extracellular matrix or adjacent cells (Figure 4b). Hence, the NTF is thought to remain on the cell surface at specialized cell-cell or cell-matrix junctions (Volynski et al. 2004) and, thus, define the sites where the LPH fragments can interact with each other. If endogenous CTF-mediated signaling requires NTF-CTF complex formation (as with α-LTX), then LPH signals may be restricted to adhesion foci.

### 4.2.5 Proposed Functions

The functions of LPH remain to be elucidated. As follows from the results of LPH1 gene knockout (Tobaben et al. 2002), this receptor is not crucial for basic neuronal survival and activity, although LPH2 and 3 may compensate for the lack of LPH1. The structure of all adhesion GPCRs indicates that these receptors can transform extracellular interactions into intracellular signaling. Being involved in some pathways leading to release of stored $Ca^{2+}$, LPH may participate in the cell contact-dependent fine tuning of $Ca^{2+}{}_i$ levels and modulation of transmitter secretion.

## *4.3 Protein Tyrosine Phosphatase σ*

In the search for a $Ca^{2+}$ independent α-LTX receptor, affinity chromatography in the absence of $Ca^{2+}$ was used to isolate brain proteins with any affinity for α-LTX. Sequencing of all proteins in the α-LTX column eluate (Krasnoperov et al. 1996; Krasnoperov et al. 2002b) has revealed that, in addition to LPH1, PTPσ can also bind to the column as a set of two minor bands.

### 4.3.1 Structure

PTPσ is a member of the family of receptor-like PTPs that contain cell adhesion molecule-like extracellular domains, a single TMR, and cytoplasmic phosphatase domains (Tonks 2006). In PTPσ, the extracellular domain comprises three immunoglobulin-like modules and, depending on alternative splicing, from four to eight fibronectin type III-like domains (Figure 5a). The cytosolic portion of the protein contains a catalytically active phosphatase domain and a pseudo-phosphatase.

Reminiscent of LPH, PTPσ is proteolytically cleaved 125 residues upstream of the TMR into two fragments, which remain non-covalently associated (Figure 5a) (Yan et al. 1993; Aicher et al. 1997). The proteolysis occurs inside the cell but, unlike in LPH, is not necessary for cell surface delivery of PTPs (Serra-Pages et al. 1994). On the cell surface, the P-subunit of PTPσ can be further cleaved by the metalloprotease TACE (Ruhe et al. 2006) six residues upstream of the TMR (Aicher et al. 1997). This process, stimulated by treating cells with $Ca^{2+}$ ionophores or phorbol esters (Aicher et al. 1997), or by interaction of PTPσ with the epidermal growth factor receptor (EGFR) (Ruhe et al. 2006), leads to the shedding of the E-subunit and sequestration of the P-subunit (Figure 5b), thus downregulating the PTPσ function.

### 4.3.2 Distribution

The large splice variant of PTPσ is found in most tissues, while the shorter variant is mostly brain-specific (Yan et al. 1993; Pulido et al. 1995) but, similar to NRX Iα and LPH1, is also found in small amounts in some other tissues (Yan et al. 1993), which may be due to autonomic innervation.

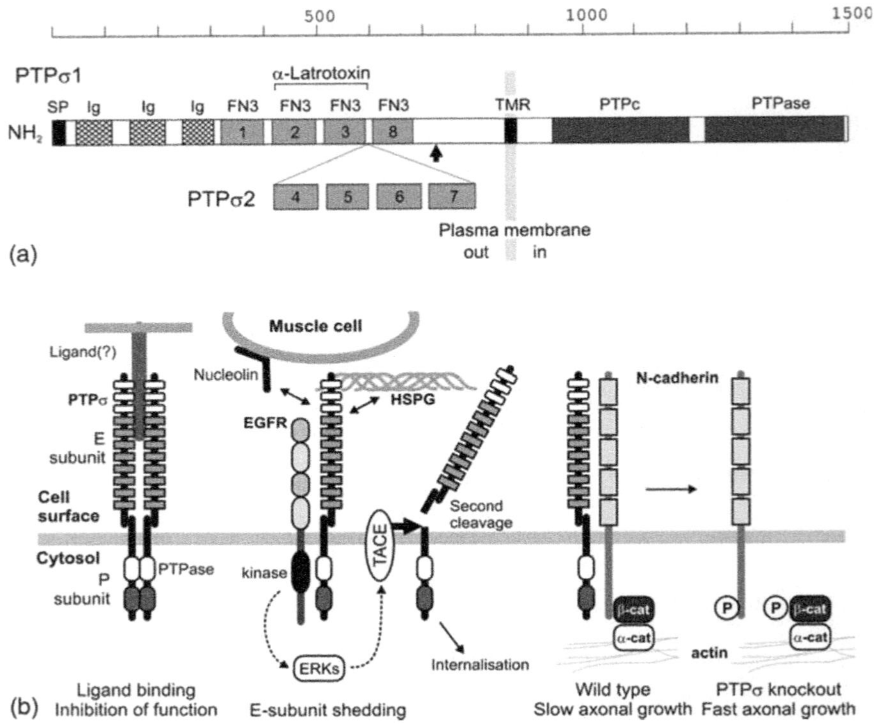


**Fig. 5** PTPσ structure and functions. (**a**) Generalized domain structure of receptor-like PTPσ. FN3, fibronectin type III-homology domains; Ig, immunoglobulin-like domains; PTPc, catalytic PTP domain; PTPase, pseudo-phosphatase domain. Arrow, proteolytic cleavage site; bracket, α-LTX binding site. (**b**) Cellular functions and interactions of PTPσ. Ligand binding leads to dimerization and inhibition of the intracellular domains. PTPσ interaction with EGFR activates the ERK-mediated signalling and upregulates the TACE metalloprotease which cleaves the P subunit and releases E subunit. PTPσ may interact with HSPG of the basement membrane and nucleolin on muscle cells. PTPσ binds N-cadherin and dephosphorylates it, causing increased adhesion and inhibition of axonal growth. In mice lacking PTPσ, N-cadherin is hyper-phosphorylated; this reduces adhesion and increases neuronal growth. EGFR, epidermic growth factor receptor; ERKs, MAP kinases ERK1 and ERK2; HSPG, heparin sulphate proteoglycan; cat, catenin.

### 4.3.3 Binding of α-LTX

Despite the very low binding of this protein to toxin columns, the α-LTX-PTPσ interaction is specific and maps onto the fibronectin domains 2 and 3 (Krasnoperov et al. 2002b). These are present in both PTPσ1 and PTPσ2, but on overexpression of these splice variants in pancreatic β-cells, only PTPσ1 mediates the secretory action of α-LTX (Lajus and Lang 2006). This may be because the toxin binds PTPσ2 too far from the membrane and is then unable to insert into the plasma membrane.

### 4.3.4 Proposed Functions

As phosphorylation of proteins localized in focal adhesions controls cell-cell and cell-matrix interactions, dephosphorylation is also important for the regulation of cell adhesion. Indeed, PTPσ is involved in negative regulation of axonal growth (McLean et al. 2002; Thompson et al. 2003) and growth cone pathfinding (Meathrel et al. 2002). Apart from EGFR (see Section 4.3.1), PTPσ interacts with N-cadherin and dephosphorylates it, augmenting its catenin-mediated attachment to the cytoskeleton and inhibiting axonal growth (Figure 5b). In PTPσ knockout mice, N-cadherin becomes over-phosphorylated (Siu et al. 2006) and unable to interact with catenins and the cytoskeleton, leading to stimulated axonal growth (McLean et al. 2002).

The structure of PTPσ is consistent with its role in binding ligands on the surface of other cells or in the extracellular matrix. It interacts with heparin sulphate proteoglycan in the basement membrane (Sajnani-Perez et al. 2003). It may also bind to the C-terminal domain of cell surface-exposed nucleolin, a normally nuclear protein that is presented on the surface of developing muscle cells (Alete et al. 2006). However, these complexes seem to mediate structural or long-term regulatory interactions rather than short-term signaling to the neuronal secretory machinery.

## 4.4 *α-LTX Receptors and Signaling*

### 4.4.1 NRX Iα

Can NRXs transduce intracellular an exocytotic signal? All currently known cytoplasmic interactions of α- and β-NRXs appear to be stable, nonenzymatic, structural connections that are essential for neuronal development, synapse organization, and long-term regulation but are unlikely to underlie a massive and relatively fast response induced by α-LTX. However, toxin binding to NRXs might cause conformational changes in Mint, which, by interacting with munc18, might regulate the availability of docking sites for synaptic vesicles. On the other hand, even a truncated NRX Iα that lacks the cytoplasmic tail can mediate the sensitivity of PC12 cells to wild-type α-LTX (Sugita et al. 1999). The most likely explanation of this is that toxin inserts into the membrane on binding any membrane-attached form of NRX Iα (Volynski et al. 2000; Hlubek et al. 2000), and because NRX experiments require $Ca^{2+}{}_{e}$, influx of this cation through the toxin pore can trigger maximal exocytosis irrespective of any NRX Iα-mediated signaling. With this in mind, there is no doubt that NRXs (full-size or C-terminally truncated) constitute functional receptors for α-LTX (Sugita et al. 1999), but only in the sense that they provide binding sites for the toxin, which subsequently exerts its own actions.

### 4.4.2 LPH1

LPH1 is the only receptor capable of G-protein-mediated signaling induced by α-LTX in central synapses (Davletov et al. 1998; Ashton et al. 2001; Section 1.4). Furthermore, LPH1 is directly involved in α-$LTX^{N4C}$-induced signaling in LPH-transfected neuroblastoma cells (Volynski et al. 2004). This mutant activates PLC also in neurons, causing release of $Ca^{2+}{}_{i}$ (Ashton et al. 2001; Volynski et al. 2003; Volynski et al. 2004) and leading to neurotransmitter release, which is blocked when the Gq cascade is inhibited by various drugs (Davletov et al. 1998; Ashton et al. 2001; Capogna et al. 2003). LPH1 also co-purifies with Gq (Rahman et al. 1999).

The endogenous ligand of LPH1 is currently unknown, and it is interesting to consider an invertebrate LPH orthologue from *Caenorhabditis elegans* (Mee et al. 2004; Willson et al. 2004). The nematode protein LAT-1 (B0457.1) lacks the olfactomedin region but contains the domains required for α-LTX binding (HRM, Stalk and GPS). Although its identity to vertebrate LPH1 is not too high (18%–45% in the area of toxin binding), it mediates the lethal effect of an α-LTX homologue, ε-latroinsectotoxin (Mee et al. 2004). *C. elegans* also expresses NRX and PTPσ homologues, but only the knockdown of LPH renders the worms resistant to ε-latroinsectotoxin (Mee et al. 2004). Interestingly, the worm LPH mediates the action of the anthelmintic octadepsipeptide emodepside, which causes paralysis in nematodes (Willson et al. 2004). By the coupling to LAT-1, emodepside activates Gαq and PLC-β (Willson et al. 2004). The product of the PLC activity, $IP_3$ stimulates release of $Ca^{2+}$ from intracellular stores, known to be important for α-LTX activity in mammals (Davletov et al. 1998; Capogna et al. 2003). In addition, the effect of emodepside involves UNC-13, a protein implicated in control of vesicular exocytosis and activated by diacyl glycerol, another product of PLC. It is possible that such different agonists as emodepside and $LTX^{N4C}$ can stimulate similar signaling from LPH orthologues, suggesting that at least part of the α-LTX effect on exocytosis is due to LPH signal transduction.

Is LPH involved only in one type of signaling, to $Ca^{2+}$ stores? In fact, α-$LTX^{N4C}$, acting via LPH, can activate $Ca^{2+}$ and inhibit $K^{+}$ channels in pancreatic β-cells (Lajus and Lang 2006), raising the possibility that, similar to many other GPCRs, LPH may be linked to ion channels. In addition, in endocrine cells, α-$LTX^{N4C}$ triggers $Ca^{2+}{}_{e}$-independent release of stored calcium and also activates PKC (Liu et al. 2005; Bittner et al. 2005). Thus, α-LTX can both provide an exocytotic stimulus (by releasing stored $Ca^{2+}$) and sensitize the secretory apparatus to it (by activating PKC).

### 4.4.3 PTPσ

Only LTX binding to PTPσ1, but not to PTPσ2, leads to secretory response in β-cells (Lajus and Lang 2006). This indicates that LTX apparently does not evoke any signaling through PTPσ, which probably simply aids toxin pore formation. Indeed, the multimeric toxin must induce the dimerization of PTPσ, leading to reciprocal inhibition of the catalytic domains (Tonks 2006). Furthermore, while all three receptors are found naturally occurring in pancreatic β-cells, the distribution of PTPσ does not correlate with the endogenous sensitivity of these cells to α-LTX (Lang et al. 1998; Lajus and Lang 2006).

### 4.4.4 α-LTX Mutants

As α-LTX$^{N4C}$ (see Section 2.3) does not form pores, it probably stimulates exocytosis by receptor mediated signaling. Therefore, the activity of α-LTX$^{N4C}$ has been investigated in $Ca^{2+}$-dependent and $Ca^{2+}$-independent exocytosis from neurons and $Ca^{2+}$-dependent exocytosis from neuroendocrine cells.

In $Ca^{2+}$-free media, the mutant toxin does not stimulate substantial exocytosis in neurons (Ashton et al. 2001; Volynski et al. 2003; Capogna et al. 2003). This strongly indicates that $Ca^{2+}$-independent exocytosis, so characteristic of wild-type α-LTX, requires more than receptor activation. However, α-LTX$^{N4C}$, strongly stimulates transmitter release from neurons in the presence of $Ca^{2+}{}_e$, presumably due to its ability to activate LPH1 (Ashton et al. 2001; Capogna et al. 2003; Volynski et al. 2003). The specific details of how receptor activation triggers exocytosis remain to be established, but it is clear that thapsigargin-sensitive release of $Ca^{2+}{}_i$ plays a pivotal role (Ashton et al. 2001; Capogna et al. 2003).

α-LTX$^{N4C}$ is also able to stimulate $Ca^{2+}$-dependent exocytosis from endocrine cells: it is active in chromaffin cells in the presence of $Ca^{2+}$ or $Sr^{2+}$ (Volynski et al. 2003) and in pancreatic β-cells (Liu et al. 2005; Lajus and Lang 2006). On the other hand, bacterially expressed α-LTX$^{N4C}$ appears unable to stimulate $Ca^{2+}$-dependent exocytosis from PC12 cells (Li et al. 2005), but this may be due to mutant's misfolding or by specific characteristics of the PC12 cell line used.

### 4.4.5 Receptor Knockouts

The genes for all three α-LTX receptors have been knocked out in mice, and these mutations are not lethal. NRX Iα knockout mice show no obvious behavioral phenotype (Geppert et al. 1998). LPH1 knockout causes social problems in mice (Tobaben et al. 2002), consistent with LPH1 involvement in schizophrenia (Chen and Chen 2005). PTPσ knockout leads to developmental abnormalities in the nervous system and outside it (Meathrel et al. 2002; Batt et al. 2002).

In mice lacking the NRX Iα gene, $Ca^{2+}$-dependent α-LTX binding to brain membranes is inhibited by 60% (Geppert et al. 1998), although the $Ca^{2+}$-independent binding seems to be affected as well (Geppert et al. 1998). LPH1 knockout leads to the loss of 75% of $Ca^{2+}$-independent binding of α-LTX but also affects the $Ca^{2+}$-dependent binding (Tobaben et al. 2002). The NRX-LPH1 double knockout still binds about 25% of α-LTX compared to wild-type mice, both in the presence and absence of $Ca^{2+}$ (Tobaben et al. 2002). $Ca^{2+}$-dependent release of glutamate induced by wild-type α-LTX in synaptosomes from the NRX Iα, LPH1 and NRX

Iα –LPH1 knockout mice is similarly inhibited, while $Ca^{2+}$-independent secretion is most grossly affected in the LPH1 single knockout, but not abolished. Surprisingly, the inactivation of both LPH1 and NRX Iα is less detrimental for the toxin-stimulated release than the LPH1 knockout alone (Tobaben et al. 2002). These data suggest that the two α-LTX receptors, NRX Iα and LPH1, may functionally interact (Südhof 2001), e.g., by reciprocally influencing each other's expression or synaptic delivery and, thus, altering the actions of α-LTX. The removal of both NRX Iα and LPH1 from synaptic cleft may allow the pore-forming toxin to access PTPσ more freely, with subsequent membrane insertion and $Ca^{2+}$-independent effects. It would be interesting to test non-pore-forming α-LTX mutants on neuronal preparations from NRX Iα, LPH1 and PTPσ knockout mice.

## 5 Overview of Mechanisms

α-LTX causes $Ca^{2+}{}_{e}$-dependent and –independent release of neurotransmitters (Figure 6). Dense core vesicles require $Ca^{2+}{}_{e}$, while synaptic vesicles, containing amino acids or acetylcholine, can be stimulated in its absence. However, in the presence of $Ca^{2+}$, transmitter is stronger overall due to initially unperturbed vesicle recycling.

All α-LTX actions in biological systems require receptors, which provide binding sites for the toxin on the cell surface. Once the toxin is bound, part of its $Ca^{2+}$-dependent action is due to pore formation and influx of $Ca^{2+}$. This mechanism

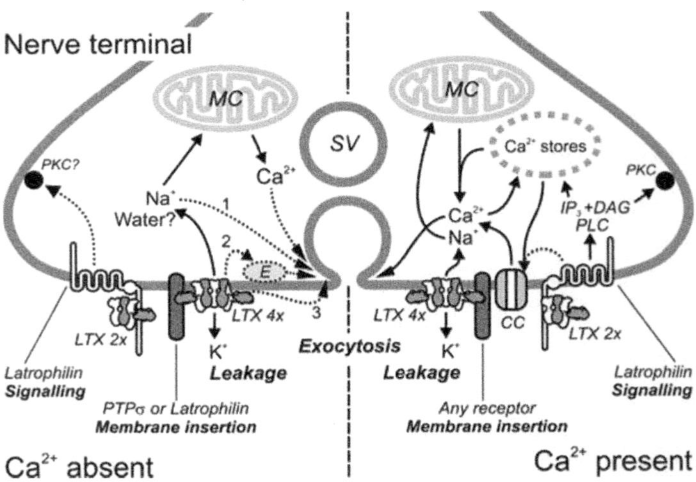


**Fig. 6** Diverse mechanisms of α-LTX action. *Right*, $Ca^{2+}$ is present in the medium. The pathways shown are described in the text. CC; $Ca^{2+}$ channels; DAG, diacyl glycerol; LTX 4x, α-LTX tetramers; MC, mitochondria. *Left*, $Ca^{2+}$-free conditions. For main comments, see text. The possible pathways for $Ca^{2+}$-independent exocytosis shown include: (1) high concentrations of $Na^+$ mimicking $Ca^{2+}$; (2) the internalised domains of α-LTX interacting with components of the exocytotic machinery (E); (3) α-LTX exerting direct fusogenic action.

triggers the release of both readily releasable and reserve pools of vesicles (Ashton et al. 2001). Another action is based on receptor-mediated signaling, which involves stimulation of PLC, production of $IP_3$ and diacyl glycerol, with respective release of stored $Ca^{2+}$ and activation of PKC. This mechanism, most likely mediated by LPH1, affects readily releasable vesicles only. Both the pore- and receptor-mediated signals can be amplified by the release of $Ca^{2+}{}_i$ and influx of $Ca^{2+}{}_e$, producing $Ca^{2+}$ waves. Mitochondria can contribute to the increase in $[Ca^{2+}]_i$. The pore-mediated component of the $Ca^{2+}$-dependent action can be blocked by $La^{3+}$, whereas the receptor-mediated action cannot (Ashton et al. 2001; Capogna et al. 2003).

In the absence of $Ca^{2+}{}_e$, α-LTX only binds to LPH1 and PTPσ. $Ca^{2+}$-independent exocytosis requires the presence of $Mg^{2+}$ and toxin insertion into the plasma membrane, but these conditions also induce formation of α-LTX channels. Influx of $Na^+$ and efflux of $K^+$ through these channels and associated efflux of small molecules and influx/efflux of water may cause secretion. In addition, transmitter release can be caused by membrane perturbation or direct interaction with secretory machinery. Some secretion may be nonvesicular. Receptor-mediated signaling can cause the activation of PKC in some cells. However, $Ca^{2+}$-independent release is blocked by $La^{3+}$, indicating that toxin pores play a crucial role in this release.

Finally, although α-LTX receptors (at least LPH1) can transduce a signal, it is obvious that this signaling cannot explain all the effects of α-LTX. So, while the use of the toxin has brought about the discovery of neuronal cell adhesion protein families (NRX and LPH) and new knowledge about PTPσ, the toxin's effect in the absence of $Ca^{2+}{}_e$ remains unclear, and its study may throw light onto the very basic mechanisms of exocytosis and different levels of its regulation.

## References


Adam-Vizi V, Deri Z, Bors P et al (1993) Lack of involvement of $[Ca^{2+}]_i$ in the external $Ca^{2+}$-independent release of acetylcholine evoked by veratridine, ouabain and α-latrotoxin: possible role of $[Na^+]_i$. J Physiol Paris 87:43–50

Aicher B, Lerch MM, Muller T et al (1997) Cellular redistribution of protein tyrosine phosphatases LAR and PTPσ by inducible proteolytic processing. J Cell Biol 138:681–96

Alete DE, Weeks ME, Hovanession AG et al (2006) Cell surface nucleolin on developing muscle is a potential ligand for the axonal receptor protein tyrosine phosphatase-σ. FEBS J 273:4668–81

Andrade MA, Perez-Iratxeta C, Ponting CP (2001) Protein repeats: structures, functions, and evolution. J Struct Biol 134:117–31

Aristotle (350 B.C.) History of Animals. In Barnes J (ed) The complete works of Aristotle: The revised Oxford translation, 1984, Princeton, Princeton University Press

Ashton AC, Rahman MA, Volynski KE et al (2000) Tetramerisation of α-latrotoxin by divalent cations is responsible for toxin-induced non-vesicular release and contributes to the $Ca^{2+}$-dependent vesicular exocytosis from synaptosomes. Biochimie 82:453–68

Ashton AC, Volynski KE, Lelianova VG et al (2001) α-Latrotoxin, acting via two $Ca^{2+}$-dependent pathways, triggers exocytosis of two pools of synaptic vesicles. J Biol Chem 276:44695–703

Auger C, Marty A (1997) Heterogeneity of functional synaptic parameters among single release sites. Neuron 19:139–50

Augustine GJ, Charlton MP, Smith SJ (1987) Calcium action in synaptic transmitter release. Annu Rev Neurosci 10:633–93

Barnett DW, Liu J, Misler S (1996) Single-cell measurements of quantal secretion induced by α-latrotoxin from rat adrenal chromaffin cells: dependence on extracellular $Ca^{2+}$. Pflugers Arch 432:1039–46

Batt J, Asa S, Fladd C et al (2002) Pituitary, pancreatic and gut neuroendocrine defects in protein tyrosine phosphatase-sigma-deficient mice. Mol Endocrinol 16:155–69

Biederer T, Südhof TC (2000) Mints as adaptors. Direct binding to neurexins and recruitment of munc18. J Biol Chem 275:39803–6

Bittner MA, Holz RW (2000) Latrotoxin stimulates secretion in permeabilized cells by regulating an intracellular $Ca^{2+}$- and ATP-dependent event: a role for protein kinase C. J Biol Chem 275:25351–7

Boehm S, Huck S (1998) Presynaptic inhibition by concanavalin A: are α-latrotoxin receptors involved in action potential-dependent transmitter release? J Neurochem 71:2421–30

Boucard AA, Chubykin AA, Comoletti D et al (2005) A splice code for trans-synaptic cell adhesion mediated by binding of neuroligin 1 to α- and β-neurexins. Neuron 48:229–36

Capogna M, Gahwiler BH, Thompson SM (1996) Calcium-independent actions of α-latrotoxin on spontaneous and evoked synaptic transmission in the hippocampus. J Neurophysiol 76:3149–58

Capogna M, Volynski KE, Emptage NJ et al (2003) The α-latrotoxin mutant $LTX^{N4C}$ enhances spontaneous and evoked transmitter release in CA3 pyramidal neurons. J Neurosci 23:4044–53

Cavalieri M, Corvaja N, Grasso A (1990) Immunocytological localization by monoclonal antibodies of α-latrotoxin in the venom gland of the spider *Latrodectus tredecimguttatus*. Toxicon 28:341–6

Ceccarelli B, Grohovaz F, Hurlbut WP (1979) Freeze-fracture studies of frog neuromuscular junctions during intense release of neurotransmitter. I. Effects of black widow spider venom and $Ca^{2+}$-free solutions on the structure of the active zone. J Cell Biol 81:163–77

Ceccarelli B, Hurlbut WP (1980) $Ca^{2+}$-dependent recycling of synaptic vesicles at the frog neuromuscular junction. J Cell Biol 87:297–303

Chanturia AN, Lishko VK (1992) Potential-dependent α-latrotoxin interaction with black lipid membranes. Toxicon 30:1059–64

Chen ML, Chen CH (2005) Microarray analysis of differentially expressed genes in rat frontal cortex under chronic risperidone treatment. Neuropsychopharmacology 30:268–77

Chih B, Gollan L, Scheiffele P (2006) Alternative splicing controls selective trans-synaptic interactions of the neuroligin-neurexin complex. Neuron 51:171–8

Comoletti D, De JA, Jennings LL et al (2004) The Arg451Cys-neuroligin-3 mutation associated with autism reveals a defect in protein processing. J Neurosci 24:4889–93

Comoletti D, Flynn RE, Boucard AA et al (2006) Gene selection, alternative splicing, and post-translational processing regulate neuroligin selectivity for β-neurexins. Biochemistry 45:12816–27

D'Amour F, Becker FE, van Riper W (1936) The black widow spider. Q Rev Biol 11:123–60

Davletov BA, Krasnoperov V, Hata Y et al (1995) High affinity binding of α-latrotoxin to recombinant neurexin Iα. J Biol Chem 270:23903–5

Davletov BA, Meunier FA, Ashton AC et al (1998) Vesicle exocytosis stimulated by α-latrotoxin is mediated by latrophilin and requires both external and stored $Ca^{2+}$. EMBO J 17:3909–20

Davletov BA, Shamotienko OG, Lelianova VG et al (1996) Isolation and biochemical characterization of a $Ca^{2+}$-independent α-latrotoxin-binding protein. J Biol Chem 271:23239–45

Dean C, Scholl FG, Choih J et al (2003) Neurexin mediates the assembly of presynaptic terminals. Nat Neurosci 6:708–16

Deri Z, Adam-Vizi V (1993) Detection of intracellular free $Na^+$ concentration of synaptosomes by a fluorescent indicator, $Na^+$-binding benzofuran isophthalate: the effect of veratridine, ouabain, and α-latrotoxin. J Neurochem 61:818–25

Deri Z, Bors P, Adam-Vizi V (1993) Effect of α-latrotoxin on acetylcholine release and intracellular $Ca^{2+}$ concentration in synaptosomes: $Na^+$-dependent and $Na^+$-independent components. J Neurochem 60:1065–72

Dresbach T, Neeb A, Meyer G et al (2004) Synaptic targeting of neuroligin is independent of neurexin and SAP90/PSD95 binding. Mol Cell Neurosci 27 227–35

Duan ZG, Yan XJ, He XZ et al (2006) Extraction and protein component analysis of venom from the dissected venom glands of *Latrodectus tredecimguttatus*. Comp Biochem Physiol B Biochem Mol Biol 145:350–7

Dudanova I, Sedej S, Ahmad M et al (2006) Important contribution of α-neurexins to $Ca^{2+}$-triggered exocytosis of secretory granules. J Neurosci 26:10599–613

Fesce R, Segal JR, Ceccarelli B et al (1986) Effects of black widow spider venom and $Ca^{2+}$ on quantal secretion at the frog neuromuscular junction. J Gen Physiol 88:59–81

Filippov AK, Tertishnikova SM, Alekseev AE et al (1994) Mechanism of α-latrotoxin action as revealed by patch-clamp experiments on *Xenopus* oocytes injected with rat brain messenger RNA. Neuroscience 61:179–89

Finkelstein A, Rubin LL, Tzeng M-C (1976) Black widow spider venom: effect of purified toxin on lipid bilayer membranes. Science 193:1009–11

Fredriksson R, Schioth HB (2005) The repertoire of G-protein-coupled receptors in fully sequenced genomes. Mol Pharmacol 67:1414–25

Frontali N, Ceccarelli B, Gorio A et al (1976) Purification from black widow spider venom of a protein factor causing the depletion of synaptic vesicles at neuromuscular junctions. J Cell Biol 68:462–79

Geppert M, Khvotchev M, Krasnoperov V et al (1998) Neurexin Iα is a major α-latrotoxin receptor that cooperates in α-latrotoxin action. J Biol Chem 273:1705–10

Grasso A, Alema S, Rufini S et al (1980) Black widow spider toxin-induced calcium fluxes and transmitter release in a neurosecretory cell line. Nature 283:774–6

Grasso A, Mercanti-Ciotti MT (1993) The secretion of amino acid transmitters from cerebellar primary cultures probed by α-latrotoxin. Neuroscience 54:595–604

Grasso A, Rufini S, Senni I (1978) Concanavalin A blocks black widow spider toxin stimulation of transmitter release from synaptosomes. FEBS Lett 85:241–4

Grasso A, Senni MI (1979) A toxin purified from the venom of black widow spider affects the uptake and release of radioactive γ-amino butyrate and N-epinephrine from rat brain synaptosomes. Eur J Biochem 102:337–44

Gray JX, Haino M, Roth MJ et al (1996) CD97 is a processed, seven-transmembrane, heterodimeric receptor associated with inflammation. J Immunol 157:5438–47

Grishin EV (1998) Black widow spider toxins: the present and the future. Toxicon 36:1693–1701

Grishin EV, Himmelreich NH, Pluzhnikov KA et al (1993) Modulation of functional activities of the neurotoxin from black widow spider venom. FEBS Lett 336:205–7

Gutman GA, Chandy KG, Grissmer S et al (2005) International Union of Pharmacology. LIII. Nomenclature and molecular relationships of voltage-gated potassium channels. Pharmacol Rev 57:473–508

Harmar AJ (2001) Family-B G-protein-coupled receptors. Genome Biol 2:3013.1–3013.10

Hata Y, Butz S, Südhof TC (1996) CASK: a novel dlg/PSD95 homolog with an N-terminal calmodulin-dependent protein kinase domain identified by interaction with neurexins. J Neurosci 16:2488–94

Hayflick JS (2000) A family of heptahelical receptors with adhesion-like domains: a marriage between two super families. J Recept Signal Transduct Res 20:119–31

Hlubek M, Tian D, Stuenkel EL (2003) Mechanism of α-latrotoxin action at nerve endings of neurohypophysis. Brain Res 992:30–42

Hlubek MD, Stuenkel EL, Krasnoperov VG et al (2000) Calcium-independent receptor for α-latrotoxin and neurexin Iα facilitate toxin-induced channel formation: evidence that channel formation results from tethering of toxin to membrane. Mol Pharmacol 57:519–28

Hurlbut WP, Ceccarelli B (1979) Use of black widow spider venom to study the release of neurotransmitters. Adv Cytopharmacol 3:87–115:87–115

Hurlbut WP, Chieregatti E, Valtorta F et al (1994) α-Latrotoxin channels in neuroblastoma cells. J Membr Biol 138:91–102

Ichtchenko K, Bittner MA, Krasnoperov V et al (1999) A novel ubiquitously expressed α-latrotoxin receptor is a member of the CIRL family of G-protein-coupled receptors. J Biol Chem 274:5491–8

Ichtchenko K, Hata Y, Nguyen T et al (1995) Neuroligin 1: a splice site-specific ligand for β-neurexins. Cell 81:435–43

Ichtchenko K, Khvotchev M, Kiyatkin N et al (1998) α-Latrotoxin action probed with recombinant toxin: receptors recruit α-latrotoxin but do not transduce an exocytotic signal. EMBO J 17:6188–99

Ichtchenko K, Nguyen T, Südhof TC (1996) Structures, alternative splicing, and neurexin binding of multiple neuroligins. J Biol Chem 271:2676–82

Kattenstroth G, Tantalaki E, Südhof TC et al (2004) Postsynaptic N-methyl-D-aspartate receptor function requires α-neurexins. Proc Natl Acad Sci U S A 101:2607–2612

Khvotchev M, Lonart G, Südhof TC (2000) Role of calcium in neurotransmitter release evoked by α-latrotoxin or hypertonic sucrose. Neuroscience 101: 793–802

Kiyatkin NI, Dulubova IE, Chekhovskaya IA et al (1990) Cloning and structure of cDNA encoding α-latrotoxin from black widow spider venom. FEBS Lett 270:127–31

Kiyatkin NI, Kulikovskaya IM, Grishin EV et al (1995) Functional characterization of black widow spider neurotoxins synthesised in insect cells. Eur J Biochem 230:854–9

Krasil'nikov OV, Sabirov RZ, Chanturiia AN et al (1988) [Conductivity and diameter of latrotoxin channels in lipid bilayers]. Ukr Biokhim Zh 60:67–71

Krasil'nikov OV, Ternovskii VI, Tashmukhamedov BA (1982) [Channel formation properties of black widow venom]. Biofizika 27:72–5

Krasil'nikov OV, Sabirov RZ (1992) Comparative analysis of latrotoxin channels of different conductance in planar lipid bilayers. Evidence for cluster organization. Biochim Biophys Acta 1112:124–8

Krasnoperov V, Bittner MA, Holz RW et al (1999) Structural requirements for α-latrotoxin binding and α-latrotoxin-stimulated secretion. A study with calcium-independent receptor of α-latrotoxin (CIRL) deletion mutants. J Biol Chem 274:3590–6

Krasnoperov V, Lu Y, Buryanovsky L et al (2002a) Post-translational proteolytic processing of the calcium-independent receptor of α-latrotoxin (CIRL), a natural chimera of the cell adhesion protein and the G protein-coupled receptor. Role of the G protein-coupled receptor proteolysis site (GPS) motif. J Biol Chem 277:46518–26

Krasnoperov VG, Beavis R, Chepurny OG et al (1996) The calcium-independent receptor of α-latrotoxin is not a neurexin. Biochem Biophys Res Commun 227:868–75

Krasnoperov VG, Bittner MA, Beavis R et al (1997) α-Latrotoxin stimulates exocytosis by the interaction with a neuronal G-protein-coupled receptor. Neuron 18:925–37

Krasnoperov VG, Bittner MA, Mo W et al (2002b) Protein tyrosine phosphatase-σ is a novel member of the functional family of α-latrotoxin receptors. J Biol Chem 277:35887–95

Kreienkamp HJ, Zitzer H, Gundelfinger ED et al (2000) The calcium-independent receptor for α-latrotoxin from human and rodent brains interacts with members of the ProSAP/SSTRIP/Shank family of multidomain proteins. J Biol Chem 275:32387–90

Lajus S, Lang J (2006) Splice variant 3, but not 2 of receptor protein-tyrosine phosphatase σ can mediate stimulation of insulin-secretion by α-latrotoxin. J Cell Biochem 98:1552–9

Lajus S, Vacher P, Huber D et al (2006) α-Latrotoxin induces exocytosis by inhibition of voltage-dependent $K^+$ channels and by stimulation of L-type $Ca^{2+}$ channels via latrophilin in β-cells. J Biol Chem 281:5522–31

Lang J, Ushkaryov Y, Grasso A et al (1998) $Ca^{2+}$-independent insulin exocytosis induced by α-latrotoxin requires latrophilin, a G protein-coupled receptor. EMBO J 17:648–57

Lelianova VG, Davletov BA, Sterling A et al (1997) α-Latrotoxin receptor, latrophilin, is a novel member of the secretin family of G protein-coupled receptors. J Biol Chem 272:21504–8

Levinson JN, Chery N, Huang K et al (2005) Neuroligins mediate excitatory and inhibitory synapse formation: involvement of PSD-95 and neurexin-1β in neuroligin-induced synaptic specificity. J Biol Chem 280:17312–19

Li G, Lee D, Wang L et al (2005) N-terminal insertion and C-terminal ankyrin-like repeats of α-latrotoxin are critical for $Ca^{2+}$-dependent exocytosis. J Neurosci 25:10188–97

Lindsay AR, Tinker A, Williams AJ (1994) How does ryanodine modify ion handling in the sheep cardiac sarcoplasmic reticulum $Ca^{2+}$-release channel? J Gen Physiol 104:425–47

Lishko VK, Terletskaya YT, Trikash IO (1990) Fusion of negatively charged phospholipid-vesicles by α-latrotoxin. FEBS Letters 266:99–101

Liu J, Misler S (1998) α-Latrotoxin alters spontaneous and depolarization-evoked quantal release from rat adrenal chromaffin cells: evidence for multiple modes of action. J Neurosci 18:6113–25

Liu J, Wan Q, Lin X et al (2005) α-Latrotoxin modulates the secretory machinery via receptor-mediated activation of protein kinase C. Traffic 6:756–65

Long SB, Campbell EB, Mackinnon R (2005) Crystal structure of a mammalian voltage-dependent *Shaker* family $K^+$ channel. Science 309:897–903

Longenecker HE, Hurlbut WP, Mauro A et al (1970) Effects of black widow spider venom on the frog neuromuscular junction. Effects on end-plate potential, miniature end-plate potential and nerve terminal spike. Nature 225:701–3

Loria PM, Hodgkin J, Hobert O (2004) A conserved postsynaptic transmembrane protein affecting neuromuscular signalling in *Caenorhabditis elegans*. J Neurosci 24 2191–2201

Lunev AV, Demin VV, Zaitsev OI et al (1991) [Electron microscopy of α-latrotoxin from the venom of the black widow spider *Latrodectus mactans tredecimguttatus*]. Bioorg Khim 17:1021–6

Matsushita H, Lelianova VG, Ushkaryov YA (1999) The latrophilin family: multiply spliced G protein-coupled receptors with differential tissue distribution. FEBS Lett 443:348–52

Matteoli M, Haimann C, Torri-Tarelli F et al (1988) Differential effect of α-latrotoxin on exocytosis from small synaptic vesicles and from large dense-core vesicles containing calcitonin gene-related peptide at the frog neuromuscular junction. Proc Natl Acad Sci U S A 85:7366–70

McLean J, Batt J, Doering LC et al (2002) Enhanced rate of nerve regeneration and directional errors after sciatic nerve injury in receptor protein tyrosine phosphatase sigma knock-out mice. J Neurosci 22:5481–91

McMahon HT, Rosenthal L, Meldolesi J et al (1990) α-Latrotoxin releases both vesicular and cytoplasmic glutamate from isolated nerve terminals. J Neurochem 55:2039–47

Meathrel K, Adamek T, Batt J et al (2002) Protein tyrosine phosphatase sigma-deficient mice show aberrant cytoarchitecture and structural abnormalities in the central nervous system. J Neurosci Res 70:24–35

Mee CJ, Tomlinson SR, Perestenko PV et al (2004) Latrophilin is required for toxicity of black widow spider venom in Caenorhabditis elegans. Biochem J 378:185–91

Meiri H, Erulkar SD, Lerman T et al (1981) The action of the sodium ionophore, monensin, on transmitter release at the frog neuromuscular junction. Brain Res 204:204–8

Meldolesi J, Huttner WB, Tsien RY et al (1984) Free cytoplasmic $Ca^{2+}$ and neurotransmitter release: studies on PC12 cells and synaptosomes exposed to α-latrotoxin. Proc Natl Acad Sci U S A 81:620–4

Meldolesi J, Madeddu L, Torda M et al (1983) The effect of α-latrotoxin on the neurosecretory PC12 cell line: studies on toxin binding and stimulation of transmitter release. Neuroscience 10:997–1009

Michaely P, Tomchick DR, Machius M et al (2002) Crystal structure of a 12 ANK repeat stack from human ankyrinR. EMBO J 21:6387–96

Michelena P, de la Fuente MT, Vega T et al (1997) Drastic facilitation by α-latrotoxin of bovine chromaffin cell exocytosis without measurable enhancement of $Ca^{2+}$ entry or $[Ca^{2+}]_i$ J Physiol (Lond) 502:481–96

Mironov SL, Sokolov Y, Chanturiya AN et al (1986) Channels produced by spider venoms in bilayer lipid membrane: mechanisms of ion transport and toxic action. Biochim Biophys Acta 862:185–98

Misler S, Falke LC (1987) Dependence on multivalent cations of quantal release of transmitter induced by black widow spider venom. Am J Physiol 253:C469–76

Misler S, Hurlbut WP (1979) Action of black widow spider venom on quantized release of acetylcholine at the frog neuromuscular junction: dependence upon external $Mg^{2+}$. Proc Natl Acad Sci U S A 76:991–5

Missler M, Fernandez-Chacon R, Südhof TC (1998a) The making of neurexins. J Neurochem 71:1339–47

Missler M, Hammer RE, Südhof TC (1998b) Neurexophilin binding to α-neurexins. A single LNS domain functions as an independently folding ligand-binding unit. J Biol Chem 273:34716–723
Missler M, Südhof TC (1998) Neurexins: three genes and 1001 products. Trends Genet 14:20–6
Missler M, Zhang W, Rohlmann A et al (2003) α-Neurexins couple $Ca^{2+}$ channels to synaptic vesicle exocytosis. Nature 423:939–48
Nguyen T, Südhof TC (1997) Binding properties of neuroligin 1 and neurexin 1β reveal function as heterophilic cell adhesion molecules. J Biol Chem 272:26032–9
Nicholls DG, Rugolo M, Scott IG et al (1982) α-Latrotoxin of black widow spider venom depolarizes the plasma membrane, induces massive calcium influx, and stimulates transmitter release in guinea pig brain synaptosomes. Proc Natl Acad Sci U S A 79:7924–8
Nishimura W, Iizuka T, Hirabayashi S et al (2000) Localization of BAI-associated protein1/membrane-associated guanylate kinase-1 at adherens junctions in normal rat kidney cells: polarized targeting mediated by the carboxyl-terminal PDZ domains. J Cell Physiol 185:358–65
Occhi G, Rampazzo A, Beffagna G et al (2002) Identification and characterization of heart-specific splicing of human neurexin 3 mRNA (NRXN3). Biochem Biophys Res Commun 298:151–5
Orlova EV, Rahman MA, Gowen B et al (2000) Structure of α-latrotoxin oligomers reveals that divalent cation-dependent tetramers form membrane pores. Nat Struct Biol 7:48–53
Pescatori M, Bradbury A, Bouet F et al (1995) The cloning of a cDNA encoding a protein (latrodectin) which co-purifies with the α-latrotoxin from the black widow spider *Latrodectus tredecimguttatus* (Theridiidae). Eur J Biochem 230:322–8
Petrenko AG, Kovalenko VA, Shamotienko OG et al (1990) Isolation and properties of the α-latrotoxin receptor. EMBO J 9:2023–7
Petrenko AG, Lazaryeva VD, Geppert M et al (1993) Polypeptide composition of the α-latrotoxin receptor. High affinity binding protein consists of a family of related high molecular weight polypeptides complexed to a low molecular weight protein. J Biol Chem 268:1860–7
Petrenko AG, Ullrich B, Missler M et al (1996) Structure and evolution of neurexophilin. J Neurosci 16:4360–9
Pettersen EF, Goddard TD, Huang CC et al (2004) UCSF Chimera – a visualization system for exploratory research and analysis. J Comput Chem 25:1605–12
Picotti GB, Bondiolotti GP, Meldolesi J (1982) Peripheral catecholamine release by α-latrotoxin in the rat. Naunyn Schmiedebergs Arch Pharmacol 320:224–9
Pulido R, Serra-Pages C, Tang M et al (1995) The LAR/PTPδ/PTPσ subfamily of transmembrane protein-tyrosine-phosphatases: multiple human LAR, PTPδ, and PTPσ isoforms are expressed in a tissue-specific manner and associate with the LAR-interacting protein LIP.1. Proc Natl Acad Sci U S A 92:11686–90
Rahman MA, Ashton AC, Meunier FA et al (1999) Norepinephrine exocytosis stimulated by α-latrotoxin requires both external and stored $Ca^{2+}$ and is mediated by latrophilin, G proteins and phospholipase C. Phil Trans R Soc Lond B 354:379–86
Robello M (1989) Dependence of the conductance of the α-latrotoxin channel on applied potential and potassium concentration. Biochim Biophys Acta 978:179–84
Robello M, Fresia M, Maga L et al (1987) Permeation of divalent cations through α-latrotoxin channels in lipid bilayers: steady-state current-voltage relationships. J Membr Biol 95:55–62
Robello M, Rolandi R, Alema S et al (1984) Trans-bilayer orientation and voltage-dependence of α-latrotoxin-induced channels. Proceedings of the Royal Society of London Series B-Biological Sciences 220:477–87
Rosenthal L, Meldolesi J (1989) α-Latrotoxin and related toxins. Pharmacol Ther 42:115–34
Rosenthal L, Zacchetti D, Madeddu L et al (1990) Mode of action of α-latrotoxin: role of divalent cations in $Ca^{2+}$-dependent and $Ca^{2+}$-independent effects mediated by the toxin. Mol Pharmacol 38:917–23
Ruhe JE, Streit S, Hart S et al (2006) EGFR signalling leads to downregulation of PTP-LAR via TACE-mediated proteolytic processing. Cell Signal 18:1515–27
Sajnani-Perez G, Chilton JK, Aricescu AR et al (2003) Isoform-specific binding of the tyrosine phosphatase PTPσ to a ligand in developing muscle. Mol Cell Neurosci 22:37–48

Scheer H, Prestipino G, Meldolesi J (1986) Reconstitution of the purified α-latrotoxin receptor in liposomes and planar lipid membranes. Clues to the mechanism of toxin action. EMBO J 5:2643–8
Scheer HW (1989) Interactions between α-latrotoxin and trivalent cations in rat striatal synaptosomal preparations. J Neurochem 52:1590–7
Scheiffele P, Fan J, Choih J et al (2000) Neuroligin expressed in nonneuronal cells triggers presynaptic development in contacting axons. Cell 101:657–69
Sedgwick SG, Smerdon SJ (1999) The ankyrin repeat: a diversity of interactions on a common structural framework. Trends Biochem Sci 24:311–16
Serra-Pages C, Saito H, Streuli M (1994) Mutational analysis of proprotein processing, subunit association, and shedding of the LAR transmembrane protein tyrosine phosphatase. J Biol Chem 269:23632–41
Serysheva II, Hamilton SL, Chiu W et al (2005) Structure of $Ca^{2+}$ release channel at 14 Å resolution. J Mol Biol 345:427–31
Silva AM, Liu-Gentry J, Dickey AS et al (2005) α-Latrotoxin increases spontaneous and depolarization-evoked exocytosis from pancreatic islet β-cells. J Physiol 565:783–99
Siu R, Fladd C, Rotin D (2006) N-cadherin is an *in vivo* substrate for PTPσ and participates in PTPσ-mediated inhibition of axon growth. Mol Cell Biol 27:208–19
Smith DS, Russell FE (1966) Structure of the venom gland of the black widow spider *Latrodectus mactans*. A preliminary light and electron microscopic study. In Russell FE, Saunders PR (eds) Animal Toxins, Oxford, Pergamon, pp 1–15
Snyder DA, Rivers AM, Yokoe H et al (1991) Olfactomedin: purification, characterization, and localization of a novel olfactory glycoprotein. Biochemistry 30:9143–53
Sokolov YV, Chanturia AN, Lishko VK (1987) Latrotoxin-induced fusion of liposomes with bilayer phospholipid-membranes. Biochimica et Biophysica Acta 900:295–9
Song JY, Ichtchenko K, Südhof TC et al (1999) Neuroligin 1 is a postsynaptic cell-adhesion molecule of excitatory synapses. Proc Natl Acad Sci U S A 96:1100–5
Stacey M, Lin HH, Gordon S et al (2000) LNB-TM7, a group of seven-transmembrane proteins related to family-B G-protein-coupled receptors. Trends Biochem Sci 25:284–9
Storchak LG, Pashkov VN, Pozdnyakova NG et al (1994) α-Latrotoxin-stimulated GABA release can occur in $Ca^{2+}$-free, $Na^{+}$-free medium. FEBS Lett 351:267–70
Südhof TC (2001) α-Latrotoxin and its receptors: neurexins and CIRL/latrophilins. Annu Rev Neurosci 24:933–62
Sugita S, Ichtchenko K, Khvotchev M et al (1998) α-Latrotoxin receptor CIRL/latrophilin 1 (CL1) defines an unusual family of ubiquitous G-protein-linked receptors. G-protein coupling not required for triggering exocytosis. J Biol Chem 273:32715–24
Sugita S, Khvochtev M, Südhof TC (1999) Neurexins are functional α-latrotoxin receptors. Neuron 22:489–96
Thompson KM, Uetani N, Manitt C et al (2003) Receptor protein tyrosine phosphatase sigma inhibits axonal regeneration and the rate of axon extension. Mol Cell Neurosci 23:681–92
Tobaben S, Südhof TC, Stahl B (2002) Genetic analysis of α-latrotoxin receptors reveals functional interdependence of CIRL/latrophilin 1 and neurexin Iα. J Biol Chem 277:6359–65
Tobaben S, Südhof TC, Stahl B (2000) The G protein-coupled receptor CL1 interacts directly with proteins of the Shank family. J Biol Chem 275:36204–10
Tonks NK (2006) Protein tyrosine phosphatases: from genes, to function, to disease. Nat Rev Mol Cell Biol 7:833–46
Tsang CW, Elrick DB, Charlton MP (2000) α-Latrotoxin releases calcium in frog motor nerve terminals. J Neurosci 20 8685–92
Tse FW, Tse A (1999) α-Latrotoxin stimulates inward current, rise in cytosolic calcium concentration, and exocytosis in pituitary gonadotropes. Endocrinology 140:3025–33
Tzeng MC, Cohen RS, Siekevitz P (1978) Release of neurotransmitters and depletion of synaptic vesicles in cerebral cortex slices by α-latrotoxin from black widow spider venom. Proc Natl Acad Sci U S A 75:4016–20

Tzeng MC, Siekevitz P (1979) The binding interaction between α-latrotoxin from black widow spider venom and a dog cerebral cortex synaptosomal membrane preparation. J Neurochem 33:263–74

Ullrich B, Ushkaryov YA, Südhof TC (1995) Cartography of neurexins: more than 1000 isoforms generated by alternative splicing and expressed in distinct subsets of neurons. Neuron 14:497–507

Umbach JA, Grasso A, Zurcher SD et al (1998) Electrical and optical monitoring of α-latrotoxin action at *Drosophila* neuromuscular junctions. Neuroscience 87:913–24

Ushkaryov YA, Hata Y, Ichtchenko K et al (1994) Conserved domain structure of β-neurexins. Unusual cleaved signal sequences in receptor-like neuronal cell-surface proteins. J Biol Chem 269:11987–92

Ushkaryov YA, Petrenko AG, Geppert M et al (1992) Neurexins: synaptic cell surface proteins related to the α-latrotoxin receptor and laminin. Science 257:50–6

Ushkaryov YA, Südhof TC (1993) Neurexin IIIα: extensive alternative splicing generates membrane-bound and soluble forms. Proc Natl Acad Sci U S A 90:6410–14

Ushkaryov YA, Volynski KE, Ashton AC (2004) The multiple actions of black widow spider toxins and their selective use in neurosecretion studies. Toxicon 43:527–42

Van Renterghem C, Iborra C, Martin-Moutot N et al (2000) α-Latrotoxin forms calcium-permeable membrane pores via interactions with latrophilin or neurexin. Eur J Neurosci 12:3953–62

Varoqueaux F, Aramuni G, Rawson RL et al (2006) Neuroligins determine synapse maturation and function. Neuron 51:741–54

Varoqueaux F, Jamain S, Brose N (2004) Neuroligin 2 is exclusively localized to inhibitory synapses. Eur J Cell Biol 83:449–56

Verhage M, Maia AS, Plomp JJ et al (2000) Synaptic assembly of the brain in the absence of neurotransmitter secretion. Science 287 864–9

Vicentini LM, Meldolesi J (1984) α Latrotoxin of black widow spider venom binds to a specific receptor coupled to phosphoinositide breakdown in PC12 cells. Biochem Biophys Res Commun 121:538–44

Volkova TM, Pluzhnikov KA, Woll PG et al (1995) Low-molecular-weight components from black-widow spider venom. Toxicon 33:483–9

Volynski KE, Capogna M, Ashton AC et al (2003) Mutant α-latrotoxin ($LTX^{N4C}$) does not form pores and causes secretion by receptor stimulation. This action does not require neurexins. J Biol Chem 278:31058–66

Volynski KE, Nosyreva ED, Ushkaryov YA et al (1999) Functional expression of α-latrotoxin in baculovirus system. FEBS Lett 442:25–8

Volynski KE, Silva JP, Lelianova VG et al (2004) Latrophilin fragments behave as independent proteins that associate and signal on binding of $LTX^{N4C}$. EMBO J 23:4423–33

Volynski KV, Meunier FA, Lelianova VG et al (2000) Latrophilin, neurexin and their signalling-deficient mutants facilitate α-latrotoxin insertion into membranes but are not involved in pore formation. J Biol Chem 275:41175–83

Wanke E, Ferroni A, Gattanini P et al (1986) α-Latrotoxin of the black widow spider venom opens a small, non-closing cation channel. Biochem Biophys Res Commun 134:320–5

Watanabe O, Meldolesi J (1983) The effects of α-latrotoxin of black widow spider venom on synaptosome ultrastructure. A morphometric analysis correlating its effects on transmitter release. J Neurocytol 12:517–31

Willson J, Amliwala K, Davis A et al (2004) Latrotoxin receptor signalling engages the UNC-13-dependent vesicle-priming pathway in *C. elegans*. Curr Biol 14:1374–9

Yan H, Grossman A, Wang H et al (1993) A novel receptor tyrosine phosphatase-σ that is highly expressed in the nervous system. J Biol Chem 268:24880–6

Zhang W, Rohlmann A, Sargsyan V et al (2005) Extracellular domains of α-neurexins participate in regulating synaptic transmission by selectively affecting N- and P/Q-type $Ca^{2+}$ channels. J Neurosci 25:4330–42

# Presynaptic Signaling by Heterotrimeric G-Proteins

**David A. Brown(✉) and Talvinder S. Sihra**



**Abstract** G-proteins (guanine nucleotide-binding proteins) are membrane-attached proteins composed of three subunits, α, β, and γ. They transduce signals from G-protein coupled receptors (GPCRs) to target effector proteins. The agonist-activated receptor induces a conformational change in the G-protein trimer so that the α-subunit binds GTP in exchange for GDP and α-GTP, and βγ-subunits separate to interact with the target effector. Effector-interaction is terminated by the α-subunit GTPase activity, whereby bound GTP is hydrolyzed to GDP. This is accelerated in situ by RGS proteins, acting as GTPase-activating proteins (GAPs). Gα-GDP and Gβγ then reassociate to form the Gαβγ trimer. G-proteins primarily involved in the modulation of neurotransmitter release are $G_o$, $G_q$ and $G_s$. $G_o$ mediates the widespread presynaptic auto-inhibitory effect of many neurotransmitters (e.g., via M2/M4 muscarinic receptors, $\alpha_2$ adrenoreceptors, $\mu/\delta$ opioid receptors, $GABA_B$ receptors). The $G_o$ βγ-subunit acts in two ways: first, and most ubiquitously, by direct binding to $Ca_V2$ $Ca^{2+}$ channels, resulting in a reduced

David A. Brown
Department of Pharmacology, University College London, Gower Street London, WC1E 6BT, UK
d.a.brown@ucl.ac.uk

T.C. Südhof, K. Starke (eds.), *Pharmacology of Neurotransmitter Release. Handbook of Experimental Pharmacology 184.*

sensitivity to membrane depolarization and reduced $Ca^{2+}$ influx during the terminal action potential; and second, through a direct inhibitory effect on the transmitter release machinery, by binding to proteins of the SNARE complex. $G_s$ and $G_q$ are mainly responsible for receptor-mediated facilitatory effects, through activation of target enzymes (adenylate cyclase, AC and phospholipase-C, PLC respectively) by the GTP-bound α-subunits. AC generates cyclic AMP which activates protein kinase A (PKA) and PLC hydrolyzes membrane phosphatidylinositol-4,5-bisphosphate ($PIP_2$) to form diacylglycerol (DAG) to activate protein kinase C (PKC). PKC phosphorylates $Ca^{2+}$ channel proteins to oppose $G_o\beta\gamma$-mediated inhibition, and both PKA and PKC phosphorylate various components of the release machinery to enhance exocytosis. cAMP and DAG can themselves facilitate release by direct, phosphorylation-independent, regulation through second messenger binding proteins, including cAMP-GEFs and Munc-13 (DAG-binding). Finally, membrane levels of $PIP_2$ play a signaling role throughout the stimulus-secretion cascade.

## 1 Overview of G Protein Signaling

Heterotrimeric G-proteins are guanine nucleotide-binding membrane-associated proteins that directly intermediate between the G-protein-coupled (heptahelical) receptor and the target effector protein. They are composed of α, β and γ subunits. The trimer is anchored in the membrane via palmitoyl or myristoyl fatty acids at the N-terminus of the α subunit and a prenyl moiety at the C-terminus of the γ-subunit (see Gilman 1987; Oldham and Hamm 2006 for details).

The basic cycle of G-protein activation and inactivation is illustrated in Figure 1. In the basal state (1) the α-subunit of the trimer binds guanosine diphosphate (GDP). On activation by agonist (Ag) the receptor "docks on" to the G-protein (2) and induces a conformational change in the latter that exposes a high-affinity guanosine triphosphate (GTP) binding site on the α-subunit. Since the concentration of GTP in the cytosol exceeds that of GDP, this promotes a GTP:GDP nucleotide exchange (the rate-limiting process for activation). This conformational change also induces a partial (see below) or complete dissociation of the α and βγ subunits, though both usually stay anchored in the membrane (except the α-subunit of rod transducin) (Calvert et al. 2006; Rosenzweig et al. 2007). The α and βγ subunits then separately (or sometimes conjointly) interact with one or more effector proteins, to activate or inhibit them (3). The cycle is completed by the hydrolysis of α-attached GTP to GDP by the GTPase activity of the α-subunit; the GDP-bound α-subunit then reassociates with the βγ-subunit to restore the ground state. Clear dissociation of α and βγ subunits occurs in the visual transduction pathway, and has also been shown for the $G_o$ trimer (Tesmer et al. 2005); however some evidence (Bunemann et al. 2003; Gales et al. 2006; Levitzki and Klein 2002; Rosenzweig et al. 2007) suggests that the α and βγ subunits do not necessarily dissociate in other systems but stay

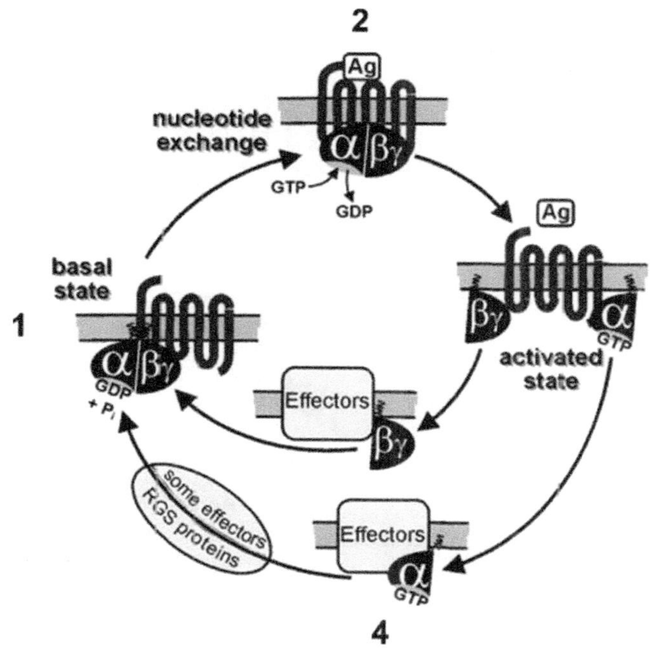


**Fig. 1** Diagram of G-protein cycle following receptor activation. See text for description. Ag = agonist. (Adapted from Wettschureck & Offermanns 2005, Physiol Rev 85:1159–1204, with permission of the American Physiological Society).

conjoined and partially separate to open up a pathway for rapid GDP exit from the trimer (Figure 2). Dissociation or otherwise may be receptor-dependent (Digby et al. 2006).

Hydrolysis of GTP is frequently accelerated by GTPase-activating proteins (GAPs). There are two classes of GAPs – effector proteins (e.g., phospholipase Cβ1 (PLCβ1) for $G_q$; phosphodiesterase-γ (PDEγ) for $G_t$ and adenylate cyclase V (ACV) for $G_s$); and, more ubiquitously, RGS proteins (Regulators of G-protein Signaling; see further below). Several other ancillary proteins can also modify G-protein activity, either by accelerating GDP:GTP exchange (guanosine nucleotide exchange factors, GEFs) or by inhibiting GDP release (GDI proteins) (Sato et al. 2006).

## 1.1 Types of G-Protein

Ignoring splice-variants, 16 species of α-subunit have been described, plus 5 species of β-subunit and 12 species of γ-subunit (see Table 2 in Wettschureck and Offermanns 2005). Trimeric G-proteins are characterized by their α-subunit, primarily in

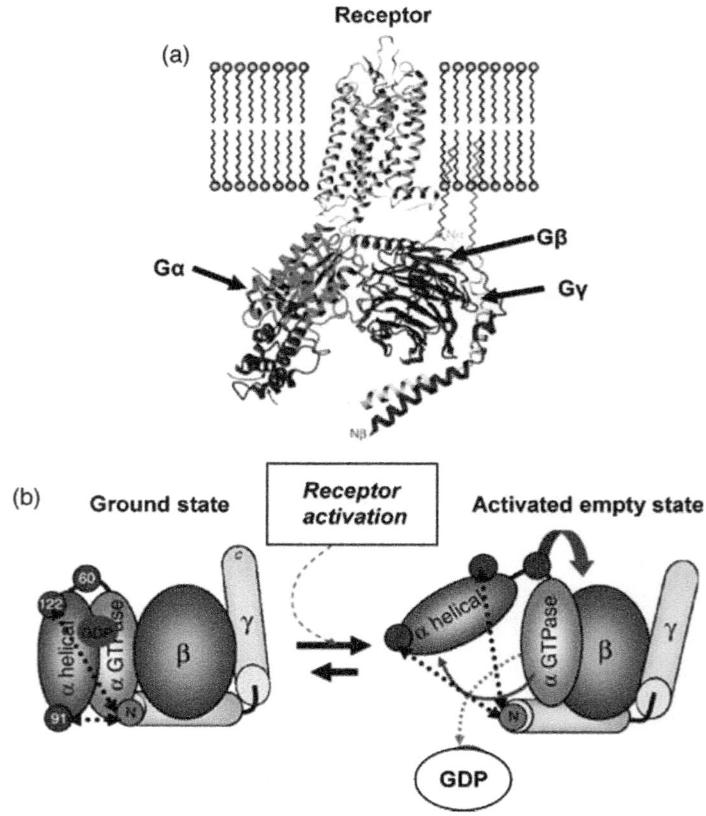


**Fig. 2** (**a**) Projection diagram of the complex between a receptor (rhodopsin) and a G-protein trimer ($G_i$). (**b**) Schematic of the structural rearrangement of the G-protein following receptor activation. Opening of the Gα GTPase and helical domains creates an exit route for GDP. (Adapted from Gales et al. 2006, Nat Struct Mol Biol 13:778–786, with permission from Macmillan Publishers, Ltd.).

terms of their principal effectors and sensitivity to *Pertussis toxin* (Ptx, which ADP-ribosylates a C-terminal cysteine present in certain α-subunits and prevents receptor coupling):

(1) $G_s/G_{olf}$-activate adenylate cyclase; not inhibited by Ptx;
(2) $G_i/G_o$ – inhibit adenylate cyclase and/or modulate certain $Ca^{2+}$ and $K^+$ channels (via βγ-subunits); inhibited by Ptx;
(3) $G_q/G_{11}$, $G_{14}$, $G_{15/16}$-activate PLCβ (β1,3,4); not inhibited by Ptx;
(4) $G_{12/13}$–multiple downstream targets; not inhibited by Ptx.

*Those most relevant to the regulation of transmitter release are $G_s$, $G_i$, $G_o$ and $G_q/G_1 1$ (and potentially $G_{12}/G_{13}$); these are widely distributed in the nervous system and are present in nerve terminals. Intraterminal localization of G-protein an-*

*tibodies has been studied in large calyciform nerve endings in chick and mammal. Isolated calyx terminals stain strongly for $G_o/G_i$ (e.g., Mirotznik et al. 2000); $G_o$ co-localizes therein with $Ca_V2.2$ $Ca^{2+}$ channels (Li et al. 2004). In the mammalian calyx of Held, $G_o$ is most abundant and co-localizes with synapsin, whereas $G_i1$-3 are predominantly postsynaptic (Kajikawa et al. 2001).*

## 1.2 Receptor Coupling and Selectivity

Upon receptor activation, the GPCR undergoes a conformational change that opens up a pocket on its inner face; the C-terminus of the G-protein α-subunit then binds within this pocket (Bourne 1997; Oldham and Hamm 2006). Thus, the Gα-subunit C-terminus forms the primary determinant of receptor–G-protein selectivity. For example, substitution of the terminal five amino acids from $G\alpha_i$ or $G\alpha_o$ into $G\alpha_q$ enables the $\alpha_2$-adrenoceptor (which normally couples to $G\alpha_i$ or $G\alpha_o$) to activate $G_q$ and stimulate PLC (Conklin et al. 1993). However, the Gβγ-subunits also contribute to binding, and can modify selectivity. For example, in GH3 pituitary tumor cells, activation of M4 muscarinic receptors preferentially inhibits the $Ca^{2+}$ current via $G\alpha_o\beta(3)\gamma(4)$, whereas somatostatin uses $G\alpha_o\beta(1)\gamma(3)$ for the same effector (Schneider et al. 1997). Although receptor–G-protein coupling is normally interpreted along diffusion-limited collision-coupling lines (e.g., Lamb and Pugh 1992), there is some old (Rimon et al. 1978) and more recent (Gales et al. 2006; Nobles et al. 2005) evidence suggesting some degree of precoupling between inactive receptors and G-proteins (but see also Hein et al. 2005; Lober et al. 2006). Certainly, some form of restricted access is required to explain why, for example, M2 and M4 muscarinic acetylcholine receptors, which readily couple to both $G_i$ and $G_o$ G-proteins in reconstituted systems, selectively activate only $G_i$ or $G_o$, respectively, in sympathetic neurons (Fernandez-Fernandez et al. 1999; Fernandez-Fernandez et al. 2001), since the density of both G-proteins in these neurons is likely to exceed that of the receptors by an order of magnitude.

## 1.3 G-Protein–Effector Coupling

As with receptor coupling, specificity of effector protein coupling is primarily determined by the α-subunit. Indeed, where the α-subunit itself acts as the messenger, effector selection is highly specific, and the three-dimensional structures for α-subunit–effector complexes have been solved for $G\alpha_s$–adenylate cyclase, $G\alpha_t$–phosphodiesterase-γ (PDEγ)–RGS9, and $G\alpha_q$–receptor kinase 2 (GRK2)–Gβγ (see Oldham and Hamm 2006). In contrast, there is relatively little selectivity among different β or γ subunits when tested as free β and/or γ combinations on effectors that respond to Gβγ such as $K_{ir}3.1/3.4$ (GIRK) channels (Wickman and Clapham

1995) and $Ca_V2.2$ $Ca^{2+}$ channels (Garcia et al. 1998; Jeong and Ikeda, 2000a; Ruiz-Velasco and Ikeda 2000; Zhou et al. 2000); such selectivity as occurs appears to depend on the β-subunit (Garcia et al. 1998; Ruiz-Velasco and Ikeda 2000). Notwithstanding, Gβγ-responsive effectors do not normally respond to βγ-subunits released indiscriminately from different G-protein trimers but only from receptor-specific αβγ-trimers. Thus, in sympathetic neurons, while $K_{ir}3.1/3.2$ channels can be activated by βγ-subunits associated with either $G\alpha_i$ or $G\alpha_o$ when these two α-subunits are overexpressed, under physiological conditions they can only be activated via $G\alpha_i$, whereas $Ca_V2.2$ channels are only inhibited by βγ-subunits associated with $G\alpha_o$ (Fernandez-Fernandez et al. 2001). Indeed, $K_{ir}3$ channels seem to show a unique response to different receptor–G-protein combinations (Benians et al. 2005). Further, the α-subunit of $G_i$ can directly interact with the channel to modify its response to Gβγ (Peleg et al. 2002). These effects may be best explained in terms of a receptor–G-protein–effector ternary complex (Clancy et al. 2005; Fowler et al. 2007; Huang et al. 1995) with restricted dissociation of the Gαβy trimer (see above). There may be similar G-protein–effector complexes involving $Ca^{2+}$ channels since these closely co-localize with $G_o$ in chick calyx terminals (Li et al. 2004) and also for $G_q$ – G-protein receptor kinase2 (GRK2) (Tesmer et al. 2005).

## *1.4 Rates of G-Protein–Mediated Events*

G-protein-mediated responses can be quite rapid. The minimal latency for physiological Gβγ activation of $K_{ir}3$ channels by a GPCR (e.g., activation of GIRK channels by M2 muscarinic receptors in cardiac sinoatrial node cells) is around 30–50 msec, with an onset time-constant for channel opening of ~250 msec and an offset time-constant of ~800 msec (Trautwein et al. 1980). Similar values have been obtained for the GPCR-activation of $K_{ir}3$ channels in neurons ($\alpha_2$-adrenoceptors on enteric neurons: latency 33 msec after afferent adrenergic nerve stimulation; rise time 85 msec; decay time-constant 290 msec [Surprenant and North 1988]; $GABA_B$ receptors on hippocampal neurons: latency 29 msec; rise-time ~110 msec; duration ~723 msec [Davies et al. 1990]). Corresponding values for Gβγ-mediated $\alpha_2$-adrenergic inhibition of $Ca_V2.2$ channels in sympathetic neurons are ~10 msec latency, 390 msec onset time-constant and ~5.6 sec offset time-constant (Zhou et al. 2000). Agonist-receptor binding appears not to be rate-limiting; instead, the rate-limiting step is the activation of the G-protein since the time-constant for free Gβγ–calcium channel interaction (measured by the time for reinhibition by the Gβγ-subunits following their dissociation from the channel by a large voltage step) was <100 msec (Zhou et al. 2000). Other measures of reinhibition kinetics following adrenergic inhibition in these neurons have yielded values of 103 msec (Ehrlich and Elmslie 1995) and 37 msec (Delmas et al. 1998b). The $GABA_B$-agonist baclofen inhibits calcium currents in cultured hippocampal neurons with comparable kinetics (delay ~80 msec; onset time-constant 220 msec; offset time-constant 2.2 sec (Pfrieger et al. 1994). This was virtually identical to the kinetics for baclofen-induced inhibition of excitatory synaptic transmission in this preparation (delay ~80 msec; onset time-constant 170 msec; offset time-constant 3.4 sec (Pfrieger et al. 1994). As pointed out by Zhou et al. (2000), such rapid-onset kinetics require that the receptors, G-proteins, and ion channel effectors be within <1 μm of each other. Indeed, there is a close co-localization of $GABA_B$ receptors and $K_{ir}3.2$ channels in hippocampal neurons (Kulik et al. 2006), and FRET analysis indicates an association between expressed $GABA_B$ receptors, $K_{ir}$ channels and RGS4 in HEK293 cells (Fowler et al. 2007). Also, as noted above, $Ca^{2+}$ channels and $G_o$ show a close co-localization at nerve terminals (Li et al. 2004), which may facilitate their rapid inhibition.

The offset rate is determined, in the first instance, by the GTPase activity of the α-subunit. In situ, this is accelerated by GTPase-activating proteins (GAPs) – sometimes by the effector itself, but more widely by RGS proteins (see further below). Thus, when $K_{ir}3.1 + 3.4$ channels are reconstituted in CHO cells with M2-muscarinic receptors, the offset of the current following removal of agonist (acetylcholine) is >10 times slower that that of the natural cardiac current, but is accelerated to match the native current kinetics on co-expressing RGS4 (Doupnik et al. 1997); similar effects on dopamine D2-receptor activation of $K_{ir}3.1 + 3.2$ channels in oocytes were observed following co-expression of RGS8 (Saitoh et al. 1997). In fact, as noted above, RGS4 appears to participate in a preassembled signaling complex with $K_{ir}3$ channels (see also Fujita et al. 2000; Tinker 2006). RGS proteins also accelerate recovery of $Ca^{2+}$ channels from G-protein-mediated inhibition (Ding et al. 2006; Jeong and Ikeda, 2000b; Mark et al. 2000; Melliti et al. 2001) and – importantly – accelerate recovery from the adenosine-mediated presynaptic inhibition at hippocampal synapses that results from $Ca^{2+}$ current inhibition (Chen and Lambert 2000) (see further below). As with $K_{ir}$ channels, individual RGS proteins may form signaling complexes with $Ca^{2+}$ channels (Richman et al. 2005). However, some RGS proteins are principally cytoplasmic or nuclear and only translocate to the membrane on activating the cognate G-protein, where they bind to the activated α-subunit (Druey et al. 1998; Heximer et al. 2001; Masuho et al. 2004; Saitoh et al. 2001).

## *1.5 Further Notes on RGS Proteins*

RGS proteins (see Abramow-Newerly et al. 2006; De Vries et al. 2000; Druey 2001; Neitzel and Hepler 2006; Tinker 2006; Willars 2006 for reviews) comprise a wide and rather diverse family, with 30+ members, but with a common 120-residue RGS domain that enables binding to the G-protein α-subunit. The effect of this is to lower the energy requirement for GTP hydrolysis and hence to accelerate the GTPase activity of the α-subunit. This is reflected in the faster offset of Gα or Gβγ-mediated effector responses noted above. Because the effective dissociation equilibrium constant for G-protein–effector interaction ($K_{diss}$) is determined by $k_{off}/k_{on} \times G^*$ (where $G^*$ is the concentration of activated G-protein), RGS-accelerated GTPase activity reduces effector response to agonist, effectively "antagonizing" agonist-action. For example, expression of an RGS-insensitive $G_o\alpha$ mutant shifts the concentration-response curve for the inhibition of the $Ca^{2+}$ current in sympathetic neurons to noradrenaline by ~tenfold to the left (Jeong and Ikeda, 2000b), implying that the endogenous RGS proteins in these cells reduce noradrenaline sensitivity by an equivalent amount. More importantly, perhaps, RGS proteins also protect the release process from *tonic* G-protein-mediated inhibition that can arise as a result of the spontaneous turnover of the $G_o$ trimer in the absence of GPCR activation (Han et al. 2006). However, superimposed on this uniform GAP activity, RGS proteins show a number of other properties.

### 1.5.1 Note 1

RGS proteins also accelerate the onset kinetics of $G_i$-mediated $K_{ir}3$ activation (e.g., Doupnik et al. 1997; Saitoh et al. 1997) and $G_o$-mediated $Ca^{2+}$ channel inhibition (e.g., Jeong and Ikeda 2000b). This is not predicted from their GAP activity; and indeed, in RGS8, accelerated $K_{ir}3$ activation is driven by a different (amino-terminal) domain from the RGS GAP-domain that determines deactivation (Jeong and Ikeda 2001). This may result from the additional binding of the RGS proteins with the effector ($K_{ir}3$ [Fujita et al. 2000] or $Ca_V2.2$ [Richman et al. 2005]), effectively reducing the diffusion time required for $G\beta\gamma$ to interact with the channel after GTP/GDP exchange on the $\alpha$-subunit.

### 1.5.2 Note 2

The binding of RGS proteins to G-protein $\alpha$-subunits can hinder binding of effectors to the activated $\alpha$-subunit, and hence antagonize $G\alpha$ (GTP)–effector coupling independently of any GAP activity (e.g., Anger et al. 2004; Hepler et al. 1997; Yan et al. 1997). By this mechanism RGS2 acts as a somewhat specific antagonist to $G\alpha_q$ (GTP)-activated phospholipase C (Hao et al. 2006; Heximer et al. 1997, 1999) – a useful property in view of the absence of Gq-inactivating toxins. Thus, RGS2 expression prevents the inhibition of the M-type (Kv7.2/7.3) $K^+$ current and the "slow" Pertussis (Ptx) toxin-insensitive inhibition of the N-type ($Ca_V2.2$)$Ca^{2+}$ current produced by expressed metabotropic mGluR1 glutamate receptors in sympathetic neurons (Kammermeier and Ikeda 1999) and the inhibition of expressed $Ca_V2.2$ calcium channels by M1-muscarinic acetylcholine receptors (Melliti et al. 2001). It also reduces the inhibitory effect of P2Y1 nucleotide receptors on expressed $K_{ir}3.1/3.2$ currents in sympathetic neurons (Filippov et al. 2004).

### 1.5.3 Note 3

RGS proteins can also modify GPCR–G-protein coupling in an agonist-selective manner. For example, in pancreatic acinar cells, RGS4 inhibits muscarinic signaling more strongly than that of bombesin or cholecystokinin (Xu et al. 1999b). Of more relevance to presynaptic regulation, RGS9-2 reduces dopamine-D2 inhibition of $Ca^{2+}$ currents without effect on M2-muscarinic inhibition (Cabrera-Vera et al. 2004). Like effector antagonism, GPCR regulation appears to be determined by the N-terminal region (Saitoh et al. 2002), so is dissociated from GAP activity or Gα-binding. This confers an ability to bind directly to the third inner loop of selective GPCRs, e.g., M1-muscarinic receptors (Bernstein et al. 2004), and $\alpha_{1B}$-adrenoceptors (Wang et al. 2005). However, binding may also be facilitated by other scaffolding proteins (see Neitzel and Hepler 2006; Willars 2006), and it has been suggested that the selectivity of RGS–receptor–G-protein modulation may depend on the formation of a ternary complex with the RGS protein, GPCR, and Gα subunit (Benians et al. 2005).

#### 1.5.4 Note 4

Importantly, RGS protein expression is subject to variable physiological and pathological regulation (Willars 2006). Thus, the modulatory effects of RGS proteins on G-protein kinetics, and on GPCR–G-protein and G-protein–effector coupling discussed above may be subject to substantial variation. One example: upregulation of RGS4 in the striatum following dopamine depletion compromises M4-muscarinic autoregulation of acetylcholine release (Ding et al. 2006), which may be relevant to Parkinson's disease.

## 2 G-Proteins and Transmitter Release

Here we consider the role of G-proteins in regulating transmitter release. We divide it into two stages – effects mediated through changes in $Ca^{2+}$ entry, and effects on the release process downstream from $Ca^{2+}$ entry (see Figure 3)

### *2.1 Effects on* $Ca^{2+}$ *Entry*

This divides into two parts – direct effects on presynaptic $Ca^{2+}$ currents, and indirect effects mediated through changes in presynaptic nerve ending properties.

#### 2.1.1 Direct Effects on $Ca^{2+}$ Channels

The $Ca^{2+}$ charge required for transmitter release results principally from $Ca^{2+}$ influx through N-, P and/or Q-type $Ca^{2+}$ channels ($Ca_V2.2$, $Ca_V2.1$, and $Ca_V2.3$ channels, respectively). The modulation of these channels by G-proteins has been

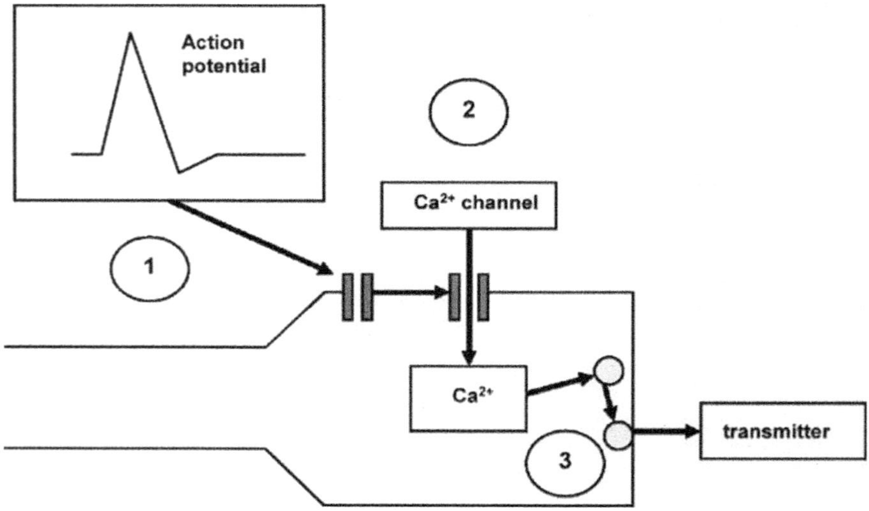


**Fig. 3** Some sites for modulation of transmitter release at nerve endings. (1) Invading action potential. (2) $Ca^{2+}$ channel / $Ca^{2+}$ influx. (3) Release mechanism.

studied intensively in both reconstituted systems and in some primary neuron somata, most notably sympathetic and sensory neurons (see DeWaard et al. 2005; Dolphin 2003; Strock and Diverse-Pierluissi 2004; Tedford and Zamponi 2006 and chapter by Zamponi, this volume). Effects on these different channel types are qualitatively similar, though differing quantitatively. Effects of GPCRs or G-protein activation on $Ca^{2+}$ channels in nerve endings have been studied directly in the large calyciform endings in the chick ciliary ganglion (Mirotznik et al. 2000; Yawo and Chuhma 1993) and in the calyx of Held in the mammalian trapezoid body (Kajikawa et al. 2001; Takahashi et al. 1996, 1998), and inferred indirectly from biophysical release parameters or biochemical studies (see below) in other preparations.

## G-Protein-Mediated $Ca^{2+}$ Channel Inhibition

Two principal forms of inhibition have been described: (1) "fast" voltage-dependent inhibition, and (2) 'slow" voltage-independent inhibition; the former is sensitive to Ptx, the latter (normally) not (Hille 1994) (see Figure 4). The former is overwhelmingly the most important and widespread, and most pertinent to the feedback auto-inhibition of transmitter release by neurotransmitters.

### *Fast voltage-dependent inhibition*

Voltage-dependent inhibition of neuronal $Ca^{2+}$ currents was first reported by Marchetti et al. (1986) and Tsunoo et al. (1986). It results from the binding of

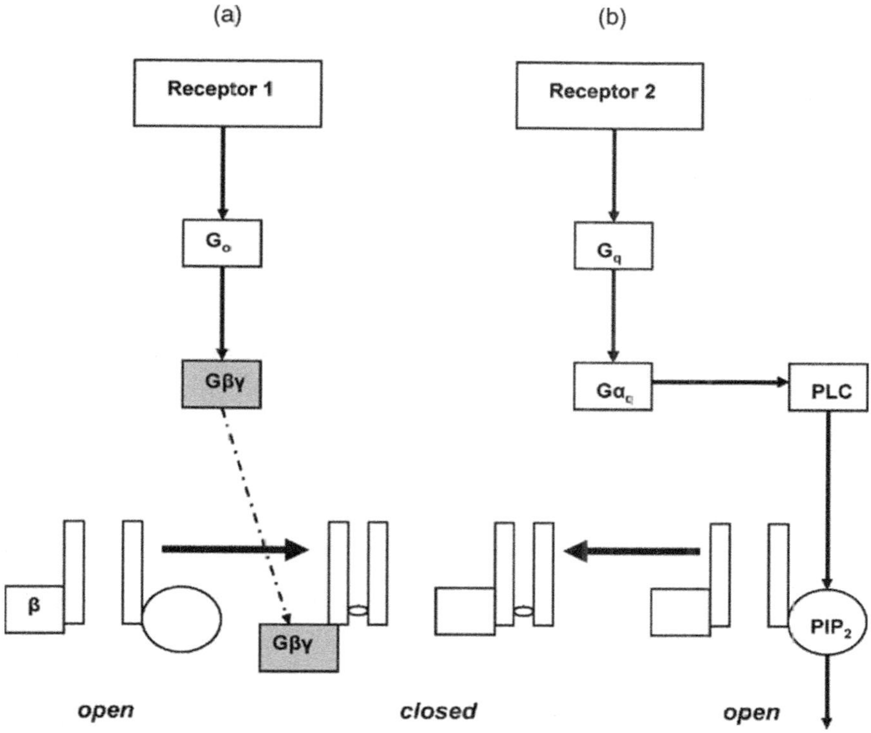


**Fig. 4** Two mechanisms for G-protein inhibition of $Ca^{2+}$ channels. (**a**) Gβγ subunits from activated $G_o\alpha\beta\gamma$ bind to channel and displace channel β-subunit, thereby reducing channel sensitivity to membrane depolarization. (**b**) Activated $G\alpha_q$ activates phospholipase C (PLC) so hydrolyzing membrane phosphatidylinositol-4,5-bisphosphate ($PIP_2$), normally required to stabilize the channel open state.

Gβy-subunits to the $\alpha_1$-subunit of the $Ca_V$ channel (see Tedford and Zamponi 2006). Inhibition is fast (latency <100 msec) because of close apposition of G-protein and $Ca^{2+}$-channel (see above). Voltage-dependence arises because depolarization favors the dissociation of the βγ-subunits from the channel, whereas hyperpolarization favors their association (they are "reluctant" to open at hyperpolarized potentials but become more "willing" at depolarized potentials) (Bean 1989; Kasai and Aosaki 1989; Patil et al. 1996). This voltage-dependence has the important physiological consequence that, during continuous GPCR-induced inhibition, the inhibited $Ca^{2+}$-current shows a frequency-dependent recovery (Park and Dunlap 1998), with a consequent partial recovery of synaptic transmission (Brody and Yue 2000). Voltage-dependence is obtunded in a splice-variant of $Ca_V2.2$channels expressing a tyrosine-phosphorylatable form of exon 37 that is expressed in a subpopulation of sensory neurons, so that their response to GPCR-inhibition is more sustained (Raingo et al. 2007).

There is relatively limited selectivity among different βγ-subunits for $Ca_V2.2$ interaction: under different circumstances, βγ-subunits associated with $G_o$, $G_i$, $G_s$, $G_q$ and $G_z$ are capable of inducing voltage-dependent inhibition (see Tedford and Zamponi 2006), but under normal conditions, GPCR-induced inhibition is primarily mediated by βγ-subunits associated with $G_o$ (Delmas et al. 1999; Fernandez-Fernandez et al. 2001). (βγ-subunits associated with $G\alpha_i$ can also contribute to fast noradrenergic inhibition of $Ca^{2+}$ current in both chick embryonic (Diverse-Pierluissi et al. 1995) and rat (Delmas et al. 1999) sympathetic neurons but this component is not voltage-sensitive so is mediated by a different mechanism to that produced by $G_o$-derived βγ-subunits.)

*Slow voltage-independent inhibition*

Receptors that couple to $G_q$, such as M1-muscarinic receptors, initiate a slower, voltage-independent inhibition of Cav2.2 currents (Beech et al. 1992; Hille 1994). This is mediated by the α-subunit of $G_q$, not the βγ-subunit (Delmas et al. 1998a; Haley et al. 2000). It results primarily from the hydrolysis of phosphatidylinositol-4,5-bisphosphate ($PIP_2$) by $G\alpha_q$-activated phospholipase-C and the consequent reduction of membrane $PIP_2$ (Gamper et al. 2004), which is required for $Ca_V2$ channel opening (see Delmas et al. 2005; Wu et al. 2002). Thus, the "slowness" compared with $G_o\beta\gamma$-mediated inhibition results in the first instance from the presence of an intermediate enzymatic step, while recovery is determined by the rate of $PIP_2$ resynthesis rather than the GTPase activity of $G\alpha_q$ (Suh et al. 2004). The time-constant for M1-receptor inhibition is ~14 seconds (Beech et al. 1992), which probably reflects the time required for $PIP_2$ to fall to a sufficient level to cause channel closure (Suh et al. 2004). However, the latency to onset of inhibition following a pulse-application of agonist may be only around 250 msec or so, as judged from the latency to closure of $PIP_2$-activated M-type $(K_V7.2/7.3)K^+$ channels following synaptically-released acetylcholine (Brown et al. 1995) – i.e., about 10 times longer than that for Gβγ-mediated inhibition. Other mechanisms further downstream from $PIP_2$ hydrolysis, such as the production of arachidonic acid from diacylglycerols, formed from $PIP_2$ hydrolysis (see Delmas et al. 2005) or phosphorylation processes (Strock and Diverse-Pierluissi 2004), may also contribute to $G_q$-mediated $Ca^{2+}$-current reduction, though the latter would be on a much slower time-scale than $PIP_2$-depletion.

*The mechanism of Ca-current inhibition at nerve endings*

Which (if either) of these processes occurs at nerve endings? Direct recording from mammalian trapezoid calyces of Held, which possess primarily $Ca_V2.1$ (P/Q-type) $Ca^{2+}$ channels (Schneggenburger and Forsythe 2006; Takahashi et al. 1996), revealed that the $GABA_B$ agonist, baclofen, strongly reduced the $Ca^{2+}$ current in a voltage-dependent manner (Takahashi et al. 1998) and that this was imitated, and occluded, by intracellular delivery of free Gβγ-subunits (Kajikawa et al. 2001). Further, baclofen totally occluded the effect of stimulating the endogenous metabotropic mGluR4 and mGluR7 receptors present at these terminals (Kajikawa et al. 2001). Stimulation of the adenosine A1 receptor produced similar (though weaker) effects, which were also occluded by baclofen (Kimura et al. 2003). Thus, these terminals show the same direct Gβγ-mediated inhibition in response to GPCR stimulation described in (2.1.1) above, with apparently no evidence for the slower $G_q$-mediated slow inhibition in (2.1.1). The same appears to be true for the calyx endings in the chick ciliary ganglion, even though they appear to contain other G-proteins α-subunits such as $\alpha_s$, $\alpha_{q/11}$. $\alpha_{12/13}$ and $\alpha_z$, which might be expected to couple through different pathways (Mirotznik et al. 2000). Recently, direct recordings from mossy fiber terminals in the hippocampus have revealed a very similar action of adenosine, acting on A1 receptors (Gundlfinger et al. 2007): $Ca^{2+}$ current inhibition (of N and P/Q-type currents) showed characteristic Gβγ-type voltage-dependence, and the reduced $Ca^{2+}$ influx was sufficient to account for the reduced transmitter release.

This may well apply to sympathetic nerve terminals, since the effects of stimulating adrenergic, muscarinic, somatostatin, PGE, and secretin receptors on the release of noradrenaline from the processes of cultured sympathetic neurons evoked by a brief pulse of $K^+$ (recorded amperometrically) were all blocked by Ptx (Koh and Hille 1997). Hence, the slow Ptx-insensitive $G_q$-mediated form of inhibition recorded in the soma of these cells on stimulating muscarinic (M1) receptors (2.1.1) did not appear to contribute significantly to muscarinic inhibition of release from their processes. Noradrenergic inhibition of transmitter release in these terminals is clearly due to Gβγ-mediated $Ca^{2+}$ current inhibition since, in autaptic connexion made in culture, transmission was blocked by instillation of Gβγ-subunits down the processes and noradrenergic inhibition was prevented by expressing the Gβγ-sequestering protein, transducin (Stephens and Mochida 2005). Further evidence that this form of $Ca^{2+}$-current inhibition underlies the presynaptic inhibitory action of many $G_o/G_i$-coupled receptors derives from experiments showing a clear reduction in intra-terminal $Ca^{2+}$ transients at both peripheral and central synapses (Dittman and Regehr 1996; Qian et al. 1997; e.g., Toth et al. 1993; Wu and Saggau 1994, 1995).

Nevertheless, several GPCRs that couple to $G_q$ rather than to Ptx-sensitive $G_o$ G-proteins are clearly able to reduce transmitter release (see, e.g., Boehm and Kubista 2002; Kubista and Boehm 2006; Miller 1998). In one case at least – that of bradykinin – it seems highly probable that this results from an inhibition of the $Ca^{2+}$-current by the $PIP_2$-depletion mechanism referred to in (**??**) above (cf. Gamper et al. 2004; Lechner et al. 2005). However, it appears from these experiments that the effects of bradykinin on $Ca^{2+}$ currents – and possibly other $G_q$-coupled receptors such as P2Y nucleotide receptors – may vary depending on the resting amount of $PIP_2$ in the membrane and the degree to which they are able to deplete $PIP_2$ to a level that results in significant $Ca^{2+}$ channel closure (Brown et al. 2007); this might vary between different neurons or under different experimental conditions.

*In sum: fast, voltage-dependent Gβγ-mediated $Ca^2+$ current inhibition triggered by activation of (primarily) $G_o$ makes a major and widespread contribution* *to the presynaptic inhibitory and auto-inhibitory effects of neurotransmitters. In contrast, the contribution of the "slow," voltage-independent inhibition produced by Gq-coupled receptors appears more variable. Nevertheless, for both forms of GPCR – G-protein pairing, additional effects downstream of the* $Ca^{2}+$ *channel may also be significant (see further below).*

## G-Protein-Mediated $Ca^{2+}$ Channel Enhancement

### *Effects mediated by protein kinase C (PKC)*

Stimulation of $G_q$-coupled receptors can also enhance $Ca_V2$ channel activity (e.g., mGluR1 Stea et al. 1995, M1, M3, M5-muscarinic receptors Bannister et al. 2004). This is mediated, not through a direct interaction of $G\alpha_q$ or $\beta\gamma$ subunits with the channel but through the formation of diacylglycerol by PLC hydrolysis and activation of PKC (Bannister et al. 2004). It results from the PKC-mediated phosphorylation of two specific serine and threonine residues in the I-II linker of the $Ca_V2\alpha_1$-subunit; the specific subtype of PKC involved is PKC$\varepsilon$, which is anchored to the C-terminus of $Ca_V2.2$ channels through a PKC-binding protein (see Tedford and Zamponi 2006). The principal effect of this is to antagonize $G\beta\gamma$-induced inhibition, so providing for an interaction between $G_q$- and $G_o$-coupled receptors (Cox et al. 2000; Queiroz et al. 2003; Gonçalves and Queiroz, this book). This may itself result in overt enhancement of resting currents because the latter are subject to a varying degree of tonic $G\beta\gamma$-inhibition as a result of the spontaneous $G_o$ trimer dissociation, but an enhancement of noninhibited currents by PKC has also been noted (Bannister et al. 2004; Stea et al. 1995), probably as a result of dual phosphorylation of the two residues (Hamid et al. 1999). Such effects may underly the enhancement of noradrenaline release from sympathetic neurons by bradykinin, which also involves a $Ca^{2+}$-independent isoform of PKC (Scholze et al. 2002).

### *Effects mediated by protein kinase A (PKA)*

The principal target for cAMP produced by GPCR activation of AC is the cognate cAMP-dependent protein kinase (PKA), which can alter presynaptic function through protein phosphorylation of multiple substrates at the level of nerve terminal excitability and $Ca^{2+}$-influx, and at loci downstream of $Ca^{2+}$-entry. Evidence for regulation at proximal loci comes from synaptosomal studies showing that PKA-mediated phosphorylation underlies the $\beta$-adrenoreceptor-stimulated facilitation of glutamate release from isolated nerve terminals (synaptosomes), an effect mimicked by cAMP phosphodiesterase inhibition (Herrero and Sanchez-Prieto 1996; Wang et al. 2002; Wang and Sihra 2003). This facilitation of neurotransmitter release is attributable to an increase in voltage-dependent $Ca^{2+}$-influx through P/Q- and N-type channels (Millan and Sanchez-Prieto 2002). The presynaptic enhancement of P-type $Ca^{2+}$-currents by $\beta$-adrenoreceptor activation has also been observed in acutely isolated amygdala neurons (Huang et al. 1996, 1998). Additionally, AC/PKA activation results in an abrogation of the effects of inhibitory GPCRs, e.g., $5HT_{1A}$ (Wang et al. 2002) and adenosine A1 receptors, in synaptosomes (Wang and Sihra 2003). These latter findings reflect a feedback regulation of Gβγ-mediated inhibition by indirect second-messenger-driven signaling in presynaptic terminals. Conversely, inhibitory mGluRs (mGluR7) stimulated by L-AP4 and negatively coupled to AC suppress glutamate release by decreasing voltage-dependent $Ca^{2+}$-entry (Millan et al. 2002, 2003). In addition to these proximal targets for GPCR/cAMP/PKA signaling, several key loci are posited distal to $Ca^{2+}$-entry (section 2.2).

### 2.1.2 Indirect Effects on $Ca^{2+}$ Entry through Changes in Presynaptic Excitability or Action Potential Waveform?

The integral of the presynaptic $Ca^{2+}$ current (and hence amount of $Ca^{2+}$ entry) is strongly dependent on the amplitude and width of the action potential (e.g., Borst and Sakmann 1999; Park and Dunlap 1998)). Because of the well-known power function between the amount of $Ca^{2+}$ entry and the amount of transmitter released ($n = 2$–$4$: motor nerve terminal [Dodge, Jr., and Rahamimoff, 1967], squid giant axon [Augustine and Charlton 1986; Llinas et al. 1982], calyx of Held [Borst and Sakmann 1999; Lou et al. 2005]), small changes in the amplitude or duration of the invading action potential might have substantial effects on transmitter release. Action potential repolarization (and hence duration) in nerve terminals such as the calyx of Held is determined by $K_V1$ and $K_V3$ voltage-gated $K^+$ channels, and possibly by $Ca^{2+}$-activated and $Na^+$-activated $K^+$-channels (Schneggenburger and Forsythe 2006). These channels are not directly controlled by G-proteins, though $K_V3$ channels can be regulated by PKC-mediated phosphorylation (Macica et al. 2003), so might be affected by $G_q$-coupled receptors (see further below). Related $K_V3.2$ channels are modulated by receptor-mediated PKA activation, but these are restricted to interneuron somata and dendrites, not terminals (Atzori et al. 2000) Thus, effects of adenosine and baclofen on evoked glutamate release at cerebellar synapses were not associated with any change in presynaptic action potential waveform (Dittman and Regehr 1996).

The most widely distributed G-protein-regulated $K^+$ channels are the G-protein-gated inwardly rectifying $K^+$-channels (GIRK channels, composed in neurons of $K_{ir}3.1 + K_{ir}3.2$ subunits), which are activated by βγ-subunits derived from $G_i$ (see section 1 and Stanfield et al. 2002). Though primarily concentrated in neuron somata and dendrites, they appear to be present in some central terminals at least (Ponce et al. 1996). These channels are also present in some sympathetic (Wang and McKinnon 1995), and parasympathetic (Surprenant and North 1988) neurons, though not necessarily in their terminals. When activated, they generate a hyperpolarization and can suppress spike generation. Since they are activated by the same presynaptic receptors that so widely activate $G_o$ to inhibit $Ca^{2+}$-currents, they could exert a profound presynaptic inhibitory effect if functionally expressed in nerve terminals. However, this appears to be very rarely (if ever) the case. Thus, there is no change in $K^+$ current in calyx endings on activating receptors that inhibit $Ca^{2+}$

currents (Takahashi et al. 1996); Kir3.2 gene-deletion eliminated the postsynaptic effects in the hippocampus produced by activating $GABA_B$, $5HT_{1A}$ or adenosine-A1 receptors without affecting their presynaptic inhibitory actions (Luscher et al. 1997); and $GABA_B$-mediated presynaptic inhibition in the retino-hypothalamic tract is not prevented by tertiapin, a $K_{ir}3$-channel blocker (Moldavan et al. 2006). On the other hand, tertiapin does reduce cannabinoid-induced inhibition of parallel fiber transmission in the cerebellum (Daniel et al. 2004), so the question may not be totally resolved.

In isolated nerve terminals, activation of type 1 mGluRs leads to activation of PKC (Coffey et al. 1994a) and facilitation of glutamate release (Coffey et al. 1993; Coffey et al. 1994a; Sanchez-Prieto et al. 1996). Based on measurements of membrane potential (Barrie et al. 1991), $^{42}K^+$ fluxes (Chaki et al. 1994; Colby and Blaustein 1988) and observations that PKC-mediated modulation is not observed when high-$[K^+]$ is used as the secretagogue (Barrie et al. 1991; Breukel et al. 1998; Cousin et al. 1999), modulation is hypothesized to occur through the inhibition of a $K^+$ channel. Electrophysiological studies have indeed shown PKC phosphorylation-dependent effects on delayed rectifier (Augustine 1990) and inward rectifier $K^+$-channels (Takano et al. 1995) – inhibition of which would tend to broaden of the action potential, and thus enhance voltage-dependent $Ca^{2+}$ entry, and thereby facilitate neurotransmitter release.

## *2.2 GPCR-Mediated Presynaptic Regulation of the Release Process Distal to* $Ca^{2+}$ *Entry*

Although the principal, and most widespread, effect of stimulating presynaptic $G_o$-coupled receptors is to inhibit transmitter release by reducing the presynaptic $Ca^{2+}$ current (see Section 2.1), there is abundant evidence that, at some synapses, such inhibitors as adenosine (at adenosine-A1 receptors) and baclofen (at $GABA_B$ receptors) exert additional effects on the release process beyond the point of $Ca^{2+}$ entry (see Thompson et al. 1993; Wu and Saggau 1997). Distal to the entry of $Ca^{2+}$, alteration of synaptic strength by modulating the exocytotic apparatus would seem a logical molecular locus for presynaptic GPCR function to regulate neurotransmitter release. Although the debate continues as to the relative contributions of voltage-dependent $Ca^{2+}$ influx inhibition (as discussed above) and downstream suppression of the exocytosis itself (Wu and Saggau 1997), both adenosine and GABA reduce the frequency of spontaneous miniature excitatory synaptic currents (mEPSC) (Scanziani et al. 1992; Scholz and Miller 1992), or the probability of such release (Prince and Stevens 1992), in a number of studies pointing to a post-$Ca^{2+}$ entry locus of action for the inhibitory presynaptic GPCRs.

Starting from the classical frog neuromuscular preparation where spontaneous acetycholine release is suppressed by purinergic adenosine receptor activation (Silinsky 1984), studies extending to mammalian central synapses contend a role for GPCR modulation of exocytosis *per se*. Thus, adenosine receptor activation reduces spontaneous glutamate release, evinced by mEPSC frequency, at synapses on dentate granule cells in hippocampal slices (Prince and Stevens 1992) and pyramidal cells in hippocampal cell cultures (Scholz and Miller 1992). At synapses on hippocampal CA3 cells, both adenosine and $GABA_B$ (baclofen) receptor activation suppresses mEPSCs in slices and this modulation is unaffected by blanket inhibition of any voltage-dependent $Ca^{2+}$ entry by $Cd^{2+}$ (Scanziani et al. 1992) A similar effect of $GABA_B$ (baclofen) receptor (but not adenosine A1 receptor) activation on the mEPSC frequency is observed in cerebellar granule cell-Purkinje cell synapses (Dittman and Regehr 1996; Harvey and Stephens 2004). Baclofen also reduces the tetrodotoxin-resistant release of GABA at hippocampal pyramidal cells (Jarolimek and Misgeld 1997). In most of these studies, the issue of VDCC modulation by the GPCR is obviated, given that during spontaneous transmitter release, there is nominally no change in intraterminal $Ca^{2+}$ concentrations. (However, it is worth noting that spontaneous and evoked transmitter release may derive from different pools of synaptic vesicles (Sara et al. 2005)). Clearly, because evoked release is contingent on voltage-dependent $Ca^{2+}$-entry, demonstration of a "$Ca^{2+}$-independent" action of inhibitory GPCR is difficult. Notwithstanding, using protocols obviating voltage-dependent $Ca^{2+}$ channel involving, evoked release is also inhibited by adenosine and $GABA_B$ receptor activation at hippocampal CA3 synapses (Capogna et al. 1996). Notably, baclofen and adenosine inhibited the release of glutamate induced by ionomycin (which raises intracellular $Ca^{2+}$ independently of $Ca^{2+}$ channel activity) (Capogna et al. 1996). Finally, methacholine and *trans*-aminocyclopentane-1,3-decarboxylic acid (*t*-ACPD)-mediated stimulation of muscarinic acetylcholine and metabotropic glutamate receptors, respectively, inhibited spontaneous mEPSCs (Scanziani et al. 1995). At cerebellar synapses, even though baclofen also reduced the $Ca^{2+}$ current, its post-$Ca^{2+}$ entry action contributed a significant component of its presynaptic inhibitory effect because it produced more synaptic inhibition than that resulting from an equivalent inhibition of the $Ca^{2+}$ current by $Ca^{2+}$ channel blocking drugs or by adenosine (Dittman and Regehr 1996; see Wu and Saggau 1997 for further examples). In contrast, the reduced $Ca^{2+}$ influx in mossy fiber terminals is sufficient of itself to account for the presynaptic inhibitory action of adenosine at this site (Gundlfinger et al. 2007). Thus, in general, it appears that post-$Ca^{2+}$ entry effects are primarily limited to the presynaptic action of GABA.

Where post-$Ca^{2+}$ entry modulation does occur, the question arises: what are the molecular targets of this GPCR-mediated regulation of exocytosis? Elucidation of the components of the exocytotic core complex, as well as upstream and downstream players, has opened the way to defining potential targets for regulation by GPCRs (Südhof 1995, 2000, 2004 and see Lang & Jahn, this volume). Thus, there are now good examples of GPCR regulation of neurotransmitter release by action at the level of recruitment of synaptic vesicles from a reserve pool (RP) into a readily-releasable pool (RRP), as well as docking, priming, exocytosis, and recovery processes (Figure 5a). In congruence with the heterotrimeric G-protein-mediated modulation exacted at the level of presynaptic excitability and $Ca^{2+}$-influx, control of the release process can be considered with respect to the direct effects of $G\beta\gamma$ subunits, and the indirect, $G\alpha$-instigated, regulation-mediated through second messengers activating protein phosphorylation pathways.

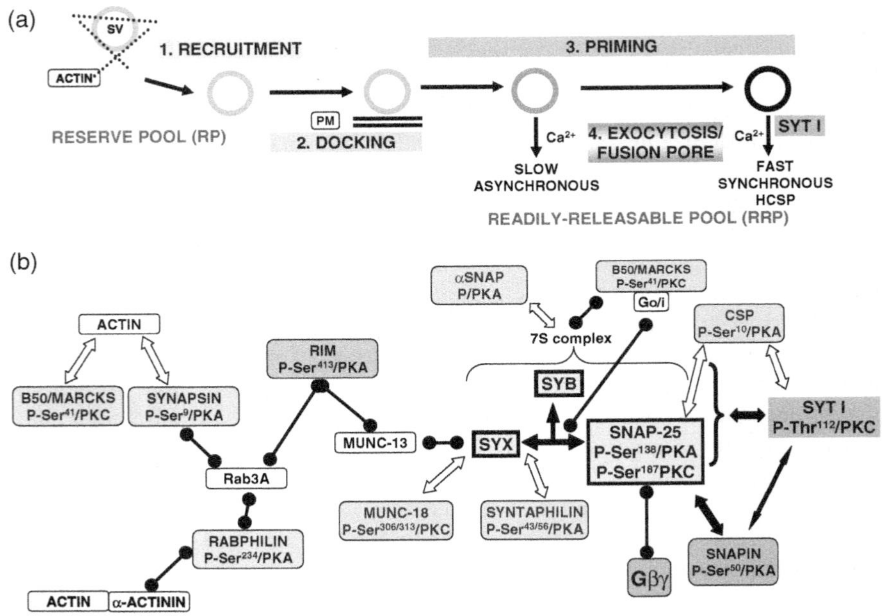


**Fig. 5** GPCR regulation of exocytosis downstream of $Ca^{2+}$-entry. (**a**) Sequence of steps leading from recruitment to maturation of synaptic vesicles from a reserve pool (RP) to a readily-releasable pool (RRP) displaying slow (asynchronous) and fast (synchronous; highly $Ca^{2+}$-sensitive pool, HCSP; synaptotagmin 1 (SYT 1) supported) components. (**b**) Protein-protein interactions of SNARES (SYX, syntaxin; SYB, synaptobrevin and SNAP-2s-7S complex) and major putative regulatory proteins. Phosphoproteins are shown in shaded boxes (phosphorylation sites for PKA and PKC are indicated where known) with phosphorylation-dependent interactions depicted by arrows (increase indicated by filled arrows; decrease indicated by open arrows). Circle-end connectors indicate a phosphorylation-independent or as yet unspecified interaction. Potential effects of interactions at various points of the sequence in A are discussed in the text.

### 2.2.1 Actions of Go-Coupled GPCRs: Direct Gβγ–Mediated Regulation of SNARE Complex Assembly and Exocytosis

Based on the effects of a number of inhibitory GPCRs on spontaneous neurotransmitter release, where the involvement of nerve terminal excitability and voltage-dependent $Ca^{2+}$-influx is obviated, direct effects of GPCRs on the exocytotic apparatus have long been posited (see Wu and Saggau 1997). Supporting the layer of control afforded by Gβγ subunits in the control of presynaptic excitability, there is now accumulating evidence that the Gβγ limb of inhibitory GPCR activation can influence the function of the SNARE complex. An initial study at the turn of the century found that the serotonin (5HT)-mediated inhibition of neurotransmission at the lamprey reticulospinal/motoneuron synapse could be mimicked by injection of Gβγ subunits (Blackmer et al. 2001). Crucially, the effects of the GPCR ligand serotonin itself is abrogated by a Gβγ scavenger based on the C-terminal binding sequence of G protein-coupled receptor kinase 2. All of the regulation observed occurs in the absence of any changes in presynaptic excitability or $Ca^{2+}$ influx, and indeed, is independent of protein phosphorylation changes.

This exciting first report, unequivocally showing a Gβγ effect on exocytosis, stimulated a hunt for the molecular target for the G-protein. Two more recent studies clarified this issue somewhat in a series of elegant experiments. Firstly, Blackmer et al. (2005) showed that the very rapid Gβγ-mediated inhibition of exocytosis seen in the lamprey synapse could be recapitulated in the permeablized PC12 cell model in which ion-channel activities play no part. In vitro studies had shown that Gβγ binds the target SNAREs, SNAP-25 and syntaxin individually, and the SNARE complex as a whole (Figure 5b). Intriguingly, specific cleavage of SNAP-25 by the prior action of the $Zn^{2+}$-protease botulinum toxin (BoNT A), substantially reduces the Gβγ binding and, in toxin-treated cells, the inhibitory effect of Gβγ subunit introduction on neurotransmitter release is attenuated. Finally, the demonstration that the binding site for Gβγ on SNAP-25 overlaps with that of the $Ca^{2+}$-sensor synaptotagmin, and that the binding of the latter is also abolished by BoNT A cleavage of the C-terminus of SNAP-25, has invoked the hypothesis that Gβγ inhibition occurs through the uncoupling of SNARE complex-synaptotagmin interaction. The effects of BoNT A-mediated cleavage of the C-terminus of SNAP-25 is even more dramatic in lamprey reticulospinal/motoneuron synapse, where inhibitory modulation by 5HT is abolished (Gerachshenko et al. 2005). Perhaps most tellingly, the 5HT-mediated modulation could be abrogated by injected peptides based on the C-terminal sequence of SNAP-25 that would be expected to scavenge the Gβγ liberated by GPCR activation. The strength of the yin-yan superposition of inhibitory Gβγ and stimulatory synaptotagmin binding to the SNAP-25 is that, while the former will bind the target in the absence of $Ca^{2+}$, the binding of the latter is $Ca^{2+}$-dependent. Gβγ competition might therefore be expected to shift the $Ca^{2+}$-dependence of the exocytotic event triggered by synaptotagmin. Interestingly, two recent studies suggest that the Gβγ produced by inhibitory GPCR activation at the lamprey reticulospinal synapse reduces the quantal size (Photowala et al. 2006; Schwartz et al. 2007). The implication from this is that inhibitory GPCRs may shift the fusion mode from “full-fusion” to a transient fusion-pore opening typified by a “kiss and run” mode of fusion.

### 2.2.2 Indirect Gα – and Gq-Mediated, Second-Messenger-Dependent Regulation of Vesicle Recruitment, Docking, Priming, and Exocytosis

As discussed above, two of the established effectors of the Gα limb of heterotrimeric G-proteins are adenylate cyclase (AC) and phospholipase C (PLCβ), which, on stimulation, lead to the generation of the second-messenger molecules, cAMP and $DAG/IP_3$, respectively. Through the respective activation of cAMP-dependent protein kinase (PKA) and $Ca^{2+}$/phospholipid-dependent protein kinase (PKC), GPCRs coupled to AC and PLCβ have the potential to effect indirect modulation of neurotransmitter release by phosphorylation of substrate proteins involved in the recruitment of synaptic vesicles and the exocytosis process itself. In the case of AC, reciprocal regulation is mediated by stimulatory $G\alpha_s$ and inhibitory $G\alpha_{i/o}$ subunits liberated by distinct GPCRs, while $G\alpha_{q/11}$ provides a monotonic stimulation of PLC$\beta$ on the activation of facilitatory GPCRs. In addition to effects mediated by cognate protein kinases, cAMP and DAG (and $PIP_2$) bind other protein effectors to each produce presynaptic modulation independent of PKA or PKC activity.

## Regulation of Adenylate Cyclase and Neurotransmitter Release – Protein Phosphorylation Mechanisms

Presynaptic coincidence and interaction of GPCR-mediated signaling cascades involving second messengers potentially form the basis for the modulation of neurotransmitter release underlying synaptic plasticity. Compelling evidence for the role of a GPCR regulated signaling pathways in synaptic plasticity has been obtained for GPCRs coupled to AC activation and production of the prototypic second-messenger, cyclic-3′,5′-adenosine monophosphate (cAMP). The demonstration that mouse knockouts of AC lead to impairment of synaptic potentiation (Storm et al. 1998; Villacres et al. 1998) certainly implicates $G\alpha_s/G\alpha_{i/o}$ mediated stimulation/inhibition of AC in the modulation of synaptic transmission and plasticity (Hansel et al. 2001; Nguyen and Woo 2003; Nicoll and Malenka 1995). Initial pharmacological evidence of a role for cAMP in presynaptic function arose from experiments utilizing the diterpine forskolin to directly stimulate AC, or the use of membrane-permeable analogues of cAMP. Subsequently, GPCRs such as $\beta$-adrenoreceptors, that are positively coupled to AC through Gs, have been shown to enhance synaptic transmission in slice preparations (Gereau and Conn 1994; Huang et al. 1996; Saitow et al. 2005). In direct support of a presynaptic mode of action, a $\beta$-adrenoreceptor/cAMP/PKA-dependent pathway has been ascribed a facilitatory role in the release of glutamate from cerebrocortical nerve terminals (Herrero and Sanchez-Prieto 1996; Wang et al. 2002). More recently, a potential role of the AC/PKA cascade has also been identified in the modulatory effects of facilitatory presynaptic kainate-type glutamate receptors at mossy fiber terminals in the hippocampus (Rodriguez-Moreno and Sihra 2004). The major effector for the second messenger cAMP is PKA. In addition to the AC/PKA cascade, there is evidence that cAMP, acting as a second messenger to stimulate GTP exchange factors (GEFs), can thereby activate small GTP-binding proteins like Rap1 and its incumbent signaling cascades. In either case, to elucidate the mechanisms by which a pleiotropic cAMP production leads to the specific regulation of presynaptic function, it is essential to delineate nerve terminal-specific target proteins for the respective signaling cascades.

### *PKA and neurotransmitter release–potential loci of action*

Although many of the protein targets for PKA are evidently ion channels involved in regulating presynaptic excitability, some of the effects of the AC activator forskolin on neurotransmitter release, under paradigms where regulation of presynaptic ion channels is obviated, point to targets for PKA downstream of $Ca^{2+}$ entry (Trudeau et al. 1996a, b, 1998). The release process distal to $Ca^{2+}$ entry can be nominally divided into four steps, viz. (1) synaptic vesicle recruitment/replenishment, (2) docking/tethering, (3) priming, and (4) exocytosis/fusion (Figure 5a). We will outline some of the best characterized presynaptic players (Figure 5b) and discuss the potential mechanisms by which their phosphorylation by the GPCR initiated AC/PKA signaling cascade, might regulate neurotransmitter release.

*PKA-synapsins: synaptic vesicle recruitment*

Functionally distinct pools of synaptic vesicles have been postulated to underpin the regulated release of neurotransmitter and presynaptic plasticity (Zucker 1993). Most proximally, $Ca^{2+}$-dependent excitation-secretion coupling is recognized to involve a "readily-releasable pool" (RRP) in which the synaptic vesicles are initially morphologically docked, to then subsequently be primed to instill fusion-competence, prior to the fusion event itself (Gillis et al. 1996; Parsons et al. 1995; Scheller 1995). Within the RRP, a kinetically slower component of release depends on supply from a pool of docked vesicles not yet competent for fusion (Rosenmund and Stevens 1996) and intimately linked with a large distal cluster of synaptic vesicles representing a "reserve pool" (RP) postulated to be mobilized during periods of intense stimulation (Koenig et al. 1993; Kuromi and Kidokoro 1998; Pieribone et al. 2002) (Figure 5a). Protein phosphorylation/dephosphorylation is a key regulatory influence on presynaptic function at the level of synaptic vesicle recruitment/mobilization from the RP (Greengard et al. 1993; Hilfiker et al. 1999a; Turner et al. 1999) (Figure 5a, step 1). In a logical strategy identifying presynaptic phosphoprotein substrates, synapsin I was among the first protein characterized (De Camilli et al. 1979). It is found associated with the cytoplasmic face of synaptic vesicles (De Camilli et al. 1983a, b; Hirokawa et al. 1989; Pieribone et al. 1995; Valtorta et al. 1988), and is present exclusively in the presynaptic cytomatrix (De Camilli et al. 1983b; Gotow et al. 1991; Landis et al. 1988; Mandell et al. 1990b; Takei et al. 1995), thereby invoking a role for the protein and its family members (Mandell et al. 1990a; Mandell et al. 1992; Stone et al. 1994; Südhof et al. 1989) in the regulation of synaptic vesicle pools and neurotransmitter release (De Camilli et al. 1990; De Camilli and Greengard, 1986a).

Synapsin I is a brain- and terminal-specific phosphoprotein which has long been shown to be a target for GPCR signaling (Dolphin et al. 1980; Dolphin and Greengard, 1981a, b). In vitro, it is a substrate for PKA on $Ser^9$ (Site 1) (Dunkley and Robinson, 1986; Huttner and Greengard, 1979), and $Ca^{2+}$/calmodulin-dependent protein kinases (CaMK) I and IV also on $Ser^9$, and CaMKII and CaMKIV on $Ser^{566}$ and $Ser^{603}$ (site 2&3) (Burke and DeLorenzo 1981; Czernik et al. 1987; DeLorenzo et al. 1979; Dunkley et al. 1986; Gower et al. 1986; Sieghart et al. 1980). All of the canonical phosphorylation sites, 1, 2, and 3, are therefore potentially subject to regulation by GPCRs that either elevate presynaptic second messenger cAMP levels or $[Ca^{2+}]_i$ (Coffey et al. 1994b; Dunkley and Robinson 1986; Krueger et al. 1977;

Robinson and Dunkley 1985; Robinson and Dunkley 1983, 1987; Sihra et al. 1989, 1992; Sihra 1993; Walaas et al. 1983). Additional phosphosites on $Ser^{62}$ (site 4), $Ser^{67}$ (site 5), and $Ser^{549}$ (site 6) respond to a noncanonical pathway instigated by neurotrophic factors (Romano et al. 1987a, b) that stimulate cognate receptor tyrosine kinase receptors to initiate signaling cascades leading to activation of proline-directed kinases mitogen-activated protein kinases (MAPK1&2; also termed extracellular signal-regulated protein kinases, ERK1&2) and cyclin-dependent protein kinase 5 (cdk5) (Jovanovic et al. 2001; Jovanovic et al. 1996; Jovanovic et al. 2000; Matsubara et al. 1996).

High concentrations of synapsin I are present in nerve terminals, about 10μM (De Camilli et al. 1990) or 10–30 molecules synapsin I/vesicle (Ho et al. 1991), and its amphipathic properties impart phospholipid affinity along with protein binding to synaptic vesicles. These interactions are under dynamic control by phosphorylation, so that phosphorylation of site 1 and sites 2/3 (Benfenati et al. 1989a, b, 1992b; Esser et al. 1998; Hosaka et al. 1999; Schiebler et al. 1986) leads to dissociation of molecule from the vesicle membrane. Likewise, synapsin I avidity for the actin cytoskeleton (Bahler and Greengard 1987; Petrucci et al. 1991; Petrucci and Morrow 1987) is altered by phosphorylation by PKA at site 1 (decrease in the actin nucleation and bundling) and by CaMKII at sites 2 and 3 (abolition of actin-nucleation and bundling but no effect on Kd) (Bahler et al. 1989; Benfenati et al. 1992a). Consistent with these alterations of synapsin I interactions, the molecule dissociates from synaptic vesicles and the cytoskeleton following in situ phosphorylation produced by forskolin or during depolarization of isolated nerve terminals (Sihra et al. 1989), or activity-dependent phosphorylation in the frog neuromuscular junction (Torri-Tarelli et al. 1990) and hippocampal neurons (Chi et al. 2001b, 2003b). These observations have therefore collectively contributed to the working hypothesis that synapsin I controls the exocytotic availability of synaptic vesicles from the RP by reversibly tethering them to the actin-based "cytonet" of the nerve terminal, in a phosphorylation-dependent manner (Chi et al. 2001a, 2003a; De Camilli et al. 1990; De Camilli and Greengard, 1986b) (Figure 5b). Although opinions differ as to the exact consequence of synapsin phosphorylation, or even its precise function (Hosaka and Südhof 1998, 1999), given its importance in presynaptic plasticity intimated from knockout studies (Li et al. 1995; Rosahl et al. 1993; Rosahl et al. 1995), it is a irrefutable candidate for modulation of presynaptic function via by GPCR activation.

### *PKA – Rab3A and transport partners: vesicle recruitment/docking*

Rab3A is a vesicle-associated small G-protein involved in the targeting and docking/tethering of synaptic vesicles to the plasma membrane prior to priming and eventual exocytosis (Geppert et al. 1994, 1997; Schluter et al. 2004). It associates in a GTP-dependent manner with two synaptic vesicle active zone effector proteins, rabphilin (Li et al. 1994; Shirataki et al. 1993) and Rab3A-interacting molecule (RIM) (Schoch et al. 2002; Wang et al. 1997) (Figure 5b).

Although Rab3A is not a substrate for PKA, both rabphilin and RIM are phosphorylated by PKA (Fykse et al. 1995; Lonart et al. 2003; Lonart and Südhof 1998) and hence represent potential targets for signaling initiated by presynaptic GPCRs coupled to AC. Interestingly, in knockouts of Rab3A (Castillo et al. 1997) and RIM (Castillo et al. 2002), but not rabphilin (Schluter et al. 1999), long-term potentiation (LTP) at mossy fiber-CA3 pyramidal cell synapses is abolished. LTP at this synapse has long been posited to be expressed presynaptically through an AC/cAMP/PKA signaling cascade (Huang et al. 1994, 1996; Weisskopf et al. 1994). In the cerebellar parallel fiber-Purkinje cell model, where plasticity is also underpinned by presynaptic AC/cAMP/PKA signaling (Salin et al. 1996), LTP, which is eliminated in cultures from RIM1α knockout mice, is rescued by wild-type RIM1α but not by RIM1α rendered phospho-null (S413A) at the N-terminal of its two PKA phosphorylation sites (Lonart et al. 2003) (Figure 5b). Together, these data point to RIM as a key player in the mediation of presynaptic plasticity under GPCR control. At the level of synaptic vesicle priming (Figure 5a, step 3), given that Rab3A and Munc13 possibly compete for (Betz et al. 2001), but certainly interact during (Dulubova et al. 2005) RIM binding (at nested N-terminal α-helical and zinc-finger domains, respectively), PKA phosphorylation of RIM may have the potential to regulate the equilibrium of its respective binding to these two partners (Betz et al. 2001). Functionally, this could mediate GPCR effects on release probability and plasticity, particularly as Munc13 is found so intimately involved in syntaxin function and its availability for SNARE formation in an open conformation (Betz et al. 1997, 1998) and in vesicle priming (Augustin et al. 1999b; Brose et al. 2000; see Section 2.2.2 herein). However, judging by the numerous interactions that RIM1α displays with components of the presynaptic active zone, the involvement of RIM1α in the control of neurotransmitter release is likely to manifest at multiple levels in the secretory cascade leading to exocytosis (Calakos et al. 2004), perhaps even through a major reorganization/restructuring of the active zone (Lonart 2002; Lonart et al. 2003) by virtue of interactions with proteins such as liprins (Zhen and Jin 1999).

Intriguingly, synapsin has recently been shown to bind to Rab3A (Giovedi et al. 2004b) and operate as novel Rab effector (Giovedi et al. 2004a) (Fig. 5b). As with RIM, this is another case of a Rab3A partnering with a PKA substrate, while itself not being phosphorylated. Rab3A and synapsin interaction results in reciprocal regulation of the respective functions of the two proteins. Thus synapsin stimulates the GTP-binding and GTPase activity of Rab3A, while Rab3A inhibits the actin binding/bundling and vesicle clustering abilities of synapsin I. Interestingly, the latter activity of Rab3A activity parallels that of one of its other effectors, rabphilin, in its interaction with α-actinin (Baldini et al. 2005; Kato et al. 1996) (Figure 5b), but differs in several other ways. Synapsin functions to recruit GDP-bound Rab3A to the synaptic vesicle and participates in the subsequent GTP binding of the small G-protein. Given that Rab3A would then be expected to reduce the actin-vesicle cross-linking ability of synapsin and thereby increase recruitment of vesicles into the RRP, it would be interesting to see whether Rab3A and PKA-dependent phosphorylation of synapsin synergize at this point to facilitate neurotransmitter release (Figure 5b).

*PKA – SNAP-25/α-SNAP/Syntaphilin: SNARE complex formation/dissolution*

At the level of SNARE complex formation and dissolution, synaptosome-associated protein (SNAP-25) (Hepp et al. 2002; Risinger and Bennett 1999), α-SNAP (Hirling and Scheller 1996) and syntaphilin (Hirling and Scheller 1996) are phosphorylated by PKA. These are therefore candidate regulators of exocytosis by GPCR/AC signaling.

Phosphorylation of SNAP-25 occurs on $Thr^{138}$, and in the chromaffin cell model, the maintenance of secretory vesicles in a primed state by high endogenous PKA activity is mediated by phosphorylation at this site (Nagy et al. 2004) (Figure 5b). Although pharmacological and mutational modulation of the PKA phosphorylation of SNAP-25 clearly impinges on defined kinetic components of release, it remains to be seen whether a similar phosphorylation of the t-SNARE affects the size of the releasable pool of small synaptic vesicles from nerve terminals and thereby alters synaptic efficacy following activation of GPCRs coupled to AC. Notwithstanding the lack of direct evidence of the role of SNAP-25 in the PKA-mediated modulation of exocytosis, indirect support has been forthcoming from studies using hippocampal neurons. Thus, following the cleavage of the C-terminus (containing phosphorylation site) of SNAP-25 by botulinum toxin, while the response to ruthenium red (acting as a $Ca^{2+}$-independent secretagogue) (Trudeau et al. 1996a) is unchanged, the facilitatory effect of forskolin seen in control conditions is abrogated (Trudeau et al. 1998).

Downstream of trans-SNARE complex formation and membrane fusion, the soluble NSF-associated protein, α-SNAP, collaborates with NSF to dissociate the resultant cis-SNARE complex. This post-fusion event promotes priming of secretory vesicles as seen by the augmentation of the RRP by overexpression of α-SNAP (Xu et al. 1999a), and begs the question as to whether the function of this protein is regulatable by GPCRs effecting increases in cAMP levels. The observation that PKA phosphorylation of α-SNAP leads to a ten-fold decrease in affinity of the protein for the 7S SNARE complex is interesting yet enigmatic (Figure 5b). On the one hand, this would imply that PKA mediates a decrease in NSF activity, a notion contrary to the recycling of the SNARES and to the stimulatory effects of PKA on neurotransmitters release. On the other hand, if, in situ, the phosphorylation-dependent decrease in association occurs following NSF function, then this would accord with the apparent docking/priming behavior of α-SNAP that has been reported (Xu et al. 1999a).

At the proximal end of the SNARE cycle, the involvement of syntaphilin as a PKA target appears less paradoxical than that of α-SNAP. Discovered as a protein that competes with SNAP-25 for syntaxin-1A binding, syntaphilin is a putative brake on SNARE complex formation and thus exocytosis (Lao et al. 2000). It is interesting therefore that PKA phosphorylation of syntaphilin at $Ser^{43}$ and $Ser^{56}$ dramatically decreases binding to syntaxin-1A, an effect mimicked by the phosphomimetic (S43D) mutant (Figure 5b). Thus, although overexpression of wild-type (dephospho-) syntaphilin inhibits exocytosis, the phosphomimetic mutant has no effect (Boczan et al. 2004). The implication is that dephospho-syntaphilin operates as negative regulator of exocytosis and thus GPCRs leading to PKA activation may well operate to abrogate this control and thereby facilitate exocytosis.

*PKA – Snapin and CSP:* $Ca^{2+}$ *triggering of exocytosis*

The final step in the exocytosis is the stimulation of fusion itself initiated by $Ca^{2+}$ binding to a $Ca^{2+}$-sensor or trigger Figure 5a. Prior regulation of the trigger by phosphorylation would be an effective mean of controlling presynaptic function given the rate-limiting nature of this step. In this context, the facilitatory effect of PKA reported to occur via a $Ca^{2+}$ independent stimulation of exocytosis (Trudeau et al. 1996b) is not attributable to increased vesicle docking or recruitment, or addition of terminals, but rather attests to an increase in release probability. The latter may potentially be imparted by a functional increase in the affinity of $Ca^{2+}$ for the release apparatus or enhanced $Ca^{2+}$ co-operativity in the release process. A mediator for this type of PKA-mediated enhancement appears to be the protein, snapin, found associated with synaptic vesicle membranes (Ilardi et al. 1999). Snapin binds directly to SNAP-25 and is thereby recruited to SNARE complex (Figure 5b). Disruption of this interaction with the C-terminal half of snapin, or the snapin-binding sequence on SNAP-25, blocks interaction of the SNARE complex with the $Ca^{2+}$ trigger synaptotagmin, and this inhibits synaptic transmission. PKA-dependent phosphorylation of snapin on $Ser^{50}$ results in a substantial increase in its affinity for SNAP-25 and enhances the binding of synaptotagmin, as does introduction of a phosphomimetic mutation (S50D) into the full-length protein (Chheda et al. 2001) (Figure 5b). In adrenal chromaffin cells, this increases the levels of release competent secretory granules and enhances the probability of release in response to $Ca^{2+}$ influx. Enhancement of the probability of release also occurs in hippocampal neuron autapses with the phosphomimetic mutant of snapin, but here the increase in release is thought to contribute to a decrease in the number of vesicles in the RRP and therefore a more rapid synaptic depression (Thakur et al. 2004). Together, these studies suggest that snapin, as a SNARE complex associating protein, could well underpin a major component of the potentiation of release seen with of stimulation of GPCR/AC/PKA activity.

Cysteine string protein (CSP) is one of a family of proteins found on secretory vesicles, being distinguished by the presence of a 14-cysteine residue string sequence at the C-terminal end, responsible for membrane association through palmitoylation, and a J-domain at the N-terminal, responsible for interaction with chaperone molecules. CSP appear to have multiple functions, including roles in vesicular uptake of neurotransmitter, RabA function, G-protein modulation of voltage-dependent $Ca^{2+}$ channels, and vesicular exocytosis (Evans and Morgan 2003). The latter effect occurs by virtue of the interaction of CSP with syntaxin and synaptotagmin. The relevance of CSP in GPCR modulation of exocytosis arises from the observation that the $Ser^{10}$ residue, just N-terminal to the J-domain, is phosphorylated by PKA (Evans et al. 2001; Evans and Morgan 2002) (Figure 5b). Crucially, this phosphorylation reduces the affinity of CSP for both syntaxin and synaptotagmin (cf snapin above). Judging by the lack of such effect by a CSP phosphonull (S10A) mutant, PKA-dependent phosphorylation of CSP would appear to be slowing the kinetics of release catecholamine release from chromaffin cells. In a phosphorylation-independent context, overexpression of CSP also inhibits exocytotic levels of catecholamine. While it remains to be seen whether CSP phosphorylation has a similar kinetic effect on exocytosis in neurons, GPCR/AC/PKA action on CSP would be predicted to decrease release and synaptic transmission, contrary to the generally accepted view that PKA is facilitatory in the presynaptic terminal.

## Regulation of Adenylate Cyclase and Neurotransmitter Release – PKA-Independent Regulation by cAMP Through HCN Channels and cAMP-GEF

While PKA constitutes the major target for the "original" second messenger cAMP, it is increasingly apparent that the nucleotide targets many other regulatory proteins in neuronal and non-neuronal cells. Beginning with cyclic nucleotide gated channels (Kaupp and Seifert 2002), the list of partners for cAMP has expanded considerably in recent years (Seino and Shibasaki 2005; Springett et al. 2004). Two key, non-kinase cAMP effectors of note, from the point of view of GPCR-mediated exocytotic control being considered here, are the *hyperpolarization-activated cyclic nucleotide-gated* (HCN) *channels* (Biel et al. 2002) and *cAMP-regulated guanine nucleotide exchange factors* (cAMP-GEFs or Epacs; *exchange proteins directly activated by cAMP*) (Bos 2003; Springett et al. 2004).

HCN or $I_h$ channels were originally characterized in cardiac muscle in the control of the pacemaking activity in the sinoatrial node. Subsequently, these members of the voltage-dependent $K^+$-channel family have been found in many cell types, including neurons. The HCN channels are nonselective to cations ($Na^+/K^+$), respond to negatively directed voltage steps around the resting membrane potential, and are regulated by cAMP by a C-terminus interaction site. The channels took the stage from the exocytotic viewpoint because their activation following GPCR stimulation with 5HT effects a facilitation of neurotransmitter release from motor terminals (Beaumont and Zucker 2000). Crucially, this occurs in the absence of any changes in intracellular $Ca^{2+}$-levels, indicating action at a distal locus in the stimulus-secretion coupling cascade. This enhancement, which lasts (tens of minutes) beyond any time-course expected of an ionotropic action of a channel, appears to arise from an increase in the mobilization of vesicle into the RRP following HCN channel activation (Wang and Zucker 1998), in the absence of any prior $Ca^{2+}$ -influx (Zhong et al. 2004). HCN channel activation also underlies cAMP-dependent regulation of release at cerebellar basket cell terminals (Southan et al. 2000) and hippocampal mossy fiber LTP, which is contingent on presynaptic modulation by cAMP (Mellor et al. 2002). The mechanism underpinning the cAMP/HCN regulation of exocytosis remains to be fully elucidated but, interestingly, the involvement of the actin cytoskeleton in the interaction of HCN channels with the release machinery has been posited (Zhong and Zucker 2004).

Following the identification of cAMP-GEFs as cAMP-binding proteins that activate the small G-protein Rap1 (De Rooij et al. 1998; Kawasaki et al. 1998), a role for mouse homologue of cAMP-GEFII in exocytosis was explicitly demonstrated in studies expressing the cAMP-sensor in PC12 cells, which lack the protein endogenously (Ozaki et al. 2000). cAMP-GEFII expression enhances forskolin facilitation of secretion, and this increase is insensitive to PKA inhibition by H89. All the diagnostics with mutant proteins indicate that cAMP binding is essential for the modulation. Furthermore, antisense oligonucleotides against cAMP-GEFII significantly diminish the exocytotic response to cyclic nucleotides in insulin-secreting cells, indicating that a PKA-independent pathway via cAMP-GEFII is of physiological significance. Indeed, in some cases, this PKA-independent route is exclusive. Thus in the calyx of Held synapse, the obligate involvement of cAMP-GEFs in cAMP-mediated facilitation is intimated from studies showing that PKA inhibitors are entirely ineffective (Sakaba and Neher 2001). On the other hand, cAMP analogues like 8-pCPT-2′-O-Me-cAMP, which preferentially activate cAMP-GEF/Epac, enhance the release of neurotransmitter in the same model (Kaneko and Takahashi 2004; Sakaba and Neher 2003).

How then does cAMP-GEF/Epac mediate effects on exocytosis? Some clues come from the observation that cAMP-GEFII co-precipitates with RIM (Ozaki et al. 2000), which has been discussed above (section 2.2.2) with respect to its putative roles in GPCR/AC-mediated facilitation of exocytosis. Curiously though, despite its enzymatic activity, cAMP-GEFII does not exchange nucleotides on the cognate binding partner for RIM, Rab3A (Ozaki et al. 2000). Clearly therefore, despite the molecular interaction of cAMP-GEFII with RIM/Rab3A, this is partitioned from the exchange activity of former, perhaps unsurprisingly given that cAMP-GEF/Epac was discovered as exchange factor for Rap1. One possibility is that RIM interaction with cAMP-GEF indirectly localizes Rap-1 to the active zone where RIM resides. Since activated Rap1 has been suggested to directly or indirectly feed into B-Raf (see Dumaz and Marais 2005), and B-Raf leads to the activation of ERKs, Rap-1 coordination by Rim could effectively activate the ERK signaling cascade in nerve terminals (Jovanovic et al. 1996). ERK activation has been shown to facilitate glutamate release from nerve terminals as a result of synapsin phosphorylation at ERK-specific sites (Jovanovic et al. 2000; Jovanovic et al. 2001). This reflects important intraterminal cross-talk between GPCRs linked to AC and neurotrophin Trk receptors linked to ERK signaling.

## Regulation of Phospholipase C and Neurotransmitter Release – Protein Phosphorylation Mechanisms

A number of presynaptic receptors are coupled to the activation of phospholipase C (PLC) through $G_q$ (Majewski and Iannazzo 1998). PLC activation results in the production of two second messengers, viz. diacylglycerol (DAG) and inositol triphosphate ($IP_3$). The former stimulates PKC together with $Ca^{2+}$, while the latter initiates the release of $Ca^{2+}$ from intracellular stores. As with PKA, PKC is pleiotropic in its effects, thus the specific regulation of presynaptic function necessitates the delineation of nerve-terminal-specific target proteins. Additionally, it has become increasingly apparent that DAG plays a fundamental role as a direct activator of some presynaptic proteins, without invoking the prior activation of PKC. Furthermore, the very same PLC activity that produces DAG and $IP_3$ also effectively reduces the levels of $PIP_2$. As with some ion channels (see Section 2.1.1), this in itself can have signaling function for membrane resident/associated proteins, as is the general norm for players involved in exocytosis. Here we will consider the role of PKC, DAG, and $PIP_2$ in affecting the function of specific proteins associated with the release process distal to $Ca^{2+}$ entry, during synaptic vesicle recruitment/replenishment, docking, priming, and fusion steps (Figure 5a).

*PKC and neurotransmitter release–potential loci of action*

In the foregoing discussion (Sections 2.1.1 and 2.1.2), we have considered the effects of PKC activation in the regulation of neurotransmitter release by modulation of nerve terminal excitability. There are numerous instances, however, of regulation of neurotransmitter by PKC downstream of $Ca^{2+}$ entry, evinced by studies using $Ca^{2+}$ ionophores to bypass ion-channel involvement (see Majewski and Iannazzo 1998). In these and other studies, the usual way to stimulate PKC is through the use of phorbol esters, which mimic DAG binding to one or two cysteine-rich domains (CRDs) found near the N-terminus of conventional and novel (but not atypical) forms of PKC. Because CRDs are found in proteins other than PKCs, the use of phorbol esters alone is not sufficient to define PKC-dependent effects, as had been assumed in early studies. To distinguish de facto regulation by PKC, therefore, additional evidence from the use of PKC inhibitors targeted to the ATP-binding site (as distinct from the DAG-binding site with which some inhibitors interact), increased PKC levels (by introduction of purified protein kinase or overexpression), or introduction/expression of phosphomimetic mutants of the PKC substrate (if known), is now deemed obligatory.

Notwithstanding, but bearing in mind the technical caveats in defining PKC-dependent regulation, there are good examples of PKC-dependent facilitation of neurotransmitter release (Majewski and Iannazzo 1998), with increasing evidence of PKC-mediated phosphorylation affecting several steps leading up to synaptic vesicle exocytosis downstream of $Ca^{2+}$ entry (Takahashi et al. 2003). These studies used bisindolylmaleimide type PKC inhibitors, which specifically target the ATP-binding site of PKC, to delineate the phosphorylation-dependent component of regulation, and verified earlier studies where the role of PKC in secretion was defined with use of pseudosubstrate peptide inhibitors of PKC (Terbush and Holz 1986, 1990) or purified PKC (Morgan and Burgoyne 1992; Naor et al. 1989) in permeablized cell systems. Indeed PKC$\alpha$ and calmodulin are found to be proteins essential for triggering $Ca^{2+}$-stimulated exocytosis in a permeablized PC12 model (Chen et al. 1999), and PKC activation underlies the increase in the RRP evinced by hypertonic sucrose-induced release in synaptosomal preparations (Khvotchev et al. 2000; Lonart and Südhof 2000).

At a proximal locus, recruitment/translocation of secretory vesicles (Figure 5a, step 1) is under the control of PKC in that its activation variously affects the RRP in both neuronal (Berglund et al. 2002; Stevens and Sullivan 1998) and non-neuronal (Gillis et al. 1996; Smith 1999; Smith et al. 1998) models. Thus phorbol ester treatment enhanced the size of the RRP (Berglund et al. 2002; Gillis et al. 1996) as well as its recovery following depletion (Smith et al. 1998; Stevens and Sullivan 1998), indicating an increase in recruitment or refilling from the RP (Figure 5a). PKC stimulation may therefore contribute to the activity-dependent facilitation of release following $Ca^{2+}$ entry (Smith 1999).

Several studies have noted an effect of PKC at a more distal locus. Thus, rather than an effect on the size or refilling of the RRP, phorbol esters effect an increase in $Ca^{2+}$-sensitivity of exocytosis in the chick ciliary ganglion (Yawo 1999a, b) (Figure 5a, step 4). So, there is a leftward shift in the exocytosis versus $[Ca^{2+}]$ relationship but without any effect on: $Ca^{2+}$-influx/buffering/extrusion, the number of $Ca^{2+}$-binding (release) site, or the size or refilling rate of the RRP. This effect on the priming of a highly $Ca^{2+}$ sensitive pool HCSP, Figure 5a) is blocked by bisindolylmaleimide inhibition of PKC activity. The inhibitor, being designed to bind the ATP-binding site on PKC, does not affect the interaction of phorbol esters with Munc13, which invokes PKC-independent priming. PKC similarly affects the $Ca^{2+}$-sensitivity of exocytosis in mammalian neuronal, calyx of Held synapse (Lou et al. 2005; Wu and Wu 2001) and non-neuronal pituitary cell (Zhu et al. 2002) release models.

At a very late point in the exocytotic process (Figure 5a, step 4), a necessary stage is the formation of a fusion pore prior to the complete coalescences of the vesicle and plasma membrane to produce full vesicular fusion as posited by the quantal hypothesis of neurotransmitter release. There is, however, experimental evidence to suggest that the fusion pore formation may be transient in some circumstances, so that sub-quantal neurotransmitter release may occur as proposed by the "kiss-and-run" hypothesis (Fernandez-Peruchena et al. 2005). The dilation and contraction of the fusion pore is therefore suggested to be under dynamic control to determine the mode of exocytosis. In the calyx of Held terminal, larger fusion pore conductances are recorded prior to full-fusion events compared to kiss-and-run exocytosis (He et al. 2006). In chromaffin cells, activity-dependent dilation of the fusion pore is suggested to produce a size-exclusion filter underlying the differential release of co-packaged small (catecholamines) and large (neuropeptides) molecule transmitters; low activity (narrow pore) releasing the former, with high activity (large pore) releasing both (Fulop et al. 2005). Furthermore, this dilation of the fusion pore is controlled by PKC activity and therefore represents an important potential target for regulation by GPCRs coupled to PLC activation (Burgoyne et al. 2001; Fulop and Smith 2006; Graham et al. 2000).

There are evidently several potential loci for the action of PKC in facilitating neurotransmitter release. The question is: what are the protein targets/substrates for PKC that mediate these multiple mechanisms of regulation of exocytosis?

*PKC-B50: regulation of secretory vesicle recruitment*

In the chromaffin and PC12 cell models, where the cortical actin cytoskeleton restricts the availability of secretory granules (Figure 5a, step 1), PKC activation promotes release (Shoji-Kasai et al. 2002; Tsuboi et al. 2001; Vitale et al. 1995) by the disruption of the actin cytoskeleton (Vitale et al. 1995). At one level, this occurs by the activity of scinderin, an actin-severing protein, which is competitively bound by membrane $PIP_2$ and PS under basal condition, but by actin when activated. Thus stimulation may be determined by PLC activation altering levels of cellular $PIP_2$ (Trifaro et al. 2002). Disruption of the actin cytoskeleton also occurs by de facto second messenger stimulation of PKC and phosphorylation of substrates such as the myristoylated-alanine-rich C-kinase substrate (MARCKS) (Rose et al. 2001) in response to stimulation of PLC by histamine GPCRs, for instance (see Trifaro et al. 2002). The actin cytoskeleton plays a similar role in neuronal transmitter release by multiple mechanisms (Doussau and Augustine 2000). MARCKS phosphorylation (Robinson 1991; Wang et al. 1989) and the phosphorylation of another PKC substrate, B50 (GAP-43 or neuromodulin, phosphorylated on $ser^{41}$) (Spencer et al. 1992), is involved in the regulation of neurotransmitter release (Dekker et al. 1991; Hens et al. 1995, 1998). In the case of B50, although a PKC-dependent modulation of the interaction with F-actin is generally acknowledged, there is no consensus as to whether B50 affects the polymerization of actin (He et al. 1997; Hens et al. 1993; Iannazzo 2001). Interestingly, B50 phosphorylation does decrease its binding of calmodulin, an effect which, in the locale of actin-bound synaptic vesicle pool, may stimulate the activation of vesicle-bound $Ca^{2+}$/calmodulin kinase II (Benfenati et al. 1992b). This would lead to the phosphorylation and thus dissociation of synapsin I from synaptic vesicles and the actin-cytoskeleton, and thereby mediate the mobilization of the reserve pool of synaptic vesicles (Iannazzo 2001). Given the observed facilitation of glutamate release by type I mGluRs (see Sanchez-Prieto et al. 1996), stimulation of these GPCRs may instigate PKC phosphorylation of B50 (and MARCKS) to support presynaptic plasticity processes (Coffey et al. 1994b; Ramakers et al. 1999).

### *PKC –* $G\alpha_o$ *and B50: regulation of vesicle priming*

Downstream of vesicle recruitment, PKC-dependent phosphorylation may affect vesicle priming (but see PKC-independent regulation of the same in Section 2.2.2) (Figure 5a, step 3). The discovery of $G\alpha_o$ and/or $G\alpha_i$ on secretory granules and synaptic vesicles (Ahnert-Hilger et al. 1994; Ahnert-Hilger and Wiedenmann 1994; Toutant et al. 1987) begs the question as to their function therein. Functional studies using mastoparan, an activator of the $G\alpha_{o/i}$, show that that heterotrimeric G-protein activation inhibits exocytosis by inhibiting an ATP-dependent priming step (see Vitale et al. 2000 for review). One possibility is that $G\alpha_o$ prevents the dissociation of SNAP-25 from VAMP, thought to be necessary for priming (Misonou et al. 1997). Given the intracellular localization of the secretory vesicle associated pool of $G\alpha_{o/i}$, physiological stimulation by extracellular signals is necessarily indirect. One potential segue between plasma membrane signaling and vesicular G-protein is the PKC substrate B50 (GAP-43), which interacts with and stimulates $G\alpha_o$ (Strittmatter et al. 1990) (Figure 5b). Therefore as B50 (GAP-43), and PKC phosphorylation thereof, supports neurotransmitter release (Dekker et al. 1991; Hens et al. 1995, 1998), and the phosphorylated protein promotes ATP-dependent priming, one possibility is that

B50 (GAP-43) phosphorylation by PKC uncouples the inhibitory tone imposed by stimulation of $G\alpha_o$, (Misonou et al. 1998; Ohara-Imaizumi et al. 1992), and, in so doing, effects facilitation of exocytosis. Interestingly in this regard, B50 (GAP-43) has been shown to interact with components of the SNARE complex in a PKC phosphorylation-dependent manner (Haruta et al. 1997).

*PKC – SNAP-25: regulation of secretory vesicle recruitment*

SNAP-25 is phosphorylated in response to phorbol esters in neuronal (Genoud et al. 1999), neuroendocrine (Iwasaki et al. 2000; Kataoka et al. 2000; Nagy et al. 2002; Shimazaki et al. 1996), and non-neuronal (Gonelle-Gispert et al. 2002) cells at $Ser^{187}$. Phosphorylation at $Thr^{138}$ by PKC is additionally intimated, this also being the target for PKA (Hepp et al. 2002 see above; Risinger and Bennett 1999) (Figure 5b). In vitro phosphorylation at the $Ser^{187}$ site using purified substrate and PKC appears to reduce the interaction of the protein with syntaxin (Shimazaki et al. 1996), an effect that is not easily reconcilable with a stimulatory effect of PKC phosphorylation on exoytosis and given that it is an increase in SNAP-25-syntaxin heterodimerization that likely represents a rate-limiting step in SNARE complex formation. However, more recently, a phosphomimetic mutation of SNAP25 at $Ser^{187}$ (S187E) was shown to cause an increase in SNAP-25 and syntaxin association (Yang et al. 2007). Overexpression of phosphomimetic (S187E) results in an increase in the kinetics of refilling (but not size) of the RRP (Nagy et al. 2002) and the size of a highly $Ca^{2+}$-sensitive pool (HCSP, Figure 5a) (Yang et al. 2007). This suggests causality between PKC-mediated phosphorylation of $Ser^{187}$ on SNAP-25, increased SNARE complex formation, and facilitation of exocytosis. This neat conclusion is, however, tempered by observations in hippocampal neurons that clostridial toxin-mediated elimination of SNAP-25 is equally well rescued by expression of toxin-insensitive exogenous wild-type, phospho-null (S187A) and phospho-mimetic (S187E ) SNAP-25, with phorbol esters still stimulating exocytosis in the phospho-null conditions (Finley et al. 2003). However, this study, using mild stimulation, may not have assayed refilling kinetics noted to be enhanced by S187E SNAP-25 phosphomimicry (Nagy et al. 2002), or the HCSP modulated by SNAP-25 (Yang et al. 2007) may be a specific feature of the chromaffin cell model, or at least not easily amenable in neuronal studies to date.

*PKC – SNAP-25: regulation of* $Ca^{2+}$*-sensitivity*

SNAP-25 phosphorylation by PKC may also underlie the increase in $Ca^{2+}$-sensitivity of exocytosis discussed above (Figure 5a, step 4). The C-terminus of SNAP-25 is known to be involved in the $Ca^{2+}$-triggering of exocytosis based on the essential $Ca^{2+}$-dependent binding of synaptotagmin to this negatively charged region (Zhang et al. 2002). Cleavage of the C-terminus 26 amino-acids of SNAP-25 with botulinum toxin E (BoNT E) abolishes binding of synaptotagmin and regulated exocytosis in PC12 cells, while removal of the C-terminus 9-amino acids by botulinum toxin A (BoNT A) increases the $Ca^{2+}$-concentration needed for binding of synaptotagmin and reduces the $Ca^{2+}$-sensitivity of exocytosis (Gerona et al. 2000; Xu et al. 1998). A similar effect to the latter is obtained if the negatively charged acidic amino-acid tail of SNAP-25 is replaced with neutral residues (Sorensen et al. 2002). The $Ser^{187}$ phosphorylation site which is targeted by PKC lies between the BoNT E and BoNT A cleavage sites and promotes the expression of the HSCP component of the RRP (Figure 5a). The HCSP, but not the RRP in general, is eliminated on cleavage with BoNT A (Yang et al. 2002). Thus, evidently, PKC phosphorylation in relation to SNAP-25 appears to be responsible, at least in part, for the enhancement of $Ca^{2+}$-sensivity of exocytosis observed following PKC activation. Clearly though, SNAP-25 phosphorylation as singular target for the DAG (phorbol ester)/PKC-induced facilitation of synaptic release is an oversimplification of the physiological situation following GPCR activation of PLC.

*PKC – Munc18: regulation of the fusion pore*

Members of the Sec1/Munc18 (SM) group of proteins are found in multiple membrane compartments and are responsible for providing the specificity in transport and docking underpinning the vesicular fusion process mediated by the SNARES. Although Munc-18 was prototypically defined by its ability to bind syntaxin with high affinity (Garcia et al. 1994; Hata et al. 1993; Pevsner et al. 1994a, b; Pevsner and Scheller 1994), sequestering it in a closed conformation before liberation in a primed/open-state (in tandem with Munc13 association, for instance), there is increasing evidence that Munc18 and other SM protein are involved at multiple loci during membrane trafficking and fusion events, independently of syntaxin interaction (Burgoyne and Morgan 2007; Toonen and Verhage 2003). Three mammalian homologues of Munc18 (18-1/2/3/) have been identified with Munc18-1 being involved in regulated exocytosis in contrast to the others being appropriated in intracellular membrane fusion (Halachmi and Lev 1996). Munc18-1 is uniquely phosphorylated by PKC at $Ser^{306}$ and $Ser^{313}$ in vitro (Fujita et al. 1996), with the latter phosphorylation, at least, also shown to occur in situ following treatment with phorbol esters (Barclay et al. 2003; Craig et al. 2003; De Vries et al. 2000) or type 1 metabotropic glutamate receptor activation (Craig et al. 2003). Crucially, phosphorylation of Munc18-1 reduces its affinity for syntaxin in vitro (Fujita et al. 1996), this effect being reproduced by phosphomimicking mutations S306E and S313E in Munc18 (Liu et al. 2004) (Figure 5b). The phosphorylation-dependent interaction of Munc18 with syntaxin could therefore represent the basis for the GPCR/PLC/PKC or phorbol ester facilitation of exocytosis. Phorbol ester stimulation and expression of phosphomimetic Munc18 mutants have mutually occlusive effects in accelerating the kinetics of exocytosis and reducing the quantal size of events (Barclay et al. 2003). The former effect is interpreted as being due to Munc18 controlling fusion pore dilation (Fisher et al. 2001; Graham et al. 2000) (Figure 5a, step 4), with the latter effect perhaps reflecting a switch to a "kiss-and-run" mode of release thought to be promoted by PKC activation among other stimuli (Cousin and Robinson 2000).

Munc18-1 likely has multiple modes of action (Burgoyne and Morgan 2007), both dependent on or independent of direct syntaxin binding, with a distal mode involving binding to preassembled SNARE complexes (Shen et al. 2007). PKC-dependent facilitation of exocytosis could therefore occur by a relief of the inhibitory constraint of Munc18-1 bound to syntaxin together with stimulation of a late stage of fusion perhaps involving the kinase regulating fusion pore dynamics (Fulop et al. 2005; Fulop and Smith 2006). Certainly Munc18-1 appears to be an essential component of PKC-mediated facilitation of exocytosis (Wierda et al. 2007).

*PKC – synaptotagmin?: regulation of the* $Ca^{2+}$*-sensitivity of exocytosis*

Given the reports showing that phorbol esters cause an increase in the $Ca^{2+}$ sensivity of exocytosis at some neuronal synapses, the question arises as to whether the $Ca^{2+}$-sensor is regulated by phosphorylation. Synaptotagmins are the established $Ca^{2+}$-sensors and triggers for exocytosis, functioning by virtue of two C2A and C2B domains, which each bind phospholipids in a $Ca^{2+}$-dependent and $Ca^{2+}$-independent manner respectively (Augustine 2001; Bai and Chapman 2004; Chapman 2002; Koh and Bellen 2003). Synaptotagmin 1, the trigger for rapid/fast exocytosis (Figure 5a, step 4), is phosphorylated by both PKC and $Ca^{2+}$/CaM kinase II at the same $Thr^{112}$ residue, both in vitro and in situ, in response to nerve terminal stimulation (Hilfiker et al. 1999b). Interestingly, phosphorylation of synaptotagmin increases its interaction with the t-SNARES, syntaxin and SNAP-25, particularly in the presence of $Ca^{2+}$ (Verona et al. 2000) (Figure 5b). This therefore represents a potential mechanism by which PKC might increase the $Ca^{2+}$-sensitivity of exocytosis. However, this hypothesis awaits confirmation in the neuronal models which display an increase in release probability in response to PKC activation (i.e. increase in release rate without an increase in RRP), because in the chromaffin cell model where PKC affects the size of the RRP, expression of phospho-mimetic or phospho-null mutants of synaptotagmin 1 is of no consequence on phorbol ester induced facilitation of exocytosis (Nagy et al. 2006).

## Regulation of Phospholipase C and Neurotransmitter Release – PKC-Independent Regulation by DAG through Munc13

Phorbol esters are commonly used as pharmacological analogues of the endogenous second messenger, DAG, produced by GPCR activation of PLC. DAG activates PKCs by binding to C1 domains (cysteine, rich $Zn^{2+}$-finger domains) in the regulatory domain of the kinase (Silinsky and Searl 2003). There are, however, other proteins in the cell that contain C1 domains, which could act as receptors for DAG (or phorbol esters) and behave as effectors, independently of any PKC-dependent phosphorylation. Munc13-1 and its homologues represent one such receptor (Betz et al. 1998), which was originally discovered as the mutation (UNC13) in *Caenorhabditis elegans,* causing a neural defect leading to uncoordinated movements. Munc13-1 is an essential brain-specific protein (Augustin et al. 1999a, b), whose overexpression leads to an enhancement of neurotransmitter release contingent on its binding to β-phorbol esters. This suggests that, in the absence of specific inhibitors of PKC, the faciliatory effects of phorbol esters arise from activation of Munc13-1 together with PKC-dependent pathways, with arguably a predominance of the former influence (Rhee et al. 2002; Searl and Silinsky 1998). However, recent evidence points to an interdependence of the PKC-dependent and PKC-independent effects leading to the potentiation of exocytosis in response to DAG/phorbol ester (Wierda et al. 2007).

Munc13-1 binds the N-terminus of syntaxin (Betz et al. 1997) (Figure 5b), Doc2 (Orita et al. 1997), RIM1 (Betz et al. 2001) and calmodulin (Junge et al. 2004). Munc13-1 plays a priming role in the preparation of vesicles for exocyotosis (Betz et al. 2001; Brose et al. 2000; Silinsky and Searl 2003; Brose, this volume). Augmentation of exocytosis pursuant to this may invoke GPCR/PLC synaptic plasticity (Junge et al. 2004; Wierda et al. 2007).

### Regulation of Phospholipase C and Neurotransmitter Release – PKC-Independent Regulation by $PIP_2$ Modification of Exocytotic Protein Function

As discussed above in relation to ion channels activity, GPCR/PLC-mediated changes in $PIP_2$ concentrations provide potent means of regulation. Part of the ATP dependence of exocytosis certainly results from the requirement of phosphtidylinositol (PI) phosphorylation by PI 4-kinase and PI(4)P 5-kinase to produce $PIP_2$ necessary for vesicle priming (Klenchin and Martin 2000 for review). Alteration of the activity of these lipid kinases and consequent changes in $PIP_2$ levels regulate exocytosis in nerve terminals (Wiedemann et al. 1998) and neuroendocrine cell models (Hay et al. 1995; Holz et al. 2000; Milosevic et al. 2005). In mice in which the type 1 PIP$\gamma$ (PI(4)P 5-kinase) is genetically ablated, levels of $PIP_2$ are decreased in nerve terminals and GTP-$\gamma$S- or depolarization-stimulated elevation of $PIP_2$ is abrogated (DiPaolo et al. 2004; DiPaolo and DeCamilli 2006). Neuronal exocytosis is decreased in these mice following a decrease in the RRP. This, together with an enhancement of rapid depression following intense nerve terminal activity, indicates a lesion in vesicle priming. A similar involvement of $PIP_2$ in priming is indicated by experiments overexpressing PI(4)P 5-kinase$\gamma$ in chromaffin cells where the RRP is enhanced (Milosevic et al. 2005). Notably, there is no apparent compensatory change with respect to exocytosis following alterations of $PIP_2$ levels, suggesting that $PIP_2$ very much sets the gain of the exocytotic system, this therefore being potentially regulatable by GPCR activation of PLC.

Several potential candidates exist that might mediate the effects of $PIP_2$ on exocytosis, including rabphilin (Chung et al. 1998), Munc18-interacting proteins (Mints) (Okamoto and Südhof 1997), $Ca^{2+}$-dependent activator protein for exocytosis (CAPS) (Loyet et al. 1998), and synaptotagmin (Schiavo et al. 1996). Given the putative roles of these binding partners at multiple loci in the exocytotic process, regulation through $PIP_2$ is likely to be multifactorial, with effects on an individual step only being decernible if it is rate-limiting. Thus, for example, while the interaction of synaptotagmin with $PIP_2$ accelerates/aids the insertion of the protein into membranes, and might therefore be predicted to increase the rate of fusion (Bai and Chapman 2004), this is not the primary effect observed in experiments upregulating $PIP_2$ concentrations, possibly because upstream regulation dominates (Milosevic et al. 2005).

Regulation of Phospholipase C and Neurotransmitter Release – Inositol-1,4,5-Trisphosphate($IP_3$) and Intracellular $Ca^{2+}$ Release

GPCRs coupled to $G_q$ can, in many neurons and non-neuronal cells, induce the formation of $IP_3$ and a rise in intracellular $Ca^{2+}$ through release from intracellular $Ca^{2+}$ stores. In some neurons, presynaptic injection of $IP_3$ or a homologue can facilitate evoked transmitter release or enhance spontaneous release (e.g., cholinergic NG108-15 neuroblastoma hybrid cells, Higashida 1988; Aplysia neurons, Chameau et al. 2001; frog neuromuscular junction, Brailoiu et al. 2003, Brailoiu and Dun 2003; crayfish neuromuscular junction, Dropic et al. 2005). Notwithstanding, there is little direct evidence to date that $IP_3$ mediates the facilitatory effects of GPCRs on transmitter release from mammalian nerve endings. Thus, Higashida (1988) suggested that $IP_3$ might be responsible for the enhanced spontaneous release of acetylcholine from processes of NG108-15 neuroblastoma cells by bradykinin (which strongly increases both $IP_3$ and $Ca^{2+}$ in these cells). However, spontaneous release was also enhanced by activating PKC, and it is this latter effect that appears to underlie the enhancement of transmitter release from homologous mammalian sympathetic neurons (Scholze et al. 2002). Probably the most convincing evidence for a role for intracellular $Ca^{2+}$ in transducing the action of a GPCR on transmitter release is for the presynaptic auto-facilitatory action of glutamate on transmission from reticulospinal axons onto motoneurons in the lamprey spinal cord, mediated through activation of type 1 metabotropic glutamate receptors (mGluRs) (Cochilla and Alford 1998). Here, stimulation of these receptors enhanced nerve-evoked presynaptic $Ca^{2+}$ transients and increased evoked transmitter release, and both were attenuated by ryanodine. This can be interpreted to suggest that $IP_3$-induced $Ca^{2+}$ release was amplified by $Ca^{2+}$-induced $Ca^{2+}$ release (CICR) from ryanodine-sensitive stores, and that this amplification was necessary for the facilitation of transmitter release, though no direct evidence for a role for $IP_3$ was advanced. CICR is widely present in presynaptic terminals and is known to play a role in amplifying transmitter release from several types of neuron, and in various forms of synaptic plasticity (Collin et al. 2005), but evidence for its participation in G-protein mediated events remains otherwise sparse.

## 3 Conclusions

**3.1** Trimeric G-proteins (composed of $\alpha$, $\beta$ and $\gamma$ subunits) intermediate between activation of G-protein coupled receptors (GPCRs) and subsequent changes in transmitter release. The most significant G-proteins in this context are $G_o$, $G_q$ and $G_s$.

**3.2** The auto-inhibitory action of most neurotransmitters on transmitter release is mediated primarily by $G_o$; the $\beta\gamma$-subunits of $G_o$ interact directly with voltage-gated $Ca^{2+}$ channels to produce a transient reduction of their voltage-response and hence of the $Ca^{2+}$ influx into the terminals following the action potential. This response is facilitated by a close association between $G_o$, the $Ca^{2+}$ channel and the receptor.

**3.3** Downstream effects of $G\beta\gamma$ on the release machinery itself – specifically, the SNARE complex – also play an important part in the response to some $G_o$-coupled GPCRs.

**3.4** Although GPCRs that activate $G_o$ also activate $G_i$, and thereby open inward-rectifier (GIRK or $K_{ir}3.2$) $K^+$ channels, this appears to play little or no part in their effect on transmitter release – probably because of the paucity of $G_i$ in nerve terminals.

**3.5** Activation of $G_q$ and $G_s$ leads primarily to facilitation of transmitter release, along a slower and more prolonged time-scale than $G_o$-mediated inhibition. This results predominantly from downstream products of $G_q$ and $G_s$ enzyme activation – diacylglycerol (DAG) from $G_q$-activated phospholipase C (PLC) and cyclicAMP (cAMP) from $G_s$-activated adenylate cyclase (AC).

**3.6** Subsequent activation of the kinases, protein kinase C (PKC) and protein kinase A (PKA) and consequent protein phosphorylation can enhance $Ca^{2+}$ channel function, opposing $G_o$-mediated inhibition, or can modify $K^+$ channel activity, so indirectly altering action potential-induced $Ca^{2+}$-entry.

**3.7** However, most of the effects of PKC and PKA activation result from phosphorylation of multiple components of the release machinery itself.

**3.8** DAG can also affect release independently of PKC activation, through an interaction with Munc13. cAMP also has non-PKA-mediated actions (e.g., on hyperpolarization-activated cation channels and cAMP-GEFs), which affect release.

**3.9** Since many of the proteins involved in transmitter release are also sensitive to membrane phosphatidylinositol-4,5-trisposphate ($PIP_2$), the reduction of $PIP_2$ consequent on $G_q$-PLC activation may also play a part in $G_q$-mediated changes in transmitter release.

**3.10** On the other hand, the release of intracellular $Ca^{2+}$ by the other $PIP_2$ product, inositol-1,4,5-trisphosphate ($IP_3$), seems rarely to mediate effects of $G_q$-coupled GPCRs.

## References

Abramow-Newerly M, Roy AA, Nunn C et al (2006) RGS proteins have a signalling complex: interactions between RGS proteins and GPCRs, effectors, and auxiliary proteins. Cell Signal 18:579–91

Ahnert-Hilger G, Wiedenmann B (1994) Requirements for exocytosis in permeabilized neuroendocrine cells. Possible involvement of heterotrimeric G proteins associated with secretory vesicles. AnnNYAcad Sci 733:298–305

Ahnert-Hilger G, Schafer T, Spicher K et al (1994) Detection of G-protein heterotrimers on large dense core and small synaptic vesicles of neuroendocrine and neuronal cells. Eur J Cell Biol 65:26–38
Anger T, Zhang W, Mende U (2004) Differential contribution of GTPase activation and effector antagonism to the inhibitory effect of RGS proteins on Gq-mediated signaling in vivo. J Biol Chem 279:3906–15
Atzori M, Lau D, Tansey EP et al (2000) H2 histamine receptor-phosphorylation of Kv3.2 modulates interneuron fast spiking. Nat Neurosci 3:791–8
Augustin I, Betz A, Herrmann C et al (1999a) Differential expression of two novel Munc13 proteins in rat brain. Biochem J 337:363–71
Augustin I, Rosenmund C, Südhof TC et al (1999b) Munc13-1 is essential for fusion competence of glutamatergic synaptic vesicles. Nature 400:457–61
Augustine GJ (1990) Regulation of transmitter release at the squid giant synapse by presynaptic delayed rectifier potassium current. J Physiol 431:343–64
Augustine GJ (2001) How does calcium trigger neurotransmitter release? Curr Opin Neurobiol 11:320–6
Augustine GJ, Charlton MP (1986) Calcium dependence of presynaptic calcium current and post-synaptic response at the squid giant synapse. J Physiol 381:619–40
Bahler M, Greengard P (1987) Synapsin I bundles F-actin in a phosphorylation-dependent manner. Nature 326:704–7
Bahler M, Benfenati F, Valtorta F et al (1989) Characterization of synapsin I fragments produced by cysteine-specific cleavage: a study of their interactions with F-actin. J Cell Biol 108:1841–9
Bai J, Chapman ER (2004) The C2 domains of synaptotagmin–partners in exocytosis. Trends Biochem Sci 29:143–51
Baldini G, Martelli AM, Tabellini G et al (2005) Rabphilin localizes with the cell actin cytoskeleton and stimulates association of granules with F-actin cross-linked by {alpha}-actinin. J Biol Chem 280:34974–84
Bannister RA, Melliti K, Adams BA (2004) Differential modulation of CaV2.3 Ca2+ channels by Galphaq/11-coupled muscarinic receptors. Mol Pharmacol 65:381–8
Barclay JW, Craig TJ, Fisher RJ et al (2003) Phosphorylation of Munc18 by protein kinase C regulates the kinetics of exocytosis. J Biol Chem 278:10538–45
Barrie AP, Nicholls DG, Sanchez-Prieto J et al (1991) An ion channel locus for the protein kinase C potentiation of transmitter glutamate release from guinea pig cerebrocortical synaptosomes. J Neurochem 57:1398–1404
Bean BP (1989) Neurotransmitter inhibition of neuronal calcium currents by changes in channel voltage dependence. Nature 340:153–6
Beaumont V, Zucker RS (2000) Enhancement of synaptic transmission by cyclic AMP modulation of presynaptic Ih channels. Nat Neurosci 3:133–41
Beech DJ, Bernheim L, Hille B (1992) Pertussis toxin and voltage dependence distinguish multiple pathways modulating calcium channels of rat sympathetic neurons. Neuron 8:97–106
Benfenati F, Bahler M, Jahn R et al (1989a) Interactions of synapsin I with small synaptic vesicles: distinct sites in synapsin I bind to vesicle phospholipids and vesicle proteins. J Cell Biol 108:1863–72
Benfenati F, Greengard P, Brunner J et al (1989b) Electrostatic and hydrophobic interactions of synapsin I and synapsin I fragments with phospholipid bilayers. J Cell Biol 108:1851–62
Benfenati F, Valtorta F, Chieregatti E et al (1992a) Interaction of free and synaptic vesicle-bound synapsin I with F-actin. Neuron 8:377–86
Benfenati F, Valtorta F, Rubenstein JL et al (1992b) Synaptic vesicle-associated Ca2+/calmodulin-dependent protein kinase II is a binding protein for synapsin I. Nature 359:417–20
Benians A, Nobles M, Hosny S et al (2005) Regulators of G-protein signaling form a quaternary complex with the agonist, receptor, and G-protein. A novel explanation for the acceleration of signaling activation kinetics. J Biol Chem 280:13383–94

Berglund K, Midorikawa M, Tachibana M (2002) Increase in the pool size of releasable synaptic vesicles by the activation of protein kinase C in goldfish retinal bipolar cells. J Neurosci 22:4776–85

Bernstein LS, Ramineni S, Hague C et al (2004) RGS2 binds directly and selectively to the M1 muscarinic acetylcholine receptor third intracellular loop to modulate Gq/11alpha signaling. J Biol Chem 279:21248–56

Betz A, Okamoto M, Benseler F et al (1997) Direct interaction of the rat unc-13 homologue Munc13-1 with the N terminus of syntaxin. J Biol Chem 272:2520–6

Betz A, Ashery U, Rickmann M et al (1998) Munc13-1 is a presynaptic phorbol ester receptor that enhances neurotransmitter release. Neuron 21:123–36

Betz A, Thakur P, Junge HJ et al (2001) Functional interaction of the active zone proteins Munc13-1 and RIM1 in synaptic vesicle priming. Neuron 30:183–96

Biel M, Schneider A, and Wahl C (2002) Cardiac HCN channels: structure, function, and modulation. Trends Cardiovasc Med 12:206–12

Blackmer T, Larsen EC, Takahashi M et al (2001) G protein betagamma subunit-mediated presynaptic inhibition: regulation of exocytotic fusion downstream of Ca2+ entry. Science 292:293–7

Blackmer T, Larsen EC, Bartleson C et al (2005) G protein betagamma directly regulates SNARE protein fusion machinery for secretory granule exocytosis. Nat Neurosci 8:421–5

Boczan J, Leenders AG, Sheng ZH (2004) Phosphorylation of syntaphilin by cAMP-dependent protein kinase modulates its interaction with syntaxin-1 and annuls its inhibitory effect on vesicle exocytosis. J Biol Chem 279:18911–19

Boehm S, Kubista H (2002) Fine tuning of sympathetic transmitter release via ionotropic and metabotropic presynaptic receptors. Pharmacol Rev 54:43–99

Borst JG, Sakmann B (1999) Effect of changes in action potential shape on calcium currents and transmitter release in a calyx-type synapse of the rat auditory brainstem. Philos Trans R Soc Lond B Biol Sci 354:347–55

Bos JL (2003) Epac: a new cAMP target and new avenues in cAMP research. Nat Rev Mol Cell Biol 4:733–8

Bourne HR (1997) How receptors talk to trimeric G proteins. Curr Opin Cell Biol 9:134–42

Brailoiu E, Dun NJ (2003) Extra- and intracellular sphingosylphosphorylcholine promote spontaneous transmitter release from frog motor nerve endings. Mol Pharmacol 63:1430–6

Brailoiu E, Miyamoto MD, Dun NJ (2003) Inositol derivatives modulate spontaneous transmitter release at the frog neuromuscular junction. Neuropharmacology 45:691–701

Breukel AI, Wiegant VM, Lopes da Silva FH et al (1998) Presynaptic modulation of cholecystokinin release by protein kinase C in the rat hippocampus. J Neurochem.70:341–8

Brody DL, Yue DT (2000) Relief of G-protein inhibition of calcium channels and short-term synaptic facilitation in cultured hippocampal neurons. J Neurosci 20:889–98

Brose N, Rosenmund C, Rettig J (2000) Regulation of transmitter release by Unc-13 and its homologues. Curr Opin Neurobiol 10:303–11

Brown DA, Buckley NJ, Caulfield MP et al (1995) Coupling of muscarinic acetylcholine receptors to neural ion channels: closure of K+ channels in Molecular Mechanisms of Muscarinic Acetylcholine Receptor Function, ed. J. Wess (Austin, Tx: R.G. Landes Co.)

Brown DA, Hughes SA, Marsh SJ et al (2007) Regulation of M(Kv7.2/7.3) channels in neurons by PIP2 and products of PIP2 hydrolysis: significance for receptor-mediated inhibition. J. Physiol 582: 917–25

Bunemann M, Frank M, and Lohse MJ (2003) Gi protein activation in intact cells involves subunit rearrangement rather than dissociation. Proc Natl Acad Sci USA 100:16077–82

Burgoyne RD, Morgan A (2007) Membrane trafficking: three steps to fusion. Curr Biol 17:R255–R258

Burgoyne RD, Fisher RJ, Graham ME et al (2001) Control of membrane fusion dynamics during regulated exocytosis. Biochem Soc Trans 29:467–72

Burke BE, DeLorenzo RJ (1981) Ca2+- and calmodulin-stimulated endogenous phosphorylation of neurotubulin. Proc Natl Acad Sci USA 78:991–5

Cabrera-Vera TM, Hernandez S, Earls LR et al (2004) RGS9-2 modulates D2 dopamine receptor-mediated Ca2+ channel inhibition in rat striatal cholinergic interneurons. Proc Natl Acad Sci USA 101:16339–44

Calakos N, Schoch S, Südhof TC et al (2004) Multiple roles for the active zone protein RIM1alpha in late stages of neurotransmitter release. Neuron 42:889–96

Calvert PD, Strissel KJ, Schiesser WE et al (2006) Light-driven translocation of signaling proteins in vertebrate photoreceptors. Trends Cell Biol 16:560–8

Capogna M, Gahwiler BH, Thompson SM (1996) Presynaptic inhibition of calcium-dependent and -independent release elicited with ionomycin, gadolinium, and alpha-latrotoxin in the hippocampus. J Neurophysiol 75:2017–28

Castillo PE, Janz R, Südhof TC et al (1997) Rab3A is essential for mossy fibre long-term potentiation in the hippocampus. Nature 388:590–3

Castillo PE, Schoch S, Schmitz F et al (2002) RIM1alpha is required for presynaptic long-term potentiation. Nature 415:327–30

Chaki S, Muramatsu M, Otomo S (1994) Involvement of protein kinase C activation in regulation of acetylcholine release from rat hippocampal slices by minaprine. Neurochem Int 24:37–41

Chameau P, Van d, V, Fossier P et al (2001) Ryanodine-, IP3- and NAADP-dependent calcium stores control acetylcholine release. Pflugers Arch 443:289–96

Chapman ER (2002) Synaptotagmin: a Ca(2+) sensor that triggers exocytosis? Nat Rev Mol Cell Biol 3:498–508

Chen H, Lambert NA (2000) Endogenous regulators of G protein signaling proteins regulate presynaptic inhibition at rat hippocampal synapses. Proc Natl Acad Sci USA 97:12810–15

Chen YA, Duvvuri V, Schulman H et al (1999) Calmodulin and protein kinase C increase Ca(2+)-stimulated secretion by modulating membrane-attached exocytic machinery. J Biol Chem 274:26469–76

Chheda MG, Ashery U, Thakur P et al (2001) Phosphorylation of Snapin by PKA modulates its interaction with the SNARE complex. Nat Cell Biol 3:331–8

Chi P, Greengard P, Ryan TA (2001a) Synapsin dispersion and reclustering during synaptic activity. Nat Neurosci 4:1187–93

Chi P, Greengard P, Ryan TA (2001b) Synapsin dispersion and reclustering during synaptic activity. Nat Neurosci 4:1187–93

Chi P, Greengard P, Ryan TA (2003a) Synaptic vesicle mobilization is regulated by distinct synapsin I phosphorylation pathways at different frequencies. Neuron 38:69–78

Chi P, Greengard P, Ryan TA (2003b) Synaptic vesicle mobilization is regulated by distinct synapsin I phosphorylation pathways at different frequencies. Neuron 38:69–78

Chung SH, Song WJ, Kim K et al (1998) The C2 domains of Rabphilin3A specifically bind phosphatidylinositol 4,5-bisphosphate containing vesicles in a Ca2+-dependent manner. In vitro characteristics and possible significance. J Biol Chem 273:10240–8

Clancy SM, Fowler CE, Finley M et al (2005) Pertussis-toxin-sensitive Galpha subunits selectively bind to C-terminal domain of neuronal GIRK channels: evidence for a heterotrimeric G-protein-channel complex. Mol Cell Neurosci 28:375–89

Cochilla AJ, Alford S (1998) Metabotropic glutamate receptor-mediated control of neurotransmitter release. Neuron 20:1007–16

Coffey ET, Sihra TS, Nicholls DG (1993) Protein kinase C and the regulation of glutamate exocytosis from cerebrocortical synaptosomes. J Biol Chem 268:21060–5

Coffey ET, Herrero I, Sihra TS et al (1994a) Glutamate exocytosis and MARCKS phosphorylation are enhanced by a metabotropic glutamate receptor coupled to a protein kinase C synergistically activated by diacylglycerol and arachidonic acid [published erratum appears in J Neurochem 1995 Jan; 64(1):471]. J Neurochem 63:1303–10

Coffey ET, Sihra TS, Nicholls DG et al (1994b) Phosphorylation of synapsin I and MARCKS in nerve terminals is mediated by $Ca^{2+}$ entry via an Aga-GI sensitive $Ca^{2+}$ channel which is coupled to glutamate exocytosis. FEBS Lett 353:264–8

Colby KA, Blaustein MP (1988) Inhibition of voltage-gated K channels in synaptosomes by sn-1,2-dioctanoylglycerol, an activator of protein kinase C. J Neurosci 8:4685–92

Collin T, Marty A, Llano I (2005) Presynaptic calcium stores and synaptic transmission. Curr Opin Neurobiol 15:275–81
Conklin BR, Farfel Z, Lustig KD et al (1993) Substitution of three amino acids switches receptor specificity of Gq alpha to that of Gi alpha. Nature; 363:274–6
Cousin MA, Robinson PJ (2000) Two mechanisms of synaptic vesicle recycling in rat brain nerve terminals. J Neurochem 75:1645–53
Cousin MA, McLaughlin M, Nicholls DG (1999) Protein kinase C modulates field-evoked transmitter release from cultured rat cerebellar granule cells via a dendrotoxin-sensitive K+ channel. Eur J Neurosci 11:101–9
Cox SL, Schelb V, Trendelenburg AU, Starke K (2000) Enhancement of noradrenaline release by angiotensin II and bradykinin in mouse atria: evidence for cross-talk between $G_{q/11}$ protein and $G_{i/o}$ protein-coupled receptors. Br J Pharmacol 129:1095–1102
Craig TJ, Evans GJ, Morgan A (2003) Physiological regulation of Munc18/nSec1 phosphorylation on serine-313. J Neurochem 86:1450–7
Czernik AJ, Pang DT, and Greengard P (1987) Amino acid sequences surrounding the cAMP-dependent and calcium/calmodulin-dependent phosphorylation sites in rat and bovine synapsin I. Proc Natl Acad Sci USA 84:7518–22
Daniel H, Rancillac A, Crepel F (2004) Mechanisms underlying cannabinoid inhibition of presynaptic Ca2+ influx at parallel fibre synapses of the rat cerebellum. J Physiol 557:159–74
Davies CH, Davies SN, Collingridge GL (1990) Paired-pulse depression of monosynaptic GABA-mediated inhibitory postsynaptic responses in rat hippocampus. J Physiol 424:513–31
De Camilli P, Greengard P (1986a) Synapsin I: a synaptic vesicle-associated neuronal phosphoprotein. Biochem Pharmacol 35:4349–57
De Camilli P, Greengard P (1986b) Synapsin I: a synaptic vesicle-associated neuronal phosphoprotein. Biochem.Pharmacol 35:4349–57
De Camilli P, Ueda T, Bloom FE et al (1979) Widespread distribution of protein I in the central and peripheral nervous systems. Proc Natl Acad Sci USA 76:5977–81
De Camilli P, Cameron R, Greengard P (1983a) Synapsin I (protein I), a nerve terminal-specific phosphoprotein. I. Its general distribution in synapses of the central and peripheral nervous system demonstrated by immunofluorescence in frozen and plastic sections. J Cell Biol 96:1337–54
De Camilli P, Harris SM, Jr., Huttner WB et al (1983b) Synapsin I (Protein I), a nerve terminal-specific phosphoprotein. II. Its specific association with synaptic vesicles demonstrated by immunocytochemistry in agarose-embedded synaptosomes. J Cell Biol 96:1355–73
De Camilli P, Benfenati F, Valtorta F et al (1990) The synapsins. Annu Rev Cell Biol 6:433–60
De Vries KJ, Geijtenbeek A, Brian EC et al (2000) Dynamics of Munc18-1 phosphorylation/dephosphorylation in rat brain nerve terminals. Eur J Neurosci 12:385–90
De Rooij J, Zwartkruis FJ, Verheijen MH et al (1998) Epac is a Rap1 guanine-nucleotide-exchange factor directly activated by cyclic AMP. Nature 396:474–7
De Vries L, Zheng B, Fischer T et al (2000) The regulator of G protein signaling family. Annu Rev Pharmacol Toxicol 40:235–71
Dekker LV, De Graan PN, Pijnappel P et al (1991) Noradrenaline release from streptolysin O-permeated rat cortical synaptosomes: effects of calcium, phorbol esters, protein kinase inhibitors, and antibodies to the neuron-specific protein kinase C substrate B-50 (GAP-43). J Neurochem 56:1146–53
Delmas P, Abogadie FC, Dayrell M et al (1998a) G-proteins and G-protein subunits mediating cholinergic inhibition of N-type calcium currents in sympathetic neurons. Eur J Neurosci 10:1654–66
Delmas P, Brown DA, Dayrell M et al (1998b) On the role of endogenous G-protein beta gamma subunits in N-type Ca2+ current inhibition by neurotransmitters in rat sympathetic neurones. J Physiol 506(Pt 2):319–29
Delmas P, Abogadie FC, Milligan G et al (1999) betagamma dimers derived from Go and Gi proteins contribute different components of adrenergic inhibition of Ca2+ channels in rat sympathetic neurones. J Physiol 518:23–36

Delmas P, Coste B, Gamper N et al (2005) Phosphoinositide lipid second messengers: new paradigms for calcium channel modulation. Neuron 47:179–82

DeLorenzo RJ, Freedman SD, Yohe WB et al (1979) Stimulation of Ca2+-dependent neurotransmitter release and presynaptic nerve terminal protein phosphorylation by calmodulin and a calmodulin-like protein isolated from synaptic vesicles. Proc Natl Acad Sci USA 76:1838–42

DeWaard M, Hering J, Weiss N et al (2005) How do G proteins directly control neuronal Ca2+ channel function? Trends Pharmacol Sci 26:427–36

Digby GJ, Lober RM, Sethi PR et al (2006) Some G protein heterotrimers physically dissociate in living cells. Proc Natl Acad Sci USA 103:17789–94

Ding J, Guzman JN, Tkatch T et al (2006) RGS4-dependent attenuation of M4 autoreceptor function in striatal cholinergic interneurons following dopamine depletion. Nat Neurosci 9:832–42

DiPaolo G, DeCamilli P (2006) Phosphoinositides in cell regulation and membrane dynamics. Nature 443:651–7

DiPaolo G, Moskowitz HS, Gipson K et al (2004) Impaired PtdIns(4,5)P2 synthesis in nerve terminals produces defects in synaptic vesicle trafficking. Nature 431:415–22

Dittman JS, Regehr WG (1996) Contributions of calcium-dependent and calcium-independent mechanisms to presynaptic inhibition at a cerebellar synapse. J Neurosci 16:1623–33

Diverse-Pierluissi M, Goldsmith PK, Dunlap K (1995) Transmitter-mediated inhibition of N-type calcium channels in sensory neurons involves multiple GTP-binding proteins and subunits. Neuron 14:191–200

Dodge FA, Jr., Rahamimoff R (1967) Co-operative action a calcium ions in transmitter release at the neuromuscular junction. J Physiol 193:419–32

Dolphin AC (2003) G protein modulation of voltage-gated calcium channels. Pharmacol Rev 55:607–27

Dolphin AC, Greengard P (1981a) Neurotransmitter- and neuromodulator-dependent alterations in phosphorylation of protein I in slices of rat facial nucleus. J Neurosci 1:192–203

Dolphin AC, Greengard P (1981b) Serotonin stimulates phosphorylation of protein I in the facial motor nucleus of rat brain. Nature 289:76–9

Dolphin AC, Goelz SE, Greengard P (1980) Neuronal protein phosphorylation: recent studies concerning protein I, a synapse-specific phosphoprotein. Pharmacol Biochem Behav 13 Suppl 1 169–74:169–74

Doupnik CA, Davidson N, Lester HA et al (1997) RGS proteins reconstitute the rapid gating kinetics of gbetagamma-activated inwardly rectifying K+ channels. Proc Natl Acad Sci USA 94:10461–6

Doussau F, Augustine GJ (2000) The actin cytoskeleton and neurotransmitter release: an overview. Biochimie 82:353–63

Dropic AJ, Brailoiu E, Cooper RL (2005) Presynaptic mechanism of action induced by 5-HT in nerve terminals: possible involvement of ryanodine and IP3 sensitive 2+ stores. Comp Biochem.Physiol A Mol Integr Physiol 142:355–61

Druey KM (2001) Bridging with GAPs: receptor communication through RGS proteins. Sci STKE 2001:RE14

Druey KM, Sullivan BM, Brown D et al (1998) Expression of GTPase-deficient Gialpha2 results in translocation of cytoplasmic RGS4 to the plasma membrane. J Biol Chem 273:18405–10

Dulubova I, Lou X, Lu J et al (2005) A Munc13/RIM/Rab3 tripartite complex: from priming to plasticity? EMBO J 24:2839–50

Dumaz N, Marais R (2005) Integrating signals between cAMP and the RAS/RAF/MEK/ERK signalling pathways. Based on the anniversary prize of the Gesellschaft fur Biochemie und Molekularbiologie Lecture delivered on 5 July 2003 at the Special FEBS Meeting in Brussels. FEBS J 272:3491–3504

Dunkley PR, Robinson PJ (1986) Depolarization-dependent protein phosphorylation in synaptosomes: mechanisms and significance. Prog Brain Res 69:273–93

Dunkley PR, Baker CM, Robinson PJ (1986) Depolarization-dependent protein phosphorylation in rat cortical synaptosomes: characterization of active protein kinases by phosphopeptide analysis of substrates. J Neurochem 46:1692–1703

Ehrlich I, Elmslie KS (1995) Neurotransmitters acting via different G proteins inhibit N-type calcium current by an identical mechanism in rat sympathetic neurons. J Neurophysiol 74:2251–7

Esser L, Wang CR, Hosaka M et al (1998) Synapsin I is structurally similar to ATP-utilizing enzymes. EMBO J 17:977–84

Evans GJ, Morgan A (2002) Phosphorylation-dependent interaction of the synaptic vesicle proteins cysteine string protein and synaptotagmin I. Biochem J 364:343–7

Evans GJ, Morgan A (2003) Regulation of the exocytotic machinery by cAMP-dependent protein kinase: implications for presynaptic plasticity. Biochem Soc Trans 31:824–7

Evans GJ, Wilkinson MC, Graham ME et al (2001) Phosphorylation of cysteine string protein by protein kinase A. Implications for the modulation of exocytosis. J Biol Chem 276:47877–85

Fernandez-Fernandez JM, Wanaverbecq N, Halley P et al (1999) Selective activation of heterologously expressed G protein-gated K+ channels by M2 muscarinic receptors in rat sympathetic neurones. J Physiol 515:631–7

Fernandez-Fernandez JM, Abogadie FC, Milligan G et al (2001) Multiple pertussis toxin-sensitive G-proteins can couple receptors to GIRK channels in rat sympathetic neurons when expressed heterologously, but only native G(i)-proteins do so in situ. Eur J Neurosci 14:283–92

Fernandez-Peruchena C, Navas S, Montes MA et al (2005) Fusion pore regulation of transmitter release. Brain Res Brain Res Rev 49:406–15

Filippov AK, Fernandez-Fernandez JM, Marsh SJ et al (2004) Activation and inhibition of neuronal G protein-gated inwardly rectifying K(+) channels by P2Y nucleotide receptors. Mol Pharmacol 66:468–77

Finley MF, Scheller RH, Madison DV (2003) SNAP-25 Ser187 does not mediate phorbol ester enhancement of hippocampal synaptic transmission. Neuropharmacology 45:857–62

Fisher RJ, Pevsner J, Burgoyne RD (2001) Control of fusion pore dynamics during exocytosis by Munc18. Science 291:875–8

Fowler CE, Aryal P, Suen KF et al (2007) Evidence for association of GABA(B) receptors with Kir3 channels and regulators of G protein signalling (RGS4) proteins. J Physiol 580:51–65

Fujita S, Inanobe A, Chachin M et al (2000) A regulator of G protein signalling (RGS) protein confers agonist-dependent relaxation gating to a G protein-gated K+ channel. J Physiol 526:341–7

Fujita Y, Sasaki T, Fukui K et al (1996) Phosphorylation of Munc-18/n-Sec1/rbSec1 by protein kinase C: its implication in regulating the interaction of Munc-18/n-Sec1/rbSec1 with syntaxin. J Biol Chem 271:7265–8

Fulop T, Smith C (2006) Physiological stimulation regulates the exocytic mode through calcium activation of protein kinase C in mouse chromaffin cells. Biochem J 399:111–19

Fulop T, Radabaugh S, Smith C (2005) Activity-dependent differential transmitter release in mouse adrenal chromaffin cells. J Neurosci 25:7324–32

Fykse EM, Li C, and Südhof TC (1995) Phosphorylation of rabphilin-3A by Ca2+/calmodulin- and cAMP-dependent protein kinases in vitro. J Neurosci 15:2385–95

Gales C, Van Durm JJ, Schaak S et al (2006) Probing the activation-promoted structural rearrangements in preassembled receptor-G protein complexes. Nat Struct Mol Biol 13:778–86

Gamper N, Reznikov V, Yamada Y et al (2004) Phosphatidylinositol [correction] 4,5-bisphosphate signals underlie receptor-specific Gq/11-mediated modulation of N-type Ca2+ channels. J Neurosci 24:10980–92

Garcia DE, Li B, Garcia-Ferreiro RE et al (1998) G-protein beta-subunit specificity in the fast membrane-delimited inhibition of Ca2+ channels. J Neurosci 18:9163–70

Garcia EP, Gatti E, Butler M et al (1994) A rat brain Sec1 homologue related to Rop and UNC18 interacts with syntaxin. Proc Natl Acad Sci USA 91:2003–7

Genoud S, Pralong W, Riederer BM et al (1999) Activity-dependent phosphorylation of SNAP-25 in hippocampal organotypic cultures. J Neurochem 72:1699–1706

Geppert M, Bolshakov VY, Siegelbaum SA et al (1994) The role of Rab3A in neurotransmitter release. Nature 369:493–7

Geppert M, Goda Y, Stevens CF et al (1997) The small GTP-binding protein Rab3A regulates a late step in synaptic vesicle fusion. Nature 387:810–14

Gerachshenko T, Blackmer T, Yoon EJ et al (2005) Gbetagamma acts at the C terminus of SNAP-25 to mediate presynaptic inhibition. Nat Neurosci 8:597–605

Gereau RW, Conn PJ (1994) Presynaptic enhancement of excitatory synaptic transmission by beta-adrenergic receptor activation. J Neurophysiol 72:1438–42

Gerona RR, Larsen EC, Kowalchyk JA et al (2000) The C terminus of SNAP25 is essential for Ca(2+)-dependent binding of synaptotagmin to SNARE complexes. J Biol Chem 275:6328–36

Gillis KD, Mossner R, Neher E (1996) Protein kinase C enhances exocytosis from chromaffin cells by increasing the size of the readily releasable pool of secretory granules. Neuron 16:1209–20

Gilman AG (1987) G proteins: transducers of receptor-generated signals. Annu Rev Biochem 56:615–49

Giovedi S, Darchen F, Valtorta F et al (2004a) Synapsin is a novel Rab3 effector protein on small synaptic vesicles. II. Functional effects of the Rab3A-synapsin I interaction. J Biol Chem 279:43769–79

Giovedi S, Vaccaro P, Valtorta F et al (2004b) Synapsin is a novel Rab3 effector protein on small synaptic vesicles. I. Identification and characterization of the synapsin I-Rab3 interactions in vitro and in intact nerve terminals. J Biol Chem 279:43760–8

Gonelle-Gispert C, Costa M, Takahashi M et al (2002) Phosphorylation of SNAP-25 on serine-187 is induced by secretagogues in insulin-secreting cells, but is not correlated with insulin secretion. Biochem J 368:223–32

Gotow T, Miyaguchi K, Hashimoto PH (1991) Cytoplasmic architecture of the axon terminal: filamentous strands specifically associated with synaptic vesicles. Neuroscience 40:587–98

Gower H, Rodnight R, Brammer MJ (1986) Ca2+ sensitivity of Ca2+-dependent protein kinase activities toward intrinsic proteins in synaptosomal membrane fragments from rat cerebral tissue. J Neurochem 46:440–7

Graham ME, Fisher RJ, Burgoyne RD (2000) Measurement of exocytosis by amperometry in adrenal chromaffin cells: effects of clostridial neurotoxins and activation of protein kinase C on fusion pore kinetics. Biochimie 82:469–79

Greengard P, Valtorta F, Czernik AJ et al (1993) Synaptic vesicle phosphoproteins and regulation of synaptic function. Science 259:780–5

Gundlfinger A, Bischofberger J, Johenning FW et al (2007) Adenosine modulates transmission at the hippocampal mossy fibre synapse via direct inhibition of presynaptic calcium channels. J Physiol 582:263–77

Halachmi N, Lev Z (1996) The Sec1 family: a novel family of proteins involved in synaptic transmission and general secretion. J Neurochem 66:889–97

Haley JE, Delmas P, Offermanns S et al (2000) Muscarinic inhibition of calcium current and M current in Galpha q-deficient mice. J Neurosci 20:3973–9

Hamid J, Nelson D, Spaetgens R et al (1999) Identification of an integration center for cross-talk between protein kinase C and G protein modulation of N-type calcium channels. J Biol Chem 274:6195–6202

Han J, Mark MD, Li X et al (2006) RGS2 determines short-term synaptic plasticity in hippocampal neurons by regulating Gi/o-mediated inhibition of presynaptic Ca2+ channels. Neuron 51:575–86

Hansel C, Linden DJ, D'Angelo E (2001) Beyond parallel fiber LTD: the diversity of synaptic and non-synaptic plasticity in the cerebellum. Nat Neurosci 4:467–75

Hao J, Michalek C, Zhang W et al (2006) Regulation of cardiomyocyte signaling by RGS proteins: differential selectivity towards G proteins and susceptibility to regulation. J Mol Cell Cardiol 41:51–61

Haruta T, Takami N, Ohmura M et al (1997) Ca2+-dependent interaction of the growth-associated protein GAP-43 with the synaptic core complex. Biochem J 325:455–63

Harvey VL, Stephens GJ (2004) Mechanism of GABA receptor-mediated inhibition of spontaneous GABA release onto cerebellar Purkinje cells. Eur J Neurosci 20:684–700

Hata Y, Slaughter CA, Südhof TC (1993) Synaptic vesicle fusion complex contains unc-18 homologue bound to syntaxin. Nature. 366:347–51

Hay JC, Fisette PL, Jenkins GH et al (1995) ATP-dependent inositide phosphorylation required for Ca(2+)-activated secretion. Nature 374:173–7

He L, Wu XS, Mohan R et al (2006) Two modes of fusion pore opening revealed by cell-attached recordings at a synapse. Nature 444:102–5

He Q, Dent EW, Meiri KF (1997) Modulation of actin filament behavior by GAP-43 (neuromodulin) is dependent on the phosphorylation status of serine 41, the protein kinase C site. J Neurosci 17:3515–24

Hein P, Frank M, Hoffmann C et al (2005) Dynamics of receptor/G protein coupling in living cells. EMBO J 24:4106–14

Hens JJ, De WM, Dekker LV et al (1993) Studies on the role of B-50 (GAP-43) in the mechanism of Ca(2+)-induced noradrenaline release: lack of involvement of protein kinase C after the Ca2+ trigger. J Neurochem 60:1264–73

Hens JJ, De WM, Boomsma F et al (1995) N-terminal-specific anti-B-50 (GAP-43) antibodies inhibit Ca(2+)-induced noradrenaline release, B-50 phosphorylation and dephosphorylation, and calmodulin binding. J Neurochem 64:1127–36

Hens JJ, Ghijsen WE, Weller U et al (1998) Anti-B-50 (GAP-43) antibodies decrease exocytosis of glutamate in permeated synaptosomes. Eur J Pharmacol 363:229–40

Hepler JR, Berman DM, Gilman AG et al (1997) RGS4 and GAIP are GTPase-activating proteins for Gq alpha and block activation of phospholipase C beta by gamma-thio-GTP-Gq alpha. Proc Natl Acad Sci USA 94:428–32

Hepp R, Cabaniols JP, Roche PA (2002) Differential phosphorylation of SNAP-25 in vivo by protein kinase C and protein kinase A. FEBS Lett 532:52–56

Herrero I, Sanchez-Prieto J (1996) cAMP-dependent facilitation of glutamate release by beta-adrenergic receptors in cerebrocortical nerve terminals. J Biol Chem 271:30554–60

Heximer SP, Watson N, Linder ME et al (1997) RGS2/G0S8 is a selective inhibitor of Gqalpha function. Proc Natl Acad Sci USA 94:14389–93

Heximer SP, Srinivasa SP, Bernstein LS et al (1999) G protein selectivity is a determinant of RGS2 function. J Biol Chem 274:34253–9

Heximer SP, Lim H, Bernard JL et al (2001) Mechanisms governing subcellular localization and function of human RGS2. J Biol Chem 276:14195–203

Higashida H (1988) Acetylcholine release by bradykinin, inositol 1,4,5-trisphosphate and phorbol dibutyrate in rodent neuroblastoma cells. J Physiol 397:209–22

Hilfiker S, Pieribone VA, Czernik AJ et al (1999a) Synapsins as regulators of neurotransmitter release. Philos Trans R Soc Lond B Biol Sci 354:269–79

Hilfiker S, Pieribone VA, Nordstedt C et al (1999b) Regulation of synaptotagmin I phosphorylation by multiple protein kinases. J Neurochem 73:921–32

Hille B (1994) Modulation of ion-channel function by G-protein-coupled receptors. Trends Neurosci 17:531–6

Hirling H, Scheller RH (1996) Phosphorylation of synaptic vesicle proteins: modulation of the alpha SNAP interaction with the core complex. Proc Natl Acad Sci USA 93:11945–9

Hirokawa N, Sobue K, Kanda K et al (1989) The cytoskeletal architecture of the presynaptic terminal and molecular structure of synapsin 1. J Cell Biol 108:111–26

Ho MF, Bahler M, Czernik AJ et al (1991) Synapsin I is a highly surface-active molecule. J Biol Chem 266:5600–7

Holz RW, Hlubek MD, Sorensen SD et al (2000) A pleckstrin homology domain specific for phosphatidylinositol 4, 5-bisphosphate (PtdIns-4,5-P2) and fused to green fluorescent protein identifies plasma membrane PtdIns-4,5-P2 as being important in exocytosis. J Biol Chem 275:17878–85

Hosaka M, Südhof TC (1998) Synapsins I and II are ATP-binding proteins with differential Ca2+ regulation. J Biol Chem 273:1425–9

Hosaka M, Südhof TC (1999) Homo- and heterodimerization of synapsins. J Biol Chem 274:16747–53

Hosaka M, Hammer RE, Südhof TC (1999) A phospho-switch controls the dynamic association of synapsins with synaptic vesicles. Neuron 24:377–87

Huang CC, Hsu KS, Gean PW (1996) Isoproterenol potentiates synaptic transmission primarily by enhancing presynaptic calcium influx via P- and/or Q-type calcium channels in the rat amygdala. J Neurosci 16:1025–33

Huang CC, Wang SJ, Gean PW (1998) Selective enhancement of P-type calcium currents by isoproterenol in the rat amygdala. J Neurosci 18:2276–82

Huang CL, Slesinger PA, Casey PJ et al (1995) Evidence that direct binding of G beta gamma to the GIRK1 G protein-gated inwardly rectifying K+ channel is important for channel activation. Neuron 15:1133–43

Huang YY, Li XC, and Kandel ER (1994) cAMP contributes to mossy fiber LTP by initiating both a covalently mediated early phase and macromolecular synthesis-dependent late phase. Cell 79:69–79

Huttner WB, Greengard P (1979) Multiple phosphorylation sites in protein I and their differential regulation by cyclic AMP and calcium. Proc Natl Acad Sci USA 76:5402–6

Iannazzo L (2001) Involvement of B-50 (GAP-43) phosphorylation in the modulation of transmitter release by protein kinase C. Clin Exp Pharmacol Physiol 28:901–4

Ilardi JM, Mochida S, Sheng ZH (1999) Snapin: a SNARE-associated protein implicated in synaptic transmission. Nat Neurosci 2:119–24

Iwasaki S, Kataoka M, Sekiguchi M et al (2000) Two distinct mechanisms underlie the stimulation of neurotransmitter release by phorbol esters in clonal rat pheochromocytoma PC12 cells. J Biochem (Tokyo) 128:407–14

Jarolimek W, Misgeld U (1997) GABAB receptor-mediated inhibition of tetrodotoxin-resistant GABA release in rodent hippocampal CA1 pyramidal cells. J Neurosci 17:1025–32

Jeong SW, Ikeda SR (2000a) Effect of G protein heterotrimer composition on coupling of neurotransmitter receptors to N-type Ca(2+) channel modulation in sympathetic neurons. Proc Natl Acad Sci USA 97:907–12

Jeong SW, Ikeda SR (2000b) Endogenous regulator of G-protein signaling proteins modify N-type calcium channel modulation in rat sympathetic neurons. J Neurosci 20:4489–96

Jeong SW, Ikeda SR (2001) Differential regulation of G protein-gated inwardly rectifying K(+) channel kinetics by distinct domains of RGS8. J Physiol 535:335–47

Jovanovic JN, Benfenati F, Siow YL et al (1996) Neurotrophins stimulate phosphorylation of synapsin I by MAP kinase and regulate synapsin I-actin interactions. Proc Natl Acad Sci USA 93:3679–83

Jovanovic JN, Czernik AJ, Fienberg AA et al (2000) Synapsins as mediators of BDNF-enhanced neurotransmitter release. Nat Neurosci 3:323–9

Jovanovic JN, Sihra TS, Nairn AC et al (2001) Opposing changes in phosphorylation of specific sites in synapsin I during Ca2+-dependent glutamate release in isolated nerve terminals. J Neurosci 21:7944–53

Junge HJ, Rhee JS, Jahn O et al (2004) Calmodulin and Munc13 form a Ca2+ sensor/effector complex that controls short-term synaptic plasticity. Cell. 118:389–401

Kajikawa Y, Saitoh N, Takahashi T (2001) GTP-binding protein beta gamma subunits mediate presynaptic calcium current inhibition by GABA(B) receptor. Proc Natl Acad Sci USA 98 8054–8

Kammermeier PJ, Ikeda SR (1999) Expression of RGS2 alters the coupling of metabotropic glutamate receptor 1a to M-type K+ and N-type Ca2+ channels. Neuron 22:819–29

Kaneko M, Takahashi T (2004) Presynaptic mechanism underlying cAMP-dependent synaptic potentiation. J Neurosci 24:5202–8

Kasai H, Aosaki T (1989) Modulation of Ca-channel current by an adenosine analog mediated by a GTP-binding protein in chick sensory neurons. Pflugers Arch 414:145–9

Kataoka M, Kuwahara R, Iwasaki S et al (2000) Nerve growth factor-induced phosphorylation of SNAP-25 in PC12 cells: a possible involvement in the regulation of SNAP-25 localization. J Neurochem 74:2058–66

Kato M, Sasaki T, Ohya T et al (1996) Physical and functional interaction of rabphilin-3A with alpha-actinin. J Biol Chem 271:31775–8

Kaupp UB, Seifert R (2002) Cyclic nucleotide-gated ion channels. Physiol Rev 82:769–824

Kawasaki H, Springett GM, Mochizuki N et al (1998) A family of cAMP-binding proteins that directly activate Rap1. Science 282:2275–9
Khvotchev M, Lonart G, Südhof TC (2000) Role of calcium in neurotransmitter release evoked by alpha-latrotoxin or hypertonic sucrose. Neuroscience 101:793–802
Kimura M, Saitoh N, Takahashi T (2003) Adenosine A(1) receptor-mediated presynaptic inhibition at the calyx of Held of immature rats. J Physiol 553:415–26
Klenchin VA, Martin TF (2000) Priming in exocytosis: attaining fusion-competence after vesicle docking. Biochimie 82:399–407
Koenig JH, Yamaoka K, Ikeda K (1993) Calcium-induced translocation of synaptic vesicles to the active site. J Neurosci 13:2313–22
Koh DS, Hille B (1997) Modulation by neurotransmitters of catecholamine secretion from sympathetic ganglion neurons detected by amperometry. Proc Natl Acad Sci USA 94:1506–11
Koh TW, Bellen HJ (2003) Synaptotagmin I, a Ca2+ sensor for neurotransmitter release. Trends Neurosci 26:413–22
Krueger BK, Forn J, Greengard P (1977) Depolarization-induced phosphorylation of specific proteins, mediated by calcium ion influx, in rat brain synaptosomes. J Biol Chem 252:2764–73
Kubista H, Boehm S (2006) Molecular mechanisms underlying the modulation of exocytotic noradrenaline release via presynaptic receptors. Pharmacol Ther 112:213–42
Kulik A, Vida I, Fukazawa Y et al (2006) Compartment-dependent colocalization of Kir3.2-containing K+ channels and GABAB receptors in hippocampal pyramidal cells. J Neurosci 26:4289–97
Kuromi H, Kidokoro Y (1998) Two distinct pools of synaptic vesicles in single presynaptic boutons in a temperature-sensitive Drosophila mutant, shibire. Neuron 20:917–25
Lamb TD, Pugh EN, Jr. (1992) A quantitative account of the activation steps involved in phototransduction in amphibian photoreceptors. J Physiol 449:719–58
Landis DM, Hall AK, Weinstein LA et al (1988) The organization of cytoplasm at the presynaptic active zone of a central nervous system synapse. Neuron 1:201–9
Lao G, Scheuss V, Gerwin CM et al (2000) Syntaphilin: a syntaxin-1 clamp that controls SNARE assembly. Neuron 25:191–201
Lechner SG, Hussl S, Schicker KW et al (2005) Presynaptic inhibition via a phospholipase C- and phosphatidylinositol bisphosphate-dependent regulation of neuronal Ca2+ channels. Mol Pharmacol 68:1387–96
Levitzki A, Klein S (2002) G-protein subunit dissociation is not an integral part of G-protein action. Chembiochem 3:815–18
Li C, Takei K, Geppert M et al (1994) Synaptic targeting of rabphilin-3A, a synaptic vesicle Ca2+/phospholipid-binding protein, depends on rab3A/3C. Neuron 13:885–98
Li L, Chin LS, Shupliakov O et al (1995) Impairment of synaptic vesicle clustering and of synaptic transmission, and increased seizure propensity, in synapsin I-deficient mice. Proc Natl Acad Sci USA 92:9235–9
Li Q, Lau A, Morris TJ et al (2004) A syntaxin 1, Galpha(o), and N-type calcium channel complex at a presynaptic nerve terminal: analysis by quantitative immunocolocalization. J Neurosci 24:4070–81
Liu J, Ernst SA, Gladycheva SE et al (2004) Fluorescence resonance energy transfer reports properties of syntaxin1a interaction with Munc18-1 in vivo. J Biol Chem 279:55924–36
Llinas R, Sugimori M, Simon SM (1982) Transmission by presynaptic spike-like depolarization in the squid giant synapse. Proc Natl Acad Sci USA 79:2415–19
Lober RM, Pereira MA, Lambert NA (2006) Rapid activation of inwardly rectifying potassium channels by immobile G-protein-coupled receptors. J Neurosci 26:12602–8
Lonart G (2002) RIM1: an edge for presynaptic plasticity. Trends Neurosci 25:329–32
Lonart G, Südhof TC (1998) Region-specific phosphorylation of rabphilin in mossy fiber nerve terminals of the hippocampus. J Neurosci 18:634–40
Lonart G, Südhof TC (2000) Assembly of SNARE core complexes prior to neurotransmitter release sets the readily releasable pool of synaptic vesicles. J Biol Chem 275:27703–7

Lonart G, Schoch S, Kaeser PS et al (2003) Phosphorylation of RIM1alpha by PKA triggers presynaptic long-term potentiation at cerebellar parallel fiber synapses. Cell 115:49–60

Lou X, Scheuss V, Schneggenburger R (2005) Allosteric modulation of the presynaptic Ca2+ sensor for vesicle fusion. Nature 435:497–501

Loyet KM, Kowalchyk JA, Chaudhary A et al (1998) Specific binding of phosphatidylinositol 4,5-bisphosphate to calcium-dependent activator protein for secretion (CAPS), a potential phosphoinositide effector protein for regulated exocytosis. J Biol Chem 273:8337–43

Luscher C, Jan LY, Stoffel M et al (1997) G protein-coupled inwardly rectifying K+ channels (GIRKs) mediate postsynaptic but not presynaptic transmitter actions in hippocampal neurons. Neuron 19:687–95

Macica CM, von Hehn CA, Wang LY et al (2003) Modulation of the kv3.1b potassium channel isoform adjusts the fidelity of the firing pattern of auditory neurons. J Neurosci 23:1133–41

Majewski H, Iannazzo L (1998) Protein kinase C: a physiological mediator of enhanced transmitter output. Prog Neurobiol 55:463–75

Mandell JW, Townes-Anderson E, Czernik AJ et al (1990a) Synapsins in the vertebrate retina: absence from ribbon synapses and heterogeneous distribution among conventional synapses. Neuron 5:19–33

Mandell JW, Townes-Anderson E, Czernik AJ et al (1990b) Synapsins in the vertebrate retina: absence from ribbon synapses and heterogeneous distribution among conventional synapses. Neuron 5:19–33

Mandell JW, Czernik AJ, De Camilli P et al (1992) Differential expression of synapsins I and II among rat retinal synapses. J Neurosci 12:1736–49

Marchetti C, Carbone E, and Lux HD (1986) Effects of dopamine and noradrenaline on Ca channels of cultured sensory and sympathetic neurons of chick. Pflugers Arch 406:104–11

Mark MD, Wittemann S, Herlitze S (2000) G protein modulation of recombinant P/Q-type calcium channels by regulators of G protein signalling proteins. J Physiol 528 Pt 1:65–77

Masuho I, Itoh M, Itoh H et al (2004) The mechanism of membrane-translocation of regulator of G-protein signaling (RGS) 8 induced by Galpha expression. J Neurochem 88:161–8

Matsubara M, Kusubata M, Ishiguro K et al (1996) Site-specific phosphorylation of synapsin I by mitogen-activated protein kinase and Cdk5 and its effects on physiological functions. J Biol Chem 271:21108–13

Melliti K, Meza U, Adams BA (2001) RGS2 blocks slow muscarinic inhibition of N-type Ca(2+) channels reconstituted in a human cell line. J Physiol 532:337–47

Mellor J, Nicoll RA, Schmitz D (2002) Mediation of hippocampal mossy fiber long-term potentiation by presynaptic Ih channels. Science 295:143–7

Millan C, Sanchez-Prieto J (2002) Differential coupling of N- and P/Q-type calcium channels to glutamate exocytosis in the rat cerebral cortex. Neurosci Lett 330:29–32

Millan C, Lujan R, Shigemoto R et al (2002) The inhibition of glutamate release by metabotropic glutamate receptor 7 affects both [Ca2+]c and cAMP: evidence for a strong reduction of Ca2+ entry in single nerve terminals. J Biol Chem 277:14092–101

Millan C, Castro E, Torres M et al (2003) Co-expression of metabotropic glutamate receptor 7 and N-type Ca(2+) channels in single cerebrocortical nerve terminals of adult rats. J Biol Chem 278:23955–62

Miller RJ (1998) Presynaptic receptors. Annu Rev Pharmacol Toxicol 38:201–27

Milosevic I, Sorensen JB, Lang T et al (2005) Plasmalemmal phosphatidylinositol-4,5-bisphosphate level regulates the releasable vesicle pool size in chromaffin cells. J Neurosci 25:2557–65

Miroznik RR, Zheng X, Stanley EF (2000) G-Protein types involved in calcium channel inhibition at a presynaptic nerve terminal. J Neurosci 20:7614–21

Misonou H, Ohara-Imaizumi M, Kumakura K (1997) Regulation of the priming of exocytosis and the dissociation of SNAP-25 and VAMP-2 in adrenal chromaffin cells. Neurosci Lett 232:182–4

Misonou H, Ohara-Imaizumi M, Murakami T et al (1998) Protein kinase C controls the priming step of regulated exocytosis in adrenal chromaffin cells. Cell Mol Neurobiol 18:379–90

Moldavan MG, Irwin RP, Allen CN (2006) Presynaptic GABA(B) receptors regulate retinohypothalamic tract synaptic transmission by inhibiting voltage-gated Ca2+ channels. J Neurophysiol 95:3727–41
Morgan A, Burgoyne RD (1992) Interaction between protein kinase C and Exo1 (14-3-3 protein) and its relevance to exocytosis in permeabilized adrenal chromaffin cells. Biochem J 286:807–11
Nagy G, Matti U, Nehring RB et al (2002) Protein kinase C-dependent phosphorylation of synaptosome-associated protein of 25 kDa at Ser187 potentiates vesicle recruitment. J Neurosci 22:9278–86
Nagy G, Reim K, Matti U et al (2004) Regulation of releasable vesicle pool sizes by protein kinase A-dependent phosphorylation of SNAP-25. Neuron 41:417–29
Nagy G, Kim JH, Pang ZP et al (2006) Different effects on fast exocytosis induced by synaptotagmin 1 and 2 isoforms and abundance but not by phosphorylation. J Neurosci 26:632–43
Naor Z, Dan-Cohen H, Hermon J et al (1989) Induction of exocytosis in permeabilized pituitary cells by alpha- and beta-type protein kinase C. Proc Natl Acad Sci USA 86:4501–4
Neitzel KL, Hepler JR (2006) Cellular mechanisms that determine selective RGS protein regulation of G protein-coupled receptor signaling. Semin Cell Dev Biol 17:383–9
Nguyen PV, Woo NH (2003) Regulation of hippocampal synaptic plasticity by cyclic AMP-dependent protein kinases. Prog Neurobiol 71:401–37
Nicoll RA, Malenka RC (1995) Contrasting properties of two forms of long-term potentiation in the hippocampus. Nature 377:115–18
Nobles M, Benians A, Tinker A (2005) Heterotrimeric G proteins precouple with G protein-coupled receptors in living cells. Proc Natl Acad Sci USA 102:18706–11
Ohara-Imaizumi M, Kameyama K, Kawae N et al (1992) Regulatory role of the GTP-binding protein, G(o), in the mechanism of exocytosis in adrenal chromaffin cells. J Neurochem 58:2275–84
Okamoto M, Südhof TC (1997) Mints, Munc18-interacting proteins in synaptic vesicle exocytosis. J Biol Chem 272:31459–64
Oldham WM, Hamm HE (2006) Structural basis of function in heterotrimeric G proteins. Q Rev Biophys 39:117–66
Orita S, Naito A, Sakaguchi G et al (1997) Physical and functional interactions of Doc2 and Munc13 in Ca2+-dependent exocytotic machinery. J Biol Chem 272:16081–4
Ozaki N, Shibasaki T, Kashima Y et al (2000) cAMP-GEFII is a direct target of cAMP in regulated exocytosis. Nat Cell Biol 2:805–11
Park D, Dunlap K (1998) Dynamic regulation of calcium influx by G-proteins, action potential waveform, and neuronal firing frequency. J Neurosci 18:6757–66
Parsons TD, Coorssen JR, Horstmann H et al (1995) Docked granules, the exocytic burst, and the need for ATP hydrolysis in endocrine cells. Neuron 15:1085–96
Patil PG, de LM, Reed RR et al (1996) Elementary events underlying voltage-dependent G-protein inhibition of N-type calcium channels. Biophys J 71:2509–21
Peleg S, Varon D, Ivanina T et al (2002) G(alpha)(i) controls the gating of the G protein-activated K(+) channel, GIRK. Neuron 33:87–99
Petrucci TC, Morrow JS (1987) Synapsin I: an actin-bundling protein under phosphorylation control. J Cell Biol 105:1355–63
Petrucci TC, Macioce P, Paggi P (1991) Axonal transport kinetics and posttranslational modification of synapsin I in mouse retinal ganglion cells. J Neurosci 11:2938–46
Pevsner J, Scheller RH (1994) Mechanisms of vesicle docking and fusion: insights from the nervous system. Curr Opin Cell Biol 6:555–60
Pevsner J, Hsu SC, Braun JE et al (1994a) Specificity and regulation of a synaptic vesicle docking complex. Neuron 13:353–61
Pevsner J, Hsu SC, Scheller RH (1994b) n-Sec1: a neural-specific syntaxin-binding protein. Proc Natl Acad Sci USA 91:1445–9

Pfrieger FW, Gottmann K, Lux HD (1994) Kinetics of GABAB receptor-mediated inhibition of calcium currents and excitatory synaptic transmission in hippocampal neurons in vitro. Neuron 12:97–107

Photowala H, Blackmer T, Schwartz E et al (2006) G protein betagamma-subunits activated by serotonin mediate presynaptic inhibition by regulating vesicle fusion properties. Proc Natl Acad Sci USA 103:4281–6

Pieribone VA, Shupliakov O, Brodin L et al (1995) Distinct pools of synaptic vesicles in neurotransmitter release. Nature 375:493–7

Pieribone VA, Porton B, Rendon B et al (2002) Expression of synapsin III in nerve terminals and neurogenic regions of the adult brain. J Comp Neurol 454:105–14

Ponce A, Bueno E, Kentros C et al (1996) G-protein-gated inward rectifier K+ channel proteins (GIRK1) are present in the soma and dendrites as well as in nerve terminals of specific neurons in the brain. J Neurosci 16:1990–2001

Prince DA, Stevens CF (1992) Adenosine decreases neurotransmitter release at central synapses. Proc Natl Acad Sci USA 89:8586–90

Qian J, Colmers WF, Saggau P (1997) Inhibition of synaptic transmission by neuropeptide Y in rat hippocampal area CA1: modulation of presynaptic Ca2+ entry. J Neurosci 17:8169–77

Queiroz G, Talaia C, Gonçalves J (2003) Adenosine $A_{2A}$ receptor-mediated facilitation of noradrenaline release involves protein kinase C activation and attenuation of presynaptic inhibitory receptor-mediated effects in the rat vas deferens. J Neurochem 85:740–8

Raingo J, Castiglioni AJ, Lipscombe D (2007) Alternative splicing controls G protein-dependent inhibition of N-type calcium channels in nociceptors. Nat Neurosci 10:285–92

Ramakers GM, McNamara RK, Lenox RH et al (1999) Differential changes in the phosphorylation of the protein kinase C substrates myristoylated alanine-rich C kinase substrate and growth-associated protein-43/B-50 following Schaffer collateral long-term potentiation and long-term depression. J Neurochem 73:2175–83

Rhee JS, Betz A, Pyott S et al (2002) Beta phorbol ester- and diacylglycerol-induced augmentation of transmitter release is mediated by Munc13s and not by PKCs. Cell 108:121–33

Richman RW, Strock J, Hains MD et al (2005) RGS12 interacts with the SNARE-binding region of the Cav2.2 calcium channel. J Biol Chem 280:1521–8

Rimon G, Hanski E, Braun S et al (1978) Mode of coupling between hormone receptors and adenylate cyclase elucidated by modulation of membrane fluidity. Nature 276:394–6

Risinger C, Bennett MK (1999) Differential phosphorylation of syntaxin and synaptosome-associated protein of 25 kDa (SNAP-25) isoforms. J Neurochem 72:614–24

Robinson PJ (1991) The role of protein kinase C and its neuronal substrates dephosphin, B-50, and MARCKS in neurotransmitter release. Mol Neurobiol 5:87–130

Robinson PJ, Dunkley PR (1983) Depolarisation-dependent protein phosphorylation in rat cortical synaptosomes: factors determining the magnitude of the response. J Neurochem 41:909–18

Robinson PJ, Dunkley PR (1985) Depolarisation-dependent protein phosphorylation and dephosphorylation in rat cortical synaptosomes is modulated by calcium. J Neurochem 44:338–48

Robinson PJ, Dunkley PR (1987) Altered protein phosphorylation in intact rat cortical synaptosomes after in vivo administration of fluphenazine. Biochem. Pharmacol 36:2203–8

Rodriguez-Moreno A, Sihra TS (2004) Presynaptic kainate receptor facilitation of glutamate release involves protein kinase A in the rat hippocampus. J Physiol 557:733–45

Romano C, Nichols RA, Greengard P (1987a) Synapsin I in PC12 cells. II. Evidence for regulation by NGF of phosphorylation at a novel site. J Neurosci 7:1300–6

Romano C, Nichols RA, Greengard P et al (1987b) Synapsin I in PC12 cells. I. Characterization of the phosphoprotein and effect of chronic NGF treatment. J Neurosci 7:1294–9

Rosahl TW, Geppert M, Spillane D et al (1993) Short-term synaptic plasticity is altered in mice lacking synapsin I. Cell 75:661–70

Rosahl TW, Spillane D, Missler M et al (1995) Essential functions of synapsins I and II in synaptic vesicle regulation [see comments]. Nature 375:488–93

Rose SD, Lejen T, Zhang L et al (2001) Chromaffin cell F-actin disassembly and potentiation of catecholamine release in response to protein kinase C activation by phorbol esters is mediated through myristoylated alanine-rich C kinase substrate phosphorylation. J Biol Chem 276:36757–63

Rosenmund C, Stevens CF (1996) Definition of the readily releasable pool of vesicles at hippocampal synapses. Neuron 16:1197–1207

Rosenzweig DH, Nair KS, Wei J et al (2007) Subunit dissociation and diffusion determine the subcellular localization of rod and cone transducins. J Neurosci 27:5484–94

Ruiz-Velasco V, Ikeda SR (2000) Multiple G-protein betagamma combinations produce voltage-dependent inhibition of N-type calcium channels in rat superior cervical ganglion neurons. J Neurosci 20:2183–91

Saitoh O, Kubo Y, Miyatani Y et al (1997) RGS8 accelerates G-protein-mediated modulation of K+ currents. Nature 390:525–9

Saitoh O, Masuho I, Terakawa I et al (2001) Regulator of G protein signaling 8 (RGS8) requires its NH2 terminus for subcellular localization and acute desensitization of G protein-gated K+ channels. J Biol Chem 276:5052–8

Saitoh O, Murata Y, Odagiri M et al (2002) Alternative splicing of RGS8 gene determines inhibitory function of receptor type-specific Gq signaling. Proc Natl Acad Sci USA 99:10138–43

Saitow F, Suzuki H, Konishi S (2005) beta-Adrenoceptor-mediated long-term up-regulation of the release machinery at rat cerebellar GABAergic synapses. J Physiol 565:487–502

Sakaba T, Neher E (2001) Preferential potentiation of fast-releasing synaptic vesicles by cAMP at the calyx of Held. Proc Natl Acad Sci U S A 98:331–6

Sakaba T, Neher E (2003) Direct modulation of synaptic vesicle priming by GABA(B) receptor activation at a glutamatergic synapse. Nature 424:775–8

Salin PA, Malenka RC, Nicoll RA (1996) Cyclic AMP mediates a presynaptic form of LTP at cerebellar parallel fiber synapses. Neuron 16:797–803

Sanchez-Prieto J, Budd DC, Herrero I et al (1996) Presynaptic receptors and the control of glutamate exocytosis. Trends Neurosci 19:235–9

Sara Y, Virmani T, Deak F et al (2005) An isolated pool of vesicles recycles at rest and drives spontaneous neurotransmission. Neuron 45:563–73

Sato M, Blumer JB, Simon V et al (2006) Accessory proteins for G proteins: partners in signaling. Annu Rev Pharmacol Toxicol 46:151–87

Scanziani M, Capogna M, Gahwiler BH et al (1992) Presynaptic inhibition of miniature excitatory synaptic currents by baclofen and adenosine in the hippocampus. Neuron 9:919–27

Scanziani M, Gahwiler BH, Thompson SM (1995) Presynaptic inhibition of excitatory synapic transmission by muscarinic and metabotropic glutamate receptor activation in the hippocampus: are Ca2+ channels involved? Neuropharmacology 34:1549–1775

Scheller RH (1995) Membrane trafficking in the presynaptic nerve terminal. Neuron 14:893897

Schiavo G, Gu QM, Prestwich GD et al (1996) Calcium-dependent switching of the specificity of phosphoinositide binding to synaptotagmin. Proc Natl Acad Sci USA 93:13327–32

Schiebler W, Jahn R, Doucet JP et al (1986) Characterization of synapsin I binding to small synaptic vesicles [published erratum appears in J Biol Chem 1986 Sep 15; 261(26):12428]. J Biol Chem 261:8383–90

Schluter OM, Schnell E, Verhage M et al (1999) Rabphilin knock-out mice reveal that rabphilin is not required for rab3 function in regulating neurotransmitter release. J Neurosci 19:5834–46

Schluter OM, Schmitz F, Jahn R et al (2004) A complete genetic analysis of neuronal Rab3 function. J Neurosci 24:6629–37

Schneggenburger R, Forsythe ID (2006) The calyx of Held. Cell Tissue Res 326:311–37

Schneider T, Igelmund P, Hescheler J (1997) G protein interaction with K+ and Ca2+ channels. Trends PharmacolSci 18:8–11

Schoch S, Castillo PE, Jo T et al (2002) RIM1alpha forms a protein scaffold for regulating neurotransmitter release at the active zone. Nature 415:321–6

Scholz KP, Miller RJ (1992) Inhibition of quantal transmitter release in the absence of calcium influx by a G protein-linked adenosine receptor at hippocampal synapses. Neuron 8:1139–50

Scholze T, Moskvina E, Mayer M et al (2002) Sympathoexcitation by bradykinin involves Ca2+-independent protein kinase C. J Neurosci 22:5823–32
Schwartz EJ, Blackmer T, Gerachshenko T et al (2007) Presynaptic G-protein-coupled receptors regulate synaptic cleft glutamate via transient vesicle fusion. J Neurosci 27:5857–68
Searl TJ, Silinsky EM (1998) Increases in acetylcholine release produced by phorbol esters are not mediated by protein kinase C at motor nerve endings. J Pharmacol Exp Ther 285:247–51
Seino S, Shibasaki T (2005) PKA-dependent and PKA-independent pathways for cAMP-regulated exocytosis. Physiol Rev 85:1303–42
Shen J, Tareste DC, Paumet F et al (2007) Selective activation of cognate SNAREpins by Sec1/Munc18 proteins. Cell 128:183–95
Shimazaki Y, Nishiki T, Omori A et al (1996) Phosphorylation of 25-kDa synaptosome-associated protein. Possible involvement in protein kinase C-mediated regulation of neurotransmitter release. J Biol Chem 271:14548–53
Shirataki H, Kaibuchi K, Sakoda T et al (1993) Rabphilin-3A, a putative target protein for smg p25A/rab3A p25 small GTP-binding protein related to synaptotagmin. Mol Cell Biol 13:2061–8
Shoji-Kasai Y, Itakura M, Kataoka M et al (2002) Protein kinase C-mediated translocation of secretory vesicles to plasma membrane and enhancement of neurotransmitter release from PC12 cells. Eur J Neurosci 15:1390–4
Sieghart W, Schulman H, Greengard P (1980) Neuronal localization of Ca2+-dependent protein phosphorylation in brain. J Neurochem 34:548–53
Sihra TS (1993) Glutamate release from isolated nerve terminals: modulatory role of protein phosphorylation and dephosphorylation. Biochem Soc Trans 21:410–16
Sihra TS, Wang JK, Gorelick FS et al (1989) Translocation of synapsin I in response to depolarization of isolated nerve terminals. Proc Natl Acad Sci USA 86:8108–12
Sihra TS, Bogonez E, Nicholls DG (1992) Localized $Ca^{2+}$ entry preferentially effects protein dephosphorylation, phosphorylation, and glutamate release. J Biol Chem 267:1983–9
Silinsky EM (1984) On the mechanism by which adenosine receptor activation inhibits the release of acetylcholine from motor nerve endings. J Physiol 346:243–56
Silinsky EM, Searl TJ (2003) Phorbol esters and neurotransmitter release: more than just protein kinase C? Br J Pharmacol 138:1191–1201
Smith C (1999) A persistent activity-dependent facilitation in chromaffin cells is caused by Ca2+ activation of protein kinase C. J Neurosci 19:589–98
Smith C, Moser T, Xu T et al (1998) Cytosolic Ca2+ acts by two separate pathways to modulate the supply of release-competent vesicles in chromaffin cells. Neuron 20:1243–53
Sorensen JB, Matti U, Wei SH et al (2002) The SNARE protein SNAP-25 is linked to fast calcium triggering of exocytosis. Proc Natl Acad Sci USA 99:1627–32
Southan AP, Morris NP, Stephens GJ et al (2000) Hyperpolarization-activated currents in presynaptic terminals of mouse cerebellar basket cells. J Physiol 526 Pt 1:91–7
Spencer SA, Schuh SM, Liu WS et al (1992) GAP-43, a protein associated with axon growth, is phosphorylated at three sites in cultured neurons and rat brain. J Biol Chem 267:9059–64
Springett GM, Kawasaki H, Spriggs DR (2004) Non-kinase second-messenger signaling: new pathways with new promise. Bioessays 26:730–8
Stanfield PR, Nakajima S, Nakajima Y (2002) Constitutively active and G-protein coupled inward rectifier K+ channels: Kir2.0 and Kir3.0. Rev Physiol Biochem Pharmacol 145:47–179
Stea A, Soong TW, Snutch TP (1995) Determinants of PKC-dependent modulation of a family of neuronal calcium channels. Neuron 15:929–40
Stephens GJ, Mochida S (2005) G protein {beta}{gamma} subunits mediate presynaptic inhibition of transmitter release from rat superior cervical ganglion neurones in culture. J Physiol 563:765–76
Stevens CF, Sullivan JM (1998) Regulation of the readily releasable vesicle pool by protein kinase C. Neuron 21:885–93
Stone LM, Browning MD, Finger TE (1994) Differential distribution of the synapsins in the rat olfactory bulb. J Neurosci 14:301–9

Storm DR, Hansel C, Hacker B et al (1998) Impaired cerebellar long-term potentiation in type I adenylyl cyclase mutant mice. Neuron 20:1199–1210
Strittmatter SM, Valenzuela D, Kennedy TE et al (1990) G0 is a major growth cone protein subject to regulation by GAP-43. Nature 344:836–41
Strock J, Diverse-Pierluissi MA (2004) Ca2+ channels as integrators of G protein-mediated signaling in neurons. Mol Pharmacol 66:1071–6
Südhof TC (1995) The synaptic vesicle cycle: a cascade of protein-protein interactions. Nature 375:645–53
Südhof TC (2000) The synaptic vesicle cycle revisited. Neuron 28:317–20
Südhof TC (2004) The synaptic vesicle cycle. Annu Rev Neurosci 27:509–47
Südhof TC, Czernik AJ, Kao HT et al (1989) Synapsins: mosaics of shared and individual domains in a family of synaptic vesicle phosphoproteins. Science 245:1474–80
Suh BC, Horowitz LF, Hirdes W et al (2004) Regulation of KCNQ2/KCNQ3 current by G protein cycling: the kinetics of receptor-mediated signaling by Gq. J Gen Physiol 123:663–83
Surprenant A, North RA (1988) Mechanism of synaptic inhibition by noradrenaline acting at alpha 2-adrenoceptors. Proc R Soc Lond B Biol Sci 234:85–114
Takahashi T, Forsythe ID, Tsujimoto T et al (1996) Presynaptic calcium current modulation by a metabotropic glutamate receptor. Science 274:594–7
Takahashi T, Kajikawa Y, Tsujimoto T (1998) G-Protein-coupled modulation of presynaptic calcium currents and transmitter release by a GABAB receptor. J Neurosci 18:3138–46
Takahashi M, Itakura M, Kataoka M (2003) New aspects of neurotransmitter release and exocytosis: regulation of neurotransmitter release by phosphorylation. J Pharmacol Sci 93:41–5
Takano K, Stanfield PR, Nakajima S et al (1995) Protein kinase C-mediated inhibition of an inward rectifier potassium channel by substance P in nucleus basalis neurons. Neuron 14:999–1008
Takei Y, Harada A, Takeda S et al (1995) Synapsin I deficiency results in the structural change in the presynaptic terminals in the murine nervous system. J Cell Biol 131:1789–1800
Tedford HW, Zamponi GW (2006) Direct G protein modulation of Cav2 calcium channels. Pharmacol Rev 58:837–62
Terbush DR, Holz RW (1986) Effects of phorbol esters, diglyceride, and cholinergic agonists on the subcellular distribution of protein kinase C in intact or digitonin-permeabilized adrenal chromaffin cells. J Biol Chem 261:17099–106
Terbush DR, Holz RW (1990) Activation of protein kinase C is not required for exocytosis from bovine adrenal chromaffin cells. The effects of protein kinase C(19-31), Ca/CaM kinase II(291–317), and staurosporine. J Biol Chem 265:21179–84
Tesmer VM, Kawano T, Shankaranarayanan A et al (2005) Snapshot of activated G proteins at the membrane: the Galphaq-GRK2-Gbetagamma complex. Science 310:1686–90
Thakur P, Stevens DR, Sheng ZH et al (2004) Effects of PKA-mediated phosphorylation of Snapin on synaptic transmission in cultured hippocampal neurons. J Neurosci 24:6476–81
Thompson SM, Capogna M, Scanziani M (1993) Presynaptic inhibition in the hippocampus. Trends Neurosci 16:222–7
Tinker A (2006) The selective interactions and functions of regulators of G-protein signalling. Semin Cell Dev Biol 17:377–82
Toonen RF, Verhage M (2003) Vesicle trafficking: pleasure and pain from SM genes. Trends Cell Biol 13:177–86
Torri-Tarelli F, Villa A, Valtorta F et al (1990) Redistribution of synaptophysin and synapsin I during alpha-latrotoxin-induced release of neurotransmitter at the neuromuscular junction. J Cell Biol 110:449–59
Toth PT, Bindokas VP, Bleakman D et al (1993) Mechanism of presynaptic inhibition by neuropeptide Y at sympathetic nerve terminals. Nature 364:635–9
Toutant M, Aunis D, Bockaert J et al (1987) Presence of three pertussis toxin substrates and Go alpha immunoreactivity in both plasma and granule membranes of chromaffin cells. FEBS Lett 215:339–44

Trautwein W, Osterrieder W, and Noma A (1980) Potassium channels and the muscarinic receptor in the sino-atrial node of the heart in Drug Receptors and their Effectors, ed. N.J.M. Birdsall (New York: Macmillan) pp 5–22

Trifaro JM, Lejen T, Rose SD et al (2002) Pathways that control cortical F-actin dynamics during secretion. Neurochem Res 27:1371–85

Trudeau LE, Doyle RT, Emery DG et al (1996a) Calcium-independent activation of the secretory apparatus by ruthenium red in hippocampal neurons: a new tool to assess modulation of presynaptic function. J Neurosci 16:46–54

Trudeau LE, Emery DG, Haydon PG (1996b) Direct modulation of the secretory machinery underlies PKA-dependent synaptic facilitation in hippocampal neurons. Neuron 17:789–97

Trudeau LE, Fang Y, and Haydon PG (1998) Modulation of an early step in the secretory machinery in hippocampal nerve terminals. Proc Natl Acad Sci USA 95:7163–8

Tsuboi T, Kikuta T, Warashina A et al (2001) Protein kinase C-dependent supply of secretory granules to the plasma membrane. Biochem Biophys Res Commun 282:621–8

Tsunoo A, Yoshii M, Narahashi T (1986) Block of calcium channels by enkephalin and somatostatin in neuroblastoma-glioma hybrid NG108-15 cells. Proc Natl Acad Sci USA 83:9832–6

Turner KM, Burgoyne RD, Morgan A (1999) Protein phosphorylation and the regulation of synaptic membrane traffic. Trends Neurosci 22:459–64

Valtorta F, Villa A, Jahn R et al (1988) Localization of synapsin I at the frog neuromuscular junction. Neuroscience 24:593–603

Verona M, Zanotti S, Schafer T et al (2000) Changes of synaptotagmin interaction with t-SNARE proteins in vitro after calcium/calmodulin-dependent phosphorylation. J Neurochem 74:209–21

Villacres EC, Wong ST, Chavkin C et al (1998) Type I adenylyl cyclase mutant mice have impaired mossy fiber long-term potentiation. J Neurosci 18:3186–94

Vitale ML, Seward EP, Trifaro JM (1995) Chromaffin cell cortical actin network dynamics control the size of the release-ready vesicle pool and the initial rate of exocytosis. Neuron 14:353–63

Vitale N, Gasman S, Caumont AS et al (2000) Insight in the exocytotic process in chromaffin cells: regulation by trimeric and monomeric G proteins. Biochimie. 82:355–73

Walaas SI, Nairn AC, Greengard P (1983) Regional distribution of calcium- and cyclic adenosine 3′ : 5′-monophosphate-regulated protein phosphorylation systems in mammalian brain. I. Particulate systems. J Neurosci 3:291–301

Wang C, Zucker RS (1998) Regulation of synaptic vesicle recycling by calcium and serotonin. Neuron 21:155–67

Wang HS, McKinnon D (1995) Potassium currents in rat prevertebral and paravertebral sympathetic neurones: control of firing properties. J Physiol 485 (Pt 2):319–35

Wang JK, Walaas SI, Sihra TS et al (1989) Phosphorylation and associated translocation of the 87-kDa protein, a major protein kinase C substrate, in isolated nerve terminals. Proc Natl Acad Sci USA 86:2253–6

Wang SJ, Sihra TS (2003) Opposing facilitatory and inhibitory modulation of glutamate release elicited by cAMP production in cerebrocortical nerve terminals (synaptosomes). Neuropharm 44:686–97

Wang SJ, Coutinho V, Sihra TS (2002) Presynaptic cross-talk of beta-adrenoreceptor and 5-hydroxytryptamine receptor signalling in the modulation of glutamate release from cerebrocortical nerve terminals. Br J Pharmacol 137:1371–9

Wang X, Zeng W, Soyombo AA et al (2005) Spinophilin regulates Ca2+ signalling by binding the N-terminal domain of RGS2 and the third intracellular loop of G-protein-coupled receptors. Nat Cell Biol 7:405–11

Wang Y, Okamoto M, Schmitz F et al (1997) Rim is a putative Rab3 effector in regulating synaptic-vesicle fusion. Nature 388:593–8

Weisskopf MG, Castillo PE, Zalutsky RA et al (1994) Mediation of hippocampal mossy fiber long-term potentiation by cyclic AMP. Science 265:1878–82

Wettschureck N, Offermanns S (2005) Mammalian G proteins and their cell type specific functions. Physiol Rev 85:1159–1204

Wickman K, Clapham DE (1995) Ion channel regulation by G proteins. Physiol Rev 75:865–85

Wiedemann C, Schäfer T, Burger MM et al (1998) An essential role for a small synaptic vesicle-associated phosphatidylinositol 4-linase in meurotransmitter release. J Neurosci 18:5594–5602
Wierda KD, Toonen RF, de WH et al (2007) Interdependence of PKC-dependent and PKC-independent pathways for presynaptic plasticity. Neuron 54:275–90
Willars GB (2006) Mammalian RGS proteins: multifunctional regulators of cellular signalling. Semin Cell Dev Biol 17:363–76
Wu L, Bauer CS, Zhen XG et al (2002) Dual regulation of voltage-gated calcium channels by PtdIns(4,5)P2. Nature 419:947–52
Wu LG, Saggau P (1994) Adenosine inhibits evoked synaptic transmission primarily by reducing presynaptic calcium influx in area CA1 of hippocampus. Neuron 12:1139–48
Wu LG, Saggau P (1995) GABAB receptor-mediated presynaptic inhibition in guinea-pig hippocampus is caused by reduction of presynaptic Ca2+ influx. J Physiol 485:649–57
Wu LG, Saggau P (1997) Presynaptic inhibition of elicited neurotransmitter release. Trends Neurosci 20:204–12
Wu XS, Wu LG (2001) Protein kinase c increases the apparent affinity of the release machinery to Ca2+ by enhancing the release machinery downstream of the Ca2+ sensor. J Neurosci 21:7928–36
Xu T, Binz T, Niemann H et al (1998) Multiple kinetic components of exocytosis distinguished by neurotoxin sensitivity. Nat Neurosci 1:192–200
Xu T, Ashery U, Burgoyne RD et al (1999a) Early requirement for alpha-SNAP and NSF in the secretory cascade in chromaffin cells. EMBO J 18:3293–3304
Xu X, Zeng W, Popov S et al (1999b) RGS proteins determine signaling specificity of Gq-coupled receptors. J. Biol. Chem. 274:3549–56
Yan Y, Chi PP, Bourne HR (1997) RGS4 inhibits Gq-mediated activation of mitogen-activated protein kinase and phosphoinositide synthesis. J Biol Chem 272:11924–7
Yang Y, Udayasankar S, Dunning J et al (2002) A highly Ca2+-sensitive pool of vesicles is regulated by protein kinase C in adrenal chromaffin cells. Proc Natl Acad Sci USA 99:17060–5
Yang Y, Craig TJ, Chen X et al (2007) Phosphomimetic mutation of Ser-187 of SNAP-25 increases both syntaxin binding and highly Ca2+-sensitive exocytosis. J Gen Physiol 129:233–44
Yawo H (1999a) Protein kinase C potentiates transmitter release from the chick ciliary presynaptic terminal by increasing the exocytotic fusion probability. J Physiol 515:169–80
Yawo H (1999b) Two components of transmitter release from the chick ciliary presynaptic terminal and their regulation by protein kinase C. J Physiol 516:461–70
Yawo H, Chuhma N (1993) Preferential inhibition of omega-conotoxin-sensitive presynaptic Ca2+ channels by adenosine autoreceptors. Nature 365:256–8
Zhang X, Kim-Miller MJ, Fukuda M et al (2002) Ca2+-dependent synaptotagmin binding to SNAP-25 is essential for Ca2+-triggered exocytosis. Neuron 34:599–611
Zhen M, Jin Y (1999) The liprin protein SYD-2 regulates the differentiation of presynaptic termini in C. elegans. Nature 401:371–5
Zhong N, Zucker RS (2004) Roles of Ca2+, hyperpolarization and cyclic nucleotide-activated channel activation, and actin in temporal synaptic tagging. JNeurosci 24:4205–12
Zhong N, Beaumont V, Zucker RS (2004) Calcium influx through HCN channels does not contribute to cAMP-enhanced transmission. J Neurophysiol 92:644–7
Zhou JY, Siderovski DP, Miller RJ (2000) Selective regulation of N-type Ca channels by different combinations of G-protein beta/gamma subunits and RGS proteins. J Neurosci 20:7143–8
Zhu H, Hille B, Xu T (2002) Sensitization of regulated exocytosis by protein kinase C. Proc Natl Acad Sci USA 99:17055–9
Zucker RS (1993) Calcium and transmitter release. J Physiol (Paris) 87:25–36

# Presynaptic Metabotropic Receptors for Acetylcholine and Adrenaline/Noradrenaline

**Ralf Gilsbach and Lutz Hein(✉)**



**Abstract** Presynaptic metabotropic receptors for acetylcholine and adrenaline/noradrenaline were first described more than three decades ago. Molecular cloning has resulted in the identification of five G protein-coupled muscarinic receptors ($M_1 - M_5$) which mediate the biological effects of acetylcholine. Nine adrenoceptors ($\alpha_{1ABD}, \alpha_{2ABC}, \beta_{123}$) transmit adrenaline/noradrenaline signals between cells. The lack of sufficiently subtype-selective ligands has prevented identification of the physiological role and therapeutic potential of these receptor subtypes for a long time. Recently, mouse lines with targeted deletions for all muscarinic and adrenoceptor genes have been generated. This review summarizes the results from these gene-targeting studies with particular emphasis on presynaptic auto- and heteroreceptor functions of muscarinic and adrenergic receptors. Specific knowledge about the function of receptor subtypes will enhance our understanding of the physiological role of the cholinergic and adrenergic nervous system and open new avenues for subtype-selective therapeutic strategies.

Lutz Hein
Institute of Experimental and Clinical Pharmacology and Toxicology, University of Freiburg, Albertstrasse 25, 79104 Freiburg, Germany
lutz.hein@pharmakol.uni-freiburg.de

T.C. Südhof, K. Starke (eds.), *Pharmacology of Neurotransmitter Release.* *Handbook of Experimental Pharmacology 184.*

# 1 Introduction

More than three decades ago, the first presynaptic metabotropic receptors regulating the release of acetylcholine, GABA, and noradrenaline were discovered (for reviews, see Langer 1997; Langer et al. 1985; Starke 2001; Starke et al. 1989). Presynaptic receptors are termed "autoreceptors" if they are activated by their neuron's own transmitter. In contrast, "heteroreceptors" are regulated by transmitters or modulators from neighboring cells. The present review will focus on recent studies which have investigated presynaptic metabotropic receptors for acetylcholine and adrenaline/noradrenaline in gene-targeted or transgenic mouse models.

# 2 Presynaptic Receptors for Acetylcholine

## 2.1 Acetylcholine Receptor Subtypes

Acetylcholine (ACh) is a widely distributed neurotransmitter in the central and peripheral nervous systems. As a classical neurotransmitter, acetylcholine is stored in intracellular vesicles and, after release into the extracellular space (i.e., the synaptic cleft), it may activate two classes of plasma membrane receptors: muscarinic ACh receptors (mAChR) and nicotinic ACh receptors (nAChR). While muscarinic receptors belong to the superfamily of seven transmembrane spanning G protein-coupled receptors (GPCRs), nicotinic receptors are ligand-gated ion channels. For a detailed description of nicotinic AChR and presynaptic functions, the reader is referred to the chapter by Dorostkar and Boehm in this *Handbook* and to recently published reviews focusing on insight derived from gene-targeted mouse models (Champtiaux et al. 2003; Drago et al. 2003).

## 2.2 Muscarinic Acetylcholine Receptors

By molecular cloning, five subtypes of G protein-coupled mAChR have been identified and are termed $M_1$-$M_5$ (Caulfield and Birdsall 1998; Wess 1996). Based on their coupling to specific groups of heterotrimeric G proteins, mAChR may be divided into two functional classes: $M_2$ and $M_4$ receptors are preferentially coupled to $G_{i/o}$ signal transduction pathways, whereas $M_1$, $M_3$, and $M_5$ subtypes activate $G_q$ proteins and downstream effectors (Wess 1996).

### 2.2.1 Transgenic and Gene-Targeted Models

In the absence of sufficiently subtype-selective ligands, mouse lines carrying targeted deletions in the muscarinic AChR genes have been generated by several laboratories (Fisahn et al. 2002; Gerber et al. 2001; Gomeza et al. 1999a,b; Hamilton et al. 1997; Matsui et al. 2000; Miyakawa et al. 2001; Wess 1996; Yamada et al. 2001a,b). All mice deficient in single mAChR subtypes were viable and fertile and were born without any obvious malformations. With few exceptions, targeted deletion of one receptor gene in the mouse genome did not result in compensation by any of the other subtypes, allowing precise assignments of receptor subtypes with specific functions (for detailed discussion, see Wess 2004).

### 2.2.2 Occurrence

Muscarinic AChR are widely expressed in many organs, tissues, and cell types of the body (Table 1) (Caulfield and Birdsall 1998; Caulfield et al. 1993; Eglen 2006; Eglen and Nahorski 2000; Levey 1993; Levey et al. 1994; Wess 2004; Wolfe and Yasuda 1995; Yasuda et al. 1993; Zhang et al. 2002a). $M_1$ AChR are abundantly expressed in the brain, including cerebral cortex, hippocampus, and striatum. $M_2$ receptors have been identified in the central nervous system, heart, and organs containing smooth muscle. The $M_3$ subtype can be found in the brain and in peripheral organs which are innervated by parasympathetic nerves. $M_4$ receptors are mostly restricted

**Table 1** Occurrence of Muscarinic AChR Subtypes

| Organ/Tissue | Muscarinic AChR Subtype | | | | |
|---|---|---|---|---|---|
| | $M_1$ | $M_2$ | $M_3$ | $M_4$ | $M_5$ |
| **Central nervous system** | | | | | |
| Cerebral cortex | ++ | ++ | + | + | (+) |
| Cerebellum | + | ++ | | + | (+) |
| Hippocampus | ++ | + | + | + | (+) |
| Hypohalamus | | | + | | |
| Limbic sytem | | | | | (−) |
| Spinal cord | | ++ | | + | |
| Striatum | + | + | + | ++ | (−) |
| **Peripheral organs** | | | | | |
| Eye | | | + | | (+) |
| Glands | + | | + | | |
| Heart | + | ++ | + | + | + |
| Lung | | + | + | + | |
| Smooth muscle | + | + | + | | |
| Sympathetic ganglia | + | + | | (+) | |
| Vascular smooth muscle | + | ++ | + | | + |

For references, see text. Symbols: ++ predominant subtype, + expressed subtype, (+) low abundant subtype.

to forebrain regions, and $M_5$ receptors are present only at low abundance in the brain, e.g., in dopaminergic neurons of the midbrain.

### 2.2.3 Presynaptic Signaling

Based on studies on gene-targeted mice, $M_2$ and $M_4$ AChR have been identified as major presynaptic inhibitory receptors (Slutsky et al. 2003; Trendelenburg et al. 2003a; Zhang et al. 2002a,b; Zhou et al. 2002). These receptor subtypes are coupled to G proteins of the $G_{i/o}$ class and may affect presynaptic signaling by a variety of mechanisms (Figure 1). Activation of inhibitory $G\alpha_{i/o}$ proteins by $M_{2/4}$ AChR leads to inhibition of adenylyl cyclase, thus resulting in decreased cellular cAMP levels. In addition, $G\beta\gamma$ subunits released from activated $G_{i/o}$ proteins are important regulators of neuronal function by inhibiting neuronal $Ca^{2+}$ channels, activating GIRK $K^+$ channels and stimulating mitogen-activated (MAP) kinases ERK1/2 (Wess 1996). Among these mechanisms, the $G\beta\gamma$-mediated inhibition of neuronal $Ca^{2+}$ channels seems to be the most important mechanism for inhibition of neurotransmitter release (Figure 1) (for reviews, see Boehm and Kubista 2002; Kubista and Boehm 2006). MAP kinase activation induced by acetylcholine was almost completely abolished in neuronal cultures from newborn mice lacking $M_1$ receptors (Berkeley et al. 2001). Similarly, $G_q$ protein activation and phosphoinositide hydrolysis was absent in $M_1$-deficient brain preparations, indicating that $M_1$

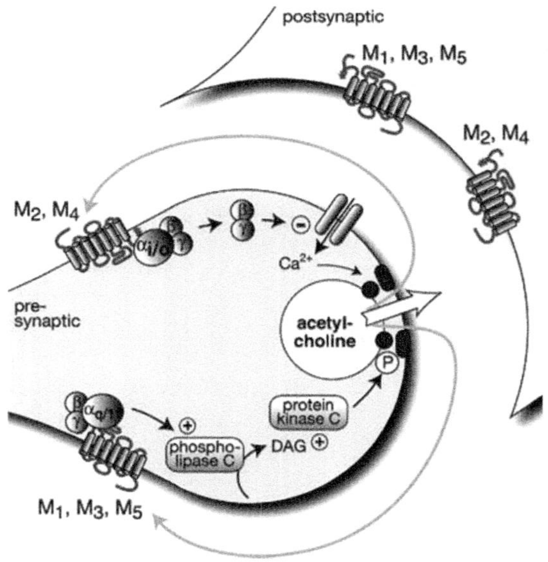


**Fig. 1** Presynaptic signaling mechanisms of muscarinic receptor subtypes ($M_1$-$M_5$). Abbreviations: $\alpha_{i/o}$, $\beta$, $\gamma$, subunits of $G_{i/o}$ heterotrimeric GTP-binding proteins; DAG, diacylglycerol.

receptors are the most abundant $G_q$-coupled muscarinic receptors in the brain (Bymaster et al. 2003a; Porter et al. 2002). Facilitory mAChR which are coupled to $G_{q/11}$ proteins may enhance neurotransmitter release primarily via protein kinase C-mediated phosphorylation of proteins of the exocytotic release machinery (Figure 1) (Boehm and Kubista 2002; Kubista and Boehm 2006).

### 2.2.4 Physiological Functions

Regulation of neurotransmitter release

$M_2$ and $M_4$ AChR have been identified as the main inhibitory presynaptic muscarinic receptors in gene-targeted mouse models (Table 2). Dependent on the type of tissue, both receptors contributed to varying degrees to presynaptic autoinhibition. In cerebral cortex and hippocampus, inhibitory $M_2$ receptors prevailed over $M_4$ receptors to inhibit release of acetylcholine in vitro. In striatal preparations, $M_4$ receptors were functionally more important than $M_2$ receptors (Zhang et al. 2002b). In the periphery, $M_4$ receptor and to a smaller degree non-$M_4$ receptors, presumably $M_2$, inhibited acetylcholine release from atria, urinary bladder, and peripheral cholinergic motor nerves innervating the diaphragm (Slutsky et al. 2003; Zhou et al. 2002).

In addition to these autoreceptors, $M_2$, $M_3$, and $M_4$ receptors have also been identified as presynaptic heteroreceptors controlling noradrenaline release in atria ($M_2$, $M_3$), urinary bladder ($M_2$, $M_4$), and vas deferens ($M_2$, $M_3$, $M_4$) (Trendelenburg et al. 2003a; Trendelenburg et al. 2005). In the striatum, activation of mAChR facilitates dopamine release. Facilitation of potassium-mediated [$^3$H]-dopamine release by the muscarinic agonist oxotremorine was absent in $M_4$-deficient striatal preparations (Zhang et al. 2002b). $M_4$ receptors may be localized on cell bodies of striatal GABAergic neurons to increase dopamine release in basal ganglia.

$M_1$ AChR functions

The $M_1$ receptor subtype is involved in a variety of CNS functions (see Table 2). Agonist-induced seizures were absent in $M_1$-deficient but not affected in mouse lines lacking $M_2$-$M_5$ receptors (Bymaster et al. 2003b; Hamilton et al. 1997). Learning and memory tasks as well as nociception, motor coordination, and anxiety-associated behavioral traits were normal in $M_1$-deficient mice (Miyakawa et al. 2001). Mild performance deficits of $M_1$-deficient mice after fear conditioning were associated with significant locomotor hyperactivity of these mice (Miyakawa et al. 2001). In addition, a small reduction in hippocampal long-term potentiation was observed in $M_1$-deficient mice (Anagnostaras et al. 2003). Furthermore, $M_1$ receptors can inhibit the "M-current" ($I_m$) through voltage-gated $K^+$ channels in sympathetic ganglia (Hamilton et al. 1997), resulting in chronotropic and inotropic stimulation of the heart (Hardouin et al. 2002). Acetylcholine-mediated, slow voltage-independent

**Table 2** Physiological Functions of Muscarinic AChR Subtypes

| Function | Phenotype of Targeted Deletion of Muscarinic AChR Subtype | | | | |
|---|---|---|---|---|---|
| | $M_1$ | $M_2$ | $M_3$ | $M_4$ | $M_5$ |
| **Central nervous system** | | | | | |
| Autoreceptor: cortex, hippocampus | | $\Downarrow\Downarrow$ | | $\Downarrow$ | |
| Autoreceptor: striatum | | $\Downarrow$ | | $\Downarrow\Downarrow$ | |
| Heteroreceptor: activation of dopamine release | = | = | $\Uparrow$ | $\Downarrow\Downarrow$ | $\Downarrow$ |
| Locomotor hyperactivity | $\Uparrow$ | = | = | $\Uparrow$ | = |
| Learning and memory | $\Downarrow$= | $\Downarrow$ | | | $\Downarrow$ |
| Agonist-induced seizures | 0 | = | = | = | = |
| Body weight, food intake | | | $\Downarrow$ | | |
| Locomotor coordination | = | | | | |
| Agonist-induced tremor | | 0 | | | |
| Agonist-induced akinesia | | 0 | | | |
| Agonist-induced hypothermia | | $\Downarrow$ | | | |
| Agonist-induced analgesia | | $\Downarrow$ | | $\Downarrow$ | |
| Anxiety-associated behavior | = | | | | |
| Reward behavior to cocaine, morphine | | | | | $\Downarrow$ |
| Agonist-induced MAPK activation | $\Downarrow\Downarrow$ | | | | |
| Agonist-induced PI hydrolysis | 0 | | | | |
| **Peripheral nervous system** | | | | | |
| Autoreceptor: atria, urinary bladder | | | | $\Downarrow\Downarrow$ | |
| Autoreceptor: phrenic nerve | | $\Downarrow$ | | | |
| Heteroreceptor: inhibition of noradrenaline release in atria, vas deferens | | $\Downarrow$ | $\Downarrow$ | $\Downarrow$ | |
| Agonist-induced bradycardia | = | $\Downarrow\Downarrow$ | = | = | = |
| Inhibition of M current ($I_m$) in sympathetic ganglia | 0 | | | | |
| Smooth muscle contraction (trachea, stomach, urinary bladder) | | $\Downarrow$ | $\Downarrow\Downarrow$ | | |
| Pupil size | | | $\Uparrow$ | | |
| Salivation | | | $\Downarrow$ | | $\Downarrow$ |
| Pepsinogen secretion from gastric chief cells | $\Downarrow$ | | $\Downarrow$ | | |
| Cholinergic vasodilatation in cerebral vessels | | | | | 0 |
| Agonist-induced dilatation of the aorta | | | $\Downarrow\Downarrow$ | | |

Assignment of functions as derived from studies in gene-targeted mice. For references, see text. Symbols: $\Uparrow$ increased function, $\Downarrow$ reduced function in receptor-deficient mice, 0 function absent, = normal function. "Agonist" refers to cholinergic agonist.

inhibition of N- and P/Q-type $Ca^{2+}$ channels was also absent in $M_1$-deficient sympathetic ganglia (Shapiro et al. 1999). In the gastrointestinal tract, secretion of pepsinogen induced by cholinergic stimulation from gastric chief cells requires a mixture of $M_1$ and $M_3$ receptors (Xie et al. 2005).

## $M_2$ AChR functions

In addition to its above mentioned functions as a presynaptic auto- and heteroreceptor, $M_2$ receptors are involved in a number of CNS functions. Agonist-induced whole-body tremor and akinesia were completely abolished, and hypothermia was significantly reduced in $M_2$-deficient mice (Gomeza et al. 1999a). $M_2$-deficient mice had problems in passive avoidance tests, demonstrating the role of this receptor subtype in learning and memory function (Tzavara et al. 2003). In the periphery, $M_2$ AChR mediate agonist-induced bradycardia (Stengel et al. 2000), contractions of smooth muscle cells in stomach, urinary bladder, and trachea (Kitazawa et al. 2007; Stengel et al. 2000; Unno et al. 2005; Unno et al. 2006), and analgesic effects (Bernardini et al. 2002; Gomeza et al. 1999a,b). Antinociception induced by the agonist oxotremorine was completely absent in mice lacking $M_2$ and $M_4$ receptors (Gomeza et al. 2001).

## $M_3$ AChR functions

Mice with deletion of the $M_3$ receptor gene were lean and consumed less food than their wild-type litter-mates (Yamada et al. 2001b). Probably $M_3$ receptors in the hypothalamus are important to control food and energy intake. $M_3$-deficient mice showed an impaired glucose tolerance and greatly reduced insulin release (Gautam et al. 2006a). Their total body energy expenditure was significantly enhanced (Gautam et al. 2006b; Kitazawa et al. 2007; Unno et al. 2005; Unno et al. 2006). In the body periphery, the $M_3$ subtype mediates a significant part of smooth muscle contractions in peripheral organs (Matsui et al. 2000) and ACh-induced salivation (Matsui et al. 2000; Yamada et al. 2001a). The vasodilatation induced by cholinergic agonists was greatly reduced in $M_3$-deficient aorta (Khurana et al. 2005).

## $M_4$ AChR functions

$M_4$ receptor-deficient mice showed a mild increase in locomotor activity. In the striatum, activation of $M_4$ AChR exerted an inhibitory effect on dopamine $D_1$ receptor-stimulated locomotor activity (Gomeza et al. 1999b). In striatal tissue slices from wild-type mice, the muscarinic agonist oxotremorine increased potassium-stimulated $[^3H]$-dopamine release (Zhang et al. 2002b). This effect of muscarinic agonists was completely abolished in $M_4$-receptor-deficient mice (Zhang et al. 2002b). In addition, $M_4$ receptors are involved in the antinociceptive effect of cholinergic agonists (Gomeza et al. 1999b).

## $M_5$ AChR functions

$M_5$ AChR represent the predominant muscarinic subtype in dopamine-containing neurons of the pars compacta of the substantia nigra and the ventral tegmental area.

Activation of these receptors facilitates dopamine release in the striatum and in the nucleus accumbens. In $M_5$-deficient striatal preparations, the potency of oxotremorine to enhance potassium-stimulated release of $[^3H]$-dopamine was reduced (Yamada et al. 2001a; Zhang et al. 2002b). In the nucleus accumbens, deletion of $M_5$ receptors selectively blunted long-term dopamine stimulation and was thus associated with reduced rewarding effects of morphine and cocaine (Araya et al. 2006; Basile et al. 2002; Fink-Jensen et al. 2003; Thomsen et al. 2005). In cerebral arteries and arterioles of $M_5$-deficient mice, cholinergic vasodilatation was significantly blunted and cerebral arteries were constricted at rest (Yamada et al. 2001a). Reduced cerebral artery perfusion was associated with learning and memory deficits in $M_5$-deficient mice (Araya et al. 2006).

### 2.2.5 Relevance in Disease and Therapy

Results obtained from studies in gene-targeted mice indicate that ligands which selectively activate or inhibit single muscarinic receptor subtypes may be beneficial for the treatment of a number of diseases including, e.g., Alzheimer's and Parkinson's disease, depression, schizophrenia, and epilepsy.

In particular, $M_1$ AChR antagonists might be useful for the treatment of Parkinson's disease, as disruption of this receptor's gene led to increased locomotor behavior (Miyakawa et al. 2001) and facilitated dopamine release in basal ganglia (Gerber et al. 2001). $M_1$ agonists could be of value to treat symptoms of schizophrenia, which has been associated with enhanced dopamine release in various forebrain areas. As $M_2$-deficient mice displayed features of reduced memory function, including reduced hippocampal long-term potentiation (Tzavara et al. 2003), selective ligands could be tested to relieve symptoms of Alzheimer's disease. $M_3$ antagonists may be useful to reduce food intake, as mice lacking this subtype were lean and hypophagic (Yamada et al. 2001b). $M_4$-specific antagonists might be beneficial to relieve extrapyramidal motor symptoms in Parkinson's disease and $M_4$ agonists may represent novel analgesic compounds. Finally, antagonists which selectively bind to $M_5$ AChR could be helpful to treat drug addiction, as $M_5$-deficient mice were less susceptible to the rewarding effects of morphine and cocaine (Basile et al. 2002; Fink-Jensen et al. 2003).

# 3 Presynaptic Receptors for Adrenaline/Noradrenaline

## *3.1 Adrenoceptor Subtypes*

Adrenoceptors (AR) are G protein-coupled receptors which mediate the biological effects of the endogenous catecholamines, adrenaline and noradrenaline. Nine adrenoceptor subtypes have been cloned which can be grouped into three different functional classes (Bylund et al. 1994). $\alpha_1$ adrenoceptors ($\alpha_{1A}$, $\alpha_{1B}$, $\alpha_{1D}$) are primarily coupled to intracelluar $G_{q/11}$-dependent signals, $\alpha_2$ receptors ($\alpha_{2A}$, $\alpha_{2B}$, $\alpha_{2C}$) activate $G_{i/o}$ proteins, and the three β receptor subtypes ($\beta_1$, $\beta_2$, $\beta_3$) stimulate $G_s$ proteins (Bylund et al. 1994). From these three receptor groups, the $\alpha_2$ receptors have been identified as presynaptic inhibitory autoreceptors and will thus be discussed first.

## 3.2 $\alpha_2$ Adrenoceptors

### 3.2.1 Transgenic and Gene-Targeted Models

The three genes encoding for $\alpha_2$AR subtypes are located on different chromosomes in the human and mouse genome and have been targeted for deletion in murine embryonic stem cells (Altman et al. 1999; Link et al. 1995; Link et al. 1996). Using gene targeting, also a "knock-in" model inserting a point mutation in the $\alpha_{2A}$AR gene ($\alpha_{2A}$-D79N) has been generated (MacMillan et al. 1996). Mice lacking $\alpha_{2A}$AR or $\alpha_{2C}$AR were born at the expected Mendelian ratios and developed normally (Altman et al. 1999; Link et al. 1995). In contrast, deficiency of the $\alpha_{2B}$ subtype was associated with decreased survival due to vascular defects, which became first apparent in mice lacking all three subtypes ($\alpha_{2ABC}$-/-) (Philipp et al. 2002b).

### 3.2.2 Occurrence

All three $\alpha_2$AR subtypes have been localized in the CNS and in peripheral tissues (Table 3) (Aoki et al. 1994; Bücheler et al. 2002; Fagerholm et al. 2004; MacDonald and Scheinin 1995; Moura et al. 2006; Nicholas et al. 1993; Nicholas et al. 1996; Rosin et al. 1996; Scheinin et al. 1994; Talley et al. 1996; Uhlen et al. 1997). In brief, the $\alpha_{2A}$AR subtype is widely expressed in the CNS, including brain stem nuclei, locus coeruleus, cerebral cortex, hypothalamus, and hippocampus. In the periphery, $\alpha_{2A}$AR have been found in lung, kidney, and spleen. $\alpha_{2B}$AR are mostly expressed in non-neuronal tissues, including heart, liver, lung, and kidney. In the CNS, the $\alpha_{2B}$ subtype shows most discrete expression in thalamic nuclei. The third subtype, $\alpha_{2C}$, is again mostly expressed in the CNS, where it was identified in the hippocampus, cortex, striatum, and olfactory tubercle (Table 3).

### 3.2.3 Presynaptic Signaling

Upon activation all three $\alpha_2$AR subtypes are coupled to their intracellular messenger systems via heterotrimeric $G_{i/o}$ proteins (Figure 2). Intracellular consequences of $\alpha_2$AR activation are reduced cAMP levels due to inhibition of adenylyl cyclase, opening of GIRK $K^+$ channels, inhibition of neuronal $Ca^{2+}$ channels, and

**Table 3** Occurrence of $\alpha_2$ Adrenoceptor Subtypes

| Organ/Tissue | $\alpha_2$ Adrenoceptor Subtype | | |
|---|---|---|---|
| | $\alpha_{2A}$ | $\alpha_{2B}$ | $\alpha_{2C}$ |
| **Central nervous system** | | | |
| Brain stem | ++ | | |
| Cerebral cortex | + | | + |
| Hypothalamus | + | | |
| Hippocampus | + | | + |
| Spinal cord | + | + | + |
| Striatum | | | ++ |
| Thalamus | | (+) | |
| **Peripheral organs** | | | |
| Adrenal medulla | + | + | + |
| Arteries | | + | |
| Heart | | + | |
| Kidney | + | + | (+) |
| Liver | | + | |
| Lung | + | + | |
| Salivary glands | + | | |
| Spleen | + | | |
| Thymus | + | | |
| Vessels | | + | |

For references, see text. Symbols: ++ predominant subtype, + expressed subtype, (+) low abundant subtype.

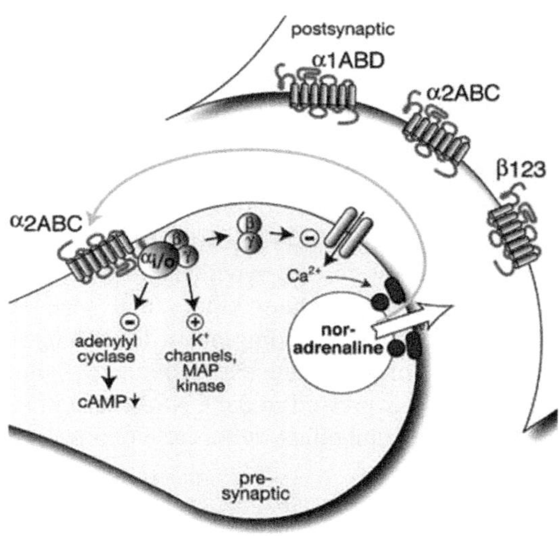


**Fig. 2** Presynaptic signaling mechanisms of $\alpha_2$ adrenoceptor subtypes ($\alpha_{2A}$, $\alpha_{2B}$, $\alpha_{2C}$). Abbreviations: $\alpha_{i/o}$, β, γ, subunits of $G_{i/o}$ heterotrimeric GTP-binding proteins; $\alpha_{1ABD}$, $\alpha_{2ABC}$, $\beta_{123}$, adrenoceptor subtypes; MAP, mitogen-activated protein kinase.

stimulation of mitogen-activated kinase ERK1/2 (Cussac et al. 2001). Several mechanism may contribute to the inhibition of presynaptic neurotransmitter release, e.g., reduced $Ca^{2+}$ channel opening, activation of $K^+$ channels, and also direct regulation of the exocytotic release machinery (for discussion see Miller 1998; Wu and Saggau 1997). Among these proposed mechanisms, the inhibitory effect of $G\beta\gamma$ subunits on presynaptic $Ca^{2+}$ channels may be the dominant effect observed in postganglionic sympathetic neurons in culture (Figure 2) (Boehm and Kubista 2002; Kubista and Boehm 2006; Stephens and Mochida 2005).

### 3.2.4 Physiological Functions

Regulation of neurotransmitter release by $\alpha_2$ autoreceptors

$\alpha_2$AR are the main inhibitory presynaptic feedback receptors which are essential to control the release of noradrenaline from adrenergic neurons (Figure 3). Pharmacological studies using ligands with partial subtype-selectivity have led to the hypothesis that the $\alpha_{2A}$AR subtype may be the major inhibitory autoreceptor in the sympathetic system (for reviews, see Starke 2001; Starke et al. 1989). This concept was later supported and extended by studies in gene-targeted mice (for detailed reviews, see Hein 2001; Philipp and Hein 2004; Philipp et al. 2002a). In brain and peripheral tissues isolated from $\alpha_{2A}$-deficient mice, maximal inhibition of [$^3$H]-noradrenaline release by exogenous $\alpha_2$-agonists was significantly blunted as compared with preparations from wild-type mice (Altman et al. 1999; Hein et al. 1999; Trendelenburg et al. 1999; Vonend et al. 2007). Deletion of the other $\alpha_2$AR subtypes, $\alpha_{2B}$ or $\alpha_{2C}$, did not affect maximal agonist-induced inhibition of noradrenaline release (Hein et al. 1999; Trendelenburg et al. 2001a), confirming pharmacological data that the $\alpha_{2A}$AR is indeed the predominant presynaptic autoreceptor in adrenergic neurons. However, the fact that $\alpha_2$ agonists still inhibited neurotransmitter release in $\alpha_{2A}$-deficient tissues indicated that non-$\alpha_{2A}$ receptors contributed to presynaptic inhibition. In the CNS, $\alpha_{2A}$ and $\alpha_{2C}$ receptors could be identified as presynaptic regulators; in peripheral organs all three $\alpha_2$AR subtypes, $\alpha_{2A}$, $\alpha_{2B}$, and $\alpha_{2C}$ operated as feedback inhibitors in vitro (Figure 3) (Trendelenburg et al. 2003b). $\alpha_2$AR subtypes may not be completely redundant in their presynaptic function, but may serve different roles in the feedback inhibition. As one example, $\alpha_2$AR showed differences in their frequency-activity relationship: In atria, $\alpha_{2C}$AR inhibited [$^3$H]-noradrenaline release better at lower frequencies of nerve stimulation, whereas the $\alpha_{2A}$AR operated at higher stimulation frequencies (Hein et al. 1999; Scheibner et al. 2001b). Furthermore, $\alpha_2$AR subtypes differed in their postnatal maturation. Whereas the $\alpha_{2A}$AR was functional immediately after birth, non-$\alpha_{2A}$AR effects could not be observed within the first postnatal days in intact tissues or in isolated sympathetic neurons (Schelb et al. 2001; Trendelenburg et al. 2001b).

In addition to the presynaptic $\alpha_2$ autoreceptor subtypes which were identified in CNS and sympathetic nerves, studies in gene-targeted mice uncovered $\alpha_{2C}$AR as autoreceptors controlling catecholamine release from chromaffin cells (Brede

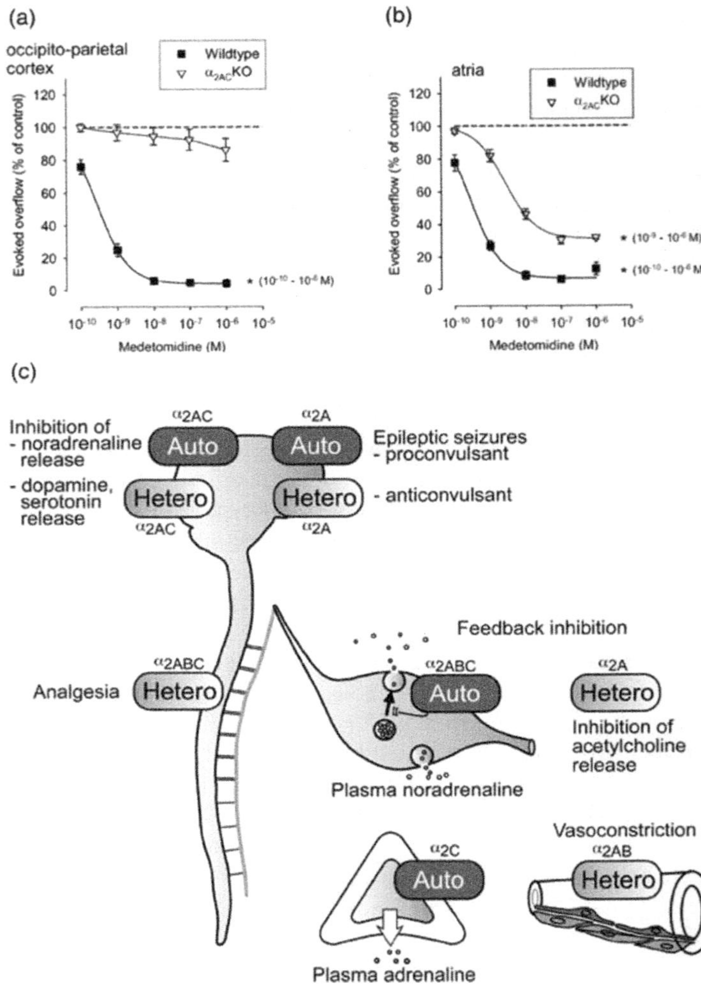


**Fig. 3** Autoreceptor versus heteroreceptor functions of $\alpha_2$-adrenoceptor subtypes. (**a** and **b**) Inhibition of electrically evoked [$^3$H]-adrenaline release by the $\alpha_2$-agonist, medetomidine, from mouse brain cortex (**a**) or heart atria (**b**). In wild-type tissue specimens, medetomidine inhibited transmitter release by >90%. In tissues from $\alpha_{2AC}$-deficient mice, the agonist effect was absent (**a**, cortex) or significantly reduced (**b**, atria). Reproduced with permission from Trendelenburg et al. 2003b. (**c**) Overview of auto- and heteroreceptor functions of $\alpha_2$-adrenoceptor subtypes. For references, see text.

et al. 2002; Brede et al. 2003). Plasma and urine adrenaline levels were >2-fold higher in $\alpha_{2C}$-deficient mice than in wild-type mice. Activation of $\alpha_{2C}$-receptors expressed in isolated mouse chromaffin cells inhibited voltage-gated $Ca^{2+}$ channels and reduced secretion of adrenaline. Inhibition of adrenaline secretion by this “presynaptic” $\alpha_{2C}$-receptor is physiologically important, as mice deficient in this receptor showed enhanced cardiac hypertrophy and heart failure in response to cardiac pressure overload (Brede et al. 2002). Interestingly, adrenal catecholamines are differentially regulated by $\alpha_2$AR subtypes. Whereas secretion of adrenaline seems to be predominantly controlled by $\alpha_{2C}$, all three $\alpha_2$AR subtypes may control noradrenaline release from adrenal preparations in vitro (Moura et al. 2005).

### $\alpha_2$ Adrenoceptors as heteroreceptors

In addition to their functions as presynaptic autoreceptors, $\alpha_2$AR can also modulate release of other neurotransmitters (Figure 3). In the CNS, $\alpha_{2A}$ and $\alpha_{2C}$ receptors inhibit dopamine release in basal ganglia (Bücheler et al. 2002) as well as serotonin secretion in mouse hippocampus and brain cortex (Scheibner et al. 2001a). In the enteric nervous system, the release of acetylcholine as determined by [$^3$H]-choline overflow from tissue slices was selectively inhibited by $\alpha_{2A}$AR (Scheibner et al. 2002).

### $\alpha_{2A}$ Adrenoceptor functions

One of the main functions of the $\alpha_{2A}$AR is to control sympathetic noradrenaline release. Dysfunction or deletion of this receptor leads to hypertension and tachycardia and accelerates the development of cardiac hypertrophy and heart failure in mice (Brede et al. 2002; Hein et al. 1999; Makaritsis et al. 1999b). Central $\alpha_{2A}$AR are essential for the hypotensive effects of $\alpha_2$ agonists, like clonidine, moxonidine, and rilmenidine (Altman et al. 1999; MacMillan et al. 1996; Zhu et al. 1999). $\alpha_{2A}$-deficient mice showed greater spontaneous blood pressure variations at rest and suppressed baroreflex gain (Niederhoffer et al. 2004). The sedative and anesthetic-sparing effects of $\alpha_2$ agonists were exclusively mediated by $\alpha_{2A}$AR (Lakhlani et al. 1997). $\alpha_{2A}$AR mediated part of the antinociceptive effect of $\alpha_2$ agonists (Fairbanks and Wilcox 1999; Hunter et al. 1997; Stone et al. 1997). In mice, $\alpha_2$ agonists may also produce pro- and anticonvulsant effects on seizure susceptibility (Janumpalli et al. 1998; Szot et al. 2004). Using dopamine β-hydroxylase-deficient mice, presynaptic $\alpha_{2A}$ autoreceptors could be identified as proconvulsant receptors, while postsynaptic $\alpha_{2A}$AR mediated anticonvulsant effects of $\alpha_2$ agonists (Figure 3) (Szot et al. 2004).

### $\alpha_{2B}$ Adrenoceptor functions

Neuronal $\alpha_{2B}$AR seems to play an important role in the antinociceptive pathway of nitrous oxide, which is used in anesthesia for its strong analgesic action (Sawamura et al. 2000). In the periphery, $\alpha_{2B}$AR elicit vasoconstriction after rapid intravascular injection of $\alpha_2$ agonists (Figure 3) (Link et al. 1996; Paris et al. 2003). Central $\alpha_{2B}$AR may be important in the development of hypertension after salt-loading and nephrectomy (Kintsurashvili et al. 2001; Makaritsis et al. 1999a; Makaritsis et al. 2000).

$\alpha_{2C}$ Adrenoceptor functions

As mentioned above, $\alpha_{2C}$AR are important feedback regulators to control adrenal adrenaline secretion (Brede et al. 2003; Moura et al. 2006). However, further studies are needed to identify the physiological relevance of the adrenal feedback mechanism. In the CNS, $\alpha_{2C}$AR affect a number of behavioral functions (for a detailed review, see Scheinin et al. 2001). Activation of $\alpha_{2C}$AR interferes with spatial working memory in mice, whereas stimulation of $\alpha_{2A}$AR and/or $\alpha_{2B}$AR may actually improve spatial working memory in mice (Bjorklund et al. 1999; Bjorklund et al. 2000; Bjorklund et al. 2001).

### 3.2.5 Relevance in Disease and Therapy

Agonists with sufficient selectivity for $\alpha_{2A}$AR or $\alpha_{2C}$AR may be helpful to prevent the detrimental cardiovascular effects of enhanced sympathetic activity (Brede et al. 2002). The failure of the $\alpha_2$ agonist, moxonidine, to produce a beneficial effect in patients with chronic heart failure in the MOXCON trial has been attributed to excessive sympatholytic effects of the drug (Cohn et al. 2003). Experimental studies in mice have clearly demonstrated that dysfunction of $\alpha_{2A}$AR and/or $\alpha_{2C}$AR contribute to the development of cardiac hypertrophy and failure (Brum et al. 2002; Hein et al. 1999). Interestingly, sedation which has been a frequent side effect of clinical $\alpha_2$ agonists may be avoided by using partial $\alpha_{2A}$ agonists, as hypotensive but not sedative $\alpha_2$AR have a high number of spare receptors (Tan et al. 2002). Furthermore, $\alpha_{2A}$ activation may limit the detrimental effects of hypoxia, ischemia, and injury in the brain (Ma et al. 2004; Paris et al. 2006).

Due to the absence of major CNS phenotypes in $\alpha_{2B}$-deficient mice, selective $\alpha_{2B}$ activation in the spinal cord may represent a novel target for the treatment of chronic pain (Sawamura et al. 2000; Zhang et al. 1999). Drugs acting via $\alpha_{2C}$AR may gain therapeutic value in disorders associated with enhanced startle responses and sensorimotor gating deficits, such as schizophrenia, attention deficit disorder, posttraumatic stress disorder, and drug withdrawal (Scheinin et al. 2001).

## *3.3 $\alpha_1$ Adrenoceptors*

### 3.3.1 Transgenic and Gene-Targeted Models

Despite the fact that noradrenergic neurons innervate many regions of the CNS, the contribution of $\alpha_1$AR to neuronal functions of catecholamines is largely unknown. Recently, gene-targeting and transgenic experiments in mice have provided novel insight in this area. Unexpectedly, transgenic expression of wild-type or constitutively active $\alpha_{1B}$AR under control of their own promotor resulted in a phenotype which was characterized by severe neurological degeneration with Parkinson-like

**Table 4** Occurrence of $\alpha_1$ Adrenoceptor Subtypes

| Organ/Tissue | $\alpha_1$ Adrenoceptor Subtype | | |
|---|---|---|---|
| | $\alpha_{1A}$ | $\alpha_{1B}$ | $\alpha_{1D}$ |
| **Central nervous system** | | | |
| Brain stem | + | | + |
| Cerebral cortex | + | + | + |
| Hypothalamus | + | | |
| Hippocampus | | | + |
| Pineal gland | | + | |
| Spinal cord | + | + | ++ |
| Thalamus | | + | + |
| **Peripheral organs** | | | |
| Bladder | + | | + |
| Heart | ++ | ++ | (+) |
| Liver | + | + | |
| Prostate | + | | |
| Vessels | + | + | + |

For references, see text. Symbols: ++ predominant subtype, + expressed subtype, (+) low abundant subtype.

locomotor deficits and epileptic seizures (Zuscik et al. 2000). Furthermore, target deletions have been integrated into the three $\alpha_1$AR genes of the mouse (Cavalli et al. 1997; Rokosh and Simpson 2002; Tanoue et al. 2002).

### 3.3.2 Occurrence

$\alpha_1$AR are abundantly expressed in the CNS (Table 4) (Drouin et al. 2002; Michelotti et al. 2000; Piascik and Perez 2001; Xiao et al. 2006). The $\alpha_{1A}$ subtype has been primarily localized in the olfactory system, hypothalamus, brain stem, and spinal cord (Day et al. 1997). $\alpha_{1B}$AR were identified in the pineal gland, in thalamic nuclei, the lateral nucleus of the amygdala, intermediate and deep layers of the cortex, as well as dorsal and median raphe nuclei (Day et al. 1997; McCune et al. 1993; Pieribone et al. 1994). Among the three $\alpha_1$AR subtypes, the $\alpha_{1D}$AR showed most discrete CNS expression in the olfactory bulb, reticular thalamic nucleus, hippocampus, layers II-V of the cerebral cortex, regions of the amygdala, motor nuclei of the brainstem, and in the spinal cord (Day et al. 1997; McCune et al. 1993; Pieribone et al. 1994). This pattern of expression was confirmed in mice with targeted expression of $\beta$-galactosidase in the $\alpha_{1D}$AR gene locus. Cortex, hippocampus, olfactry bulb, dorsal geniculate, and ventral posterolateral nuclei of the thalamus were densely labeled with $\beta$-galactosidase (Sadalge et al. 2003).

### 3.3.3 Presynaptic Signaling

Despite their abundant expression in the CNS (see Section 3.3.2), no studies have yet been performed in gene-targeted mice to address the precise subcellular localization and signal transduction of $\alpha_1$AR subtypes.

### 3.3.4 Physiological Functions

In the past, cardiovascular functions of $\alpha_1$AR have generated the greated interest and research effort. However, cardiovascular phenotypes which have been observed in gene-targeted mouse models deficient in $\alpha_1$AR are beyond the scope of this review and have been reviewed elsewhere (Chen and Minneman 2005; Koshimizu et al. 2006; Piascik and Perez 2001; Tanoue et al. 2003).

$\alpha_{1A}$ Adrenoceptor functions

Despite the fact that $\alpha_{1A}$AR are expressed in the CNS, no studies addressing the function of this receptor in gene-targeted mice have appeared yet.

$\alpha_{1B}$ Adrenoceptor functions

A number of behavioral phenotypes have been reported in $\alpha_{2B}$-deficient mice, including altered responses to *d*-amphetamine (Auclair et al. 2002; Drouin et al. 2002). Amphetamine and other psychostimulants cause drug addiction in humans and induce locomotor stimulant effects in rodents. Single drug doses enhance locomotor activity and repeated drug administration leads to behavioral sensitization with greatly enhanced activity. In $\alpha_{1B}$-deficient mice, locomotor hyperactivity and behavioral sensitization to amphetamine, cocaine, and morphine were significantly decreased (Drouin et al. 2002). Basal and amphetamine-stimulated extracellular dopamine levels were lower in $\alpha_{1B}$-deficient mice than in wild-type control animals (Auclair et al. 2004). However, basal dopaminergic transmission and locomotor activity after administration of a dopamine D1 agonist were similar in wild-type and $\alpha_{1B}$ mutant mice (Drouin et al. 2002). $\alpha_{1B}$ mutant mice showed an exaggerated biochemical and behavioral response to the serotonin releasing drug, *p*-chloroamphetamine (Salomon et al. 2006). Salomon et al. suggest that repeated amphetamine administration leads to behavioral sensitization due to reduced reciprocal inhibition of serotonin and noradrenaline signaling via 5-$HT_{2A}$ and $\alpha_{1B}$ receptors, respectively. Additional studies indicate that $\alpha_{1B}$AR are involved in modulation of memory consolidation and fear-motivated exploratory activity (Knauber and Muller 2000) as well as in spatial learning (Spreng et al. 2001).

$\alpha_{1D}$ Adrenoceptor functions

This receptor subtype has been associated with the control of pain perception (Harasawa et al. 2003), motor coordination (Mishima et al. 2004), working memory, and attention (Mishima et al. 2004), as well as exploratory behavior (Sadalge et al. 2003). Furthermore, $\alpha_{1D}$-deficient mice showed altered functions of the lower urinary tract (Chen et al. 2005).

### 3.3.5 Relevance in Disease and Therapy

As mentioned above, transgenic expression of $\alpha_{1B}$AR under control of its native promotor led to a complex phenotype with multiple system atrophy and overt neurodegeneration (Zuscik et al. 2000). This transgene expressed neurodegenerative disease that resembled Parkinson's syndrome. Neurons and oligodendocytes were filled with cytoplasmic inclusion bodies which were positive for α-synuclein and ubiquitin and therefore may be classified as a synucleinopathy (Papay et al. 2002). Other phenotypes characteristic for multiple system atrophy, such as Purkinje cell loss in the cerebellum and degeneration of the intermediolateral cell columns of the spinal cord, also existed in transgenic $\alpha_{1B}$AR mice (Papay et al. 2002). Better understanding of this complex phenotype may identify the $\alpha_{1B}$AR as a target for the treatment of neurodegenerative diseases.

## *3.4 β Adrenoceptors*

### 3.4.1 Transgenic and Gene-Targeted Models

β-Adrenoceptors are best known for their role in the regulation of cardiovascular, airway, uterine, and peripheral metabolic functions. Mice lacking $\beta_1$, $\beta_2$ and $\beta_3$ receptors have been generated (Chruscinski et al. 1999; Rohrer et al. 1996; Susulic et al. 1995), but only few studies have used these models to identify subtype-specific functions of β-adrenoceptors in neurons.

### 3.4.2 Occurrence

β-Adrenoceptor subtypes are widely expressed throughout the body (Table 5) (Arch 2002; Arner 2005; Brodde and Michel 1999; Dennedy et al. 2002; Guimaraes and Moura 2001; Michel and Vrydag 2006; Ordway et al. 1987, 1988; Rohrer et al. 1999). The brain is densely innervated by noradrenaline- and adrenaline-containing neurons, most of which originate in the locus coeruleus or in the reticular formation (Foote et al. 1983). β-adrenoceptors have been identified in the target regions of brain stem catecholaminergic neurons. By *in situ* hybridization of rat brains, $\beta_1$-adrenoceptors were identified in cerebral cortex, thalamus, pineal gland, and sympathetic ganglia (Nicholas et al. 1996). In contrast, the $\beta_2$AR subtype was most densely expressed in the olfactory bulb, cerebral cortex, hippocampus, thalamus and hypothalamus, pineal gland, and spinal cord. In the pineal gland, noradrenaline released from sympathetic nerves controls the circadian rhythm of melatonin synthesis via $\beta_1$AR (for review, see Simonneaux and Ribelayga 2003).

**Table 5** Occurrence of β Adrenoceptor Subtypes

| Organ/Tissue | β Adrenoceptor Subtype | | |
|---|---|---|---|
| | $\beta_1$ | $\beta_2$ | $\beta_3$ |
| **Central nervous system** | | | |
| Cerebral cortex | + | + | |
| Cortex | + | + | |
| Hippocampus | + | + | |
| Hypothalamus | | + | |
| Pineal gland | + | + | |
| Spinal cord | | + | |
| Thalamus | + | + | |
| Sympathetic ganglia | + | + | |
| **Peripheral organs** | | | |
| Smooth muscle | + | ++ | (+) |
| Endothelium | + | ++ | (+) |
| Juxtaglomerular apparatus | + | | |
| Heart | ++ | + | (+) |
| Pancreas | | + | |
| Adipocyte | + | + | ++ |
| Lung | + | ++ | |
| Urinary bladder | | + | + |
| Uterus | | + | + |
| Skeletal muscle | + | ++ | + |
| Kidney | + | | |

For references, see text. Symbols: ++ predominant subtype, + expressed subtype, (+) low abundant subtype.

### 3.4.3 Presynaptic Signaling

βAR subtypes are primarily coupled to stimulatory $G_s$ proteins, which activate adenylyl cyclase to increase intracellular cAMP levels. Based on pharmacological studies, presynaptic βAR belong mostly to the $\beta_2$AR subtype which can facilitate noradrenaline release from sympathetic nerves (Trendelenburg et al. 2000). Most likely, cAMP-dependent activation of protein kinase A resulting in phosphorylation of protein components of the vesicle exocytosis machinery is responsible for the enhanced transmitter secretion (Figure 4) (see discussion in Boehm and Kubista 2002; Kubista and Boehm 2006).

### 3.4.4 Physiological Functions

In addition to their peripheral functions, including presynaptic facilitation of noradrenaline release, β-adrenoceptors play a role in regulating numerous functions of the central nervous system, including sympathetic tone, learning and memory, mood, and food intake. Mice lacking dopamine β-hydroxylase or β-adrenoceptor subtypes were used to study the involvement in memory functions. Noradrenaline acting at $\beta_1$-receptors was found to be essential for retrieval of contextual and spatial memory, but it was not necessary for retrieval of emotional memories (Murchison

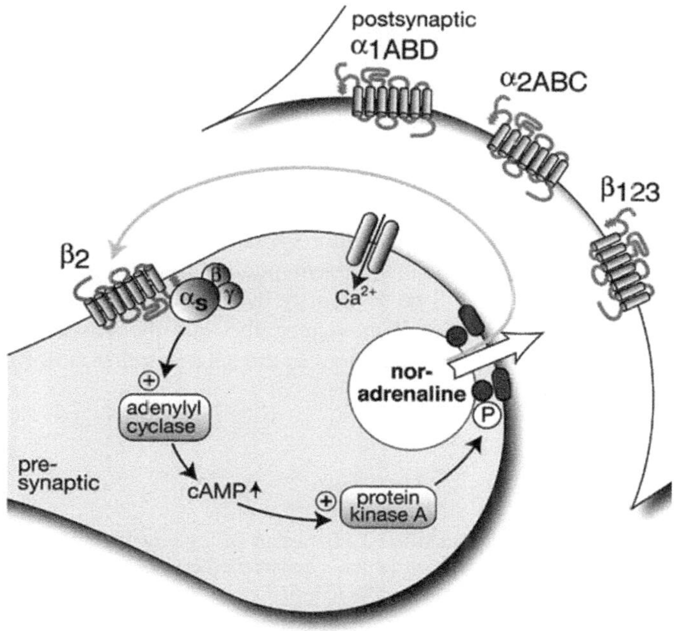


**Fig. 4** Presynaptic signaling mechanisms of $\beta_2$ adrenoceptors. Abbreviations: $\alpha_{i/o}$, β, γ, subunits of $G_{i/o}$ heterotrimeric GTP-binding proteins; $\alpha_{1ABD}$, $\alpha_{2ABC}$, $\beta_{123}$, adrenoceptor subtypes.

et al. 2004; Winder et al. 1999). These studies suggest that memory retrieval has separate pathways from memory acquisition, consolidation, or storage and that separate signaling pathways may control these two aspects of memory function.

### 3.4.5 Relevance in Disease and Therapy

While the initial studies in gene-targeted mice suggest that βAR in the CNS may play an important role in memory function, further studies are required to better understand the role of central βAR and their potential role as therpeutic targets. However, due to the prominent role of $\beta_1$AR and $\beta_2$AR in the cardiovascular system, it may be difficult to identify ligands for CNS therapies which lack significant side effects.

## 4 Conclusions

Molecular cloning has led to the identification of a great complexity of receptor subtypes in the cholinergic and adrenergic systems. Five muscarinic GPCRs and altogether nine adrenoceptors mediate the biological effects of acetylcholine and adrenaline/noradrenaline, respectively. Despite the fact that these receptors were the first presynaptic receptors to be described more than three decades ago, their precise physiological function and therapeutic potential has remained elusive until gene-targeted mouse models became available. These mouse models have greatly helped to identify specific roles for each of these receptors. In the context of this overview, at least two muscarinic receptors ($M_2$ and $M_4$), as well as three adrenoceptors ($\alpha_{2A}$, $\alpha_{2B}$, $\alpha_{2C}$), have been demonstrated to serve as presynaptic inhibitory autoreceptors in their respective transmitter systems. Additional receptor subtypes act as facilitory receptors to enhance presynaptic neurotransmitter secretion. The next challenges in this field will be to identify the molecular basis of the complex intracellular signaling properties of these receptors and their differential roles in neurons and thus to validate these receptors as therapeutic target molecules.

## References

Altman JD, Trendelenburg AU, MacMillan L, Bernstein D, Limbird L, Starke K, Kobilka BK, Hein L (1999) Abnormal regulation of the sympathetic nervous system in $\alpha_{2A}$-adrenergic receptor knockout mice. Mol Pharmacol 56:154–61

Anagnostaras SG, Murphy GG, Hamilton SE, Mitchell SL, Rahnama NP, Nathanson NM, Silva AJ (2003) Selective cognitive dysfunction in acetylcholine $M_1$ muscarinic receptor mutant mice. Nature Neurosci 6:51–8

Aoki C, Go CG, Venkatesan C, Kurose H (1994) Perikaryal and synaptic localization of $\alpha_{2A}$-adrenergic receptor-like immunoreactivity. Brain Res 650:181–204

Araya R, Noguchi T, Yuhki M, Kitamura N, Higuchi M, Saido TC, Seki K, Itohara S, Kawano M, Tanemura K, Takashima A, Yamada K, Kondoh Y, Kanno I, Wess J, Yamada M (2006) Loss of $M_5$ muscarinic acetylcholine receptors leads to cerebrovascular and neuronal abnormalities and cognitive deficits in mice. Neurobiol Dis 24:334–44

Arch JR (2002) $\beta_3$-adrenoceptor agonists:potential, pitfalls and progress. Eur J Pharmacol 440:99–107

Arner P (2005) Human fat cell lipolysis:biochemistry, regulation and clinical role. Best Pract Res Clin Endocrinol Metab 19:471–82

Auclair A, Cotecchia S, Glowinski J, Tassin JP (2002) D-amphetamine fails to increase extracellular dopamine levels in mice lacking $\alpha_{1B}$-adrenergic receptors:relationship between functional and nonfunctional dopamine release. J Neurosci 22:9150–4

Auclair A, Drouin C, Cotecchia S, Glowinski J, Tassin JP (2004) 5-$HT_{2A}$ and $\alpha_{1B}$-adrenergic receptors entirely mediate dopamine release, locomotor response and behavioural sensitization to opiates and psychostimulants. Eur J Neurosci 20:3073–84

Basile AS, Fedorova I, Zapata A, Liu X, Shippenberg T, Duttaroy A, Yamada M, Wess J (2002) Deletion of the $M_5$ muscarinic acetylcholine receptor attenuates morphine reinforcement and withdrawal but not morphine analgesia. Proc Natl Acad Sci USA 99:11452–7

Berkeley JL, Gomeza J, Wess J, Hamilton SE, Nathanson NM, Levey AI (2001) $M_1$ muscarinic acetylcholine receptors activate extracellular signal-regulated kinase in CA1 pyramidal neurons in mouse hippocampal slices. Mol Cell Neurosci 18:512–24

Bernardini N, Roza C, Sauer SK, Gomeza J, Wess J, Reeh PW (2002) Muscarinic $M_2$ receptors on peripheral nerve endings:a molecular target of antinociception. J Neurosci 22:RC229.

Bjorklund M, Sirvio J, Sallinen J, Scheinin M, Kobilka BK, Riekkinen P, Jr. (1999) $\alpha_{2C}$-adrenoceptor overexpression disrupts execution of spatial and non-spatial search patterns. Neuroscience 88:1187–98

Bjorklund M, Sirvio J, Riekkinen M, Sallinen J, Scheinin M, Riekkinen P, Jr. (2000) Overexpression of $\alpha_{2C}$-adrenoceptors impairs water maze navigation. Neuroscience 95:481–7
Bjorklund M, Siverina I, Heikkinen T, Tanila H, Sallinen J, Scheinin M, Riekkinen P, Jr. (2001) Spatial working memory improvement by an $\alpha_{2C}$-adrenoceptor agonist dexmedetomidine is not mediated through $\alpha_{2C}$-adrenoceptor. Prog Neuropsychopharmacol Biol Psychiatry 25:1539–54
Boehm S, Kubista H (2002) Fine tuning of sympathetic transmitter release via ionotropic and metabotropic presynaptic receptors. Pharmacol Rev 54:43–99
Brede M, Wiesmann F, Jahns R, Hadamek K, Arnolt C, Neubauer S, Lohse MJ, Hein L (2002) Feedback inhibition of catecholamine release by two different $\alpha_2$-adrenoceptor subtypes prevents progression of heart failure. Circulation 106:2491–6
Brede M, Nagy G, Philipp M, Sorensen JB, Lohse MJ, Hein L (2003) Differential control of adrenal and sympathetic catecholamine release by $\alpha_2$-adrenoceptor subtypes. Mol Endocrinol 17:1640–6.
Brodde OE, Michel MC (1999) Adrenergic and muscarinic receptors in the human heart. Pharmacol Rev 51:651–90
Brum PC, Kosek J, Patterson A, Bernstein D, Kobilka B (2002) Abnormal cardiac function associated with sympathetic nervous system hyperactivity in mice. Am J Physiol Heart Circ Physiol 283:H1838–45
Bücheler M, Hadamek K, Hein L (2002) Two $\alpha_2$-adrenergic receptor subtypes, $\alpha_{2A}$ and $\alpha_{2C}$, inhibit transmitter release in the brain of gene-targeted mice. Neuroscience 109:819–26
Bylund DB, Eikenberg DC, Hieble JP, Langer SZ, Lefkowitz RJ, Minneman KP, Molinoff PB, Ruffolo RR, Jr., Trendelenburg U (1994) International Union of Pharmacology nomenclature of adrenoceptors. Pharmacol Rev 46:121–36
Bymaster FP, Carter PA, Yamada M, Gomeza J, Wess J, Hamilton SE, Nathanson NM, McKinzie DL, Felder CC (2003a) Role of specific muscarinic receptor subtypes in cholinergic parasympathomimetic responses, in vivo phosphoinositide hydrolysis, and pilocarpine-induced seizure activity. The Eur J Neurosci 17:1403–10
Bymaster FP, McKinzie DL, Felder CC, Wess J (2003b) Use of $M_1$-$M_5$ muscarinic receptor knockout mice as novel tools to delineate the physiological roles of the muscarinic cholinergic system. Neurochem Res 28:437–42
Caulfield MP, Birdsall NJ (1998) International Union of Pharmacology. XVII. Classification of muscarinic acetylcholine receptors. Pharmacol Rev 50:279–90
Caulfield MP, Robbins J, Higashida H, Brown DA (1993) Postsynaptic actions of acetylcholine:the coupling of muscarinic receptor subtypes to neuronal ion channels. Progr Brain Res 98:293–301
Cavalli A, Lattion AL, Hummler E, Nenniger M, Pedrazzini T, Aubert JF, Michel MC, Yang M, Lembo G, Vecchione C, Mostardini M, Schmidt A, Beermann F, Cotecchia S (1997) Decreased blood pressure response in mice deficient of the $\alpha_{1B}$-adrenergic receptor. Proc Natl Acad Sci USA 94:11589–94
Champtiaux N, Gotti C, Cordero-Erausquin M, David DJ, Przybylski C, Lena C, Clementi F, Moretti M, Rossi FM, Le Novere N, McIntosh JM, Gardier AM, Changeux JP (2003) Subunit composition of functional nicotinic receptors in dopaminergic neurons investigated with knock-out mice. J Neurosci 23:7820–9
Chen Q, Takahashi S, Zhong S, Hosoda C, Zheng HY, Ogushi T, Fujimura T, Ohta N, Tanoue A, Tsujimoto G, Kitamura T (2005) Function of the lower urinary tract in mice lacking $\alpha_{1D}$-adrenoceptor. J Urol 174:370–4
Chen ZJ, Minneman KP (2005) Recent progress in $\alpha_1$-adrenergic receptor research. Acta Pharmacol Sin 26:1281–7
Chruscinski AJ, Rohrer DK, Schauble E, Desai KH, Bernstein D, Kobilka BK (1999) Targeted disruption of the $\beta_2$ adrenergic receptor gene. J Biol Chem 274:16694–700
Cohn JN, Pfeffer MA, Rouleau J, Sharpe N, Swedberg K, Straub M, Wiltse C, Wright TJ (2003) Adverse mortality effect of central sympathetic inhibition with sustained-release moxonidine in patients with heart failure (MOXCON). Eur J Heart Fail 5:659–67

Cussac D, Schaak S, Gales C, Flordellis C, Denis C, Paris H (2001) $\alpha_{2B}$-adrenergic receptors activate MAPK and modulate the proliferation of primary cultured proximal tubule cells. Am J Physiol Renal Physiol 8:8

Day HE, Campeau S, Watson SJ, Jr., Akil H (1997) Distribution of $\alpha_{1A}$-, $\alpha_{1B}$- and $\alpha_{1D}$-adrenergic receptor mRNA in the rat brain and spinal cord. Journal of chemical neuroanatomy 13:115–39

Dennedy MC, Houlihan DD, McMillan H, Morrison JJ (2002) $\beta_2$- and $\beta_3$-adrenoreceptor agonists:human myometrial selectivity and effects on umbilical artery tone. Am J Obstet Gynecol 187:641–7

Drago J, McColl CD, Horne MK, Finkelstein DI, Ross SA (2003) Neuronal nicotinic receptors:insights gained from gene knockout and knockin mutant mice. Cell Mol Life Sci 60:1267–80

Drouin C, Darracq L, Trovero F, Blanc G, Glowinski J, Cotecchia S, Tassin JP (2002) $\alpha_{1B}$-adrenergic receptors control locomotor and rewarding effects of psychostimulants and opiates. J Neurosci 22:2873–24

Eglen RM (2006) Muscarinic receptor subtypes in neuronal and non-neuronal cholinergic function. Auton Autacoid Pharmacol 26:219–33

Eglen RM, Nahorski SR (2000) The muscarinic $M_5$ receptor:a silent or emerging subtype? Br J Pharmacol 130:13–21

Fagerholm V, Philipp M, Hein L, Scheinin M (2004) [Ethyl-$^3$H]RS-79948-197 $\alpha_2$-adrenoceptor autoradiography validation in $\alpha_2$-adrenoceptor knockout mice. Eur J Pharmacol 497:301–9

Fairbanks CA, Wilcox GL (1999) Spinal antinociceptive synergism between morphine and clonidine persists in mice made acutely or chronically tolerant to morphine. J Pharmacol Exp Ther 288:1107–16

Fink-Jensen A, Fedorova I, Wortwein G, Woldbye DP, Rasmussen T, Thomsen M, Bolwig TG, Knitowski KM, McKinzie DL, Yamada M, Wess J, Basile A (2003) Role for $M_5$ muscarinic acetylcholine receptors in cocaine addiction. J Neurosci Res 74:91–6

Fisahn A, Yamada M, Duttaroy A, Gan JW, Deng CX, McBain CJ, Wess J (2002) Muscarinic induction of hippocampal gamma oscillations requires coupling of the $M_1$ receptor to two mixed cation currents. Neuron 33:615–24

Foote SL, Bloom FE, Aston-Jones G (1983) Nucleus locus ceruleus:new evidence of anatomical and physiological specificity. Physiol Rev 63:844–914

Gautam D, Gavrilova O, Jeon J, Pack S, Jou W, Cui Y, Li JH, Wess J (2006a) Beneficial metabolic effects of $M_3$ muscarinic acetylcholine receptor deficiency. Cell Metabol 4:363–75

Gautam D, Han SJ, Hamdan FF, Jeon J, Li B, Li JH, Cui Y, Mears D, Lu H, Deng C, Heard T, Wess J (2006b) A critical role for $\beta$ cell $M_3$ muscarinic acetylcholine receptors in regulating insulin release and blood glucose homeostasis in vivo. Cell Metabol 3:449–61

Gerber DJ, Sotnikova TD, Gainetdinov RR, Huang SY, Caron MG, Tonegawa S (2001) Hyperactivity, elevated dopaminergic transmission, and response to amphetamine in $M_1$ muscarinic acetylcholine receptor-deficient mice. Proc Natl Acad Sci USA 98:15312–17

Gomeza J, Shannon H, Kostenis E, Felder C, Zhang L, Brodkin J, Grinberg A, Sheng H, Wess J (1999a) Pronounced pharmacologic deficits in $M_2$ muscarinic acetylcholine receptor knockout mice. Proc Natl Acad Sci USA 96:1692–7

Gomeza J, Zhang L, Kostenis E, Felder C, Bymaster F, Brodkin J, Shannon H, Xia B, Deng C, Wess J (1999b) Enhancement of D1 dopamine receptor-mediated locomotor stimulation in $M_4$ muscarinic acetylcholine receptor knockout mice. Proc Natl Acad Sci USA 96:10483–8

Gomeza J, Zhang L, Kostenis E, Felder CC, Bymaster FP, Brodkin J, Shannon H, Xia B, Duttaroy A, Deng CX, Wess J (2001) Generation and pharmacological analysis of $M_2$ and $M_4$ muscarinic receptor knockout mice. Life Sci 68:2457–66

Guimarães S, Moura D (2001) Vascular adrenoceptors:an update. Pharmacol Rev 53:319–56

Hamilton SE, Loose MD, Qi M, Levey AI, Hille B, McKnight GS, Idzerda RL, Nathanson NM (1997) Disruption of the $M_1$ receptor gene ablates muscarinic receptor-dependent M current regulation and seizure activity in mice. Proc Natl Acad Sci USA 94:13311–16

Harasawa I, Honda K, Tanoue A, Shinoura H, Ishida Y, Okamura H, Murao N, Tsujimoto G, Higa K, Kamiya HO, Takano Y (2003) Responses to noxious stimuli in mice lacking $\alpha_{1D}$-adrenergic receptors. Neuroreport 14:1857–60

Hardouin SN, Richmond KN, Zimmerman A, Hamilton SE, Feigl EO, Nathanson NM (2002) Altered cardiovascular responses in mice lacking the $M_1$ muscarinic acetylcholine receptor. J Exp Pharmacol Ther 301:129–37

Hein L (2001) Transgenic models of $\alpha_2$-adrenergic receptor subtype function. Rev Physiol Biochem Pharmacol 142:161–85

Hein L, Altman JD, Kobilka BK (1999) Two functionally distinct $\alpha_2$-adrenergic receptors regulate sympathetic neurotransmission. Nature 402:181–4

Hunter JC, Fontana DJ, Hedley LR, Jasper JR, Lewis R, Link RE, Secchi R, Sutton J, Eglen RM (1997) Assessment of the role of $\alpha_2$-adrenoceptor subtypes in the antinociceptive, sedative and hypothermic action of dexmedetomidine in transgenic mice. Br J Pharmacol 122:1339–44

Janumpalli S, Butler LS, MacMillan LB, Limbird LE, McNamara JO (1998) A point mutation (D79N) of the $\alpha_{2A}$ adrenergic receptor abolishes the antiepileptogenic action of endogenous norepinephrine. J Neurosci 18:2004–8

Khurana S, Yamada M, Wess J, Kennedy RH, Raufman JP (2005) Deoxycholyltaurine-induced vasodilation of rodent aorta is nitric oxide- and muscarinic $M_3$ receptor-dependent. Eur J Pharmacol 517:103–10

Kintsurashvili E, Gavras I, Johns C, Gavras H (2001) Effects of antisense oligodeoxynucleotide targeting of the $\alpha_{2B}$-adrenergic receptor messenger RNA in the central nervous system. Hypertension 38:1075–80

Kitazawa T, Hashiba K, Cao J, Unno T, Komori S, Yamada M, Wess J, Taneike T (2007) Functional roles of muscarinic $M_2$ and $M_3$ receptors in mouse stomach motility:studies with muscarinic receptor knockout mice. Eur J Pharmacol 554:212–22

Knauber J, Muller WE (2000) Decreased exploratory activity and impaired passive avoidance behaviour in mice deficient for the $\alpha_{1B}$-adrenoceptor. Eur Neuropsychopharmacol 10:423–7

Koshimizu TA, Tanoue A, Tsujimoto G (2006) Clinical implications from studies of $\alpha_1$ adrenergic receptor knockout mice. Biochem Pharmacol.

Kubista H, Boehm S (2006) Molecular mechanisms underlying the modulation of exocytotic noradrenaline release via presynaptic receptors. Pharmacol Ther 112:213–42

Lakhlani PP, MacMillan LB, Guo TZ, McCool BA, Lovinger DM, Maze M, Limbird LE (1997) Substitution of a mutant $\alpha_{2A}$-adrenergic receptor via "hit and run" gene targeting reveals the role of this subtype in sedative, analgesic, and anesthetic-sparing responses in vivo. Proc Natl Acad Sci USA 94:9950–5

Langer SZ (1997) 25 years since the discovery of presynaptic receptors:present knowledge and future perspectives. Trends Pharmacol Sci 18:95–9

Langer SZ, Duval N, Massingham R (1985) Pharmacologic and therapeutic significance of $\alpha$-adrenoceptor subtypes. J Cardiovasc Pharmacol 7:S1–8

Levey AI (1993) Immunological localization of $M_1$-$M_5$ muscarinic acetylcholine receptors in peripheral tissues and brain. Life Sci 52:441–8

Levey AI, Edmunds SM, Heilman CJ, Desmond TJ, Frey KA (1994) Localization of muscarinic $M_3$ receptor protein and $M_3$ receptor binding in rat brain. Neuroscience 63:207–21

Link RE, Stevens MS, Kulatunga M, Scheinin M, Barsh GS, Kobilka BK (1995) Targeted inactivation of the gene encoding the mouse $\alpha_{2C}$-adrenoceptor homolog. Mol Pharmacol 48:48–55

Link RE, Desai K, Hein L, Stevens ME, Chruscinski A, Bernstein D, Barsh GS, Kobilka BK (1996) Cardiovascular regulation in mice lacking $\alpha_2$-adrenergic receptor subtypes b and c. Science 273:803–5

Ma D, Hossain M, Rajakumaraswamy N, Arshad M, Sanders RD, Franks NP, Maze M (2004) Dexmedetomidine produces its neuroprotective effect via the $\alpha_{2A}$-adrenoceptor subtype. Eur J Pharmacol 502:87–97

MacDonald E, Scheinin M (1995) Distribution and pharmacology of $\alpha_2$-adrenoceptors in the central nervous system. J Physiol Pharmacol 46:241–58

MacMillan LB, Hein L, Smith MS, Piascik MT, Limbird LE (1996) Central hypotensive effects of the $\alpha_{2A}$-adrenergic receptor subtype. Science 273:801–3
Makaritsis KP, Handy DE, Johns C, Kobilka B, Gavras I, Gavras H (1999a) Role of the $\alpha_{2B}$-adrenergic receptor in the development of salt–induced hypertension. Hypertension 33:14–17
Makaritsis KP, Johns C, Gavras I, Altman JD, Handy DE, Bresnahan MR, Gavras H (1999b) Sympathoinhibitory function of the $\alpha_{2A}$-adrenergic receptor subtype. Hypertension 34:403–7
Makaritsis KP, Johns C, Gavras I, Gavras H (2000) Role of $\alpha_2$-adrenergic receptor subtypes in the acute hypertensive response to hypertonic saline infusion in anephric mice. Hypertension 35:609–13
Matsui M, Motomura D, Karasawa H, Fujikawa T, Jiang J, Komiya Y, Takahashi S, Taketo MM (2000) Multiple functional defects in peripheral autonomic organs in mice lacking muscarinic acetylcholine receptor gene for the $M_3$ subtype. Proc Natl Acad Sci USA 97:9579–84
McCune SK, Voigt MM, Hill JM (1993) Expression of multiple $\alpha$ adrenergic receptor subtype messenger RNAs in the adult rat brain. Neuroscience 57:143–51
Michel MC, Vrydag W (2006) $\alpha_1$-, $\alpha_2$- and $\beta$-adrenoceptors in the urinary bladder, urethra and prostate. Br J Pharmacol 147 Suppl 2:S88–119
Michelotti GA, Price DT, Schwinn DA (2000) $\alpha_1$-adrenergic receptor regulation:basic science and clinical implications. Pharmacol Ther 88:281–309
Miller RJ (1998) Presynaptic receptors. Annu Rev Pharmacol Toxicol 38:201–27
Mishima K, Tanoue A, Tsuda M, Hasebe N, Fukue Y, Egashira N, Takano Y, Kamiya HO, Tsujimoto G, Iwasaki K, Fujiwara M (2004) Characteristics of behavioral abnormalities in $\alpha_{1D}$-adrenoceptors deficient mice. Behav Brain Res 152:365–73
Miyakawa T, Yamada M, Duttaroy A, Wess J (2001) Hyperactivity and intact hippocampus-dependent learning in mice lacking the $M_1$ muscarinic acetylcholine receptor. J Neurosci 21:5239–50
Moura E, Afonso J, Hein L, Vieira-Coelho MA (2006) $\alpha_2$-adrenoceptor subtypes involved in the regulation of catecholamine release from the adrenal medulla of mice. Br J Pharmacol 149:1049–58
Murchison CF, Zhang XY, Zhang WP, Ouyang M, Lee A, Thomas SA (2004) A distinct role for norepinephrine in memory retrieval. Cell 117:131–43
Nicholas AP, Pieribone V, Hokfelt T (1993) Distributions of mRNAs for $\alpha_2$ adrenergic receptor subtypes in rat brain:an in situ hybridization study. J Comp Neurol 328:575–94
Nicholas AP, Hokfelt T, Pieribone VA (1996) The distribution and significance of CNS adrenoceptors examined with in situ hybridization. Trends Pharmacol Sci 17:245–55
Niederhoffer N, Hein L, Starke K (2004) Modulation of the baroreceptor reflex by $\alpha_{2A}$-adrenoceptors:a study in $\alpha_{2A}$ knockout mice. Br J Pharmacol 141:851–9
Ordway GA, O'Donnell JM, Frazer A (1987) Effects of clenbuterol on central $\beta_1$ and $\beta_2$ adrenergic receptors of the rat. J Pharmacol Exp Ther 241:187–95
Ordway GA, Gambarana C, Frazer A (1988) Quantitative autoradiography of central beta adrenoceptor subtypes:comparison of the effects of chronic treatment with desipramine or centrally administered l-isoproterenol. J Pharmacol Exp Ther 247:379–89
Papay R, Zuscik MJ, Ross SA, Yun J, McCune DF, Gonzalez-Cabrera P, Gaivin R, Drazba J, Perez DM (2002) Mice expressing the $\alpha_{1B}$-adrenergic receptor induces a synucleinopathy with excessive tyrosine nitration but decreased phosphorylation. J Neurochem 83:623–34
Paris A, Philipp M, Tonner PH, Steinfath M, Lohse M, Scholz J, Hein L (2003) Activation of $\alpha_{2B}$-adrenoceptors mediates the cardiovascular effects of etomidate. Anesthesiol 99:889–95
Paris A, Mantz J, Tonner PH, Hein L, Brede M, Gressens P (2006) The effects of dexmedetomidine on perinatal excitotoxic brain injury are mediated by the $\alpha_{2A}$-adrenoceptor subtype. Anesth Analg 102:456–61
Philipp M, Brede M, Hein L (2002a) Physiological significance of $\alpha_{2A}$-adrenergic receptor subtype diversity:one receptor is not enough. Am J Physiol Regul Integr Comp Physiol 283:R287–95
Philipp M, Brede ME, Hadamek K, Gessler M, Lohse MJ, Hein L (2002b) Placental $\alpha_2$-adrenoceptors control vascular development at the interface between mother and embryo. Nat Genet 31:311–15

Philipp M, Hein L (2004) Adrenergic receptor knockout mice:distinct functions of 9 receptor subtypes. Pharmacol Ther 101:65–74

Piascik MT, Perez DM (2001) $\alpha_1$-adrenergic receptors:new insights and directions. J Exp Pharmacol Ther 298:403–10

Pieribone VA, Nicholas AP, Dagerlind A, Hokfelt T (1994) Distribution of $\alpha_1$ adrenoceptors in rat brain revealed by in situ hybridization experiments utilizing subtype-specific probes. J Neurosci 14:4252–68

Porter AC, Bymaster FP, DeLapp NW, Yamada M, Wess J, Hamilton SE, Nathanson NM, Felder CC (2002) $M_1$ muscarinic receptor signaling in mouse hippocampus and cortex. Brain Res 944:82–9

Rohrer DK, Chruscinski A, Schauble EH, Bernstein D, Kobilka BK (1999) Cardiovascular and metabolic alterations in mice lacking both $\beta_1$- and $\beta_2$-adrenergic receptors. J Biol Chem 274:16701–8

Rohrer DK, Desai KH, Jasper JR, Stevens ME, Regula DP, Jr., Barsh GS, Bernstein D, Kobilka BK (1996) Targeted disruption of the mouse $\beta_1$-adrenergic receptor gene:developmental and cardiovascular effects. Proc Natl Acad Sci USA 93:7375–80

Rokosh DG, Simpson PC (2002) Knockout of the $\alpha_{1A/C}$-adrenergic receptor subtype:the $\alpha_{1A/C}$ is expressed in resistance arteries and is required to maintain arterial blood pressure. Proc Natl Acad Sci USA 99:9474–9

Rosin DL, Talley EM, Lee A, Stornetta RL, Gaylinn BD, Guyenet PG, Lynch KR (1996) Distribution of $\alpha_{2C}$-adrenergic receptor-like immunoreactivity in the rat central nervous system. J Comp Neurol 372:135–65

Sadalge A, Coughlin L, Fu H, Wang B, Valladares O, Valentino R, Blendy JA (2003) $\alpha_{1D}$ Adrenoceptor signaling is required for stimulus induced locomotor activity. Mol Psychiatry 8:664–72

Salomon L, Lanteri C, Glowinski J, Tassin JP (2006) Behavioral sensitization to amphetamine results from an uncoupling between noradrenergic and serotonergic neurons. Proc Natl Acad Sci USA 103:7476–81

Sawamura S, Kingery WS, Davies MF, Agashe GS, Clark JD, Kobilka BK, Hashimoto T, Maze M (2000) Antinociceptive action of nitrous oxide is mediated by stimulation of noradrenergic neurons in the brainstem and activation of $\alpha_{2B}$ adrenoceptors. J Neurosci 20:9242–51

Scheibner J, Trendelenburg AU, Hein L, Starke K (2001a) $\alpha_2$-adrenoceptors modulating neuronal serotonin release:a study in $\alpha_2$-adrenoceptor subtype-deficient mice. Br J Pharmacol 132:925–33

Scheibner J, Trendelenburg AU, Hein L, Starke K (2001b) Stimulation frequency-noradrenaline release relationships examined in $\alpha_{2A}$-, $\alpha_{2B}$- and $\alpha_{2C}$-adrenoceptor-deficient mice. Naunyn Schmiedeberg's Arch Pharmacol 364:321–8

Scheibner J, Trendelenburg AU, Hein L, Starke K, Blandizzi C (2002) $\alpha_2$-adrenoceptors in the enteric nervous system:a study in $\alpha_{2A}$-adrenoceptor-deficient mice. Br J Pharmacol 135:697–704

Scheinin M, Lomasney JW, Hayden-Hixson DM, Schambra UB, Caron MG, Lefkowitz RJ, Fremeau RT, Jr. (1994) Distribution of $\alpha_2$-adrenergic receptor subtype gene expression in rat brain. Brain Res Mol Brain Res 21:133–49

Scheinin M, Sallinen J, Haapalinna A (2001) Evaluation of the $\alpha_{2C}$-adrenoceptor as a neuropsychiatric drug target studies in transgenic mouse models. Life Sci 68:2277–85

Schelb V, Gobel I, Khairallah L, Zhou H, Cox SL, Trendelenburg AU, Hein L, Starke K (2001) Postnatal development of presynaptic receptors that modulate noradrenaline release in mice. Naunyn Schmiedeberg's Arch Pharmacol 364:359–71

Shapiro MS, Loose MD, Hamilton SE, Nathanson NM, Gomeza J, Wess J, Hille B (1999) Assignment of muscarinic receptor subtypes mediating G-protein modulation of $Ca^{2+}$ channels by using knockout mice. Proc Natl Acad Sci USA 96:10899–904

Simonneaux V, Ribelayga C (2003) Generation of the melatonin endocrine message in mammals:a review of the complex regulation of melatonin synthesis by norepinephrine, peptides, and other pineal transmitters. Pharmacol Rev 55:325–95

Slutsky I, Wess J, Gomeza J, Dudel J, Parnas I, Parnas H (2003) Use of knockout mice reveals involvement of $M_2$-muscarinic receptors in control of the kinetics of acetylcholine release. J Neurophysiol 89:1954–67

Spreng M, Cotecchia S, Schenk F (2001) A behavioral study of $\alpha_{1B}$ adrenergic receptor knockout mice:increased reaction to novelty and selectively reduced learning capacities. Neurobiology of learning and memory 75:214–29

Starke K (2001) Presynaptic autoreceptors in the third decade:focus on $\alpha_2$-adrenoceptors. J Neurochem 78:685–93

Starke K, Gothert M, Kilbinger H (1989) Modulation of neurotransmitter release by presynaptic autoreceptors. Physiol Rev 69:864–89

Stengel PW, Gomeza J, Wess J, Cohen ML (2000) $M_2$ and $M_4$ receptor knockout mice:muscarinic receptor function in cardiac and smooth muscle in vitro. J Exp Pharmacol Ther 292:877–85

Stephens GJ, Mochida S (2005) G protein $\beta\gamma$ subunits mediate presynaptic inhibition of transmitter release from rat superior cervical ganglion neurones in culture. J Physiol 563:765–76

Stone LS, MacMillan LB, Kitto KF, Limbird LE, Wilcox GL (1997) The $\alpha_{2A}$ adrenergic receptor subtype mediates spinal analgesia evoked by $\alpha_2$ agonists and is necessary for spinal adrenergic-opioid synergy. J Neurosci 17:7157–65

Susulic VS, Frederich RC, Lawitts J, Tozzo E, Kahn BB, Harper ME, Himms-Hagen J, Flier JS, Lowell BB (1995) Targeted disruption of the $\beta_3$-adrenergic receptor gene. J Biol Chem 270:29483–92

Szot P, Lester M, Laughlin ML, Palmiter RD, Liles LC, Weinshenker D (2004) The anticonvulsant and proconvulsant effects of $\alpha_2$-adrenoreceptor agonists are mediated by distinct populations of $\alpha_{2A}$-adrenoreceptors. Neuroscience 126:795–803

Talley EM, Rosin DL, Lee A, Guyenet PG, Lynch KR (1996) Distribution of $\alpha_{2A}$-adrenergic receptor-like immunoreactivity in the rat central nervous system. J Comp Neurol 372:111–34

Tan CM, Wilson MH, MacMillan LB, Kobilka BK, Limbird LE (2002) Heterozygous $\alpha_{2A}$-adrenergic receptor mice unveil unique therapeutic benefits of partial agonists. Proc Natl Acad Sci USA 99:12471–6

Tanoue A, Koba M, Miyawaki S, Koshimizu TA, Hosoda C, Oshikawa S, Tsujimoto G (2002) Role of the $\alpha_{1D}$-adrenergic receptor in the development of salt-induced hypertension. Hypertension 40:101–6

Tanoue A, Koshimizu TA, Shibata K, Nasa Y, Takeo S, Tsujimoto G (2003) Insights into $\alpha_1$ adrenoceptor function in health and disease from transgenic animal studies. Trends Endocrinol Metab 14:107–13

Thomsen M, Woldbye DP, Wortwein G, Fink-Jensen A, Wess J, Caine SB (2005) Reduced cocaine self-administration in muscarinic $M_5$ acetylcholine receptor-deficient mice. J Neurosci 25:8141–9

Trendelenburg AU, Hein L, Gaiser EG, Starke K (1999) Occurrence, pharmacology and function of presynaptic $\alpha_2$-autoreceptors in $\alpha_{2A/D}$-adrenoceptor-deficient mice. Naunyn Schmiedeberg's Arch Pharmacol 360:540–51

Trendelenburg AU, Cox SL, Schelb V, Klebroff W, Khairallah L, Starke K (2000) Modulation of $^3$H-noradrenaline release by presynaptic opioid, cannabinoid and bradykinin receptors and beta-adrenoceptors in mouse tissues. Br J Pharmacol 130:321–30

Trendelenburg AU, Klebroff W, Hein L, Starke K (2001a) A study of presynaptic $\alpha_2$-autoreceptors in $\alpha_{2A/D}$-, $\alpha_{2B}$- and $\alpha_{2C}$-adrenoceptor-deficient mice. Naunyn Schmiedeberg's Arch Pharmacol 364:17–30.

Trendelenburg AU, Norenberg W, Hein L, Meyer A, Starke K (2001b) $\alpha_2$-adrenoceptor-mediated inhibition of cultured sympathetic neurons:changes in $\alpha_{2A/D}$-adrenoceptor-deficient mice. Naunyn Schmiedeberg's Arch Pharmacol 363:110–19

Trendelenburg AU, Gomeza J, Klebroff W, Zhou H, Wess J (2003a) Heterogeneity of presynaptic muscarinic receptors mediating inhibition of sympathetic transmitter release:a study with $M_2$- and $M_4$-receptor-deficient mice. Br J Pharmacol 138:469–80

Trendelenburg AU, Philipp M, Meyer A, Klebroff W, Hein L, Starke K (2003b) All three $\alpha_2$-adrenoceptor types serve as autoreceptors in postganglionic sympathetic neurons. Naunyn Schmiedeberg's Arch Pharmacol 368:504–12

Trendelenburg AU, Meyer A, Wess J, Starke K (2005) Distinct mixtures of muscarinic receptor subtypes mediate inhibition of noradrenaline release in different mouse peripheral tissues, as studied with receptor knockout mice. Br J Pharmacol 145:1153–9

Tzavara ET, Bymaster FP, Felder CC, Wade M, Gomeza J, Wess J, McKinzie DL, Nomikos GG (2003) Dysregulated hippocampal acetylcholine neurotransmission and impaired cognition in $M_2$, $M_4$ and $M_2/M_4$ muscarinic receptor knockout mice. Mol Psychiatry 8:673–9

Uhlen S, Lindblom J, Johnson A, Wikberg JE (1997) Autoradiographic studies of central $\alpha_{2A}$- and $\alpha_{2C}$-adrenoceptors in the rat using [$^3$H]MK912 and subtype-selective drugs. Brain Res 770:261–6

Unno T, Matsuyama H, Sakamoto T, Uchiyama M, Izumi Y, Okamoto H, Yamada M, Wess J, Komori S (2005) $M_2$ and $M_3$ muscarinic receptor-mediated contractions in longitudinal smooth muscle of the ileum studied with receptor knockout mice. Br J Pharmacol 146:98–108

Unno T, Matsuyama H, Izumi Y, Yamada M, Wess J, Komori S (2006) Roles of $M_2$ and $M_3$ muscarinic receptors in cholinergic nerve-induced contractions in mouse ileum studied with receptor knockout mice. Br J Pharmacol 149:1022–30

Vonend O, Habbel S, Stegbauer J, Roth J, Hein L, Rump LC (2007) $\alpha_{2A}$-adrenoceptors regulate sympathetic transmitter release in mice kidneys. Br J Pharmacol 150:121–7

Wess J (1996) Molecular biology of muscarinic acetylcholine receptors. Crit Rev Neurobiol 10:69–99

Wess J (2004) Muscarinic acetylcholine receptor knockout mice:novel phenotypes and clinical implications. Ann Rev Pharmacol Toxicol 44:423–50

Winder DG, Martin KC, Muzzio IA, Rohrer D, Chruscinski A, Kobilka B, Kandel ER (1999) ERK plays a regulatory role in induction of LTP by theta frequency stimulation and its modulation by β-adrenergic receptors. Neuron 24:715–26

Wolfe BB, Yasuda RP (1995) Development of selective antisera for muscarinic cholinergic receptor subtypes. Ann NY Acad Sci 757:186–93

Wu LG, Saggau P (1997) Presynaptic inhibition of elicited neurotransmitter release. Trends Neurosci 20:204–12

Xiao RP, Zhu W, Zheng M, Cao C, Zhang Y, Lakatta EG, Han Q (2006) Subtype-specific $\alpha_1$- and β-adrenoceptor signaling in the heart. Trends Pharmacol Sci 27:330–7

Xie G, Drachenberg C, Yamada M, Wess J, Raufman JP (2005) Cholinergic agonist-induced pepsinogen secretion from murine gastric chief cells is mediated by $M_1$ and $M_3$ muscarinic receptors. Am J Physiol 289:G521–9

Yamada M, Lamping KG, Duttaroy A, Zhang W, Cui Y, Bymaster FP, McKinzie DL, Felder CC, Deng CX, Faraci FM, Wess J (2001a) Cholinergic dilation of cerebral blood vessels is abolished in $M_5$ muscarinic acetylcholine receptor knockout mice. Proc Natl Acad Sci USA 98:14096–101

Yamada M, Miyakawa T, Duttaroy A, Yamanaka A, Moriguchi T, Makita R, Ogawa M, Chou CJ, Xia B, Crawley JN, Felder CC, Deng CX, Wess J (2001b) Mice lacking the $M_3$ muscarinic acetylcholine receptor are hypophagic and lean. Nature 410:207–12

Yasuda RP, Ciesla W, Flores LR, Wall SJ, Li M, Satkus SA, Weisstein JS, Spagnola BV, Wolfe BB (1993) Development of antisera selective for m4 and m5 muscarinic cholinergic receptors:distribution of $M_4$ and $M_5$ receptors in rat brain. Mol Pharmacol 43:149–57

Zhang C, Davies MF, Guo TZ, Maze M (1999) The analgesic action of nitrous oxide is dependent on the release of norepinephrine in the dorsal horn of the spinal cord. Anesthesiology 91:401–7

Zhang W, Basile AS, Gomeza J, Volpicelli LA, Levey AI, Wess J (2002a) Characterization of central inhibitory muscarinic autoreceptors by the use of muscarinic acetylcholine receptor knock-out mice. J Neurosci 22:1709–17

Zhang W, Yamada M, Gomeza J, Basile AS, Wess J (2002b) Multiple muscarinic acetylcholine receptor subtypes modulate striatal dopamine release, as studied with $M_1$-$M_5$ muscarinic receptor knock-out mice. J Neurosci 22:6347–52

Zhou H, Meyer A, Starke K, Gomeza J, Wess J, Trendelenburg AU (2002) Heterogeneity of release-inhibiting muscarinic autoreceptors in heart atria and urinary bladder:a study with $M_2$- and $M_4$-receptor-deficient mice. Naunyn-Schmiedeberg's Arch Pharmacol 365:112–22
Zhu QM, Lesnick JD, Jasper JR, MacLennan SJ, Dillon MP, Eglen RM, Blue DR, Jr. (1999) Cardiovascular effects of rilmenidine, moxonidine and clonidine in conscious wild-type and D79N $\alpha_{2A}$-adrenoceptor transgenic mice. Br J Pharmacol 126:1522–30
Zuscik MJ, Sands S, Ross SA, Waugh DJ, Gaivin RJ, Morilak D, Perez DM (2000) Overexpression of the $\alpha_{1B}$-adrenergic receptor causes apoptotic neurodegeneration:multiple system atrophy. Nature Med 6:1388–94

# Presynaptic Receptors for Dopamine, Histamine, and Serotonin

**Thomas J. Feuerstein**



Thomas J. Feuerstein
Neurochirurgische Universitätsklinik Breisacherstrasse 64 D - 79106 Freiburg, Germany
thomas.feuerstein@uniklinik-freiburg.de

T.C. Südhof, K. Starke (eds.), *Pharmacology of Neurotransmitter Release.*
*Handbook of Experimental Pharmacology 184.*

**Abstract** Presynaptic receptors for dopamine, histamine and serotonin that are located on dopaminergic, histaminergic and sertonergic axon terminals, respectively, function as autoreceptors. Presynaptic receptors also occur as heteroreceptors on other axon terminals. Auto- and heteroreceptors mainly affect $Ca^{2+}$-dependent exocytosis from the receptor-bearing nerve ending. Some additionally subserve other presynaptic functions.

Presynaptic dopamine, histamine and serotonin receptors are involved in various (patho)physiological conditions. Examples are the following:

Dopamine autoreceptors play a role in Parkinson's disease, schizophrenia and drug addiction. Dopamine heteroreceptors affecting the release of acetylcholine and of amino acid neurotransmitters in the basal ganglia are also relevant for Parkinson's disease. Peripheral dopamine heteroreceptors on postganglionic sympathetic terminals influence heart rate and vascular resistance through modulation of noradrenaline release.

Blockade of histamine autoreceptors increases histamine synthesis and release and may support higher CNS functions such as arousal, cognition and learning. Peripheral histamine heteroreceptors on C fiber and on postganglionic sympathetic fiber terminals diminish neuropeptide and noradrenaline release, respectively. Both inhibititory effects are beneficial in myocardial ischemia. The inhibition of neuropeptide release also explains the antimigraine effects of some agonists of presynaptic histamine receptors.

Upregulation of presynaptic serotonin autoreceptors is probably involved in the pathogenesis of major depression. Correspondingly, antidepressant treatments can be linked with a reduced density of 5-HT autoreceptors. 5-HT Heteroreceptor activation diminishes acetylcholine and GABA release and may therefore increase anxiety. In the periphery, presynaptic 5-HT heteroreceptor agonists shorten migraine attacks by inhibition of the release of neuropeptides from trigeminal afferents, apart from their constrictive action on meningeal vessels.

# 1 Introduction

Our knowledge of presynaptic dopamine and serotonin receptors dates back to the 1970s (Farnebo and Hamberger 1971). Presynaptic histamine receptors were discovered in 1983 (Arrang et al. 1983). Presynaptic dopamine receptors occur as autoreceptors, i.e., on dopaminergic axon terminals, and as heteroreceptors on nondopaminergic axon terminals. By analogy the same holds true for presynaptic histamine and serotonin receptors. The early days of the dopamine autoreceptors were stormy, but the controversies were finally solved (see Starke et al. 1989). The main function that presynaptic receptors affect is transmitter release, which in this article means $Ca^{2+}$-dependent exocytosis. However, some receptors discussed in this chapter also modify other presynaptic functions, such as transmitter synthesis, using signal transduction cascades somewhat different from those leading to modulation of exocytosis.

**Dopamine** is one of the main neurotransmitters in the central nervous system (CNS), but it also plays a role in the periphery, where, for example, it acts as an autocrine/paracrine substance in the kidneys, the mesenteric organs, and the lungs (Goldstein et al. 1995). In the CNS, cell bodies of dopamine neurons are to be found in midbrain areas, hypothalamus, olfactory bulb, and retina. Cells responsive to dopamine express five receptor types, all G-protein-coupled and known as $D_1$, $D_2$, $D_3$, $D_4$ and $D_5$. Through alternative splicing the $D_2$ receptor gene can generate two forms of the receptor, $D_{2long}$ and $D_{2short}$ (Dal Toso et al. 1989).

The $D_1$-like receptors ($D_1$, $D_5$) couple predominantly to Gs and thus can stimulate adenylyl cyclase, yielding cAMP. The $D_2$-like receptors ($D_2$, $D_3$, $D_4$) couple to Gi/o proteins and may inhibit adenylyl cyclase or modulate many other different signalling molecules and pathways. $D_3$ receptors may also couple to Gs (Obadiah et al. 1999; Ilani et al. 2002).

The receptor types occurring presynaptically are mainly $D_2$-like. Of these, the $D_2$ type prevails, whereas $D_3$ and $D_4$ are uncommon. $D_1$-like receptors were rarely further differentiated as $D_1$ or $D_5$.

Apart from its role as a major mediator of inflammation and allergic reactions and as physiological regulator of gastric acid secretion, **histamine** is also a neurotransmitter in the CNS. Central histaminergic cell bodies are located in the posterior hypothalamus and project diffusely to almost all brain regions and to the spinal cord. There are four types of histamine receptors, all G-protein-coupled, $H_1$, $H_2$, $H_3$ and $H_4$. $H_1$ receptors couple to Gq/11 proteins. $H_2$ receptors couple to Gs. $H_3$ and $H_4$ receptors couple to Gi/o.

The histamine $H_3$ receptor was first identified as an autoreceptor (Arrang et al. 1983), negatively regulating the synthesis and release of histamine in the CNS. Release-inhibiting $H_3$ receptors have also been identified on terminals of various other neurons in the CNS and the periphery. $H_1$, $H_2$ and $H_4$ receptors function as presynaptic receptors rarely, if ever (Schwartz et al. 1986; Tanaka and Ichikawa 2006).

$H_3$ receptors are hetereogeneous for three reasons: species differences in amino acid sequence and, hence, pharmacology (Zaragoza et al. 2004); receptor oligomerization (Bakker 2004); and the existence of, so far, six splice variants, $H_{3A}$, $H_{3B}$, $H_{3C}$, $H_{3D}$, $H_{3E}$, and $H_{3F}$ (Bakker et al. 2006). The latter three variants, $H_{3D}$, $H_{3E}$, and $H_{3F}$, are retained intracellulary and may control the cell surface expression of $H_{3A}$, $H_{3B}$, and $H_{3C}$. Since most reports on histamine receptor modulation of transmitter release only refer to the $H_3$ receptor, not to the splice variants, I do not differentiate between these in the following.

The main source of **serotonin** (5-HT) is the enterochromaffin cells, which release it as tissue hormone either into the gut lumen or the portal vein circulation. In the CNS, cell bodies of 5-HT-containing neurons are concentrated along the raphé nuclei of the brainstem and are also found in more lateral regions and in the reticular formation. The neurons terminate in almost all brain regions and in the spinal cord.

5-HT interacts with a large diversity of G-protein-coupled receptors, namely the 5-$HT_{1,2,4,5,6}$ and $_7$ families, and in addition with a ligand-gated cationic channel, 5-$HT_3$. The diversity has been explained by the fact that serotonin is one of the oldest neurotransmitters in evolution. The 5-$HT_1$ receptor family (5-$HT_{1A,1B,1D,1E \text{ and} 1F}$) couples mainly to Gi/o. 5-$HT_{2A}$, 5-$HT_{2B}$ and 5-$HT_{2C}$ receptors couple to Gq/11. 5-$HT_4$, 5-$HT_6$, and 5-$HT_7$ receptors couple to Gs. The 5-$HT_{5A}$ receptor couples principally to Gi/o. The 5-$HT_{5B}$ receptor seems to occur only in rodents, and no transduction mechanism has been identified.

Members of all serotonin receptor families except the 5-$HT_6$ receptor (Hamon et al. 1999; Marcos et al. 2006) occur presynaptically. Presynaptic 5-$HT_3$ receptors are discussed in this *Handbook* in the chapter on ionotropic receptors by Dorostkar and Boehm.

With dopamine receptors on terminals with seven different transmitters, histamine receptors on terminals with eight different transmitters, serotonin receptors on terminals with eight different transmitters, and several of these receptors occurring in more than a single subtype, presynaptic modulation is manifold. The question always arises, of course – and I shall try to answer it in each case – whether a particular receptor receives an endogenous agonist input or lacks it and, hence, belongs to those presynaptic receptors that are "vestiges of evolution which continue to exist because they do us no harm" (Starke 1981).

# 2 Presynaptic Receptors for Dopamine

Presynaptic dopamine receptors modulate the release of a variety of neurotransmitters. When looking at Table 1 one may be impressed by the number of studies and the number of neuron systems studied. It is clear that, overall, $D_1$-like presynaptic receptors are facilitatory, whereas $D_2$-like presynaptic receptors are inhibitory; in the case of the few exceptions, the reason often remains unclear.

A final general observation on Table 1 is that, in several cases, both facilitation through $D_1$-like and inhibition by $D_2$-like receptors have been reported. This holds true for noradrenaline release in rat nucleus accumbens, acetylcholine release in rat striatum, GABA release in rat striatum, nucleus accumbens, and ventral tegmental area (VTA), and glutamate release in rat substantia nigra pars reticulata.

## *2.1 Presynaptic Dopamine Autoreceptors: Modulation of Dopamine Release*

The main effect mediated by presynaptic dopamine autoreceptors is modulation of exocytotic release of dopamine (Table 1); however, the receptors in addition subserve other functions (Table 2).

**Table 1** Tissues in which neurotransmitter release is increased (↑) or decreased (↓) by presynaptic dopamine receptor activation. Receptor classification (except with superscript[a]) as by the authors cited. Designation as $D_1$-like or $D_2$-like where no differentiation between $D_1$ and $D_5$, or between $D_2$, $D_3$ and $D_4$, was made

| Release of Transmitter | Species and Tissue | Receptor Type | References |
|---|---|---|---|
| Dopamine (see 2.1) | Rat neocortex | $D_2$-like↓ | 1; Fedele et al. 1999 |
| | striatum | $D_2$-like↓ | 1; see Goldstein et al. 1990 |
| | nucleus accumbens | $D_3$↓ | Joseph et al. 2002 |
| | hypothalamus | $D_2$-like↓ | 1 |
| | median eminence | $D_2$-like↓ | 1 |
| | pituitary neurointermediate lobe | $D_2$-like↓ | 1 |
| | retina | $D_2$↓ | see Nguyen-Legros et al. 1999 |
| | Mouse striatum | $D_2$-like↓ | Bull et al. 1991 |
| | cultured mesencephalic cells | $D_2$↓, $D_3$↓ | Tang et al. 1994 |
| | Guinea pig spinal cord | $D_2$-like↓ | 1 |
| | retina | $D_2$-like↓ | Weber et al. 2001 |
| | Rabbit neocortex | $D_2$-like↓ | 1 |
| | nucleus caudatus | $D_2$-like↓ | 1 |
| | tuberculum olfactorium | $D_2$-like↓ | 1 |
| | retina | $D_2$-like↓ | 1; see Nguyen-Legros et al. 1999 |
| | Cat nucleus caudatus | $D_2$-like↓ | 1 |
| | Human neocortex | $D_2$↓ | Fedele et al. 1999; Löffler et al. 2006 |
| Nor-adrenaline (see 2.4) | Rat medial prefrontal cortex | $D_1$-like↑ | Pan et al. 2004 |
| | hippocampus | $D_2$-like↓ | Monnet 2002 |
| | nucleus accumbens | $D_1$-like↑, $D_2$-like↓ | Vanderschuren et al. 1999 |
| | kidney | $D_2$-like↓ | 2 |
| | portal vein | $D_2$-like↓ | 2 |
| | Guinea pig mesenteric artery | $D_2$-like↓ | 2 |
| | Rabbit hippocampus | $D_2$-like↓ | Jackisch et al. 1985 |
| | hypothalamus | $D_2$-like↓ | Galzin et al. 1982 |
| | ear artery | $D_2$-like↓ | 2 |
| | heart | $D_2$-like↓ | 2 |
| | lung | $D_2$-like↓ | 2 |
| | kidney | $D_2$-like↓ | 2 |
| | Cat spleen | $D_2$-like↓ | 2 |
| | nictitating membrane | $D_2$-like↓ | 2 |
| | atria | $D_2$-like↓ | 2 |
| | Dog gracilis muscle | $D_2$-like↓ | 2 |
| | saphenous vein | $D_2$-like↓ | 2 |
| | kidney | $D_2$-like↓ | 2 |
| | Cattle kidney | $D_2$-like↓ | 2 |
| | Human peripheral vessels | $D_2$-like↓ | 2; Haeusler et al. 1992; Mannelli et al. 1997 |
| | peripheral tissues | $D_2$-like↓ | Francis 1995; Murphy 2000 |
| | kidney | $D_2$-like↓ | Rump et al. 1993 |
| | right atrium | $D_2$↓ | Rump et al. 1995 |
| | gastric and uterine arteries | $D_2$-like↓ | Morgadinho et al. 1999 |

*(Continued)*

**Table 1** Continued

| Release of Transmitter | Species and Tissue | Receptor Type | References |
|---|---|---|---|
| Serotonin (see 2.5) | Rat hypothalamus | $D_2$-like↑[a] | Cox et al. 1980 |
| | substantia nigra | $D_1$-like↓ | Benkirane et al. 1987 |
| Acetyl-choline (see 2.6) | Rat striatum | $D_1$-like↑ | Consolo et al. 1987 |
| | striatum | $D_2$-like↓ | 1; 4; Gorell and Czarnecki 1986; Stoof et al. 1982; Enz et al. 1990 |
| | striatum | $D_2$↓ | Yan et al. 1997 |
| | tuberculum olfactorium | $D_2$-like↓ | 1; 4 |
| | nucleus acumbens | $D_2$-like↓ | 1; 4 |
| | heart | $D_2$↓ | 3 |
| | Mouse striatum | $D_2$↓ | Wang et al. 2006 |
| | Guinea pig cerebral cortex | $D_2$-like↓ | 1; 4 |
| | caudate nucleus | $D_2$-like↓ | 1; 4 |
| | tuberculum olfactorium | $D_2$-like↓ | 1; 4 |
| | brain stem | $D_2$-like↓ | 1; 4 |
| | stomach | $D_2$↓ | 3 |
| | ileum | $D_2$-like↓ | 3 |
| | Rabbit hippocampus | $D_2$-like↓ | 1 |
| | striatum | $D_2$-like↓ | 1; 4; Dolezal et al. 1992 |
| | retina | $D_1$-like↑ | 3 |
| | Cat caudate nucleus | $D_2$-like↓ | 1 |
| GABA (see 2.7) | Rat prefrontal cortex | $D_1$-like↓ | Gonzalez-Islas and Hablitz 2001 |
| | striatum | $D_2$-like↓ | 4; Delgado et al. 2000; Pisani et al. 2000; Momiyama and Koga 2001 |
| | striatum | $D_1$-like↑, $D_2$-like↓ | Guzman et al. 2003 |
| | striatum | $D_1$-like↑ | Arias-Montano et al. 2001; 2007; Floran et al. 1990 |
| | nucleus accumbens | $D_1$-like↑ | 4 |
| | nucleus accumbens | $D_5$↓[a] | Nicola and Malenka 1997 |
| | cultured n. accumbens cells | $D_1$-like↑ or ↓[b], $D_2$-like↓ | Geldwert et al. 2006 |
| | dorsolateral septal nucleus | $D_4$↓ | Asaumi et al. 2006 |
| | supraoptic nucleus | $D_4$↓ | Azdad et al. 2003 |
| | paraventricular nucleus | $D_4$↓ | Baimoukhametova et al. 2004 |
| | substantia nigra | $D_1$-like↑ | 4 |
| | substantia nigra pars reticulata | $D_1$-like↑ | Floran et al. 1990; Garcia et al. 1997; Radnikow and Misgeld 1998 |
| | entopeduncular nucleus | $D_1$-like↑ | Floran et al. 1990 |
| | globus pallidus | $D_1$-like↑ | Floran et al. 1990 |
| | globus pallidus | $D_2$-like↓ | Cooper et al. 2001; Floran et al. 1997 |
| | VTA | $D_1$↑ | Cameron and Williams 1993 |
| | VTA | $D_2$-like↓ | Koga and Momiyama 2000 |

*(Continued)*

**Table 1** Continued

| Release of Transmitter | Species and Tissue | Receptor Type | References |
|---|---|---|---|
| Glutamate (see 2.8) | Rat striatum | $D_2$-like↓ | 4 |
| | striatum | $D_2$↓ | Bamford et al. 2004 |
| | globus pallidus medialis | $D_2$↓, $D_3$↓ | Hernandez et al. 2006 |
| | substantia nigra pars reticulata | $D_1$-like↑, $D_2$-like↓ | Ibanez-Sandoval et al. 2006 |
| | olfactory bulb | $D_2$-like↓ | Davila et al. 2003 |
| | supraoptic nucleus | $D_4$↓ | Price and Pittman 2001 |
| | Mouse nucleus accumbens | $D_5$↓[a] | Nicola et al. 1996 |
| | Turtle olfactory bulb | $D_2$-like↓ | Berkowicz and Trombley 2000 |
| Cholecysto-kinin (see 2.9) | Rat neo cortex | $D_1$-like↑ | Brog and Beinfeld 1992 |
| | hippocampus | $D_1$-like↑ | Brog and Beinfeld 1992 |
| | striatum | $D_2$-like↑ | Meyer and Krauss 1983 |
| | striatum | $D_1$-like↑ | Brog and Beinfeld 1992 |

Table summarizes studies in which release was determined either as overflow (Starke 1977) or by means of the postsynaptic response. References to reviews: 1, Starke et al. 1989; 2, Willems et al. 1985; 3, Fuder and Muscholl 1995; 4, Starke 1981.

[a] Retrospective classification by T.J.F.

[b] Electrophysiological measurements; some neurons responded to the $D_1$ agonist SKF38393 with inhibition and some with enhancement of GABA release.

**Table 2** Functions of presynaptic dopamine autoreceptors other than modulation of release (↑ increase, ↓ decrease in function). Receptor classification as by the authors cited Designation as $D_1$-like or $D_2$-like where nc differentiation between $D_1$ and $D_5$, or between $D_2$, $D_3$ and $D_4$, was made.

| Modulation of | Species and Tissue | Receptor Type | References |
|---|---|---|---|
| Dopamine synthesis | Rat striatum | $D_2$-like↓ | Booth et al. 1990; Lindgren et al. 2001; Thompson and Certain 2005; see Starke et al. 1989 |
| | retina | $D_2$ | see Nguyen-Legros et al. 1999 |
| | Mouse striatum | $D_2$↓ | Hakansson et al. 2004 |
| | striatum | $D_{2short}$↓ | Lindgren et al. 2003 |
| | cultured mesencephalic cells | $D_2$↓, $D_3$↓ | O'Hara et al. 1996 |
| | Guinea pig striatum | $D_2$-like↓ | Johnson et al. 1992a |
| Dopamine uptake | Rat striatum | $D_2$-like↑ | Meiergerd et al. 1993; Wieczorek and Kruk 1994; Rothblat and Schneider 1997; Batchelor and Schenk 1998 |
| | nucleus accumbens | $D_2$-like↑ | Wieczorek and Kruk 1994; Rothblat and Schneider 1997 |
| | Mouse striatum | $D_2$↑ | Dickinson et al. 1999 |
| | striatum | $D_2$↓ | Schmitz et al. 2002 |
| Neurotensin release | Rat prefrontal cortex | $D_2$-like↑ | Bean et al. 1990; Bean and Roth 1991 |

The autoreceptors modulating dopamine release are exclusively $D_2$-like and inhibitory (Table 1). They operate in many tissues, including the retina, and in various species, including humans (Table 1). Differences between species, however, exist: $[^3H]$-dopamine release in slices of mouse neocortex was not inhibited by the $D_2/D_3$ preferring agonist quinpirole, whereas $[^3H]$-dopamine release evoked identically in slices of human neocortex clearly was (Löffler et al. 2006). When activated by endogenous released dopamine, the receptors mediate a physiological autoinhibition, as shown by the uniform increase in dopamine release caused by antagonists. In human neocortex slices *in vitro*, the concentration of endogenous dopamine at the autoreceptors, the so-called biophase concentration, has been estimated. At an action potential frequency of 3 Hz it was 3.6 nM, which is close to the estimated dissociation constant of the dopamine-autoreceptor complex of 5 nM (Löffler et al. 2006).

Where the $D_2$-like inhibitory receptor was identifed further it was mostly $D_2$, $D_3$ in rat nucleus accumbens, and never $D_4$ (Table 1). The $D_3$ receptor was thought to play a minor role also in the mouse corpus striatum (Joseph et al. 2002). However, this suggestion was refuted by the finding that no autoreceptor function remained in mouse striatum after genetic deletion of the $D_2$ receptor (Schmitz et al. 2002).

$D_2$-like receptors couple mainly to Gi/o proteins, as mentioned above. However, there is no direct evidence to support this coupling for the release-modulating autoreceptors. Moreover, the subsequent intracellular signal transduction has never been studied directly in axon terminals. Mouse AtT-20 pituitary cells, which release acetylcholine and adrenocorticotropic hormone, have been used as a model for axon terminals. When expressed in these cells, $D_3$ receptors mediated inhibition of P/Q-type calcium channels and activation of G protein-coupled inward rectifier potassium channels (Kuzhikandathil et al. 1998; Kuzhikandathil and Oxford 1999). Both would explain the autoreceptor-mediated inhibition of dopamine exocytosis.

Presynaptic dopamine autoreceptors are rapidly desensitized by dopamine but not other agonists auch as apomorphine or roxindole (Arbilla et al. 1985; Seyfried and Bartoszyk 1994; see also Kim et al. 2005). The reason for the difference is unknown.

A large receptor reserve – i.e., receptors in excess of those necessary to produce a maximum response – has been described for release-modulating dopamine autoreceptors in rat striatum (Yokoo et al. 1988). No reserve exists for the autoreceptors in human neocortex (Löffler et al. 2006).

## 2.2 Presynaptic Dopamine Autoreceptors: Additional Functions

As mentioned above, presynaptic dopamine autoreceptors subserve functions in addition to modulation of dopamine release. They are summarized in Table 2.

One such function is inhibition of tyrosine hydroxylase (TH), independently of the inhibition of release. In analogy to release, it results in a feedback inhibition of the synthesis of dopamine (Table 2). Like the release-inhibiting autoreceptors, those inhibiting dopamine synthesis are $D_2$-like and, where identified further, $D_2$ or $D_3$

(Table 2). In mouse cultured mesencephalic cells transfected with either $D_2$, $D_3$, or $D_4$ receptors, the former two but not the $D_4$ receptors mediated inhibition of TH (O'Hara et al. 1996). The $D_2$ receptor effect was maintained for at least 60 min, whereas the $D_3$ receptor effect desensitized rapidly (O'Hara et al. 1996). It has been suggested that $D_2$ autoreceptors are the $D_{2short}$ splice variant, whereas postsynaptic $D_2$ receptors are $D_{2long}$ (Usiello et al. 2000). This remains an hypothesis for the release-inhibitory, but has been borne out experimentally for the TH-inhibiting autoreceptors (Lindgren et al. 2003). Signal transduction behind the autoreceptor starts with pertussis toxin-sensitive Gi/o proteins (O'Hara et al. 1996), inhibition of adenylyl cyclase, and a fall in cAMP levels. The cAMP-dependent phosphorylation of TH at serine 40 is then diminished and enzyme activity reduced due a reinforcement of end product inhibition (see, for instance, Daubner et al. 1992).

Starting in 1993, reports have appeared that in rat striatum and nucleus accumbens $D_2$-like autoreceptors, when activated, enhanced removal of dopamine from the extracellular space by means of the dopamine transporter (DAT) as studied by voltammetry (Table 2). $D_2$ receptor activation also increased [$^3$H]-dopamine uptake in *Xenopus* oocytes transfected with both the human $D_2$ receptor and the human DAT (Mayfield and Zahniser 2001), as well as the uptake of a fluorescent DAT substrate in human embryonic kidney cells equally transfected with both the human $D_2$ receptor and the human DAT (Bolan et al. 2007). In accord with this autoreceptor-transporter connection, clearence of dopamine from the striatal extracellular space *in vivo* was reduced in $D_2$ receptor knockout mice (Dickinson et al. 1999). Unfortunately, however, the data in the literature are by no means consistent. No effect of $D_2$ receptor activation on the DAT was observed in another rat striatum study (Prasad and Amara 2001), in synaptosomes from rat and human neocortex (Feuerstein, unpublished observations), and in PC12 cells that possess $D_2$ receptors and the DAT (Pothos et al. 1998). In a second study with $D_2$ receptor knockout mice, uptake of dopamine in the striatum was increased rather than decreased (Schmitz et al. 2002). The reason for these discrepancies is not clear. It has been suggested that the clearance measurements by means of voltammetry were not appropriate for assessment of the function of the DAT (Prasad and Amara 2001).

Dopamine and neurotensin are co-transmitters in neurons projecting from the VTA to the prefrontal cortex. Dopamine release in rat prefrontal cortex, like the neocortex generally (Table 1), is inhibited through $D_2$-like autoreceptors. It is surprising, therefore, that prefrontal release of neurotensin was enhanced through the autoreceptors (Table 2). This is not the only example of opposite presynaptic modulation of the release of two cotransmitters. Presynaptic $\beta$-adrenoceptors increase release of noradrenaline from some postganglionic sympathetic neurons but diminish release of its cotransmitter ATP (Driessen et al. 1996). It has been suggested that in such cases some vesicles contain mainly transmitter A and others mainly transmitter B and that presynaptic receptor activation leads to inhibition of the release of one and facilitation of the release of the other vesicle population (Goncalves et al. 1996).

## 2.3 Dopamine Autoreceptors in Disease and Therapy

Dopamine neurons are involved in several neuropsychiatric disorders such as Parkinson's disease, schizophrenia, and psychotropic drug abuse. It seems likely that dopamine autoreceptor function is affected in these disorders as well as by their drug treatment.

Levodopa and dopamine receptor agonists improve the motor symptoms of Parkinson's disease. A severe potential side effect of the agonists but not of levodopa is impairment of cognitive functions such as learning. The reason for this difference probably is the mode of activation of dopamine receptors. Levodopa increases presynaptic dopamine availability and thereby action potential-induced dopamine release. It leaves the pattern of dopamine release (phasic versus tonic) and the pattern of presynaptic as well as postsynaptic dopamine receptor activation (phasic versus tonic) unaffected. Agonists such as pergolide, in contrast, activate pre- and postsynaptic dopamine receptors continuously, in an action potential-independent manner. Animal as well as human experiments indicate that both the continuous activation of postsynaptic receptors and the continuous autoreceptor-mediated attenuation of dopamine release contribute to the deterioration of cognitive functions during treatment with dopamine agonists (Arnsten et al. 1995; Breitenstein et al. 2006). There may be an additional component. The agonists clinically used are $D_2$-selective. By inhibition of dopamine release as well as their postsynaptic action they favor postsynaptic $D_2$ receptor activation at the expense of $D_1$.

Increased dopaminergic neurotransmission is a component of psychostimulant addiction, Tourette's syndrome, and schizophrenia. Chlorpromazine-like neuroleptic drugs are thought to help these patients by blockade of postsynaptic dopamine (and other) receptors. An alternative therapeutic principle would be selective activation of dopamine autoreceptors. The idea led to the development of a group of drugs (e.g., the thiazoloazepine derivative B-HT 920) allegedly displaying selective $D_2$ autoreceptor agonist activity. However, pharmacological differences between $D_2$ autoreceptors and postsynaptic $D_2$ receptors have remained elusive (e.g., Starke et al. 1989; Drukarch and Stoof 1990). It has been thought that a large autoreceptor reserve might make $D_2$ agonists more potent at the autoreceptors than at postsynaptic receptors (see Goldstein et al. 1990). However, as mentioned above, no autoreceptor reserve exists in human neocortex, an important projection field of dopamine neurons in these disorders. There is an additional reason for skepticism: dopamine autoreceptors may be nonfunctional in states of chronically increased dopaminergic transmission. DAT knockout mice are characterized by such "hyperdopaminergia." In these animals, any dopamine autoreceptor function, soma-dendritically as well as presynaptically, was almost lost (Jones et al. 1999).

As opposed to the disorders of the preceding paragraph, a *decrease* in dopaminergic transmission may be one of the neurochemical alterations in depression (Dailly et al. 2004). The selective antagonists at $D_2$-like receptors sulpiride and amisulpride, when given at low doses, reduce depression symptoms, presumably by blockade of $D_2$-autoreceptors and enhancement of dopamine release (Racagni et al. 2004). Sulpiride in fact increased the release of [$^3$H]-dopamine in human neocortex slices

(Löffler et al. 2006). Coadministration of sulpiride and fluvoxamine, a serotonin reuptake inhibitor, increased *in vivo* dopamine release in the prefrontal cortex of rats (Ago et al. 2005a). The increase probably is the neurochemical counterpart of the antidepressant-like effect of coadministration of sulpiride and fluvoxamine in mice, as measured in the tail suspension test (Ago et al. 2005b). A recent clinical trial showed superior antidepressant efficacy of the combination of sulpiride plus the serotonin reuptake inhibitor paroxetine compared with paroxetine alone and, importantly enough, an accelerated antidepressant response of the combination (Uchida et al. 2005).

A key effect of addictive psychotropic drugs is to increase the extracellular concentration of dopamine in the nucleus accumbens. Ethanol does this in part by direct excitation of the mesolimbic dopamine neurons projecting to the accumbens (Koyama et al. 2007). However, studies in alcohol-preferring rats, animals which tend to develop alcohol dependence, have indicated an additional component: a reduction of presynaptic $D_2$ autoreceptor function (Engleman et al. 2003; Thielen et al. 2004).

## 2.4 Presynaptic Dopamine Heteroreceptors on Noradrenergic Terminals

These receptors were first observed in the sympathetic nervous system: in the nictitating membrane of the cat, dopamine reduced the release of noradrenaline via a receptor distinct from the $\alpha_2$-autoreceptor (Enero and Langer 1975). Most of the heteroreceptors are $D_2$-like and inhibitory, but in rat nucleus accumbens and medial prefrontal cortex $D_1$ receptors enhance noradrenaline release (Table 1). The opposite effects mediated by $D_1$ and $D_2$ receptors in rat nucleus accumbens have been explained by location of the inhibitory $D_2$ receptors close to, and location of the facilitatory $D_1$ receptors more distant from, dopaminergic varicosities (Vandschuren et al. 1999).

Although noradrenergic terminals normally contain too little dopamine for presynaptic dopamine heteroreceptors to become activated, and despite the fact that the hippocampus is only sparsely innervated by dopaminergic fibers (Bischoff et al. 1979), the release of [$^3$H]-noradrenaline in rabbit (Jackisch et al. 1985) and rat (Monnet 2002) hippocampus was inhibited by endogenous dopamine as shown by the facilitatory effect of $D_2$ antagonists. Voltage-sensitive calcium channels seem to play a role in the dopaminergic inhibition of noradrenaline release (Monnet 2002).

Presynaptic receptors, when located on the same terminals, may interact with one another (see Schlicker and Göthert 1998; Trendelenburg et al. 2003). Such cross-talk also occurs between presynaptic $D_2$-like heteroreceptor on noradrenergic neurons and the presynaptic $\alpha_2$-autoreceptors: inhibition by one narrows the scope for inhibition by the other (Jackisch et al. 1985; see also 1.2.3, 1.2.5 and 1.3.3).

## 2.5 Presynaptic Dopamine Heteroreceptors on Serotonergic Terminals

An increase in serotonin release through a $D_2$-like receptor has been reported for rat hypothalamus and a decrease in serotonin release through a $D_1$-like receptor for rat substantia nigra (Table 1). The mechanisms of these unusual effects are not discussed in the two papers. The $D_1$ antagonist SCH 23390 increased the release of serotonin in the substantia nigra, thus revealing an endogenous dopamine input (Benkirane et al. 1987). No dopaminergic modulation of serotonin release was seen in rat and rabbit hippocampus (Cox et al. 1980; Jackisch et al. 1985; Matsumoto et al. 1996), rat neocortex (Baumann and Waldmeier 1981), rat nucleus accumbens (Ferre and Artigas 1995), rabbit striatum (Feuerstein et al. 1986). and rat septum (Rutz et al. 2007). It seems worth mentioning that in attempts to label serotonergic neurons with [$^3$H]-5-HT care has to be taken to avoid a false labelling of dopaminergic neurons (Feuerstein et al. 1986).

## 2.6 Presynaptic Dopamine Heteroreceptors on Cholinergic Terminals

Due to the dopamine-acetylcholine link in Parkinson's disease, these receptors are of particular interest, and due to this link the majority of studies have been carried out on striatal cholinergic interneurons (Table 1). They are endowed with presynaptic $D_2$-like – where further classified $D_2$ – receptors inhibiting acetylcholine release and possibly also with presynaptic $D_1$-like receptors increasing acetylcholine release (Table 1). They also possess soma-dendritic $D_1$- and $D_2$-like receptors that increase and decrease, respectively, their firing frequency (Aosaki et al. 1998; Centonze et al. 2003; Maurice et al. 2004). However, blockade of the soma-dendritic dopamine receptors failed to influence the tonic activity of cholinergic interneurons (Bennett and Wilson 1999), so they may not be activated by endogenous dopamine. The presynaptic receptors, in contrast, do receive an endogenous dopamine input to mediate tonic presynaptic inhibition (e.g., Hertting et al. 1980; Helmreich et al. 1982).

Single cell reverse transcription – polymerase chain reaction detected $D_2$ and $D_5$ but not $D_{3/4}$ and little $D_1$ receptor mRNA in rat striatal cholinergic interneurons. The $D_2$ receptors coupled to Gi/o and reduced somatic N-type $Ca^{2+}$ currents (Yan et al. 1997), thus providing a cellular mechanism for the reduction of acetylcholine release by $D_2$ receptors located on cholinergic terminals.

A $D_2$ receptor reserve of 25%–30% has been estimated for the cholinergic interneurons of rat striatum, without differentiation between presynaptic and soma-dendritic receptors (Enz et al. (1990), but the reliability of the method used to assess the receptor reserve has been questioned (Agneter et al. 1997).

Transmission from cortical glutamate neurons to striatal GABAergic medium spiny neurons is subject to long-term depression (LTD), a long-lasting decrease of transmission following repetitive activity of the glutamate neurons. Inhibitory $D_2$ receptors on cholinergic striatal interneurons are necessary for the expression of this LTD. The reasons have been elucidated by Wang et al. (2006). Figure 1 shows their explanation – one of the most explicit hypotheses on the physiological role of a presynaptic receptor. LTD is due to inhibition of presynaptic glutamate release by endocannabinoids synthesized in, and released from, the postsynaptic spines. One prerequisite for postsynaptic endocannabinoid synthesis is postsynaptic calcium entry through L-type calcium channels. Acetylcholine released from the spontaneously active cholinergic interneuron, by activating $M_1$ receptors on the spines, inhibits the L-type calcium channels and thus tends to attenuate LTD. As long as the cholinergic interneurons fire at their normal slow rate and release little

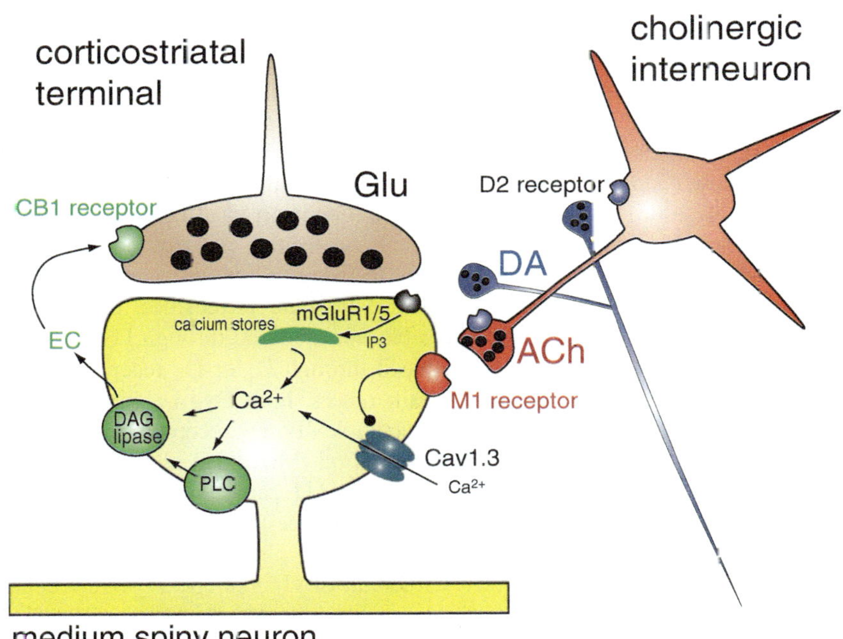


**Fig. 1** Elements controlling LTD induction in the synapse between a corticostriatal glutamate terminal and the spine of a striatal medium spiny neuron. LTD is due to activation of postsynaptic metabotropic mGlu1/5 and (not shown) ionotropic AMPA receptors, promotion of postsynaptic endocannabinoid synthesis and release and subsequent $CB_1$ receptor-mediated inhibition of glutamate release. Calcium derived from intracellular stores as well entry through Cav1.3 L-type calcium channels is necessary for endocannabinoid synthesis. A cholinergic interneuron projects to the spine, innervating an $M_1$-receptor that, when activated, inhibits the L-type calcium channel. The cholinergic neuron in turn receives dopamine input from the nigro-striatal dopamine pathway. Dopamine reduces both the firing rate of the neuron and the release of acetylcholine per pulse by activation of soma-dendritic and presynaptic $D_2$ receptors, respectively. ACh = Acetylcholine; DA = dopamine; DAG lipase = diacylglycerol lipase; EC = endocannabinoid; PLC = phospholipase C. Reproduced, with permission, from Wang et al. (2006).

acetylcholine per impulse, LTD can be induced. When, however, the cholinergic neurons are disinhibited by sulpiride blockade of their $D_2$ receptors, the massive release of acetylcholine suppresses postsynaptic calcium entry to the extent that LTD disappears. In accord with this explanation, LTD in the presence of sulpiride can be rescued when postsynaptic $M_1$ receptors are blocked by pirenzepine (Wang et al. 2006).

## *2.7 Presynaptic Dopamine Heteroreceptors on GABAergic Terminals*

Most of the observations of dual presynaptic effects of dopamine – inhibition of release through $D_2$-like receptors and facilitation (but occasionally also inhibition) of release through $D_1$-like receptors – refer to GABA release in rat striatum, nucleus accumbens and VTA (Table 1). It was thought for some time that the two receptor types were located on different neurons (Gerfen et al. 1990; Yung et al. 1995). However, co-localization of $D_1$-like and $D_2$-like receptors on the same GABA neuron has now been demonstrated immunohistochemically (Aizman et al. 2000; Geldwert et al. 2006) as well as functionally (Geldwert et al. 2006). Particularly noteworthy is the demonstration of different localization patterns on single varicosities (Mizuno et al. 2007). The authors measured $Ca^{2+}$ transients in individual varicosities in mouse nucleus accumbens. In some varicosities, the transients were enhanced through $D_1$-like receptors but not modified through $D_2$-like receptors; in other varicosities, they were reduced through $D_2$ or $D_3$ receptors but not modified through $D_1$-like receptors; in still others, both $D_1$-like facilitation and $D_{2/3}$ inhibition occurred. To complete the complexity, $D_3$ receptors in some varicosities mediated facilitation rather than inhibition (Mizuno et al. 2007). Possibly some GABAergic neurons transport $D_1$-like, $D_2$ and $D_3$ receptors for various lengths and deposit them in their terminals in a target cell-specific manner (Geldwert et al. 2006). It must be emphasized, however, that despite much ingenious research, the physiological significance of the co-localization of $D_1$-like and $D_2$-like receptors remains uncertain, especially since the two receptor groups possess similar affinity for dopamine (Schoffelmeer et al. 1994).

The results of Mizuno et al. (2007) make it likely that modulation of $Ca^{2+}$ currents underlies the various modulations of GABA release. This view is supported by other studies for both $D_2$-like inhibition (Pisani et al. 2000; Momiyama and Koga 2001) and $D_1$-like facilitation (Arias-Montano et al. 2007). The $D_2$-like inhibitory receptors presumably couple to Gi/o (Momiyama and Koga 2001), whereas the $D_1$-like facilitatory receptors presumably couple to Gs, activation of adenylyl cyclase and activation of protein kinase A (Arias-Montano et al. 2007). The riddle of the exceptional cases of inhibition through $D_1$-like receptors remains. For example, in rat prefrontal cortex $D_1$-like receptors presynaptically inhibited GABAergic transmission, an effect blocked by an inhibitor of protein kinase A (Gonzalez-Islas and Hablitz 2001) – the same enzyme involved in facilitation through $D_1$-like receptors

(Arias-Montano et al. 2007). The occasional facilitation by presynaptic $D_3$ receptors (Mizuno et al. 2007) may be due to the fact that this subtype can couple to Gs in addition to Gi/o (see Introduction).

In rat dorsolateral septal, supraoptic, and paraventricular nuclei, dopamine receptors inhibiting GABA release were pharmacologically identified as $D_4$ (Table 1). In the supraoptic and paraventricular nuclei the $D_4$ receptors reduce the GABAergic inhibition of the oxytocin- and vasopressin-secreting neurons and thus enhance neurophypophysial hormone release.

## *2.8 Presynaptic Dopamine Heteroreceptors on Glutamatergic Terminals*

Dopamine presynaptically inhibits the release of glutamate in a number of brain nuclei through $D_2$-like receptors – where identified as either $D_2$, $D_3$ or $D_4$ (Table 1). In rat substantia nigra pars reticulata it simultaneously can facilitate the release of glutamate through $D_1$-like receptors – another case of co-localization of receptors mediating opposite effects. In mouse nucleus accumbens, a $D_1$-like receptor inhibited glutamate release. Since the inhibitory effect of dopamine persisted after genetic deletion of the $D_1$ receptor, the inhibitory receptor was $D_5$ (Table 1).

An interesting presynaptic location $D_2$ receptor is on corticostriatal glutamate terminals – the terminals shown in Figure 1 as targets of retrograde endocannabinoid signaling in LTD. These glutamate terminals (and not only the cholinergic interneurons as in Figure 1) are endowed with $D_2$ receptors. Dopamine selectively inhibits release of glutamate from terminals of low activity and thus favors transmission from the more active terminals (Bamford et al. 2004). These findings of a presynaptic $D_2$ receptor-mediated control of striatal glutamate release replace an older view that $D_2$ receptors were almost exclusively located on neurons whose somata reside in the striatum and not on terminals of corticostriatal axons (Joyce and Marshall 1987).

The glomerular layer of the olfactory bulb contains a substantial population of dopaminergic neurons. Dopamine acting at $D_2$-like heteroreceptors inhibits glutamate release from terminals of the olfactory sensory neurons and hence may modulate the olfactory nerve synapse (Table 1).

The neurosecretory cells of the supraoptic nucleus receive both the GABAergic inhibitory input mentioned above (Section 2.7) and an excitatory glutamate input: Whereas the $D_4$ heteroreceptor-mediated depression of GABA release disinhibits the neurosecretory cells (Section 2.7), the $D_4$ heteroreceptor-mediated depression of glutamate release (Table 1) inhibits them and, hence, the pituitary secretion of oxytocin and vasopressin.

## 2.9 Presynaptic Dopamine Heteroreceptors on Cholecystokinergic Terminals

Since cholecystokinin peptides show a neuroleptic-like profile in several screening tests for neuroleptics, cholecystokinin-dopamine interactions are of interest, especially the modulation of cholecystokinin release by dopamine receptors (Table 1). In the rat striatum both $D_1$ and $D_2$ receptor activation was reported to increase, and both $D_1$ and $D_2$ receptor blockade to depress, the release of the peptide.

## 2.10 Presynaptic Dopamine Heteroreceptors in Disease and Therapy

$D_2$ receptors on postganglionic sympathetic axon terminals diminish noradrenaline release and in consequence heart rate and vascular resistance (Ensinger et al. 1985). Selective $D_2$ agonists have therefore been studied for the treatment of hypertension. However, an unfavorable side effect profile (largely nausea and orthostatic hypotension) precluded wide use for this indication (Murphy 2000). Orthostatic hypotension is also a side effect when $D_2$ agonists are taken for other purposes, such as to decrease prolactin levels. Moreover, there is a therapeutic situation in which the $D_2$ heteroreceptors on sympathetic varicosities receive an input of endogenous dopamine. This is when parkinsonian patients are treated with L-dopa and dopamine accumulates in postganglionic sympathetic nerves (Lokhandwala and Buckley 1978). The release of noradrenaline is then reduced for two reasons: because part of the noradrenaline is replaced by the false transmitter dopamine and because exocytosis is diminished due to presynaptic $D_2$ heteroreceptor activation. In fact, recording of electrical sympathetic nerve activity in humans has shown that L-dopa-induced orthostatic hypotension is peripheral rather than central in origin (Takeuchi et al. 1993). Note, however, that orthostatic hypotension is also a characteristic of Parkinson's disease, independent of L-dopa treatment (Goldstein et al. 2005).

The effects of presynaptic $D_1$-like and $D_2$-like receptors on the release of amino acid neurotransmitters in the basal ganglia are important for the pathophysiology and treatment of Parkinson's disease. The fact that recent reviews on antiparkinsonian drugs highlight the old drug apomorphine with the rationale that it can stimulate both $D_1$-like and $D_2$-like receptors (e.g., Subramony 2006) may reflect the growing awareness that the benefits of pure $D_2$-like receptor agonists in Parkinson therapy are limited. The dopaminergic $D_1$-like and $D_2$-like modulation of amino acid release takes place mainly in the striatum and the reciprocally innervated network of the globus pallidus and the subthalamic nucleus (see Hernandez et al. 2006). It may be assumed, therefore, that apomorphine owes its superior beneficial effects in parkinsonian patients inter alia to activation of presynaptic $D_1$-like and $D_2$-like receptors located along the axons of striatal GABAergic neurons projecting to the output nuclei of the basal ganglia.

## 3 Presynaptic Receptors for Histamine

Presynaptic histamine receptors are more uniform than presynaptic dopamine receptors. Only one type, $H_3$, has been identified with certainty. Like presynaptic dopamine receptors, presynaptic $H_3$ receptors occur as auto- and heteroreceptors

**Table 3** Tissues in which neurotransmitter release is decreased by presynaptic histamine $H_3$ receptor activation

| Release of transmitter | Species and Tissue | References |
|---|---|---|
| Histamine (see 3.1) | Rat neocortex | see Starke et al. 1989; Morisset et al. 2000; Westerink et al. 2002 |
| | hippocampus | see Starke et al. 1989 |
| | hypothalamus | see Starke et al. 1989; Yamamoto et al. 1997 |
| | striatum | see Starke et al. 1989 |
| | Mouse neocortex | Morisset et al. 2000 |
| | Guinea pig cardiac synaptosomes | Li et al. 2006 |
| | Human neocortex | see Starke et al. 1989 |
| Nor-adrenaline (see 3.3) | Rat neocortex | Schlicker et al. 1989; Smits and Mulder 1991 |
| | hypothalamus | Smits and Mulder 1991 |
| | heart | see Fuder and Muscholl 1995 |
| | blood vessels | see Fuder and Muscholl 1995 |
| | Mouse neocortex | Schlicker et al. 1993; 1994 |
| | cardiac synaptosomes | Koyama et al. 2003 |
| | Guinea pig mesenteric artery | Ishikawa and Sperelakis 1987 |
| | atria | Endou et al. 1994, see Fuder and Muscholl 1995 |
| | heart | Imamura et al. 1994; 1996b; see Fuder and Muscholl 1995 |
| | cardiac synaptosomes | Li et al. 2006; Silver et al. 2002 |
| | Cat iris | see Fuder and Muscholl 1995 |
| | nictitating membrane | see Fuder and Muscholl 1995 |
| | Human neocortex and hippocampus | Schlicker et al. 1999 |
| | saphenous vein | see Fuder and Muscholl 1995 |
| | atria | Imamura et al. 1995; Hatta et al. 1997 |
| | Pig retina | see Fuder and Muscholl 1995 |
| | Dog heart | Mazenot et al. 1999 |
| Serotonin (see 3.4) | Rat neocortex | Schlicker et al. 1988; Smits and Mulder 1991; Fink et al. 1990 |
| | hypothalamus | Smits and Mulder 1991 |
| | striatum | Smits and Mulder 1991 |
| | substantia nigra pars reticulata | Threlfell et al. 2004 |
| Dopamine (see 3.5) | Rat nucleus accumbens | Prast et al. 1999a |
| | Mouse striatum | Schlicker et al. 1993 |
| Acetyl-choline (see 3.6) | Guinea pig ileum | see Fuder and Muscholl 1995 |
| | colon | see Fuder and Muscholl 1995 |
| | trachea | see Fuder and Muscholl 1995 |

*(Continued)*

**Table 3** Continued

| Release of transmitter | Species and Tissue | References |
|---|---|---|
| GABA (see 3.7) | Rat striatum | Arias-Montano et al. 2001 |
| | substantia nigra pars reticulata | Garcia et al. 1997 |
| | medial vestibular nucleus | Bergquist et al. 2006 |
| Glutamate (see 3.8) | Rat dentate gyrus | Brown and Haas 1999 |
| | thalamus | Garduno-Torres et al. 2007 |
| | striatum | Molina-Hernandez et al. 2001; Doreulee et al. 2001 |
| | amygdala | Jiang et al. 2005 |
| Sensory neuro-peptides (see 3.9) | Rat dura mater | Matsubara et al. 1992; Dimitriadou et al. 1997 |
| | lung and spleen | Dimitriadou et al. 1994 |
| | skin | Ohkubo and Shibata 1995 |
| | Guinea pig airways | Ichinose et al. 1990 |
| | dura mater | Matsubara et al. 1992 |
| | atria | Imamura et al. 1996a |
| | Rabbit lung | Nemmar et al. 1999 |

Table summarizes studies in which release was determined either as overflow or by means of the postsynaptic response.

(Table 3). They have been detected both functionally (Table 3) and, morphologically, by comparison of the distribution of $H_3$ receptor mRNA and $H_3$ receptor protein. By the latter approach it has been shown, for example, that thalamo-cortical and hippocampal pyramidal glutamate neurons and striato-nigral as well as striato-pallidal GABA neurons possess presynaptic $H_3$ receptors (see Pillot et al. 2002; Jin and Panula 2005; see also immunohistochemical detection in Section 3.9).

Presynaptic $H_3$ receptors also are uniform in their signal transduction. They couple to Gi/o proteins and decrease the depolarization-induced release of neurotransmitters by inhibiting multiple calcium channels (e.g., Arrang et al. 1985; Schlicker et al. 1994; Endou et al. 1994; Brown and Haas 1999; see Stark et al. 2004). For comparison, the signal transduction of soma-dendritic $H_3$ autoreceptors in histaminergic neurons also involves a pertussis toxin-sensitive G-protein with subsequent inhibition of N- and P-type $Ca^{2+}$ channels (Takeshita et al. 1998). The few exceptions to this signal transduction pathway are discussed in the corresponding subsections below (see Sections 3.1, 3.3, and 3.9).

There are some reports on modulation of noradrenaline, acetylcholine, and glutamate release through presynaptic $H_1$ or $H_2$ receptors. The evidence is doubtful, however. For example, effects of drugs thought to be $H_1$- or $H_2$-selective may in fact have been due to an action on $H_3$ receptors (see page 291 of Fuder and Muscholl 1995). Morphological evidence for presynaptic $H_1$ or $H_2$ receptors similar to that mentioned for $H_3$ receptors is lacking.

### *3.1 Presynaptic Histamine Autoreceptors: Modulation of Release and Synthesis*

Like presynaptic dopamine autoreceptors, presynaptic histamine autoreceptors are activated by the released endogenous transmitter to inhibit further histamine release, as shown by the increase in histamine release caused by antagonists at $H_3$ receptors: a definite piece of physiology. Evidence has been presented recently that cardiac postganglionic sympathetic neurons of the guinea pig synthesize and release histamine as a co-transmitter (Li et al. 2003; 2006). These noradrenaline-histamine neurons possess $H_3$ autoreceptors which, when activated, depress the release of both noradrenaline and histamine – unlike the $D_2$-like autoreceptors of dopamine-neurotensin neurons which modulate the release of the two cotransmitters in opposite direction (see Section 2.2). It would be of interest to see whether, conversely, activation of $\alpha_2$-autoreceptors inhibits the release of histamine in the guinea pig heart.

Apart from exocytosis, presynaptic $H_3$ autoreceptors also inhibit the synthesis of histamine at the level of nerve endings, at least in part through pathways distinct from those leading to the inhibition of release. One pathway is inhibition of adenylyl cyclase (Gomez-Ramirez et al. 2002; Moreno-Delgado et al. 2006): activation of cAMP-dependent protein kinase A by cAMP stimulates histamine synthesis through phosphorylation of histidine decarboxylase, and this stimulation is diminished when adenylyl cyclase activity decreases following activation of $H_3$ autoreceptors and Gi/o proteins.

Presynaptic $H_3$ autoreceptors are among those G protein-coupled receptors now known to display constitutive activity, inducing an effect in the absence of an agonist. This has been shown, for example, for the release-inhibiting autoreceptors in rat and mouse neocortex (Morisset et al. 2000) and for the synthesis-inhibiting autoreceptors in rat neocortex (Moreno-Delgado et al. 2006). Thioperamide, previously considered as the classical $H_3$ receptor antagonist, is in fact an extremely potent inverse agonist. Other $H_3$ receptor antagonists also have been identified as inverse agonists (e.g., Leurs et al. 2005; Moreno-Delgado et al. 2006). Constitutive activity of presynaptic $H_3$ autoreceptors differs among splice variants and species (Coge et al. 2001). In this context it seems worth mentioning that an inverse agonist is a pure antagonist as long as the receptor under investigation is not spontaneously active, implying that the antagonist/inverse agonist character of a ligand depends on the degree of constitutive activity of the receptor under investigation.

So far, constitutive activity of presynaptic $H_3$ heteroreceptors has not been described.

## 3.2 Histamine Autoreceptors in Disease and Therapy

Wherever histamine neurons are involved in pathophysiology, their $H_3$ autoreceptors surely are involved as well. For example, histamine release in the prefrontal cortex is increased during handling stress, and this increase is potentiated when $H_3$ autoreceptors are blocked (Westerink et al. 2002).

Histaminergic neuron activity promotes arousal, attention, cognition, and learning, and $H_3$ antagonists exert procognitive effects in animals (Huang et al. 2004; Yu et al. 2006; see Stark et al. 2004). It seems likely that the effects are due to blockade of $H_3$ autoreceptors, thereby increasing histamine synthesis and release. In accord with this view, the procognitive effect of the antagonists was potentiated by histidine but attenuated by blockade of histidine decarboxylase. Released histamine acts on postsynaptic $H_1$ and, to a lesser extent, $H_2$ receptors to improve memory (above references).

Histaminergic neuron activity also seems to suppress appetite. For this reason, $H_3$ antagonists or inverse agonists, which promote histamine release, were expected to act similarly, whereas autoreceptor-selective agonists were expected to increase food intake. However, what was found was the opposite: the antagonists/inverse agonist thioperamide increased, whereas classical $H_3$ agonists decreased, food intake in diet-induced obese wild-type mice (Yoshimoto et al. 2006). To further cloud the picture, although the $H_3$ agonist-induced anorexigenic effect was associated with a reduction in histamine release, a comparable reduction in histamine release due to blockade of histidine decarboxylase increased appetite. Apparently the effects of $H_3$ receptor ligands on appetite and body weight are largely due to actions at $H_3$ receptors other than the autoreceptors.

Finally, histaminergic neuron activity seems to protect against convulsions (Zhang et al. 2003; Lintunen et al. 2005), epilepsy is a potential indication for $H_3$ receptor antagonists (Harada et al. 2004; Stark et al. 2004), and their anticonvulsive effect has been partly explained by an increase in histamine release (Yokoyama et al. 1994; Harada et al. 2004).

## 3.3 Presynaptic Histamine Heteroreceptors on Noradrenergic Terminals

They were the first presynaptic $H_3$ heteroreceptors to be discovered (in guinea pig mesenteric artery; Table 1). They mediate inhibition of transmitter release from many noradrenergic neurons, peripheral as well as central (Table 1). They may be activated by histamine from mast cells, as happens in the heart upon myocardial ischemia (see Section 3.10). In the CNS an input of endogenous histamine has not been demonstrated (see references in Table 3). For example, the release of noradrenaline in the hippocampus of freely moving rats remained unchanged upon local administration of the $H_3$ receptor antagonist/inverse agonist thioperamide (Di Carlo et al. 2000).

Quantitative differences have been observed between humans and animals regarding noradrenaline release from neocortical slices: release in human tissue was inhibited by $H_3$ receptor activation to a much lower extent than in mouse neocortex (see Stark et al. 2004).

Like $D_2$-like heteroreceptors (Section 2.4), the $H_3$ heteroreceptors on noradrenergic terminals may interact with terminal $\alpha_2$ autoreceptors: preactivation of one receptor attenuates the effect obtainable by activation of the other (see Schlicker and Göthert 1998).

As for all presynaptic $H_3$ receptors, the $H_3$ heteroreceptor-mediated inhibition of exocytotic noradrenaline release from sympathetic terminals of the heart involves Gi/o and a decrease in $Ca^{2+}$ entry (Hatta et al. 1997; Levi and Smith 2000; Silver et al. 2002). This signal transduction is also responsible for the inhibition by histamine of exocytotic noradrenaline release in *acute* myocardial ischemia (Hatta et al. 1997). A striking change, however, takes place in *protracted* myocardial ischemia. Although noradrenaline release continues, and although histamine still inhibits release, the mechanisms are totally different. In protracted myocardial ischemia, ATP depletion and intracellular pH reduction promote axoplasmic $Na^+$ accumulation upon activation of the $Na^+/H^+$ exchanger and diminish the pH gradient between cytoplasm and vesicle interior in the sympathetic terminals, a constellation that prevents the vesicular storage of noradrenaline and engenders a massive carrier-mediated, non-exocytotic, $Ca^{2+}$-independent release of noradrenaline (Imamura et al. 1994; 1996b; see Levi and Smith 2000). The inhibition of this carrier-mediated release through $H_3$ receptors involves a depression of the $Na^+/H^+$ exchanger (Imamura et al. 1996b), thereby reducing the intraterminal accumulation of $Na^+$ with an ensuing decrease in noradrenaline outward transport. The second messengers mediating the reduction of $Na^+/H^+$ exchanger activity remain unclear. In addition to the depression of the $Na^+/H^+$ exchanger, suppression of voltage-dependent $Na^+$ channels may be part of the $H_3$ heteroreceptor signal transduction in protracted cardiac ischemia (Hatta et al. 1997).

### 3.4 Presynaptic Histamine Heteroreceptors on Serotonergic Terminals

$H_3$ receptor activation inhibits the release of serotonin in several rat brain regions (Table 3). An endogenous histamine tone has been found at the heteroreceptors in rat brain cortex (Schlicker et al. 1988; Fink et al. 1990). At high micromolar concentrations histamine may enhance the release of serotonin (and of noradrenaline) via a tyramine-like mechanism (Starke and Weitzell 1978; Young et al. 1988).

## 3.5 Presynaptic Histamine Heteroreceptors on Dopaminergic Terminals

The inhibitory $H_3$ heteroreceptors on dopaminergic nerve terminals in mouse striatal slices were not activated by endogenous histamine under the experimental conditions chosen, since their blockade did not enhance the release of dopamine (Schlicker et al. 1993). In the rat nucleus accumbens *in vivo*, however, indirect evidence (an histamine-evoked increase in acetylcholine release) suggests that dopamine release is permanently inhibited by endogenous histamine (Prast et al. 1999a; see Section 3.6).

Blockade of dopamine autoreceptors increased the extent of $H_3$ receptor-mediated inhibition of dopamine release in mouse striatal slices; this another example of an autoreceptor/heteroreceptor interaction (Schlicker et al. 1993; compare Sections 2.4, 3.3, 4.3).

The effect of $H_3$ receptor activation on release of dopamine in rat corpus striatum has not been investigated, but $H_3$ receptor activation reduced the synthesis of dopamine in rat striatum (Molina-Hernandez et al. 2000).

## 3.6 Presynaptic Histamine Heteroreceptors on Cholinergic Terminals

Presynaptic $H_3$ heteroreceptors, when activated, inhibit the release of transmitter from several peripheral cholinergic neuron systems, probably preganglionic as well as postganglionic parasympathetic (Table 3). The receptors may be activated by histamine secreted from mast cells.

In the CNS, the existence of presynaptic $H_3$ heteroreceptors on cholinergic neurons is less certain. $H_3$ receptor activation depressed the release of acetylcholine in slices (Clapham and Kilpatrick 1992) but not synaptosomes (Arrang et al. 1995) of rat entorhinal cortex, suggesting location of the $H_3$ receptors outside the cholinergic terminals. Still, Prast et al. (1999a; b) in their extensive *in vivo* push-pull superfusion studies in rat nucleus accumbens suggest a minor $H_3$ receptor on cholinergic terminals (Fig. 6 of Prast et al. 1999b).

Several mechanisms for modulation of acetylcholine release through cerebral $H_3$ receptors outside cholinergic terminals have been elucidated. First, in the parietal cortex of rats *in vivo*, local $H_3$ receptor activation reduced the release of acetylcholine. This was due to a primary increase in the release of GABA (at soma-dendritic $H_3$ receptors of GABA neurons, the authors suggest) followed by $GABA_A$ receptor-mediated depression of acetylcholine release (Giorgetti et al. 1997). Second, in the corpus amygdaloideum of rats *in vivo*, local $H_3$ receptor activation also reduced, whereas $H_3$ receptor blockade increased, acetylcholine release. Here the primary sites of action were presynaptic $H_3$ autoreceptors, and the ensuing change of histamine release subsequently changed the release of acetylcholine through an $H_2$ receptor of unknown location (Cangioli et al. 2002; Passani et al. 2001). Third, in the nucleus acumbens of rats *in vivo*, local $H_3$ receptor activation *increased* the release of acetylcholine. One mechanism was primary $H_3$ receptor inhibition of dopamine release (see Section 3.5) and, hence, secondary attenuation of the dopaminergic inhibition of acetylcholine release (see Section 2.6). Another mechanism was primary $H_3$ receptor inhibition of GABA release (see Section 2.7) and secondary attenuation of a GABAergic inhibition of acetylcholine release (Prast et al. 1999a; b).

### 3.7 Presynaptic Histamine Heteroreceptors on GABAergic Terminals

As just mentioned, $H_3$ receptor-mediated inhibition of GABA release has been dissected as an intermediate step in $H_3$ agonist-induced enhancement of acetylcholine release in rat nucleus accumbens (Section 3.6; Prast et al. 1999a; b). While direct evidence for $H_3$ inhibition of GABA release in the accumbens is lacking, histamine does inhibit the release of GABA in other brain structures (Table 3). The effect in rat substantia nigra pars reticulata and corpus striatum is an interesting case of fine-tuning. In both structures, release of GABA is facilitated through dopamine $D_1$-like heteroreceptors (Table 1). Histamine and a selective $H_3$ agonist suppressed only that part of GABA release that was due to tonic $D_1$-like receptor activation (Garcia et al. 1997; Arias-Montano et al. 2001). The $D_1$ receptors increase striatal GABA release by promoting $Ca^{2+}$ entry through P/Q-type channels. The $H_3$ receptors conversely inhibit release by depression of $Ca^{2+}$ entry through these channels (Arias-Montano et al. 2007). Thus, presynaptic $D_1$ and $H_3$ heteroreceptors of striatal GABA neurons are co-localized and interact both on terminals in the target nucleus substantia nigra pars reticulata and on terminals of axon collaterals in the corpus striatum.

### 3.8 Presynaptic Histamine Heteroreceptors on Glutamatergic Terminals

$H_3$ receptor activation attenuates the release of glutmate in several brain regions (Table 3), but an endogenous input to the receptors has not been demonstrated. In the dentate gyrus adenosine, acting at $A_1$ receptors, also attenuated the release of glutamate and occluded further inhibition by histamine, presumably because a common signal transduction pathway (Brown and Haas 1999).

## 3.9 Presynaptic Histamine Heteroreceptors on Peptidergic Terminals

Sensory neuropeptides such as substance P and calcitonin gene-related peptide (CGRP) release histamine from mast cells (Dimitriadou et al. 1994; 1997; Imamura et al. 1996a). Histamine, in turn, can inhibit the release of neuropeptides through $H_3$ receptors which have been detected on peripheral sensory axon terminals by immunohistochemistry (Cannon et al. 2007). In three of the studies listed in Table 3, inhibition of substance P release (Ohkubo and Shibata 1995; Nemmar et al. 1999) and CGRP release (Imamura et al. 1996a) was measured as a decrease of overflow; in two others it was measured as attenuation of plasma extravasation (Ichinose et al. 1990; Matsubara et al. 1992). The presynaptic signal transduction may involve the stimulation of ATP-sensitive $K^+$ channels (Ohkubo and Shibata 1995). The close proximity of mast cells and peripheral sensory axon endings makes this $H_3$ effect an endogenous modulation mechanism. In the guinea pig heart, for example, the $H_3$ antagonist thioperamide increased the release of CGRP elicited by capsaicin (Imamura et al. 1996a). It is likely that $H_3$ receptors are also located on the central terminals of primary sensory neurons, where their activation inhibits transmission at the first nociceptive synapse in the spinal cord or brain stem (Cannon et al. 2003; Cannon and Hough 2005).

## 3.10 Presynaptic Histamine Heteroreceptors in Disease and Therapy

In myocardial ischemia, several of the mechanisms presented above come into play. First, neuropeptides such as CGRP are released from cardiac sensory C fibers and subsequently release histamine from mast cells as just mentioned. Histamine then can act at least at two presynaptic $H_3$ heteroreceptors: on the C fibers to attenuate further neuropeptide release (Section 3.9), and on postganglionic sympathetic fibers to attenuate exocytotic as well as carrier-mediated noradrenaline release (Section 3.3). Both presynaptic effects are potentially beneficial. The $H_3$ receptors are unique in this pattern of effects. Presynaptic adenosine $A_1$ receptors, when activated, also inhibit both exocytotic and carrier-mediated noradrenaline release, but cardiac $A_1$ receptors in addition mediate negative chronotropic and dromotropic effects. Presynaptic $\alpha_2$-adrenoceptors, when activated, reduce exocytotic noradrenaline release but *enhance* carrier-mediated noradrenaline release (due to stimulation of the $Na^+/H^+$ exchanger, Imamura et al. 1996b), which is the major mode of noradrenaline release and the major arrhythmogenic risk in protracted myocardial ischemia (see Levi and Smith 2000; Koyama et al. 2003).

$H_3$ heteroreceptors inhibiting noradrenaline release from sympathetic nerve terminals in the nasal mucosa may contribute to nasal congestion in allergic rhinitis (Varty et al. 2004).

Both the histamine neuron innervation and the density of $H_3$ receptors are increased in the substantia nigra of Parkinsonian patients (Anichtchik et al. 2000; 2001). The receptors probably are the $H_3$ heteroreceptors on striato-nigral GABA terminals, and their increase may contribute to the depression of GABA release in the substantia nigra in Parkinson's disease (Garcia-Ramirez et al. 2004).

The inhibition through $H_3$ receptors of the release of substance P and other neurotransmitters of primary sensory neurons may become highly important clinically: It explains the antinociceptive and antiinflammatory properties of some $H_3$ receptor agonists, including their antimigraine effects (McLeod et al. 1998; Cannon et al. 2003; see also Göthert et al. 1995).

# 4 Presynaptic Receptors for Serotonin

If presynaptic histamine receptors are more uniform than presynaptic dopamine receptors, the contrary holds true for presynaptic serotonin receptors: they are even more diverse than presynaptic dopamine receptors. As mentioned in the Introduction, presynaptic 5-$HT_3$ receptors, being ligand-gated ion channels, are covered in the chapter by Dorostkar and Boehm and will be mentioned here only occasionally. Presynaptic G protein-coupled 5-HT receptors inhibit the release of serotonin from serotonergic axon terminals and inhibit or enhance the release of other neurotransmitters (Table 4).

## *4.1 Presynaptic Serotonin Autoreceptors: Modulation of 5-HT Release*

All G protein-coupled presynaptic 5-HT autoreceptors – there is no consistent evidence for 5-$HT_3$ autoreceptors (Dorostkar and Boehm, this book) – are inhibitory. They belong mainly to the 5-$HT_1$ family and where subclassified were 5-$HT_{1B}$ or 5-$HT_{1D}$ (Table 4). Non-5-$HT_1$ presynaptic autoreceptors, presumably 5-$HT_5$ and/or 5-$HT_7$, replaced the 5-$HT_1$-like receptors in 5-$HT_{1B}$ knockout mice (Pineyro et al. 1995b; see page 327 of Göthert and Schlicker 1997). Note that, as opposed to dopamine and histamine autoreceptors, 5-HT autoreceptors differ in the neurons' terminal and soma-dendritic region: in the latter, they are 5-$HT_{1A}$ (Sprouse and Aghajanian 1986).

Studies, *in vitro* and *in vivo*, in which appropriate antagonists increased the release of serotonin have shown that presynaptic 5-HT autoreceptors are physiological regulators of 5-HT release in all regions of the CNS (references in Table 4; and see Table 1 of Göthert and Schlicker 1997). A study with pertussis toxin has confirmed that the 5-$HT_{1B}$ autoreceptors in rat striatum couple to Gi/o (Sarhan and Fillion 1999). Further downstream, release is inhibited by reduction of transmembrane $Ca^{2+}$ influx (see p. 934 of Starke et al. 1989; p. 329 of Göthert and Schlicker

**Table 4** Tissues in which neurotransmitter release is increased (↑) or decreased (↓) by presynaptic5-HT receptor activation. Receptor classification (except with superscript [a]) as by the authors cited. Designation as $5\text{-}HT_1$-like where no differentiation between $5\text{-}HT_{1A}$, $5\text{-}HT_{1B}$, $5\text{-}HT_{1D}$, $5\text{-}HT_{1E}$, and $5\text{-}HT_{1F}$ was made. Designation as $5\text{-}HT_2$-like where no differentiation between $5\text{-}HT_{2A}$, $5\text{-}HT_{2B}$ and $5\text{-}HT_{2C}$ was made

| Release of transmitter | Species and Tissue | Receptor Type | References |
|---|---|---|---|
| 5-HT (see 4.1) | Rat neocortex | $5\text{-}HT_{1B}$↓ | 1; Gobert et al. 1998; Rutz et al. 2006 |
| | neocortex | $5\text{-}HT_{1B}$ and $5\text{-}HT_{1D}$↓ | 2 |
| | hippocampus | $5\text{-}HT_{1B}$↓ | 1; Birthelmer et al. 2002; 2003; Rutz et al. 2006 |
| | diencephalon | $5\text{-}HT_1$-like↓ | 2 |
| | septal region | $5\text{-}HT_{1B}$↓ | Rutz et al. 2007 |
| | striatum | $5\text{-}HT_1$-like↓ | 1 |
| | striatum | $5\text{-}HT_{1B}$↓ | Sarhan and Fillion 1999 |
| | nucleus accumbens | $5\text{-}HT_1$-like↓ | 1 |
| | hypothalamus | $5\text{-}HT_{1B}$↓ | 1 |
| | cerebellum | $5\text{-}HT_{1B}$↓ | 1 |
| | brain stem | $5\text{-}HT_1$-like↓ | 1 |
| | medulla oblongata | $5\text{-}HT_1$-like↓ | 1 |
| | spinal cord | $5\text{-}HT_{1B}$↓ | 2 |
| | Mouse neocortex | $5\text{-}HT_{1B}$↓ | 2 |
| | neocortex | $5\text{-}HT_5$ and/or $5\text{-}HT_7$ ↓ | 2 |
| | hippocampus | $5\text{-}HT_{1B}$↓ | 2 |
| | hippocampus | $5\text{-}HT_5$ and/or $5\text{-}HT_7$ ↓ | 2 |
| | hypothalamus | $5\text{-}HT_1$-like↓ | 2 |
| | cerebellum | $5\text{-}HT_1$-like↓ | 1 |
| | Guinea pig neocortex | $5\text{-}HT_{1B}$↓ | 2 |
| | hippocampus | $5\text{-}HT_{1B}$↓ | 2 |
| | hypothalamus | $5\text{-}HT_1$-like↓ | 2 |
| | substantantia nigra | $5\text{-}HT_{1B}$↓ | 2 |
| | medulla oblongata | $5\text{-}HT_1$-like↓ | 2 |
| | medulla oblongata | $5\text{-}HT_{1D}$↓ | Bobker and Williams (1990) |
| | Rabbit neocortex | $5\text{-}HT_{1B}$↓ | 2 |
| | hippocampus | $5\text{-}HT_{1B}$↓ | 2 |
| | hypothalamus | $5\text{-}HT_1$-like↓ | 1 |
| | striatum | $5\text{-}HT_{1B}$↓ | 2 |
| | superior colliculus | $5\text{-}HT_1$-like↓ | 2 |
| | Cat caudate nucleus | $5\text{-}HT_1$-like↓ | 1 |
| | Pig neocortex | $5\text{-}HT_{1D}$↓ | 1 |
| | Rhesus monkey | $5\text{-}HT_{1B}$↓ | 2 |
| | Human neocortex | $5\text{-}HT_{1B}$↓ | 2 |
| | hippocampus | $5\text{-}HT_1$-like↓ | 2 |
| Nor-adrenaline (see 4.3) | Rat pithed *in vivo* | $5\text{-}HT_1$-like↓ | 4 |
| | kidney | $5\text{-}HT_1$-like↓ | 4 |
| | vena cava | $5\text{-}HT_{1B}$↓ | 4 |
| | Guinea pig mesenteric artery | $5\text{-}HT_1$-like↓ | 4 |
| | Rabbit ear artery | $5\text{-}HT_1$-like↓ | 4 |
| | Cat nictitating membrane | $5\text{-}HT_1$-like↑ | 4; Park et al. 1991 |

(*Continued*)

**Table 4** Continued

| Release of transmitter | Species and Tissue | Receptor Type | References |
|---|---|---|---|
| | Dog coronary artery | $5\text{-HT}_1$-like↓ | 4 |
| | tibial artery | $5\text{-HT}_1$-like↓ | 4 |
| | saphenous vein | $5\text{-HT}_1$-like↓ | 4 |
| | Cattle ovarian follicle | $5\text{-HT}_1$-like↓ | 4 |
| | Human atria | $5\text{-HT}_{1D}$↓ | 2; Göthert et al. 1996; Göthert 2003 |
| | saphenous vein | $5\text{-HT}_{1B}$↓ | 4; Göthert et al. 1996; 2003 |
| | dura mater | $5\text{-HT}_1$-like↓[a] | Göthert et al. 1996 |
| Dopamine (see 4.4) | Rat striatum | $5\text{-HT}_2$-like↓[a] | Ennis et al. 1981 |
| | striatum | $5\text{-HT}_2$-like↓ | Muramatsu et al. 1988a; Ng et al. 1999 |
| | striatum | $5\text{-HT}_{2C}$↓ | Alex et al. 2005 |
| | striatum | $5\text{-HT}_{2C}$↑ | Lucas and Spampinato 2000 |
| | striatum | $5\text{-HT}_{1B}$↓ | Sarhan et al. 1999; Sarhan and Fillion 1999; Sarhan et al. 2000 |
| Acetyl-choline (see 4.5) | Rat neocortex | $5\text{-HT}_4$↑ | Consolo et al. 1994 |
| | entorhinal cortex | $5\text{-HT}_{1B}$↓ | Feuerstein et al. 1996a |
| | hippocampus | $5\text{-HT}_{1B}$↓ | 3; Feuerstein et al. 1996a; Birthelmer et al. 2002; 2003; Sarhan and Fillion 1999 |
| | hippocampus | $5\text{-HT}_2$-like↓ | Muramatsu et al. 1988b |
| | hippocampus | $5\text{-HT}_4$↑ | Siniscalchi et al. 1999 |
| | striatum | $5\text{-HT}_1$- or $5\text{-HT}_2$-like↓ | 3 |
| | Guinea pig neocortex | $5\text{-HT}_4$↑ | Siniscalchi et al. 1999 |
| | n. basalis Meynert | $5\text{-HT}_4$↑ | Siniscalchi et al. 1999 |
| | hippocampus | $5\text{-HT}_1$-like↓ | 3 |
| | hippocampus | $5\text{-HT}_4$↑ | Siniscalchi et al. 1999 |
| | striatum | $5\text{-HT}_2$-like↑ | 3 |
| | ileum | $5\text{-HT}_{1A}$↓, $5\text{-HT}_4$↑ | 4; LePard et al. 2004 |
| | Human neocortex | $5\text{-HT}_{1F}$↓ | Feuerstein and Seeger 1997 |
| | stomach | $5\text{-HT}_4$↑ | Leclere and Lefebvre 2002 |
| | urinary bladder | $5\text{-HT}_1$-like↓, $5\text{-HT}_4$↑ | 4 |
| GABA (see 4.6) | Rat cingulate cortex | $5\text{-HT}_{1B}$↓ | Tanaka and North 1993 |
| | hippocampus | $5\text{-HT}_{1A}$↓ | Katsurabayashi et al. 2003 |
| | hippocampus | $5\text{-HT}_2$-like↑ | Piguet and Galvan 1994 |
| | amygdala | $5\text{-HT}_{1A}$↓ | Koyama et al. 1999 |
| | hypothalamus | $5\text{-HT}_1$-like↓[a] | Malinina et al. 2005 |
| | suprachiasmatic nucleus | $5\text{-HT}_{1B}$↓ | Bramley et al. 2005 |
| | substantia nigra | $5\text{-HT}_{1B}$↓ | Johnson et al. 1992b |
| | medulla oblongata | $5\text{-HT}_{1B}$↓ | Bobker and Williams (1989) |
| | spinal cord | $5\text{-HT}_1$-like↓ | Wu et al. 1991 |
| | Mouse suprachiasmatic nucleus | $5\text{-HT}_{1B}$↓ | Bramley et al. 2005 |
| | ventral midbrain | $5\text{-HT}_{1B}$↓ | Morikawa et al. 2000 |
| | Guinea pig VTA | $5\text{-HT}_{1D}$↓ | Cameron and Williams 1994 |
| | Human neocortex | $5\text{-HT}_{1D}$↓ | Feuerstein et al. 1996b |

(*Continued*)

**Table 4** Continued

| Release of transmitter | Species and Tissue | Receptor Type | References |
|---|---|---|---|
| Glutamate (see 4.7) | Rat neocortex | $5\text{-HT}_{2A}\uparrow$ | Aghajanian and Marek 1999; Arvanov et al. 1999; Marek et al. 2001 |
| | cingulate cortex | $5\text{-HT}_{1B}\downarrow$ | Tanaka and North 1993 |
| | entorhinal cortex | $5\text{-HT}_{1A}\downarrow$ | Schmitz et al. 1998 |
| | subicular cortex | $5\text{-HT}_{1B}\downarrow$ | Boeijinga and Boddeke 1996 |
| | nucleus accumbens | $5\text{-HT}_{1B}\downarrow$ | Muramatsu et al. 1998 |
| | amygdala | $5\text{-HT}_{1A}\downarrow$ | Cheng et al. 1998 |
| | hypothalamus | $5\text{-HT}_1\text{-like}\downarrow$[a] | Malinina et al. 2005 |
| | brainstem | $5\text{-HT}_{1B}\downarrow$ | Singer et al. 1996; Mizutani et al. 2006 |
| | cerebellum | $5\text{-HT}_1\text{-like}\downarrow$ | Raiteri et al. 1986 |
| | medulla oblongata | $5\text{-HT}_{1A}\downarrow$, $5\text{-HT}_{1B}\downarrow$ | Bobker and Williams 1989 |
| | spinal cord | $5\text{-HT}_{1A}\downarrow$, $5\text{-HT}_{1B}\downarrow$ | Wu et al. 1991 |
| | Mouse suprachiasmatic nucleus | $5\text{-HT}_{1B}\downarrow$ | Pickard et al. 1999 |
| | nucleus accumbens | $5\text{-HT}_{1B}\downarrow$ | Morikawa et al. 2000 |
| | Guinea pig spinal cord | $5\text{-HT}_{1D}\downarrow$ | Travagli and Williams 1996 |
| | Hamster suprachiasmatic nucleus | $5\text{-HT}_{1B}\downarrow$ | Pickard et al. 1999 |
| | superior colliculus | $5\text{-HT}_{1B}\downarrow$ | Mooney et al. 1994 |
| | Human neocortex | $5\text{-HT}_{1D}\downarrow$ | see Raiteri 2006 |
| Glycine (see 4.8) | Rat brain stem | $5\text{-HT}_{1B}\downarrow$ | Umemiya and Berger 1995 |
| Sensory neuro-peptides (see 4.9) | Rat trigeminal nerve endings | $5\text{-HT}_1\text{-like}\downarrow$ | Buzzi et al. 1991 |

Table summarizes studies in which release was determined either as overflow or by means of the postsynaptic response. References to reviews: 1, Starke et al. 1989; 2, Göthert and Schlicker 1997; 3, Cassel and Jeltsch 1995; 4, Fuder and Muscholl 1995.
[a] Retrospective receptor classification by T.J.F.

1997). How $5\text{-HT}_7$ receptors manage to inhibit release through $G_s$ – if they couple to $G_s$ also presynaptically – remains open.

## *4.2 Serotonin Autoreceptors in Disease and Therapy*

Major depression has been classically explained by deficiencies in the monoamine neurotransmitters 5-HT, noradrenaline, and dopamine. The importance of these monoamine systems for the vulnerablity to depression has recently been confirmed (Ruhé et al. 2007). Do presynaptic $5\text{-HT}_{1B}$ autoreceptors play any role? Two groups of findings tend to answer this question in the affirmative. The depression-like state of learned helplessness in rats is associated with a decrease in cortical 5-HT release *in vivo* (Petty et al. 1992) and an increase in the density of $5\text{-HT}_{1B}$ binding sites – presumably including presynaptic $5\text{-HT}_{1B}$ autoreceptors – in the neocortex, hippocampus, and septum (Edwards et al. 1991). Conversely, chronic treatments that are clinically effective – with tricyclic antidepressants, selective serotonin reuptake inhibitors, monoamine oxidase inhibitors, and electroconvulsive shock – increased the release of 5-HT in various brain regions in animal experiments, although not in all studies (see Table 4 of Göthert and Schlicker 1997). Chronic treatment with selective serotonin reuptake inhibitors also reduced the $5\text{-HT}_{1B}$ mRNA in the dorsal raphé nucleus of the rat, an effect that disappeared after discontinuation of the drugs; the mRNA decrease probably reflected a decrease in the density of presynaptic autoreceptors (see Sari 2004).

$5\text{-HT}_{1A}$ and $5\text{-HT}_{1B}$ receptors are probably involved in aggressive behavior (see Popova 2006). The postsynaptic $5\text{-HT}_{1A}$ receptor density is decreased in the frontal cortex, hypothalamus, and amygdala of highly aggressive rats. Moreover, aggressive behavior is increased in $5\text{-HT}_{1B}$ receptor knockout mice, and conversely $5\text{-HT}_{1B}$ receptor agonists dampen aggression (see Sari 2004; Popova 2006). The common denominator of these findings may be an increase in serotonin release in aggressive animals, caused by lack of $5\text{-HT}_{1B}$ autoreceptors in the knockout mice and leading to downregulation of postsynaptic $5\text{-HT}_{1A}$ receptors in the rat experiments. Importantly, a linkage of aggressive behavior with the $5\text{-HT}_{1B}$ receptor gene was revealed in two human populations (see Popova 2006).

### *4.3 Presynaptic Serotonin Heteroreceptors on Noradrenergic Terminals*

5-HT heteroreceptors occur on postganglionic sympathetic (Table 4) but not central (Allgaier et al. 1995) noradrenergic terminals. All are $5\text{-HT}_1$ and, where subclassified, $5\text{-HT}_{1B}$ or $5\text{-HT}_{1D}$. Interestingly, human postganglionic sympathetic nerve terminals possess $5\text{-HT}_{1D}$ receptors in the heart but $5\text{-HT}_{1B}$ receptors in the saphenous vein (see Göthert 2003). The case is reminiscent of GABAergic neurons, which seem to transport $D_1$, $D_2$ and $D_3$ receptors from the cell bodies to the terminals in a target cell-specific manner (see Section 2.7). An endogenous serotonin input to the receptors has not so far been noticed and seems unlikely. The mechanism of inhibition involves a decrease of $Ca^{2+}$ influx through N-type $Ca^{2+}$ channels (see Göthert 2003). For a final time in this review, an heteroreceptor/autoreceptor interaction has to be mentioned: presynaptic $5\text{-HT}_{1B/D}$ heteroreceptors and presynaptic $\alpha_2$-autoreceptors mutually narrow one another's scope for inhibition (compare Sections 2.4, 3.3, and 3.5). The facilitation of noradrenaline release through $5\text{-HT}_1$-like receptors in the cat nictitating membrane (Table 4; Park et al. 1991) remains a somewhat doubtful exception (see Table 2 of Fuder and Muscholl 1995).

### 4.4 Presynaptic Serotonin Heteroreceptors on Dopaminergic Terminals

The corpus striatum is innervated both by the nigrostriatal dopamine neurons and by serotonin neurons from the raphé nuclei, and serotonin modifies dopamine release in rat striatum (Table 4). The way it does so, however, remains uncertain. A majority of studies indicate inhibition through 5-$HT_2$, probably 5-$HT_{2C}$, receptors, but one study has reported facilitation through 5-$HT_{2C}$ receptors, possibly due to a different region of the striatum being examined. Alex et al. (2005) showed by microdialysis *in vivo* that local administration of a 5-$HT_{2C}$ antagonist/inverse agonist increased dopamine release in rat striatum, indicating either an endogenous serotonin input or constitutive activity of the receptors. They assumed location of the 5-$HT_{2C}$ receptors outside dopaminergic terminals, specifically on GABA neurons. On the other hand, location of 5-$HT_2$-like receptors on striatal dopaminergic terminals is supported by several findings, as a decrease in 5-$HT_2$-like receptor binding after 6-hydroxydopamine lesioning of the nigrostriatal pathway (Ng et al. 1999), and preservation of the inhibitory effect of agonists on high potassium-evoked dopamine release in the presence of tetrodotoxin (Ennis et al. 1981).

The already mentioned false labelling of striatal dopaminergic neurons by 5-HT (see Section 2.5) also works in the opposite direction, i.e., false labelling of 5-HT terminals by dopamine (Berger and Glowinski 1978; Feuerstein et al. 1986). The question arises of whether some of the studies in the dopamine section of Table 4 in which [$^3$H]-dopamine was used (Ennis et al. 1981; Muramatsu et al. 1988a; Ng et al. 1999; Sarhan et al. 1999; Sarhan and Fillion 1999; Sarhan et al. 2000) inadvertently measured [$^3$H]-dopamine release from 5-HT neurons. In none of these studies were measures taken against false labelling. So were the presumed serotonin heteroreceptors in fact serotonin autoreceptors? This seems unlikely in the case of the 5-$HT_2$-like receptors, which do not occur as presynaptic 5-HT autoreceptors (Section 4.1). The three reports on 5-$HT_{1B}$ heteroreceptors are more "suspicious" in this respect. The latter couple to Gi/o (Sarhan and Fillion 1999).

Although striatal 5-$HT_4$ receptors, when activated, facilitate dopamine release, the receptors probably are located on striatal GABA neurons, so the receptors have not been entered in Table 4 (see Eglen et al. 1995).

5-$HT_{1A}$ heteroreceptors depress, whereas 5-$HT_{2A}$ heteroreceptors enhance, the synthesis of dopamine in rat corpus striatum. The former effect (Johnson et al. 1993) occurs even in the absence of any dopamine release, is shared by endogenous serotonin, and may involve adenylyl cyclase inhibition and a diminution of TH phosphorylation as discussed above for dopamine $D_2$ autoreceptors (Section 2.2). The 5-$HT_{2A}$ increase in synthesis, in contrast (Lucas and Spampinato 2000), occurs only when dopamine neurons are activated by blockade of their $D_2$ autoreceptors.

### *4.5 Presynaptic Serotonin Heteroreceptors on Cholinergic Terminals*

Of all neurons, the cholinergic ones possess the greatest variety of presynaptic serotonin receptors: inhibition of acetylcholine release through 5-$HT_1$-like receptors, inhibition or facilitation through 5-$HT_2$ receptors, inhibition through 5-$HT_3$ receptors, and facilitation through 5-$HT_4$ receptors have all been reported (Table 4). I leave out the 5-$HT_3$ receptors here for reasons given; they are doubtful anyway (Boehm and Dorostkar, this book). As reviewed in detail by Cassel and Jeltsch (1995), the serotonergic modulation of brain cholinergic systems has cognitive implications. For this reason, a number of studies have been carried out in the hippocampus and the entorhinal cortex.

The literature agrees that, in rat and guinea pig hippocampus and rat entorhinal cortex, acetylcholine release is inhibited through 5-$HT_1$-like – where examined 5-$HT_{1B}$ – receptors (Table 4). However, this is not the only serotonergic modulation. When Feuerstein et al. (1996a) applied serotonin to hippocampal or entorhinal cortex slices, it failed to change the release of [$^3$H]-acetylcholine, although the 5-$HT_{1B}$-selective agonist Ru24969 caused the expected inhibition. Serotonin inhibited release only when either (1) 5-$HT_2$ receptors were blocked, or (2) release was elicited by high $K^+$ concentrations in the presence of tetrodotoxin to interrupt action potential conduction, or (3) substance P was depleted by treatment of capsaicin, or (4) neurokinin$_1$ ($NK_1$) receptors were blocked. In the same study 5-$HT_{2A}$ receptors were located to GABA/substance P neurons by in situ hybridization (Feuerstein et al. 1996a). The ensuing hypothesis is depicted in Figure 2. In the hippocampus and the entorhinal cortex, cholinergic terminals carrying 5-$HT_{1B}$ inhibitory and $NK_1$ facilitatory presynaptic receptors are innervated by GABA/substance P interneurons. Serotonergic agonists may modulate this release by two mechanisms: (1) direct inhibition through the presynaptic 5-$HT_{1B}$ receptor or (2) indirect facilitation through a soma-dendritic 5-$HT_{2A}$ receptor on the GABA/substance P neuron followed by substance P release and activation of the presynaptic $NK_1$ receptor. In the case of serotonin itself as agonist, the two opposing mechanisms cancel each other.

A very similar twofold mechanism modulates acetylcholine release in the human neocortex, with one exception: the inhibitory presynaptic serotonin receptor is 5-$HT_{1F}$ (Feuerstein and Seeger 1997).

5-$HT_4$ agonists enhance the release of acetylcholine in several peripheral tissues and several brain areas (Table 4). 5-$HT_4$ antagonists counteract the effect but, in the studies published so far, caused no change when given alone. The co-location of facilitatory 5-$HT_4$ and inhibitory 5-$HT_1$ receptors on hippocampal cholinergic terminals is reminiscent of the co-location of $D_1$ and $D_2$ receptors on, for example, GABAergic terminals (Section 2.7). In contrast to the two dopamine receptors,

mediated by 5-$HT_1$-like – where identified 5-$HT_{1A}$, 5-$HT_{1B}$ or 5-$HT_{1D}$ – receptors. No evidence was found for a serotonergic modulation of [$^3$H]-glutamate release in rat hippocampal slices (Sehmisch et al. 2001).

In the calyx of Held synapse in the auditory brainstem of rats the 5-$HT_{1B}$ receptor inhibition of glutamate release was fully explained by inhibition of $Ca^{2+}$ entry through voltage-dependent channels (Mizutani et al. 2006). In rat nucleus accumbens, in contrast, the analogous 5-$HT_{1B}$ effect did not involve inhibition of $Ca^{2+}$ channels (Muramatsu et al. 1998). In rat amygdala, the 5-$HT_{1A}$ inhibition operated via a cyclic AMP pathway and apparently inhibition of $Ca^{2+}$ entry; an unexplained observation was the failure of pertussis toxin to abolish the 5-$HT_{1A}$ inhibition (Cheng et al. 1998).

In the preceding section (4.6), the inhibition by cocaine of GABA release in the VTA was explained by reinforcement of serotonergic, not dopaminergic, presynaptic inhibition. An interesting parallel, but with the opposite explanation, has now to be addressed. In the nucleus accumbens, glutamate terminals synapse onto the medium spiny GABA neurons. Another feature of the effect of cocaine is inhibition of the release of glutamate from these terminals. In mouse nucleus accumbens slices, the inhibition by cocaine was the same, irrespective of whether the slices had been cut from wild-type mice or animals in which the 5-$HT_{1B}$ receptor gene had been deleted (Morikawa et al. 2000). Cocaine operated by a dopaminergic, not a serotonergic mechanism: blockade of the reuptake of dopamine released from nigro-striatal axons followed by reinforcement of $D_5$ receptor-mediated inhibition of glutamate release (Nicola et al. 1996; Table 1).

In contrast to the inhibitory actions of 5-$HT_1$-like receptors, 5-$HT_{2A}$ receptors facilitate glutamate release. This facilitation may be mediated through protein kinase C activation (Arvanov et al. 1999).

## *4.8 Presynaptic Serotonin Heteroreceptors on Glycinergic Terminals*

Glycine is the major inhibitory neurotransmitter in the spinal cord and brain stem. 5-HT presynaptically inhibits glycinergic synaptic transmission to rat brain stem hypoglossal motoneurons through 5-$HT_{1B}$ receptors (Umemiya and Berger 1995). The inhibition did not involve $Ca^{2+}$ channel modulation.

## *4.9 Presynaptic Serotonin Heteroreceptors on Peptidergic Terminals*

Inhibitory 5-HT receptors on sensory trigeminal fibers of blood vessels in the meninges inhibit the release of neuropeptides such as substance P and CGRP (compare Section 3.10). This leads to inhibition of the sterile neurogenic inflammation induced by these peptides.

## 4.10 Presynaptic Serotonin Heteroreceptors in Disease and Therapy

An increase in serotonergic transmission in the CNS induces anxiety – hence the therapeutic effect of the $5\text{-HT}_{1A}$ receptor partial agonist buspirone, which probably acts by activation of the inhibitory soma-dendritic autoreceptors of serotonin neurons (see Göthert and Starke 2005). Anxiogenic effects have been obtained by activation of $5\text{-HT}_{1B}$ receptors in rodents (see Sari 2004). The mechanism may involve presynaptic inhibition of the release of several neurotransmitters. One is acetylcholine. In the hippocampus acetylcholine suppresses anxiety through both nicotinic and muscarinic receptors, and inhibition of its release will produce anxiety (see Sari 2004). Another is GABA. Its anxiolytic role is established (e.g., Roy-Byrne 2005), and again inhibition of its release, as through $5\text{-HT}_{1D}$ receptors in the human neocortex (Feuerstein et al. 1996b), will be anxiogenic. In particular, the combination of a depressed GABAergic and an enhanced serotonergic neurotransmission seems to be anxiogenic (Kahn et al. 1988; Green 1991).

As pointed out above (Section 4.2), $5\text{-HT}_{1B}$ agonists may dampen aggression via activation of $5\text{-HT}_{1B}$ autoreceptors. However, $5\text{-HT}_{1B}$ heteroreceptors may contribute (see Sari 2004). One hypothesis implicates inhibitory soma-dendritic $5\text{-HT}_{1B}$ receptors on vasopressin neurons in the anterior hypothalamus. Activity of these neurons promotes attacks and biting in rodents, hence inhibition of their activity leads to the opposite effect (Ferris et al. 1997).

Serotonin also plays an important role in the pathogenesis and therapy of migraine. The triptanes are $5\text{-HT}_{1B/D}$ receptor agonists. They shorten migraine attacks by two mechanisms, meningeal vasoconstriction through smooth muscle $5\text{-HT}_{1B}$ receptors and inhibition of the release of sensory neuropeptides from trigeminal afferents through presynaptic $5\text{-HT}_{1D}$ receptors (see Göthert and Starke 2005).

Serotonergic/cholinergic interactions influence cognition (see Buhot et al. 1995; Cassel and Jeltsch 1995). For example, 5-HT controls subcortical cholinergic projections to the hippocampus and the neocortex, cooperating in this function with substance P-containing interneurons (Sections 4.5 and Figure 2). The cooperation has led to therapeutic speculations, as follows (Feuerstein and Seeger 1997). In Alzheimer's disease, serotonin uptake inhibitors are used to relieve depression. This may have consequences for acetylcholine release and, hence, cognition; it may detrimentally impair (through $5\text{-HT}_{1B}$ or $5\text{-HT}_{1F}$ heteroreceptors at cholinergic terminals) and beneficially enhance (through soma-dendritic $5\text{-HT}_2$ receptors of the substance P interneurons) hippocampal and cortical acetylcholine release. Therapeutic improvement of cognition may be expected from a (pure) $5\text{-HT}_{1F}$ receptor antagonist, a $5\text{-HT}_2$ receptor agonist, and an $NK_1$ receptor agonist or from their combination.

Apart from 5-HT and acetylcholine, glutamate is involved in most aspects of cognition, for instance in learning. 5-$HT_{1B}$ heteroreceptors depress the release of glutamate from the terminals, in the subiculum, of glutamatergic CA1 hippocampal pyramidal cells (Boeijinga and Boddeke 1996; Table 4). 5-$HT_{1B}$ agonists impair learning by this mechanism, whereas 5-$HT_{1B}$ antagonists enhance learning (page 573 of Sari 2004).

Finally, serotonin contributes to the pharmacology of cocaine, although less prominently than dopamine. The 5-$HT_{1B}$ heteroreceptor mechanism of the inhibition by cocaine of GABA release in the VTA has been explained in Section 4.6. The inhibition of GABA release in turn disinhibits VTA dopamine neurons, thus enhancing dopamine release in the terminal region of the nucleus accumbens, a key structure of addiction (compare Section 2.3).

## References


Aghajanian GK, Marek GJ (1999) Serotonin, via 5-$HT_{2A}$ receptors, increases EPSCs in layer V pyramidal cells of prefrontal cortex by an asynchronous mode of glutamate release. Brain Res 825:161–71

Agneter E, Singer EA, Sauermann W, Feuerstein TJ (1997) The slope parameter of concentration-response curves used as a touchstone for the existence of spare receptors. Naunyn-Schmiedeberg's Arch Pharmacol 356:283–92

Ago Y, Nakamura S, Baba A, Matsuda T (2005a) Sulpiride in combination with fluvoxamine increases *in vivo* dopamine release selectively in rat prefrontal cortex. Neuropsychopharmacology 30:43–51

Ago Y, Harasawa T, Itoh S, Nakamura S, Baba A, Matsuda T (2005b) Antidepressant-like effect of coadministration of sulpiride and fluvoxamine in mice. Eur J Pharmacol 520:86–90

Aizman O, Brismar H, Uhlen P, Zettergren E, Levey AI, Forssberg H, Greengard P, Aperia A (2000) Anatomical and physiological evidence for $D_1$ and $D_2$ dopamine receptor colocalization in neostriatal neurons. Nat Neurosci 3:226–30

Alex KD, Yavanian GJ, McFarlane HG, Pluto CP, Pehek EA (2005) Modulation of dopamine release by striatal 5-$HT_{2C}$ receptors. Synapse 55:242–51

Allgaier C, Warnke P, Stangl AP, Feuerstein TJ (1995) Effects of 5-HT receptor agonists on depolarization-induced [$^3$H]-noradrenaline release in rabbit hippocampus and human neocortex. Br J Pharmacol 116:1769–74

Anichtchik OV, Rinne JO, Kalimo H, Panula P (2000) An altered histaminergic innervation of the substantia nigra in Parkinson's disease. Exp Neurol 163:20–30

Anichtchik OV, Peitsaro N, Rinne JO, Kalimo H, Panula P (2001) Distribution and modulation of histamine $H_3$ receptors in basal ganglia and frontal cortex of healthy controls and patients with Parkinson's disease. Neurobiol Dis 8:707–16

Aosaki T, Kiuchi K, Kawaguchi Y (1998) Dopamine $D_1$-like receptor activation excites rat striatal large aspiny neurons *in vitro*. J Neurosci 18:5180–90

Arias-Montano JA, Floran B, Garcia M, Aceves J, Young JM (2001) Histamine $H_3$ receptor-mediated inhibition of depolarization-induced, dopamine $D_1$ receptor-dependent release of [$^3$H]-gamma-aminobutryic acid from rat striatal slices. Br J Pharmacol 133:165–71

Arias-Montano JA, Floran B, Floran L, Aceves J, Young JM (2007) Dopamine $D_1$ receptor facilitation of depolarization-induced release of gamma-amino-butyric acid in rat striatum is mediated by the cAMP/PKA pathway and involves P/Q-type calcium channels. Synapse 61:310–19

Arbilla S, Nowak JZ, Langer SZ (1985) Rapid desensitization of presynaptic dopamine autoreceptors during exposure to exogenous dopamine. Brain Res 337:11–17

Arnsten AF, Cai JX, Steere JC, Goldman-Rakic PS (1995) Dopamine $D_2$ receptor mechanisms contribute to age-related cognitive decline: the effects of quinpirole on memory and motor performance in monkeys. J Neurosci 15:3429–39

Arrang JM, Garbarg M, Schwartz JC (1983) Auto-inhibition of brain histamine release mediated by a novel class ($H_3$) of histamine receptor. Nature 302:832–7

Arrang JM, Garbarg M, Schwartz JC (1985) Autoregulation of histamine release in brain by presynaptic $H_3$-receptors. Neuroscience 15:553–62

Arrang JM, Drutel G, Schwartz JC (1995) Characterization of histamine $H_3$ receptors regulating acetylcholine release in rat entorhinal cortex. Br J Pharmacol 114:1518–22

Arvanov VL, Liang X, Magro P, Roberts R, Wang RY (1999) A pre- and postsynaptic modulatory action of 5-HT and the 5-$HT_{2A}$, 2C receptor agonist DOB on NMDA-evoked responses in the rat medial prefrontal cortex. Eur J Neurosci 11:2917–34

Asaumi Y, Hasuo H, Akasu T (2006) Dopamine presynaptically depresses fast inhibitory synaptic transmission via D4 receptor-protein kinase A pathway in the rat dorsolateral septal nucleus. J Neurophysiol 96:591–601

Azdad K, Piet R, Poulain DA, Oliet SH (2003) Dopamine $D_4$ receptor-mediated presynaptic inhibition of GABAergic transmission in the rat supraoptic nucleus. J Neurophysiol 90:559–65

Baimoukhametova DV, Hewitt SA, Sank CA, Bains JS (2004) Dopamine modulates use-dependent plasticity of inhibitory synapses. J Neurosci 24:5162–71

Bakker RA (2004) Histamine $H_3$-receptor isoforms. Inflamm Res 53:509–16

Bakker RA, Lozada AF, van Marle A, Shenton FC, Drutel G, Karlstedt K, Hoffmann M, Lintunen M, Yamamoto Y, van Rijn RM, Chazot PL, Panula P, Leurs R (2006) Discovery of naturally occurring splice variants of the rat histamine $H_3$ receptor that act as dominant-negative isoforms. Mol Pharmacol 69:1194–1206

Bamford NS, Zhang H, Schmitz Y, Wu NP, Cepeda C, Levine MS, Schmauss C, Zakharenko SS, Zablow L, Sulzer D (2004) Heterosynaptic dopamine neurotransmission selects sets of corticostriatal terminals Neuron 42:653–63

Batchelor M, Schenk JO (1998) Protein kinase A activity may kinetically upregulate the striatal transporter for dopamine. J Neurosci 18:10304–9

Baumann PA, Waldmeier PC (1981) Further evidence for negative feedback control of serotonin release in the central nervous system. Naunyn-Schmiedeberg's Arch Pharmacol 317:36–43

Bean AJ, During MJ, Roth RH (1990) Effects of dopamine autoreceptor stimulation on the release of colocalized transmitters: *in vivo* release of dopamine and neurotensin from rat prefrontal cortex. Neurosci Lett 108:143–8

Bean AJ, Roth RH (1991) Extracellular dopamine and neurotensin in rat prefrontal cortex *in vivo*: effects of median forebrain bundle stimulation frequency, stimulation pattern, and dopamine autoreceptors. J Neurosci 11:2694–2702

Benkirane S, Arbilla S, Langer SZ (1987) A functional response to $D_1$ dopamine receptor stimulation in the central nervous system: inhibition of the release of $[^3H]$-serotonin from the rat substantia nigra. Naunyn-Schmiedeberg's Arch Pharmacol 335:502–7

Bennett BD, Wilson CJ (1999) Spontaneous activity of neostriatal cholinergic interneurons *in vitro*. J Neurosci 19:5586–96

Berger B, Glowinski J (1978) Dopamine uptake in serotoninergic terminals *in vitro*: a valuable tool for the histochemical differentiation of catecholaminergic and serotoninergic terminals in rat cerebral structures. Brain Res 147:29–45

Berkowicz DA, Trombley PQ (2000) Dopaminergic modulation at the olfactory nerve synapse. Brain Res 855:90–9

Bergquist F, Ruthven A, Ludwig M, Dutia MB (2006) Histaminergic and glycinergic modulation of GABA release in the vestibular nuclei of normal and labyrinthectomised rats. J Physiol 577:857–68

Bianchi C, Rodi D, Marino S, Beani L, Siniscalchi A (2002) Dual effects of 5-$HT_4$ receptor activation on GABA release from guinea pig hippocampal slices. Neuroreport 13:2177–80

Birthelmer A, Schweizer T, Jeltsch H, Jackisch R, Cassel JC (2002) 5,7-Dihydroxytryptamine lesions enhance and serotonergic grafts normalize the evoked overflow of acetylcholine in rat hippocampal slices. Eur J Neurosci Nov;16(10):1839–49

Birthelmer A, Ehret A, Amtage F, Forster S, Lehmann O, Jeltsch H, Cassel JC, Jackisch R (2003) Neurotransmitter release and its presynaptic modulation in the rat hippocampus after selective damage to cholinergic or/and serotonergic afferents. Brain Res Bull 59:371–81

Bischoff S, Scatton B, Korf JB J (1979) Biochemical evidence for a transmitter role of dopamine in the rat hippocampus. Brain Res 165:161–5

Bobker DH, Williams JT (1989) Serotonin agonists inhibit synaptic potentials in the rat locus ceruleus *in vitro* via 5-hydroxytryptamine$_{1A}$ and 5-hydroxytryptamine$_{1B}$ receptors. J Pharmacol Exp Ther 250:37–43

Bobker DH, Williams J (1990) Serotonin-mediated inhibitory postsynaptic potential in guinea pig prepositus hypoglossi and feedback inhibition by serotonin. J Physiol 422:447–42

Boeijinga PH, Boddeke HW (1996) Activation of 5-HT$_{1B}$ receptors suppresses low but not high frequency synaptic transmission in the rat subicular cortex *in vitro*. Brain Res 721:59–65

Bolan EA, Kivell B, Jaligam V, Oz M, Jayanthi LD, Han Y, Sen N, Urizar E, Gomes I, Devi LA, Ramamoorthy S, Javitch JA, Zapata A, Shippenberg TS (2007) $D_2$ receptors regulate dopamine transporter function via an extracellular signal-regulated kinases 1 and 2-dependent and phosphoinositide 3 kinase-independent mechanism. Mol Pharmacol 71:1222–32

Booth RG, Baldessarini RJ, Kula NS, Gao Y, Zong R, Neumeyer JL (1990) Presynaptic inhibition of dopamine synthesis in rat striatal tissue by enantiomeric mono- and dihydroxyaporphines. Mol Pharmacol 38:92–101

Bramley JR, Sollars PJ, Pickard GE, Dudek FE (2005) 5-HT$_{1B}$ receptor-mediated presynaptic inhibition of GABA release in the suprachiasmatic nucleus. J Neurophysiol 93:3157–64

Breitenstein C, Korsukewitz C, Floel A, Kretzschmar T, Diederich K, Knecht S (2006) Tonic dopaminergic stimulation impairs associative learning in healthy subjects. Neuropsychopharmacology 31:2552–64

Brog JS, Beinfeld MC (1992) Cholecystokinin release from the rat caudate-putamen, cortex and hippocampus is increased by activation of the $D_1$ dopamine receptor. J Pharmacol Exp Ther 260:343–8

Brown RE, Haas HL (1999) On the mechanism of histaminergic inhibition of glutamate release in the rat dentate gyrus. J Physiol 515:777–86

Buhot MC, Patra SK, Naili S (1995) Spatial memory deficits following stimulation of hippocampal 5-HT$_{1B}$ receptors in the rat. Eur J Pharmacol 285:221–8

Bull DR, Bakhtiar R, Sheehan MJ (1991) Characterization of dopamine autoreceptors in the amygdala: a fast cyclic voltammetric study *in vitro*. Neurosci Lett 134:41–4

Buzzi MG, Carter WB, Shimizu T, Heath H 3[rd], Moskowitz MA (1991) Dihydroergotamine and sumatriptan attenuate levels of CGRP in plasma in rat superior sagittal sinus during electrical stimulation of the trigeminal ganglion. Neuropharmacology 30:1193–1200

Cameron DL, Williams JT (1993) Dopamine $D_1$ receptors facilitate transmitter release. Nature 366:344–7

Cameron DL, Williams JT (1994) Cocaine inhibits GABA release in the VTA through endogenous 5-HT. J Neurosci 14:6763–7

Cangioli I, Baldi E, Mannaioni PF, Bucherelli C, Blandina P, Passani MB (2002) Activation of histaminergic $H_3$ receptors in the rat basolateral amygdala improves expression of fear memory and enhances acetylcholine release. Eur J Neurosci 16:521–8

Cannon KE, Nalwalk JW, Stadel R, Ge P, Lawson D, Silos-Santiago I, Hough LB (2003) Activation of spinal histamine H3 receptors inhibits mechanical nociception. Eur J Pharmacol 470:139–47

Cannon KE, Hough LB (2005) Inhibition of chemical and low-intensity mechanical nociception by activation of histamine $H_3$ receptors. J Pain 6:193–200

Cannon KE, Chazot PL, Hann V, Shenton F, Hough LB, Rice FL (2007) Immunohistochemical localization of histamine $H_3$ receptors in rodent skin, dorsal root ganglia, superior cervical ganglia, and spinal cord: potential antinociceptive targets. Pain 129:76–92

Cassel JC, Jeltsch H (1995) Serotonergic modulation of cholinergic function in the central nervous system: cognitive implications. Neuroscience 69:1–41

Centonze D, Grande C, Usiello A, Gubellini P, Erbs E, Martin AB, Pisani A, Tognazzi N, Bernardi G, Moratalla R, Borrelli E, Calabresi P(2003) Receptor subtypes involved in the presynaptic and postsynaptic actions of dopamine on striatal interneurons. J Neurosci 23 6245–54

Cheng LL, Wang SJ, Gean PW (1998) Serotonin depresses excitatory synaptic transmission and depolarization-evoked $Ca^{2+}$ influx in rat basolateral amygdala via 5-$HT_{1A}$ receptors. Eur J Neurosci 10:2163–72

Clapham J, Kilpatrick GJ (1992) Histamine $H_3$ receptors modulate the release of [$^3$H]-acetylcholine from slices of rat entorhinal cortex: evidence for the possible existence of $H_3$ receptor subtypes. Br J Pharmacol 107:919–23

Coge F, Guenin SP, Audinot V, Renouard-Try A, Beauverger P, Macia C, Ouvry C, Nagel N, Rique H, Boutin JA, Galizzi JP (2001) Genomic organization and characterization of splice variants of the human histamine $H_3$ receptor. Biochem J 355:279–88

Consolo S, Wu CF, Fusi R (1987) $D_1$ receptor-linked mechanism modulates cholinergic neurotransmission in rat striatum. J Pharmacol Exp Ther 242:300–5

Consolo S, Arnaboldi S, Giorgi S, Russi G, Ladinsky H (1994) 5-$HT_4$ receptor stimulation facilitates acetylcholine release in rat frontal cortex. Neuroreport 5:1230–2

Cooper AJ, Stanford IM (2001) Dopamine $D_2$ receptor mediated presynaptic inhibition of striatopallidal $GABA_A$ IPSCs *in vitro*. Neuropharmacol 41:62–71

Cox B, Kerwin RW, Lee TF, Pycock CJ (1980) A dopamine-5-hydroxytryptamine link in the hypothalamic pathways which mediate heat loss in the rat. J Physiol 303:9–21

Dailly E, Chenu F, Renard CE, Bourin M (2004) Dopamine, depression and antidepressants. Fundam Clin Pharmacol 18:601–7

Dal Toso R, Sommer B, Ewert M, Herb A, Pritchett DB, Bach A, Shivers BD, Seeburg PH (1989) The dopamine $D_2$ receptor: two molecular forms generated by alternative splicing. EMBO J 8:4025–34

Daubner SC, Lauriano C, Haycock JW, Fitzpatrick PF (1992) Site-directed mutagenesis of serine 40 of rat tyrosine hydroxylase. Effects of dopamine and cAMP-dependent phosphorylation on enzyme activity. J Biol Chem 267:12639–46

Davila NG, Blakemore LJ, Trombley PQ (2003) Dopamine modulates synaptic transmission between rat olfactory bulb neurons in culture. J Neurophysiol 90:395–404

Delgado A, Sierra A, Querejeta E, Valdiosera RF, Aceves J (2000) Inhibitory control of the GABAergic transmission in the rat neostriatum by $D_2$ dopamine receptors. Neuroscience 95:1043–8

Di Carlo G, Ghi P, Orsetti M (2000) Effect of R-(−)-alpha-methylhistamine and thioperamide on *in vivo* release of norepinephrine in the rat hippocampus. Prog Neuropsychopharmacol Biol Psychiatry 24:275–84

Dickinson SD, Sabeti J, Larson GA, Giardina K, Rubinstein M, Kelly MA, Grandy DK, Low MJ, Gerhardt GA, Zahniser NR (1999) Dopamine $D_2$ receptor-deficient mice exhibit decreased dopamine transporter function but no changes in dopamine release in dorsal striatum. J Neurochem 72:148–56

Dimitriadou V, Rouleau A, Dam Trung Tuong M, Newlands GJ, Miller HR, Luffau G, Schwartz JC, Garbarg M (1994) Functional relationship between mast cells and C-sensitive nerve fibres evidenced by histamine $H_3$-receptor modulation in rat lung and spleen. Clin Sci (Lond) 87:151–63

Dimitriadou V, Rouleau A, Trung Tuong MD, Newlands GJ, Miller HR, Luffau G, Schwartz JC, Garbarg M (1997) Functional relationships between sensory nerve fibers and mast cells of dura mater in normal and inflammatory conditions. Neuroscience 77:829–39

Dolezal V, Jackisch R, Hertting G, Allgaier C (1992) Activation of dopamine $D_1$ receptors does not affect $D_2$ receptor-mediated inhibition of acetylcholine release in rabbit striatum. Naunyn Schmiedeberg's Arch Pharmacol 345:16–20

Doreulee N, Yanovsky Y, Flagmeyer I, Stevens DR, Haas HL, Brown RE (2001) Histamine $H_3$ receptors depress synaptic transmission in the corticostriatal pathway. Neuropharmacology 40:106–13

Driessen B, Bultmann R, Goncalves J, Starke K (1996) Opposite modulation of noradrenaline and ATP release in guinea pig vas deferens through prejunctional β-adrenoceptors: evidence for the $\beta_2$ subtype. Naunyn-Schmiedeberg's Arch Pharmacol 353:564–71

Drukarch B, Stoof JC (1990) $D_2$ dopamine autoreceptor selective drugs: do they really exist? Life Sci 47:361–76

Edwards E, Harkins K, Wright G, Henn FA (1991) 5-$HT_{1B}$ receptors in an animal model of depression. Neuropharmacology 30:101–5

Eglen RM, Wong EH, Dumuis A, Bockaert J (1995) Central 5-$HT_4$ receptors. Trends Pharmacol Sci 16:391–8

Endou M, Poli E, Levi R (1994) Histamine $H_3$-receptor signaling in the heart: possible involvement of Gi/Go proteins and N-type $Ca^{++}$ channels. J Pharmacol Exp Ther 269:221–9

Enero MA, Langer SZ (1975) Inhibition by dopamine of $^3$H-noradrenaline release elicited by nerve stimulation in the isolated cat's nictitating membrane. Naunyn Schmiedeberg's Arch Pharmacol 289:179–203

Engleman EA, McBride WJ, Li TK, Lumeng L, Murphy JM (2003) Ethanol drinking experience attenuates (−)sulpiride-induced increases in extracellular dopamine levels in the nucleus accumbens of alcohol-preferring (P) rats. Alcohol Clin Exp Res 27:424–31

Ennis C, Kemp JD, Cox B (1981) Characterisation of inhibitory 5-hydroxytryptamine receptors that modulate dopamine release in the striatum. J Neurochem 36:1515–20

Ensinger H, Majewski H, Hedler L, Starke K (1985) Neuronal and postjunctional components in the blood pressure effects of dopamine and bromocriptine in rabbits. J Pharmacol Exp Ther 234:681–90

Enz A, Goldstein M, Meller E (1990) Dopamine agonist-induced elevation of striatal acetylcholine: relationship between receptor occupancy and response in normal and denervated rat striatum. Mol Pharmacol 37:560–5

Farnebo LO, Hamberger B (1971) Drug-induced changes in the release of $^3$H-monoamines from field stimulated rat brain slices. Acta Physiol Scand Suppl 371:35–44

Fedele E, Fontana G, Munari C, Cossu M, Raiteri M (1999) Native human neocortex release-regulating dopamine $D_2$ type autoreceptors are dopamine $D_2$ subtype. Eur J Neurosci 11:2351–8

Ferre S, Artigas F (1995) Clozapine decreases serotonin extracellular levels in the nucleus accumbens by a dopamine receptor-independent mechanism. Neurosci Lett 187:61–4

Ferris CF, Melloni RH Jr, Koppel G, Perry KW, Fuller RW, Delville Y (1997) Vasopressin/ serotonin interactions in the anterior hypothalamus control aggressive behavior in golden hamsters. J Neurosci 17:4331–40

Feuerstein TJ, Seeger W (1997) Modulation of acetylcholine release in human cortical slices: possible implications for Alzheimer's disease. Pharmacol Ther 74:333–47

Feuerstein TJ, Hertting G, Lupp A, Neufang B (1986) False labelling of dopaminergic terminals in the rabbit caudate nucleus: uptake and release of [$^3$H]-5-hydroxytryptamine. Br J Pharmacol 88:677–84

Feuerstein TJ, Gleichauf O, Landwehrmeyer GB (1996a) Modulation of cortical acetylcholine release by serotonin: the role of substance P interneurons. Naunyn-Schmiedeberg's Arch Pharmacol 354:618–26

Feuerstein TJ, Hüring H, van Velthoven V, Lücking CH, Landwehrmeyer GB (1996b) 5-$HT_{1D}$-like receptors inhibit the release of endogenously formed [$^3$H]GABA in human, but not in rabbit, neocortex. Neurosci Lett 209:210–14

Fink K, Schlicker E, Neise A, Göthert M (1990) Involvement of presynaptic $H_3$ receptors in the inhibitory effect of histamine on serotonin release in the rat brain cortex. Naunyn-Schmiedeberg's Arch Pharmacol 342:513–19

Floran B, Aceves J, Sierra A, Martinez-Fong D (1990) Activation of $D_1$ dopamine receptors stimulates the release of GABA in the basal ganglia of the rat. Neurosci Lett 116:136–40

Floran B, Floran L, Sierra A, Aceves J (1997) $D_2$ receptor-mediated inhibition of GABA release by endogenous dopamine in the rat globus pallidus. Neurosci Lett 237:1–4

Francis GS (1995) Receptor systems involved in norepinephrine release in heart failure: focus on dopaminergic systems. Clin Cardiol 18(3 Suppl I):I13–I16

Fuder H, Muscholl E (1995) Heteroreceptor-mediated modulation of noradrenaline and acetylcholine release from peripheral nerves. Rev Physiol Biochem Pharmacol 126:265–412

Galzin AM, Dubocovich ML, Langer SZ (1982) Presynaptic inhibition by dopamine receptor agonists of noradrenergic neurotransmission in the rabbit hypothalamus. J Pharmacol Exp Ther 221:461–71

Garcia M, Floran B, Arias-Montano JA, Young JM, Aceves J (1997) Histamine $H_3$ receptor activation selectively inhibits dopamine $D_1$ receptor-dependent [$^3$H]GABA release from depolarization-stimulated slices of rat substantia nigra pars reticulata. Neuroscience 80:241–9

Garcia-Ramirez M, Aceves J, Arias-Montano JA (2004) Intranigral injection of the $H_3$ agonist immepip and systemic apomorphine elicit ipsilateral turning behaviour in naive rats, but reduce contralateral turning in hemiparkinsonian rats. Behav Brain Res 154:409–15

Garduno-Torres B, Trevino M, Gutierrez R, Arias-Montano JA (2007) Pre-synaptic histamine $H_3$ receptors regulate glutamate, but not GABA release in rat thalamus. Neuropharmacology 52:527–35

Geldwert D, Norris JM, Feldman IG, Schulman JJ, Joyce MP, Rayport S (2006) Dopamine presynaptically and heterogeneously modulates nucleus accumbens medium-spiny neuron GABA synapses *in vitro*. BMC Neurosci 7:53

Gerfen CR, Engber TM, Mahan LC, Susel Z, Chase TN, Monsma FJ Jr, Sibley DR (1990) $D_1$ and $D_2$ dopamine receptor-regulated gene expression of striatonigral and striatopallidal neurons. Science 250:1429–32

Giorgetti M, Bacciottini L, Bianchi L, Giovannini MG, Cecchi M, Blandina P(1997) GABAergic mechanism in histamine $H_3$ receptor inhibition of $K^+$-evoked release of acetylcholine from rat cortex *in vivo*. Inflamm Res 46:S33–S34

Gobert A, Rivet JM, Audinot V, Newman-Tancredi A, Cistarelli L, Millan MJ (1998) Simultaneous quantification of serotonin, dopamine and noradrenaline levels in single frontal cortex dialysates of freely-moving rats reveals a complex pattern of reciprocal auto- and heteroreceptor-mediated control of release. Neurosci 84:413–29

Göthert M (2003) Modulation of noradrenaline release in human cardiovascular tissues. Pharmacol Toxicol 92:156–9

Göthert M, Schlicker E (1997) Regulation of 5-HT release in the CNS by presynaptic 5-HT autoreceptors and by 5-HT heteroreceptors. In: Baumgarten HG, Göthert M (eds) Serotonergic Neurons and 5-HT Receptors in the CNS. Springer, Berlin Heidelberg New York, pp 307–50

Göthert M, Starke K (2005) Pharmakologie des Serotonins. Pharmakotherapie primärer Kopfschmerzen. In: Allgemeine und spezielle Pharmakologie und Toxikologie. Eds. K. Aktories, U. Förstermann, F. Hofmann, K. Starke, 9. edition, Urban & Fischer, München Jena, pp 213–19

Göthert M, Garbarg M, Hey JA, Schlicker E, Schwartz JC, Levi R (1995) New aspects of the role of histamine in cardiovascular function: identification, characterization, and potential pathophysiological importance of $H_3$ receptors. Can J Physiol Pharmacol 73:558–64

Göthert M, Fink K, Frolich D, Likungu J, Molderings G, Schlicker E, Zentner J (1996) Presynaptic 5-HT auto- and heteroreceptors in the human central and peripheral nervous system. Behav Brain Res 73:89–92

Goldstein M, Harada K, Meller E, Schalling M, Hökfelt T (1990) Dopamine autoreceptors. Biochemical, pharmacological, and morphological studies. Ann N Y Acad Sci 604:169–75

Goldstein DS, Mezey E, Yamamoto T, Aneman A, Friberg P, Eisenhofer G (1995) Is there a third peripheral catecholaminergic system? Endogenous dopamine as an autocrine/paracrine substance derived from plasma DOPA and inactivated by conjugation. Hypertens Res Suppl 1:S93–9

Goldstein DS, Eldadah BA, Holmes C, Pechnik S, Moak J, Saleem A, Sharabi Y (2005) Neurocirculatory abnormalities in Parkinson disease with orthostatic hypotension: independence from levodopa treatment. Hypertension 46:1333–9

Gomez-Ramirez J, Ortiz J, Blanco I (2002) Presynaptic $H_3$ autoreceptors modulate histamine synthesis through cAMP pathway. Mol Pharmacol 61:239–45

Goncalves J, Bultmann R, Driessen B (1996) Opposite modulation of cotransmitter release in guinea pig vas deferens: increase of noradrenaline and decrease of ATP release by activation of prejunctional beta-adrenoceptors. Naunyn-Schmiedeberg's Arch Pharmacol 353:184–92

Gonzalez-Islas C, Hablitz J (2001) Dopamine inhibition of evoked IPSCs in rat prefrontal cortex. J Neurophysiol 86:2911–18

Gorell JM, Czarnecki B (1986) Pharmacologic evidence for direct dopaminergic regulation of striatal acetylcholine release. Life Sci 38:2239–46

Green S (1991) Benzodiazepines, putative anxiolytics and animal models of anxiety. Trends Neurosci 14:101–4

Guzman JN, Hernandez A, Galarraga E, Tapia D, Laville A, Vergara R, Aceves J, Bargas J (2003) Dopaminergic modulation of axon collaterals interconnecting spiny neurons of the rat striatum. J Neurosci 23:8931–40

Haeusler G, Lues I, Minck KO, Schelling P, Seyfried CA (1992) Pharmacological basis for antihypertensive therapy with a novel dopamine agonist. Eur Heart J 13 Suppl D:129–35

Hakansson K, Pozzi L, Usiello A, Haycock J, Borrelli E, Fisone G (2004) Regulation of striatal tyrosine hydroxylase phosphorylation by acute and chronic haloperidol. Eur J Neurosci 20:1108–12

Hamon M, Doucet E, Lefevre K, Miquel MC, Lanfumey L, Insausti R, Frechilla D, Del Rio J, Verge D (1999) Antibodies and antisense oligonucleotide for probing the distribution and putative functions of central 5-$HT_6$ receptors. Neuropsychopharmacology 21:68S–76S

Harada C, Fujii Y, Hirai T, Shinomiya K, Kamei C (2004) Inhibitory effect of iodophenpropit, a selective histamine H3 antagonist, on amygdaloid kindled seizures. Brain Res Bull 63:143–6

Hatta E, Yasuda K, Levi R (1997) Activation of histamine $H_3$ receptors inhibits carrier-mediated norepinephrine release in a human model of protracted myocardial ischemia. J Pharmacol Exp Ther 283:494–500

Helmreich I, Reimann W, Hertting G, Starke K (1982) Are presynaptic dopamine autoreceptors and postsynaptic dopamine receptors in the rabbit caudate nucleus pharmacologically different? Neuroscience 7:1559–66

Hernandez A, Ibanez-Sandoval O, Sierra A, Valdiosera R, Tapia D, Anaya-Martinez V, Galarraga E, Bargas J, Aceves J (2006) Control of the subthalamic innervation of the rat globus pallidus by $D_{2/3}$ and $D_4$ dopamine Receptors. J Neurophysiol 96:2877–88

Hertting G, Zumstein A, Jackisch R, Hoffmann I, Starke K (1980) Modulation by endogenous dopamine of the release of acetylcholine in the caudate nucleus of the rabbit. Naunyn-Schmiedeberg's Arch Pharmacol 315:111–17

Huang YW, Hu WW, Chen Z, Zhang LS, Shen HQ, Timmerman H, Leurs R, Yanai K (2004) Effect of the histamine $H_3$-antagonist clobenpropit on spatial memory deficits induced by MK-801 as evaluated by radial maze in Sprague-Dawley rats. Behav Brain Res 151:287–93

Ibanez-Sandoval O, Hernandez A, Floran B, Galarraga E, Tapia D, Valdiosera R, Erlij D, Aceves J, Bargas J (2006) Control of the subthalamic innervation of substantia nigra pars reticulata by $D_1$ and $D_2$ dopamine receptors. J Neurophysiol 95:1800–11

Ichinose M, Belvisi MG, Barnes PJ (1990) Histamine $H_3$-receptors inhibit neurogenic microvascular leakage in airways. J Appl Physiol 68:21–5

Ilani T, Fishburn CS, Levavi-Sivan B, Carmon S, Raveh L, Fuchs S (2002) Coupling of dopamine receptors to G proteins: studies with chimeric $D_2/D_3$ dopamine receptors. Cell Mol Neurobiol 22:47–56

Imamura M, Poli E, Omoniyi AT, Levi R (1994) Unmasking of activated histamine $H_3$-receptors in myocardial ischemia: their role as regulators of exocytotic norepinephrine release. J Pharmacol Exp Ther 271:1259–66

Imamura M, Seyedi N, Lander HM, Levi R (1995) Functional identification of histamine $H_3$-receptors in the human heart. Circ Res 77:206–10
Imamura M, Smith NC, Garbarg M, Levi R (1996a) Histamine $H_3$-receptor-mediated inhibition of calcitonin gene-related peptide release from cardiac C fibers. A regulatory negative-feedback loop. Circ Res 78:863–9
Imamura M, Lander HM, Levi R (1996b) Activation of histamine $H_3$-receptors inhibits carrier-mediated norepinephrine release during protracted myocardial ischemia. Comparison with adenosine $A_1$-receptors and $\alpha_2$-adrenoceptors. Circ Res 78:475–81
Ishikawa S, Sperelakis N (1987) A novel class ($H_3$) of histamine receptors on perivascular nerve terminals. Nature 327:158–60
Jackisch R, Moll S, Feuerstein TJ, Hertting G (1985) Dopaminergic modulation of hippocampal noradrenaline release. Evidence for $\alpha_2$-antagonistic effects of some dopamine receptor agonists and antagonists. Naunyn-Schmiedeberg's Arch Pharmacol 330:105–13
Jiang X, Chen A, Li H (2005) Histaminergic modulation of excitatory synaptic transmission in the rat basolateral amygdala. Neuroscience 131:691–703
Jin CY, Panula P (2005) The laminar histamine receptor system in human prefrontal cortex suggests multiple levels of histaminergic regulation. Neuroscience 132:137–49
Johnson EA, Tsai CE, Lucci J, Harrison-Shahan Y, Azzaro AJ (1992a) Dopamine $D_2$ synthesis-modulating receptors are present in the striatum of the guinea pig. Neuropharmacology 31:95–101
Johnson SW, Mercuri NB, North RA (1992b) 5-hydroxytryptamine$_{1B}$ receptors block the $GABA_B$ synaptic potential in rat dopamine neurons. J Neurosci 12:2000–5
Johnson EA, Tsai CE, Shahan YH, Azzaro AJ (1993) Serotonin 5-$HT_{1A}$ receptors mediate inhibition of tyrosine hydroxylation in rat striatum. J Pharmacol Exp Ther 266:133–41
Jones SR, Gainetdinov RR, Hu XT, Cooper DC, Wightman RM, White FJ, Caron MG (1999) Loss of autoreceptor functions in mice lacking the dopamine transporter. Nat Neurosci 2 649–55
Joseph JD, Wang YM, Miles PR, Budygin EA, Picetti R, Gainetdinov RR, Caron MG, Wightman RM (2002) Dopamine autoreceptor regulation of release and uptake in mouse brain slices in the absence of $D_3$ receptors. Neuroscience 112:39–49
Joyce JN, Marshall JF (1987) Quantitative autoradiography of dopamine $D_2$ sites in rat caudate-putamen: localization to intrinsic neurons and not to neocortical afferents. Neuroscience 20:773–95
Kahn RS, van Praag HM, Wetzler S, Asnis GM, Barr G (1988) Serotonin and anxiety revisited. Biol Psychiatry 23:189–208
Katsurabayashi S, Kubota H, Tokutomi N, Akaike N (2003) A distinct distribution of functional presynaptic 5-HT receptor subtypes on GABAergic nerve terminals projecting to single hippocampal CA1 pyramidal neurons. Neuropharmacology 44:1022–30
Kim KM, Gainetdinov RR, Laporte SA, Caron MG, Barak LS (2005) G protein-coupled receptor kinase regulates dopamine $D_3$ receptor signaling by modulating the stability of a receptor-filamin-beta-arrestin complex. A case of autoreceptor regulation. J Biol Chem 280:12774–80
Koga E, Momiyama T (2000) Presynaptic dopamine $D_2$-like receptors inhibit excitatory transmission onto rat ventral tegmental dopaminergic neurones. J Physiol 523:163–73
Koyama S, Kubo C, Rhee JS, Akaike N (1999) Presynaptic serotonergic inhibition of GABAergic synaptic transmission in mechanically dissociated rat basolateral amygdala neurons. J Physiol 518:525–38
Koyama M, Heerdt PM, Levi R (2003) Increased severity of reperfusion arrhythmias in mouse hearts lacking histamine $H_3$-receptors. Biochem Biophys Res Commun 306:792–6
Koyama S, Brodie MS, Appel SB (2007) Ethanol inhibition of M-current and ethanol-induced direct excitation of ventral tegmental area dopamine neurons. J Neurophysiol 97:1977–85
Kuzhikandathil EV, Yu W, Oxford GS (1998) Human dopamine $D_3$ and $D_{2L}$ receptors couple to inward rectifier potassium channels in mammalian cell lines. Mol Cell Neurosci 12:390–402
Kuzhikandathil EV, Oxford GS (1999) Activation of human $D_3$ dopamine receptor inhibits P/Q-type calcium channels and secretory activity in AtT-20 cells. J Neurosci 19:1698–1707

Leclere PG Lefebvre RA (2002) Presynaptic modulation of cholinergic neurotransmission in the human proximal stomach. Br J Pharmacol 135:135–42

LePard KJ, Ren J, Galligan J (2004) Presynaptic modulation of cholinergic and non-cholinergic fast synaptic transmission in the myenteric plexus of guinea pig ileum. Neurogastroenterol Motil 16:355–64

Leurs R, Bakker RA, Timmerman H, de Esch IJ (2005) The histamine $H_3$ receptor: from gene cloning to $H_3$ receptor drugs. Nat Rev Drug Discov 4:107–20

Levi R, Smith NC (2000) Histamine $H_3$-receptors: a new frontier in myocardial ischemia. J Pharmacol Exp Ther 292:825–30

Li M, Luo X, Chen L, Zhang J, Hu J, Lu B (2003) Co-localization of histamine and dopamine-beta-hydroxylase in sympathetic ganglion and release of histamine from cardiac sympathetic terminals of guinea pig. Auton Autacoid Pharmacol 23:327–33

Li M, Hu J, Chen Z, Meng J, Wang H, Ma X, Luo X (2006) Evidence for histamine as a neurotransmitter in the cardiac sympathetic nervous system. Am J Physiol Heart Circ Physiol 291:45–51

Lindgren N, Xu ZQ, Herrera-Marschitz M, Haycock J, Hokfelt T, Fisone G (2001) Dopamine $D_2$ receptors regulate tyrosine hydroxylase activity and phosphorylation at Ser40 in rat striatum. Eur J Neurosci 13:773–80

Lindgren N, Usiello A, Goiny M, Haycock J, Erbs E, Greengard P, Hokfelt T, Borrelli E, Fisone G (2003) Distinct roles of dopamine $D_{2L}$ and $D_{2S}$ receptor isoforms in the regulation of protein phosphorylation at presynaptic and postsynaptic sites. Proc Natl Acad Sci USA 100:4305–9

Lintunen M, Sallmen T, Karlstedt K, Panula P (2005) Transient changes in the limbic histaminergic system after systemic kainic acid-induced seizures. Neurobiol Dis 20:155–69

Löffler M, Bubl B, Huethe F, Hubbe U, McIntosh JM, Jackisch R, Feuerstein TJ (2006) Dopamine release in human neocortical slices: characterization of inhibitory autoreceptors and of nicotinic acetylcholine receptor-evoked release. Brain Res Bull 68:361–73

Lokhandwala MF, Buckley JP (1978) The effect of L-dopa on peripheral sympathetic nerve function: role of presynaptic dopamine receptors. J Pharmacol Exp Ther 204:362–71

Lucas G, Spampinato U (2000) Role of striatal serotonin$_{2A}$ and serotonin$_{2C}$ receptor subtypes in the control of *in vivo* dopamine outflow in the rat striatum. J Neurochem 74:693–701

Malinina E, Druzin M, Johansson S (2005) Fast neurotransmission in the rat medial preoptic nucleus. Brain Res 1040:157–68

Mannelli M, Lazzeri C, Ianni L, La Villa G, Pupilli C, Bellini F, Serio M, Franchi F (1997) Dopamine and sympathoadrenal activity in man. Clin Exp Hypertens 19:163–79

Marcos B, Gil-Bea FJ, Hirst WD, Garcia-Alloza M, Ramirez MJ (2006) Lack of localization of 5-$HT_6$ receptors on cholinergic neurons: implication of multiple neurotransmitter systems in 5-$HT_6$ receptor-mediated acetylcholine release. Eur J Neurosci 24:1299–1306

Marek GJ, Wright RA, Gewirtz JC, Schoepp DD (2001) A major role for thalamocortical afferents in serotonergic hallucinogen receptor function in the rat neocortex. Neuroscience 105:379–92

Matsubara T, Moskowitz MA, Huang Z (1992) UK-14,304, R(-)-alpha-methyl-histamine and SMS 201–995 block plasma protein leakage within dura mater by prejunctional mechanisms. Eur J Pharmacol 224:145–50

Matsumoto M, Yoshioka M, Togashi H, Ikeda T, Saito H (1996) Functional regulation by dopamine receptors of serotonin release from the rat hippocampus: *in vivo* microdialysis study. Naunyn-Schmiedeberg's Arch Pharmacol 353:621–9

Maurice N, Mercer J, Chan CS, Hernandez-Lopez S, Held J, Tkatch T, Surmeier DJ (2004) $D_2$ dopamine receptor-mediated modulation of voltage-dependent $Na^+$ channels reduces autonomous activity in striatal cholinergic interneurons. J Neurosci 24:10289–301

Mayfield RD, Zahniser NR (2001) Dopamine $D_2$ receptor regulation of the dopamine transporter expressed in Xenopus laevis oocytes is voltage-independent. Mol Pharmacol 59:113–21

Mazenot C, Ribuot C, Durand A, Joulin Y, Demenge P, Godin-Ribuot D (1999) *In vivo* demonstration of $H_3$-histaminergic inhibition of cardiac sympathetic stimulation by R-alpha-methyl-histamine and its prodrug BP 2.94 in the dog. Br J Pharmacol 126:264–8

McLeod RL, Aslanian R, del Prado M, Duffy R, Egan RW, Kreutner W, McQuade R, Hey JA (1998) Sch 50971, an orally active histamine H3 receptor agonist, inhibits central neuro-

genic vascular inflammation and produces sedation in the guinea pig. J Pharmacol Exp Ther 287:43–50
Meiergerd SM, Patterson TA, Schenk JO (1993) $D_2$ receptors may modulate the function of the striatal transporter for dopamine: kinetic evidence from studies *in vitro* and *in vivo*. J Neurochem 61:764–7
Meyer DK, Krauss J (1983) Dopamine modulates cholecystokinin release in neostriatum. Nature 301:338–40
Mizuno T, Schmauss C, Rayport S (2007) Distinct roles of presynaptic dopamine receptors in the differential modulation of the intrinsic synapses of medium-spiny neurons in the nucleus accumbens. BMC Neurosci 8:8–23
Mizutani H, Hori T, Takahash T (2006) 5-$HT_{1B}$ receptor-mediated presynaptic inhibition at the calyx of Held of immature rats. Eur J Neurosci 24:1946–54
Molina-Hernandez A, Nunez A, Arias-Montano JA (2000) Histamine $H_3$-receptor activation inhibits dopamine synthesis in rat striatum. Neuroreport 11:163–6
Molina-Hernandez A, Nunez A, Sierra JJ, Arias-Montano JA (2001) Histamine $H_3$ receptor activation inhibits glutamate release from rat striatal synaptosomes. Neuropharmacology 41:928–34
Momiyama T, Koga E (2001) Dopamine $D_2$-like receptors selectively block N-type $Ca^{2+}$ channels to reduce GABA release onto rat striatal cholinergic interneurones. J Physiol 533:479–92
Monnet FP (2002) Functional cooperation between neurosteroids and $D_2$ dopamine antagonists on KCl-evoked [$^3$H]noradrenaline release: modulation by calcium channel blockers. J Neuroendocrinol 14:955–62
Mooney RD, Shi MY, Rhoades RW (1994) Modulation of retinotectal transmission by presynaptic 5-$HT_{1B}$ receptors in the superior colliculus of the adult hamster. J Neurophysiol 72:3–13
Moreno-Delgado D, Torrent A, Gomez-Ramirez J, de Esch I, Blanco I, Ortiz J (2006) Constitutive activity of $H_3$ autoreceptors modulates histamine synthesis in rat brain through the cAMP/PKA pathway. Neuropharmacology 51:517–23
Morgadinho MT, Fontes Ribeiro CA, Macedo TR (1999) Presynaptic dopamine receptors involved in the inhibition of noradrenaline and dopamine release in the human gastric and uterine arteries. Fundam Clin Pharmacol 13:662–70
Morikawa H, Manzoni OJ, Crabbe JC, Williams JT (2000) Regulation of central synaptic transmission by 5-$HT_{1B}$ auto- and heteroreceptors. Mol Pharmacol 58:1271–8
Morisset S, Rouleau A, Ligneau X, Gbahou F, Tardivel-Lacombe J, Stark H, Schunack W, Ganellin CR, Schwartz JC, Arrang JM (2000) High constitutive activity of native $H_3$ receptors regulates histamine neurons in brain. Nature 408:860–4
Muramatsu M, Tamaki-Ohashi J, Usuki C, Araki H, Chaki S, Aihara H (1988a) 5-$HT_2$ antagonists and minaprine block the 5-HT-induced inhibition of dopamine release from rat brain striatal slices. Eur J Pharmacol 153:89–95
Muramatsu M, Tamaki-Ohashi J, Usuki C, Araki H, Aihara H (1988b) Serotonin-2 receptor-mediated regulation of release of acetylcholine by minaprine in cholinergic nerve terminal of hippocampus of rat. Neuropharmacology 27:603–9
Muramatsu M, Lapiz MD, Tanaka E, Grenhoff J (1998) Serotonin inhibits synaptic glutamate currents in rat nucleus accumbens neurons via presynaptic 5-$HT_{1B}$ receptors. Eur J Neurosci 10:2371–9
Murphy MB (2000) Dopamine: a role in the pathogenesis and treatment of hypertension. J Hum Hypertens 14 Suppl 1:S47–S50
Nemmar A, Delaunois A, Beckers JF, Sulon J, Bloden S, Gustin P (1999) Modulatory effect of imetit, a histamine $H_3$ receptor agonist, on C-fibers, cholinergic fibers and mast cells in rabbit lungs *in vitro*. Eur J Pharmacol 371:23–30
Nicola SM, Kombian SB, Malenka RC (1996) Psychostimulants depress excitatory synaptic transmission in the nucleus accumbens via presynaptic $D_1$-like dopamine receptors. J Neurosci 16:1591–604
Nicola SM, Malenka RC (1997) Dopamine depresses excitatory and inhibitory synaptic transmission by distinct mechanisms in the nucleus accumbens. J Neurosci 17:5697–5710

Ng NK, Lee HS, Wong PT (1999) Regulation of striatal dopamine release through 5-HT$_1$ and 5-HT$_2$ receptors. J Neurosci Res 55:600–7
Nguyen-Legros J, Versaux-Botteri C, Vernier P (1999) Dopamine receptor localization in the mammalian retina. Mol Neurobiol 19:181–204
Obadiah J, Avidor-Reiss T, Fishburn CS, Carmon S, Bayewitch M, Vogel Z, Fuchs S, Levavi-Sivan B (1999) Adenylyl cyclase interaction with the D$_2$ dopamine receptor family; differential coupling to Gi, Gz, and Gs. Cell Mol Neurobiol 19:653–64
O'Hara CM, Uhland-Smith A, O'Malley KL, Todd RD (1996) Inhibition of dopamine synthesis by dopamine D2 and D3 but not D4 receptors. J Pharmacol Exp Ther 277:186–92
Ohkubo T, Shibata M (1995) ATP-sensitive $K^+$ channels mediate regulation of substance P release via the prejunctional histamine H$_3$ receptor. Eur J Pharmacol 277:45–9
Pan WH, Yang SY, Lin SK (2004) Neurochemical interaction between dopaminergic and noradrenergic neurons in the medial prefrontal cortex. Synapse 53:44–52
Park KH, Long JP, Cannon JG (1991) Effects of serotonin$_1$-like receptor agonists on autonomic neurotransmission. Can J Physiol Pharmacol 69:1855–60
Passani MB, Cangioli I, Baldi E, Bucherelli C, Mannaioni PF, Blandina P (2001) Histamine H$_3$ receptor-mediated impairment of contextual fear conditioning and in-vivo inhibition of cholinergic transmission in the rat basolateral amygdala. Eur J Neurosci 14:1522–32
Petty F, Kramer G, Wilson L (1992) Prevention of learned helplessness: *in vivo* correlation with cortical serotonin. Pharmacol Biochem Behav 43:361–7
Pillot C, Heron A, Cochois V, Tardivel-Lacombe J, Ligneau X, Schwartz JC, Arrang JM (2002) A detailed mapping of the histamine H$_3$ receptor and its gene transcripts in rat brain. Neuroscience 114:173–93
Pickard GE, Smith BN, Belenky M, Rea MA, Dudek FE, Sollars P (1999) 5-HT$_{1B}$ receptor-mediated presynaptic inhibition of retinal input to the suprachiasmatic nucleus. J Neurosci 19:4034–45
Piguet P, Galvan M (1994) Transient and long-lasting actions of 5-HT on rat dentate gyrus neurones *in vitro*. J Physiol 481:629–39
Pineyro G, de Montigny C, Blier P (1995a) 5-HT$_{1D}$ receptors regulate 5-HT release in the rat raphe nuclei. *In vivo* voltammetry and *in vitro* superfusion studies. Neuropsychopharmacology 13:249–60
Pineyro G, Castanon N, Hen R, Blier P (1995b) Regulation of [$^3$H]5-HT release in raphe, frontal cortex and hippocampus of 5-HT$_{1B}$ knock-out mice. Neuroreport 7:353–9
Pisani A, Bonsi P, Centonze D, Calabresi P, Bernardi G (2000) Activation of D$_2$-like dopamine receptors reduces synaptic inputs to striatal cholinergic interneurons. J Neurosci 20:RC69 (1–6)
Popova NK (2006) From genes to aggressive behavior: the role of serotonergic system. Bioessays 28:495–503
Pothos EN, Przedborski S, Davila V, Schmitz Y, Sulzer D (1998) D$_2$-Like dopamine autoreceptor activation reduces quantal size in PC12 cells. J Neurosci 18:5575–85
Prasad BM, Amara SG (2001) The dopamine transporter in mesencephalic cultures is refractory to physiological changes in membrane voltage. J Neurosci 21:7561–7
Prast H, Tran MH, Fischer H, Kraus M, Lamberti C, Grass K, Philippu A (1999a) Histaminergic neurons modulate acetylcholine release in the ventral striatum: role of H$_3$ histamine receptors. Naunyn Schmiedebergs Arch Pharmacol 360:558–64
Prast H, Tran MH, Lamberti C, Fischer H, Kraus M, Grass K, Philippu A (1999b) Histaminergic neurons modulate acetylcholine release in the ventral striatum: role of H$_1$ and H$_2$ histamine receptors. Naunyn Schmiedebergs Arch Pharmacol 360:552–7
Price CJ, Pittman QJ (2001) Dopamine D$_4$ receptor activation inhibits presynaptically glutamatergic neurotransmission in the rat supraoptic nucleus. J Neurophysiol 86:1149–55
Racagni G, Canonico PL, Ravizza L, Pani L, Amore M (2004) Consensus on the use of substituted benzamides in psychiatric patients. Neuropsychobiology 50:134–43
Radnikow G, Misgeld U (1998) Dopamine D$_1$ receptors facilitate GABA$_A$ synaptic currents in the rat substantia nigra pars reticulata. J Neurosci 18:2009–16
Raiteri M (2006) Functional pharmacology in human brain. Pharmacol Rev 58:162–93

Raiteri M, Maura G, Bonanno G, Pittaluga A (1986) Differential pharmacology and function of two 5-$HT_1$ receptors modulating transmitter release in rat cerebellum. J Pharmacol Exp Ther 237:644–8

Rothblat DS, Schneider JS (1997) Regionally specific effects of haloperidol and clozapine on dopamine reuptake in the striatum. Neurosci Lett 228:119–22

Roy-Byrne PP (2005) The GABA-benzodiazepine receptor complex: structure, function, and role in anxiety. J Clin Psychiatry 66 Suppl 2:14–20

Ruhé HG, Mason NS, Schene AH (2007) Mood is indirectly related to serotonin, norepinephrine and dopamine levels in humans: a meta-analysis of monoamine depletion studies. Mol Psychiatry 12:331–59

Rump LC, Schwertfeger E, Schaible U, Schuster MJ, Frankenschmidt A, Schollmeyer P (1993) Dopamine receptor modulation of noradrenaline release by carmoxirole in human cortical kidney slices. Eur J Clin Pharmacol 44 Suppl 1:S47–S49

Rump LC, Riera-Knorrenschild G, Schwertfeger E, Bohmann C, Spillner G, Schollmeyer P (1995) Dopaminergic and alpha-adrenergic control of neurotransmission in human right atrium. J Cardiovasc Pharmacol 26:462–70

Rutz S, Riegert C, Rothmaier AK, Buhot MC, Cassel JC, Jackisch R (2006) Presynaptic serotonergic modulation of 5-HT and acetylcholine release in the hippocampus and the cortex of 5-$HT_{1B}$-receptor knockout mice. Brain Res Bull 70:81–93

Rutz S, Riegert C, Rothmaier AK, Jackisch R (2007) Presynaptic modulation of 5-HT release in the rat septal region. Neurosci 146:643–58

Sarhan H, Cloez-Tayarani I, Massot O, Fillion MP, Fillion G (1999) 5-$HT_{1B}$ receptors modulate release of [$^3$H] dopamine from rat striatal synaptosomes. Naunyn Schmiedeberg's Arch Pharmacol 359:40–7

Sarhan H, Fillion G (1999) Differential sensitivity of 5-$HT_{1B}$ auto and heteroreceptors. Naunyn Schmiedeberg's Arch Pharmacol 360:382–90

Sarhan H, Grimaldi B, Hen R, Fillion G (2000) 5-$HT_{1B}$ receptors modulate release of [$^3$H] dopamine from rat striatal synaptosomes: further evidence using 5-HT moduline, polyclonal 5-$HT_{1B}$ receptor antibodies and 5-$HT_{1B}$ receptor knock-out mice. Naunyn Schmiedebergs Arch Pharmacol 361:12–18

Sari Y (2004) Serotonin$_{1B}$ receptors: from protein to physiological function and behavior. Neurosci Biobehav Rev 28:565–82

Sehmisch S, Blauth E, Thorn D, Cassel JC, Kelche C, Feuerstein TJ, Jackisch R (2001) Electrically evoked release of glutamate in rat hippocampal slices: effects of various drugs and fimbria-fornix lesions. Naunyn-Schmiedeberg's Arch Pharmacol 363:481–90

Schlicker E, Göthert M (1998) Interactions between the presynaptic $\alpha_2$-autoreceptor and presynaptic inhibitory heteroreceptors on noradrenergic neurones. Brain Res Bull 47:129–32

Schlicker E, Betz R, Göthert M (1988) Histamine $H_3$ receptor-mediated inhibition of serotonin release in the rat brain cortex. Naunyn-Schmiedeberg's Arch Pharmacol 337:588–90

Schlicker E, Fink K, Hinterthaner M, Göthert M (1989) Inhibition of noradrenaline release in the rat brain cortex via presynaptic $H_3$ receptors. Naunyn Schmiedeberg's Arch Pharmacol 340:633–8

Schlicker E, Fink K, Detzner M, Göthert M (1993) Histamine inhibits dopamine release in the mouse striatum via presynaptic $H_3$ receptors. J Neural Transm Gen Sect 93:1–10

Schlicker E, Kathmann M, Detzner M, Exner HJ, Göthert M (1994) $H_3$ receptor-mediated inhibition of noradrenaline release: an investigation into the involvement of $Ca^{2+}$ and $K^-$ ions, G protein and adenylate cyclase. Naunyn Schmiedeberg's Arch Pharmacol 350:34–41

Schlicker E, Werthwein S, Zentner J (1999) Histamine $H_3$ receptor-mediated inhibition of noradrenaline release in the human brain. Fundam Clin Pharmacol 13:120–2

Schmitz D, Gloveli T, Empson RM, Draguhn A, Heinemann U (1998) Serotonin reduces synaptic excitation in the superficial medial entorhinal cortex of the rat via a presynaptic mechanism. J Physiol 508:119–29

Schmitz Y, Schmauss C, Sulzer D (2002) Altered dopamine release and uptake kinetics in mice lacking $D_2$ receptors. J Neurosci 22:8002–9

Schoffelmeer AN, Hogenboom F, Mulder AH, Ronken E, Stoof JC, Drukarch B (1994) Dopamine displays an identical apparent affinity towards functional dopamine $D_1$ and $D_2$ receptors in rat striatal slices: possible implications for the regulatory role of $D_2$ receptors. Synapse. 17:190–5

Schwartz JC, Arrang JM, Garbarg M, Korner M (1986) Properties and roles of the three subclasses of histamine receptors in brain. J Exp Biol 124:203–24

Seyfried CA, Bartoszyk GD (1994) Sensitivity of dopamine $D_2$ receptors following long-term treatment with roxindole. Eur J Pharmacol 257:67–72

Silver RB, Poonwasi KS, Seyedi N, Wilson SJ, Lovenberg TW, Levi R (2002) Decreased intracellular calcium mediates the histamine $H_3$-receptor-induced attenuation of norepinephrine exocytosis from cardiac sympathetic nerve endings. Proc Natl Acad Sci USA 99:501–6

Singer JH, Bellingham MC, Berger AJ (1996) Presynaptic inhibition of glutamatergic synaptic transmission to rat motoneurons by serotonin. J Neurophysiol 76:799–807

Siniscalchi A, Badini I, Beani L, Bianchi C (1999) 5-$HT_4$ receptor modulation of acetylcholine outflow in guinea pig brain slices. Neuroreport 10:547–51

Smits RPJM, Mulder AH (1991) Inhibitory effects of histamine on the release of serotonin and noradrenaline from rat brain slices. Neurochem Int 18:215–20

Sprouse JS, Aghajanian GK (1986) (−)-Propranolol blocks the inhibition of serotonergic dorsal raphe cell firing by 5-$HT_{1A}$ selective agonists. Eur J Pharmacol 128:295–8

Stark H, Kathmann M, Schlicker E, Schunack W, Schlegel B, Sippl W (2004) Medicinal chemical and pharmacological aspects of imidazole-containing histamine $H_3$ receptor antagonists. Mini Rev Med Chem 4:965–77

Starke K (1977) Regulation of noradrenaline release by presynaptic receptor systems. Rev Physiol Biochem Pharmacol 77:1–124

Starke K (1981) Presynaptic receptors. Annu Rev Pharmacol Toxicol 21:7–30

Starke K, Weitzell R (1978) Is histamine involved in the sympathomimetic effect of nicotine? Naunyn-Schmiedeberg's Arch Pharmacol 304:237–48

Starke K, Göthert M, Kilbinger H (1989) Modulation of neurotransmitter release by presynaptic autoreceptors. Physiol Rev 69:864–89

Stoof JC, De Boer T, Sminia P, Mulder AH (1982) Stimulation of $D_2$-dopamine receptors in rat neostriatum inhibits the release of acetylcholine and dopamine but does not affect the release of gamma-aminobutyric acid, glutamate or serotonin. Eur J Pharmacol 84:211–14

Subramony JA (2006) Apomorphine in dopaminergic therapy. Mol Pharmaceut 3:380–5

Takeshita Y, Watanabe T, Sakata T, Munakata M, Ishibashi H, Akaike N (1998) Histamine modulates high-voltage-activated calcium channels in neurons dissociated from the rat tuberomammillary nucleus. Neuroscience 87:797–805

Takeuchi S, Sugiyama Y, Mano T, Iwase S, Matsukawa T (1993) Effect of L-dopa on human muscle sympathetic nerve activity. Environ Med 37:99–102

Tanaka E, North RA (1993) Actions of 5-hydroxytryptamine on neurons of the rat cingulate cortex. J Neurophysiol 69:1749–57

Tanaka S, Ichikawa A (2006) Recent advances in molecular pharmacology of the histamine systems: immune regulatory roles of histamine produced by leukocytes. J Pharmacol Sci 101:19–23

Tang L, Todd RD, O'Malley KL (1994) Dopamine $D_2$ and $D_3$ receptors inhibit dopamine release. J Pharmacol Exp Ther 270:475–9

Thielen RJ, Engleman EA, Rodd ZA, Murphy JM, Lumeng L, Li TK, McBride WJ (2004) Ethanol drinking and deprivation alter dopaminergic and serotonergic function in the nucleus accumbens of alcohol-preferring rats. J Pharmacol Exp Ther 309:216–25

Thompson TL, Certain ME (2005) Estrogen mediated inhibition of dopamine transport in the striatum: regulation by G alpha i/o. Eur J Pharmacol 511:121–6

Threlfell S, Cragg SJ, Kallo I, Turi GF, Coen CW, Greenfield SA (2004) Histamine $H_3$ receptors inhibit serotonin release in substantia nigra pars reticulata. J Neurosci 24:8704–10

Travagli RA, Williams J T (1996) Endogenous monoamines inhibit glutamate transmission in the spinal trigeminal nucleus of the guinea pig. J Physiol 491:177–85

Trendelenburg AU, Meyer A, Klebroff W, Guimaraes S, Starke K (2003) Cross-talk between presynaptic angiotensin receptors, bradykinin receptors and $\alpha_2$-autoreceptors in sympathetic neurons: a study in $\alpha_2$-adrenoceptor-deficient mice. Br J Pharmacol 138:1389–1402

Uchida H, Takeuchi H, Suzuki T, Nomura K, Watanabe K, Kashima H (2005) Combined treatment with sulpiride and paroxetine for accelerated response in patients with major depressive disorder. J Clin Psychopharmacol 25:545–51

Umemiya M, Berger AJ (1995) Presynaptic inhibition by serotonin of glycinergic inhibitory synaptic currents in the rat brain stem. J Neurophysiol 73:1192–1201

Usiello A, Baik JH, Rouge-Pont F, Picetti R, Dierich A, LeMeur M, Piazza PV, Borrelli E (2000) Distinct functions of the two isoforms of dopamine D2 receptors. Nature 408:199–203

Vanderschuren LJ, Wardeh G, De Vries TJ, Mulder AH, Schoffelmeer AN (1999) Opposing role of dopamine $D_1$ and $D_2$ receptors in modulation of rat nucleus accumbens noradrenaline release. J Neurosci 19:4123–31

Varty LM, Gustafson E, Laverty M, Hey JA (2004) Activation of histamine $H_3$ receptors in human nasal mucosa inhibits sympathetic vasoconstriction. Eur J Pharmacol 484:83–9

Wang Z, Kai L, Day M, Ronesi J, Yin HH, Ding J, Tkatch T, Lovinger DM, Surmeier DJ (2006) Dopaminergic control of corticostriatal long-term synaptic depression in medium spiny neurons is mediated by cholinergic interneurons. Neuron 50:443–52

Weber B, Schlicker E, Sokoloff P, Stark H (2001) Identification of the dopamine autoreceptor in the guinea pig retina as $D_2$ receptor using novel subtype-selective antagonists. Br J Pharmacol 133:1243–8

Westerink BH, Cremers TI, De Vries JB, Liefers H, Tran N, De Boer P (2002) Evidence for activation of histamine $H_3$ autoreceptors during handling stress in the prefrontal cortex of the rat. Synapse 43:238–43

Wieczorek WJ, Kruk ZL (1994) A quantitative comparison on the effects of benztropine, cocaine and nomifensine on electrically evoked dopamine overflow and rate of re-uptake in the caudate putamen and nucleus accumbens in the rat brain slice. Brain Res 657:42–50

Willems JL, Buylaert WA, Lefebvre RA, Bogaert MG (1985) Neuronal dopamine receptors on autonomic ganglia and sympathetic nerves and dopamine receptors in the gastrointestinal system. Pharmacol Rev 37:165–216

Wu SY, Wang MY, Dun NJ (1991) Serotonin via presynaptic 5-$HT_1$ receptors attenuates synaptic transmission to immature rat motoneurons *in vitro*. Brain Res 554:111–21

Yamamoto Y, Mochizuki T, Okakura-Mochizuki K, Uno A, Yamatodani A (1997) Thioperamide, a histamine $H_3$ receptor antagonist, increases GABA release from the rat hypothalamus. Methods Find Exp Clin Pharmacol 19:289–98

Yan Z, Song WJ, Surmeier J (1997) $D_2$ dopamine receptors reduce N-type $Ca^{2+}$ currents in rat neostriatal cholinergic interneurons through a membrane-delimited, protein-kinase-C-insensitive pathway. J Neurophysiol 77:1003–15

Yokoo H, Goldstein M, Meller E (1988) Receptor reserve at striatal dopamine receptors modulating the release of [$^3$H] dopamine. Eur J Pharmacol 155:323–7

Yokoyama H, Onodera K, Maeyama K, Sakurai E, Iinuma K, Leurs R, Timmerman H, Watanabe T (1994) Clobenpropit (VUF-9153), a new histamine $H_3$ receptor antagonist, inhibits electrically induced convulsions in mice. Eur J Pharmacol 260:23–8

Yoshimoto R, Miyamoto Y, Shimamura K, Ishihara A, Takahashi K, Kotani H, Chen AS, Chen HY, Macneil DJ, Kanatani A, Tokita S (2006) Therapeutic potential of histamine $H_3$ receptor agonist for the treatment of obesity and diabetes mellitus. Proc Natl Acad Sci USA, in press

Young CS, Mason R, Hill SJ (1988) Studies on the mechanism of histamine-induced release of noradrenaline and 5-hydroxytryptamine from slices of rat cerebral cortex. Biochem Pharmacol 37:2799–2805

Yung KK, Bolam JP, Smith AD, Hersch SM, Ciliax BJ, Levey AI (1995) Immunocytochemical localization of $D_1$ and $D_2$ dopamine receptors in the basal ganglia of the rat: light and electron microscopy. Neuroscience 65:709–30

Yu C, Shen Y, Xu L, Zhu Y, Zhuge Z, Huang Y, Henk T, Rob L, Wei E, Chen Z (2006) Effect of endogenous histamine in the ventral hippocampus on fear memory deficits induced by scopolamine as evaluated by step-through avoidance response in rats. Physiol Behav 87:687–93

Zhang LS, Chen Z, Huang YW, Hu WW, Wei EQ, Yanai K (2003) Effects of endogenous histamine on seizure development of pentylenetetrazole-induced kindling in rats. Pharmacology 69:27–32

Zaragoza F, Stephensen H, Knudsen SM, Pridal L, Wulff BS, Rimvall K (2004) 1-alkyl-4-acylpiperazines as a new class of imidazole-free histamine $H_3$ receptor antagonists. J Med Chem 47:2833–8

# Presynaptic Adenosine and P2Y Receptors

**Jorge Gonçalves(✉) and Glória Queiroz**



**Abstract** Adenine-based purines, such as adenosine and ATP, are ubiquitous molecules that, in addition to their roles in metabolism, act as modulators of neurotransmitter release through activation of presynaptic P1 purinoceptors or adenosine receptors (activated by adenosine) and P2 receptors (activated by nucleotides). Of the latter, the P2Y receptors are G protein-coupled, whereas the P2X receptors are ligand-gated ion channels and not covered in this review.

Jorge Gonçalves
Department of Pharmacology, Faculty of Pharmacy, University of Porto, Porto, Portugal
jgoncalves@reit.up.pt

T.C. Südhof, K. Starke (eds.), *Pharmacology of Neurotransmitter Release.*
*Handbook of Experimental Pharmacology 184.*

The adenosine receptor family comprises the subtypes $A_1$, $A_{2A}$, $A_{2B}$ and $A_3$; the P2Y receptor family includes eight subtypes: $P2Y_1$, $P2Y_2$, $P2Y_4$, $P2Y_6$, $P2Y_{11}$, $P2Y_{12}$, $P2Y_{13}$ and $P2Y_{14}$. Presynaptic $A_1$ receptors and the $P2Y_{1,2,4,12}$ subtypes depress neurotransmitter release, whereas presynaptic $A_{2A}$ and $A_{2B}$ receptors facilitate neurotransmitter release. $A_3$ receptors and the other P2Y subtypes do not seem to modulate neurotransmitter release. Activation of $A_1$ or the $P2Y_{1,2,4,12}$ receptor subtypes may inhibit calcium entry through voltage-sensitive N-type calcium channels and decrease the affinity of the release apparatus for $Ca^{2+}$ with a subsequent inhibition of transmitter release. Facilitation of neurotransmitter release caused by activation of $A_2$ receptors involves an increase in cyclic AMP and/or activation of protein kinases A and C. These kinases may directly augment $Ca^{2+}$-currents or act directly at the exocytotic machinery, increasing the probability of vesicle-fusion and/or release or they may act indirectly, preventing an ongoing inhibition of voltage-sensitive $Ca^{2+}$ channels.

Different types of adenosine and P2Y receptors may be present in the same nerve terminal. The effect that prevails may be determined by the functional interaction, in the receptor biophase, between purine release, uptake and metabolism.

The adenosine modulation of synaptic transmission exceeds the simple direct control of neurotransmitter release. Adenosine may have a permissive effect on responses to other presynaptic modulators, may regulate the rate of desensitization of other presynaptic receptors, or prevent an excessive influence of inhibitory receptors. In this respect it is often viewed as a "modulator of modulators" involved in the fine-tuning of neurotransmitter release.

## 1 The Beginning—Purines and Modulation of Neurotransmitter Release

Adenine-based purines such as adenosine and ATP are ubiquitous molecules that, in addition to their roles in metabolism, function as extracellular signalling messengers. Effects of adenosine on neurotransmitter release were first described by Ginsborg and Hirst (1972), based on a study carried out in the rat phrenic nerve-diaphragm preparation. In that study adenosine reduced the quantum content of end-plate potentials and the frequency of miniature end-plate potentials, an effect interpreted as a depression of acetylcholine release. In the same preparation, ATP and ADP also depressed release, raising the hypothesis that all these substances were acting by a common mechanism (Ribeiro and Walker 1975). The role of adenosine as inhibitory modulator of neurotransmitter release was confirmed for several other cholinergic synpases, in the peripheral nervous system (PNS) as well as in the central nervous system (CNS; Harms et al. 1979; Pedata et al. 1983; Jackisch et al. 1984), and was extended to other transmitters. At the end of the 1980s, adenosine was accepted to be a general inhibitory neuromodulator, depressing release of excitatory and inhibitory neurotransmitters in the CNS and PNS by activation of

$A_1$ receptors. Subsequently, the detection of release-facilitating effects of adenosine, acting at $A_2$ receptors, and of effects of nucleotides mediated by P2 receptors complicated the picture.

# 2 Adenosine Receptor-Mediated Inhibition of Neurotransmitter Release

## *2.1 The Receptors Involved*

Effects of adenosine were initially explained by its role in intracellular signalling, as an intermediary of cyclic nucleotide production (Londos and Preston 1977). The antagonism of methylxanthines against adenosine (Ther et al. 1957) was also ascribed to an intracellular action (Sattin and Rall 1970). Later, the use of derivatives of adenosine and of theophylline covalently linked to an oligosaccharide and large enough to prevent their access to the intracellular compartment (Olsson et al. 1976) provided evidence for an involvement of cell membrane receptors in the effects of extracellular adenosine and methylxanthines. The existence of different types of membrane receptors for endogenous purines was demonstrated by differences in the order of potency of agonists, in the type of influence they exerted on cyclic AMP (cAMP) levels, and in the ability of methylxanthines to antagonize the responses (Burnstock 1978). Adenosine was the most potent endogenous agonist at receptors that were blocked by methylxanthines; therefore these were named adenosine receptors or P1 purinoceptors (Burnstock 1978). Nucleotides were more potent than adenosine on other receptors (see Section 4.1). Adenosine receptors were initially divided into $A_1$ and $A_2$ subtypes based on inhibitory and stimulatory effects on adenylate cyclase, respectively (van Calker et al. 1978). They were later further divided into $A_1$, $A_{2A}$, $A_{2B}$ and $A_3$ subtypes, each presenting distinct pharmacological properties (Table 1; see Fredholm et al. 2001).

Adenosine analogues that do not easily cross the cell membrane were shown to depress neurotransmitter release either when release was measured as postsynaptic response (mechanical: Muller and Paton 1979; Paton 1981; electrophysiological: Silinsky 1980; Ribeiro and Sebastião 1985; Dunwiddie et al. 1986) or as overflow (Fredholm et al. 1983; Jackisch et al. 1984; 1985; Feuerstein et al. 1985). The target, hence, was a membrane receptor, and the order of potency of the various agonists led to the assumption that the presynaptic inhibitory adenosine receptors were of the $A_1$ subtype. However, differences in the order of potency of agonists raised some doubts. In the neuromuscular junction of the frog (Ribeiro and Sebastião 1985) and rat (Sebastião and Ribeiro 1988), 5′-N-ethylcarboxamidoadenosine (NECA) inhibited the release of neurotransmitter more potently than 2-chloroadenosine, while the opposite was true in the heart. A retrospective analysis of the results obtained by others was also used to claim the involvement of a xanthine-sensitive receptor distinct

**Table 1** Pharmacology of Adenosine Receptor Subtypes Expressed in Mammalian Tissues

| Type | Principal Agonists | Antagonists | References |
|---|---|---|---|
| $A_1$ | CPA | DPCPX | Moos et al. 1985; Lohse et al. 1987 |
| $A_{2A}$ | CGS 21680 | ZM 241385 | Hutchison et al. 1989; Jarvis et al. 1989; Lupica et al. 1990 |
| | | SCH 58261 | Poucher et al. 1995; Ongini et al. 1999 |
| $A_{2B}$ | (*) | Enprofylline | Auchampach et al. 1997 |
| | | MRS 1754 | Feoktistov and Biaggioni 1997; Kim et al. 2000 |
| $A_3$ | AB-MECA | MRE-3008-F20 | van Galen et al. 1994 |
| | Cl-IB-MECA | VUF5574 | van Schaick et al. 1996; Varani et al. 2000; Van Muijlwijk-Koezen et al. 2000 |

AB-MECA: $N^6$-(iodo-4-amino-benzyl)-5′-N-methyl-carboxamidoadenosine; CGS 21680: 2-p- (2-carboxyethyl) phenylamino)-5′-N-carboxamidoadenosine; Cl-IB-MECA: 2-chloro-$N^6$-(3-iodobenzyl)adenosine-5′-N-methylcarboxamide; CPA: $N^6$-cyclopentyladenosine; DPCPX: 1,3-dipropyl-8-cyclopentylxanthine; MRE-3008-F20: 5-[[(4-methoxyphenyl)amino]carbonyl]amino-8-ethyl-2-(2-furyl)-pyrazolo[4,3-e]1,2,4-triazolo[1,5-c]pyrimidine; MRS 1754: 8-[4-[((4-cyanophenyl)carbamoylmethyl) oxy]phenyl]-1,3-di(n-propyl)-xanthine; SCH 58261: 7-(2-phenylethyl)-5-amino-2-(2-furyl)-pyrazolo-[4,3-e]-1,2,4-triazolo[1,5-c]pyrimidine; VUF5574: N-(2-methoxyphenyl)-N-(2(3-pyridyl)quinazolin-4-yl)urea; ZM 241385: 4-(2-[7-amino-2-(2-furyl)[1,2,4]triazolo[2,3-a][1,3,5]triazin-5-ylamino]ethyl)phenol.

(*) NECA, although nonselective, remains the most potent $A_{2B}$ agonist so far (Brackett and Daly 1994; Feoktistov and Biaggioni 1997) and thus, characterization of $A_{2B}$ receptor-mediated effects is based mainly on negative results, i.e. the failure of CGS 21680 or of CPA to reproduce NECA elicited responses (Feoktistov and Biaggioni 1997).

from the $A_1$ receptor (Ribeiro and Sebastião 1986). In addition to the discrepancies in the order of potency of the agonists, the observations obtained in some experimental models (Ribeiro et al. 1979; Dunwiddie and Fredholm 1984; Ribeiro and Sebastião 1986; Fredholm and Lindgren 1987)—that presynaptic inhibitory adenosine receptors were not negatively coupled to adenylate cyclase—were taken to support the claim that presynaptic inhibitory adenosine receptors were distinct from $A_1$. The controversy was later resolved by the demonstration that the same inhibitory receptor could couple to different intracellular targets to cause inhibition of neurotransmitter release (Fredholm and Dunwiddie 1988; Sebastião and Ribeiro 1988). The availability of selective $A_1$ adenosine receptor antagonists (Table 1) and their use in release experiments (Sebastião et al. 1990) combined with immunohistochemical and molecular biology methods (Rivkees et al. 1995; Ruiz et al. 2000; Schindler et al. 2001) provided definite confirmation that $A_1$ is the subtype involved in the inhibition of neurotransmitter release by adenosine.

## 2.2 Intracellular Events in $A_1$ Receptor-Mediated Presynaptic Inhibition

The proteins that regulate calcium entry seem to be the main targets of the transduction mechanism triggered by presynaptic $A_1$ adenosine receptors. Adenosine was shown to inhibit calcium uptake by synaptosomes (Ribeiro et al. 1979) and presynaptic calcium currents in rat hippocampal slices (Reddington et al. 1982; Schubert et al. 1986) and dorsal root ganglion cells of rat (Dolphin et al. 1986) and mouse (Gross et al. 1989). The main effector candidates are the voltage-sensitive calcium channels (VSCC), mainly of the N-type (Gross et al. 1989; Yawo and Chuhma 1993; Mynlieff and Bean 1994) but also of the P/Q-type (Ambrosio et al. 1997; Silinsky 2004).

Coupling of presynaptic $A_1$ adenosine receptors to VSCC occurs via $G_{i/o}$ proteins (Munshi et al. 1991), since the adenosine-induced inhibition of neurotransmitter release was blocked by N-ethylmaleimide (Fredholm et al. 1986; Allgaier et al. 1987; Fredholm and Lindgren 1987) and by pertussis toxin (PTX; Gross et al. 1989; Jeong and Ikeda 2000), two compounds that uncouple $G_{i/o}$ proteins from receptors (Hertting et al. 1990). Coupling seems to occur via a direct interaction between $G_{i/o}$ protein $\beta\gamma$ subunits and VSCC without requiring an intermediate component (Yasuda et al. 1996; Lim et al. 2001). Coupling of $\beta\gamma$ subunits to VSCC may be influenced by the composition of the heterotrimeric G protein. As far as N-type VSCC are concerned, coupling efficiency is higher when $\alpha_{i2}$ is the G protein $\alpha$ subunit (Jeong and Ikeda 2000).

The possibility that $A_1$ receptors influence targets downstream of calcium entry has also been proposed because, at the amphibian neuromuscular junction, adenosine inhibited acetylcholine release by a mechanism that did not depend on calcium entry into the presynaptic terminal (Hunt et al. 1994; Robitaille et al. 1999). This downstream pathway may also involve G-protein $\beta\gamma$ subunits (Blackmer et al. 2001).

Figure 1 summarizes the different transduction mechanisms. Interestingly, adenylate cyclase, a common target of $G_{i/o}$ proteins in other biological systems, does not seem to be involved in the intracellular effects triggered by activation of presynaptic $A_1$ adenosine receptors (Fredholm et al. 1986; Fredholm and Lindgren 1987).

## 2.3 Distribution of Presynaptic $A_1$ Receptors

The distribution of the $A_1$ adenosine receptor and its mRNA has been investigated by autoradiography, in situ hybridization, and immunohistochemistry in several animal species (Weber et al. 1988; Rivkees et al. 1995; Dixon et al. 1996) including humans (Fastbom et al. 1986; Glass et al. 1996; Lynge and Hellsten 2000; Schindler et al. 2001). The receptor occurs in the CNS and PNS, without marked differences between species. In the CNS, the highest densities are found in the hippocampus and dentate gyrus, cerebellum, thalamic nuclei, and neocortex (see

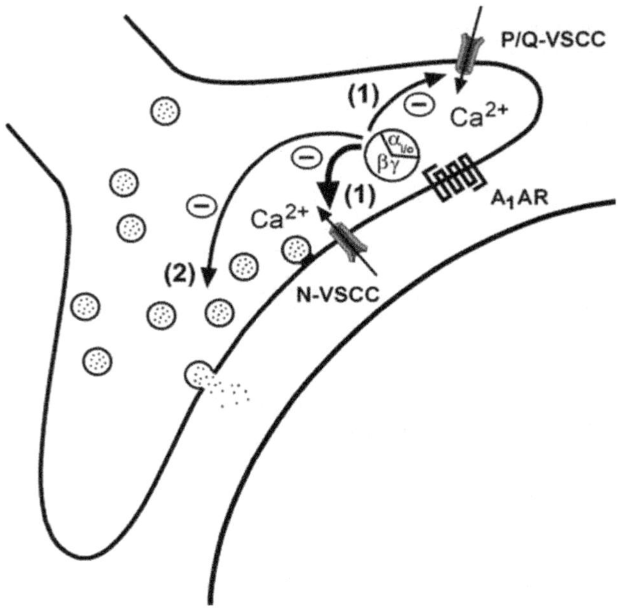


**Fig. 1** Mechanisms involved in presynaptic inhibition through $A_1$ adenosine receptors ($A_1AR$). $A_1AR$ couple to PTX-sensitive $G_{i/o}$ proteins. Two pathways are then mediated by $G_{i/o}\beta\gamma$ subunits: (1) to N-type voltage-sensitive calcium channels (VSCC) and, to a smaller degree, P/Q-type VSCC, leading to voltage -dependent inhibition, and (2) to the exocytotic machinery, leading to inhibition by a mechanism independent of calcium currents.

Fredholm et al. 2001). However, such levels of expression reflect the presence of both pre- and postsynaptic receptors. Thus, the distribution of the presynaptic $A_1$ receptors must be evaluated by functional approaches rather than be inferred from the total level of receptor expression.

Inhibition of neurotransmitter release mediated by $A_1$ adenosine receptors is a widespread phenomenon. As mentioned in Section 1, it was first described for cholinergic neurons. But presynaptic $A_1$ receptors also inhibit the release of several other neurotransmitters both in CNS and PNS. Table 2 summarizes early relevant studies. As to postganglionic parasympathetic neurons, there is only one study to our knowledge. In that study, carried out in guinea pig atria, no $A_1$ receptor-mediated modulation of acetylcholine release was found (Nakatsuka et al. 1995).

## 2.4 Endogenous Activation of Presynaptic $A_1$ Receptors

The extracellular levels of adenosine have been estimated to be approximately 30 to 500 nM, but may dramatically increase during hypoxia (Decking et al. 1997; MacLean et al. 1998; see Fredholm et al. 2001; Latini and Pedata 2001). Adenosine

**Table 2** Inhibition by Neurotransmitter Release through $A_1$ Adenosine Receptors

| Neurotransmitter | Tissue | Preparation (species; receptor) | References |
|---|---|---|---|
| *Acetylcholine* | Striatum | Slices (rat; **) | Harms et al. 1979 |
| | Hippocampus | Slices (rabbit; $A_1$) | Jackisch et al. 1984 |
| | Brain cortex | Slices (rat; **) | Pedata et al. 1983 |
| | Neuromuscular junction | Phrenic nerve-diaphragm preparation (rat; xanthine-sensitive receptor; $A_1$) | Ginsborg and Hirst 1972<br>Ribeiro and Walker 1975; Sebastião et al. 1990; Ribeiro and Sebastião 1985 |
| | | Phrenic nerve-diaphragm preparation (frog; xanthine-sensitive receptor) | |
| *Noradrenaline* | Striatum | Slices (rat; **) | Harms et al. 1979 |
| | Brain cortex | Synaptosomes (guinea pig; $A_1$) | Ebstein and Daly 1982 |
| | Hippocampus | Synaptosomes (guinea pig; $A_1$) | Ebstein and Daly 1982 |
| | | Slices (rat; $A_1$) | Jonzon and Fredholm 1984 |
| | | Slices (rabbit; $A_1$) | Jackisch et al. 1985 |
| | Postganglionic sympathetic nerve terminals | Kidney (rabbit; **) | Hedqvist and Fredholm 1976 |
| | | Subcutaneous adipose tissue (dog; **) | Hedqvist and Fredholm 1976 |
| | | Vas deferens (guinea pig; **) | Hedqvist and Fredholm 1976 |
| *Dopamine* | Striatum | Slices (rat; **) | Harms et al. 1979 |
| | | Synaptosomes (rat; **) | Michaelis et al. 1979 |
| | | Synaptosomes (guinea pig; ** $A_1$) | Ebstein and Daly 1982 |
| *Serotonin* | Hippocampus | Slices (rabbit; $A_1$) | Feuerstein et al. 1985 |
| *Glutamate* | Hippocampus | Slices (rat; **) | Corradetti et al. 1984; Dolphin and Archer 1983 |
| *Aspartate* | Hippocampus | Slices (rat; **) | Corradetti et al. 1984; Bowker and Chapman 1986 |
| *GABA* | Cerebral cortex | Slices (rat; **) | Hollins and Stone 1980 |

Receptor type according to authors' conclusions; ** receptor not identified.

activates $A_1$ receptors in concentrations ranging from 3–30 nM (Fredholm et al. 2001). Thus, the concentration of endogenous adenosine is sufficient to activate $A_1$ receptors either under physiological or pathophysiological conditions.

Three experimental observations prove tonic effects mediated by presynaptic $A_1$ adenosine receptors: effects of $A_1$ receptor antagonists, of exogenous adenosine deaminase (ADA; the enzyme that metabolizes adenosine), and of ADA inhibitors. $A_1$ receptor antagonists were shown to enhance neurotransmitter release estimated by electrophysiological responses (Dunwiddie et al. 1981; Ribeiro and Sebastião 1987) or by overflow methods (Katsuragi and Su 1982; Jackisch et al. 1984; Feuerstein et al. 1985; Gonçalves and Queiroz 1993). Incubation of tissue preparations with exogenous ADA also enhanced neurotransmitter release, irrespectively of the method used to estimate release (electrophysiology: Sebastião and Ribeiro 1985; Haas and Green 1988; overflow: Michaelis et al. 1979; Jackisch et al. 1984; Feuerstein et al. 1985; Gonçalves and Queiroz 1996). Inhibition of ADA exerted the expected opposite effect: inhibition of transmitter release (Enero and Saidman 1977).

# 3 Adenosine Receptor-Mediated Facilitation of Neurotransmitter Release

## *3.1 The Receptors Involved and Their Distribution*

The possibility that adenosine could influence neurotransmitter release by a mechanism other than the common $A_1$ receptor-mediated inhibition was raised by the observation that, when $A_1$ receptor activation was prevented, adenosine analogues enhanced release of catecholamines (brain synaptosomes: Ebstein and Daly 1982; guinea pig pulmonary artery: Wiklund et al. 1989; striatal dialysates: Zetterstrom and Fillenz 1990; rat iris: Fuder et al. 1992) and acetylcholine (rat striatal immunoaffinity-purified cholinergic synaptosomes: Brown et al. 1990). The number of studies reporting adenosine receptor-mediated facilitation increased after the discovery of the selective and high-affinity $A_{2A}$ receptor agonist CGS 21680 (Table 1). CGS 21680 facilitated acetylcholine release at the rat neuromuscular junction without the need to block $A_1$ adenosine receptors (Correia-de-Sá et al. 1991) and made it possible to demonstrate the facilitation of acetylcholine, glutamate, noradrenaline, dopamine, serotonin, glycine, taurine, and GABA release (see Table 3). Interestingly, in the prostatic portion of rat vas deferens, where neuronal $A_{2A}$ adenosine receptors seem to be absent (Diniz et al. 2003), the nonselective adenosine receptor agonist NECA, but not CGS 21680, enhanced noradrenaline release when $A_1$ receptors were blocked (Queiroz et al. 2002; Queiroz et al. 2004), an effect prevented by the $A_{2B}$ adenosine receptor antagonists enprofylline and MRS 1754 (Table 1), indicating the involvement of $A_{2B}$ receptors. This is as yet the only observation of an $A_{2B}$ receptor-mediated facilitation of neurotransmitter release, but the possibility that it may operate in other tissues that lack $A_{2A}$ adenosine receptors should be investigated as soon as more selective $A_{2B}$ adenosine receptor agonists become available.

## *3.2 Intracellular Events in $A_2$ Receptor-Mediated Presynaptic Facilitation*

The $A_{2A}$ receptors modulating neurotransmitter release are known to couple to $G_s$, causing stimulation of adenylate cyclase, an increase in cyclic AMP formation and

**Table 3** Facilitation Neurotransmitter Release Through $A_{2A}$ Adenosine Receptors

| Neurotransmitter | Tissue | Preparation (species; receptor) | References |
|---|---|---|---|
| *Acetylcholine* | Striatum | Synaptosomes (rat; $A_{2A}$) | Kirk and Richardson 1994; Gubitz et al. 1996 |
| | | *In vivo* microdialysis (rat; $A_{2A}$) | Kurokawa et al. 1996 |
| | Hippocampus | Synaptosomes (rat; $A_{2A}$) | Cunha et al. 1994a; 1994b |
| | | Slices (rat; $A_{2A}$) | Jin and Fredholm 1997 |
| | | Slices of CA3 region and dentate gyrus (rat; $A_{2A}$) | Cunha et al. 1994a |
| | Enteric nervous system | Myenteric motoneurons (rat; $A_{2A}$) | Duarte-Araujo et al. 2004 |
| | Neuromuscular junction | Phrenic nerve-diaphragm preparation (rat; $A_2$) | Correia-de-Sá et al. 1991 |
| | | Neurons innervating transversus abdominis and lumbrical muscles (mouse; $A_{2A}$) | Baxter et al. 2005 |
| *Noradrenaline* | Brain stem | Nucleus tractus solitarii (rat; $A_{2A}$) | Barraco et al. 1995 |
| | Postganglionic sympathetic nerve terminals | Epididymal portion of vas deferens (rat; $A_{2A}$) | Gonçalves and Queiroz 1993 |
| | | Tail artery (rat; $A_{2A}$) | Gonçalves and Queiroz 1996 |
| *Dopamine* | Striatum | *In vivo* microdialysis (rat; $A_2$) | Okada et al. 1996 |
| | Nucleus accumbens | *In vivo* microdialysis (rat; $A_{2A}$) | Quarta et al. 2004 |
| *Serotonin* | Brain stem | Nucleus tractus solitarii (rat; $A_{2A}$) | Helfman et al. 1996; Barraco et al. 1996 |
| *Glutamate* | Cerebral cortex | *In vivo* cortical cup technique (rat; $A_2$) | O'Regan et al. 1992 |
| | | Synaptosomes (rat; $A_{2A}$) | Marchi et al. 2002 |
| | Brain stem | Nucleus tractus solitarii (rat; $A_{2A}$) | Castillo-Melendez et al. 1994 |
| | Striatum | *In vivo* microdialysis (rat; $A_{2A}$) | Popoli et al. 1995 |
| | Hippocampus | Slices (rat; $A_{2A}$) | Lopes et al. 2002 |
| *GABA* | Striatum | Sriatopallidal slices (rat; $A_{2A}$) | Mayfield et al. 1993 |
| | | Globus pallidus (rat; $A_{2A}$) | Shindou et al. 2001 |
| | Hippocampus | Synaptosomes (rat; $A_{2A}$) | Cunha and Ribeiro 2000a |
| *Glycine* | Brain stem | Brainstem slices (rat; $A_2$) | Umemiya and Berger 1994 |
| *Taurine* | Hippocampus | *In vivo* microdialysis (rat; $A_{2A}$) | Hada et al. 1998 |

Receptor type according to authors' conclusions. CGS 21680 was the agonist used.

subsequent activation of protein kinase A (PKA; Correia-de-Sá and Ribeiro 1994c; Gubitz et al. 1996; Okada et al. 2001; Rebola et al. 2002; Oliveira and Correia-de-Sá

2005). The same holds true for the $A_{2B}$ receptors in the prostatic portion of the rat vas deferens (Queiroz et al. 2004). Possible targets of PKA are SNARE (soluble *N*-ethyl-maleimide-sensitive factor attachment protein [SNAP] receptor) proteins. PKA may enhance transmitter release by favoring the formation of the SNARE complex, increasing the readily of the releasable pool of synaptic vesicles (Nagy et al. 2004; Saitow et al. 2005). PKA may also influence $Ca^{2+}$-currents through P-VSCC. In rat hippocampus, both pathhways (to SNAREs and to P-VSCC) seem to operate, since the $A_2$ receptor-mediated enhancement of $K^+$-evoked serotonin release was attenuated by inhibiton of PKA, by botulinum toxin inhibition of synaptobrevin, a SNARE protein, and by inhibition of P-VSCC with ω-agatoxin IVA (Okada et al. 2001).

In addition to $G_s$, however, presynaptic $A_{2A}$ adenosine receptors may couple to $G_q$, causing subsequent activation of a protein kinase C (PKC; Gubitz et al. 1996; Lopes et al. 1999; Cunha and Ribeiro 2000a, 2000b; Lopes et al. 2002; Queiroz et al. 2003a). In analogy to PKA, possible targets of PKC are SNAREs and VSCC, in this case N-VSCC. PKC may regulate binding of the SNARE protein syntaxin and of synaptotagmin to N-VSCC (Vaughan et al. 1998; Shoji-Kasai et al. 2002). In rat hippocampus, the $A_2$ receptor-mediated enhancement of basal serotonin release was attenuated by PKC inhibition, botulinum toxin inhibition of syntaxin, and inhibition of N-VSCC with ω-conotoxin GVIA (Okada et al. 2001). In the epididymal portion of rat vas deferens, $A_{2A}$ receptor-mediated facilitation of noradrenaline release was prevented by PKC inhibition (Queiroz et al. 2003a). The PKC and the PKA pathways may interact as shown in rat tail artery, where the $A_{2A}$ receptor-mediated facilitation of noradrenaline release required activation of both PKC and PKA, with PKA activation occurring downstream of PKC activation (Fresco et al. 2004). In some tissues, activation of PKC enhances neurotransmitter release only in the presence of an ongoing inhibition of neurotransmitter release mediated by $G_{i/o}$ proteins. $A_{2A}$ receptor coupling to a PKC pathway seems to operate here in order to prevent excessive presynaptic inhibition (see Section 7). Figure 2 summarizes these mechanisms.

### 3.3 Endogenous Activation of Presynaptic $A_2$ Receptors

Presynaptic $A_2$ receptors may be activated by endogenous adenosine as revealed by the effects of $A_2$ receptor antagonists and exogenous ADA. For instance, in the rat tail artery, under conditions that no evidence of tonic activation of $A_1$ adenosine receptors was obtained, the nonselective $A_2$ receptor antagonist 3,7-dimethyl-1-propargylxanthine (DMPX) and exogenous ADA depressed noradrenaline release (Gonçalves and Queiroz 1996). In the rat vas deferens the effect of ADA was concentration-dependent: a low concentration depressed noradrenaline release by removing a tonic activation of adenosine $A_{2A}$ receptors, whereas a higher concentration facilitated release likely due to removal of the remaining tonic $A_1$ receptor-mediated inhibition (Gonçalves and Queiroz 1993). Similar approaches

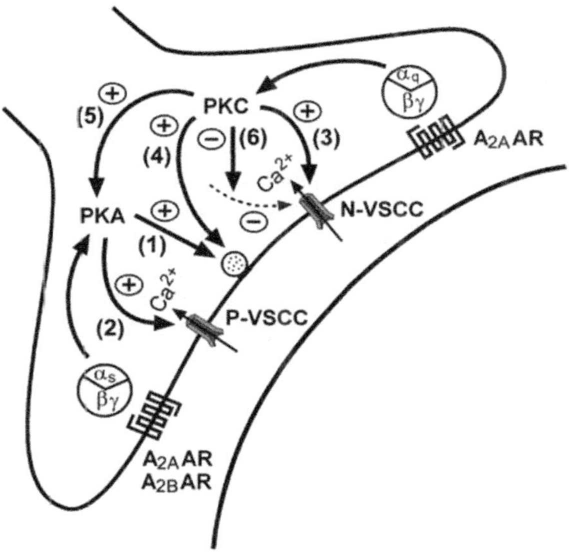


**Fig. 2** Mechanisms involved in presynaptic facilitation through $A_2$ adenosine receptors. $A_{2A}$ and $A_{2B}$ adenosine receptors ($A_{2A}$AR, $A_{2B}$AR), by coupling to $G_s$, activate adenylate cyclase and protein kinase A (PKA). This may (1) influence SNARE proteins or (2) enhance calcium currents through P-type voltage-sensitive calcium channels (P-VSCC). $A_{2A}$AR may also couple to $G_q$, leading to activation of a protein kinase C (PKC) pathway. This may (3) enhance calcium currents through N-VSCC, (4) influence SNARE proteins, (5) promote the PKA pathways or (6) remove an ongoing $G_{i/o}$ mediated inhibition of release.

provided evidence for an endogenous activation of facilitatory $A_2$ adenosine receptors in the rat neuromuscular junction (Correia-de-Sá and Ribeiro 1994a, 1994b; Correia-de-Sá et al. 1996) and hippocampus (Sebastião et al. 2000).

# 4 P2Y Receptor-Mediated Inhibition of Neurotransmitter Release

## *4.1 The Receptors Involved*

The first suggestion that ATP could act as an extracellular messenger was raised by the observation that antidromic stimulation of sensory nerves elicited ATP release (Holton and Holton 1953) and that application of ATP to various regions of the brain produced biochemical or electrophysiological changes (Phillis et al. 1975). The observation that autonomic nerves could release ATP (Su et al. 1971) raised the hypothesis that it might act as neurotransmitter or as cotransmitter (Burnstock

1976), a possibility that later was confirmed for both PNS and CNS synapses (Burnstock 2004).

At some receptors for endogenous purines, nucleotides were more potent than adenosine, and responses to ATP were not antagonized by methylxanthines. These receptors were named P2 receptors (Burnstock 1978) and later subdivided in two groups, P2X and P2Y, according to pharmacological profiles, tissue distribution, and function (Burnstock and Kennedy 1985). It was also early recognized that some P2Y receptors responded to pyrimidine as well as purine nucleotides (von Kügelgen et al. 1987; von Kügelgen and Starke 1987). Studies of transduction mechanisms showed that P2X receptors were ligand-gated cationic channels, whereas P2Y receptors were coupled to G proteins. At present, the P2Y receptor subfamily includes the eight subtypes $P2Y_{1,2,4,6,11,12,13,14}$ (Abbracchio et al. 2006). According to their sensitivity to nucleotides they fall into four groups: (1) selective purinoceptors ($P2Y_1$, $P2Y_{12}$ and $P2Y_{13}$), (2) selective pyrimidinoceptors ($P2Y_6$), (3) receptors with mixed sensitivity for purine and pyrimidine nucleotides ($P2Y_2$, $P2Y_4$, $P2Y_{11}$), and (4) receptors activated by nucleotide sugars such as UDP-glucose ($P2Y_{14}$). The $P2Y_1$, $P2Y_6$, $P2Y_{12}$ and $P2Y_{13}$ subtypes are preferentially activated by nucleoside-diphosphates, ADP selectively activating $P2Y_1$, $P2Y_{12}$ and $P2Y_{13}$ receptors, and UDP selectively activating $P2Y_6$ receptors. The $P2Y_2$, $P2Y_4$ and $P2Y_{11}$ subtypes are preferentially activated by nucleoside-triphosphates (von Kügelgen 2006).

Effects of adenine nucleotides on neurotransmitter release were first observed at peripheral cholinergic synapses, where ATP, ADP, and adenosine were equipotent in inhibiting transmission (Ribeiro and Walker 1975; Su 1983). ATP also inhibited electrically evoked overflow of acetylcholine (Akasu et al. 1982; Silinsky and Ginsborg 1983) and noradrenaline (Clanachan et al. 1977; Enero and Saidman 1977), but since these effects were attenuated by theophylline derivatives, they were generally attributed to breakdown of ATP to adenosine and activation of presynaptic $A_1$ receptors. Some doubts were raised against this hypothesis, however, because ATP metabolism is not instantaneous and complete (Shinozuka et al. 1988), and yet ATP and adenosine were equipotent in inhibiting neurotransmitter release in some tissue preparations (Enero and Saidman 1977; Verhaeghe et al. 1977). These doubts eventually led to the discovery that nucleotides such as ATP inhibited transmitter release through additional, separate presynaptic P2Y receptors (Wiklund et al. 1985; Wiklund and Gustafsson 1986; von Kügelgen et al. 1989, 1993, 1994).

In contrast to presynaptic P2Y receptors, presynaptic P2X receptors mediate facilitation of neurotransmitter release. They are covered in Boehm and Dorostkar's chapter in this *Handbook*.

The major problem in the identification of P2Y receptor-mediated inhibition of ATP on neurotransmitter release lay in the difficulty to clearly exclude breakdown to adenosine and an involvement at $A_1$ receptors. However, the demonstration that ATP acted *per se* on P2 receptors obtained a solid support when it was shown that breakdown-resistant nucleotides such as adenosine 5′-[γ-thio]triphosphate (ATPγS) and βγ-methyleneadenosine-5′-triphosphate (β,γ-MeATP) mimicked the effects of

ATP (Wiklund et al. 1985; Shinozuka et al. 1988, von Kügelgen et al. 1989, 1994) and that the inhibitory effect of ATP was enhanced by ectonucleotidase inhibitors and not changed by exogenous ADA (Wiklund et al. 1985; Wiklund and Gustafsson 1986) or adenosine uptake inhibitors (Shinozuka et al. 1988). Furthermore, the effects of adenine nucleotides were selectively attenuated by P2 receptor antagonists: in the mouse and rat vas deferens, the P2 receptor antagonist suramin preferentially antagonized the release-inhibiting effects of nucleotides without a major influence on the effect of adenosine (von Kügelgen et al. 1989, 1994), and the P2 receptor antagonist reactive blue 2 but not the selective $A_1$ adenosine receptor antagonist DPCPX abolished the effect of nucleotides (von Kügelgen et al. 1994; Queiroz et al. 2003b). Similar results were obtained in rat atrium and pancreas (von Kügelgen et al. 1995; Koch et al. 1998).

In most studies in which a P2Y receptor-mediated inhibition of neurotransmitter release was demonstrated, the subtypes involved were not clearly identified. However, according to their cellular expression, involvement of $P2Y_{11}$ and $P2Y_{14}$ receptors in the presynaptic inhibition of neurotransmitter release is very unlikely (Hussl and Boehm 2006) and inhibition of neurotransmitter release by UDP has never been observed, which excludes the $P2Y_6$ receptor. Recently, the P2Y receptors inhibiting sympathetic transmitter release in rat vas deferens were identified as $P2Y_{12}$ and/or $P2Y_{13}$ (Queiroz et al. 2003b). This identification was based (a) on the rank order of potency of nucleotides, 2-methylthioadenosine 5′-triphosphate (2-MeSATP) > 2-methylthioadenosine 5′-diphosphate (2-MeSADP) > ADP ≥ ATP, which is compatible with activation of $P2Y_1$, $P2Y_{12}$ or $P2Y_{13}$ receptors, and (b) on the finding that nucleotide inhibition was prevented by the antagonist of $P2Y_{12}$ and $P2Y_{13}$ receptors, 2-methylthioadenosine 5′-monphosphate (2-MeSAMP; Hollopeter et al. 2001; Zhang et al. 2002) but not by blockade of $P2Y_1$ receptors. Similarly, in rat cultured sympathetic neurons and the ontogenetically related PC12 cell line, $P2Y_{12}$ receptors mediated an inhibition of catecholamine release triggered by $K^+$ depolarization (Lechner et al. 2004): 2-MeSADP was more potent than ATP or ADP and the effects of nucleotides were blocked by 2-MeSAMP and the selective and competitive $P2Y_{12}$ receptor antagonist cangrelor (AR-C69931MX; Ingall et al. 1999).

In the CNS, other P2Y receptors may be involved in presynaptic modulation. In hippocampal pyramidal neurons of rats, a combination of single-cell PCR analysis, western blot analysis of a purified presynaptic active zone fraction, and immunocytochemical analysis revealed that $P2Y_1$, $P2Y_2$ and $P2Y_4$ were present in the active zone of glutamatergic terminals (Rodrigues et al. 2005). Functional studies in hippocampal synaptosomes supported the modulation of glutamate release through these subtypes: the inhibitory effect of ATP analogues was abolished by reactive blue 2 and attenuated by the selective $P2Y_1$ receptor antagonist 2′-deoxy-$N^6$-methyladenosine-3′,5′-bisphosphate (MRS 2179; Boyer et al. 1998); moreover, the agonist at $P2Y_2$ and $P2Y_4$ receptors P1-(uridine 5′-),P4-(inosine 5′)tetraphosphate (INS45973; Shaver et al. 2005) inhibited glutamate release (Rodrigues et al. 2005). In the rat medial habenula, UTP caused inhibition of glutamate release by activation of a $P2Y_2$-like receptor (Price et al. 2003).

## 4.2 Intracellular Events in P2Y Receptor-Mediated Presynaptic Inhibition

Studies on post-receptor mechanisms activated by P2Y receptors resulted in a division into two groups that differ in G protein coupling: the $P2Y_{1,2,4,6,11}$ subtypes, which predominantly couple to $G_{q/11}$ proteins ($P2Y_{11}$ receptors may additionally couple to $G_s$; Communi et al. 1997); and the $P2Y_{12,13,14}$ subtypes, that almost exclusively couple to members of the $G_{i/o}$ protein family (Abbracchio et al. 2006). It should be noted, however, that presynaptically $P2Y_{1,2,4}$ receptors couple to $G_{i/o}$ in addition to $G_{q/11}$ proteins. The $P2Y_{1,2,4,6,12,13}$ subtypes were shown to modulate the activity of various classes of neuronal $K^+$ channels (Filippov et al. 2004), transmitter-gated ion channels (Gerevich et al. 2005; Luthardt et al. 2003) and VSCC (Brown et al. 2000).

Two main pathways lead to the modulation of neurotransmitter release by inhibitory P2Y receptors, both involving modulation of VSCC. The first pathway, mediated by $G_{i/o}$ proteins, results in PTX-sensitive, voltage-dependent inhibition of VSCC by the G-protein $\beta\gamma$ subunits (Delmas et al. 1999; Dolphin 2003; mechanism 1 in Figure 3). Evidence for this pathway was initially obtained in adrenal chromaffin cells, where ATP and ADP reduced $Ca^{2+}$ currents through activation of receptors

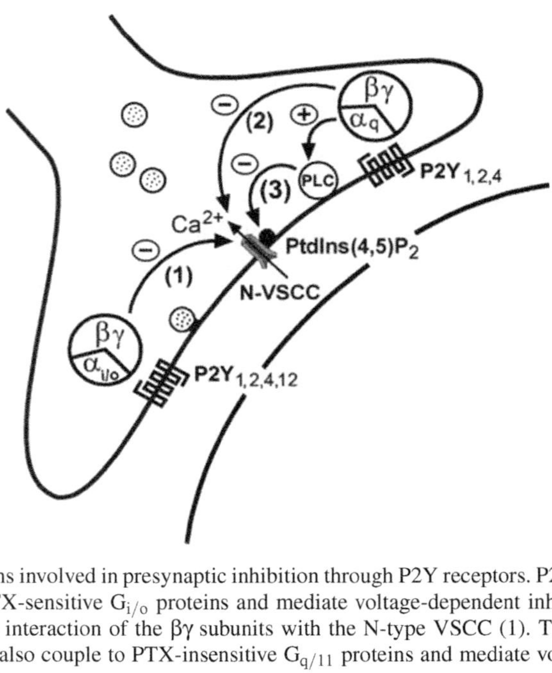


**Fig. 3** Mechanisms involved in presynaptic inhibition through P2Y receptors. $P2Y_{1,2,4,12}$ receptors may couple to PTX-sensitive $G_{i/o}$ proteins and mediate voltage-dependent inhibition of calcium currents by direct interaction of the $\beta\gamma$ subunits with the N-type VSCC (1). The $P2Y_{1,2,4}$ receptor subtypes may also couple to PTX-insensitive $G_{q/11}$ proteins and mediate voltage-independent inhibition of $Ca^{2+}$ currents by (2) direct interaction of G protein $\beta\gamma$ subunits with the channel or (3) activation of phospholipase C (PLC), causing depletion of membrane phosphatidylinositol 4,5-bisphosphate ($PtdIns(4,5)P_2$).

coupled to PTX-sensitive $G_{i/o}$ proteins (Diverse-Pierluissi et al. 1991; Lim et al. 1997), an effect later shown to be voltage-dependent and to involve an inhibition of N- and P/Q-VSCC (Currie and Fox 1996; Lim et al. 1997; Powell et al. 2000). Subsequent studies supported the involvement of this pathway in the inhibition of VSCC mediated by the $P2Y_{1,2,4}$ receptors (Rodrigues at al. 2005) that inhibit glutamate release in hippocampal neurons (Mendoza-Fernandez et al. 2000); by $P2Y_2$ receptors in hamster submandibular neurons (Abe et al. 2003); and by $P2Y_{12}$ receptors in sympathetic superior cervical ganglion neurons (Simon et al. 2002; Lechner et al. 2004) and PC12 cells (Vartian and Boehm 2001; Lechner et al. 2004).

The second pathway, mediated by $G_{q/11}$ proteins, results in PTX-resistant and voltage-independent inhibition of VSCC. In this case, both the $G\alpha_q$ and the $\beta\gamma$ subunits carry the pathway forward, the $\beta\gamma$ subunit acting directly on the VSCC, the $\alpha_q$ subunit causing a PLC$\beta$-dependent depletion of membrane phosphatidylinositol 4,5-bisphosphate as observed with other $G_{q/11}$ protein-coupled receptors (mechanisms 2 and 3 in Figure 3; Gamper et al. 2004; Gerevich et al. 2004; Lechner et al. 2005). A voltage-independent, PTX-resistant inhibition of VSCC was demonstrated in dorsal root ganglion neurons upon activation of $P2Y_1$ receptors (Gerevich et al. 2004). It was also demonstrated in sympathetic superior cervical ganglion neurons upon activation of $P2Y_{1,2,4}$ receptors, in addition to the PTX-sensitive inhibition described above (Brown et al. 2000; Filippov et al. 2003).

### *4.3 Distribution and Endogenous Activation of Presynaptic P2Y Receptors*

Presynaptic inhibition through P2Y receptors has been detected in many tissues and several species and for cholinergic, noradrenergic, dopaminergic, sertoninergic as well as vasopressin and glutamate neurons (Table 4).

Some presynaptic P2Y receptors work as autoreceptors, mediating a negative feedback mechanism to control neurotransmitter release in synapses where ATP acts as cotransmitter, since P2 receptor antagonists, when tested alone, enhanced release (Fuder and Muth 1993; von Kügelgen et al 1993). In the CNS, P2Y receptors have been shown to tonically suppress glutamatergic transmission (Zhang et al. 2003). These P2Y receptors seem to be preferentially activated by ATP released from glial cells upon activation of glutamate receptors (Queiroz et al. 1997; Zhang et al. 2003), suggesting that modulation of synaptic transmition by presynaptic inhibitory P2Y receptors may depend on astroglial activity.

## 5 Presynaptic P3 Receptors?

Presynaptic P2Y receptors are well established. So are presynaptic P2X receptors, as mentioned above (see chapter by Boehm and Dorostkar in this *Handbook*). In some studies, a third class of receptors, named P3, was proposed to mediate inhibitory

**Table 4** Inhibition of Neurotransmitter Release Through P2 Receptors

| Neurotransmitter | Tissue | Preparation (species; receptor) | References |
|---|---|---|---|
| *Acetylcholine* | Brain cortex | Slices (rat; P2Y) | Cunha et al. 1994b |
| | Ganglion | Sympathetic ganglia (frog; P2) | Silinsky and Ginsborg 1983 |
| | Neuromuscular junction | Sartorius muscle fibres (frog; P2) | Giniatullin and Sokolova 1998 |
| | | Phrenic nerve-diaphragm (mouse; P2, P2Y) | Hong and Chang 1998; De Lorenzo et al. 2006 |
| | Enteric nervous system | Ileum longitudinal muscle (guinea pig; P2) | Wiklund et al. 1985; Wiklund and Gustafsson 1986 |
| | | Submucosal neurons (guinea pig; P3) | Barajas-López et al. 1995 |
| *Noradrenaline* | Brain cortex | Slices (rat; P2Y) | von Kügelgen et al. 1994 |
| | Hippocampus | Slices (rat; P2Y) | Koch et al. 1997 |
| | Postganglionic sympathetic nerve terminals | Saphenous artery (guinea pig; P2) | Fujioka and Cheung 1987 |
| | | Tail artery (rat; P3, P2Y) | Shinozuka et al. 1988; Gonçalves and Queiroz 1996 |
| | | Mesenteric artery (rat; P3) | Shinozuka et al. 2001 |
| | | Atria (rat; P2Y) | von Kügelgen et al. 1995 |
| | | Vas deferens (mouse; P2Y) | von Kügelgen et al. 1989; Kurz et al. 1993 |
| | | Vas deferens (rat; P3, P2Y; $P2Y_{12,13}$) | Forsyth et al. 1991; von Kügelgen et al. 1993; Kurz et al. 1993; Queiroz et al. 2003b |
| | | Vas deferens (rabbit; P3) | Todorov et al. 1994 |
| | | Iris (rat; P2Y) | Fuder and Muth 1993 |
| | | Pancreas (rat; P2Y) | Koch et al. 1998 |
| | | Kidney (rat; P2Y) | Bohmann et al. 1997 |
| | Adrenals | Isolated chromaffin cells (bovine; P2Y) | Powell et al. 2000 |
| *Dopamine* | Neostriatum | Sices (rat; P2Y) | Trendelenburg and Bültmann 2000 |
| *Serotonin* | Brain cortex | Slices (rat; P2Y) | von Kügelgen et al. 1997 |
| *Vasopressin* | Hypophysis | Isolated posterior lobe (rat; P2Y) | Sperlágh et al. 1999 |
| *Glutamate* | Brain cortex | Sices (rat; P2Y) | Bennett and Boarder 2000 |
| | Hippocampus | Neuronal cultures (rat; P2Y) | Koizumi and Inoue 1997; Koizumi et al. 2003 |
| | | Slices (rat; P2Y) | Inoue et al. 1999; Zhang et al. 2003 |
| | | Pyramidal neurons (rat; P2Y, P3) | Mendoza-Fernandez et al. 2000 |
| | | Synaptossomes (rat, $P2Y_{1,2,4}$) | Rodrigues et al. 2005 |
| | Medial habenula | Slices (rat, $P2Y_2$) | Price et al. 2003 |

Receptor according to the authors' conclusions.

effects of nucleotides on neurotransmitter release in the CNS and PNS (see Table 4). They were described to be equally sensitive to ATP and adenosine, and blocked by xanthines. However, it is not clear whether the mixed P1/P2 pharmacology is due to a distinct molecular entity not yet cloned (Saitoh and Nakata 1996); to a particular association between ecto-nucleotidases and $A_1$ receptors (Sebastião et al. 1999), which might facilitate conversion of ATP into adenosine and the subsequent activation of $A_1$ receptors (see Section 6); or even to hetero-oligomerization that may occur between $A_1$ and P2Y receptors (Nakata et al. 2005).

# 6 The Diversity of the Purinergic Mechanisms in the Synaptic Cleft

The coexistence of inhibitory and facilitatory adenosine- and P2-receptors in the same nerve terminal raises the question of which factors influence the concentration of endogenous agonists in the vicinity of each receptor (i.e., in the receptor biophase) and, consequently, determine the net modulatory effect. Tissue differences in receptor expression and purine release/formation or inactivation systems create a diversity of possibilities that makes the net modulatory effect by endogenous purines quite variable from tissue to tissue.

## *6.1 Release and Inactivation of Purines*

ATP may be released by exocytosis from neurons, where it is co-stored in synaptic vesicles with other neurotransmitters (Burnstock 2004), and from astrocytes (Volknandt 2002). ATP is also released upon stimulation of membrane receptors or upon mechanical stimulation by mechanisms that mainly involve transporters of the ABC protein family or ATP-permeable anion channels (Bodin and Burnstock 2001). Under pathological conditions, massive release of ATP may occur upon damage of the cell membrane or cell lysis.

ATP is converted in the extracellular space into ADP, AMP, adenosine, and eventually into inosine by the action of a set of enzymes located at cell surfaces. These enzymes include NTPDase1, NTPDase2, NTPDase3, and NTPDase8, which are members of the ecto-nucleoside 5′-triphosphate diphosphohydrolase (E-NTPDase) family (Robson et al. 2006). They further include ecto-5′-nucleotidase (CD73; Zimmermann 1992), ecto-alkaline phosphatase (Ohkubo et al. 2000) and adenosine deaminase (ADA; Franco et al. 1986). The NTPDases1, 2, and 3 are the main E-NTPDase isoforms expressed in neurons. They metabolize tri- and diphosphonucleosides, although with different kinetics and substrate preferences: NTPDase1 hydrolyses ATP and ADP equally well and very efficiently, leading to formation of AMP; NTPDase2 hydrolyses ATP rapidly but ADP poorly, leading to accumulation of ADP. NTPDase3 hydrolyses both ATP and ADP, like NTPDase1, but less efficiently, allowing some accumulation of both ATP and ADP (Kegel et al. 1997; Kukulski et al. 2004; see also Robson et al. 2006). The chain of nucleotide degradation terminates with the hydrolysis of AMP to adenosine by ecto-5′-nucleotidase or ecto-alkaline phosphatase. Inactivation of extracellular adenosine may occur by irreversible deamination to inosine, catalyzed by ecto-ADA, a membrane-bounded form of ADA with the active site facing the extracellular space (Franco et al. 1986), or by exo-ADA, a releasable form of ADA, as recently suggested (Correia-de-Sá et al. 2006). Adenosine may also be inactivated by uptake through the concentrative nucleoside transporter family (SLC28; Baldwin et al. 1999; Gray et al. 2004) or the equilibrative nucleoside transporter family (SLC29; Baldwin et al. 1999).

The involvement of releasable enzymes in the metabolism of extracellular nucleotides has been suggested to explain an increase in the capacity to metabolize ATP observed in some tissues following nerve stimulation (Todorov et al. 1996; Mihaylova-Todorova et al. 2002). Released enzymes resembling the E-NTPDases mentioned above seem to be able to convert released ATP to AMP. In some tissues (guinea pig but not rabbit vas deferens), the released enzymes also hydrolyse AMP into adenosine (see Westfall et al. 2002).

## *6.2 Factors That Influence Purine Concentration in the Receptor Biophase*

The concentration of nucleotides in the P2 receptor biophase – including the presynaptic P2Y receptor biophase – is determined by the type of E-NTPDases present in the vicinity of the receptors (Figure 4a–4c). The nucleotide concentration in turn determines the types of P2Y receptor activated: either the dinucleoside-preferring $P2Y_{1,2,13}$ or the trinucleotides-preferring $P_{2,4}$ receptors (see Section 4.1). Because of its high efficiency for both ATP and ADP hydrolysis, NTPDase1 minimizes any nucleotide effect at P2Yreceptors (Figure 4a). Expression of NTPDase2 favors activation of the $P2Y_{1,12,13}$ diphosphonucleoside receptors (Figure 4b). Expression of NTPDase3, allowing some accumulation of both ATP and ADP, permits activation of both the diphosphonucleoside $P2Y_{1,12,13}$ and the triphosphonucleoside $P2Y_{2,4}$ receptors (Figure 4c).

Localization of ecto-5′ectonucleotidase close to NTPDases creates a very efficient system to produce adenosine from released ATP, favoring activation of adenosine receptors (Nitahara et al. 1995; Cunha et al. 1998; Sebastião et al. 1999; Duarte-Araújo et al. 2004; see Matsuoka and Ohkubo 2004). This situation seems to prevail in the rat hippocampus and may explain the inhibition of synaptic transmission caused by ATP and ATP analogues (Cunha et al. 1998).

The concentration of adenosine in the P1 receptor biophase and the balance between $A_1$ and $A_{2A}$ receptor-mediated effects is influenced by the adenosine inactivation systems (ADA and the adenosine uptake carriers; see Figure 4d). $A_1$ receptors are surface anchoring proteins for ecto-ADA, and binding of ecto-ADA to $A_1$ recep-

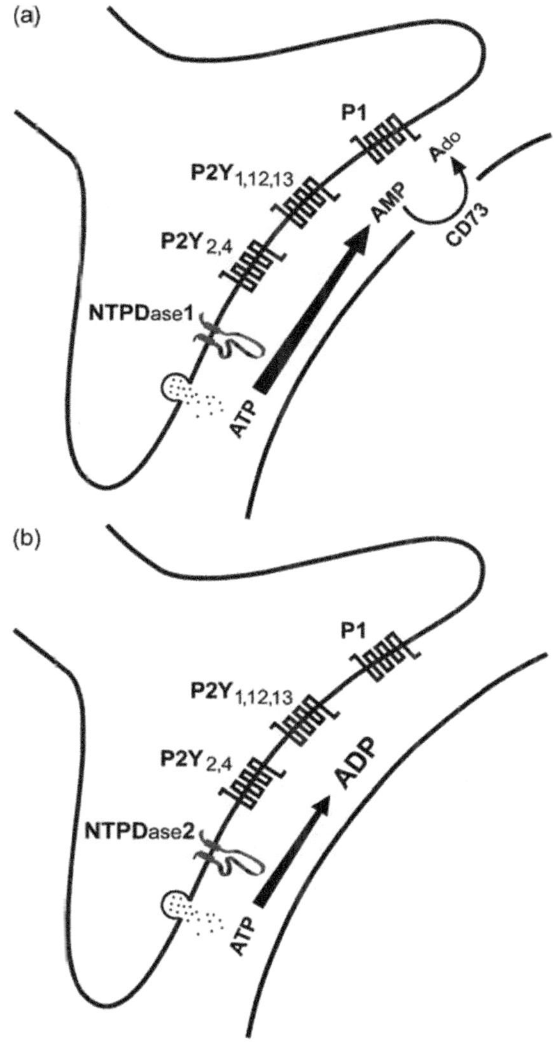


**Fig. 4** Mechanisms involved in the extracellular inactivation of nucleotides (**a**, **b** and **c**) and adenosine (**d**) and their influence on purine concentration in the P2Y and P1 receptor biophases. (**a**) NTPDase1 hydrolyses ATP and ADP very efficiently, thus preventing their action on P2Y receptors; (**b**) NTPDase2 metabolizes ATP preferentially, allowing an accumulation of ADP and thus favouring activation of $P2Y_{1,12,13}$ receptors; (**c**) NTPDase3 hydrolyses both ATP and ADP slowly, giving them time to activate both $P2Y_{2,4}$ and $P2Y_{1,12,13}$ receptors. Formation of adenosine depends on the activity of ecto 5′-nucleotidase (CD73). Adenosine inactivation systems also influence adenosine concentration in the P1 receptor biophase (**d**): the nucleoside transporters take up adenosine; adenosine deaminase (ADA) regulates both the concentration of adenosine in the $A_1$ receptor biophase and the functionality of $A_1$ receptors.

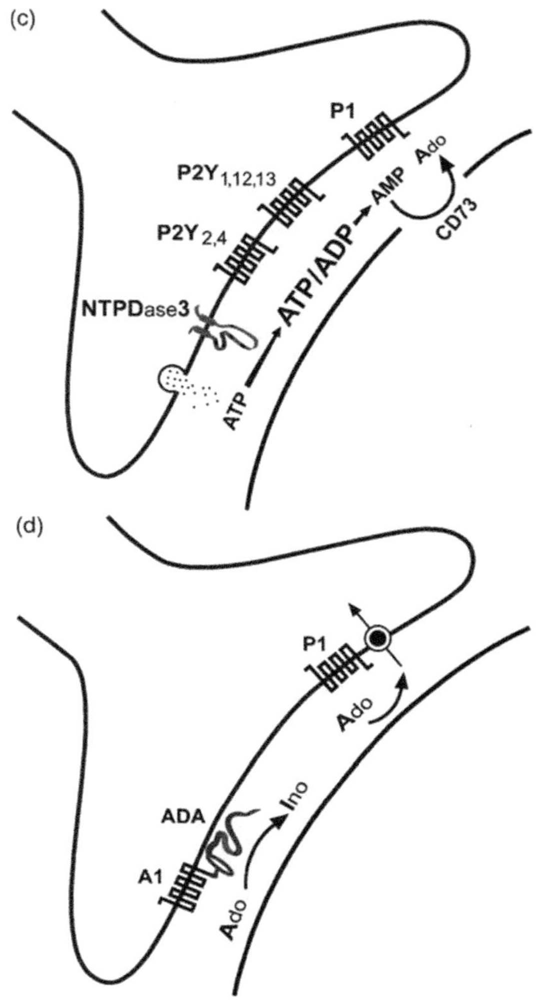


**Fig. 4** Continued.

tors increases the affinity and efficacy of $A_1$ receptor agonists (Ciruela et al. 1996; Ruiz et al. 2000) and prevents $A_1$ receptor desensitization (Saura et al. 1998) and internalization (Ginés et al. 2001). Adenosine uptake carriers also influence the adenosine concentration in the receptor biophase. In rat tail artery, inhibition of adenosine uptake decreased transmitter release (Gonçalves and Queiroz 1996), suggesting that $A_1$ adenosine receptors are located close to the uptake carriers. However, this is not a general rule, since in other tissues inhibition of adenosine uptake facilitated neurotransmitter release (rat vas deferens: Gonçalves and Queiroz 1993; Queiroz et al. 2004; rat myenteric plexus: Duarte-Araújo et al. 2004), indicating that the functional interaction between adenosine uptake carriers and presynaptic adenosine receptors may vary from tissue to tissue.

## 7 Adenosine and the Fine-Tuning of Transmitter Release

The adenosine receptor-mediated modulation of synaptic transmission exceeds the direct control of neurotransmitter release, and the concept of adenosine as a "modulator of modulators" has been developed (Sebastião and Ribeiro 2000), highlighting the potential of adenosine to fine-tune synaptic activity. Fine-tuning of transmitter release by adenosine may be exerted by (1) triggering responses to other modulators, (2) regulating the rate of desensitization of other receptors, (3) forming hetero-oligomers with other receptor types and/or (4) cross-talk of intracellular pathways they activate.

Adenosine has a permissive effect on responses to the peptides calcitonin gene-related peptide (CGRP; Correia-de-Sá and Ribeiro 1994a; Sebastião et al. 2000), vasoactive intestinal peptide (VIP; Correia-de-Sá et al. 2001), and brain derived neurotrophic factor (BDNF; Diógenes et al. 2004): the modulatory actions of these peptides depend strictly on an ongoing activation of facilitatory $A_{2A}$ receptors. Receptors for CGRP and VIP are often positively coupled to adenylate cyclase (Rosenberg and Li 1995), and the permissive effect of $A_{2A}$ receptors on CGRP and VIP responses has been ascribed to a synergic effect on adenylate cyclase (Sebastião et al. 2000).

A second mechanism by which adenosine receptors may indirectly influence release is by regulating the rate of desensitization of other receptors. For example, at rat phrenic motor nerve terminals, $A_1$ and $A_{2A}$ receptors inhibit and facilitate, respectively, acetylcholine release (Correia-de-Sá et al. 1991; Oliveira and Correia-de-Sá 2005). Additionally, they modulate nicotinic function, with $A_{2A}$ receptors accelerating and $A_1$ receptors slowing the rate of desensitization of the autofacilitatory nicotinic receptors (Correia-de-Sá and Ribeiro 1994b; Timóteo et al. 2003; Duarte-Araújo et al. 2004), which influences the global effects of adenosine on transmitter release.

Adenosine may also coactivate other release-modulatory receptors, an influence tentatively explained by receptor hetero-oligomerization and/or by a cross-talk between the intracellular pathways to which adenosine receptors and the interacting receptors are coupled. Hetero-oligomerization of adenosine receptors has been demonstrated for the pairs $A_1/P2Y_1$ (Yoshioka et al. 2002), $A_1$/mGluR1 (Ciruela et al. 2001), $A_{2A}$/mGluR5 (Ferré et al. 2002), $A_1/D_1$ (Ginés et al. 2000), $A_{2A}/D_2$ (Hillion et al. 2002), and $A_1/A_{2A}$ receptors (Ciruela et al. 2006). The receptor complexes formed have ligand binding and signalling properties distinct from their constituent receptors, but so far no evidence supports their presynaptic localization.

In most cases, mutual influences between presynaptic receptors have been explained by a cross-talk of intracellular signalling pathways. In several synapses, presynaptic actions of $A_1$ receptors and other release-inhibitory receptors are mutually occlusive (Schlicker and Göthert 1998). For example, in central (rabbit brain cortex: Limberger et al. 1988; rabbit hippocampus: Allgaier et al. 1991) and peripheral noradrenergic neurons (guinea pig atrium: Nakatsuka et al. 1995; rat tail artery: Bucher et al. 1992) activation of $\alpha_2$-autoreceptors attenuated the $A_1$ receptor-mediated inhibition of noradrenaline release. This kind of interaction has been explained by competition for the same intracellular pathway (e.g., N-type VSCC and $K^+$-channels; Schlicker and Göthert 1998). $A_1$ adenosine receptors and other release-inhibitory receptors, such as the $\alpha_2$-adrenoceptors, are coupled to $G_{i/o}$ proteins and the inhibition of transmitter release they cause is mainly a consequence of an inhibition of VSCC by $G_{i/o}$ protein $\beta\gamma$ subunits (Delmas et al. 1999; Lim et al. 2001; Dolphin 2003). Accordingly, the simultaneous activation of other inhibitory receptors may reduce the coupling of $A_1$ receptors to the $G_{i/o}$ proteins (less available) and the subsequent inhibition of neurotransmitter release.

Cross-talk between facilitatory $A_2$ receptors and presynaptic inhibitory receptors has also been described. In rat sympathetic neurons, facilitation of noradrenaline release mediated by $A_{2A}$ receptors (Fresco et al. 2002; Queiroz et al. 2003a) or by $A_{2B}$ receptors (Talaia et al. 2005) is enhanced by ongoing activation of inhibitory receptors, such as the $\alpha_2$-adrenoceptors. In rat hippocampus, $A_{2A}$ receptor-mediated facilitation of glutamate release (O'Kane and Stone 1998; Lopes et al. 1999, 2002) was shown to depend on coactivation of inhibitory $A_1$ receptors. The involvement of a diffusible second messenger in the cross-talk between $A_1$ and $A_{2A}$ receptors has been suggested based on the observation that activation of $A_{2A}$ receptors decreased $A_1$ receptor binding in hippocampal (Lopes et al. 1999) and striatal (Dixon et al. 1997) synaptosomes, but not in membrane fragments. Furthermore, in synapses where cross-talk between presynaptic $A_{2A}$ and inhibitory receptors was observed, $A_{2A}$ receptors were shown to couple to a PKC pathway (Dixon et al. 1997; Lopes et al. 2002; Queiroz et al. 2003a; see Figure 2) and facilitation of neurotransmitter release caused by direct activation of PKC with phorbol esters was also attenuated when activation of release-inhibitory receptors was prevented (Fredholm and Lindgren 1988; Queiroz et al. 2003a). Similar interactions have been reported in sympathetic neurons between release-inhibitory receptors and facilitatory angiotensin $AT_1$ (Cox et al. 2000; Trendelenburg et al. 2003; Mota and Guimarães 2003; Talaia et al. 2006) and bradykinin $B_2$ receptors (Cox et al. 2000), which also couple to the $G_{q/11}$-PLC-PKC pathway. A PKC-mediated disruption of the $G_{i/o}$ pathway to which the inhibitory receptors are coupled was the main mechanism proposed to explain interactions between presynaptic release-inhibitory and facilitatory receptors (Cox et al. 2000; Talaia et al. 2006), including $A_{2A}$ receptors (Queiroz et al. 2003a). Such disruption may occur by phosphorylation of the $G_{i/o}$ proteins (Katada et al. 1985), the receptors (Liang et al. 1998; Dixon et al. 1997; Lopes et al. 1999), or the N-type of VSCC they regulate (Hamid et al. 1999; see Figure 2, mechanism 6).

Interactions between receptors do not follow a constant pattern and involve multiple mechanisms (Kubista and Boehm 2006) that depend on the cellular organization of the interacting receptors, of the proteins involved in intracellular signalling pathways, and of the molecular targets they regulate (Hur and Kim 2002). The release-inhibitory receptors can be viewed as a local homeostatic mechanism to control the amount of neurotransmitter at the synaptic cleft. The physiological meaning of facilitatory receptors is less clear. One possibility is that they may select the more convenient cotransmitter combination (Gonçalves et al. 1996). The requirement of an ongoing activation of inhibitory receptors to observe facilitation of neurotransmitter release by $A_2$ receptors and other presynaptic facilitatory receptors supports the proposal that, in certain synapses, they may operate to prevent excessive inhibition, contributing to a fine-tuning of transmitter release.

**Acknowledgements** The work at the authors' laboratory is supported by the Fundação da Ciência e Tecnologia (FCT) Project POCI/SAU-FCF/60714/2004 and REQUIMTE, Centro de Química da Universidade do Porto (CEQUP).

## References

Abbracchio MP, Burnstock G, Boeynaems JM et al (2006) International Union of Pharmacology LVIII: update on the P2Y G protein-coupled nucleotide receptors: from molecular mechanisms and pathophysiology to therapy. Pharmacol Rev 58:281–341

Abe M, Endoh T, Suzuki T (2003) Extracellular ATP-induced calcium channel inhibition mediated by P1/P2Y purinoceptors in hamster submandibular ganglion neurons. Br J Pharmacol 138:1535–43

Akasu T, Hirai K, Koketsu K (1982) Modulatory effect of ATP on the release of acetylcholine from presynaptic nerve terminals in bullfrog sympathetic ganglia. Kurume Med J 29:75–83

Allgaier C, Hertting G, Kügelgen OV (1987) The adenosine receptor-mediated inhibition of noradrenaline release possibly involves an N-protein and is increased by $\alpha_2$-autoreceptor blockade. Br J Pharmacol 90:403–12

Allgaier C, Greber R, Hertting G (1991) Studies on the interaction between presynaptic $\alpha_2$-adrenoceptors and adenosine $A_1$ receptors located on noradrenergic nerve terminals. Naunyn-Schmiedberg's Arch Pharmacol 344:187–92

Ambrosio AF, Malva JO, Carvalho AP et al (1997) Inhibition of N-,P/Q- and other types of $Ca^{2+}$ channels in rat hippocampal nerve terminals by the adenosine $A_1$ receptor. Eur J Pharmacol 340:301–10

Auchampach JA, Jin X, Wan TC et al (1997) Canine mast cell adenosine receptors: cloning and expression of the $A_3$ receptor and evidence that degranulation is mediated by the $A_{2B}$ receptor. Mol Pharmacol 52:846–60

Baldwin SA, Mackey JR, Cass CE et al (1999) Nucleoside transporters: molecular biology and implications for therapeutic development. Mol Med Today 5:216–24

Barajas-Lopez C, Müller MJ, Prieto-Gomez B et al (1995) ATP inhibits the synaptic release of acetylcholine in submucosal neurons. J Pharmacol Exp Ther 274:1238–45

Barraco RA, Clough-Helfman C, Goodwin BP et al (1995) Evidence for presynaptic adenosine $A_{2a}$ receptors associated with norepinephrine release and their desensitization in the rat nucleus tractus solitarius. J Neurochem 65:1604–11

Barraco RA, Helfman CC, Anderson GF (1996) Augmented release of serotonin by adenosine $A_{2a}$ receptor activation and desensitization by CGS 21680 in the rat nucleus tractus solitarius. Brain Res 733:155–61

Baxter RL, Vega-Riveroll LJ, Deuchars J et al (2005) $A_{2A}$ adenosine receptors are located on presynaptic motor nerve terminals in the mouse. Synapse 57:229–34

Bennett GC, Boarder MR (2000) The effect of nucleotides and adenosine on stimulus-evoked glutamate release from rat brain cortical slices. Br J Pharmacol 131:617–23

Blackmer T, Larsen EC, Takahashi M et al (2001) G protein βγ subunit-mediated presynaptic inhibition: regulation of exocytotic fusion downstream $Ca^{2+}$ entry. Science 292:293–7

Bodin P, Burnstock G (2001) Purinergic signalling: ATP release. Neurochem Res 26:959–69

Bohmann C, von Kügelgen I, Rump LC (1997) P2-receptor modulation of noradrenergic neurotransmission in rat kidney. Br J Pharmacol 121:1255–62

Bowker HM, Chapman AG (1986) Adenosine analogues. The temperature-dependence of the anticonvulsant effect and inhibition of $^3$H-D-aspartate release. Biochem Pharmacol 35:2949–53

Boyer JL, Mohanram A, Camaioni E et al (1998) Competitive and selective antagonism of $P2Y_1$ receptors by $N^6$-methyl 2′-deoxyadenosine 3′,5′-biphosphate. Br J Pharmacol 124:1–3

Brackett LE, Daly JW (1994) Functional characterization of the $A_{2b}$ adenosine receptor in NIH 3T3 fibroblasts. Biochem Pharmacol 47:801–14

Brown DA, Pilippov AK, Branard EA (2000) Inhibition of potassium and calcium currents in neurons by molecularly-defined P2Y receptors. J Auton Nerv Sys 81:31–6

Brown SJ, James S, Reddington M et al (1990) Both $A_1$ and $A_{2a}$ purine receptors regulate striatal acetylcholine release. J Neurochem 55:31–8

Bucher B, Corriu C, Stoclet JC (1992) Prejunctional opioid μ-receptors and adenosine $A_1$-receptors on the sympathetic nerve endings of the rat-tail artery interact with the $\alpha_2$-adrenoceptors. Naunyn-Schmiedberg's Arch Pharmacol 345:37–43

Burnstock G (1976) Do some nerve cells release more than one transmitter? Neuroscience 1:239–48

Burnstock G (1978) A basis for distinguishing two types of purinergic receptor. In: Bolis L, Straub RW (eds) Cell Membrane Receptors for Drugs and Hormones. Raven Press, New York, pp 107–18

Burnstock G (2004) Cotransmission. Curr Opin Pharmacol 4:47–52

Burnstock G, Kennedy C (1985) Is there a basis for distinguishing two types of P2-purinoceptor? Gen Pharmacol 16:433–40

Castillo-Melendez M, Krstew E, Lawrence AJ et al (1994) Presynaptic adenosine $A_{2a}$ receptors on soma and central terminals of rat vagal afferent neurons. Brain Res 652:137–44

Ciruela F, Saura C, Canela EI et al (1996) Adenosine deaminase affects ligand-induced signalling by interacting with cell surface adenosine receptors. FEBS Lett 380:219–23

Ciruela F, Escriche M, Burgueño J et al (2001) Metabotropic glutamate 1α and adenosine $A_1$ receptors assemble into functionally interacting complexes. J Biol Chem 276:18345–51

Ciruela F, Casadó V, Rodrigues RJ et al (2006) Presynaptic control of striatal glutamatergic neurotransmission by adenosine $A_1$-$A_{2A}$ receptor heteromers. J Neurosci 26:2080–7

Clanachan AS, Johns A, Paton D M (1977) Presynaptic inhibitory actions of adenine nucleotides and adenosine on the neurotransmission in the rat vas deferens. Neuroscience 2:597–602

Communi D, Govaerts C, Parmentier M et al (1997) Cloning of a human purinergic P2Y receptor coupled to phospholipase C and adenylyl cyclase. J Biol Chem 272:31969–73

Corradetti R, Lo Conte G, Moroni F et al (1984) Adenosine decreases aspartate and glutamate release from rat hippocampal slices. Eur J Pharmacol 104:19–26

Correia-de-Sá P, Sebastião AM, Ribeiro JA (1991) Inhibitory and excitatory effects of adenosine receptor agonists on evoked transmitter release from phrenic nerve ending of the rat. Br J Pharmacol 103:1614–20

Correia-de-Sá P, Ribeiro JA (1994a) Potentiation by tonic $A_{2a}$-adenosine receptor activation of CGRP-facilitated [$^3$H]-ACh release from rat motor nerve endings. Br J Pharmacol 111:582–8

Correia-de-Sá P, Ribeiro JA (1994b) Tonic adenosine $A_{2A}$ receptor activation modulates nicotinic autoreceptor function at the rat neuromuscular junction. Eur J Pharmacol 271:349–55

Correia-de-Sá P, Ribeiro JA (1994c) Evidence that presynaptic $A_{2A}$-adenosine receptor of the motor nerve endings is positively coupled to adenylate cyclase. Naunyn-Schmiedberg's Arch Pharmacol 350:514–22

Correia-de-Sá P, Timóteo MA, Ribeiro JA (1996) Presynaptic $A_1$ inhibitory/$A_{2A}$ facilitatory adenosine receptor activation balance depends on motor nerve stimulation paradigm at the rat hemidiaphragm. J Neurophysiol 76:3910–19

Correia-de-Sá P, Timóteo MA, Ribeiro JA (2001) Synergism between $A_{2A}$-adenosine receptor activation and vasoactive intestinal peptide to facilitate [$^3$H]-acetylcholine release from the rat motor nerve terminals. Neurosci Lett 309:101–4

Correia-de-Sá P, Adaes S, Timóteo MA et al (2006) Fine-tuning modulation of myenteric motoneurons by endogenous adenosine: on the role of secreted adenosine deaminase. Auton Neurosci 126–127:211–24

Cox SL, Schelb V, Trendelenburg AU et al (2000) Enhancement of noradrenaline release by angiotensin II and bradykinin in mouse atria: evidence for cross talk between Gq/11 protein- and Gi/o protein-coupled receptors. Br J Pharmacol 129:1095–1102

Cunha RA, Milusheva E, Vizi ES et al (1994a) Excitatory and inhibitory effects of $A_1$ and $A_{2A}$ adenosine receptor activation on the electrically evoked [$^3$H]acetylcholine release from different areas of the rat hippocampus. J Neurochem 63:207–14

Cunha RA, Ribeiro JA, Sebastião AM (1994b) Purinergic modulation of the evoked release of [$^3$H] acetylcholine from the hippocampus and cerebral cortex of the rat: role of the ectonucleotidases. Eur J Neurosci 6:33–42

Cunha RA, Sebastião AM, Ribeiro JA (1998) Inhibition by ATP of hippocampal synaptic transmission requires localized extracellular catabolism by ecto-nucleotidases into adenosine and channelling to adenosine $A_1$ receptors. J Neurosci 18:1987–95

Cunha RA, Ribeiro JA (2000a) Purinergic modulation of [$^3$H]GABA release from rat hippocampal nerve terminals. Neuropharmacology 39:1156–67

Cunha RA, Ribeiro JA (2000b) Adenosine $A_{2A}$ receptor facilitation of synaptic transmission in the CA1 area of the rat hippocampus requires protein kinase C but not protein kinase A activation. Neurosci Lett 289:127–30

Currie KP, Fox AP (1996) ATP serves as a negative feedback inhibitor of voltage-gated $Ca^{2+}$ channel currents in cultured bovine adrenal chromaffin cells. Neuron 16:1027–36

Decking UKM, Schlieper G, Kroll K, Schrader J (1997) Hypoxia-induced inhibition of adenosine kinase potentiates cardiac adenosine release. Circ Res 81:154–64

De Lorenzo S, Veggetti M, Muchnik S et al (2006) Presynaptic inhibition of spontaneous acetylcholine release mediated by P2Y receptors at the mouse neuromuscular junction. Neuroscience. 142:71–85

Delmas P, Abogadie FC, Milligan G et al (1999) βγ Dimers derived from Go and Gi proteins contribute different components of adrenergic inhibition of $Ca^{2+}$ channels in rat sympathetic neurons. J Physiol 518:23–36

Diniz C, Leal S, Gonçalves J (2003) Regional differences in the adenosine $A_2$ receptor-mediated modulation of contractions in rat vas deferens. Eur J Pharmacol 460:191–9

Diógenes MJ, Fernandes CC, Sebastião AM et al (2004) Activation of adenosine $A_{2A}$ receptor facilitates brain-derived neurotrophic factor modulation of synaptic transmission in hippocampal slices. J Neurosci 24:2905–13

Diverse-Pierluissi M, Dunlap K, Westhead EW (1991) Multiple actions of extracellular ATP on calcium currents in cultured bovine chromaffin cells. Proc Natl Acad Sci USA 88:1261–5

Dixon AK, Gubitz AK, Sirinathsinghji DJ et al (1996) Tissue distribution of adenosine receptor mRNAs in the rat. Br J Pharmacol 118:1461–8

Dixon AK, Widdowson L, Richardson PJ (1997) Desensitization of the adenosine $A_1$ receptor by the $A_{2A}$ receptor in the rat striatum. J Neurochem 69:315–21

Dolphin AC (2003) G protein modulation of voltage-gated calcium channels. Pharmacol Rev 55:607–27

Dolphin AC, Archer ER (1983) An adenosine agonist inhibits and a cyclic AMP analogue enhances the release of glutamate but not GABA from slices of rat dentate gyrus. Neurosci Lett 43:49–54

Dolphin AC, Forda SR, Scott RH (1986) Calcium-dependent currents in cultured rat dorsal root ganglion neurones are inhibited by an adenosine analogue. J Physiol 373:47–61

Duarte-Araújo M, Nascimento C, Timóteo MA et al (2004) Dual effects of adenosine on acetylcholine release from myenteric motoneurons are mediated by junctional facilitatory $A_{2A}$ and extrajunctional inhibitory $A_1$ receptors. Br J Pharmacol 141:925–34

Dunwiddie TV, Fredholm BB (1984) Adenosine receptors mediating inhibitory electrophysiological responses in rat hippocampus are different from receptors mediating cyclic AMP accumulation. Naunyn Schmiedeberg's Arch Pharmacol 326:294–301
Dunwiddie TV, Hoffer BJ, Fredholm BB (1981) Alkylxanthines elevate hippocampal excitability. Evidence for a role of endogenous adenosine. Naunyn-Schmiedeberg's Arch Pharmacol 316:326–30
Dunwiddie TV, Worth TS, Olsson RA (1986) Adenosine analogs mediating depressant effects on synaptic transmission in rat hippocampus: structure-activity relationships for the N6 subregion. Naunyn Schmiedeberg's Arch Pharmacol 334:77–85
Ebstein RP, Daly JW (1982) Release of norepinephrine and dopamine from brain vesicular preparations: effects of adenosine analogues. Cell Mol Neurobiol 2:193–204
Enero MA, Saidman BQ (1977) Possible feed-back inhibition of noradrenaline release by purine compounds. Naunyn-Schmiedeberg's Arch Pharmacol 297:39–46
Fastbom J, Pazos A, Probst A et al (1986) Adenosine $A_1$-receptors in human brain: characterization and autoradiographic visualization. Neurosci Lett 65:127–32
Feoktistov I, Biaggioni I (1997) Adenosine $A_{2B}$ receptors. Pharmacol Rev 49:381–402
Ferré S, Karcz-Kubicha M, Hope BT et al (2002) Synergistic interaction between adenosine $A_{2A}$ and glutamate mGlu5 receptors: Implications for striatal neuronal function. Proc Natl Acad Sci USA 99:11940–5
Feuerstein TJ, Hertting G, Jackisch R (1985) Modulation of hippocampal serotonin (5-HT) release by endogenous adenosine. Eur J Pharmacol 107:233–42
Filippov AK, Simon J, Barnard EA et al (2003) Coupling of the nucleotide $P2Y_4$ receptor to neuronal ion channels. Br J Pharmacol 138:400–6
Filippov AK, Fernandez-Fernandez JM, Marsh SJ et al (2004) Activation and inhibition of neuronal G protein-gated inwardly rectifying $K^+$ channels by P2Y nucleotide receptors. Mol Pharmacol 66:468–77
Forsyth KM, Bjur RA, Westfall DP (1991) Nucleotide modulation of norepinephrine release from sympathetic nerves in the rat vas deferens. J Pharmacol Exp Ther 1256:821–6
Franco R, Canela EI, Bozal J (1986) Enzymes of the purine metabolism in rat brain microsomes. Neurochem Res 11:407–22
Fredholm BB, Dunwiddie TV (1988) How does adenosine inhibit transmitter release? Trends Pharmacol Sci 9:130–4
Fredholm BB, Lindgren E (1987) Effects of N-ethylmaleimide and forskolin on noradrenaline release from rat hippocampal slices. Evidence that prejunctional adenosine and alpha-receptors are linked to N-proteins but not to adenylate cyclase. Acta Physiol Scand 130:95–105
Fredholm BB, Lindgren E (1988) Protein kinase C activation increases noradrenaline release from rat hippocampus and modifies the inhibitory effect of $\alpha_2$-adrenoceptor and adenosine $A_1$-receptor agonists. Naunyn Schmiedeberg's Arch Pharmacol 337:477–83
Fredholm BB, Jonzon B, Lindgren E (1983) Inhibition of noradrenaline release from hippocampal slices by a stable adenosine analogue. Acta Physiol Scand Suppl 515:7–10
Fredholm BB, Fastbom J, Lindgren E (1986) Effects of N-ethylmaleimide and forskolin on glutamate release from rat hippocampal slices. Evidence that prejunctional adenosine receptors are linked to N-proteins, but not to adenylate cyclase. Acta Physiol Scand 127:381–6
Fredholm BB, Ijzerman AP, Jacobson KA et al (2001) Nomenclature and classification of adenosine receptors. Pharmacol Rev 53: 527–52
Fresco P, Diniz C, Queiroz G et al (2002) Release inhibitory receptors activation favours the $A_{2A}$-adenosine receptor-mediated facilitation of noradrenaline release in isolated rat-tail artery. Br J Pharmacol 136:230–6
Fresco P, Diniz C, Gonçalves J (2004) Facilitation of noradrenaline release by activation of adenosine $A_{2A}$ receptors triggers both phospholipase C and adenylate cyclase pathways in rat tail artery. Cardiovasc Res 63:739–46
Fuder H, Muth U (1993) ATP and endogenous agonists inhibit evoked [$^3$H]-noradrenaline release in rat iris via $A_1$ and P2Y-like purinoceptors. Naunyn-Schmiedeberg's Arch Pharmacol 348:352–7

Fuder H, Brink A, Meincke M et al (1992) Purinoceptor-mediated modulation by endogenous and exogenous agonists of stimulation-evoked [$^3$H]-noradrenaline release on rat iris. Naunyn Schmiedeberg's Arch Pharmacol 345:417–23

Fujioka M, Cheung DW (1987) Autoregulation of neuromuscular transmission in the guinea-pig saphenous artery. Eur J Pharmacol 139:147–53

Gamper N, Reznikov V, Yamada Y et al (2004) Phosphatidylinositol 4,5-bisphosphate signals underlie receptor-specific Gq/11-mediated modulation of N-type $Ca^{2+}$ channels. J Neurosci 24:10980–992

Gerevich Z, Borvendeg SJ, Schroeder W et al (2004) Inhibition of N type voltage-activated calcium channels in rat dorsal root ganglion neurons by P2Y receptors is a possible mechanism of ADP-induced analgesia. J Neurosci 24:797–807

Gerevich Z, Muller C, Illes P (2005) Metabotropic $P2Y_1$ receptors inhibit $P2X_3$ receptor-channels in rat dorsal ganglion neurons. Eur J Pharmacol 521:34–38

Ginés S, Hillion J, Torvinen M et al (2000) Dopamine $D_1$ and adenosine $A_1$ receptors from functionally interacting heteromeric complexes. Proc Natl Acad Sci USA 97:8606–11

Ginés S, Ciruela F, Burgueño J et al (2001) Involvement of caveolin in ligand-induced recruitment and internalization of $A_1$ adenosine receptor and adenosine deaminase in an epithelial cell line. Mol Pharmacol 59:1314–23

Giniatullin RA, Sokolova EM (1998) ATP and adenosine inhibit transmitter release at the frog neuromuscular junction through distinct presynaptic receptors. Br J Pharmacol 124:839–44

Ginsborg BL, Hirst GD (1972) The effect of adenosine on the release of the transmitter from the phrenic nerve of the rat. J Physiol 224:629–45

Glass M, Faull RL, Dragunow M (1996) Localisation of the adenosine uptake site in the human brain: a comparison with the distribution of adenosine $A_1$ receptors. Brain Res 710:79–91

Gonçalves J, Queiroz G (1993) Facilitatory and inhibitory modulation by endogenous adenosine of noradrenaline release in the epididymal portion of rat vas deferens. Naunyn-Schmiedeberg's Arch Pharmacol 348:367–71

Gonçalves J, Queiroz G (1996) Purinoceptor modulation of noradrenaline release in rat tail artery: tonic modulation mediated by inhibitory P2Y- and facilitatory $A_{2A}$-purinoceptors. Br J Pharmacol 117:156–60

Gonçalves J, Bültmann R, Driessen B (1996) Opposite modulation of cotransmitter release in guinea-pig vas deferens: increase of noradrenaline and decrease of ATP release by activation of prejunctional β-adrenoceptors. Naunyn-Schmiedeberg's Arch Pharmacol 353:184–92

Gray JH, Owen RP, Giacomini KM (2004) The concentrative nucleoside transporter family, SLC28. Pflugers Arch 447 728–34

Gross RA, Macdonald RL, Ryan-Jastrow T (1989) 2-Chloroadenosine reduces the N calcium current of cultured mouse sensory neurones in a pertussis toxin-sensitive manner. J Physiol 411:585–95

Gubitz AK, Widdowson L, Kurokawa M et al (1996) Dual signalling by the adenosine $A_{2a}$ receptor involves activation of both N- and P-type calcium channels by different G proteins and protein kinases in the same striatal nerve terminals. J Neurochem 67:374–81

Haas HL, Greene RW (1988) Endogenous adenosine inhibits hippocampal CA1 neurones: further evidence from extra- and intracellular recording. Naunyn Schmiedeberg's Arch Pharmacol 1988 337:561–5

Hada J, Kaku T, Morimoto K et al (1998) Activation of adenosine $A_2$ receptors enhances high $K^+$-evoked taurine release from rat hippocampus: a microdialysis study. Amino Acids15:43–52

Hamid J, Nelson D, Spaetgens R et al (1999) Identification of an integration center for cross-talk between protein kinase C and G protein modulation of N-type calcium channels. J Biol Chem 274:5195–202

Harms HH, Wardeh G, Mulder AH (1979) Effect of adenosine on depolarization-induced release of varicus radiolabelled neurotransmitters from slices of rat corpus striatum. Neuropharmacology 18:577–80

Hedqvist P, Fredholm BB (1976) Effects of adenosine on adrenergic neurotransmission; prejunctional inhibition and postjunctional enhancement. Naunyn Schmiedeberg's Arch Pharmacol 293:217–23

Helfman CC, Zhong H, Barraco RA et al (1996) The effects of 5′-N-ethylcarboxamidoadenosine on evoked release of [$^3$H]serotonin in the rat nucleus tractus solitarius. Neurosci Lett 213:61–5

Hertting G, Wurster S, Allgaier C (1990) Regulatory proteins in presynaptic function. Ann N Y Acad Sci 604:289–304

Hillion J, Canals M, Torvinen M et al (2002) Coaggregation, cointernalization, and codesensitization of adenosine $A_{2A}$ receptors and dopamine $D_2$ receptors. Biol Chem 277:18091–7

Hollins C, Stone TW (1980) Adenosine inhibition of γ-aminobutyric acid release from slices of rat cerebral cortex. Br J Pharmacol 69:107–12

Hollopeter G, Jantzen HM, Vincent D et al (2001) Identification of the platelet ADP receptor targeted by antithrombotic drugs. Nature 409:202–7

Holton FA, Holton P (1953) The possibility that ATP is a transmitter at sensory nerve endings. J Physiol 119: 50P—51P

Hong SJ, Chang CC (1998) Evaluation of intrinsic modulation of synaptic transmission by ATP in mouse fast twitch muscle. J Neurophysiol 80:2550–8

Hunt JM, Redman RS, Silinsky EM (1994) Reduction by intracellular calcium chelation of acetylcholine secretion without occluding the effects of adenosine at frog motor nerve endings. Br J Pharmacol 111:753–8

Hur EM, Kim KT (2002) G protein-coupled receptor signalling and cross-talk: achieving rapidity and specificity. Cell Signal 14:397–405

Hussl S, Boehm S (2006) Functions of neuronal P2Y receptors. Eur J Physiol 452:538–51

Hutchison AJ, Webb RL, Oei HH et al (1989) CGS 21680C, an $A_2$ selective adenosine receptor agonist with preferential hypotensive activity. J Pharmacol Exp Ther 251:47–55

Ingall AH, Dixon J, Bailey A et al (1999) Antagonists of the platelet P2T receptor: A novel approach to antithrombotic therapy. J Med Chem 42:213–20

Inoue K, Koizumi S, Ueno S et al (1999) The functions of ATP receptors in the synaptic transmission in the hippocampus. Prog Brain Res 120:193–206

Jackisch R, Strittmatter H, Kasakov L et al (1984) Endogenous adenosine as a modulator of hippocampal acetylcholine release. Naunyn-Schmiedeberg's Arch Pharmacol 327:319–25

Jackisch R, Fehr R, Hertting G (1985) Adenosine: an endogenous modulator of hippocampal noradrenaline release. Neuropharmacology 24:499–507

Jarvis MF, Schulz R, Hutchison AJ et al (1989) [$^3$H]CGS 21680, a selective $A_2$ adenosine receptor agonist directly labels $A_2$ receptors in rat brain. J Pharmacol Exp Therap 251:888–93

Jeong SW, Ikeda SR (2000) Effect of G protein heterotrimer composition on coupling of neurotransmitter receptors to N-type $Ca^{2+}$ channel modulation in sympathetic neurons. Proc Natl Acad Sci 97:907–12

Jin S, Fredholm BB (1997) Adenosine $A_{2A}$ receptor stimulation increases release of acetylcholine from rat hippocampus but not striatum, and does not affect catecholamine release. Naunyn-Schmiedeberg's Arch Pharmacol 355:48–56

Jonzon B, Fredholm BB (1984) Adenosine receptor mediated inhibition of noradrenaline release from slices of the rat hippocampus. Life Sci 35:1971–9

Katada T, Gilman AG, Watanabe Y et al (1985) Protein kinase C phosphorylates the inhibitory guanine-nucleotide-binding regulatory component and apparently suppresses its function in hormonal inhibition of adenylate cyclase. Eur J Biochem 151:431–7

Katsuragi T, Su C (1982) Augmentation by theophylline of [$^3$H]purine release from vascular adrenergic nerves: evidence for presynaptic autoinhibition. J Pharmacol Exp Ther 220:152–6

Kegel B, Braun N, Heine P et al (1997) An ecto-ATPase and an ecto-ATP diphosphohydrolase are expressed in rat brain. Neuropharmacology 36:1189–1200

Kim YC, Ji X, Melman N et al (2000) Anilide derivatives of an 8-phenylxanthine carboxylic congener are highly potent and selective antagonists at human $A_{2B}$ adenosine receptors. J Med Chem 43:1165–72

Kirk IP, Richardson PJ (1994) Adenosine $A_{2a}$ receptor-mediated modulation of striatal [$^3$H]GABA and [$^3$H]acetylcholine release. J Neurochem 62:960–6
Koch H, von Kügelgen I, Starke K (1997) P2-receptor-mediated inhibition of noradrenaline release in the rat hippocampus. Naunyn-Schmiedeberg's Arch Pharmacol 355:707–15
Koch H, von Kügelgen I, Starke K (1998) P2-receptor-mediated inhibition of noradrenaline release in the rat pancreas. Naunyn-Schmiedeberg's Arch Pharmacol 357:431–40
Koizumi S, Inoue K (1997) Inhibition by ATP of calcium oscillations in rat cultured hippocampal neurones. Br J Pharmacol 122:51–8
Koizumi S, Fujishita K, Tsuda M et al (2003) Dynamic inhibition of excitatory synaptic transmission by astrocyte-derived ATP in hippocampal cultures. Proc Natl Acad Sci USA 100:11023–8
Kubista H, Boehm S (2006) Molecular mechanisms underlying the modulation of exocytotic noradrenaline release via presynaptic receptors. Pharmacol Ther 112:213–42
Kukulski F, Sevigny J, Komoszynski M (2004) Comparative hydrolysis of extracellular adenine nucleotides and adenosine in synaptic membranes from porcine brain cortex, hippocampus, cerebellum and medulla oblongata. Brain Res 1030:49–56
Kurokawa M, Koga K, Kase H et al (1996) Adenosine $A_{2a}$ receptor-mediated modulation of striatal acetylcholine release in vivo. J Neurochem 66:1882–8
Kurz K, von Kügelgen I, Starke K (1993) Prejunctional modulation of noradrenaline release in mouse and rat vas deferens: contribution of P1- and P2-purinoceptors. Br J Pharmacol 110:1465–72
Latini S, Pedata F (2001) Adenosine in the central nervous system: release mechanisms and extracellular concentrations. J Neurochem 79:463–84
Lechner SG, Dorostkar MM, Mayer M et al (2004) Autoinhibition of transmitter release from PC12 cells and sympathetic neurons through a $P2Y_{12}$ receptor-mediated inhibition of voltage-gated $Ca^{2+}$ channels. Eur J Neurosci 20:2917–28
Lechner SG, Hussl S, Schicker KW et al (2005) Presynaptic inhibition via a phospholipase C- and phosphatidylinositol bisphosphate-dependent regulation of neuronal $Ca^{2-}$ channels. Mol Pharmacol 68:1387–96
Liang M, Eason MG, Jewell-Motz EA et al (1998) Phosphorylation and functional desensitisation of the $\alpha_{2A}$-adrenergic receptor by protein kinase C. Mol Pharmacol 54:44–9
Lim W, Kim SJ, Yan HD et al (1997) $Ca^{2+}$-channel dependent and -independent inhibition of exocytosis by extracellular ATP in voltage-clamped rat adrenal chromaffin cells. Pflugers Arch 435:34–42
Lim WK, Myung C-S, Garrison JC et al (2001) Receptor-G protein $\gamma$ specificity: $\gamma$11 shows unique potency for $A_1$ adenosine and $5\text{-}HT_{1A}$ receptors. Biochemistry 40:10532–41
Limberger N, Späth L, Starke K (1988) Presynaptic $\alpha_2$-adrenoceptor, opioid $\kappa$-receptor and adenosine A1-receptor interactions on noradrenaline release in rabbit brain cortex. Naunyn-Schmiedberg's Arch Pharmacol 338:53–61
Lohse MJ, Klotz KN, Lindenborn-Fotinos J et al (1987) 8-Cyclopentyl-1,3-dipropylxanthine (DPCPX)-a selective high affinity antagonist radioligand for $A_1$ adenosine receptors. Naunyn-Schmiedeberg's Arch Pharmacol 336:204–10
Londos C, Preston MS (1977) Regulation of glucagon and divalent cations of inhibition of hepatic adenylate cyclase by adenosine. J Biol Chem 252:5951–6
Lopes LV, Cunha RA, Ribeiro JA (1999) Cross talk between $A_1$ and $A_{2A}$ adenosine receptors in the hippocampus and cortex of young adult and old rats. J Neurophysiol 82:3196–3203
Lopes LV, Cunha RA, Kull B et al (2002) Adenosine $A_{2A}$ receptor facilitation of hippocampal synaptic transmission is dependent on tonic $A_1$ receptor inhibition. Neuroscience 112:319–29
Lupica CR, Cass WA, Zahniser NR et al (1990) Effects of the selective adenosine $A_2$ receptor agonist CGS 21680 on in vitro electrophysiology, cAMP formation and dopamine release in rat hippocampus and striatum. J Pharmacol Exp Therap 252:1134–41
Luthardt J, Borvendeg SJ, Sperlágh B et al (2003) $P2Y_1$ receptor activation inhibits NMDA receptor-channels in layer V pyramidal neurons of the rat prefrontal and parietal cortex. Neurochem Int 42:161–72

Lynge J, Hellsten Y (2000) Distribution of adenosine $A_1$, $A_{2A}$ and $A_{2B}$ receptors in human skeletal muscle. Acta Physiol Scand 169:283–90

MacLean DA, Sinoway LI, Leuenberger U (1998) Systemic hypoxia elevates skeletal muscle interstitial adenosine levels in humans. Circulation 98:1990–2

Marchi M, Raiteri L, Risso F et al (2002) Effects of adenosine $A_1$ and $A_{2A}$ receptor activation on the evoked release of glutamate from rat cerebrocortical synaptosomes. Br J Pharmacol 136:434–40

Matsuoka I, Ohkubo S (2004) ATP- and adenosine-mediated signaling in the central nervous system: adenosine receptor activation by ATP through rapid and localized generation of adenosine by ecto-nucleotidases. J Pharmacol Sci 94: 95–9

Mayfield RD, Suzuki F, Zahniser NR (1993) Adenosine $A_{2a}$ receptor modulation of electrically evoked endogenous GABA release from slices of rat globus pallidus. J Neurochem 60:2334–7

Mendoza-Fernandez V, Andrew RD, Barajas-Lopez C (2000) ATP inhibits glutamate synaptic release by acting at P2Y receptors in pyramidal neurons of hippocampal slices. J Pharmacol Exp Ther 293:172–9

Michaelis ML, Michaelis EK, Myers SL (1979) Adenosine modulation of synaptosomal dopamine release. Life Sci 24:2083–92

Mihaylova-Todorova ST, Todorov LD, Westfall DP (2002) Enzyme kinetics and pharmacological characterization of nucleotidases released from the guinea pig isolated vas deferens during nerve stimulation: evidence for a soluble ecto-nucleoside triphosphate diphosphohydrolase-like ATPase and a soluble ecto-5′ nucleotidase-like AMPase. J Pharmacol Exp Ther 302:992–1001

Moos WH, Szotek DS, Bruns RF (1985) $N^6$-Cycloalkyladenosines. Potent $A_1$-selective adenosine agonists. J Med Chem 28:1383–4

Mota A, Guimarães S (2003) Influence of $\alpha_2$-autoreceptor-stimulation on the facilitation by angiotensin II and bradykinin of noradrenaline release. Naunyn-Schmiedberg's Arch Pharmacol 368:443–7

Muller MJ, Paton DM (1979) Presynaptic inhibitory actions of 2-substituted adenosine derivatives on neurotransmission in rat vas deferens: effects of inhibitors of adenosine uptake and deamination. Naunyn-Schmiedeberg's Arch Pharmacol 306:23–8

Munshi R, Pang IH, Sternweis PC et al (1991) $A_1$ adenosine receptors of bovine brain couple to guanine nucleotide-binding proteins Gi1, Gi2, and Go. J Biol Chem 266:22285–9

Mynlieff M, Beam KG (1994) Adenosine acting at an $A_1$ receptor decreases N-type calcium current in mouse motoneurons. J Neurosci 14:3628–34

Nagy G, Reim K, Matti U et al (2004) Regulation of releasable vesicle pool sizes by protein kinase A-dependent phosphorylation of SNAP-25. Neuron 41:417–29

Nakata H, Yoshioka K, Kamiya T et al (2005) Functions of heteromeric association between adenosine and P2Y receptors. J Mol Neurosci 25:233–8

Nakatsuka H, Nagano O, Foldes FF et al (1995) Effects of adenosine on norepinephrine and acetylcholine release from guinea pig right atrium: role of $A_1$-receptors. Neurochem Int 27:345–53

Nitahara K, Kittel A, Liang SD, Vizi ES (1995) $A_1$-receptor-mediated effect of adenosine on the release of acetylcholine from the myenteric plexus: role and localization of ecto-ATPase and 5′-nucleotidase. Neuroscience 67:159–68

O'Kane EM, Stone TW (1998) Interaction between adenosine $A_1$ and $A_2$ receptor-mediated responses in the rat hippocampus in vitro. Eur J Pharmacol 362:17–25

Ohkubo S, Kimura J, Matsuoka I (2000) Ecto-alkaline phosphatase in NG108-15 cells: a key enzyme mediating P1 antagonist sensitive ATP response. Br J Pharmacol 131:1667–2

Okada M, Mizuno K, Kaneko S (1996) Adenosine $A_1$ and $A_2$ receptors modulate extracellular dopamine levels in rat striatum. Neurosci Lett 212:53–6

Okada M, Nutt DJ, Murakami T et al (2001) Adenosine receptor subtypes modulate two major functional pathways for hippocampal serotonin release. J Neurosci 21:628–40

Oliveira L, Correia-de-Sá P (2005) Protein kinase A and $Ca_v1$ (L-Type) channels are common targets to facilitatory adenosine $A_{2A}$ and muscarinic $M_1$ receptors on rat motoneurons. Neurosignals 14:262–72

Olsson RA, Davies CJ, Khouri EM et al (1976) Evidence for an adenosine receptor on the surface of dog coronary myocytes. Circ Res 39:93–8

Ongini E, Dionisotti S, Gessi S, Irnius E, Fredholm BB (1999) Comparison of CGS 15943, ZM 241385 and SCH 58261 as antagonists at human adenosine receptors. Naunyn-Schmiedeberg's Arch Pharmacol 359:7–10

O'Regan MH, Simpson RE, Perkins LM et al (1992) The selective $A_2$ adenosine receptor agonist CGS 21680 enhances excitatory transmitter amino acid release from the ischemic rat cerebral cortex. Neurosci Lett 138:169–72

Paton DM (1981) Structure-activity relations for presynaptic inhibition of noradrenergic and cholinergic transmission by adenosine: evidence for action on $A_1$ receptors. J Auton Pharmacol 1:287–90

Pedata F, Antonelli T, Lambertini L et al (1983) Effect of adenosine, adenosine triphosphate, adenosine deaminase, dipyridamole and aminophylline on acetylcholine release from electrically-stimulated brain slices. Neuropharmacol 22:609–14

Phillis JW, Kostopoulos GK, Limacher JJ (1975) A potent depressant action of adenine derivatives on cerebral cortical neurones. Eur J Pharmacol 30:125–9

Popoli P, Betto P, Reggio R et al (1995) Adenosine $A_{2A}$ receptor stimulation enhances striatal extracellular glutamate levels in rats. Eur J Pharmacol 287:215–17

Poucher SM, Keddie JR, Singh P et al (1995) The in vitro pharmacology of ZM 241385, a potent, non-xanthine $A_{2a}$ selective adenosine receptor antagonist. Br J Pharmacol 115: 1096–1102

Powell AD, Teschemacher AG, Seward EP (2000) P2Y purinoceptors inhibit exocytosis in adrenal chromaffin cells via modulation of voltage-operated calcium channels. J Neurosci 20:606–16

Price GD, Robertson SJ, Edwards FA (2003) Long-term potentiation of glutamatergic synaptic transmission induced by activation of presynaptic P2Y receptors in the rat medial habenula nucleus. Eur J Neurosci 17:844–50

Quarta D, Ferré S, Solinas M et al (2004) Opposite modulatory roles for adenosine $A_1$ and $A_{2A}$ receptors on glutamate and dopamine release in the shell of the nucleus accumbens. Effects of chronic caffeine exposure. J Neurochem 88:1151–8

Queiroz G, Gebicke-Haerter PJ, Schobert A et al (1997) Release of ATP from cultured rat astrocytes elicited by glutamate receptor activation. Neuroscience 78 1203–8

Queiroz G, Diniz C, Gonçalves J (2002) Facilitation of noradrenaline release by adenosine $A_{2A}$ receptors in the epididymal portion and adenosine $A_{2B}$ receptors in the prostatic portion of the rat vas deferens. Eur J Pharmacol 448:45–50

Queiroz G, Quintas C, Talaia C et al (2004) Coupling to protein kinases A and C of adenosine $A_{2B}$ receptors involved in the facilitation of noradrenaline release in the prostatic portion of rat vas deferens. Neuropharmacology 47:216–24

Queiroz G, Talaia C, Gonçalves J (2003a) Adenosine $A_{2A}$ receptor–mediated facilitation of noradrenaline release involves protein kinase C activation and attenuation of presynaptic inhibitory receptor-mediated effects in the rat vas deferens. J Neurochem 85:740–8

Queiroz G, Talaia C, Gonçalves J (2003b) ATP modulates noradrenaline release by activation of inhibitory P2Y receptors and facilitatory P2X receptors in the rat vas deferens. J Pharmacol Exp Ter 307:809–15

Rebola R, Oliveira CR, Cunha RA (2002) Transducing system operated by adenosine $A_{2A}$ receptors to facilitate acetylcholine release in the rat hippocampus. Eur J Pharmacol 454:31–8

Reddington M, Lee KS, Schubert P (1982) An $A_1$-adenosine receptor, characterized by [$^3$H] cyclohexyladenosine binding, mediates the depression of evoked potentials in a rat hippocampal slice preparation. Neurosci Lett 28:275–9

Ribeiro JA, Sebastião AM (1985) On the type of receptor involved in the inhibitory action of adenosine at the neuromuscular junction.Br J Pharmacol 84:911–18

Ribeiro JA, Sebastião AM (1986) Adenosine receptors and calcium: basis for proposing a third ($A_3$) adenosine receptor. Prog Neurobiol 26:179–209

Ribeiro JA, Sebastião AM (1987) On the role, inactivation and origin of endogenous adenosine at the frog neuromuscular junction. J Physiol (Lond) 384: 571–85

Ribeiro JA, Walker J (1975) The effects of adenosine triphosphate and adenosine diphosphate on transmission at the rat and frog neuromuscular junctions. Br J Pharmacol 54:213–18

Ribeiro JA, Sa-Almeida AM, Namorado JM (1979) Adenosine and adenosine triphosphate decrease $^{45}Ca$ uptake by synaptosomes stimulated by potassium. Biochem Pharmacol 28:1297–1300

Rivkees SA, Price SL, Zhou FC (1995) Immunohistochemical detection of $A_1$ adenosine receptors in rat brain with emphasis on localization in the hippocampal formation, cerebral cortex, cerebellum, and basal ganglia. Brain Res 677:193–203

Robitaille R, Thomas S, Charlton MP (1999) Effects of adenosine on $Ca^{2+}$ entry in the nerve terminal of the frog neuromuscular junction. Can J Physiol Pharmacol 77:707–14

Robson SC, Sévigny J, Zimmermann H (2006) The E-NTPDase family of ectonucleotidases: Structure function relationships and pathophysiological significance. Purinergic Signalling 2:409–30

Rodrigues RJ, Almeida T, Richardson PJ et al (2005) Dual presynaptic control by ATP of glutamate release via facilitatory $P2X_1$, $P2X_{2/3}$, and $P2X_3$ and inhibitory $P2Y_1$, $P2Y_2$, and/or $P2Y_4$ receptors in the rat hippocampus. J Neurosci 25:6286–95

Rosenberg PA, Li Y (1995) Vasoactive intestinal peptide regulates extracellular adenosine levels in rat cortical cultures. Neurosci Lett 200:93–6

Ruiz MA, Escriche M, Lluis C et al (2000) Adenosine $A_1$ receptor in cultured neurons from rat cerebral cortex: colocalization with adenosine deaminase. J Neurochem 75:656–64

Saitoh Y, Nakata H (1996) Photoaffinity labelling of a P3 purinoceptor-like protein purified from rat brain membranes. Biochem Biophys Res Commun 219: 469–74

Saitow F, Suzuki H, Konishi S (2005) Beta-adrenoceptor-mediated long term up-regulation of the release machinery at rat cerebellar GABAergic synapses. J Physiol 565:487–502

Sattin A, Rall TW (1970) The effect of adenosine and adenine nucleotides on the cyclic adenosine 3',5'-phosphate content of guinea pig cerebral cortex slices. Mol Pharmacol 6:13–23

Saura CA, Mallol J, Canela EI et al (1998) Adenosine deaminase and $A_1$ adenosine receptors internalize together following agonist-induced receptor desensitization. J Biol Chem 273:17610–17

Schindler M, Harris CA, Hayes B et al (2001) Immunohistochemical localization of adenosine $A_1$ receptors in human brain regions. Neurosci Lett 297:211–15

Schlicker E, Göthert M (1998) Interactions between the presynaptic $\alpha_2$-autoreceptor and presynaptic inhibitory heteroreceptors on noradrenergic neurones. Brain Res Bull 47:129–32

Schubert P, Heinemann U, Kolb R (1986) Differential effect of adenosine on pre- and postsynaptic calcium fluxes. Brain Res 376:382–6

Sebastião AM, Ribeiro JA (1985) Enhancement of transmission at the frog neuromuscular junction by adenosine deaminase: evidence for an inhibitory role of endogenous adenosine on neuromuscular transmission. Neurosci Lett 62:267–70

Sebastião AM, Ribeiro JA (1988) On the adenosine receptor and adenosine inactivation at the rat diaphragm neuromuscular junction. Br J Pharmacol 94:109–20

Sebastião A M, Ribeiro JA (2000) Fine-tuning neuromodulation by adenosine. Trends Pharmacol Sci 21:341–6

Sebastião AM, Stone TW, Ribeiro JA (1990) The inhibitory adenosine receptor at the neuromuscular junction and hippocampus of the rat: antagonism by 1,3,8-substituted xanthines. Br J Pharmacol 101:453–9

Sebastião AM, Cunha RA, Cascalheira JF et al (1999) Adenine nucleotides as inhibitors of synaptic transmission: role of localised ectonucleotidases. Prog Brain Res 120:183–92

Sebastião AM, Macedo MP, Ribeiro JA (2000) Tonic activation of $A_{2A}$ adenosine receptors unmasks, and of $A_1$ receptors prevents, a facilitatory action of calcitonin gene–related peptide in the rat hippocampus. Br J Pharmacol 129:374–80

Shaver SR, Rideout JL, Pendergast W et al (2005) Structure-activity relationships of dinucleotides: potent and selective agonists of P2Y receptors. Purinergic Signal 1:183–91

Shindou T, Mori A, Kase H et al (2001) Adenosine $A_{2A}$ receptor enhances $GABA_A$-mediated IPSCs in the rat globus pallidus. J Physiol 532:423–34

Shinozuka K, Bjur RA, Westfall DP (1988) Characterization of prejunctional purinoceptors on adrenergic nerves of the rat caudal artery. Naunyn Schmiedeberg's Arch Pharmacol 338:221–7

Shinozuka K, Tanioka Y, Kwon YM et al (2001) Characterization of prejunctional purinoceptors inhibiting noradrenaline release in rat mesenteric arteries. Jpn J Pharmacol 85:41–6

Shoji-Kasai Y, Itakura M, Kataoka M et al (2002) Protein kinase C-mediated translocation of secretory vesicles to plasma membrane and enhancement of neurotransmitter release from PC12 cells. Eur J Neurosci 15:1390–4

Silinsky EM (1980) Evidence for specific adenosine receptors at cholinergic nerve endings. Br J Pharmacol 71:191–4

Silinsky EM (2004) Adenosine decreases both presynaptic calcium currents and neurotransmitter release at the mouse neuromuscular junction. J Physiol 558:389–401

Silinsky EM, Ginsborg BL (1983) Inhibition of acetylcholine release from preganglionic frog nerves by ATP but not adenosine. Nature 305:327–8

Simon J, Filippov AK, Goransson S et al (2002) Characterization and channel coupling of the $P2Y_{12}$ nucleotide receptor of brain capillary endothelial cells. J Biol Chem 277:31390–400

Sperlágh B, Mergl Z, Juranyi Z et al (1999) Local regulation of vasopressin and oxytocin secretion by extracellular ATP in the isolated posterior lobe of the rat hypophysis. J Endocrinol 160:343–50

Su C (1983) Purinergic neurotransmission and neuromodulation. Pharmacol Rev 23: 397–411

Su C, Bevan JA, Burnstock G (1971) [$^3$H]adenosine triphosphate: release during stimulation of enteric nerves. Science 173:336–38

Talaia C, Queiroz G, Quintas C et al (2005) Interaction between adenosine $A_{2B}$ receptors and $\alpha_2$-adrenoceptors on the modulation of noradrenaline release in the rat vas deferens: possible involvement of a group 2 adenylyl cyclase isoform. Neurochem Int 47:418–29

Talaia C, Queiroz G, Pinheiro H et al (2006) Involvement of G-protein $\beta\gamma$ subunits on the influence of inhibitory $\alpha_2$-autoreceptors on the angiotensin $AT_1$-receptor-modulation of noradrenaline release in the rat vas deferens. Neurochem Int 49:698–707

Ther L, Muschaweck R, Hergott J (1957) Antagonism between adenosine & methylxanthine at the bundle of His of the heart. Naunyn-Schmiedeberg's Arch Exp Pathol Pharmakol 231:586–90

Todorov LD, Bjur RA, Westfall DP (1994) Inhibitory and facilitatory effects of purines on transmitter release from sympathetic nerves. J Pharmacol Exp Ther 268:985–9

Todorov LD, Mihaylova-Todorova S, Craviso GL et al (1996) Evidence for the differential release of the cotransmitters ATP and noradrenaline from sympathetic nerves of the guinea-pig vas deferens. J Physiol 496:731–48

Trendelenburg AU, Bültmann R (2000) P2 receptor-mediated inhibition of dopamine release in rat neostriatum. Neuroscience 96:249–52

Trendelenburg AU, Meyer A, Klebroff W et al (2003) Cross talk between presynaptic angiotensin receptors, bradykinin receptors and $\alpha_2$-autoreceptors in sympathetic neurons: a study in $\alpha_2$-adrenoceptor-deficient mice. Br J Pharmacol 138:1389–1402

Umemiya M, Berger AJ (1994) Activation of adenosine $A_1$ and $A_2$ receptors differentially modulates calcium channels and glycinergic synaptic transmission in rat brainstem. Neuron 13 1439–46

van Calker D, Muller M, Hamprecht B (1978) Adenosine inhibits the accumulation of cyclic AMP in cultured brain cells. Nature 276:839–41

van Galen PJ, van Bergen AH, Gallo-Rodriguez C et al (1994) A binding site model and structure-activity relationships for the rat $A_3$ adenosine receptor. Mol Pharmacol 45:1101–11

van Muijlwijk-Koezen JE, Timmerman H, van der Goot H et al (2000) Isoquinoline and quinazoline urea analogues as antagonists for the human adenosine A(3) receptor. J Med Chem 43:2227–38

van Schaick EA, Jacobson KA, Kim HO et al (1996) Hemodynamic effects and histamine release elicited by the selective adenosine $A_3$ receptor agonist 2-Cl-IB-MECA in conscious rats. Eur J Pharmacol 308:311–14

Varani K, Merighi S, Gessi S et al (2000) [$^3$H]MRE 3008F20: a novel antagonist radioligand for the pharmacological and biochemical characterization of human $A_3$ adenosine receptors. Mol Pharmacol 57:968–75

Vartian N, Boehm S (2001) P2Y receptor-mediated inhibition of voltage-activated $Ca^{2+}$ currents in PC12 cells. Eur J Neurosci 13:899–908
Vaughan PF, Walker JH, Peers C (1998) The regulation of neurotransmitter secretion by protein kinase C. Mol Neurobiol 18:125–55
Verhaeghe RH, Vanhoutte PM, Shepherd JT (1977) Inhibition of sympathetic neurotransmission in canine blood vessels by adenosine and adenine nucleotides. Circ Res. 1977 40:208–15
Volknandt W (2002) Vesicular release mechanisms in astrocytic signalling. Neurochem Int 41: 301–6
von Kügelgen I (2006) Pharmacological profiles of cloned mammalian P2Y-receptor subtypes. Pharmacol Ther 110:415–32
von Kügelgen I, Starke K (1987) Evidence for two separate vasoconstrictor-mediating nucleotide receptors, both distinct from $P_{2X}$-receptor, in rat basilar artery: a receptor for pyrimidine nucleotides and a receptor for purine nucleotides. Naunyn-Schmiedeberg's Arch Pharmacol 341:538–46
von Kügelgen I, Häussinger D, Starke K (1987) Evidence for a vasoconstriction-mediating receptor for UTP, distinct from the P2 purinoceptor, in rabbit ear artery. Naunyn Schmiedeberg's Arch Pharmacol 336:556–60
von Kügelgen I, Schöffel E, Starke K (1989) Inhibition by nucleotides acting at presynaptic P2-receptors of sympathetic neuro-efector transmission in the mouse isolated vas deferens. Naunyn-Schmiedeberg's Arch Pharmacol 340:522–32
von Kügelgen I, Kurz K, Starke K (1993) Axon terminal P2-purinoceptors in feedback control of sympathetic transmitter release. Neuroscience 56:263–67
von Kügelgen I, Kurz K, Starke K (1994) Evidence for $P_2$-purinoceptor-mediated inhibition of noradrenaline release in rat brain cortex. Br J Pharmacol 113:815–22
von Kügelgen I, Stöffel D, Starke K (1995) P2-purinoceptor-mediated inhibition of noradrenaline release in rat atria. Br J Pharmacol 115:247–54
von Kügelgen I, Koch H, Starke K (1997) P2-receptor-mediated inhibition of serotonin release in the rat brain cortex. Neuropharmacology 36:1221–7
Weber RG, Jones CR, Palacios JM et al (1988) Autoradiographic visualization of $A_1$-adenosine receptors in brain and peripheral tissues of rat and guinea pig using $^{125}$I-HPIA. Neurosci Lett 87:215–20
Westfall DP, Todorov LD, Mihaylova-Todorova ST (2002) ATP as a cotransmitter in sympathetic nerves and its inactivation by releasable enzymes. J Pharm Exp Ther 303:439–44
Wiklund NP, Gustafsson LE (1986) Neuromodulation by adenine nucleotides, as indicated by experiments with inhibitors of nucleotide inactivation. Acta Physiologica Scandinava 126:217–23
Wiklund NP, Gustafsson LE, Lundin J (1985) Pre- and postjunctional modulation of cholinergic neuroeffector transmission by adenine nucleotides. Experiments with agonist and antagonist. Acta Physiol Scand 125:681–91
Wiklund NP, Cederqvist B, Gustafsson LE (1989) Adenosine enhancement of adrenergic neuroeffector transmission in guinea-pig pulmonary artery. Br J Pharmacol 96:425–33
Yasuda H, Lindorfer MA, Woodfork KA et al (1996) Role of the prenyl group on the G protein $\gamma$ subunit in coupling trimeric G proteins to $A_1$ adenosine receptors. J Biol Chem 271: 18588–95
Yawo H, Chuhma N (1993) Preferential inhibition of omega-conotoxin-sensitive presynaptic $Ca^{2+}$ channels by adenosine autoreceptors. Nature 365:256–8
Yoshioka K, Saitoh O, Nakata H (2002) Agonist-promoted heteromeric oligomerization between adenosine $A_1$ and $P2Y_1$ receptors in living cells. FEBS Lett 523:147–51
Zetterstrom T, Fillenz M (1990) Adenosine agonists can both inhibit and enhance in vivo striatal dopamine release. Eur J Pharmacol 180:137–43
Zhang FL, Luo L, Gustafson E et al (2002) $P2Y_{13}$: identification and characterization of a novel $G_{\alpha i}$-coupled ADP receptor from human and mouse. J Pharmacol Exp Ther 301:705–13
Zhang JM, Wang HK, Ye CH et al (2003) ATP released by astrocytes mediates glutamatergic activity-dependent heterosynaptic suppression. Neuron 40:971–82
Zimmermann H (1992) 5′-Nucleotidase: molecular structure and functional aspects. Biochem J 285:345–65

# Presynaptic Metabotropic Glutamate and $GABA_B$ Receptors

**M. Raiteri**



**Abstract** Glutamate and GABA, the two most abundant neurotransmitters in the mammalian central nervous system, can act on metabotropic receptors that are structurally quite dissimilar from those targeted by most other neurotransmitters/modulators. Accordingly, metabotropic glutamate receptors (mGluRs) and $GABA_B$ receptors ($GABA_B$Rs) are classified as members of family 3 (or family C) of G protein-coupled receptors. On the other hand, mGluRs and $GABA_B$Rs exhibit pronounced and partly unresolved differences between each other. The most intriguing difference is that mGluRs exist as multiple pharmacologically as well as structurally distinct subtypes, whereas, in the case of $GABA_B$Rs, molecular biologists have so

M. Raiteri
Department of Experimental Medicine, Pharmacology and Toxicology Section, University of Genoa, 16148 Genoa, Italy
Center of Excellence for Biomedical Research, University of Genoa, 16132 Genoa, Italy
raiteri@pharmatox.unige.it

T.C. Südhof, K. Starke (eds.), *Pharmacology of Neurotransmitter Release.*
*Handbook of Experimental Pharmacology 184.*

far identified only one structurally distinct heterodimeric complex whose few variants seem unable to explain the pharmacological heterogeneity of $GABA_B$Rs observed in many functional studies. Both mGluRs and $GABA_B$Rs can be localized on axon terminals of different neuronal systems as presynaptic autoreceptors and heteroreceptors modulating the exocytosis of various transmitters.

## 1 Introduction

Glutamate and GABA mediate transmission at most central excitatory and inhibitory synapses, respectively. Besides their quick actions at ionotropic receptors, glutamate and GABA, similarly to most other chemical mediators, can activate metabotropic receptors generally endowed with modulatory functions.

The interest in mGluRs and $GABA_B$Rs has in part been stimulated by the peculiarity of their structures, which are quite different from those of all other G protein-coupled receptors (GPCRs); of note, the structure of $GABA_B$Rs could only be determined after almost two decades of unsuccessful efforts. Due to their special properties, mGluRs and $GABA_B$Rs have been included in family 3 of GPCRs (Bockaert and Pin 1999). Metabotropic receptors of family 3 possess a very large N-terminal domain in which the agonist binding sites are localized. The binding site of these GPCRs is constituted of two lobes; glutamate or GABA is supposed to bind to one lobe within the cleft between the two lobes and, according to the Venus flytrap model (O'Hara et al. 1993; Galvez et al. 1999), this leads to the closure of the two lobes.

As for other GPCRs, those of family 3 can undergo dimerization. It has been shown that mGluRs form homodimers via a disulfide bridge between the N-terminal domains (Romano et al. 1996; Pace et al. 1999). In the case of $GABA_B$Rs, heterodimerization between two subunits, termed $GABA_{B1}$ and $GABA_{B2}$, seems to be the absolute requirement for an efficient coupling to G proteins and takes place in the cytoplasm via coiled-coil interaction of the subunit C-terminal domains (White et al. 1998; Kuner et al. 1999).

It is well established that mGluRs are structurally and pharmacologically heterogeneous and exist as eight subtypes (mGluR1 to mGluR8) having discrete regional, cellular, and subcellular localization (Pin and Bockaert 1995; Conn and Pin 1997; Cartmell and Schoepp 2000; Bockaert et al. 2002; Pin et al. 2003). In contrast, $GABA_B$Rs have often been reported to exhibit pharmacological heterogeneity (Bonanno and Raiteri 1993a; Mott and Lewis 1994; Bowery et al. 2002), but this has remained so far unexplained because molecular biologists tend to exclude the existence of more than one $GABA_B$R formed by the obligatory heterodimerization of $GABA_{B1}$ and $GABA_{B2}$ subunits (see Bowery et al. 2002; Bettler et al. 2004, for reviews).

The transduction mechanisms associated with mGluR activation have been extensively reviewed (Pin and Bockaert 1995; Conn and Pin 1997; Bockaert et al. 2002). Depending on the receptor involved, the main mechanisms are activation of the phosphatidylinositol (PI) pathway with mobilization of internal $Ca^{2+}$ ions (group I mGluRs, including mGluR1 and 5) and inhibition of adenylyl cyclase (AC) activity (group II mGluRs, including mGluR2 and 3, and group III mGluRs, including mGluR4,6,7 and 8), although a series of mGluR-mediated cellular events do not seem to result from a simple activation of phospholipase C (PLC) or inhibition of cyclic AMP production (see Bockaert et al. 2002). Neuronal GABA$_B$Rs are negatively coupled via G proteins to their effector processes, namely, plasma membrane $Ca^{2+}$ and $K^+$ channels (Dolphin et al. 1990; Gage 1992) and AC (Hill et al. 1984; Karbon et al. 1984). Activation of GABA$_B$Rs decreases $Ca^{2+}$ conductance, increases conductance to $K^+$ ions, and inhibits AC activity (see Bowery et al. 2002).

PLC-coupled mGluRs are generally localized postsynaptically (Baude et al. 1993; Ottersen and Landsend 1997; Shigemoto et al. 1997), although examples of presynaptic localization have been described (see below). Conversely, mGluRs negatively coupled to AC or to voltage-sensitive $Ca^{2+}$ channels (VSCCs) are mostly localized on presynaptic axon terminals (Shigemoto et al. 1996, 1997; Ottersen and Landsend 1997; Masugi et al. 1999; Millán et al. 2002), where they modulate neurotransmitter release. Both pre- and postsynaptically located GABA$_B$Rs have been described in the mammalian CNS (Bonanno and Raiteri 1993a; Bowery 1993; Bowery et al. 2002; Bettler et al. 2004; Raiteri 2006). GABA$_B$Rs stimulation at postsynaptic sites normally causes a long-lasting neuronal hyperpolarization mediated by an increase in $K^+$ conductance. The presynaptic GABA$_B$Rs mediate negative modulation of neurotransmitter exocytosis. The interested reader can find detailed information on the localization of mGluRs and GABA$_B$Rs, as determined by morphological approaches, in several of the reviews quoted in this Introduction.

Presynaptic mGluRs and GABA$_B$Rs can be both autoreceptors regulating release of glutamate or GABA, respectively, and heteroreceptors regulating release of various transmitters. Since a review article specifically devoted to the regulation of neurotransmitter release by mGluRs occurring in various *in vitro* and *in vivo* preparations was published a few years ago (Cartmell and Schoepp 2000), this chapter will mainly focus on the modulatory roles of mGluRs presynaptically located on axon terminals releasing a given transmitter in *in vitro* conditions. I will also try to update the information available on release-regulating presynaptic GABA$_B$Rs. Although modulation of neurotransmission by mGluRs and, less so, by GABA$_B$Rs, has been the object of a great number of electrophysiological studies (see for reviews Anwyl 1999; Nicoll 2004), the present chapter will emphasize neurochemical and pharmacological evidence for release-regulating presynaptic mGluRs and GABA$_B$Rs originated from direct studies of neurotransmitter overflow.

## 2 Metabotropic Glutamate Autoreceptors

The glutamatergic autoreceptor system is by far the most complex among the brain autoreceptor systems (compare Starke et al. 1989). Glutamatergic terminals possess several autoreceptors, both ionotropic and metabotropic, some of which coexist and may interact on the same terminal. Investigations on mGlu autoreceptors have included multiple experimental approaches. The excellent review of Cartmell and Schoepp (2000) contains a table that summarizes the literature on release-regulating mGluRs, including autoreceptors. Interestingly, as many as 19 out of 48 reports on glutamate release had been carried out using synaptosomes, probably chosen as a relatively simple experimental approach offering easier interpretations of the results.

All three groups of mGluRs (I to III) have been implicated in the presynaptic modulation of excitatory synaptic transmission. Much of the electrophysiological evidence available (see Anwyl 1999), including that regarding PI-linked receptors, has shown inhibitory effects generally attributed to direct activation of mGluRs present on glutamate/aspartate axon terminals.

## *2.1 Group I Metabotropic Glutamate Autoreceptors*

Many authors believe that the simplest and best established model for studying release-regulating presynaptic receptors is the isolated nerve terminal (synaptosome) preparation. This technical approach was chosen by Herrero et al. (1992) in their fundamental work on mGlu autoreceptors: synaptosomal preparations from adult rat cerebral cortex were exposed to the broad-spectrum mGluR agonist (1S,3R)-ACPD, and glutamate release was monitored by on-line fluorometry. (1S,3R)-ACPD enhanced the release of glutamate produced by 50 μM 4-aminopyridine (4-AP) in the presence of low concentrations of arachidonic acid. This facilitation of glutamate release was paralleled by a transient elevation of diacylglycerol (DAG) and by activation of protein kinase C (PKC), suggesting the involvement of PI-linked group I mGluRs (Coffey et al. 1994). In apparent contrast to these data, some electrophysiological studies reported that presynaptic group I mGluRs mediated reduction of excitatory transmission (Gereau and Conn 1995; Manzoni and Bockaert 1995).

In the meantime, 3,5-DHPG had been introduced as a novel selective group I receptor agonist (Schoepp et al. 1994), able to activate mGluR1 and mGluR5 with consequent stimulation of PI hydrolysis and mobilization of intracellular calcium (Fagni et al. 2000). Curiously, when 3,5-DHPG was added to synaptosomes, in the presence of arachidonic acid, the release of glutamate provoked by submaximal depolarization with 4-AP (50 μM) was enhanced, but the compound failed to affect glutamate exocytosis when this was induced by maximal depolarization with 1 mM 4-AP (Herrero et al. 1998). To better understand this observation, nerve terminals were pre-exposed to 3,5-DHPG for 1 min (first stimulation) and, after washings to remove the agonist, they were rechallenged with 3,5-DHPG (second stimulation). A suppression of the release-facilitatory effect of 3,5-DHPG was observed during the second stimulation. Furthermore, the agonist became concomitantly able to *inhibit* the release induced by maximal depolarization with 4-AP (1 mM). The ability of 3,5-DHPG to potentiate the release was recovered 30 min after the first stimulation, and the inhibitory action was lost. One explanation for the switch from facilitation to inhibition in the control of glutamate release might have been that, during the desensitization of group I mGluRs, inhibitory mGluRs prevailed. However, agonists at the AC-coupled mGluRs failed to elicit release inhibition, suggesting that the receptors involved were PI-coupled group I mGluRs.

Herrero et al. (1998) then found that, while the stimulatory effect of 3,5-DHPG on the 4-AP (50 μM)-evoked glutamate release was PKC-dependent, its inhibitory effect on the 4-AP(1 mM)-evoked release was not altered by the PKC inhibitor staurosporine. PKC activity was, however, involved in the switch from release facilitation to release inhibition. To show this, nerve terminals were pre-exposed to 3,5-DHPG for 1 min to induce the switch, washed and treated with staurosporine. Under these conditions, the inhibitory effect of 3,5-DHPG on the 4-AP(1 mM)-evoked glutamate exocytosis remained unaltered. Conversely, in experiments in which the exposure to 3,5-DHPG occurred after a prior inhibition of PKC with staurosporine, the second stimulation with 3,5-DHPG failed to induce release inhibition.

The dual role of the group I mGluRs in the release of glutamate studied in nerve endings isolated from the cerebrocortex of adult rats was also identified in rat hippocampal synaptosomes (Rodriguez-Moreno et al. 1998). Moreover, to get information on the potential relevance of the mechanism in synaptic transmission, electrophysiological experiments with hippocampal slices were carried out. Excitatory postsynaptic potentials (EPSPs) were evoked by electrically stimulating the Schaffer collaterals and recorded from identified CA1 pyramidal cells. Exposure to 3,5-DHPG produced dramatic reduction (to $\sim$ 15%) in the amplitude of EPSPs, confirming previous electrophysiological findings (Gereau and Conn 1995; Manzoni and Bockaert 1995). Thus, hippocampal slices seemed to respond to 3,5-DHPG like isolated nerve terminals during the "second stimulation" with the group I mGluR agonist, suggesting that, in slices, the desensitization of the stimulatory component had already occurred, possibly due to extracellular endogenous glutamate tonically activating mGluRs. If this were the case, preincubation of the slices with an mGluR antagonist should prevent receptor desensitization. Hippocampal slices were therefore pretreated with the antagonist MCPG and then stimulated; in MCPG-treated slices, 3,5-DHPG enhanced the EPSPs amplitude. The facilitatory response was transient and was followed by an inhibition; facilitation was recovered following 30 min of renewed incubation with the antagonist (Rodriguez-Moreno et al. 1998).

A more direct approach to verify the hypothesis that, in slices, an increase in the extracellular concentrations of glutamate could promote the switch from facilitation to inhibition in response to a group I mGluR agonist was to incubate synaptosomes with a depolarizing concentration of KCl to elevate the extraterminal concentration of the natural mGluR agonist. Under basal conditions (3 mM KCl; 37° C), the extraterminal concentration of glutamate after 10 min incubation amounted to $\sim$1 μM; addition of 3,5-DHPG elicited facilitation of the 50 μM 4-AP-evoked glutamate release, while inhibition of the 1 mM 4-AP-evoked release was absent. Raising the concentration of KCl to 13 mM increased extraterminal glutamate to 3.5 μM. Under these conditions, 3,5-DHPG failed to facilitate release and became clearly inhibitory (Herrero et al. 1998; Rodriguez-Moreno et al. 1998).

It has to be noted that the electrophysiological experiments in which 3,5-DHPG exclusively caused inhibition of EPSP amplitude in hippocampal slices were carried out at room temperature (21° C–22° C). When the temperature was raised to 31° C, 3,5-DHPG first caused some facilitation of synaptic transmission, followed by the expected inhibition. The interpretation given by Rodriguez-Moreno et al. (1998) was that, at 31° C, the extracellular concentration of glutamate was maintained lower than at 21° C by the glial/neuronal uptake systems, and this could have prevented receptor desensitization. Whatever the explanation, the work of the Spanish laboratories represents one convincing example of the usefulness of the synaptosomal preparation in investigating presynaptic mechanisms. Indeed, the dual role of group I mGluRs in the modulation of glutamate release and the switch from facilitation to inhibition of this modulation, mechanisms that may have important implications in synaptic physiology and neurotoxicity, probably have been discovered thanks to the use of isolated nerve terminal preparations. These dynamic changes had in fact not been observed in electrophysiological experiments.

The ambient concentration of glutamate appears to be critical for the function of group I mGluRs. Extrasynaptic spillover of glutamate was reported to be regulated through the function of excitatory amino acid transporters, which depends, inter alia, on the temperature (Asztely et al. 1997). When extraterminal glutamate concentrations are maintained at ~1 μM in incubated synaptosomes, the dual role of group I mGluRs can be clearly observed. According to Rodriguez-Moreno et al. (1998), if, under physiological conditions, the brain extracellular concentrations of glutamate are kept low enough by the uptake systems, the facilitatory response mediated by mGluRs would not be desensitized and group I mGluRs should be ready to mediate enhancement of evoked glutamate exocytosis whenever necessary; for instance, during plasticity processes. Activity-dependent increases in extracellular glutamate will instead desensitize the group I mGluR facilitatory function, which may be particularly important in preventing undesired accumulation of extracellular excitatory amino acids consequent to repetitive activity.

Desensitization of mGluRs linked to PI hydrolysis in response to prolonged or repetitive exposure to agonists had previously been observed in cultured cerebellar neurons (Catania et al. 1991; Aronica et al. 1993). These authors reported that desensitization was reduced by inhibitors of PKC. As to presynaptic group I mGluRs autoregulating glutamate release, the functional switch from facilitation to inhibition of release is also thought to be mediated by PKC-induced phosphorylation and was found to be prolonged by inhibiting phosphatase activity. Inhibitors of phosphatase 1 and 2A indeed prevented recovery of the facilitation of evoked glutamate release lost following a first "stimulation" with group I mGluR agonists (Sistiaga and Sanchez-Prieto 2000). The targets for the phosphorylation in the PKC-mediated desensitization are likely to be the group I mGluRs themselves, based on studies performed with recombinant receptors in expression systems (see, for a review, Alagarsamy et al. 2001).

The exact identity of the autoreceptors of group I mGluR (whether mGluR1 or mGluR5 or both) undergoing the desensitization-dependent switch in glutamate exocytosis regulation has long remained elusive, mainly due to the lack of selective mGluR1 and mGluR5 ligands. Reid et al. (1999) exploited some new ligands to characterize pharmacologically these autoreceptors. Two novel group I-preferring ligands were utilized: the mGluR5 agonist CHPG (Doherty et al. 1997) and the selective mGluR1 antagonist AIDA (Pellicciari et al. 1995; Moroni et al. 1997). When tested at a concentration reported to be mGluR5 selective, CHPG failed to potentiate 4AP-evoked [$^3$H]glutamate release from cerebrocortical synaptosomes, suggesting that presynaptic mGluR5 on glutamatergic terminals are not involved. On the other hand, the selective mGluR1 antagonist AIDA abolished the 3,5-DHPG potentiation, although at high concentration (1 mM), consistent with the possibility that the group I mGluR under study was of the mGluR1 subtype (Reid et al. 1999).

This conclusion seemed in contrast, however, with the almost concomitant findings of Sistiaga et al. (1998), according to which mGluR1 is not involved in the 3,5-DHPG-mediated facilitation of endogenous glutamate release as judged by experiments in cerebrocortical nerve terminals isolated from mGluR1-deficient mice. Whether technical problems, including insufficient selectivity of the antagonists used, or compensatory changes occurring in mGluR1 knockout animals, could explain the controversy, remained to be established. Whatever the reasons, subsequent works with selective ligands provided strong evidence that mGluR5 plays a major role as a release-enhancing group I autoreceptor.

Thomas et al. (2000) used several, and in part novel, group I mGluR ligands in experiments with rat forebrain slices preloaded with [$^3$H]D-aspartate ([$^3$H]D-ASP) and stimulated electrically. The potentiation of the evoked [$^3$H]D-ASP release produced by group I mGluR agonists was completely insensitive to a number of selective mGluR1 antagonists. In addition, CHPG, a selective mGluR5 agonist, concentration-dependently increased the evoked release of [$^3$H]D-ASP. Unfortunately, during this work, selective mGluR5 antagonists were not yet available, so that the effect of CHPG could only be counteracted by (S)-MCPG, a mixed mGluR1/mGluR5 antagonist. Subsequently, Croucher's group (Fazal et al. 2003) extended the pharmacological characterization of the mGlu autoreceptor thanks to the recent availability of novel mGluR5 selective antagonists. Rat cerebrocortical minislices were prelabeled with [$^3$H]D-ASP and depolarized with high-$K^+$; the evoked release of [$^3$H]D-ASP was enhanced by 3,5-DHPG. The responses to 3,5-DHPG were largely insensitive to mGluR1 antagonists, but were potently antagonized by some selective mGluR5 blockers. Fazal et al. (2003) concluded that the group I mGluRs sited on glutamatergic terminals in the cerebral cortex are autoreceptors of the mGluR5 subtype (see Figure 1, left).

Very recently, Rodrigues et al. (2005) hypothesized that the anti-Parkinsonian (Breysse et al. 2002) and neuroprotective (Battaglia et al. 2002) properties of mGluR5 antagonists could be related in part to the presence of presynaptic mGluR5 that modulate glutamate release on striatal glutamatergic terminals and control, together with dopamine D2 and adenosine $A_{2a}$ receptors, the responsiveness of medium spiny neurons (Ferré et al. 2003). By monitoring glutamate release from superfused purified rat striatal synaptosomes, Rodrigues et al. (2005) found that the mGluR5 agonist CHPG enhanced the $K^+$-evoked release of the excitatory amino acid, an effect prevented by the mGluR5 antagonist MPEP. Furthermore, using a recent method of subsynaptic fractionation, Rodrigues et al. (2005) demonstrated that mGluR5 are localized (as expected) in the postsynaptic density fraction, but also

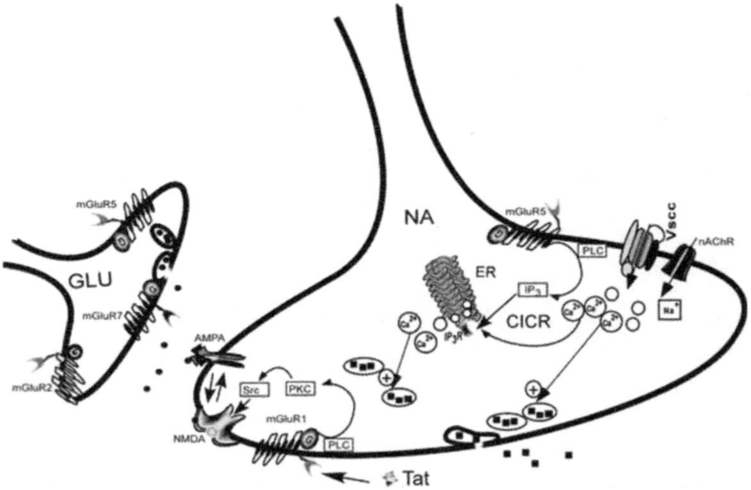


**Fig. 1** Presynaptic mGluRs: metabotropic autoreceptors, metabotropic heteroreceptors on noradrenergic terminals, and their interactions with ionotropic receptors. Group I autoreceptors mediate enhancement of glutamate (solid circles) release and are likely to be mGluR5. Release inhibitory autoreceptors of group II appear to belong to the mGluR2 subtype; they exhibit high affinity for glutamate and are situated perisynaptically. Group III autoreceptors, particularly mGluR7, are localized at the active zone of presynaptic glutamatergic terminals; they exhibit very low affinity for glutamate. Presynaptic mGluR5 on noradrenaline (NA) terminals become functional, i.e., mediate release of NA (solid squares), upon application of appropriate stimuli, like activation of nicotinic cholinergic receptors (nAChRs) co-localized on the same NA terminal (right). Activation of nAChRs evokes NA exocytosis through depolarizing $Na^+$ entry and voltage-sensitive $Ca^{2+}$ channel (VSCC) opening. This influx of $Ca^{2+}$ triggers a $Ca^{2+}$-induced $Ca^{2+}$ release (CICR) from the endoplasmic reticulum (ER) that allows mGluR5 to participate in the NA release by an $IP_3$-dependent process. Presynaptic mGluR1 on NA terminals (left) co-localize with presynaptic NMDARs. When NMDARs are activated to evoke NA exocytosis, mGluR1 contribute to the evoked release by enhancing NMDAR function through the PLC/PKC/Src pathway. The HIV-1 protein Tat is a potent agonist at mGluR1.

in the presynaptic nerve ending membrane. Moreover, by an immunocytochemistry approach, the authors showed that ~50% of the whole nerve terminal population exhibited mGluR5 immunoreactivity, while almost 60% of the glutamatergic terminals co-stained for mGluR5 and $A_{2a}$. Thus, presynaptic mGluR5 are present on striatal nerve terminals. Some of them are autoreceptors on glutamatergic terminals, while others could be located on terminals of neurons whose identity remains to be determined.

Studies with cerebrocortical synaptosomes had reported high PKC activity in unstimulated conditions and in the absence of agonists (Coffey et al. 1994), suggesting the possibility of a basal, constitutive activity of group I mGluRs. If stimulatory presynaptic group I mGluRs on glutamatergic terminals were constitutively active, glutamate release should be inhibited by selective antagonists, particularly by noncompetitive antagonists (also known as negative allosteric modulators) previously reported to inhibit constitutive receptor activity occurring in heterologous expression systems (Gasparini et al. 2002).

Wang and Sihra (2004) indeed observed that SIB1893, a noncompetitive mGluR5 antagonist, inhibited glutamate exocytosis evoked by 4-AP from neocortical synaptosomes, suggesting that SIB1893 prevented an effect mediated by mGluR5 made "constitutively" active by depolarization. According to the authors, a presynaptic mGluR5/DAG/PKC pathway mediates facilitation of glutamate release. Moreover, the reactions that follow the activation of PKC mediated by mGluR5 take place at the $Ca^{2+}$ entry step of the stimulus-exocytosis coupling cascade. Vazquez and Sanchez-Prieto (1997) had previously reported that presynaptic modulation of glutamate release targeted different VSCCs in cerebrocortical synaptosomes. In the work of Wang and Sihra (2004), the inhibition of glutamate release induced by SIB1893 was abrogated by the combined blockade of N- and P/Q-type channels brought about by ω-conotoxin GVIA and ω-agatoxin IVA, consistent with the hypothesis that mGluR5 activation leads to potentiation of glutamate release through a facilitation of N- and P/Q-type $Ca^{2+}$ channel activities. SIB1893 inhibited both the 4AP- and the high-$K^+$-evoked release of glutamate, but did not affect the exocytotic release evoked by the $Ca^{2+}$ ionophore ionomycin, which directly introduces $Ca^{2+}$ into the cytosol without VSCC activation, a finding that, according to Wang and Sihra (2004), rules out any significant effect of mGluR5 activation at steps downstream of $Ca^{2+}$ entry and at the level of the coupling of $Ca^{2+}$ to the exocytotic release machinery itself.

## *2.2 Group II and Group III Metabotropic Glutamate Autoreceptors*

Apart from the studies on the dual action of group I mGluRs on glutamate release and the switch from facilitatory to inhibitory that occurs upon receptor desensitization, there is sufficient functional evidence supporting an inhibitory role on glutamate exocytosis of group II and III mGluRs localized at presynaptic glutamatergic terminals.

Studies by Attwell et al. (1995; 1998a,b) with cerebrocortical synaptosomes prelabeled with [$^3$H]D-ASP and depolarized in superfusion with veratridine, showed that the group II selective mGluR2/3 agonists DCG-IV and 2R,4R-APDC blocked the neuronal release of the glutamate analogue, possibly through inhibition of VSCCs. A number of reports had described modulations of [$^3$H]D-ASP release from *slices* by mGluRs. Of interest, the broad-spectrum agonist 1S,3R-ACPD was found to inhibit the output of [$^3$H]D-ASP evoked by high-$K^+$ from rat striatal slices (Lombardi et al. 1993), whereas it potentiated the evoked release in cortical slices (Lombardi et al. 1994). These different effects of 1S,3R-ACPD probably reflect activation of group I mGluRs in the cortex, but not in the striatum, where inhibitory group II mGluRs might prevail. It should be considered that brain slices include abundant glial components. Astrocytes are known to express mGluR3 and mGluR5 (Winder and Conn 1996), and Bezzi et al. (1998) reported that 1S,3R-ACPD and

3,5-DHPG elicited glutamate exocytotic release from cultured astrocytes, under some particular conditions. Purified synaptosomes contain very modest "gliosome" contaminations (Stigliani et al. 2006) and seem therefore more appropriate as a model to investigate presynaptic auto- and heteroreceptors regulating glutamate release.

Most of the information available on group II mGluRs (see, for instance, Shigemoto et al. 1997) indicates that mGluR2, more than mGluR3, are localized to presynaptic nerve endings of glutamatergic neurons (Figure 1). Experiments with mGluR2 knockout mice are consistent with the view that mGluR2 is a presynaptic receptor that modulates glutamate release in the hippocampus (Yokoi et al. 1996). As determined with cloned mGluR2, the affinity for glutamate is low micromolar (Schoepp et al. 1999); following release, glutamate is thought to reach much higher concentrations in the synaptic cleft, so that mGluR2, if they were localized in the immediate vicinity of the synapse, would be almost saturated by the endogenous ligand. Indeed, although experiments on glutamate release with exogenous mGluR2/3 agonists support a location of mGluR2 on presynaptic terminals, immunocytochemical studies could not show mGluR2 near the active zone of the glutamatergic synapse. Synaptic spillover of glutamate seems therefore to be necessary for activation of presynaptic mGluR2 (see Scanziani et al. 1997). It is likely that mGluR2 plays a role in a neuronal mechanism endowed with essentially protective functions.

The selective group III agonist L-AP4 exhibits high affinity ($\leq 1\,\mu M$) at mGluR4, mGluR6 (almost exclusively expressed in retina) and mGluR8, but low affinity ($>100\,\mu M$) at mGluR7 (see Cartmell and Schoepp 2000). Inhibitory effects of L-AP4 on glutamate release from synaptosomes suggest the involvement of different subtypes of group III receptors depending on the age of the animals and on the brain region examined (East et al. 1995; Vazquez et al. 1995; Herrero et al. 1996; Rodriguez-Moreno et al. 1998; Reid et al. 1999). A recent article by Millán et al. (2002) reported that, in cerebrocortical synaptosomes from adult rats, L-AP4 inhibited 4-AP(1 mM)- or KCl(30 mM)-evoked glutamate release with an $IC_{50}$ of $\simeq 300\,\mu M$. The localization of mGluR7 within presynaptic active zones (Shigemoto et al. 1996; see Figure 1, left), together with the low affinity for glutamate, are consistent with their role as autoreceptors that, in certain synapses, mediate feedback inhibition of glutamate release under normal physiological conditions. Indeed, the low affinity of mGluR7 for glutamate appears as an appropriate characteristic if these receptors are localized within synapses in which glutamate concentrations following release likely reach hundreds of micromolar. It has to be noted that autoreceptors of the mGluR7 subtype do not seem to be present on all glutamatergic terminals in the CNS; their expression may be more concentrated in certain regions and their function could be limited to certain synapses (see below in this section).

The mechanisms underlying the inhibition of glutamate release by mGluR7 have been studied in detail by Millán et al. (2002). The L-AP4 inhibition was abolished by the group III mGluR antagonist CPPG (see Cartmell and Schoepp 2000) and largely prevented by pertussis toxin, suggesting involvement of $G_{i/o}$ proteins. Surprisingly enough, the inhibition of glutamate release by L-AP4 seemed to be unrelated to decreases in intrasynaptosomal cyclic AMP, although group III mGluRs are known to reduce AC activity.

Immunolabeling with mGluR7 antibodies and $Ca^{2+}$-imaging of single nerve terminals revealed that the population of mGluR7-expressing nerve terminals largely responded to L-AP4 by reducing their cytoplasmic $Ca^{2+}$-concentration, consistent with the idea that mGluR7 is the principal presynaptic release-inhibiting autoreceptor in the cerebral cortex of adult rats. Assuming that synaptophysin is a reliable marker for axon terminals, Millán et al. (2002) performed double immunolabeling experiments and found that, in their purified synaptosomal preparation, 25%–35% of the synaptophysin-positive particles also contained mGluR7. These results correlate with those obtained in $Ca^{2+}$-imaging experiments showing that 28% of nerve terminals responded to L-AP4 which, in turn, inhibited by 25% the $Ca^{2+}$-dependent glutamate exocytosis. According to Millán et al. (2002), the majority of the axon terminals expressing mGluR7 may be glutamatergic, other nonglutamatergic nerve endings that possess heteroreceptors of the mGluR7 subtype, i.e., GABAergic nerve endings, representing relatively minor subpopulations of nerve terminals. The apparently limited inhibition of glutamate exocytosis by L-AP4 (25% maximal effect) is probably due to the expression of mGluR7 in only a subpopulation of glutamatergic terminals.

Given that mGluR7 activation reduces the depolarization-evoked rise in cytosolic $[Ca^{2+}]$ and release of glutamate, and considering that the effect of L-AP4 on glutamate release seemed unrelated to changes in cyclic AMP levels, Millán et al. (2002) investigated the involvement of $Ca^{2+}$ channels. The results obtained suggest that presynaptic mGluR7 attenuate the function of N-type VSCCs but not that of P/Q-type channels. The effects of L-AP4 were insensitive to PKA and PKC inhibitors, suggesting a direct interaction between the $G_{i/o}$ proteins linked to mGluR7 and N-type channels.

The finding that the inhibition of the evoked glutamate release by mGluR7 activation occurred without changes in synaptosomal cyclic AMP levels was unexpected, as mentioned, considering that activation of group III mGluRs either in heterologous expression systems or in neuronal preparations had been shown to decrease cyclic AMP levels. However, Millán et al. (2002) found that mGluR7 activation could reduce intrasynaptosomal cyclic AMP levels provided that they were previously increased by forskolin. To conclude, mGluR7 mediates inhibition of glutamate release by signaling through two pathways. In the first, depression of glutamate release is consequent to reduction in the activity of N-type VSCCs and no detectable change in cyclic AMP. In the second pathway, if cyclic AMP levels are high, mGluR7 mediates decrease in the nucleotide levels, thereby counteracting the facilitation of glutamate release by PKA activation.

The presynaptic mGluR7 shows the highest evolutionary conservation within mGluRs; its characterization has however been hampered because no selective ligand was available until very recently. Mitsukawa et al. (2005) have just characterized an mGluR7-selective agonist, AMN082, which is able to activate the mGluR7 signaling via an allosteric site located in the receptor transmembrane domain. Its effects on glutamate release remains to be determined.

## 3 Metabotropic Glutamate Heteroreceptors

The review by Cartmell and Schoepp (2000) covered almost exhaustively the field of mGluRs that are localized on nonglutamatergic terminals and regulate as presynaptic heteroreceptors the release of various transmitters; this section of my chapter will therefore be essentially restricted to recent aspects of the functional pharmacology of presynaptic metabotropic glutamate heteroreceptors.

### *3.1 Modulation of Acetylcholine Release*

Marti et al. (2001) studied effects of mGluR activation on the release of acetylcholine (ACh) in corpus striatum. The authors monitored release of endogenous ACh from rat striatal synaptosomes depolarized in superfusion with high-$K^+$. The group I agonist (S)-3,5-DHPG potentiated the $K^+$-evoked exocytosis of ACh. The effect of (S)-3,5-DHPG was counteracted by a mixture of CPCCOEt, a mGluR1 selective antagonist, and MPEP, a mGluR5 selective antagonist (Gasparini et al. 1999), but not by either antagonist alone. Thus group I mGluRs (mGluR1 or mGluR5 or both) able to enhance the depolarization-evoked release of ACh are present on nerve terminals of striatal cholinergic interneurons. On the other hand, the $K^+$-evoked release of ACh from striatal synaptosomes was inhibited by the group II agonist DCG-IV, an effect prevented by the group II selective antagonist EGLU. To conclude, the study by Marti et al. (2001) supports the view that the terminals of striatal cholinergic interneurons are endowed with release-stimulating mGluRs of group I and release-inhibiting mGluRs of group II. The availability of novel ligands and knockout animals should permit to verify the possible coexistence of mGluR1 and mGluR5 in striatal cholinergic interneurons, as proposed by Pisani et al. (2001), and to identify the group II mGluR subtype involved in the inhibition of ACh release.

Changes in the function of presynaptic group II mGluRs inhibiting ACh release in experimental parkinsonism were then investigated by the Ferrara laboratory (Marti et al. 2003). The effects on ACh release of group II mGluR and NMDA receptor ligands were examined in striatal slices and synaptosomes from naive rats, 6-hydroxydopamine hemilesioned rats, and 6-hydroxydopamine hemilesioned rats chronically treated with L-DOPA plus benserazide. The inhibition of ACh release caused by the group II mGluR agonist LY354740 was reduced on the lesioned side, whereas the potentiating effect of NMDA (Morari et al. 1998) was enhanced. Interestingly, chronic L-DOPA normalized the function of both group II mGluRs and NMDA receptors. According to Marti et al. (2003), it can be predicted that the loss of presynaptic mGluR-mediated inhibition of ACh release and the enhancement of the NMDA-evoked release seen in hemiparkinsonian rats will shift the glutamatergic input to cholinergic interneurons towards enhanced excitability.

Assuming that the above group II mGluRs and NMDARs coexist on the same striatal cholinergic terminals, possible receptor-receptor interactions were investigated (Mela et al. 2006). The group II mGluR agonist LY354740 and NMDA were

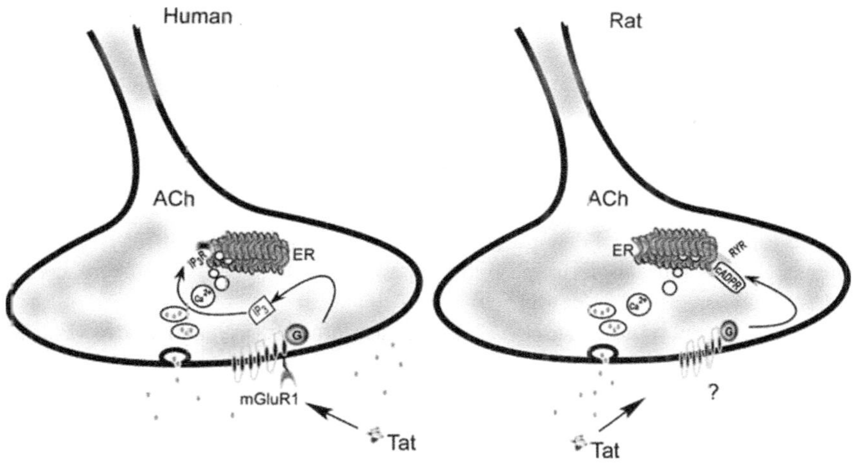


**Fig. 2** Presynaptic mGluRs on human and rat neocortical cholinergic (ACh) nerve endings and effect of the HIV-1 protein Tat. In human neocortex, Tat activates mGluR1 leading to inositol trisphosphate ($IP_3$) production, $IP_3$ receptor ($IP_3R$) activation, mobilization of $Ca^{2+}$ from the endoplasmic reticulum (ER), and vesicular ACh release. In rat neocortex ACh terminals Tat binds to an unidentified receptor whose activation also leads to ACh release. This release again is dependent on intraterminal $Ca^{2+}$, but this is mobilized by ryanodine receptor (RYR) activation via the endogenous agonist cyclic adenosine diphosphoribose (cADPR).

co-applied to striatal slices or synaptosomes. LY354740 inhibited the NMDA-evoked ACh release from both synaptosomes and slices. VSCCs of the P/Q-type were involved, as the modulation was prevented by ω-agatoxin IVA. Co-immunoprecipitation experiments excluded direct interaction between the receptors, whose coexistence is however indicated by the experiments with synaptosomes.

Group I mGluRs were found to exist also on cholinergic terminals of human neocortex (Feligioni et al. 2003). The selective group I mGluRs agonist 3,5-DHPG released [$^3$H]ACh from human neocortical synaptosomes. The effect was prevented by CPCCOEt, suggesting activation of mGluR1 (see Figure 2). Differently from the effect observed by Marti et al. (2001) in striatal cholinergic terminals, activation of the mGluR1 on human neocortical cholinergic terminals caused ACh release in absence of depolarization.

Our laboratory has long been interested in studying effects of the human immunodeficiency virus (HIV-1) proteins gp120 and Tat on neurotransmitter release and on receptors modulating neurotransmitter release (Pittaluga and Raiteri 1994; Pittaluga et al. 1996; Pattarini et al., 1998; Gemignani et al. 2000; Feligioni et al. 2003; Longordo et al. 2006). In fact, these proteins may play important roles in the neuropsychiatric symptoms (termed AIDS dementia) that infected patients frequently develop (Lawrence and Major 2002). When human neocortical synaptosomes prelabeled with [$^3$H]choline were exposed in superfusion to Tat (1 nM), a release of [$^3$H]ACh partly dependent on external $Ca^{2+}$ ions and partly on mobilization of $Ca^{2+}$ from intraterminal stores was observed (Feligioni et al. 2003). Xestospongin

C, a membrane-permeant $IP_3R$ antagonist, inhibited the Tat-induced release of [$^3$H]ACh indicating involvement of $IP_3$-sensitive internal $Ca^{2+}$ stores. Among the various possibilities, Tat might have activated group I mGluRs, possibly of the mGluR1 subtype (Figure 2). Accordingly, the effect of 1 nM Tat on the release of [$^3$H]ACh from human neocortex synaptosomes was prevented by the selective mGluR1 antagonists CPCCOEt and LY367385. Of note, the concentrations of Tat ($\leq$1 nM) able to significantly enhance the release of ACh approach those found extracellularly during HIV-1 infection (Gurwell et al. 2001). Tat may represent a potent pathological agonist at mGluR1 possibly involved in the neurotoxic events characteristic of HIV-1 infection.

Tat also caused release of [$^3$H]ACh in synaptosomes from rat neocortex (Feligioni et al. 2003). However, the mechanisms involved were different from those in human brain. In rats, the release was independent of external $Ca^{2+}$; it was largely dependent on intraterminal $Ca^{2+}$, but insensitive to xestospongin C, excluding involvement of $IP_3Rs$. The Tat effect in rat synaptosomes was blocked by dantrolene, indicating ryanodine-sensitive stores as the origin of intraterminal $Ca^{2+}$ (Figure 2). Ryanodine receptors are thought to be endogenously activated by cADP-ribose (Galione et al. 1993; Lee 2001). When rat cerebrocortical synaptosomes were exposed to Br-cADP-ribose, a competitive antagonist of the cADP-ribose binding site (Walseth and Lee 1993), the Tat-induced release of [$^3$H]ACh was prevented, an effect not observed in human neocortex synaptosomes (Feligioni et al. 2003). The differences between the mechanisms of Tat-evoked ACh release in human and rat cholinergic terminals are further supported by the observation that, in rats, Tat did not act through group I mGluRs since 3,5-DHPG was unable to elicit [$^3$H]ACh release.

In conclusion, Tat can evoke exocytotic release of ACh from human and rat cerebrocortex cholinergic axon terminals, by different mechanisms. In human brain Tat behaves as an extremely potent mGluR1 agonist mediating VSCCs activation and mobilizing intraterminal $Ca^{2+}$ through $IP_3$. Binding of Tat to rat cholinergic terminals leads to ACh release that is also dependent on intraterminal $Ca^{2+}$, which is, however, mobilized by ryanodine receptor activation, probably via cADP-ribose. In rat brain Tat appears to be a potent activator of an as yet unidentified "receptor"' linked to the second messenger cADP-ribose (Figure 2).

### *3.2 Modulation of Noradrenaline Release*

Presynaptic group I mGluRs on noradrenergic axon terminals were identified only very recently (Longordo et al. 2006; Parodi et al. 2006), because the low density of noradrenergic terminals in CNS makes immunocytochemical approaches quite difficult. Moreover, some receptors (and this is one case) may be silent under basal conditions and exhibit function only in presence of appropriate stimuli.

Several laboratories have reported that nicotinic cholinergic receptors (nAChRs) are situated on central noradrenergic terminals where they mediate enhancement of

NA release (see review by Wonnacott 1997). Parodi et al. (2006) found that the external $Ca^{2+}$-dependent overflow of [$^3$H]NA provoked by (−)-nicotine in rat hippocampal synaptosomes was significantly enhanced when the group I mGluR agonist 3,5-DHPG, inactive on its own, was added to the superfusion medium together with (−)-nicotine. This effect was reduced down to the release evoked by (−)-nicotine alone when MPEP, a selective mGluR5 antagonist was also present. Addition of CPCCOEt, a selective mGluR1 antagonist, was instead unable to affect the overflow caused by 3,5-DHPG. The nAChR antagonist mecamylamine abrogated the release of [$^3$H]NA evoked by (−)-nicotine plus 3,5-DHPG. Thus release-enhancing presynaptic mGluR5 exist on noradrenergic terminals of rat hippocampus. Exposure to a group I mGluR agonist alone is insufficient to cause release of NA; a mGluR5-mediated release can however be observed when (−)-nicotine and a group I mGluR agonist are added together, suggesting coexistence on the same noradrenergic terminal of nAChRs and mGluR5 with nAChRs exerting a permissive role on the activation of mGluR5 (Figure 1, right). The 3,5-DHPG-mediated component of the release provoked by (−)-nicotine plus 3,5-DHPG was sensitive to the $IP_3$ receptor antagonist xestospongin C (Parodi et al. 2006).

As to the mechanism of this receptor-receptor interaction, it has to be considered that nAChRs mediating NA release are non-$\alpha$7 receptors highly permeable to $Na^+$. Influx of $Na^+$ can cause membrane depolarization and consequent activation of VSCCs (Kulak et al. 2001). $Ca^{2+}$ influx through VSCCs may trigger a calcium-induced calcium release (CICR) from $Ca^{2+}$ stores primed by $IP_3$ (Berridge 1998); indeed group I mGluRs are positioned close to $IP_3$ receptors on the endoplasmic reticulum by the Homer family proteins (Brakeman et al. 1997; Fagni et al. 2000). In other words, nicotine could first trigger influx of $Na^-$ into noradrenergic terminals and activation of VSCCs; entry of $Ca^{2+}$ would then (1) directly induce NA exocytosis and (2) permit, perhaps through a CICR process, an additional release of NA mediated by mGluR5 (Figure 1, right).

The noradrenergic system is involved in stress-related and anxiety disorders (Bremner et al. 1996). *In vivo* microdialysis studies show that exposure of animals to stressful stimuli leads to increase in the release of noradrenaline in corticolimbic areas including prefrontal cortex and hippocampus (Kawahara et al. 2000; Dazzi et al. 2002). Moreover, there is increasing evidence that excessive glutamatergic transmission also plays a role in the pathophysiology of stress and anxiety conditions (see Moghaddam and Jackson 2004). Interactions between glutamatergic and noradrenergic systems are therefore likely to be enhanced during stressful stimuli. Accordingly, blockade of mGluR5 by MPEP has recently been reported to decrease stress-induced cortical noradrenaline release in rodents (Page et al. 2005) and morphine withdrawal-induced activation of locus coeruleus neurons (Rasmussen et al. 2005). Interestingly, Grilli et al. (2005) have found that nAChRs mediating enhancement of NA release from hippocampal nerve terminals are functionally upregulated soon after withdrawal from chronic nicotine, a condition that, according to the findings of Parodi et al. (2006), may favor mGluR5 activation. In rodents, MPEP was reported to decrease nicotine self-administration (Paterson et al. 2003; Tessari et al. 2004). Considering that some of the behavioral modifications experienced by smokers acutely withdrawn from nicotine represent stress symptoms, possibly related to increased noradrenergic transmission mediated by nAChR upregulation and mGluR5 activation, mGluR5 antagonists may turn out to be useful to quit smoking also by reducing withdrawal stress.

In addition to the interaction of group I mGluRs with nAChRs, an interaction with NMDA receptors on noradrenergic terminals also seems to occur. In particular, noradrenergic terminals in human and rat neocortex (Fink et al. 1992; Pittaluga et al. 1996) and in rat hippocampus (Fink et al. 1990; Pittaluga and Raiteri 1990) are endowed with release-enhancing NMDA receptors which might interact with mGluRs coexisting on the same terminals. The HIV-1 protein Tat played a role in the discovery of the interaction. In the absence of external $Mg^{2+}$, concomitant addition of NMDA and glycine to the superfusion medium elicited release of [$^3$H]NA from human neocortex synaptosomes. Interestingly, when this preparation was exposed to NMDA plus Tat (1 nM), without adding glycine, a release of [$^3$H]NA quantitatively identical to that provoked by NMDA plus 10 μM glycine was observed. This effect was entirely dependent on NMDA receptor activation since it was almost eliminated by the NMDA receptor antagonist MK-801 (Longordo et al. 2006). Results qualitatively similar to those in human neocortex were obtained with rat cortical or hippocampal synaptosomes. Based on these data, the effect of Tat would appear compatible with an extremely potent action at the glycine site of the NMDA receptors. However, differently from the HIV-1 coat protein gp120, which does mimic glycine at NMDARs (Pittaluga and Raiteri 1994; Pattarini et al. 1998), Tat failed to revert that antagonism of the NMDA/Tat effects brought about by the glycine site blocker 7-Cl-kynurenate both in human and rat synaptosomes, a result inconsistent with an action of Tat at the NMDA receptor glycine site (Longordo et al. 2006). Tat and glycine (or gp120) act therefore at different sites on NA terminals, a view also supported by the ability of Tat to further enhance the maximal effect of glycine or gp120 (see Pittaluga and Raiteri 1994; Pittaluga et al. 1996; Longordo et al. 2006).

The possibility existed that Tat could bind to a “receptor,” possibly a mGluR, coexisting with NMDA receptors on noradrenergic terminals and able to interact with the ionotropic receptor. The release of [$^3$H]NA caused by NMDA/Tat or NMDA/glycine/Tat from human or rat synaptosomes was pertussis-toxin sensitive suggesting that Tat acts through $G_{i/o}$ protein-coupled receptors. The Tat-mediated component of the NMDA/glycine/Tat-evoked release was completely blocked by the mGluR1 antagonist CPCCOEt, whereas the mGluR5 antagonist MPEP only exerted a slight nonsignificant inhibition. To confirm the involvement of an mGluR1 subtype, LY367385, a selective mGluR1 antagonist, was tested both in human neocortical and rat neocortical or hippocampal synaptosomes: LY367385 abrogated the Tat-mediated component of the NMDA/glycine/Tat-evoked [$^3$H]noradrenaline release (Longordo et al. 2006). Thus, mGluR1 seems to be localized on human and rat noradrenergic axon terminals where they interact with coexisting NMDA receptors and Tat can activate mGluR1 with extremely high potency (Figure 1). Whether, given the absence of selective mGluR1 agonists, Tat could represent a useful tool in studying mGluR1 remains to be established.

The mechanisms underlying the enhancement of NMDA receptor function by Tat acting at presynaptic mGluR1 have been in part clarified by Longordo et al. (2006). The release of [$^3$H]NA evoked by NMDA/glycine/Tat from human or rat noradrenergic terminals was significantly inhibited by the PLC inhibitor U73122, by the PKC blocker GF 109203X, and by the inhibitor of cytosolic tyrosine kinase (Src) PP2. On the contrary, PKA and $Ca^{2+}$/calmodulin-dependent protein kinase II seemed not to be involved. The inhibitions observed likely concerned the Tat-mediated component of the NMDA/glycine/Tat-evoked release since the inhibitors did not affect the release evoked by NMDA/glycine. The involvement of Src in the NMDAR activation would imply phosphorylation of tyrosine residues. Src is associated with NMDA receptors, and phosphorylation by Src was reported to enhance NMDA receptor currents (Yu et al. 1997). Notably, activation of Src family kinase can be triggered by PKC (Salter and Kalia 2004). Thus, the PLC/PKC/Src pathway seems to couple presynaptic mGluR1 activation and NMDA receptor function in brain noradrenergic terminals. NMDA receptor function is usually studied in the absence of external $Mg^{2+}$. However, NMDA was found to elicit NA release in the presence of physiological concentrations of $Mg^{2+}$ during concomitant activation of AMPA receptors, which coexist with NMDA receptors on noradrenergic terminals (Raiteri et al. 1992). The following scenario could be envisaged *in vivo* (see Figure 1, left): glutamate reaching noradrenergic terminals activates depolarizing AMPA receptors, which in turn permit $Mg^{2+}$ removal and NMDA receptor activation; if Tat mimics glutamate and binds at mGluR1, the NMDA receptor-mediated response could be strongly enhanced. In addition, Tat might also contribute by releasing ACh (Feligioni et al. 2003) onto α7 nicotinic receptors localized on glutamatergic terminals and eliciting glutamate release (Marchi et al. 2002), as well as onto non-α7 nicotinic receptors on noradrenergic terminals able to exert a permissive role on NMDA receptor (Risso et al. 2004) and mGluR5 (Parodi et al. 2006) function.

The inability of Tat (or of the group I agonist 3,5-DHPG; Parodi et al. 2006) to release NA when added alone suggests that the mere activation of mGluR1/5 stimulates the above enzymatic pathway insufficiently to trigger NA release although, alternatively, group I mGluRs might be in a desensitized state (that can be reverted by NMDA) under basal conditions (see Alagarsamy et al. 2001). Nonetheless, presynaptic mGluR1 activation can dramatically promote the functions of NMDA receptors if one considers that these receptors, almost silent when glycine was not added (only the unavoidable glycine contamination present in the solutions at ~50 nM was available), became fully responsive in the presence of a concentration of Tat as low as 1 nM. If glycine is an obligatory glutamate coagonist, a reasonable explanation for the Tat effect could be that the protein causes, through mGluR1 activation and the above phosphorylation pathway, a several-fold increase in the affinity of the glycine site for its natural ligand.

The concentrations of Tat able to activate mGluR1 and modulate directly ACh release (Feligioni et al. 2003) or, through interactions with NMDA receptors, NA release (Longordo et al. 2006) are lower than those reported to cause overt neurotoxicity (Nath and Geiger 1998). It is therefore possible that mGluR1 represents a major target through which Tat produces early impairments in neurotransmission independent of cell death. Since Tat or its active fragments cannot be removed as easily as glutamate in the CNS, mGluR1 could therefore be "pathologically" activated during HIV-1 infection. The persistent effects of Tat may provoke transmitter release outside of the physiological range and with abnormal kinetic characteristics, potentially contributing to cognitive impairments occurring in AIDS-dementia. The situation could become quite serious because gp120 also could be present: the HIV-1 coat protein is a potent agonist at the glycine site of NMDA receptors and its effects can be additive to those of Tat (Nath et al. 2000; Longordo et al. 2006). Assuming that Tat and some of its fragments can behave as pathological mGluR1 agonists, AIDS-dementia symptoms might be controlled, in part, by mGluR1 antagonists or mGluR1 negative allosteric modulators.

# 4 Presynaptic $GABA_B$ Receptors

In studies on modulation of transmitter release from CNS slices, Bowery et al. (1980) found that (-)baclofen and GABA decreased the evoked release of noradrenaline, dopamine, and serotonin by acting at a novel site. These inhibitory actions of (-)baclofen and GABA were not blocked by bicuculline, were not mimicked by the GABA analogue isoguvacine, and were not mediated by $Cl^-$ influx, all of which are characteristics of the classical ionotropic GABA receptor. The classical GABA receptor was insensitive to baclofen which, in ligand-binding studies, showed distinct attachment sites on CNS membranes. For these reasons, the term $GABA_B$ receptor was proposed for the baclofen-sensitive site to distinguish it from the bicuculline-sensitive receptor, which was, in turn, designated $GABA_A$ (Hill and Bowery 1981).

Differently from the ionotropic $GABA_A$ receptors, $GABA_B$ receptors ($GABA_B$Rs) are metabotropic GPCRs (Karbon and Enna 1985) endowed with peculiar structural and pharmacological properties. Contained in the present article is an overview of recent reports on release-modulating $GABA_B$Rs, concentrating on the mystery of their unexplained pharmacological heterogeneity. Readers desiring additional or more detailed information on particular aspects of this topic may consult other sources (e.g., Bowery 1993, 2006; Bonanno and Raiteri 1993a; Bowery and Enna 2000; Bowery et al. 2002; Bettler et al. 2004; Couve et al. 2004; Raiteri 2006).

## *4.1 Structural Characteristics and Distribution of* $GABA_B$ *Rs*

Some of the structural properties of $GABA_B$Rs appear unique. It is widely accepted that GPCRs can be expressed as a single protein or form homodimers (Bouvier 2001). Differently, $GABA_B$Rs exist as heterodimers with the subunits designated $GABA_{B1}$ and $GABA_{B2}$ (Jones et al. 1998; Kaupmann et al. 1998a; White et al.

1998; Kuner et al. 1999; Ng et al. 1999). The $GABA_{B1}$ subunit was first cloned (Kaupmann et al. 1997); however, this seven-transmembrane-spanning protein with homology to mGluRs exhibited poor binding affinities for agonists and inefficient coupling to presumed $GABA_B$ effector systems in heterologous cells. It was subsequently shown (Couve et al. 1998) that $GABA_{B1}$ needs to be transported from the endoplasmic reticulum to the plasma membrane by the $GABA_{B2}$ subunit, which not only serves to escort $GABA_{B1}$ to the cell surface, but is likely to include the receptor portion that links to the G protein, while $GABA_{B1}$ is critical for agonist activation (Margeta-Mitrovic et al. 2000; Pagano et al. 2001). It is now generally accepted that the coupling of $GABA_{B2}$ with $GABA_{B1}$ yields a full functional $GABA_BR$ in which the $GABA_{B1}$ subunit, now expressed as a part of the heterodimer, exhibits agonist affinity similar to that of the wild-type $GABA_B$ receptor (Kaupmann et al. 1998a; White et al. 1998). In other words, the binding of GABA or $GABA_BR$ agonists to $GABA_{B1}$ causes activation of G protein via $GABA_{B2}$. $GABA_BRs$ are located in the CNS both pre- and postsynaptically, where they are coupled to $Ca^{2+}$ and $K^+$ channels which mediate inhibition of transmitter release and neuronal hyperpolarization upon receptor activation.

Some splice variants of the $GABA_{B1}$ subunit have been identified (see Bettler et al. 2004 for a review). Most studies concern the two variants of the $GABA_{B1}$ gene originally identified, $GABA_{B1(a)}$ and $GABA_{B1(b)}$, which originate molecules that differ at the extreme amino terminal domain (Kaupmann et al. 1997; see Sections 4.3 and 4.4).

In the CNS, $GABA_B$ binding sites are widely distributed and particularly abundant in the thalamic nuclei, the molecular layer of the cerebellum, the cerebral cortex, and the dorsal horn of the spinal cord (Bowery et al. 1987). Based on in situ hybridization studies of mRNA for the two major splice variants of $GABA_{B1}$, $GABA_{B1(a)}$ exhibits association with presynaptic receptors in rat and human cerebellum and spinal cord, whereas $GABA_{B1(b)}$ is predominantly postsynaptic (Kaupmann et al. 1998b; Billinton et al. 1999; Bischoff et al. 1999; Princivalle et al. 2000; Towers et al. 2000). In other CNS regions, however, $GABA_{B1(b)}$ appears to be on presynaptic terminals and $GABA_{B1(a)}$ is postsynaptically localized (Benke et al. 1999; Princivalle et al. 2001), suggesting that a cellular location or function cannot be generally attributed to a given $GABA_{B1}$ splice variant.

## 4.2 *Presynaptic* $GABA_B$ *Autoreceptors and Heteroreceptors*

Compelling evidence indicates that a major function of $GABA_BRs$ is to mediate inhibition of neurotransmitter release from nerve terminals where they are localized as presynaptic auto- and heteroreceptors (see Bonanno and Raiteri 1993a; Bowery et al. 2002; Raiteri 2006, for reviews).

Electrophysiological studies in mammalian CNS suggested subtle distinctions between pre- and postsynaptic $GABA_BRs$ (see Nicoll 2004 for a review). Stronger evidence that pharmacological differences may exist between $GABA_B$ presynaptic

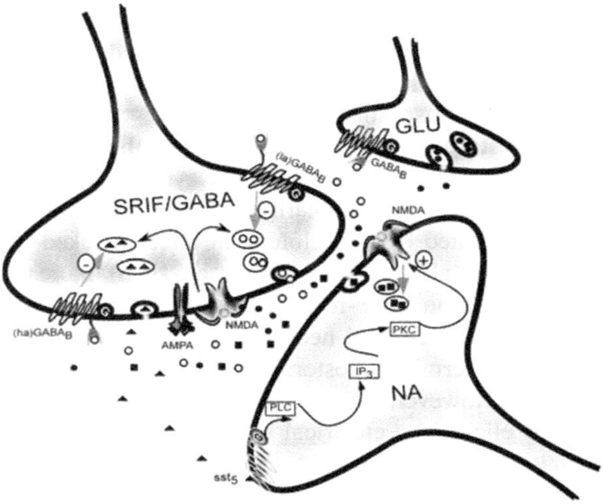


**Fig. 3** Mechanisms that may underlie the cognitive enhancing properties of the GABA$_B$R antagonist CGP36742 (now termed SGS742). GABA and somatostatin (SRIF) are often co-stored in and co-released from interneurons. Glutamate (GLU; black dots) evokes release of noradrenaline (NA; solid squares), GABA (white dots) and SRIF (solid triangles) by activating ionotropic NMDA/AMPA receptors located on the respective releasing axon terminals. GABA can inhibit its own release and SRIF release through pharmacologically distinct GABA$_B$R subtypes (see Table 1). CGP36742 (SGS742) binds with high affinity to the GABA$_B$ receptor (ha, high affinity, GABA$_B$) modulating SRIF release but with very low affinity to the GABA$_B$ receptor (la, low affinity GABA$_B$) modulating GABA release (Bonanno et al. 1999). SRIF enhances NA release by activating sst$_5$ receptors colocalized with NMDAR on noradrenergic terminals. Upon activation of NMDA receptors on GABAergic terminals, GABA is released and inhibits the release of SRIF. CGP36742 (SGS742) prevents this GABAergic inhibition, allowing the neuropeptide to reach the sst$_5$ receptors on NA terminals. These SRIF receptors trigger potentiation of coexisting NMDAR receptors through the phospholipase C (PLC)-inositol trisphosphate (IP$_3$)-protein kinase C (PKC) pathway (see Pittaluga et al. 2005). Figure redrawn from Raiteri (2006).

illustrated in Figure 3 in which SRIF and GABA are shown as cotransmitters whose release can be inhibited through GABA$_B$ autoreceptors exhibiting high affinity (ha) and low affinity (la) for CGP36742/SGS742, respectively. It has to be noted that CGP36742/SGS742 is the first GABA$_B$R antagonist entered into clinical trials as a possible treatment of mild cognitive impairments and, more recently, of more severe conditions such as Alzheimer's disease (Froestl et al. 2004).

Thus the evidence for pharmacologically distinct subtypes of the presynaptic GABA$_B$R derived from release studies with superfused synaptosomes appears to at least equate with other receptor systems which can boast the chrism of molecular biology evidence. No doubt, from the data reported in Table 1, the existence of subtypes of the GABA$_B$R seems quite obvious to a pharmacologist, with strong similarities between human and rat receptors.

### 4.3 Structural Heterogeneity Does not Reflect Pharmacological Heterogeneity

As a rule, pharmacological heterogeneity is based on structural diversities between the proteins representing receptor subtypes. This seems not to be the case for $GABA_B$ receptors, however, whose pharmacological subtypes are likely to originate through novel, unconventional, and yet unexplained strategies. The mystery of $GABA_BR$ subtypes lies in the so far unsuccessful attempts to identify subunits other than $GABA_{B1}$ and $GABA_{B2}$ and the increasing belief that $GABA_{B1}$ and $GABA_{B2}$ exclusively form all $GABA_B$ receptors. Having assumed such a peculiar structural uniqueness of the heterodimeric $GABA_B$ receptor, the origin of the observed striking pharmacological heterogeneity of release-regulating presynaptic $GABA_BRs$ remains a matter of speculation.

When it became clear that probably all $GABA_BRs$ in the CNS brain'are the sole products of the $GABA_{B1}$ and $GABA_{B2}$ genes, much attention focused on subunit variants as a potential source of the pharmacological heterogeneity observed with native $GABA_BRs$. In general, mixing and matching of isoforms did not lead to $GABA_BRs$ with distinct pharmacological profiles; only in some cases, differences were reported that remain, however, highly controversial (see Bettler et al. 2004, for references).

The most investigated $GABA_BR$ isoforms are $GABA_{B1(a)}$ and $GABA_{B1(b)}$, which differ in their extracellular domain (Kaupmann et al. 1998a). These two isoforms are the most abundant variants and the only variants highly conserved among different species (see Bettler et al. 2004). At the $NH_2$ terminus, the first 147 amino acids of the $GABA_{B1(a)}$ isoform are replaced in $GABA_{B1(b)}$ with a sequence of 18 amino acid residues. Alternative $NH_2$ termini are rather exceptional and do not occur in other families of GPCRs. One major difference between $GABA_{B1(a)}$ and $GABA_{B1(b)}$ consists in the presence of a pair of sushi repeats (see Figure 4) in the $GABA_{B1(a)}$-specific domain (Bettler et al. 1998; Hawrot et al. 1998). Sushi repeats, also known as short consensus repeats, were originally observed in complement proteins and proposed to take part in protein-protein interactions (Blein et al. 2004). Of note, sushi repeats have been rarely, if any, observed in neurotransmitter receptors, adding one more peculiarity to the $GABA_B$ receptor.

The review article of Bettler et al. (2004) reports numerous studies indicating that $GABA_{B1(a)}$ and $GABA_{B1(b)}$ exhibit differences in their spatial and temporal expression patterns. Although these spatial and temporal differences in the expression of the $GABA_{B1(a)}$ and $GABA_{B1(b)}$ subunits are consistent with their separate transcriptional regulation and suggest distinct functional roles, they have so far not been of help in explaining the heterogeneous pharmacology of $GABA_BRs$. Several additional $GABA_{B1}$ isoforms ($GABA_{B1(c\text{-}g)}$) were reported; however, in contrast to the $GABA_{B1(a)}$ and $GABA_{B1(b)}$ variants, the existence of a protein *in vivo* has not yet been demonstrated for any of the $GABA_{B1(c\text{-}g)}$ splice variants. Finally $GABA_{B2}$ splice variants do not seem to occur (Bettler et al. 2004)

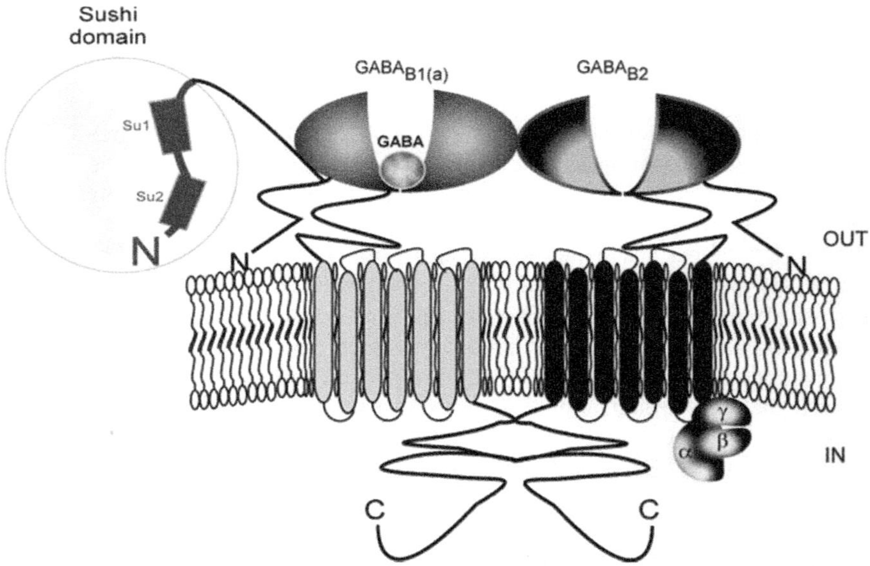


**Fig. 4** Schematic representation of the $GABA_{B1(a)}$-$GABA_{B2}$ heterodimer. The receptor represented contains the 1a variant of the $GABA_{B1}$ subunit. The 1a variant differs from the 1b variant because it contains in the N terminus an additional 120-residue-long sequence consisting of two sushi domains (Su1 and Su2). The sushi domains can interact with a variety of proteins, not only extracellular but also present in the plasma membrane. Such interactions might be neuron-specific and induce conformational changes in the GABA binding site of the $GABA_{B1}$ subunit able to determine binding selectivity towards ligand antagonists and, consequently, "pharmacological" heterogeneity of $GABA_B$ receptors.

If the only prominent molecular diversity in the $GABA_BR$ system resides in the two variants $GABA_{B1(a)}$ and $GABA_{B1(b)}$, it seems justified to concentrate on these variants in efforts aimed to find out if these structural diversities could originate functional and pharmacological $GABA_BR$ heterogeneity. Accordingly, it was recently thought that a genetic approach, i.e., producing mice with selective ablations of the $GABA_{B1(a)}$ or $GABA_{B1(b)}$ subunit to dissociate the native functions of the two subunits, could have been fruitful.

## 4.4 *Structural and Functional Diversity Between Presynaptic and Postsynaptic* $GABA_B$ *Receptors*

Two elegant studies (Pérez-Garci et al. 2006; Vigot et al. 2006) have demonstrated for the first time that $GABA_{B1(a)}$ and $GABA_{B1(b)}$ show strategically distinct subcellular localization and physiological function and may represent the starting point on which robust working hypothesis aimed to understand $GABA_BR$ diversity can be constructed.

Bettler's group (Vigot et al. 2006) used a combined genetic, physiological, and morphological approach to demonstrate that $GABA_{B1}$ isoforms localize to distinct synaptic sites and convey separate functions in the hippocampus *in vivo*. These authors generated homozygous mice with mutations in the $GABA_{B1(a)}$ (referred to as $1a^{-/-}$ mice) or $GABA_{B1(b)}$ ($1b^{-/-}$ mice). Using whole-cell patch-clamp recording in slice preparations, Vigot et al. (2006) first checked for the presence of $GABA_B$ heteroreceptors on excitatory terminals. Baclofen was effective in reducing the excitatory postsynaptic currents in CA1 pyramidal neurons induced by stimulation of the Schaffer collateral-commissural fibers, both in wild-type and $1b^{-/-}$ mice, but not in $1a^{-/-}$ mice. This indicates that $1a^{-/-}$ mice, in contrast to $1b^{-/-}$ mice, lack $GABA_B$ presynaptic heteroreceptors on Schaffer collateral terminals, consistent with the idea that heteroreceptors on glutamate terminals in wild-type animals are $GABA_{B1(a)}$-$GABA_{B2}$ receptors.

The study of Vigot et al. (2006) was then extended to the $GABA_B$ autoreceptors on GABAergic terminals by recording inhibitory postsynaptic currents in CA1 pyramidal neurons. Baclofen reduced the amplitude of the currents in the neurons of all genotypes, suggesting that both $GABA_{B1(a)}$ and $GABA_{B1(b)}$ can efficiently participate in autoreceptor function. Whether GABAergic terminals possess autoreceptors assembled with $GABA_{B1(a)}$ or $GABA_{B1(b)}$ or with both subunits remains to be clarified.

Postsynaptic $GABA_BRs$ induce a late inhibitory postsynaptic current by activating Kir3-type $K^+$ channels (Lüscher et al. 1997). In CA1 pyramidal cells, baclofen induced similar outward currents in $1a^{-/-}$ and wild-type mice; however, in cells of $1b^{-/-}$ mice, the effect of baclofen was reduced by $\sim 60\%$ compared to wild-type or $1a^{-/-}$ mice, indicating that postsynaptic inhibition is predominantly mediated by the $GABA_{B1(b)}$ isoform. Electron microscopy observation in the CA1 stratum radiatum revealed a synaptic distribution of $GABA_{B1}$ isoforms in agreement with the observed functional differences, with $GABA_{B1(b)}$ mostly found on dendritic spines in $1a^{-/-}$ mice, whereas, in $1b^{-/-}$ mice, $GABA_{B1(a)}$ was predominantly localized to axon terminals. Finally, Vigot et al. (2006) showed that transfected CA3 neurons selectively express $GABA_{B1(a)}$ in distal axons, suggesting that the sushi repeats specify heteroreceptor localization of $GABA_BRs$ on presynaptic nerve terminals.

In an accompanying report, Pérez-Garci et al. (2006) used the knockout mice generated according to Vigot et al. (2006) to perform experiments with L5 pyramidal neurons in slices of the primary somatosensory cortex. The authors conclude that postsynaptic GABA-mediated inhibition of $Ca^{2+}$ spikes involves direct blockade of dendritic $Ca^{2+}$ channels and is mediated by $GABA_{B1(b)}$ containing $GABA_BRs$. On the other hand, presynaptic inhibition of GABA release was absent in $1a^{-/-}$ but not $1b^{-/-}$ mice, suggesting that the $GABA_{B1(a)}$ subunit exclusively makes up the presynaptic autoreceptors in the inhibitory terminals of the circuit studied by Pérez-Garci et al. (2006).

The findings that the $GABA_{B1(a)}$ isoform, which possesses the pair of sushi repeats, a conserved protein interaction motif, assembles presynaptic auto- and heteroreceptors, may have important implications. In fact, the pair of sushi repeats in

$GABA_{B1(a)}$ exhibit different structural properties that would favor interactions with multiple auxiliary proteins (Blein et al. 2004), thus generating $GABA_B$ receptor heterogeneity.

## *4.5 Is Pharmacological Heterogeneity Limited to Presynaptic* $GABA_B$ *Receptors?*

The two recent studies by Vigot et al. (2006) and Pérez-Garci et al. (2006) provide convincing evidence that $GABA_{B1(a)}$ and $GABA_{B1(b)}$ isoforms localize to distinct synaptic sites, i.e., presynaptic and postsynaptic, respectively, and convey separate functions, i.e., $GABA_B$ autoreceptor and heteroreceptor-mediated inhibition of GABA and glutamate exocytosis and $GABA_B$R-mediated postsynaptic inhibition. However, there is no mention in these reports of $GABA_B$R pharmacological diversity, which remains therefore an unresolved question.

The literature on $GABA_B$Rs includes several reports regarding the existence of pharmacological receptor subtypes (see reviews by Bonanno and Raiteri 1993a; Bowery 1993, 2006; Mott and Lewis 1994; Kerr and Ong 1995; Malcangio and Bowery 1996; Marshall et al. 1999; Bowery and Enna 2000; Bowery et al. 2002; Calver et al. 2002; Bettler et al. 2004; Couve et al. 2004; Nicoll 2004; Raiteri 2006). The results obtained with various experimental approaches (ligand binding with brain homogenates, electrophysiology, tests with cloned transfected receptors, transmitter release from brain preparations) are either negative or far from conclusive and are often contradictory. In my opinion, one observation deserves particular attention: the only way to clearly show the existence of pharmacologically distinct functional $GABA_B$Rs seems to be monitoring release from superfused synaptosomes (Raiteri et al. 1974; Raiteri and Raiteri 2000), a technique best suited to study presynaptic events and that has been successful in identifying pharmacological subtypes of a variety of receptors for which structural diversity has also been demonstrated by molecular biology.

According to Vigot et al. (2006) and Pérez-Garci et al. (2006), $GABA_{B1(a)}$ assembles release-regulating presynaptic receptors at axon terminals; since only $GABA_{B1(a)}$ possesses the sushi repeats, synaptosomes would represent a subcellular fraction in which $GABA_B$Rs with sushi repeats are particularly concentrated. On the other hand, nerve terminals of different neuronal families are likely to express on their plasma membrane neuron-specific proteins (transporters and so on) which might interact differentially at the sushi repeat sites of the unique $GABA_{B1(a)}$-$GABA_{B2}$ heterodimer expressed on the terminal axons. Such interactions might induce in the cleft of the $GABA_{B1(a)}$ Venus Flytrap module neuron-specific conformational changes able to affect the binding characteristics of $GABA_B$R ligands, especially of antagonists, although changes in the heptahelical domains leading to differential binding of $GABA_B$R allosteric modulators (Chen et al. 2006) can not be ruled out. Such a scenario would be compatible with the existence of $GABA_B$R subtypes with distinct location (terminals of different neurons), function (inhibition of release of different transmitters), and pharmacology (selective blockade by different antagonists) as summarized in some review articles (Bonanno and Raiteri 1993a; Bowery et al. 2002; Raiteri 2006).

If this were the case, the $GABA_BR$ heterogeneity could be largely, though not exclusively, a presynaptic phenomenon compatible with the idea that presynaptic regulation of transmitter release is a major function of $GABA_BRs$. Postsynaptic $GABA_BRs$ mostly contain the $GABA_{B1(b)}$ isoform, which lacks the sushi repeats necessary to bind the neuron-specific proteins that would make the receptors pharmacologically distinct. Considering that, in general, the density of receptors sited on soma and dendrites is much higher than that of the receptors sited on terminal boutons, the presence of pharmacologically distinct presynaptic $GABA_BRs$ could be obscured by the abundant population of identical postsynaptic $GABA_BRs$ during experiments of binding with brain tissue homogenates, which in fact rarely showed signs of pharmacological heterogeneity. Similarly, significant pharmacological differences should not be expected from experiments with cloned heterodimeric receptors expressed in cells unable to produce the neuron-specific proteins able to bind at the sushi repeats (see for references, Bettler et al. 2004) and generate pharmacologically distinct $GABA_BRs$.

## 5 Conclusions and Future Implications

The autoreceptor modulation of glutamate release from presynaptic terminals is very complex, but since glutamate is the major excitatory transmitter in the mammalian CNS, a thorough understanding of how glutamate exocytosis is regulated is of fundamental importance. While group III mGluR7 appear to be autoreceptors relatively well characterized, group I autoreceptors, although the most studied, remain less well understood. Group I mGluRs localized on glutamatergic terminals may on the one side participate in physiological processes of synaptic plasticity; on the other side, inappropriate facilitation of autoreceptor-mediated glutamate release may contribute to a number of pathological states, including ischemia, epilepsy, and neurodegenerative disorders. Given the relevance of these implications, the definition of the relative contributions of the two group I mGluR subtypes to the facilitation of glutamate release, still a matter of debate (Moroni et al. 1998; Sistiaga et al. 1998; Reid et al. 1999; Thomas et al. 2000; Fazal et al. 2003; Rodrigues et al. 2005), is of critical importance.

The investigation on the mGluR subtypes present as presynaptic heteroreceptors on quantitatively minor families of nerve endings, but able to modulate the exocytosis of transmitters involved in important physiopathological conditions, is still in its infancy. Due to the limitations imposed to the immunocytochemical approaches by the low density of most nerve ending families, major advances regarding presynaptic heteroreceptors regulating exocytosis release of classical transmitters and neuropeptides will continue to originate, in my opinion, from functional studies of release.

Opposite to glutamate, the autoreceptor system of the other major CNS transmitter, GABA, appears surprisingly simple: one, and only one, heterodimeric metabotropic receptor, termed $GABA_B$, would subserve the function of inhibiting GABA exocytosis throughout the CNS. Even more impressive, this very same $GABA_B$ heterodimer would regulate the release of tens of transmitters as an heteroreceptor located on the corresponding families of releasing nerve terminals. In the next few years we will probably show that this cannot be the case, after having unveiled the tricks used by the $GABA_B$ receptor to play as an extraordinary quick-change actor.

**Acknowledgements** I thank and acknowledge the contribution of members of my laboratory past and present. I am very grateful to Mrs. Maura Agate for her invaluable help during the preparation of the manuscript. I also wish to thank Dr. Massimo Grilli for the production of the figures for this article.

## References

Alagarsamy S, Sorensen SD, Conn PJ (2001) Coordinate regulation of metabotropic glutamate receptors. Curr Opin Neurobiol 11:357–62

Anwyl R (1999) Metabotropic glutamate receptors: electrophysiological properties and role in plasticity. Brain Res Rev 29:83–120

Aronica E, Dell'Albani P, Condorelli DF et al (1993) Mechanisms underlying developmental changes in the expression of metabotropic glutamate receptors in cultured cerebellar granule cells: homologous desensitization and interactive effects involving N-methyl-D-aspartate receptors. Mol Pharmacol 44:981–9

Asztely F, Erdemli G, Kullmann DM (1997) Extrasynaptic glutamate spillover in the hippocampus: dependence on temperature and the role of active glutamate uptake. Neuron 18:281–93

Attwell PJE, Kaura S, Sigala G et al (1995) Blockade of both epileptogenesis and glutamate release by (1*S*,3*S*)-ACPD, a presynaptic glutamate receptor agonist. Brain Res 698:155–62

Attwell PJE, Koumentaki A, Croucher MJ, Bradford HF (1998a) Specific group II metabotropic glutamate receptor activation inhibits the development of kindled epilepsy in rats. Brain Res 787:286–91

Attwell PJE, Singh Kent N, Jane DE, et al (1998b) Anticonvulsant and glutamate release-inhibiting properties of the highly potent metabotropic glutamate receptor agonist (2*S*,2′*R*,3′*R*)-2-(2′,3′-dicarboxycyclopropyl)glycine (DCG-IV). Brain Res 805:138–43

Battaglia G, Fornai F, Busceti C L et al (2002) Selective blockade of mGlu5 metabotropic glutamate receptors is protective against methamphetamine neurotoxicity. J Neurosci 22:2135–41

Baude A, Nusser Z, Roberts JD et al (1993) The metabotropic glutamate receptor (mGluR1 alpha) is concentrated at perisynaptic membrane of neuronal subpopulations as detected by immunogold reaction. Neuron 11:771–87

Benke D, Honer M, Michel C et al (1999) Gamma-aminobutyric acid type B receptor splice variant proteins GBR1a and GBR1b are both associated with GBR2 *in situ* and display differential regional and subcellular distribution. J Biol Chem 274:27323–30

Berridge MJ (1998) Neuronal calcium signaling, Neuron 21:13–26

Bettler B, Kaupmann K, Bowery N (1998) GABA-B receptors: drugs meet clones. Curr Opin Neurobiol 8:345–50

Bettler B, Kaupmann K, Mosbacher J, Gassmann M (2004) Molecular structure and physiological functions of $GABA_B$ receptors. Physiol Rev 84:835–67

Bezzi P, Carmignoto G, Pasti L et al (1998) Prostaglandins stimulate calcium-dependent glutamate release in astrocytes. Nature 391:281–85

Billinton A, Upton N, Bowery NG (1999) GABA$_B$ receptor isoforms GBR1a and GBR1b, appear to be associated with pre- and post-synaptic elements respectively in rat and human cerebellum. Br J Pharmacol 126:1387–92
Bischoff S, Leonhard S, Reymann N et al (1999) Spatial distribution of GABA$_B$R1 receptor mRNA and binding sites in the rat brain. J Comp Neurol 412:1–16
Blein S, Ginham R, Uhrin D et al (2004) Structural analysis of the complement control protein (CCP) modules of GABA$_B$ receptor 1a. J Biol Chem 279:48292–306
Bockaert J, Pin J-P (1999) Molecular tinkering of G-protein-coupled receptors: an evolutionary success. EMBO J 18:1723–9
Bockaert J, Fagni L, Pin J-P (2002) Metabotropic glutamate receptors (mGluRs). Structure and function. In: Egebjerg J, Schousboe A, Krogsgaard-Larsen P (eds) Glutamate and GABA receptors and transporters. Structure, function and pharmacology. Taylor & Francis, London and New York, pp 121–50
Bonanno G, Raiteri M (1993a) Multiple GABA$_B$ receptors. Trends Pharmacol Sci 14:259–61
Bonanno G, Raiteri M (1993b) Gamma-aminobutyric (GABA) autoreceptors in rat cerebral cortex and spinal cord represent pharmacologically distinct subtypes of the GABA$_B$ receptor. J Pharmacol Exp Ther 265:765–70
Bonanno G, Raiteri M, Emson PC (1988) In vitro release of somatostatin from cerebral cortical slices: Characterization of electrically-evoked release. Brain Res 447:92–7
Bonanno G, Gemignani A, Schmid G et al (1996) Human brain somatostatin release from isolated cortical nerve endings and its modulation through GABA$_B$ receptors. Br J Pharmacol 118: 1441–6
Bonanno G, Fassio A, Schmid G et al (1997) Pharmacologically distinct GABA$_B$ receptors that mediate inhibition of GABA and glutamate release in human neocortex. Br J Pharmacol 120:60–4
Bonanno G, Fassio A, Sala R et al (1998) GABA$_B$ receptors as potential targets for drugs able to prevent excitatory amino acid transmission in the spinal cord. Eur J Pharmacol 362:143–8
Bonanno G, Carita F, Cavazzani P, et al (1999) Selective block of rat and human neocortex GABA$_B$ receptors regulating somatostatin release by a GABA$_B$ antagonist endowed with cognition enhancing activity. Neuropharmacology 38:1789–95
Bouvier M (2001) Oligomerization of G-protein-coupled transmitter receptors. Nat Rev Neurosci 2: 274–86
Bowery NG (1993) GABA$_B$ receptor pharmacology. Annu Rev Pharmacol Toxicol 33:109–47
Bowery NG (2006) GABA$_B$ receptor: a site of therapeutic benefit. Curr Opin Pharmacol 6:37–43
Bowery NG, Enna SJ (2000) Gamma-aminobutyric acid (B) receptors: first of the functional metabotropic heterodimers. J Pharmacol Exp Ther 292:2–7
Bowery NG, Hill DR, Hudson AL et al (1980)(–)Baclofen decreases neurotransmitter release in the mammalian CNS by an action at a novel GABA receptor. Nature 283:92–4
Bowery NG, Hudson AL and Price GW (1987) GABA$_A$ and GABA$_B$ receptor site distribution in the rat central nervous system. Neuroscience 20:365–83
Bowery NG, Bettler B, Froestl W et al (2002) International Union of Pharmacology. XXXIII. Mammalian $\gamma$-aminobutyric acid$_B$ receptors: structure and function. Pharmacol Rev 54:247–64
Brakeman PR, Lanahan AA, O'Brien R et al (1997) Homer: a protein that selectively binds metabotropic glutamate receptors. Nature 386:284–8
Bremner JD, Krystal JH, Southwick SM, Charney DS (1996) Noradrenergic mechanisms in stress and anxiety: II. Preclinical studies. Synapse 23:28–38
Breysse N, Baunez C, Spooren W et al (2002) Chronic but not acute treatment with a metabotropic glutamate 5 receptor antagonist reverses the akinetic deficits in a rat model of parkinsonism. J Neurosci 22:5669–78
Calver AR, Davies CH, Pangalos M (2002) GABA(B) receptors: from monogamy to promiscuity. Neurosignals 11:299–314
Cartmell J, Schoepp DD (2000) Regulation of neurotransmitter release by metabotropic glutamate receptors. J Neurochem 75:889–907
Catania MV, Aronica E, Sortino MA et al (1991) Desensitization of metabotropic glutamate receptors in neuronal cultures. J Neurochem 56:1329–35

Chen Y, Menendez-Roche N, Sher E (2006) Differential modulation by the $GABA_B$ receptor allosteric potentiator 2,6-Di-tert-butyl-4-(3-hydroxy-2,2-dimethylpropyl)-phenol (CGP7930) of synaptic transmission in the rat hippocampal CA1 area. J Pharmacol Exp Ther 317:1170–7
Coffey ET, Herrero I, Sihra TS et al (1994) Glutamate exocytosis and MARCKS phosphorylation are enhanced by a metabotropic glutamate receptor coupled to protein kinase C synergistically activated by diacylglycerol and archidonic acid. J Neurochem 63:1303–10
Conn PJ, Pin J-P (1997) Pharmacology and functions of metabotropic glutamate receptors. Annu Rev Pharmacol Toxicol 37:205–37
Couve A, Filippov AK, Connelly CN et al (1998) Intracellular retention of recombinant GABAB receptors. J Biol Chem 273:26361–7
Couve A, Calver AR, Fairfax B et al (2004) Unravelling the unusual signalling properties of the $GABA_B$ receptor. Biochem Pharmacol 68:1527–36
Dazzi L, Ladu S, Spiga F et al (2002) Chronic treatment with imipramine or mirtazepine antagonizes stress- and FG7142-induced increase in cortical norepinephrine output in freely moving rats. Synapse 43:70–7
Doherty AJ, Palmer MJ, Henley JM et al (1997) (RS)-2-chloro-5-hydroxyphenylglycine (CPHG) activates mGlu5, but no mGlu1, receptors expressed in CHO cells and potentiates NMDA responses in the hippocampus. Neuropharmacology 36:265–7
Dolphin AC, Huston E, Scott RH (1990) $GABA_B$-mediated inhibition of calcium currents: a possible role in presynaptic inhibition. In: Bowery NG, Bittiger H, Olpe HR (eds) $GABA_B$ receptors in mammalian function. John Wiley, Chichester, pp 259–71
East SJ, Hill MO, Brotchie JM (1995) Metabotropic glutamate receptor agonists inhibit endogenous glutamate release from rat striatal synaptosomes. Eur J Pharmacol 277:117–21
Fagni L, Chavis P, Ango F, Bockaert J (2000) Complex interactions between mGluRs, intracellular $Ca^{2+}$ stores and ion channels in neurons. Trends Neurosci 23:80–8
Fassio A, Bonanno G, Cavazzani P, Raiteri M (1994) Characterization of the GABA autoreceptor in human neocortex as a pharmacological subtype of the $GABA_B$ receptor. Eur J Pharmacol 263:311–14
Fazal A, Parker F, Palmer AM, Croucher MJ (2003) Characterisation of the actions of group I metabotropic glutamate receptor subtype selective ligands on excitatory amino acid release and sodium-dependent re-uptake in rat cerebrocortical mini slices. J Neurochem 86:1346–58
Feligioni M, Raiteri L, Pattarini R et al (2003) The human immunodeficiency virus-1 protein Tat and its discrete fragments evoke selective release of acetylcholine from human and rat cerebrocortical terminals through species-specific mechanisms. J Neurosci 23:6810–18
Ferré S, Ciruela F, Woods AS et al (2003) Glutamate $mGlu_5$-adenosine $A_{2A}$–dopamine $D_2$ receptor interactions in the striatum. Implications for drug therapy in neuro-psychiatric disorders and drug abuse. Curr Med Chem. 3:1–26
Fink K, Bönisch H, Göthert M (1990) Presynaptic NMDA receptors stimulate noradrenaline release in the cerebral cortex. Eur J Pharmacol 185:115–17
Fink K, Schultheiss R, Göthert M (1992) Stimulation of noradrenaline release in human cerebral cortex mediated by N-methyl-D-aspartate (NMDA) and non-NMDA receptors. Br J Pharmacol 106:67–72
Froestl W, Mickel SJ, Von Sprecher G et al (1995) Phosphinic acid analogues of GABA. 2. Selective, orally active $GABA_B$ antagonists. J Med Chem 38:3313–31
Froestl W, Gallagher M, Jenkins H et al (2004) SGS742: the first $GABA_B$ receptor antagonist in clinical trials. Biochem Pharmacol 68:1479–87
Gabriel SM, Biere LM, Haroutunian V et al (1993) Widespread deficits in somatostatin but not neuropeptide Y concentrations in Alzheimer's disease cerebral cortex. Neurosci Lett 155:116–20
Gage PW (1992) Activation and modulation of neuronal $K^+$ channels by GABA. Trends Neurosci 15:46–51
Galione A, White A, Willmott N et al (1993) cGMP mobilizes intracellular $Ca^{2+}$ in sea urchin eggs by stimulating cyclic ADP-ribose synthesis. Nature 365:456–9

Galvez T, Joly C, Parmentier M-L et al (1999) Mutagenesis and modeling of the GABA$_B$ receptor extracellular domain support a Venus Flytrap mechanism for ligand binding. J Biol Chem 247:13362–9

Gasparini F, Lingenhohl K, Stoehr N et al (1999) 2-Methyl-6-(phenylethynyl)-pyridine (MPEP), a potent, selective and systemically active mGlu5 receptor antagonist. Neuropharmacology 38:1493–1503

Gasparini F, Kuhn R, Pin JP (2002) Allosteric modulators of group I metabotropic glutamate receptors: novel subtype-selective ligands and therapeutic perspectives. Curr Opin Pharmacol 2:43–9

Gemignani A, Paudice P, Bonanno G, Raiteri M (1994) Pharmacological discrimination between gamma-aminobutyric acid type B receptors regulating cholecystokinin and somatostatin release from rat neocortex synaptosomes. Mol Pharmacol 46:558–62

Gemignani A, Paudice P, Pittaluga A, Raiteri M (2000) The HIV-1 coat protein gp120 and some of its fragments potently activate native cerebral NMDA receptors mediating neuropeptide release. Eur J Neurosci 12:2839–46

Gereau RW, Conn PJ (1995) Multiple presynaptic metabotropic glutamate receptors modulate excitatory and inhibitory synaptic transmission in hippocampal area CA1. J Neurosci 15:6879–89

Grilli M, Parodi M, Raiteri M, Marchi M (2005) Chronic nicotine differentially affects the function of nicotinic receptor subtypes regulating neurotransmitter release. J Neurochem 93:1353–60

Gurwell JA, Nath A, Sun Q et al (2001) Synergistic neurotoxicity of opioids and human immunodeficiency virus-1 Tat protein in striatal neurons in vitro. Neuroscience 102:555–63

Hawrot E, Xiao Y, Shi QL et al (1998) Demonstration of a tandem pair of complement protein modules in GABA-B receptor 1a. FEBS Lett 432:103–8

Hendry SHC, Jones EG, Defelipe J et al. (1984) Neuropeptide-containing neurons of the cerebral cortex are also GABAergic. Proc Natl Acad Sci USA 81:6526–30

Herrero I., Miras-Portugal M.T., Sanchez-Prieto J. (1992) Positive feedback of glutamate exocytosis by metabotropic presynaptic receptor stimulation. Nature 360 1630–6

Herrero I, Castro E, Miras-Portugal MT, Sanchez-Prieto J (1996) Two components of glutamate exocytosis differentially affected by presynaptic modulation. J Neurochem 67:2346–54

Herrero I, Miras-Portugal MT, Sanchez-Prieto J (1998) Functional switch from facilitation to inhibition in the control of glutamate release by metabotropic glutamate receptors. J Biol Chem 273:1951–8

Hill DR, Bowery NG (1981) $^3$H-baclofen and $^3$H-GABA bind to bicuculline-insensitive GABA$_B$ sites in rat brain. Nature 290:149–52

Hill DR, Bowery NG, Hudson AL (1984) Inhibition of GABA$_B$ receptor binding by guanyl nucleotides. J Neurochem 42:652–7

Jones KA, Borowsky B, Tamm JA et al (1998) GABA$_B$ receptors function as a heteromeric assembly of the subunits GABA$_B$R1 and GABA$_B$R2. Nature 396:674–9

Karbon EW, Enna SJ (1985) Characterization of the relationship between gamma-aminobutyric acid B agonists and transmitter-coupled cyclic nucleotide-generating systems in rat brain. Mol Pharmacol 27:53–9

Karbon EW, Duman RS, Enna SJ (1984) GABA$_B$ receptors and norepinephrine-stimulated cAMP production in rat brain cortex. Brain Res 306:327–32

Kaupmann K, Huggel K, Heid J et al (1997) Expression cloning of GABA$_B$ receptors uncovers similarity to metabotropic glutamate receptors. Nature 386:239–46

Kaupmann K, Malitschek B, Schuler V et al (1998a) GABA$_B$-receptor subtypes assemble into functional heteromeric complexes. Nature 396:683–7

Kaupmann K, Schuler V, Mosbacher J et al (1998b) Human gamma-aminobutyric acid type B receptors are differentially expressed and regulate inwardly rectifying $K^+$ channels. Proc Natl Acad Sci USA 95:14991–6

Kawahara H, Kawahara Y, Westerink BHC (2000) The role of afferents to the locus coeruleus in the handling stress-induced increase in the release of noradrenaline in the medial prefrontal cortex: a dual probe microdialysis study in the rat brain. Eur J Pharmacol 387:279–86

Kerr DIB, Ong J (1995) GABAB receptors. Pharmacol Ther 67:187–246

Kulak JM, McIntosh JM, Yoshikami D, Olivera BM (2001) Nicotine-evoked transmitter release from synaptosomes: functional association of specific presynaptic acetylcholine receptors and voltage-gated calcium channels. J Neurochem 77:1581–9

Kuner R, Grunewald KH, Eisenhardt G et al (1999) Role of heteromer formation in $GABA_B$ receptor function. Science 283:74–7

Lawrence DM, Major EO (2002) HIV-1 and the brain: connections between HIV-1-associated dementia, neuropathology and neuroimmunology. Microbes Infect 4:301–8

Lee HC (2001) Physiological functions of cyclic ADP-ribose and NAADP as calcium messengers. Annu Rev Pharmacol Toxicol 41:317–45

Lombardi G, Alesiani M, Leonardi P et al (1993) Pharmacological characterization of the metabotropic glutamate receptor inhibiting D-[$^3$H]-aspartate output in rat striatum. Br J Pharmacol 110:407–12

Lombardi G, Pellegrini-Giampietro DE, Leonardi P et al (1994) The depolarization-induced outflow of D-[$^3$H]aspartate from rat brain slices is modulated by metabotropic glutamate receptors. Neurochem Int 24:525–32

Longordo F, Feligioni M, Chiaramonte G et al (2006) The human immunodeficiency virus-1 protein transactivator of transcription up-regulates N-methyl-D-aspartate receptor function by acting at metabotropic glutamate receptor 1 receptors coexisting on human and rat brain noradrenergic neurones. J Pharmacol Exp Ther 317:1097–1105

Lüscher C, Jan LY, Stoffel M et al (1997) G protein-coupled inwardly rectifying $K^+$ channels (GIRKs) mediate postsynaptic but not presynaptic transmitter actions in hippocampal neurons. Neuron 19:687–95

Malcangio M, Bowery NG (1996) GABA and its receptors in the spinal cord. Trends Pharmacol Sci 17:457–62

Manzoni O, Bockaert J (1995) Metabotropic glutamate receptors inhibiting excitatory synapses in the CA1 area of rat hippocampus. Eur J Neurosci **7**:2518–23

Marchi M, Risso F, Viola C et al (2002) Direct evidence that release-stimulating $\alpha 7^*$ nicotinic cholinergic receptors are localized on human and rat brain glutamatergic axon terminals. J Neurochem 80:1071–8

Margeta-Mitrovic M, Jan YN, Jan LY (2000) A trafficking checkpoint controls $GABA_B$ receptor heterodimerization. Neuron 27:97–106

Marshall FH, Jones KA, Kaupmann K, Bettler B (1999) $GABA_B$ receptors—the first 7TM heterodimers. Trends Pharmacol Sci 20:396–9

Marti M, Paganini F, Stocchi S et al (2001) Presynaptic group-I and II metabotropic glutamate receptors oppositely modulate striatal acetylcholine release. Eur J Neurosci 14:1181–4

Marti M, Paganini F, Stocchi S et al (2003) Plasticity of glutamatergic control of striatal acetylcholine release in experimental parkinsonism: opposite changes at group-II metabotropic and NMDA receptors. J Neurochem 84:792–802

Masugi M, Yokoi M, Shigemoto R et al (1999) Metabotropic glutamate receptor subtype 7 ablation causes deficit in fear response and conditioned taste aversion. J Neurosci 19:955–63

Mela F, Marti M, Fiorentini C et al (2006) Group-II metabotropic glutamate receptors negatively modulate NMDA transmission at striatal cholinergic terminals: role of P/Q-type high voltage activated $Ca^{2+}$ channels and endogenous dopamine. Mol Cell Neurosci 31:284–92

Millán C, Luj NR, Shigemoto R, Sánchez-Prieto J (2002) The inhibition of glutamate release by metabotropic glutamate receptor 7 affects both $[Ca^{2+}]_c$ and cAMP. Evidence for a strong reduction of $Ca^{2+}$ entry in single nerve terminals. J Biol Chem 277:14092–101

Mitsukawa K, Yamamoto R, Ofner S et al (2005) A selective metabotropic glutamate receptor 7 agonist: activation of receptor signaling via an allosteric site modulates stress parameters *in vivo*. Proc Natl Acad Sci USA 102:18712–17

Moghaddam B, Jackson M (2004) Effect of stress on prefrontal cortex function. Neurotox Res 6:1–6

Mondadori C, Jaekel J, Preiswerk G (1993) CGP 36742: the first orally active $GABA_B$ blocker improves the cognitive performance of mice, rats, and rhesus monkeys. Behav Neural Biol 60:62–8

Morari M, Sbrenna S, Marti M et al (1998) NMDA and non-NMDA ionotropic glutamate receptors modulate striatal acetylcholine release via pre- and postsynaptic mechanisms. J Neurochem 71:2006–17
Moroni F, Lombardi G, Thomsen C et al (1997) Pharmacological characterization of 1-aminoindan-1, 5-dicarboxylic acid, AIDA, a potent mGluR1 antagonist. J Pharmacol Exp Ther 281:721–9
Moroni F, Cozzi A, Lombardi G et al (1998) Presynaptic $mGlu_1$ type receptors potentiate transmitter output in the rat cortex. Eur J Pharmacol 347:189–95
Mott DD, Lewis DV (1994) The pharmacology and function of central GABAB receptors. Int Rev Neurobiol 36:97–223
Nath A, Geiger JD (1998) Neurobiological aspects of HIV infections: neurotoxic mechanisms. Prog Neurobiol 54:19–3
Nath A, Haughey NJ, Jones M et al (2000) Synergistic neurotoxicity by human immunodeficiency virus proteins Tat and gp120: protection by memantine. Ann Neurol 47:186–94
Ng GYK, Clark J, Coulombe N et al (1999) Identification of a $GABA_B$ receptor subunit, gb2, required for functional $GABA_B$ receptor activity. J Biol Chem 274:7607–10
Nicoll RA (2004) My close encounter with $GABA_B$ receptors. Biochem Pharmacol 68:1667–74
O'Hara PJ, Sheppard PO, Thogersen H et al (1993) The ligand-binding domain in metabotropic glutamate receptors is related to bacterial periplasmic binding proteins. Neuron 11:41–52
Ottersen OP, Landsend AS (1997) Organization of glutamate receptors at the synapse. Eur J Neurosci 9:2219–24
Pace A, Gama L, Breitwieser GE (1999) Dimerization of the calcium-sensing receptor occurs within the extracellular domain and is eliminated by Cys-Ser mutations at Cys101 and Cys236. J Biol Chem 274:11629–34
Pagano A, Rovelli G, Mosbacher J et al (2001) C-terminal interaction is essential for surface trafficking but not for heteromeric assembly of $GABA_B$ receptors. J Neurosci 21:1189–1202
Page ME, Szeliga P, Gasparini F, Cryan JF (2005) Blockade of the mGlu5 receptor decreases basal and stress-induced cortical norepinephrine in rodents. Psychopharmacology (Berl) 179:240–6
Parodi M, Patti L, Grilli M et al (2006) Nicotine has a permissive role on the activation of metabotropic glutamate 5 receptors coexisting with nicotinic receptors on rat hippocampal noradrenergic nerve terminals. Neurochem Int 48:138–43
Paterson NE, Semenova S, Gasparini F, Marku A (2003) The mGluR5 antagonist MPEP decreased nicotine self-administration in rats and mice. Psychopharmacology 167:257–64
Pattarini R, Pittaluga A, Raiteri M (1998) The human immunodeficiency virus-1 envelope protein gp120 binds through its V3 sequence to the glycine site of N-methyl-D-aspartate receptors mediating noradrenaline release in the hippocampus. Neuroscience 87:147–57
Pellicciari R, Luneia R, Constantino G et al (1995) 1-Aminoindan-1,5-dicarboxylic acid: a novel antagonist at phospholipase C-linked metabotropic glutamate receptors. J Med Chem 38:3717–19
Pérez-Garci E, Gassmann M, Bettler B, Larkum ME (2006) The $GABA_{B1b}$ isoform mediates long-lasting inhibition of dendritic $Ca^{2+}$ spikes in layer 5 somatosensory pyramidal neurons. Neuron 50:603–16
Pin J-P, Bockaert J (1995) Get receptive to metabotropic glutamate receptors. Curr Opin Neurobiol 5:342–9
Pin J-P, Galvez T, Prézeau L (2003) Evolution, structure, and activation mechanism of family 3/C G protein-coupled receptors. Pharmacol Ther 98:325–54
Pisani A, Bonsi P, Centonze D et al (2001) Functional coexpression of excitatory mGluR1 and mGluR5 on striatal cholinergic interneurons. Neuropharmacology 40:460–3
Pittaluga A, Raiteri M (1990) Release-enhancing glycine-dependent presynaptic NMDA receptors exist on noradrenergic terminals of hippocampus. Eur J Pharmacol 191:231–4
Pittaluga A, Raiteri M (1994) HIV-1 envelope protein gp120 potentiates NMDA-evoked noradrenaline release by a direct action at rat hippocampal and cortical noradrenergic nerve endings. Eur J Neurosci 6:1743–9

Pittaluga A, Pattarini R, Severi P, Raiteri M (1996) Human brain N-methyl-D-aspartate receptors regulating noradrenaline release are positively modulated by HIV-1 coat protein gp120. AIDS 10:463–8
Pittaluga A, Bonfanti A, Raiteri M (2000) Somatostatin potentiates NMDA receptor function via activation of $InsP_3$ receptors and PKC leading to removal of the $Mg^{2+}$ block without depolarization. Br J Pharmacol 130:557–66
Pittaluga A, Feligioni M, Ghersi C et al (2001) Potentiation of NMDA receptor function through somatostatin release: a possible mechanism for the cognition-enhancing activity of $GABA_B$ receptor antagonists. Neuropharmacology 41:301–10
Princivalle A, Spreafico R, Bowery N, de Curtis M (2000) Layer-specific immunocytochemical localization of $GABA_BR1a$ and $GABA_BR1b$ receptors in the rat piriform cortex. Eur J Neurosci 12:1516–20
Princivalle AP, Pangalos MN, Bowery NG, Spreafico R (2001) Distribution of $GABA_{B(1a)}$, $GABA_{B(1b)}$ and $GABA_{B2}$ receptor protein in cerebral cortex and thalamus of adult rats. Neuroreport 12:591–95
Raiteri L, Raiteri M (2000) Synaptosomes still viable after 25 years of superfusion. Neurochem Res 25:1265–74
Raiteri M (2006) Functional pharmacology in human brain. Pharmacol Rev 58:162–93
Raiteri M, Angelini F, Levi G (1974) A simple apparatus for studying the release of synaptosomes from neurotransmitters. Eur J Pharmacol 25:411–14
Raiteri M, Garrone B, Pittaluga A (1992) N-methyl-D-aspartic acid (NMDA) and non-NMDA receptors regulating hippocampal norepinephrine release. II. Evidence for functional cooperation and for coexistence on the same axon terminal. J Pharmacol Exp Ther 260:238–42
Raiteri M, Bonanno G, Paudice P et al (1996) Human brain cholecystokinin release of cholecystokinin-like immunoreactivity (CCK-LI) from isolated cortical nerve endings and its modulation through $GABA_B$ receptors. J Pharmacol Exp Ther 228:747–51
Rasmussen K, Martin H, Berger JE, Seager MA (2005) The mGlu5 receptor antagonists MPEP and MTEP attenuate behavioral signs of morphine withdrawal and morphine-withdrawal-induced activation of locus coeruleus neurons in rats. Neuropharmacology 48:173–80
Reid ME, Toms NJ, Bedingfield JS, Roberts PJ (1999) Group I mGlu receptors potentiate synaptosomal $[^3H]$glutamate release independently of exogenously applied arachidonic acid. Neuropharmacology 38:477–85
Risso F, Grilli M, Parodi M et al (2004) Nicotine exerts a permissive role on NMDA receptor function in hippocampal noradrenergic terminals. Neuropharmacology 47:65–71
Rodrigues RJ, Alfaro TM, Rebola N et al (2005) Co-localization and functional interaction between adenosine $A_{2A}$ and metabotropic group 5 receptors in glutamatergic nerve terminals of the rat striatum. J Neurochem 92:433–41
Rodriguez-Moreno A, Sistiaga A, Lerma J, Sanchez-Prieto J (1998) Switch from facilitation to inhibition of excitatory synaptic transmission by group I mGluR desensitization. Neuron 21 1477–86
Romano C, Yang W-L, O'Malley KL (1996) Metabotropic glutamate receptor 5 is a disulfide-linked dimer. J Biol Chem 271:28612–16
Salter MW, Kalia LV (2004) Src kinases: a hub for NMDA receptor regulation. Nat Rev Neurosci 5:317–28
Scanziani M, Salin PA, Vogt KE et al (1997) Use-dependent increases in glutamate concentration activate presynaptic metabotropic glutamate receptors. Nature 385:630–4
Schoepp DD, Goldsworthy J, Johson BG et al (1994) 3,5-dihydroxyphenylglycine is a highly selective agonist for phosphoinositide linked metabotropic glutamate receptors in the rat hippocampus. J Neurochem 63:769–72
Schoepp DD, Jane DE, Monn JA (1999) Pharmacological agents acting at subtypes of metabotropic glutamate receptors. Neuropharmacology 38:1431–76
Shigemoto R, Kulik A, Roberts JDB et al (1996) Target-cell-specific concentration of a metabotropic glutamate receptor in the presynaptic active zone. Nature 381:523–5

Shigemoto R, Kinoshita A, Wada E et al (1997) Differential presynaptic localization of metabotropic glutamate receptor subtypes in the rat hippocampus. J Neurosci 17:7503–22

Sistiaga A, Sanchez-Prieto J (2000) Protein phosphatase 1 and 2° inhibitors prolong the switch in the control of glutamate release by group I metabotropic glutamate receptors: characterization of the inhibitory pathway. J Neurochem 75:1566–74

Sistiaga A, Herrero I, Conquet F, Sanchez-Prieto J (1998) The metabotropic glutamate receptor 1 is not involved in the facilitation of glutamate release in cerebrocortical nerve terminals. Neuropharmacology 37 1485–92

Starke K, Göthert M, Kilbinger H (1989) Modulation of neurotransmitter release by presynaptic autoreceptors. Physiol Rev 69:864–989

Stigliani S, Zappettini S, Raiteri L et al (2006) Glia re-sealed particles freshly prepared from adult rat brain are competent for exocytotic release of glutamate. J Neurochem 96:656–68

Tessari M, Pilla M, Andreoli M et al (2004) Antagonism at metabotropic glutamate 5 receptors inhibits nicotine- and cocaine-taking behaviours and prevents nicotine-triggered relapse to nicotine-seeking. Eur J Pharmacol 499:121–33

Thomas LS, Jane DE, Harris JR, Croucher MJ (2000) Metabotropic glutamate autoreceptors of the mGlu(5) subtype positively modulate neuronal glutamate release in the rat forebrain in vitro. Neuropharmacology 39:1554–66

Towers S, Princivalle A, Billinton A et al (2000) $GABA_B$ receptor protein and mRNA distribution in rat spinal cord and dorsal root ganglia. Eur J Neurosci 12:3201–10

Vazquez E, Sanchez-Prieto J (1997) Presynaptic modulation of glutamate release targets different calcium channels in rat cerebrocortical nerve terminals. Eur J Neurosci 9:2009–18

Vazquez E, Herrero I, Miras-Portugal MT, Sanchez-Prieto J (1995) Developmental change from inhibition to facilitation in the presynaptic control of glutamate exocytosis by metabotropic glutamate receptors. Neuroscience 68:117–24

Vigot R, Barbieri S, Bräuner-Osborne H et al (2006) Differential compartmentalization and distinct functions of $GABA_B$ receptor variants. Neuron 50:589–601

Waldmeier PC, Wicki P, Feldtrauer J-J, Baumann PA (1988) Potentiation involvement of a baclofen-sensitive autoreceptor in the modulation of the release of endogenous GABA from rat brain slices in vitro: Naunyn-Schmiedeb Arch Pharmacol 337:289–95

Walseth TF, Lee HC (1993) Synthesis and characterization of antagonists of cyclic ADP ribose-induced $Ca^{2+}$ release. Biochim Biophys Acta 1178:235–42

Wang S-J, Sihra TS (2004) Noncompetitive metabotropic glutamate5 receptor antagonist (E)-2-methyl-6-styryl-pyridine (SIB1893) depresses glutamate release through inhibition of voltage-dependent $Ca^{2+}$ entry in rat cerebrocortical nerve terminals (synaptosomes). J Pharmacol Exp Ther 309:951–8

White JH, Wise A, Main MJ et al (1998) Heterodimerization is required for the formation of a functional $GABA_B$ receptor. Nature 396:679–82

Winder DG, Conn PJ (1996) Roles of metabotropic glutamate receptors in glial function and glial-neuronal communication. J Neurosci Res 46:131–7

Wonnacott S (1997) Presynaptic nicotinic ACh receptors. Trends Neurosci 20:92–8

Yokoi M, Kobayashi K, Manabe T et al (1996) Impairment of hippocampal mossy fiber LTD in mice lacking mGluR2. Science 273:645–7

Yu XM, Askalan R, Keil GJ 2$^{nd}$, Salter MW (1997) NMDA channel regulation by channel-associated protein tyrosine kinase SRc. Science 275:74–8

# Presynaptic Neuropeptide Receptors

**E. Schlicker(✉) and M. Kathmann**



**Abstract** Presynaptic receptors for four families of neuropeptides will be discussed: opioids, neuropeptide Y, adrenocorticotropic hormone (ACTH), and orexins. Presynaptic receptors for the opioids ($\mu$, $\delta$, $\kappa$, and $ORL_1$) and neuropeptide Y ($Y_2$) inhibit transmitter release from a variety of neurones, both in the peripheral and central nervous systems. These receptors, which were also identified in human tissue, are coupled to $G_{i/o}$ proteins and block voltage-dependent $Ca^{2+}$ channels, activate voltage-dependent $K^+$ channels, and/or interfere with the vesicle release machinery. Presynaptic receptors for ACTH ($MC_2$ receptors) have so far been identified almost exclusively in cardiovascular tissues from rabbits, where they facilitate noradrenaline release; they are coupled to $G_s$ protein and act via stimulation of adenylyl cyclase. Presynaptic receptors for orexins (most probably $OX_2$ receptors) have so far almost exclusively been identified in the rat and mouse brain, where they facilitate the release of glutamate and $\gamma$-aminobutyric acid (GABA); they are most probably linked to $G_q$ and directly activate the vesicle release machinery or act via a transduction mechanism upstream of the release process. Agonists and antagonists at opioid receptors owe at least part of their therapeutic effects to actions on presynaptic receptors. Therapeutic drugs targeting neuropeptide Y and orexin receptors and presynaptic ACTH receptors so far are not available.

E. Schlicker
Institut für Pharmakologie und Toxikologie, Rheinische Friedrich-Wilhelms-Universität, Reuterstrasse 2b, 53113 Bonn, Germany
e.schlicker@uni-bonn.de

T.C. Südhof, K. Starke (eds.), *Pharmacology of Neurotransmitter Release.*
*Handbook of Experimental Pharmacology 184.*

# 1 Introduction

One reason to describe presynaptic neuropeptide receptors in a separate chapter is the chemistry of their endogenous ligands – peptides with up to about 50 amino acids. However, peptides have more properties that discriminate them from endogenous ligands at other types of presynaptic receptors. First, peptide transmitters, unlike transmitters of low molecular weight (e.g., noradrenaline, acetylcholine, GABA, or glutamate), are not formed in the axon terminals but are synthetized in the perikaryon of the neurone and are transported to the axon terminals; they are stored there in separate vesicles characterized by a large diameter and a dense core. Second, peptide transmitters frequently are cotransmitters, which are released in addition to one or even two low-molecular-weight transmitter(s). Third, endogenous peptides do not act via ionotropic receptors as opposed to many small transmitters, e.g., GABA (via $GABA_{A/C}$ receptors), glutamate (via NMDA or AMPA receptors), acetylcholine (via nicotinic receptors), or serotonin (via 5-$HT_3$ receptors). In other words, peptide transmitters are slower to act than some transmitters of low molecular weight.

There are dozens of presynaptic receptors for endogenous peptides. This review will focus on receptors for four peptides or peptide families: the opioid peptides, neuropeptide Y and related peptides, adrenocorticotropic hormone (corticotropin, ACTH), and the orexins. This choice is representative, since part of the receptors inhibit and part of them facilitate transmitter release; moreover, the receptors under consideration are coupled to the major G proteins, namely $G_{i/o}$, $G_s$ and $G_q$ (Table 1).

**Table 1** Synopsis of Receptors Activated by the Four Peptide Families

| Peptide Family | Endogenous Ligands | Receptors | | G Protein |
|---|---|---|---|---|
| | | Total | With Presynaptic Location | |
| Opioid | Endorphins, enkephalins, dynorphins, nociceptin | μ, δ, κ, $ORL_1$ | μ, δ, κ, $ORL_1$ | $G_{i/o}$ |
| Neuropeptide Y | Neuropeptide Y, peptide YY, pancreatic polypeptide | $Y_1$, $Y_2$, $Y_3$, $Y_5$, $Y_6$[1] | $Y_2$ | $G_{i/o}$ |
| Melanocortin | ACTH, α-, β-, γ-melanocyte-stimulating hormone | $MC_1$, $MC_2$, $MC_3$, $MC_4$, $MC_5$ | $MC_2$ | $G_s$ |
| Orexin | Orexin A and B | $OX_1$, $OX_2$ | $OX_2$? | $G_q$ |

[1] This subtype is functional in the mouse only.

This review is based mainly on four types of experiments in which exocytotic, vesicular, ATP-dependent transmitter release was determined directly or indirectly. Other types of transmitter release including "basal" release or carrier-mediated release will not be considered. (1) In superfusion studies transmitter release is determined directly (as "overflow"). (2) In other studies on isolated tissues transmitter release is not determined directly but via the endorgan response. (3) In electrophysiological studies, spontaneously occurring or electrically evoked currents or potentials elicited by the transmitter are studied at the postsynaptic membrane. (4) A small part of the experiments stem from pithed animals in which the noradrenaline spillover or the increase in blood pressure or heart rate elicited by electrical sympathetic nerve stimulation is studied. Apart from experiments on pithed animals, in vivo studies were usually not considered in this review since their interpretation bears difficulties. In the studies under (2), (3) and (4), control experiments were carried out in which direct effects of the drug under study at the postsynaptic receptor were excluded. In the electrophysiological studies a special type of experiment was performed to demonstrate presynaptic receptor location: spontaneously occurring currents or potentials were determined in the presence of tetrodotoxin in order to block impulse flow along the axon. If the drug affects the frequency of the remaining currents or potentials without affecting their amplitude, a presynaptic site of action can be assumed. Although transmitter release was not determined directly in numerous of the studies considered here, the term "transmitter release" will be used for this type of investigation as well.

For each of the four presynaptic receptors or receptor families, the occurrence in the autonomic and/or central nervous system will be described. Next, the signal transduction will be considered. Subsequently we will discuss their physiological role and their possible role in disease and therapy.

## 2 Presynaptic Opioid Receptors

Although an example of a presynaptic opioid receptor, activated by morphine, was discovered 90 years ago (Trendelenburg 1917), the endogenous peptides acting on opioid receptors have been described only since 1975 (Figure 1). There are four types of opioid receptors, termed μ, δ, and κ opioid and opioid receptor-like$_1$ (ORL$_1$). Although other nomenclatures have been proposed (e.g., OP$_3$, OP$_1$, OP$_2$ and OP$_4$, respectively; Alexander et al. 2006), the traditional designations will be used in the present review. The term "opioid receptors" will be used in a broad sense covering not only the opioid receptors *sensu stricto* (μ, δ, and κ) but also the ORL$_1$ receptor. There are at least four types of endogenous peptides acting on opioid receptors: endorphins, enkephalins, dynorphins, and nociceptin. These peptides are not synthetized as such but are cleaved from precursors (e.g., proopiomelanocortin), which in turn have been processed from pre-precursors (e.g., pre-proopiomelanocortin). ß-Endorphin activates μ, the enkephalins activate δ, and the dynorphins activate κ opioid receptors, whereas nociceptin activates

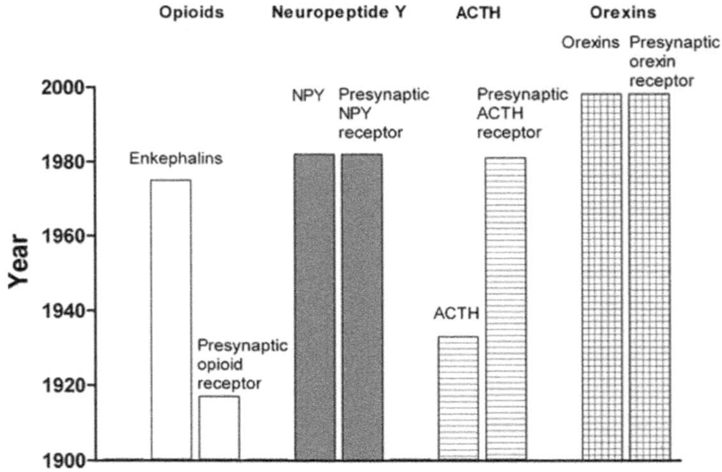


**Fig. 1** First description of the four peptides (or peptide families) and their respective presynaptic receptors. The papers are (from left to right): Trendelenburg (1917); Hughes et al. (1975); Tatemoto et al. (1982); Allen et al. (1982); Collip et al. (1933); Göthert (1981); de Lecea et al. (1998); van den Pol et al. (1998).

$ORL_1$ receptors. Although nociceptin is selective for $ORL_1$ receptors, the other three peptides show a marked degree of promiscuity with regard to the classical opioid receptors (μ, δ, and κ). Endorphins occur in the brain (almost exclusively in neurones which have their somata in the infundibular nucleus) and in the adenohypophysis. Enkephalins are found on many sites in the CNS but also occur in the adrenal medulla and in the gut wall. Dynorphins are also found in the CNS and in the gut (for review, see Gutstein and Akil 2006).

Figure 2 gives an example how presynaptic opioid receptors can be identified in a superfusion model. We examined the effect of subtype-selective agonists on noradrenaline release in mouse brain cortex slices. Figure 2a shows that noradrenaline release was inhibited by a μ and an $ORL_1$ receptor agonist but was not affected by high concentrations of a δ and κ opioid receptor agonist. In further experiments we showed that the effects of the μ agonist (Figure 2b) and of the $ORL_1$ agonist (Figure 2c) indeed involve μ and $ORL_1$ receptors since the concentration-response curves were shifted to the right by antagonists with apparent $pA_2$ values close to their potency values in other functional studies for each receptor subtype (Guerrini et al. 1998; Berger et al. 2006).

Using superfusion experiments, electrophysiological techniques, pithed animal preparations, and experiments in which transmitter release was determined indirectly via the end-organ response (e.g., twitch response of vas deferens preparations), numerous presynaptic opioid receptors have been identified (Table 2). For the identification of the receptors, classical drug tools were used; for future studies, knockout mice (now available for each of the four opioid receptor subtypes) and special nucleotides (e.g., antisense oligodeoxynucleotides or short interfering

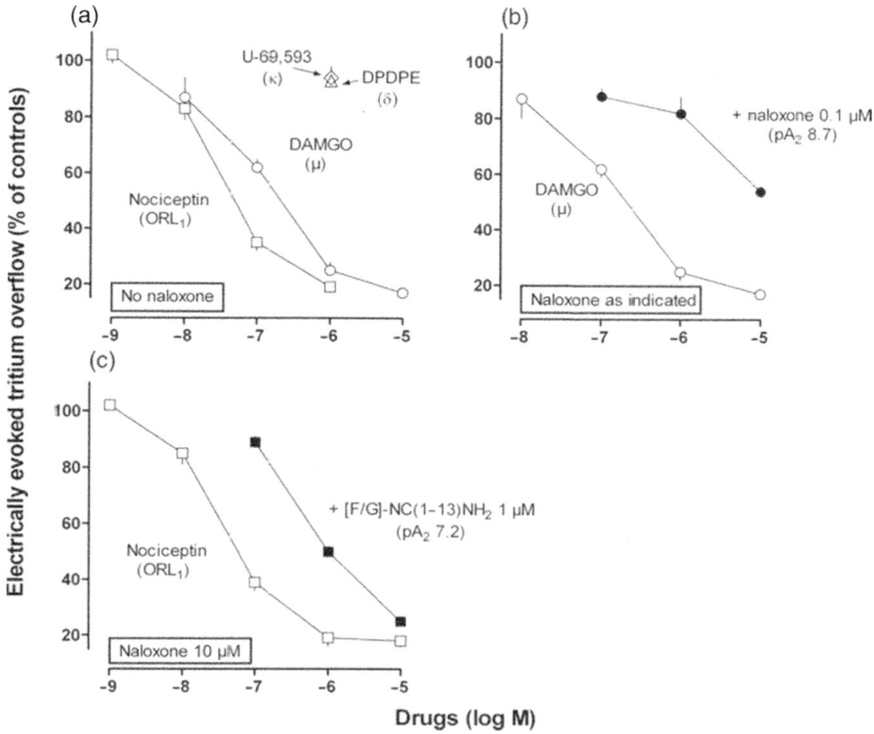


**Fig. 2** Effect of opioid receptor agonists on the electrically (0.3 Hz) evoked tritium overflow from superfused mouse brain cortex slices preincubated with $^{3}$H-noradrenaline. The evoked overflow represents quasi-physiological exocytotic noradrenaline release. (**a**) shows that noradrenaline release was inhibited by agonists at the μ and the $ORL_1$ receptor but not affected by high concentrations of a δ and κ receptor agonist. The effect of DAMGO in fact involved μ receptors since it was antagonized by a low concentration of naloxone, which possesses some preference for μ over δ, κ and $ORL_1$ receptors (**b**). On the other hand, the effect of nociceptin was indeed related to the activation of $ORL_1$ receptors since it was antagonized by the $ORL_1$ antagonist [$Phe^1\Psi(CH_2$-NH)$Gly^2$]-nociceptin(1–13)$NH_2$ ([F/G]-NC(1–13)$NH_2$) ([F/G]-NC(1–13)$NH_2$) (**c**). Data with nociceptin and its antagonist from Schlicker et al. (1998; redrawn); other data published in abstract form only (Schlicker and Kathmann 2000). Similar data were published by Trendelenburg et al. (2000).

RNA) are promising new approaches. In all examples listed in Table 2 transmitter release was inhibited. Each of the four opioid receptor subtypes can serve as an inhibitory presynaptic receptor (Table 1), and each has been identified also in human tissue (Table 2). In a few papers (not further considered here) a facilitatory rather than inhibitory effect of opioids on transmitter release was reported. In such cases, the possibility has to be considered that the opioid receptor was located presynaptically at an inhibitory interneurone projecting to the neurone under study (and not at the latter itself).

**Table 2** Synopsis of presynaptic inhibitory opioid receptors

| | $\mu$ opioid receptors | | | |
|---|---|---|---|---|
| | Transmitter | Tissue | Species | References |
| Sympathetic nervous system | *preganglionic* Acetylcholine | Superior cervical ganglion | Cat, rabbit | 2 |
| | | Hypogastric ganglion | Mouse | Rogers and Henderson 1990 |
| | | Superior and inferior mesenteric ganglion | Guinea-pig | 2 |
| | *postganglionic* Noradrenaline | Tail artery | Rat | 1 |
| | | Nictitating membrane | Cat | 1 |
| | | Colon | Guinea-pig | 1 |
| | | Vas deferens | Rat, mouse | 1 |
| Parasympathetic/ enteric nervous system | Acetylcholine | Heart | Rabbit | 2 |
| | | Lung | Rat | Yu et al. 2006 |
| | | Ciliary ganglion | Chick | Endo and Yawo 2000 |
| | | Small intestine myenteric plexus | Guinea pig | Cherubini and North 1985 |
| | | Ileum | Guinea-pig, rat | 2, Storr et al. 2002 |
| | | Colon | Cat | 2 |
| | | Gall bladder | Guinea-pig | Guarraci et al. 2002 |
| Central nervous system | Noradrenaline | Cerebral cortex, hippocampus, cerebellum | Guinea-pig | 3 |
| | | Cerebral cortex, hippocampus, amygdala, nucleus tractus solitarii, periaqueductal grey, cerebellum | Rat | 3 |
| | | Cerebral cortex | Mouse | see Fig. 2 |
| | Dopamine | Striatum | Rat | Schlosser et al. 1995 |
| | Serotonin | Cerebral cortex | Human | Berger et al. 2006 |
| | | Cerebral cortex, hippocampus | Rat | 3, Berger et al. 2006 |
| | Acetylcholine | Nucleus accumbens, olfactory tubercle, hippocampus, amygdala | Rat | 3 |
| | Glutamate | Cerebral cortex | Rat | Ostermeier et al. 2000 |
| | | Amygdala | Rat | *Zhu and Pan 2005** |
| | | Arcuate nucleus, ventromedial hypothalamus | Rat | Emmerson and Miller 1999 |
| | | Subthalamic nucleus | Rat | Shen and Johnson 2002 |
| | | Hypothalamus (supraoptic nucleus) | Rat, mouse | *Liu et al. 1999*; *Honda et al. 2004* |
| | | Periaqueductal grey | Rat | Vaughan and Christie 1997 |
| | | Raphe pallidus | Rat | Bouryi and Lewis 2004 |
| | | Dorsal motor nucleus | Rat | Browning et al. 2002 |
| | | Nucleus tractus solitarii | Rat | Glatzer and Smith 2005 |
| | | Rostral ventrolateral medulla | Rat | *Hayar and Guyenet 1998* |
| | | Spinal cord | Rat | Glaum et al. 1994 |

(*Continued*)

**Table 2** Continued

| | μ opioid receptors | | | |
|---|---|---|---|---|
| | Transmitter | Tissue | Species | References |
| | GABA | Hippocampus | Rat | Capogna et al. 1993 |
| | | Amygdala | Rat | *Finnegan et al. 2005; 2006* |
| | | Globus pallidus | Rat | Stanford and Cooper 1999 |
| | | Hypothalamus (supraoptic nucleus) | Mouse | *Honda et al. 2004* |
| | | Subthalamic nucleus | Rat | Shen and Johnson 2002 |
| | | Ventral tegmental area | Rat | Bergevin et al. 2002 |
| | | Locus coeruleus | Rat | *Pan et al. 2004* |
| | | Periaqueductal gray | Rat | Kishimoto et al. 2001 |
| | | Rostral ventrolateral medulla | Rat | Hayar and Guyenet 1998 |
| | | Spinal cord (substantia gelatinosa) | Rat | Grudt and Henderson 1998 |
| | | Spinal cord (dorsal horn) | Rat | *Kerchner and Zhuo 2002* |
| | Glycine | Spinal cord (substantia gelatinosa) | Rat | *Grudt and Henderson 1998* |
| | Substance P | Hypothalamus, spinal cord | Cat | 3, 4 |
| | | Spinal cord | Rat | Aimone and Yaksh 1989 |
| | ß-Endorphin, dynorphin | Hypothalamus | Rat | Nikolarakis et al. 1989 |
| | Cholecystokinin | Hypothalamus | Cat | 3, 4 |

| | δ opioid receptors | | | |
|---|---|---|---|---|
| | Transmitter | Tissue | Species | References |
| Sympathetic nervous system | *preganglionic* Acetylcholine | Hypogastric ganglion | Mouse | Rogers and Henderson 1990 |
| | *postganglionic* Noradrenaline | Atrium | Guinea-pig, mouse | 1 |
| | | Arteries and veins | Rabbit | 1 |
| | | Colon | Guinea-pig | 1 |
| | | Spleen | Cat | 1 |
| | | Vas deferens | Hamster, rat, mouse | 1 |
| | | Cultured sympathetic neurones | Chicken | 1 |
| Parasympathetic nervous system | Acetylcholine | Heart | Rabbit | 2 |
| | | Ciliary ganglion | Chick | Endo and Yawo 2000 |
| | | Ileum | Mouse | 2 |
| | | Colon | Cat | 2 |
| | | Gall bladder | Guinea-pig | Guarraci et al. 2002 |
| | | Sacral colonic ganglia | Cat | 2 |

(*Continued*)

**Table 2** Continued

| | $\delta$ opioid receptors | | | |
|---|---|---|---|---|
| | Transmitter | Tissue | Species | References |
| Central nervous system | Noradrenaline | Cerebral cortex | Human | Berger et al. 2006 |
| | Dopamine | Striatum | Rat | Schlosser et al. 1995 |
| | Serotonin | Hippocampus | Rat | 3 |
| | Acetylcholine | Striatum, nucleus accumbens, olfactory tubercle | Rat | 3 |
| | Glutamate | Cerebral cortex | Rat | Ostermeier et al. 2000 |
| | | Raphe pallidus | Rat | Bouryi and Lewis 2004 |
| | | Spinal cord | Rat | *Glaum et al. 1994* |
| | GABA | Globus pallidus | Rat | *Stanford and Cooper 1999* |
| | | Subthalamic nucleus | Rat | Shen and Johnson 2002 |
| | | Locus coeruleus | Rat | Pan et al. 2002 |
| | Substance P | Hypothalamus, spinal cord | Cat | 3, 4 |
| | | Spinal cord | Rat | 3, 4, Aimone and Yaksh 1989 |
| | Metenkephalin, ß-endorphin, dynorphin | Hypothalamus | Rat | Nikolarakis et al. 1989 |
| | $\kappa$ opioid receptors | | | |
| | Transmitter | Tissue | Species | References |
| Sympathetic nervous system | Noradrenaline | Heart | Rabbit, guinea-pig | 1 |
| | | Vasculature | Rabbit | 1 |
| | | Colon | Guinea-pig | 1 |
| | | Vas deferens | Rabbit, mouse | 1 |
| | | Cultured sympathetic neurones | Chicken | 1 |
| Parasympathetic nervous system | Acetylcholine | Small intestine (myenteric plexus) | Guinea-pig | Cherubini and North 1985 |
| | | Ileum | Guinea-pig, mouse | 2 |
| | | Colon | Cat | 2 |
| | | Gall bladder | Guinea-pig | Guarraci et al. 2002 |
| Sensory neurones | Calcitonin gene-related peptide | Skin | Rat | Averbeck et al. 2001 |
| Central nervous system | Noradrenaline | Cerebral cortex, hippocampus, cerebellum | Guinea-pig | 3 |

(*Continued*)

**Table 2** Continued

| | κ opioid receptors | | | |
|---|---|---|---|---|
| | Transmitter | Tissue | Species | References |
| | Dopamine | Striatum, frontal cortex, olfactory tubercle, nucleus accumbens, mediobasal hypothalamus, amygdala | Rat | 3 |
| | | Striatum | Cat, guinea-pig | 3 |
| | Serotonin | Cerebral cortex | Human, rat | Berger et al. 2006 |
| | | Superior colliculus | Rabbit | 3 |
| | Histamine | Cerebral cortex | Rat | 3 |
| | Acetylcholine | Cerebral cortex | Human | Feuerstein et al. 1996 |
| | Glutamate | Hippocampus | Guinea-pig | Gannon and Terrian 1991; Simmons et al. 1994 |
| | | Hypothalamus (arcuate nucleus) | Rat | *Emmerson and Miller 1999* |
| | | Nucleus accumbens (shell) | Rat | *Hjelmstadt and Fields 2001* |
| | | Rostral ventromedial medulla | Rat | Ackley et al. 2001 |
| | GABA | Globus pallidus | Rat | Ogura and Kita 2000 |
| | Glycine | Nucleus ambiguus | Rat | *Wang et al. 2004* |
| | Metenkephalin | Brainstem | Rat | Ueda et al. 1987 |
| | Dynorphin | Hippocampus | Guinea-pig | Gannon and Terrian 1991 |
| | | Hypothalamus | Rat | Nikolarakis et al. 1989 |

| | $ORL_1$ receptors | | | |
|---|---|---|---|---|
| | Transmitter | Tissue | Species | References |
| Sympathetic nervous system | Noradrenaline | Heart | Rat | 1 |
| | | Atrium | Mouse | 1 |
| | | Tail artery | Rat | 1 |
| | | Resistance vessels | Rat | 1 |
| | | Anococcygeus muscle | Rat | 1 |
| | | Vas deferens | Rabbit, rat mouse | 1 |
| Parasympathetic/enteric nervous system | Acetylcholine | Small intestine (myenteric plexus) | Guinea-pig | Nicholson et al. 1998; Liu et al. 2001 |
| | Non-adrenergic-non-cholinergic transmitter | Small intestine (myenteric plexus) | Guinea-pig | Nicholson et al. 1998 |

(*Continued*)

**Table 2** Continued

| | *ORL*$_1$ receptors | | | |
|---|---|---|---|---|
| | Transmitter | Tissue | Species | References |
| Central nervous system | Noradrenaline | Cerebral cortex, cerebellum, hippocampus, hypothalamus | Mouse | 5 |
| | | Cerebral cortex | Human | Rominger et al. 2002 |
| | Serotonin | Cerebral cortex | Human, rat, mouse | 5; Berger et al. 2006 |
| | Glutamate | Hippocampus | Rat | Yu et al. 1997 |
| | | Hypothalamus (arcuate nucleus) | Rat | Emmerson and Miller 1999 |
| | | Periaqueductal grey | Rat | *Vaughan et al. 1997* |
| | GABA | Periaqueductal grey | Rat | *Vaughan et al. 1997* |

1 Reviewed by Boehm and Kubista (2002)
2 Reviewed by Illes (1989)
3 Reviewed by Mulder and Schoffelmeer (1993)
4 Reviewed by Jackisch (1991)
5 Reviewed by Schlicker and Morari (2000)
* In the studies given in *italics* opioids decreased postsynaptic currents or potentials also in the presence of tetrodotoxin.

Presynaptic opioid receptors were found in all parts of the autonomic nervous system (sympathetic, parasympathetic, and enteric) and on many sites of the central nervous system. The release of more than 10 transmitters can be inhibited via opioid receptors. The latter serve as presynaptic heteroreceptors in most instances, i.e., the opioid modulating release differs from the transmitter subject to modulation. However, there are also examples of presynaptic opioid autoreceptors. For example, in the study of Nikolarakis et al. (1989) on rat hypothalamus slices, antagonists of the μ, δ, and κ opioid receptor increased the release of endogenous opioids, suggesting that the latter decreased their own release. Another example of an autoreceptor has been shown in the guinea pig hippocampus, where the release of dynorphin (which mainly activates κ opioid receptors) was inhibited by the κ opioid receptor agonist U-69,593 (Gannon and Terrian 1991). With respect to the distribution of presynaptic opioid receptors, marked species differences exist. For example, noradrenaline release in the cerebral cortex is inhibited via δ and ORL$_1$ receptors in humans, via κ and ORL$_1$ receptors in the guinea pig, and via μ and ORL$_1$ receptors in the rat and mouse (Table 2, Figure 2). These examples show that there are nerve terminals in which more than one opioid receptor can be identified. Transmitter release from the postganglionic sympathetic neurone in the mouse vas deferens is even subject to inhibition via each of the four opioid subtypes (Table 2).

Which mechanisms are involved in the opioid receptor-mediated inhibition of transmitter release? In general, the four types of opioid receptor are $G_{i/o}$ protein-coupled receptors (Alexander et al. 2006), but theoretically another transduction pathway might exist for presynaptic opioid receptors. However, the coupling of presynaptic opioid receptors to $G_{i/o}$ (labelled with **1** in Figure 3) has been

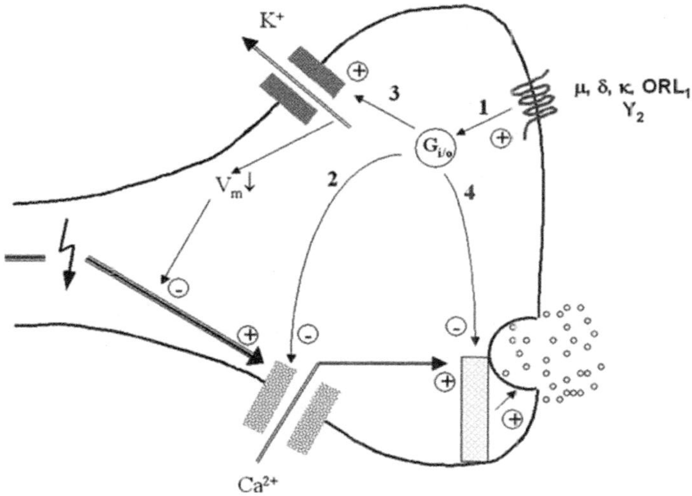


**Fig. 3** Mechanisms involved in the opioid (μ, δ, κ, $ORL_1$) and neuropeptide Y ($Y_2$) receptor-mediated inhibition of exocytotic transmitter release. Following activation of the respective receptor and $G_{i/o}$ (**1**), three signal transduction pathways are possible, namely inhibition of voltage-dependent $Ca^{2+}$ channels (**2**), opening of $K^+$ channels (**3**), and a direct inhibitory effect on the vesicle release machinery (**4**). *Glossary*: − > - leading to; => - ion flux; ≡> - action potential; $V_m$ - membrane potential; "(+)" and "(−)" mean stimulatory and inhibitory effect, respectively.

demonstrated in several studies, using pertussis toxin or the somewhat less selective agent N-ethylmaleimide. In this manner, coupling to $G_{i/o}$ has been shown (1) for presynaptic μ opioid receptors inhibiting the release of noradrenaline in rat and guinea pig hippocampus (Werling et al. 1989) and of GABA in rat hippocampus (Capogna et al. 1993) and periaqueductal grey (Kishimoto et al. 2001; Hahm et al. 2004) and (2) for presynaptic κ opioid receptors inhibiting noradrenaline release in rabbit hippocampus (Allgaier et al. 1989).

In order to better understand how the events downstream of G protein activation are influenced by opioid receptors, one should briefly recall the major steps of the electrosecretory coupling. When an action potential invades the axon terminal, voltage-dependent $Ca^{2+}$ channels open. The $Ca^{2+}$ ions entering the axoplasma finally lead to fusion of vesicles with the plasma membrane and release of the transmitter into the synaptic cleft. $Ca^{2+}$ influx is crucially dependent on the membrane potential; e.g., hyperpolarization will impair $Ca^{2+}$ influx Hyperpolarization can be elicited by the efflux of $K^+$ ions. In general, $G_{i/o}$-coupled receptors can decrease $Ca^{2+}$ influx via voltage-dependent $Ca^{2+}$ channels (labelled with **2** in Figure 3), increase $K^+$ efflux via voltage-dependent $K^+$ channels (**3** in Figure 3), and/or directly interfere with the release process (**4** in Figure 3) (Miller 1998). Each of the three mechanisms, alone or in combination, has been reported to play a role in opioid receptor-mediated inhibition of transmitter release. To show the involvement of ion channels, (more or less) selective drug tools have been used. The direct influence on the release machinery has been shown in electrophysiological studies in which impulse flow was abolished by tetrodotoxin or $Ca^{2+}$ influx through voltage-dependent $Ca^{2+}$ channels was blocked by appropriate agents. If a drug inhibits the frequency of postsynaptic currents or potentials under this scenario, a direct interference with the release process can be assumed (Miller 1998).

Voltage-dependent *$Ca^{2+}$ channels* play a role in the μ receptor-mediated inhibition of noradrenaline release in the rat cerebral cortex (Mulder and Schoffelmeer 1993), of glutamate release in the chick ciliary ganglion (Endo and Yawo 2000), and of GABA release in the rat striatum (Barral et al. 2003). Voltage-dependent *$K^+$ channels* are involved in the μ receptor-mediated inhibition of the release of glutamate in the rat subthalamic nucleus (Shen and Johnson 2002) and periaqueductal grey (Vaughan and Christie 1997) and of GABA in the rat striatum (Barral et al. 2003), amygdala (Finnegan et al. 2006), subthalamic nucleus (Shen and Johnson 2002), ventral tegmental area (Bergevin et al. 2002), and periaqueductal grey (Vaughan and Christie 1997). They are also involved in the κ receptor-mediated inhibition of glutamate release in the guinea pig dentate gyrus (Simmons et al. 1994). A *direct effect on the release machinery* has been shown for each of the four opioid receptors in at least one region of the central nervous system, and both for fast excitatory (glutamate) and inhibitory transmission (GABA, glycine) (references given in italics in Table 2).

Are the presynaptic opioid receptors described here activated by endogenous opioids? If so, one would expect that an opioid receptor antagonist like naloxone would alter transmitter release in the direction opposite to that elicited by opioid receptor agonists, i.e., would increase transmitter release. In most experimental models, an effect of naloxone per se has not been found, suggesting that the corresponding opioid receptors are not subject to an endogenous tone (listed in the reviews by Starke 1977; Jackisch 1991; Boehm and Kubista 2002). In a series of models, however, naloxone did increase transmitter release. This has been shown for the opioid receptors (most probably μ, see Table 2) modulating noradrenaline release in the rat cerebral cortex (Taube et al. 1976) and for the opioid receptors modulating acetylcholine release in the guinea pig thalamus (Beani et al. 1982). It has also been shown for the μ and δ opioid receptors inhibiting substance P release in the cat spinal cord (Go and Yaksh 1987) and the δ opioid receptors inhibiting acetylcholine release in the cat parasympathetic colonic ganglia (Kennedy and Krier 1987) and the rat striatum (Sándor et al. 1991). Using subtype-selective opioid receptor antagonists, an endogenous tone has furthermore been shown for the μ, δ, and κ opioid autoreceptors causing inhibition of ß-endorphin, metenkephalin, and dynorphin release in the rat hypothalamus (Nikolarakis et al. 1989), for the μ and κ opioid receptors inhibiting noradrenaline and acetylcholine release in the guinea pig myenteric plexus (Cosentino et al. 1995), for the κ opioid autoreceptors inhibiting metenkephalin in the rat brainstem (Ueda et al. 1987), and for the $ORL_1$ receptors inhibiting noradrenaline release in the rat anococcygeus muscle (Ho et al. 2000) and serotonin release in the human cerebral cortex (Berger et al. 2006). In a final study on human cerebral cortex (Feuerstein et al. 1996), an endogenous tone at δ opioid receptors was shown to be likely. This is suggested by the fact that the selective δ opioid receptor antagonist naltrindole facilitated acetylcholine release in the presence (but not absence) of a cocktail of peptidase inhibitors and that the cocktail per se inhibited acetylcholine release. The simplest explanation is that the concentration of endogenous opioids in the synaptic cleft normally is relatively low but can reach a critical level if degradation is blocked. (The situation in the study is complicated since the δ opioid receptor is not located at the cholinergic axon terminals but on an unidentified interneurone; Feuerstein et al. 1996.)

The observation that an endogenous tone at presynaptic opioid receptors is rarely detectable fits well to the fact that an endogenous tone is rare in opioid receptor models in general (see Table 21-2 in the review by Gutstein and Akil 2006). Endogenous opioid systems come into play under stressful situations. For example, the slight euphoria occurring after physical exercise or the suppression of pain that sometimes occurs in severely injured persons is believed to be caused by a marked increase in the release of endogenous opioids (Gutstein and Akil 2006).

Drugs acting at opioid receptors are among the oldest and continue to be among the most important medicines. There is no doubt that at least some targets for their desired, as well as undesired, effects are presynaptic receptors. This holds true for analgesia. Presynaptic μ receptors involved in the analgesic effect of, e.g., morphine are located both spinally and supraspinally. Spinal presynaptic μ opioid receptors inhibit the release of transmitters like glutamate, substance P, and calcitonin gene-related peptide from the terminals of the primary afferent neurones conveying pain from the periphery to the central nervous system (Zöllner and Stein 2007). One supraspinal site is in the periaqueductal grey, from where the descending pain-suppressing tract originates. This tract is under the control of tonically active GABAergic neurones, the axon terminals of which are equipped with presynaptic μ receptors. Activation of these receptors increases the pain-suppressing effect of the descending pathway (Vaughan and Christie 1997). Finally, presynaptic μ opioid receptors on tonically active GABAergic interneurones in the ventral tegmental area are associated with the rewarding properties of opioids. These GABAergic interneurones synapse with dopaminergic neurones projecting to the nucleus accumbens; activation of the presynaptic μ receptors leads to an increase in dopamine release and, hence, euphoria (Bergevin et al. 2002).

## 3 Presynaptic Neuropeptide Y Receptors

Neuropeptide Y and its presynaptic receptors were discovered 25 years ago (Figure 1). There are four subtypes of neuropeptideY receptors, $Y_1$, $Y_2$, $Y_4$ and $Y_5$ (Table 1, Alexander et al. 2006); a fifth subtype, $Y_6$, is functional in the mouse, whereas in primates the related gene is nonfunctional due to a frame-shift mutation (Alexander et al. 2006). Neuropeptide Y receptors are also activated by another two peptides, peptide YY (PYY) and pancreatic polypeptide (PP). The three peptides are not synthetized as such but are processed from larger precursors via two steps, i.e., an initial cleavage of the N-terminal signal peptide and the subsequent cleavage of the C-terminal part (Tatemoto 2004). The affinity of PYY is very similar to that of neuropeptide Y at each of the five receptor subtypes; PP has about the same affinity as neuropeptide Y (or PYY) at $Y_5$ receptors, a lower affinity at $Y_1$, $Y_2$ and $Y_6$ receptors and a higher affinity at $Y_4$ receptors (Alexander et al. 2006). Neuropeptide Y is a neurotransmitter in sympathetic, parasympathetic, enteric, and sensory neurones and in neurones of the CNS. PYY and PP are hormones; the former is stored in the pancreas and the latter in the terminal intestine (Brain and Cox 2006).

Figure 4 shows how a presynaptic neuropeptide Y receptor was identified in a superfusion model. In mouse brain cortex slices, serotonin release was concentration-dependently inhibited by neuropeptide Y and this effect was potently mimicked by neuropeptide Y-(13-36). Since the latter is selective for $Y_2$ over $Y_1$ and $Y_5$ receptors (Alexander et al. 2006) one may conclude that neuropeptide Y acts via $Y_2$ receptors.

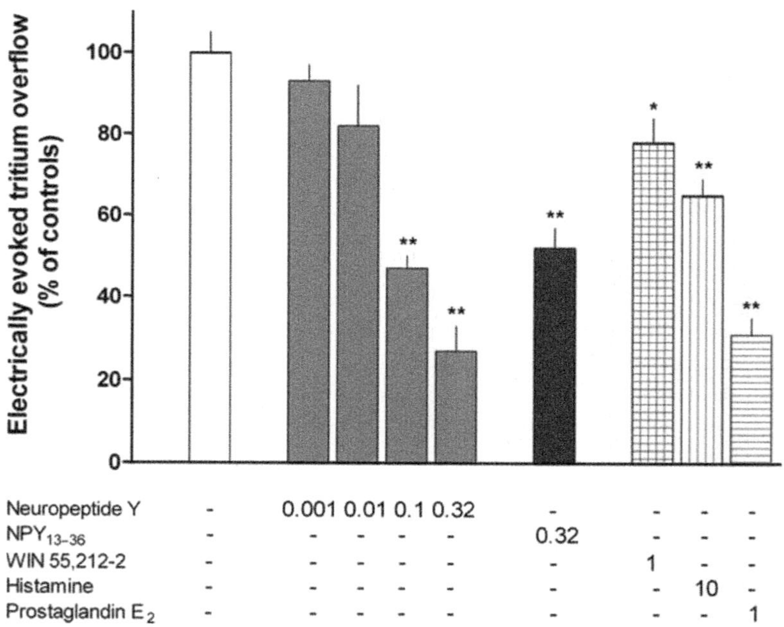


**Fig. 4** Effect of various peptides and nonpeptides on the electrically (3 Hz) evoked tritium overflow from superfused mouse brain cortex slices preincubated with $^3$H-serotonin. The evoked overflow represents quasi-physiological exocytotic serotonin release. In all experiments, serotonin autoreceptors were blocked by metitepine. The figure shows that human neuropeptide Y concentration-dependently inhibited serotonin release and that this effect was mimicked by human neuropeptide Y (13–36) ($NPY_{13-36}$), which has a high affinity for $Y_2$ but a very low affinity for $Y_1$ receptors. These results are compatible with the view that neuropeptide Y acts via $Y_2$ receptors in the present model. For the sake of comparison, the figure also shows the inhibitory effects of another three agonists, acting via cannabinoid $CB_1$, histamine $H_3$ and prostaglandin $EP_3$ receptors and used at concentrations causing the maximum or near-maximum effect at their respective receptors. Drug concentrations in μM. $^*P < 0.05$, $^{**}P < 0.003$, compared to the control (from Nakazi et al. 2000 and Nakazi 2001; redrawn).

Figure 4 also shows that the extent of inhibition obtained with neuropeptide Y was very marked and comparable to that obtained with prostaglandin $E_2$ (acting via $EP_3$ receptors), which was examined for the sake of comparison. The maximum inhibition obtained with WIN 55,212-2 (acting via cannabinoid $CB_1$ receptors) and histamine (acting via $H_3$ receptors ) was far lower.

Many examples of inhibitory presynaptic neuropeptide Y receptors have been found in tissues from experimental animals and humans (Table 3), based on experiments with each of the four techniques described in the Introduction (overflow experiments, electrophysiological techniques, pithed animals, electrically induced twitches in isolated tissues). For the identification of the receptors, the three peptides and modified (e.g., truncated) congeners (Figure 4) were the only tools for many years. More recently, selective antagonists (e.g., in the study by Schwertfeger et al. [2004] on the human heart) and knockout mice (e.g., in the study by Smith-White et al. [2002] on the mouse heart) have been used. According to most studies, only $Y_2$ receptors serve as presynaptic neuropeptide Y receptors.

Presynaptic neuropeptide Y receptors have been identified in the sympathetic, parasympathetic and enteric nervous system, on sensory neurones, and on many sites in the CNS; they inhibit the release of 11 different transmitters (Table 3). Most presynaptic neuropeptide Y receptors are heteroreceptors, but some serve as autoreceptors, both in the sympathetic nervous system (pig spleen and kidney) and the brain (rat hypothalamus) (Table 3). Marked species differences occur. For example, $Y_2$ receptors inhibit the sympathetically mediated tachycardia, the vagally mediated bradycardia, and the vagally mediated bronchoconstriction in the guinea pig but not in the rabbit (Serone et al. 1999; Abrahamsson 2000). Moreover, $Y_2$ receptors inhibit noradrenaline release in the guinea pig carotid but not femoral artery, and conversely inhibit noradrenaline release in the rat femoral but not carotid artery (Potter and Tripovic 2006).

Like opioid receptors, neuropeptide Y receptors are $G_{i/o}$ protein-coupled, and for this reason the marked similarities in signal transduction are not surprising. The coupling to $G_{i/o}$ was proven for the presynaptic $Y_2$ receptors inhibiting the substance P-mediated plasma extravasation in the rat dura mater (Yu and Moskowitz 1996) and the release of noradrenaline in the rat pineal gland (Simmoneaux et al. 1994) (step **1** in Figure 3). With respect to the downstream pathways, a coupling to voltage-dependent $Ca^{2-}$ channels (**2** in Figure 3) has been shown for the presynaptic neuropeptide Y receptors inhibiting noradrenaline release in the dog splenic artery (Yang and Chiba 2002), glutamate release in the rat hippocampus (Klapstein and Colmers 1992; Qian et al. 1997), and GABA release in the rat thalamus (Sun et al. 2001). Relatively little information is available on whether presynaptic neuropeptide Y receptors can couple to voltage-dependent $K^+$ channels (step **3** in Figure 3). This may hold true for the neuropeptide Y receptor involved in inhibition of ATP release in the mouse vas deferens (Stjärne et al. 1989). For three presynaptic neuropeptide Y receptors in the rat spinal cord a direct effect on the vesicle release machinery (step **4** in Figure 3) can be assumed, since agonists inhibited the frequency of the miniature inhibitory or excitatory postsynaptic currents (Moran et al. 2004; italics in Table 3).

**Table 3** Synopsis of presynaptic neuropeptide Y receptors

| | Transmitter | Tissue | Species | References |
|---|---|---|---|---|
| Sympathetic nervous system | Noradrenaline | Heart | Human, guinea-pig | 1 |
| | | Submandibular artery | Human | 1 |
| | | Splenic artery | Dog | Yang and Chiba 2002 |
| | | Gracilis muscle vasculature | Dog | 1 |
| | | Ear artery | Rabbit | 1 |
| | | Carotid artery | Guinea-pig | Potter and Tripovic 2006 |
| | | Mesenteric, renal, femoral, basilar artery, resistance vessels | Rat | 1; Potter and Tripovic 2006 |
| | | Saphenous vein | Dog | 1 |
| | | Portal vein | Rat | 1 |
| | | Kidney | Human, pig, rabbit | 1 |
| | | Spleen | Pig | 1 |
| | | Urinary bladder | Guinea-pig | 1 |
| | | Vas deferens | Rat, mouse | 1 |
| | | Uterus | Rat | 1 |
| | | Oviduct | Rabbit | 1 |
| | ATP | Vas deferens | Guinea-pig, mouse | 1 |
| | Neuropeptide Y | Spleen | Pig | Modin et al. 1994 |
| | | Kidney | Pig | 1 |
| Para-sympathetic nervous system | Acetylcholine | Heart | Human | Schwertfeger et al. 2004 |
| | | Heart | Dog, guinea-pig, rat, mouse | 1; Abrahamsson 2000 |
| | | Bronchial arterioles | Dog | Mahns et al. 1998 |
| | | Nasal vessels | Dog, cat | 1 |
| | | Trachea, bronchial muscle | Guinea-pig | Abrahamsson 2000; 1 |
| | | Uterus | Rat | 1 |
| Sensory nerves | Substance P | Dura mater vasculature | Rat | Yu and Moskowitz 1996 |
| | Calcitonin-gene-related peptide | Mesenteric artery | Rat | 1 |
| | | Small intestine | Guinea-pig | 1 |
| CNS | Noradrenaline | Cerebral cortex, hippocampus, hypothalamus, pineal gland, medulla oblongata | Rat | 1 |
| | Dopamine | Striatum | Rat | 1 |
| | Serotonin | Cerebral cortex | Rat, mouse | 1; Nakazi et al. 2000; Nakazi 2001 |
| | Glutamate | Cerebral cortex | Rat | Wang 2005 |
| | | Olfactory bulb | Rat | Blakemore et al. 2006 |
| | | Hippocampus | Rat | 1 |
| | | Striatum | Rat | 1 |

(*Continued*)

**Table 3** Continued

| Transmitter | Tissue | Species | References |
|---|---|---|---|
| | Hypothalamus | Rat, mouse | Rhim et al. 1997; Fu et al. 2004 |
| | Spinal cord | Rat | *Moran et al. 2004** |
| GABA | Thalamus | Rat | 1 |
| | Hypothalamus | Rat | 1 |
| | Spinal cord | Rat | *Moran et al. 2004* |
| Glycine | Spinal cord | Rat | *Moran et al. 2004* |
| Neuropeptide Y | Hypothalamus | Rat | 1 |

1 Reviewed by Westfall (2004)
*In this study neuropeptide Y decreased the frequency of postsynaptic currents also in the presence of tetrodotoxin.

To address the question of whether presynaptic neuropeptide Y receptors are also activated by endogenous neuropeptide Y or related peptides, the selective $Y_2$ receptor antagonist BIIE 0246 was used. Based on these experiments, an endogenous tone can be assumed for the presynaptic $Y_2$ receptors inhibiting noradrenaline release in the perfused mesenteric arterial bed of the rat (Westfall 2004), noradrenaline and ATP release in the dog splenic artery (Yang and Chiba 2002), and the release of neuropeptide Y itself in the rat hypothalamus (King et al. 2000). The number of tonically activated presynaptic $Y_2$ receptors may be much higher since many sites were examined earlier than 1999 when BIIE 0246 became available.

There is now good evidence for a role of presynaptic neuropeptide Y receptors in disease states. As to the cardiovascular system, plasma neuropeptide Y levels are slightly increased in hypertensive patients and to a greater extent in cardiac failure (Morris 2004). An increase in neuropeptide Y levels is even an important prognostic marker for cardiovascular death in hemodialysis patients (Odar-Cederlof et al. 2003). The exact role played by the presynaptic neuropeptide Y receptor, however, cannot easily be pinpointed for at least two reasons. First, presynaptic inhibitory $Y_2$ receptors occur both on sympathetic and parasympathetic nerve endings of the human heart (Table 3); i.e., they inhibit two opposing systems. Second, within the sympathetic nervous system postsynaptic neuropeptide Y receptors ($Y_1$) lead to vasoconstriction and increase the effect elicited by other vasoconstrictors (Morris 2004), whereas presynaptic $Y_2$ receptors inhibit the release of noradrenaline, ATP, and neuropeptide and thereby impair vasoconstriction.

The pathophysiological role played by another presynaptic $Y_2$ receptor is better understood. In humans suffering from temporal lobe epilepsy (with Ammon's horn sclerosis) and in rats with chemically induced epilepsy, the density of $Y_2$ receptors on hippocampal glutamatergic mossy fibers is greatly increased. The increase is believed to be a protective mechanism against epileptic seizures since the $Y_2$ receptor-mediated inhibition of the release of the excitatory transmitter glutamate is also enhanced (Vezzani and Sperk 2004). In an electrophysiological study glutamate release from the mossy fibers was inhibited by exogenous neuropeptide Y in epileptic but not in control rats. In the epileptic animals there was even a tonic inhibition by endogenous neuropeptide Y since the $Y_2$ antagonist BIIE 0246 increased glutamate release (Tu et al. 2005).

Clinical trials with neuropeptide Y receptor ligands have so far not been successful (Brain and Cox 2006).

## 4 Presynaptic ACTH Receptors

Although ACTH has been known since the 1930s, presynaptic ACTH receptors were identified much later (Figure 1) and have attracted the attention of investigators for only a few years. ACTH consists of 39 amino acid residues and is processed from proopiomelanocortin, i.e., the same precursor molecule giving rise to ß-endorphin (see section 2). Besides ACTH and endogenous opioids, a series of melanocyte-stimulating hormones are formed from proopiomelanocortin (Table 1). The latter activate $MC_1$, $MC_3$, $MC_4$ and $MC_5$ receptors. ACTH activates the $MC_2$ receptor (i.e., the receptor involved in its corticotropic effect) but also has a high affinity for some of the other melanocortin receptors (Gantz and Fong 2003; Alexander et al. 2006). ACTH is not only one of the hormones of the adenohypophysis but serves also as a neurotransmitter in hypothalamic neurones (Bloch et al. 1979; Bugnon et al. 1979).

Unlike the opioids and neuropeptide Y, ACTH increases transmitter release. As shown in superfusion studies and in pithed and anesthetized animal preparations, ACTH increases noradrenaline release in cardiovascular tissues from the rabbit; in an electrophysiological study it increased acetylcholine release in frog skeletal muscle (Table 4). Thus, according to our present knowledge, presynaptic ACTH receptors occur as heteroreceptors only. No effect of ACTH on noradrenaline release was found in rat atrium and rat and guinea pig pulmonary artery (Costa and Majewski 1988). Human tissues have not been examined. The studies summarized in Table 4 are based on experiments with the full-length ACTH molecule and its active sequence $ACTH_{1-24}$. The effect of ACTH was not mimicked by $ACTH_{4-10}$, a

**Table 4** Synopsis of presynaptic facilitatory ACTH receptors

| | Transmitter | Tissue | Species | References |
|---|---|---|---|---|
| Sympathetic nervous system | Noradrenaline | Heart | Rabbit | Costa and Majewski 1988; Szabo et al. 1988 |
| | | Aorta | Rabbit | Göthert 1984 |
| | | Pulmonary artery | Rabbit | Göthert 1981; 1984; Göthert and Hentrich 1984; Costa and Majewski 1988 |
| | | Various sympathetically innervated tissues | Rabbit | Szabo et al. 1987; 1989 |
| Motor nerves | Acetylcholine | Cutaneous pectoris and sartorius muscle | Frog | Johnston et al. 1983 |

behaviorally active peptide devoid of a corticotropic effect, and was antagonized by an appropriate concentration of the ACTH receptor antagonist $ACTH_{7-38}$ (Göthert 1981). The data suggest that the facilitatory effect is mediated via the $MC_2$ receptor.

In general, the $MC_2$ receptor is $G_s$ protein-coupled (Table 1), and two studies in the rabbit pulmonary artery indicate that this is also true for the presynaptic receptor. The evidence is, indirect, however, in that it suggests activation of adenylyl cyclase, the typical transduction step downstream from $G_s$. In the study by Göthert and Hentrich (1984), the facilitatory effect of ACTH was increased by simultaneous administration of forskolin, an activator of the catalytic subunit of adenylyl cyclase, and AH 21–132, a phosphodiesterase inhibitor. In the study by Costa and Majewski (1988), the facilitatory effect of ACTH was occluded when the vessel was superfused with a lipid-soluble cAMP analogue.

The presynaptic $MC_2$ receptor plays no physiological role, since the concentration of ACTH found in the blood is far too low for its activation. This view is supported by the finding that the antagonist $ACTH_{7-38}$, when given alone, did not affect noradrenaline release in the rabbit heart (Szabo et al. 1989) or pulmonary artery (Göthert 1981). The situation may change under pathophysiological conditions when ACTH concentrations are increased. ACTH may then increase noradrenaline release in cardiovascular tissues and help to maintain cardiovascular function (Szabo et al. 1989). However, at presynaptically effective concentrations ACTH has additional postsynaptic cardiovascular effects: vasodilation and an increase in heart rate (Szabo et al. 1987; 1989).

Although $ACTH_{1-24}$ is used for diagnostic purposes and for the treatment of epilepsies in babies and infants, drugs specifically targeting *presynaptic* $MC_2$ receptors are not available.

## 5 Presynaptic Orexin Receptors

Orexins (also known as hypocretins) and presynaptic receptors activated by orexins were first described in 1998 (Figure 1). Orexin-A (hypocretin-1) and orexin-B (hypocretin-2) consist of 33 and 28 amino acid residues, respectively, and are derived from a common precursor molecule (prepro-orexin). They act on two receptors, $OX_1$ and $OX_2$. Orexinergic neurones have their perikarya in the lateral and posterior part of the hypothalamus and project to many parts of the brain, including the cerebral cortex, thalamus, limbic system, locus coeruleus and raphe nuclei, and to the spinal cord. Orexin-like immunoreactive neurones also occur in the small intestine (for review, see Smart and Jerman 2002).

Orexins increase transmitter release. This has been shown in electrophysiological studies for acetylcholine release in the myenteric plexus of the ileum and for GABA and glutamate release in a series of locations of the central nervous system (Table 5); in other words, presynaptic orexin receptors serve as heteroreceptors. Experiments on human tissue are so far lacking. In some of the experimental models listed in Table 5 orexin-A and orexin-B were studied and orexin-B was at least as

**Table 5** Synopsis of presynaptic facilitatory orexin receptors

| | Transmitter | Tissue | Species | References |
|---|---|---|---|---|
| Parasympathetic nervous system | Acetylcholine | Ileum | Guinea-pig | Katayama et al. 2003; 2005 |
| Central nervous system | GABA | Medial hypothalamus | Rat | *van den Pol et al. 1998** |
| | | Dorsal vagal complex | Rat | *Davis et al. 2003* |
| | Glutamate | Prefrontal cortex | Rat | Lambe and Aghajanian 2003 |
| | | Medial hypothalamus | Rat | *van den Pol et al. 1998* |
| | | Lateral hypothalamus | Mouse | *Li et al. 2002* |
| | | Laterodorsal tegmentum | Mouse | Burlet et al. 2002 |
| | | Caudal nucleus tractus solitarii | Rat | *Smith et al. 2002* |

*In the studies given in *italics* orexins facilitated postsynaptic currents or potentials also in the presence of tetrodotoxin.

potent as orexin-A (Burlet et al. 2002; Li et al. 2002; Davis et al. 2003; Lambe and Aghajanian 2003; Katayama et al. 2005). Since both orexins are equally potent at $OX_2$ receptors, whereas orexin-A is 10 times more potent than orexin-B at $OX_1$ receptors (Alexander et al. 2006), the presynaptic effects of orexins may involve $OX_2$ receptors or perhaps a mixture of $OX_1$ and $OX_2$. For final proof, selective antagonists would be necessary, but $OX_2$ receptor antagonists are so far not available and the $OX_1$ receptor antagonists have a relatively weak potency; the agonist peptide [$Ala^{11}$, D-$Leu^{15}$] orexin-B, which possesses a high preference for $OX_2$ receptors, may be helpful (Alexander et al. 2006). Experiments with knockout mice would be valuable as well.

Since the orexin receptors are $G_q$ protein-coupled (Alexander et al. 2006), one may assume that this also holds true for the presynaptic orexin receptor(s), but so far no data are available. Nonetheless, the six studies carried out in central nervous preparations permit some conclusions on the post-G protein mechanisms. In all instances, the orexins increased the frequency of spontaneous inhibitory or excitatory postsynaptic potentials or currents. The results differed, however, with respect to the influence of tetrodotoxin. In the medial and lateral hypothalamus (van den Pol et al. 1998; Li et al. 2002), dorsal vagal complex (Davis et al. 2003), and caudal nucleus tractus solitarii (Smith et al. 2002), orexins increased the frequency of the miniature potentials or currents also in the presence of tetrodotoxin, suggesting that they directly influenced the vesicle release machinery (references in italics in Table 5). On the other hand, in the prefrontal cortex (Lambe and Aghajanian 2003) and laterodorsal tegmentum (Burlet et al. 2002), the orexins did not retain their facilitatory effect in the presence of tetrodotoxin, suggesting an effect further upstream e.g., on $Ca^{2+}$ and/or $K^+$ channels.

The physiological and pathophysiological role of presynaptic orexin receptors is incompletely understood, and drugs targeting orexin receptors so far have not been developed. Orexins play an important role in the control of sleep and wakefulness, as highlighted by the findings that the knockout of preproorexin in mice (Chemelli et al. 1999) or a mutation of the $OX_2$ gene in dogs (Lin et al. 1999) produces a disturbance resembling narcolepsy in humans. Indeed a decreased number of orexin neurones was found in humans suffering from that disease (Thannickal et al. 2000). There are two presynaptic facilitatory sites that may be involved in the control of sleep and wakefulness by the orexins. The first one is on glutamatergic neurones in the laterodorsal tegmentum, and the second one is on glutamatergic neurones in the prefrontal cortex (i.e., at the final synapse in the ascending arousal pathway). One may assume that the increase in glutamate release excites ascending cholinergic neurones involved in the arousal reaction in the first instance (Burlet et al. 2002; Table 5) and glutamatergic output neurones in the second model (Lambe and Aghajanian 2003; Table 5).

## References


Abrahamsson C (2000) Neuropeptide Y1- and Y2-receptor-mediated cardiovascular effects in the anaesthetized guinea pig, rat, and rabbit. J Cardiovasc Pharmacol 36:451–8

Ackley MA, Hurley RW, Virnich DE et al (2001) A cellular mechanism for the antinociceptive effect of a kappa opioid receptor agonist. Pain 91:377–88

Aimone LD, Yaksh TL (1989) Opioid modulation of capsaicin-evoked release of substance P from rat spinal cord in vivo. Peptides 10:1127–31

Alexander SPH, Mathie A, Peters JA (2006) Guide to receptors and channels, 2nd edition. Br J Pharmacol 147 (Suppl 3):S1–S180

Allen JM, Adrian TE, Tatemoto K et al (1982) Two novel related peptides, neuropeptide Y (NPY) and peptide YY (PYY) inhibit the contraction of the electrically stimulated mouse vas deferens. Neuropeptides 3:71–7

Allgaier C, Daschmann B, Sieverling J et al (1989) Presynaptic κ-opioid receptors on noradrenergic nerve terminals couple to G-proteins and interact with $\alpha_2$-adrenoceptors. J Neurochem 53:1629–35

Averbeck B, Reeh PW, Michaelis M (2001) Modulation of CGRP and PGE2 release from isolated rat skin by alpha-adrenoceptors and kappa-opioid-receptors. Neuroreport 12:2097–2100

Barral J, Mendoza E, Galarraga E et al (2003) The presynaptic modulation of corticostriatal afferents by μ-opioids is mediated by $K^+$ conductances. Eur J Pharmacol 462:91–8

Beani L, Bianchi C, Siniscalchi A (1982) The effect of naloxone on opioid-induced inhibition and facilitation of acetylcholine release in brain slices. Br J Pharmacol 76:393–401

Berger B, Rothmaier AK, Wedekind F et al (2006) Presynaptic opioid receptors on noradrenergic and serotonergic neurons in the human as compared to the rat neocortex. Br J Pharmacol 148:795–806

Bergevin A, Girardot D, Bourque MJ et al (2002) Presynaptic μ-opioid receptors regulate a late step of the secretory process in rat ventral tegmental area GABAergic neurons. Neuropharmacology 42:1065–78

Blakemore LJ, Levenson CW, Trombley PQ (2006) Neuropeptide Y modulates excitatory synaptic transmission in the olfactory bulb. Neuroscience 138:663–74

Bloch B, Bugnon C, Fellmann D, Lenys D, Gouget A (1979) Neurons of the rat hypothalamus reactive with antisera against endorphins, ACTH, MSH and ß-LPH. Cell Tissue Res 204:1–15

Boehm S, Kubista H (2002) Fine tuning of sympathetic transmitter release via ionotropic and metabotropic presynaptic receptors. Pharmacol Rev 54:43–99

Bouryi VA, Lewis DI (2004) Enkephalinergic inhibition of raphe pallidus inputs to rat hypoglossal motoneurones in vitro. Neuroscience 129:55–64

Brain SD, Cox HM (2006) Neuropeptides and their receptors: innovative science providing novel therapeutic targets. Br J Pharmacol 147:S202–S211

Browning KN, Kalyuzhny AE, Travagli RA (2002) Opioid peptides inhibit excitatory but not inhibitory synaptic transmission in the rat dorsal motor nucleus of the vagus. J Neurosci 22:2998–3004

Bugnot C, Bloch B, Lenys D, Fellmann D (1979) Infundibular neurons of the human hypothalamus simultaneously reactive with antisera against endorphins, ACTH, MSH, and beta-LPH. Cell Tissue Res 199:177–96

Burlet S, Tyler CJ, Leonard CS (2002) Direct and indirect excitation of laterodorsal temental neurons by hypocretin/orexin peptides: implications for wakefulness and narcolepsy. J Neurosci 22: 2862–72

Capogna M, Gahwiler BH, Thompson SM (1993) Mechanism of mu-opioid receptor-mediated presynaptic inhibition in the rat hippocampus in vitro. J Physiol 470:539–58

Chemelli RM, Willie JT, Sinton CM et al (1999) Narcolepsy in orexin knockout mice: molecular genetics of sleep regulation. Cell 98:437–51

Cherubini E, North RA (1985) Mu and kappa opioids inhibit transmitter release by different mechanisms. Proc Natl Acad Sci 82:1860–3

Collip JB, Anderson EM, Thomson DL (1933) The adrenotropic hormone of the anterior pituitary lobe. Lancet 222:347–8

Cosentino M, Marino F, De Ponti F et al (1995) Tonic modulation of neurotransmitter release in the guinea-pig myenteric plexus: effect of mu and kappa opioid receptor blockade and of chronic sympathetic denervation. Neurosci Lett 194:185–8

Costa M, Majewski H (1988) Facilitation of noradrenaline release from sympathetic nerves through activation of ACTH receptors, ß-adrenoceptors and angiotensin II receptors. Br J Pharmacol 95:993–1001

Davis SF, Williams KW, Xu W (2003) Selective enhancement of synaptic inhibition by hypocretin (orexin) in rat vagal motor neurons: implications for autonomic regulation. J Neurosci 23: 3844–54

de Lecea L, Kilduff T, Peyron C et al (1998) The hypocretins: hypothalamus-specific peptides with neuroexcitatory activity. Proc Natl Acad Sci 95:322–7

Emmerson PJ, Miller RJ (1999) Pre- and postsynaptic actions of opioid and orphan opioid agonists in the rat arcuate nucleus and ventromedial hypothalamus in vitro. J Physiol 517:431–45

Endo K, Yawo H (2000) mu-Opioid receptor inhibits N-type $Ca^{2+}$ channels in the calyx presynaptic terminal of the embryonic chick ciliary ganglion. J Physiol 524:769–81

Feuerstein TJ, Gleichauf O, Peckys D et al (1996) Opioid receptor-mediated control of acetylcholine release in human neocortex tissue. Naunyn-Schmiedeberg's Arch Pharmacol 354:586–92

Finnegan TF, Chen SR, Pan HL (2005) Effect of the μ opioid on excitatory and inhibitory synaptic inputs to periqueductal gray-projecting neurons in the amygdala. J Pharmacol Exp Ther 312:441–8

Finnegan TF, Chen SR, Pan HL (2006) Mu opioid receptor activation inhibits GABAergic inputs to basolateral amygdala neurons through Kv1.1/1.2 channels. J Neurophysiol 95:2032–41

Fu LY, Acuna-Goycolea C, van den Pol (2004) Neuropeptide Y inhibits hypocretin/orexin neurons by multiple presynaptic and postsynaptic mechanisms: tonic depression of the hypothalamic arousal system. J Neurosci 24:8741–51

Gannon RL, Terrian DM (1991) U-50,488H inhibits dynorphin and glutamate release from guinea pig hippocampal mossy fiber terminals. Brain Res 548:242–7

Gantz I, Fong JM (2003) The melanocortin system. Am J Physiol Endocrinol Metabol 284:E468–E474

Glatzer NR, Smith BN (2005) Modulation of synaptic transmission in the rat nucleus of the solitary tract by endomorphin-1. J Neurophysiol 93:2530–40

Glaum SR, Miller RJ, Hammond DL (1994) Inhibitory actions of delta 1-, delta 2-, and mu-opioid receptor agonists on excitatory transmission in lamina II neurons of adult rat spinal cord. J Neurosci 14:4965–71

Go VLW, Yaksh TL (1987) Release of substance P from the cat spinal cord. J Physiol (Lond) 391:141–67

Göthert M (1981) $ACTH_{1-24}$ increases stimulation-evoked noradrenaline release from sympathetic nerves by acting on presynaptic ACTH receptors. Eur J Pharmacol 76:295–6

Göthert M (1984) Facilitatory effect of adrenocorticotropic hormone and related peptides on $Ca^{2+}$-dependent noradrenaline release from sympathetic nerves. Neuroscience 11:1001–9

Göthert M, Hentrich F (1984) Role of cAMP for regulation of impulse-evoked noradrenaline release from the rabbit pulmonary artery and its possible relationship to presynaptic ACTH receptors. Naunyn-Schmiedeberg's Arch Pharmacol 328:127–34

Grudt TJ, Henderson G (1998) Glycine and $GABA_A$ receptor-mediated synaptic transmission in rat substantia gelatinosa: inhibition by mu-opioid and $GABA_B$ agonists. J Physiol 507:473–83

Guerraci FA, Pozo MJ, Palomares SM et al (2002) Opioid agonists inhibit excitatory neurotransmission in ganglia and at the neuromuscular junction in guinea pig gallbladder. Gastroenterology 122:340–51

Guerrini R, Calo G, Rizzi A et al (1998) A new selective antagonist of the nociceptin receptor. Brit J Pharmacol 123:163–5

Gutstein HB, Akil H (2006) Opioid analgesics. In: Brunton LL (ed) Goodman and Gilman's The Pharmacological Basis of Therapeutics. McGraw-Hill, New York, pp 547–90

Hahm ET, Lee JJ, Min BI et al (2004) Opioid inhibition of GABAergic neurotransmission in mechanically isolated rat periqueductal gray neurons. Neurosci Res 50:343–54

Hayar A, Guyenet PG (1998) Pre- and postsynaptic inhibitory actions of methonine-enkephalin on identified bulbospinal neurons of the rat RVL. J Neurophysiol 80:2003–14

Hjelmstadt GO, Fields HL (2001) Kappa opioid receptor inhibition of glutamatergic transmission in the nucleus accumbens shell. J Neurophysiol 85:1153–8

Ho M, Corbett AD, McKnight AT (2000) Characterization of the $ORL_1$ receptor on adrenergic nerves in the rat anccoccygeus muscle. Br J Pharmacol 131:349–55

Honda E, Ono K, Inenaga K (2004) DAMGO suppresses both excitatory and inhibitory synaptic transmission in supraoptic neurones of mouse hypothalamic slice preparations. J Neuroendocrinol 16:198–207

Hughes J, Smith TW, Kosterlitz HW et al (1975) Identification of two related pentapeptides from the brain with potent opiate agonist activity. Nature 258:577–9

Illes P (1989) Modulation of transmitter and hormone release by multiple neuronal opioid receptors. Rev Physiol Biochem Pharmacol 112:141–233

Jackisch R (1991) Regulation of neurotransmitter release by opiates and opioid peptides in the central nervous system. In: Feigenbaum J, Hanani M (eds) Presynaptic regulation of neurotransmitter release: a handbook. Freund Publishing House, Tel Aviv, pp 551–92

Johnston MF, Kravitz EA, Meiri H et al (1983) Adrenocorticotropic hormone causes long-lasting potentiation of transmitter release from frog motor nerve terminals. Science 220:1071–2

Katayama Y, Homma T, Honda K et al (2003) Actions of orexin A in the myenteric plexus of the guinea-pig small intestine. Neuroreport 14:1515–18

Katayama Y, Hirai K, Homma T et al (2005) Actions of orexins on individual myenteric neurons of the guinea-pig ileum: orexin A or B? Neuroreport 16:745–9

Kennedy C, Krier J (1987) δ-Opioid receptors mediate inhibition of fast excitatory postsynaptic potentials in cat parasympathetic colonic ganglia. Br J Pharmacol 92:437–43

Kerchner GA, Zhuo M (2002) Presynaptic suppression of dorsal horn inhibitory transmission by mu-opioid receptors. J Neurophysiol 88:520–2

King PJ, Williams G, Doods H et al (2000) Effect of a selective neuropeptide Y $Y_2$ receptor antagonist BIIE 0246 on neuropeptide Y release. Eur J Pharmacol 396:1–3

Kishimoto K, Koyama S, Akaike N (2001) Synergistic mu-opioid and $5\text{-}HT_{1A}$ presynaptic inhibition of GABA release in rat periaqueductal gray neurons. Neuropharmacology 41:529–38

Klapstein GJ, Colmers WF (1992) 4-Aminopyridine and low $Ca^{2+}$ differentiate presynaptic inhibition mediated by neuropeptide Y, baclofen and 2-chloroadenosine in rat hippocampal CA1 in vitro. Br J Pharmacol 105:470–4

Lambe E, Aghajanian GK (2003) Hypocretin (orexin) induces calcium transients in single spines postsynaptic to identified the boutons in prefrontal slice. Neuron 40:139–50

Li Y, Gao XB, Sakurai T et al (2002) Hypocretin/orexin excites hypocretin neurons via a local glutamate neuron – a potential mechanism for orchestrating the hypothalamic arousal system. Neuron 36:1169–81

Lin L, Faraco J, Li R et al (1999) The sleep disorder canine narcolepsy is caused by a mutation in the hypocretin (orexin) receptor 2 gene. Cell 98:365–76

Liu QS, Han S, Jia YS et al (1999) Selective modulation of excitatory transmission by mu-opioid receptor activation in rat supraoptic neurons. J Neurophysiol 82:3000–5

Liu S, Hu HZ, Ren J et al (2001) Pre- and postsynaptic inhibition by nociceptin in guinea pig small intestinal myenteric plexus in vitro. Am J Physiol Gastrointest Liver Physiol 281:G237–G246

Mahns DA, Lacroix JS, Potter EK (1998) Inhibition of vagal vasodilatation by a selective neuropeptide YY2 receptor agonist in the bronchial circulation of anaesthetised dogs. J Auton Nerv Syst 73:80–5

Miller RJ (1998) Presynaptic receptors. Annu Rev Pharmacol Toxicol 38:201–27

Modin A, Pernow J, Lundberg JM (1994) Repeated renal and splenic sympathetic nerve stimulation in anaesthetized pigs. J Auton Nerv Syst 49:123–34

Moran TD, Colmers WF, Smith PA (2004) Opioid-like actions of neuropeptide Y in rat substantia gelatinosa: Y1 suppression of inhibition and Y2 suppression of excitation. J Neurophysiol 92:3266–75

Morris MJ (2004) Neuropeptide Y and cardiovascular function. In: Michel MC (ed) Neuropeptide Y and related peptides, *Handbook of Experimental Pharmacology*, vol 162. Springer, Berlin, pp 327–59

Mulder AH, Schoffelmeer ANM (1993) Multiple opioid receptors and presynaptic modulation of neurotransmitter release in the brain. In: Herz A (ed) Opioids I, *Handbook of Experimental Pharmacology*, vol 104/I. Springer, Berlin, pp 125–44

Nakazi M (2001) Hemmung der Serotoninfreisetzung im Mäusegehirn durch Cannabinoid-$CB_1$-Rezeptoren und andere Rezeptorsysteme. Thesis, University of Bonn

Nakazi M, Bauer U, Nickel T et al (2000) Inhibition of serotonin release in the mouse brain via presynaptic cannabinoid $CB_1$ receptors. Naunyn-Schmiedeberg's Arch Pharmacol 361:19–24

Nicholson JR, Paterson SJ, Menzies JRW et al (1998) Pharmacological studies on the "orphan" opioid receptor in central and peripheral sites. Can J Physiol Pharmacol 76:304–13

Nikolarakis KE, Almeida OF, Yassouridis A et al (1989) Presynaptic auto- and allelo-receptor regulation of hypothalamic opioid peptide release. Neuroscience 31:269–73

Odar-Cederlof I, Ericsson F, Theodorsson E et al (2003) Neuropeptide-Y and atrial natriuretic peptide as prognostic markers in patients on hemodialysis. ASAIO J 49:74–80

Ogura M, Kita H (2000) Dynorphin exerts both postsynaptic and presynaptic effects in the Globus pallidus of the rat. J Neurophysiol 83:3366–76

Ostermeier AM, Schlosser B, Schwender D et al (2000) Activation of mu- and delta-opioid receptors causes presynaptic inhibition of glutamatergic excitation in neocortical neurons. Anesthesiology 93:1053–63

Pan YZ, Li DP, Chen SR et al (2002) Activation of delta-opioid receptors excites spinally projecting locus coeruleus neurons through inhibition of GABAergic inputs. J Neurophysiol 88:2675–83

Pan YZ, Li DP, Chen SR et al (2004) Activation of mu-opioid receptors excites a population of locus coeruleus-spinal neurons through presynaptic disinhibition. Brain Res 997:67–78

Potter EK, Tripovic D (2006) Modulation of sympathetic neurotransmission by neuropeptide Y Y2 receptors in rats and guinea pigs. Exp Brain Res 173:346–52

Qian J, Colmers WF, Saggau P (1997) Inhibition of synaptic transmission by neuropeptide Y in rat hippocampal area CA1: modulation of presynaptic $Ca^{2+}$ entry. J Neurosci 17:8169–77

Rhim H, Kinney GA, Emmerson PJ et al (1997) Regulation of neurotransmission in the arcuate nucleus of the rat by different neuropeptide Y receptors. J Neurosci 17:2980–9

Rogers H, Henderson G (1990) Activation of mu- and delta-opioid receptors present on the same nerve terminals depresses transmitter release in the mouse hypogastric ganglion. Brit J Pharmacol 101:505–12

Rominger A, Förster S, Zentner J et al (2002) Comparison of the ORL1 receptor-mediated inhibition of noradrenaline release in human and rat neocortical slices. Br J Pharmacol 135:800–6
Sándor NT, Kiss J, Sándor A et al (1991) Naloxone enhances the release of acetylcholine from cholinergic interneurons of the striatum if the dopaminergic input is impaired. Brain Res 552:343–5
Schlicker E, Kathmann M (2000) Modulation of the release of noradrenaline and other transmitters via cannabinoid and opioid receptors. Fundam Clin Pharmacol 14:64
Schlicker E, Morari M (2000) Nociceptin/orphanin FQ and neurotransmitter release in the central nervous system. Peptides 21:1023–9
Schlicker E, Werthwein S, Kathmann M et al (1998) Nociceptin inhibits noradrenaline release in the mouse brain cortex via presynaptic $ORL_1$ receptors. Naunyn-Schmiedeberg's Arch Pharmacol 358:418–22
Schlosser B, Kudernatsch MB, Sutor B et al (1995) Delta, mu and kappa opioid receptor agonists inhibit dopamine overflow in rat neostriatal slices. Neurosci Lett 191:126–30
Schwertfeger E, Klein T, Vonend O et al (2004) Neuropeptide Y inhibits acetylcholine release in human heart atrium by activation of $Y_2$-receptors. Naunyn-Schmiedeberg's Arch Pharmacol 369:455–61
Serone AP, Wright CE, Angus JA (1999) Heterogeneity of prejunctional NPY receptor-mediated inhibition of cardiac neurotransmission. Brit J Pharmacol 127:99–108
Shen KZ, Johnson SW (2002) Presynaptic modulation of synaptic transmission by opioid receptor in rat subthalamic nucleus in vitro. J Physiol 541:219–30
Simmoneaux V, Ouichou A, Craft C et al (1994) Presynaptic and postsynaptic effects of neuropeptide Y in the rat pineal gland. J Neurochem 62:2464–71
Simmons ML, Terman GW, Drake CT et al (1994) Inhibition of glutamate release by presynaptic kappa 1-opioid receptors in the guinea pig dentate gyrus. J Neurophysiol 72:1697–1705
Smart D, Jerman JC (2002) The physiology and pharmacology of the orexins Pharmacol Ther 94:51–61
Smith BN, Davis SF, van den Pol AN et al (2002) Selective enhancement of excitatory synaptic activity in the rat nucleus tractus solitarius by hypocretin 2. Neuroscience 115:707–14
Smith-White MA, Herzog H, Potter EK (2002) Role of neuropeptide Y $Y_2$ receptors in modulation of cardiac parasympathetic neurotransmission. Regul Peptides 103:105–11
Stanford IM, Cooper AJ (1999) Presynaptic mu and delta opioid receptor modulation of $GABA_A$ IPSCs in the rat globus pallidus in vitro. J Neurosci 19:4796–4803
Starke K (1977) Regulation of noradrenaline release by presynaptic receptor systems. Rev Physiol Biochem Pharmacol 77:1–124
Stjärne L, Stjärne E, Mshgina M (1989) Does clonidine or neuropeptide Y-mediated inhibition of ATP secretion from sympathetic nerves operate primarily by increasing a potassium conductance? Acta Physiol Scand 136:137–8
Storr M, Hahn A, Gaffal E et al (2002) Effects of endomorphin-1 and -2 on on mu-opioid receptors in myenteric neurons and in the peristaltic reflex in rat small intestine. Clin Exp Pharmacol Physiol 29:428–34
Sun QQ, Akk G, Huguenard JR et al (2001) Differential regulation of GABA release and neuronal excitability mediated by neuropeptide Y1 and Y2 receptors in rat thalamic neurons. J Physiol 531:81–94
Szabo B Hedler L, Lichtwald K et al (1987) ACTH increases noradrenaline release in pithed rabbits with electrically stimulated sympathetic outflow. Eur J Pharmacol 136:391–9
Szabo B, Hedler L, Schurr C et al (1988) ACTH increases noradrenaline release in the rabbit heart. Naunyn-Schmiedeberg's Arch Pharmacol 338:368–72
Szabo B, Hedler L, Starke K (1989) Facilitation of the release of noradrenaline: an extra-adrenal effect of adrenocorticotropic hormone. Resuscitation 18:229–42
Tatemoto K (2004) Neuropeptide Y: history and overview. In: Michel MC (ed) Neuropeptide Y and related peptides, *Handbook of Experimental Pharmacology*, vol 162. Springer, Berlin, pp 1–21
Tatemoto K, Carlquist M, Mutt V (1982) Neuropeptide Y – a novel brain peptide with structural similarities to peptide YY and pancreatic polypeptide. Nature 296:659–60

Taube HD, Borowski E, Endo T et al (1976) Enkephalin: a potential modulator of noradrenaline release in rat brain. Eur J Pharmacol 38:377–80
Thannickal TC, Moore RY, Nienhuis R (2000) Reduced number of hypocretin neurons in human narcolepsy. Neuron 27:469–74
Trendelenburg P (1917) Physiologische und pharmakologische Versuche über die Dünndarmperistaltik. Arch Exp Pathol Pharmakol 81:55–129
Trendelenburg AU, Cox SL, Schelb V et al (2000) Modulation of $^{3}$H-noradrenaline release by presynaptic opoid, cannabinoid and bradykinin and ß-adrenoceptors in mouse tissues. Brit J Pharmacol 130:321–30
Tu B, Timofeeva O, Jiao Y et al (2005) Spontaneous release of neuropeptide Y tonically inhibits recurrent mossy fiber synaptic transmission in epileptic brain. J Neurosci 25:1718–29
Ueda H, Fukushima N, Ge M et al (1987) Presynaptic opioid kappa-receptor and regulation of the release of Met-enkephalin in the rat brainstem. Neurosci Lett 81:309–13
van den Pol AN, Gao XB, Obrietan K et al (1998) Presynaptic and postsynaptic actions and modulation of neuroendocrine neurons by a new hypothalamic peptide, hypocretin/orexin. J Neurosci 18:7962–71
Vaughan CW, Christie MJ (1997) Presynaptic inhibitory action of opioids on synaptic transmission in the rat periaqueductal grey in vitro. J Physiol 498:463–72
Vaughan CW, Ingram SL, Christie MJ (1997) Actions of the ORL1 receptor ligand nociceptin on membrane properties of rat periaqueductal gray neurons in vitro. J Neurosci 17:996–1003
Vezzani A, Sperk G (2004) Overexpression of NPY and Y2 receptors in epileptic brain tissue: an endogenous neuroprotective mechanism in temporal lobe epilepsy ? Neuropeptides 38:245–52
Wang SJ (2005) Activation of neuropeptide Y Y1 receptors inhibits glutamate release through reduction of voltage-dependent Ca2+ entry in the rat cerebral cortex nerve terminals: suppression of this inhibitory effect by the protein kinase C-dependent facilitatory pathway. Neuroscience 134:987–1000
Wang X, Dergacheva O, Griffioen KJ et al (2004) Action of kappa and delta opioid agonists on premotor cardiac vagal neurons in the nucleus ambiguus. Neuroscience 129:235–41
Werling LL, McMahon PN, Cox BM (1989) Effects of pertussis toxin on opioid regulation of catecholamine release from rat and guinea pig brain slices. Naunyn-Schmiedeberg's Arch Pharmacol 339:509–13
Westfall TC (2004) Prejunctional effects of of neuropeptide Y and its role as a cotransmitter. In: Michel MC (ed) Neuropeptide Y and related peptides, *Handbook of Experimental Pharmacology*, vol 162. Springer, Berlin, pp 137–83
Yang XP, Chiba S (2002) Antagonistic interaction between BIIE 0246, a neuropeptide Y Y2-receptor antagonist, and omega-conotoxin GVIA, a Ca2+ channel antagonist, in presynaptic transmitter releases in dog splenic arteries. Jpn J Pharmacol 89:188–91
Yu XJ, Moskowitz MA (1996) Neuropeptide Y $Y_2$ receptor-mediated attenuation of neurogenic plasma extravasation acting through pertussis toxin-sensitive mechanisms. Br J Pharmacol 119:229–32
Yu TP, Fein J, Phan T et al (1997) Orphanin FQ inhibits synaptic transmission and long-term potentiation in rat hippocampus. Hippocampus 7:88–94
Yu Y, Wang X, Cui Y et al (2006) Abnormal modulation of cholinergic neurotransmission by endomorphin 1 and endomorphin 2 in isolated bronchus of type 1 diabetic rats. Peptides 27:2770–7
Zhu W, Pan ZZ (2005) Mu-opioid-mediated inhibition of glutamate synaptic transmission in rat central amygdala neurons. Neuroscience 133:97–103
Zöllner C, Stein C (2007) Opioids. In: Stein C (ed) Analgesia, *Handbook of Experimental Pharmacology*, vol 177. Springer, Berlin, pp 31–63

# Presynaptic Modulation by Endocannabinoids

**David M. Lovinger**



**Abstract** Modulation of neurotransmitter release by G-protein-coupled receptors (GPCRs) is a prominent presynaptic mechanism for regulation of synaptic transmission. Activation of GPCRs located at the presynaptic terminal can decrease the probability of neurotransmitter release. This presynaptic depression involves activation

David M. Lovinger
Laboratory for Integrative Neuroscience, National Institute on Alcohol Abuse and Alcoholism, National Institutes of Health, 5625 Fishers Lane, Bethesda, MD 20892-9411, USA
lovindav@mail.nih.gov

T.C. Südhof, K. Starke (eds.), *Pharmacology of Neurotransmitter Release*. *Handbook of Experimental Pharmacology 184*.

of Gi/o-type G-proteins that mediate different inhibitory mechanisms, including inhibition of voltage-gated calcium channels, activation of potassium channels, and direct inhibition of the vesicle fusion process. A variety of neurotransmitters and modulatory agents can activate GPCRs that produce presynaptic depression. Among these are lipid metabolites that serve as agonists for GPCRs. The discovery of endocannabinoids and their cognate receptors, including the CB1 receptor, has stimulated intense investigation into the neurophysiological roles of these lipid metabolites. It is now clear that presynaptic depression is the major physiological role for the CB1 receptor. Endocannabinoids activate this receptor mainly via a retrograde signaling process in which these compounds are synthesized in and released from postsynaptic neuronal elements, and travel back to the presynaptic terminal to act on the CB1 receptor. This retrograde endocannabinoid modulation has been implicated in short-term synaptic depression, including suppression of excitatory or inhibitory transmission induced by postsynaptic depolarization and transient synaptic depression induced by activation of postsynaptic GPCRs during agonist treatment or synaptic activation. Endocannabinoids and the CB1 receptor also play a key role in one form of long-term synaptic depression (LTD) that involves a long-lasting decrease in neurotransmitter release. Endocannabinoid-dependent synaptic modulation and plasticity have now been implicated in a variety of brain processes, including learning and memory. The present chapter discusses our current knowledge about the cellular and molecular mechanisms involved in endocannabinoid-mediated presynaptic modulation and endocannabinoid-dependent LTD.

## 1 Introduction

Presynaptic neurotransmitter release is crucial to the function of chemical synapses, as it provides a rapid surge of the chemical needed to activate postsynaptic processes that insure effective intercellular communication. Neurotransmitter release is subject to control from a variety of molecules, as the chapters in this volume will attest. One prominent source of such control is modulation by presynaptic G protein-coupled receptors (GPCRs) that inhibit transmitter release. A variety of receptors for a host of neuromodulatory agents have been identified (see Dunwiddie and Lovinger 1993; Miller 1998; Wu and Saggau 1997 for review). Such inhibition can be either homosynaptic, inhibition of release of a neurotransmitter by receptors for that same transmitter, or heterosynaptic, inhibition of release of a neurotransmitter by receptors for a second neurotransmitter. These neuromodulatory mechanisms allow for control of the amount and timing of neurotransmitter release. Ultimately this form of presynaptic modulation can regulate the pattern and timing of activity in neural circuits. Presynaptic GPCRs are also targets for therapeutic agents. For example, morphine and other opioid drugs produce their analgesic actions, at least in part, through activation of presynaptic opiate receptors (see Reisine and Pasternak 1996). Baclofen, an agonist at presynaptic GABAB receptors, is effective in treatment of spasticity and other neuromuscular disorders (c.f. Orsnes et al. 2000). The therapeutic efficacy of these compounds likely stems from the fact that they can produce subtle adjustments in neurotransmitter release without fully blocking synaptic transmission.

There are several common features of GPCR-mediated presynaptic depression. First, the receptors mediating such effects are coupled to Gi/o-type G-proteins that are known to engage intracellular signaling pathways, including inhibition of adenylyl cyclase, inhibition of voltage-gated calcium channels, and activation of potassium channels (Dunwiddie and Lovinger 1993; Miller 1998). Second, receptors and downstream signaling can be activated on a subsecond timescale, providing rapid feedback for intercellular alterations in secretion (Mintz and Bean 1993; Heinbockel et al. 2005). Third, regulation of intracellular calcium, due mainly to inhibition of the aforementioned calcium channels, appears to play a prominent role in receptor-mediated decreases in neurotransmitter release (Wu and Saggau 1997). Given the strong calcium sensitivity of excitation/secretion coupling, regulation of presynaptic calcium levels is certainly an attractive mechanism for presynaptic modulation. Fourth, presynaptic receptors have been shown to inhibit neurotransmitter release via mechanisms downstream from calcium channel activity, and likely independent of ion channel activity (see Dunwiddie and Lovinger 1993; Wu and Saggau 1997 for review). Receptor-mediated alterations in vesicle exocytosis have been proposed to account for these effects. The preponderance of data suggests that inhibition of voltage-gated calcium channels and inhibition of exocytosis are the major mechanisms through which GPCRs inhibit transmitter release (Wu and Saggau 1997).

The many neurotransmitters and modulatory agents involved in GPCR-mediated presynaptic depression appear to come from a variety of cellular sources. Traditional neurotransmitters and many neuropeptides are released from presynaptic terminals to act on GPCRs on their own terminals or on those of nearby neurons (Dunwiddie and Lovinger 1993; Miller 1998). Neuromodulators such as adenosine and glutamate might be released from glia as well as neurons (Volterra and Meldolesi 2005), and the constant presence of low levels of adenosine is thought to have a "tonic" modulatory effect at brain presynaptic terminals (Dunwiddie and Masino 2001). Yet other substances appear to be made in, or released from, postsynaptic neuronal elements and can act on presynaptic terminals via signaling in a "retrograde" fashion (Stanton et al. 2004). Among the molecules thought to serve such retrograde signaling roles are adenosine (Brundege and Dunwiddie 1996), integrin proteins (Sargent Jones 2005), small gaseous molecules (Kyrozis et al. 2004), neurotrophins (Bramham and Messaoudi 2004), and various lipid metabolites (Feinmark et al. 2003; Chevaleyre et al. 2006). Signaling by certain members of this latter class of compounds is the subject of this chapter.

Long-lasting decreases in the efficacy of synaptic transmission (so-called long-term depression, or LTD) are observed at synapses throughout the brain (Malenka and Bear 2004). The initiation of LTD often begins with repetitive activation of a given set of synapses, and LTD is expressed as decreased synaptic function at the same set of synapses or another synaptic input impinging on the same postsynaptic neuron (homosynaptic versus heterosynaptic LTD, respectively). The maintenance of certain subtypes of LTD appears to involve decreased presynaptic neurotransmitter release (Choi and Lovinger 1997a,b; Feinmark et al. 2003; Castellucci et al. 1978; Mato et al. 2005). Presynaptic GPCRs, including receptors for glutamate and opioid peptides, have been implicated in presynaptically expressed LTD (Kahn et al. 2001; Kobayashi et al. 1996; Weisskopf et al. 1993).

In this chapter, the roles of cannabinoid-type GPCRs in regulating neurotransmitter release wiil be discussed. In addition, the role of retrograde endocannabinoid signaling and cannabinoid receptors in short-term and long-term synaptic depression will be discussed. This latter line of research has its origins in the discovery of the phenomena termed depolarization-induced suppression of excitation and inhibition (DSE and DSI, respectively). These forms of short-lasting synaptic depression were first discovered in the early 1990s at synapses in the cerebellum and hippocampus (Llano et al. 1991; Pitler and Alger 1992). The experimental paradigm for eliciting DSE/I starts with depolarization of a postsynaptic neuron. This depolarization leads to a decrease in excitatory glutamatergic or inhibitory GABAergic transmission onto that same neuron. The subsequent discovery that retrograde signaling involving endocannabinoids and their cognate receptors is necessary for DSE and DSI will be discussed in detail later in the chapter. A short time after the characterization of DSE and DSI, investigators studying long-term synaptic plasticity in the striatum observed a form of LTD that was later found to involve postsynaptic induction and presynaptic expression mechanisms (Calabresi et al. 1992a; Lovinger et al. 1993; Walsh 1993; Choi et al. 1997a,b). Endocannabinoid retrograde signaling and cannabinoid receptor activation play key roles in this and related forms of LTD, and this long-lasting endocannabinoid-dependent plasticity will be discussed and compared with DSE/I and other shorter-lasting endocannabinoid actions. Prior to discussing endocannabinoid-dependent synaptic plasticity it is necessary to review our current knowledge of endocannabinoids and their receptors.

## 2 Endocannabinoids and their Receptors

### *2.1 Endocannabinoid Synthesis and Degradation*

The so-called endocannabinoids are metabolites of membrane lipids that are part of a host of modified lipid and fatty acid molecules that have now been implicated in intercellular signaling across a wide spectrum of biological organisms. Within the mammalian body, paracrine effects of lipid metabolites are widespread (see Bazan 2005 for review). The endocannabinoids were so named because of their ability to activate the GPCRs known to be targets for the active ingredient, Δ9-tetrahydrocannabinol, in marijuana, hashish, and other drugs generated from *Cannabis sativa*. Two metabolites of arachidonic acid-containing phospholipids, arachidonoyl ethanolamide (AEA or anandamide) and 2-arachidonoyl glycerol (2-AG) are the most prominent endocannabinoids in the brain and periphery.

Anandamide was the first endocannabinoid to be discovered (Devane et al. 1992). This compound can be formed by transacylation of phosphatidylinositol and subsequent degradation by the so-called N-acetyl-phospatidylethanolamine phospholipase D (NAPE-PLD) (Di Marzo et al. 1994; Schmid et al. 1990), as well as via a newly characterized pathway involving phospholipase C (PLC) and dephosphorylation mediated by protein tyrosine phosphatase N22 (PTPN22) and possibly other phosphatases (Liu et al. 2006). It should be noted that AEA is only one of a class of fatty acid ethanolamides, including oleaoyl ethanolamide (OEA) and palmitoylethanolamide (PEA) made from lipids containing other fatty acid moieties (Bachur et al. 1965). These other FA ethanolamides do not act on the CB1 receptor, and the molecular targets of these compounds are just beginning to be elucidated (LoVerme et al. 2005a,b). Both OEA and PEA are substrates for the AEA catabolic enzyme fatty acid amide hydrolase (FAAH), and it is possible that these compounds could slow AEA degradation competing for FAAH activity, an action that has been dubbed the entourage effect (Ben-Shabat et al. 1998). However, on balance it is inaccurate to refer to this class of compounds as endocannabinoids, as they form a group of juxtacrine/autocrine-acting FA ethanolamides with several molecular targets. Anandamide can be released into the extracellular environment (Di Marzo et al. 1994; Giuffrida et al. 1999) and can participate in cell-cell signaling (Chevaleyre et al. 2006; Ralevic 2003). The FAAH enzyme is widely distributed throughout the body (Cravatt et al. 1996). Two FAAH isoforms have been cloned, but the subtype identified more recently is not expressed in mouse and rat (Wei et al. 2006). FAAH activity appears to be a key step in termination of AEA signaling (Glaser et al. 2003).

The endocannabinoid 2-AG is produced via a two-step process: degradation of arachidonate-containing phospholipids to diacylglycerol (DAG) by PLC followed by DAG lipase-catalyzed degradation to 2-AG (Mechoulam et al. 1995; Sugiura et al. 1995), although an alternative synthesis pathway has been proposed (Bisogno et al. 1999). Release of 2-AG from cells has also been measured (Bisogno et al. 1997), and in general brain tissue levels of 2-AG are higher than those of AEA (Jung et al. 2005, but see Giuffrida et al. 1999). This may reflect the presence of a metabolic as well as a signaling pool of 2-AG, and therefore some refinement will be needed to obtain accurate measures of the 2-AG involved in synaptic modulation. Inactivation of 2-AG proceeds through monoglycerol lipase (MGL) (Dinh et al. 2002), although cyclooxygenase-2 (Cox-2) may also participate in 2-AG catabolism (Kozak et al. 2002). Studies examining the synaptic localization of enzymes involved in 2-AG synthesis and degradation have generally concluded that DAG lipase is mainly present in postsynaptic elements, while MGL is localized to presynaptic terminals (Bisogno et al. 2003; Gulyas et al. 2004; Katona et al. 2006). Interestingly, the AEA-degrading enzyme FAAH is often found in a postsynaptic neuronal location, where it has been associated with endoplasmic reticulum and other intracellular membranous structures (Gulyas et al. 2004). Thus, the steps involved in AEA degradation could be quite complex (Mckinney and Cravatt 2005). Other compounds have been suggested to act as endocannabinoids, including virodhamine (Porter et al. 2002) and arachidonoyl dopamine (Bisogno et al. 2000), but little evidence has accumulated for actions of these compounds in vivo. Table 1 lists the compounds that inhibit endocannabinoid synthetic and degradative enzymes and known nonspecific targets of these drugs.

## 2.2 The CB1 Receptor

The neural actions of endocannabinoids are mediated mainly through activation of the CB1 class of cannabinoid receptors (Freund et al. 2003; Mackie 2005; Matsuda et al. 1990), although recent reports indicate the presence of another major CB receptor, CB2, in glial cells as well as in neurons in some brain regions (Gong et al. 2006; Van Sickle et al. 2005). Cannabinoid receptors are classical Gi/o-coupled GPCRs (Howlett et al. 1986; Matsuda et al. 1990). As such, activation of CBs by their cognate ligands produces typical responses such as inhibition of adenylyl cyclase, phosphorylation/activation of extracellular signal-related kinase (ERK), activation of G-protein-activated inwardly rectifying potassium channels (GIRKs), and inhibition of voltage-gated calcium channels (VGCCs) (Clapham and Neer 1997; Freund et al. 2003; Davis et al. 2003; Wartmann et al. 1995). Additional signaling pathways for CBs have been noted (e.g., Glass and Felder 1997), but it is not clear if these responses are widespread. Table 1 lists the most commonly used CB1 receptor ligands, and compares their actions at CB2 receptors.

Within the central nervous system CB1 receptors are found mainly on presynaptic terminals (Freund et al. 2003; Mackie 2005), and within a given brain region selected terminals express the receptor (Figure 1). Postsynaptic CB1 receptors and CB1 signaling has been seen as well (Tsou et al. 1997; Bacci et al. 2004), but appear to be much rarer than presynaptic receptors. Activation of CB1 has been demonstrated to decrease release of a wide variety of neurotransmitters, including ACh, GABA, glutamate, NE, and others at synapses throughout the CNS and periphery (Alger 2002; Mackie 2006; Szabo and Schlicker 2005). These presynaptic inhibitory effects proceed through the sorts of mechanisms described above, namely Gi/o-mediated alterations in ion channel function and other mechanisms involved in neurotransmitter release. More will be said about the role of particular mechanisms later in this chapter.

## 2.3 Endocannabinoid Release and Reuptake

Endocannabinoids are lipophilic biomolecules, which complicates our understanding of mechanisms of release of these substances. Most neurotransmitters are released from cells via vesicular mechanisms, and such a mechanism has even been suggested to underlie dopamine release from postsynaptic elements in ventral tegmental area (VTA) dopaminergic neurons (Beckstead et al. 2004). Endocannabinoids would not be packaged into the lumen of a vesicle like hydrophilic molecules.

**Table 1** Drugs that Act on the Endocannabinoid System

| Target | Compound | Action | Affinity/Potency |
|---|---|---|---|
| **CB1 Receptor** | ACEA | Agonist (AEA analog) | 1.4 nM (Ki) |
| | ACPA | Agonist (AEA analog) | 2.2 nM (Ki) |
| | (*R*)-(+)-Methanandamide | Agonist (AEA analog) | 20 nM (Ki) |
| | Noladin ether | Agonist (endogenous) | 21.2 nM (Ki) |
| | NADA | Agonist (endogenous) | 250 nm (Ki) |
| | DEA | Agonist (endogenous) | 34.4 nM (Ki) |
| | Leelamine | Agonist | 2.9 μM (Ki) |
| | AM 251 | Antagonist/inverse agonist | 7.5 nM (Ki) |
| | AM 281 | Antagonist/Inverse Agonist | 14 nM (Ki) |
| | LY 320135 | Antagonist/Inverse Agonist | 141 nM (Ki) |
| | O-2050 | Antagonist | 2.5 nM (Ki) |
| | SR 141716A | Antagonist/Inverse Agonist | 1.2 nM (Kd) |
| **CB2 Receptor** | JWH 133 | Agonist | 34 nM (Ki) |
| | JWH 015 | Agonist | 13.8 nM (Ki) |
| | L-759,633 | Agonist | 6.4 nM (Ki) |
| | L-759,656 | Agonist | 11.8 nM (Ki) |
| | AM 630 | Antagonist/Inverse Agonist | 31.2 nM (Ki) |
| | JTE 907 | Antagonist/Inverse Agonist | 0.38–35.9 nM (Ki) |
| **CB1 and CB2 Receptors** | Anandamide (AEA) | Agonist (endogenous) | 89, 371 nM; CB1, CB2 (Ki) |
| | 2-AG | Agonist (endogenous) | 0.47, 1.4 μM; CB1, CB2 (Ki) |
| | Virodhamine | Agonist (endogenous) | 1.9 , 1.4 μM; CB1, CB2 ($EC_{50}$) |
| | CP 55,940 | Agonist | 0.6, 0.7 nM. CB1, CB2 (Ki) |
| | HU 210 | Agonist | 0.06, 0.52 nM; CB1, CB2 (Ki) |
| | WIN 55,212-2 | Agonist | 62, 3.3 nM; CB1, CB2 ($K_i$) |
| | Bay 59–3074 | Partial Agonist | 48.3, 45, 5 nM; CB1, CB2 (Ki) |
| | CAY10508 | Antagonist/Inverse Agonist | 243 nM CB1 (Ki) |
| | Δ9-THC | Agonist/Partial Agonist | 37-199 nM (Ki) |
| **Endocannabinoid Uptake** | AM 404 | Transport/FAAH inhibitor | 1 μM at transporter ($IC_{50}$) |
| | LY 2183240 | Transport inhibitor | 270 pM at transporter ($IC_{50}$) |
| | O-2093 | Transport inhibitor | 17.3 μM($IC_{50}$) |
| | OMDM-2 | Transport inhibitor | 3 μM (Ki) |
| | UCM 707 | Transport inhibitor | 0.8 μM($IC_{50}$) |
| | VDM 11 | Transport/FAAH inhibitor | 4–11 μM at transporter ($IC_{50}$) |
| **AEA and 2-AG Hydolyzing Enzymes (FAAH and MGL)** | Palmitoyliso-propylamide | FAAH Inhibitor | ~10 μM($IC_{50}$) |
| | MAFP | Inhibitor (FAAH and MGL) | 2.5 nM at FAAH ($IC_{50}$) |
| | ATFMK | Inhibitor (FAAH and MGL) | low μM($IC_{50}$) |
| | URB 597 | FAAH Inhibitor | 4.6 nM ($IC_{50}$) |
| **Diacylglycerol Lipase (2-AG Synthetic Enzyme)** | Tetrahydrolip-statin (THL) | Inhibitor | 2–6 nM (IC50) |
| | RHC 80267 | Inhibitor | 1.1–4 μM($IC_{50}$) |

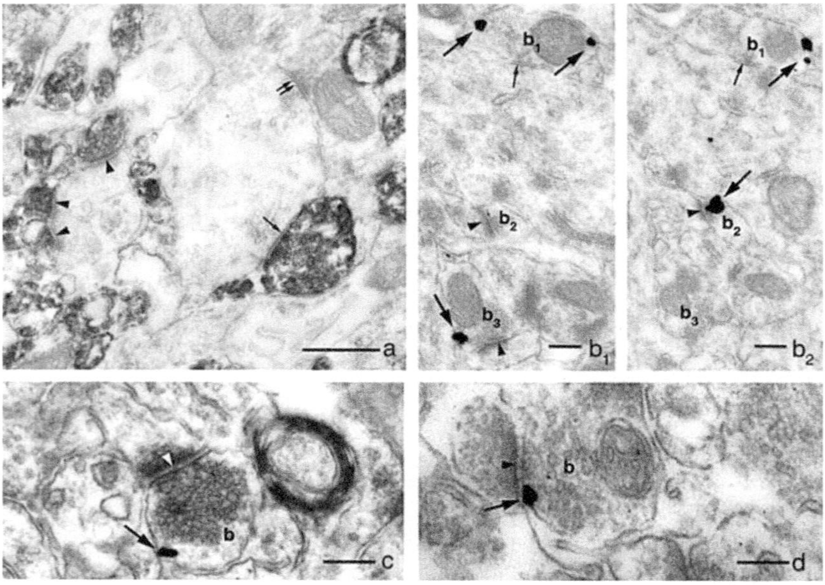


**Fig. 1** Presynaptic expression of CB1 receptors. (**a**) Electron micrograph showing CB1 labeling on both asymmetrical, presumed glutamatergic, excitatory synaptic terminals (arrowheads), and symmetrical, presumed GABAergic, inhibitory terminals (single arrow). Note that some terminals at symmetrical, presumed GABAergic, synapses do not show CB1 labeling (double arrows). No evidence of postsynaptic labeling is observed. These images are from the molecular layer of the dentate gyrus. (**b1**, **b2**) Labeling of CB1 receptors (denoted by large arrows) in both types of axon terminals (b1-b3) in the molecular layer of the dentate gyrus using the EM-immunogold technique. Note that receptors are localized to the intracellular surface of the plasma membrane in both excitatory (arrowheads) and inhibitory (small arrows) synaptic terminals. Similar CB1 labeling is also seen in stratum oriens of the CA3 hippocampal subfield (**c**), and stratum oriens of the CA1 subfield (**d**) (terminals and receptors identified as in **b1** and **b2**). Scale bars: **a**, 0,5 μM, **b-d**, 0.2 μM. (Reprinted with permission from Katona et al. 2006, copyright 2006, by the Society for Neuroscience).

Although ECs might reside in vesicular membranes, there is no reason to believe that this would constitute a better delivery system than having the molecules exit the cell via the plasma membrane itself. Furthermore, evidence from studies of EC signaling in the hippocampus suggests that botulinum toxin E light chain does not block postsynaptic EC release (Wilson and Nicoll 2001). These findings support the idea that vesicular fusion is not necessary for EC release. Two potential EC release mechanisms have been suggested. The first is simple diffusion through the plasma membrane followed by liberation into the extracellular space. This is a simple and elegant model, and of course there is ample biophysical evidence that ECs can move through lipid membranes (Bojesen and Hansen 2005; Glaser et al. 2003; Alexander and Cravatt 2006; Makriyannis et al. 2005b; McFarland and Barker 2004). The second idea involves a transporter or carrier-mediated system that facilitates the diffusion of ECs from one side of the membrane to the other (Fegley et al. 2004; Hillard and Jarrahian 2000; Maccarrone et al. 2002; Moore et al. 2005; Ronesi et al. 2004). At first glance, it would appear that an EC transport system is needlessly complex for a lipophilic molecule. However, a system for increasing the rate of transmembrane EC movement might help to localize the most rapid influx and efflux of these molecules near synthetic and degradative enzymes, thus ensuring rapid handling of the largest EC pools.

Evidence for and against the existence of an EC transporter (called the anandamide membrane transporter or endocannabinoid membrane transporter, AMT or EMT alternatively) has accumulated, and this has been a topic of fierce debate in recent papers (Bojesen and Hansen 2005; Glaser et al. 2003; Alexander and Cravatt 2006; Makriyannis et al. 2005b; McFarland and Barker 2004; Fegley et al. 2004; Hillard and Jarrahian 2000; Maccarrone et al. 2002; Mechoulam and Deutsch 2005; Moore et al. 2005). This topic can only be summarized briefly at present. Pharmacological experiments have indicated that cellular uptake and release of AEA and 2-AG can be disrupted by various agents that appear to act independently of other EC-interacting proteins (Hillard and Jarrahian 2000; Fegley et al. 2004; Moore et al. 2005). However, the protein or other molecules involved in EC transport has yet to be identified. Furthermore, the specificity of the putative AMT blockers has been called into question, and in particular it appears that many of these compounds also inhibit the FAAH enzyme (Glaser et al. 2003, Alexander and Cravatt 2006). Thus, it is possible that the slowing of EC uptake produced by these enzymes is secondary to disruption of an EC gradient maintained by FAAH activity (although it is not so clear how this would disrupt 2-AG uptake if this compound is not metabolized by FAAH; Lichtman et al. 2002; Saario et al. 2004). It is clear that FAAH plays a large role in regulating AEA uptake, as disruption of this enzyme greatly slows the uptake process (Glaser et al. 2003). Indeed, FAAH itself may even serve as a conduit for entry of fatty acids into the cell (Mckinney and Cravatt 2005). Thus, the field awaits either creation of a cleaner AMT blocker (Moore et al. 2005; Alexander and Cravatt 2006) or definitive molecular identification of a transporter molecule(s). Table 1 lists the drugs that are currently used as AMT inhibitors along with their known nonspecific targets.

Regardless of the mechanism of transmembrane endocannabinoid movement, investigators interested in the mechanisms of intercellular communication by ECs must explain how these lipophilic compounds are induced to leave the plasma membrane of one cell and travel to receptors on another cell. Fatty acids and their derivatives are known to interact with carrier proteins such as albumin, and anandamide interacts with such proteins as well (Hsu and Storch 1996; Sacchettini and Gordon 1993; Bojesen and Hansen 2003; Makriyannis et al. 2005a). Fatty acid binding proteins also mediate movement of these lipophilic molecules within the cell (Bass 1998). Thus, the idea of an FA-binding carrier protein mediating the extraction from the cell membrane and the intercellular movement of ECs is attractive (Makriyannis et al. 2005a). However, no such carrier protein has been identified to date. It is possible that EC movement across the synapse involves a lipophilic component of the molecular matrix within the synaptic cleft itself, but that hypothesis has not yet been fully explored. Thus, there is still much to learn about how ECs participate in cell-cell communication.

## 3 Cannabinoids, Endocannabinoids, and Presynaptic Depression

### *3.1 Actions of Synthetic Cannabinoids*

The most widely reported effect of CB1 agonists is reduction in synaptic transmission or the extracellular concentration of neurotransmitters (Alger 2002; Chevaleyre et al. 2006; Freund et al. 2003; Mackie 2006; Szabo and Schlicker 2005). This synaptic depression appears to result predominantly from reductions in neurotransmitter release produced by presynaptic CB1 receptors. The presynaptic depressant effect of CB1 agonists is most often blocked by CB1 antagonists and is absent in gene-targeted CB1-/- mice (Domenici et al. 2006; Gerdeman et al. 2002; Hajos et al. 2001; Katona et al. 2006; Kawamura et al. 2006; Takahashi and Castillo 2006; Wilson et al. 2001; Yoshida et al. 2002), with notable exceptions (Hajos et al. 2001; Hoffman et al. 2005).

The mechanisms underlying CB1-mediated presynaptic depression are similar to those implicated in presynaptic depression by other presynaptic GPCRs. Direct demonstrations of CB1-mediated inhibition of VGCCs and activation of GIRK and A-type potassium channels have provided convincing evidence that this receptor can activate molecular pathways implicated in presynaptic inhibition (Bacci et al. 2004; Henry and Chavkin 1995; Mackie and Hille 1992; Mackie et al. 1995, Pan et al. 1996; Guo and Ikeda 2004). Inhibition of VGCCs almost certainly plays a role in presynaptic CB1 actions (Brown et al. 2004; Foldy et al. 2006; Kushmerick et al. 2004). The study by Kushmerick et al. (2004) is particularly convincing, as the authors took advantage of their ability to perform electrophysiological recordings in the large presynaptic terminals at the calyx of Held to provide direct evidence for decreased presynaptic calcium current induced by CB1 activation. The study by Brown et al. (2004) also provides compelling evidence for a calcium channel role from experiments using presynaptic calcium imaging. Activation of potassium channels may also be involved in CB1-mediated presynaptic depression (Daniel and Crepel 2001; Diana and Marty 2003; Varma et al. 2002), but it is often difficult to disentangle potassium and calcium channel function when examining presynaptic modulation with indirect methods (Brown et al. 2004). At synapses in the amygdala and nucleus accumbens CB1 activation does not appear to alter the frequency or other properties of miniature synaptic responses, indicating that the receptor has little effect on mechanisms downstream from VGCCs (Hoffman and Lupica 2001; Katona et al. 2001; Zhu and Lovinger 2005). However, agonist-induced decreases in miniature frequency have been detected, suggesting a more proximal action on neurotransmitter release mechanisms at some synapses (Chevaleyre et al. 2006; Gerdeman and Lovinger 2001).

Cannabinoid and endocannabinoid-induced synaptic depression is observed in both the peripheral nervous system and the CNS. Indeed, Δ9-THC inhibition of transmitter release was first demonstrated in mouse vas deferens (Graham et al. 1974), and further evidence for presynaptic inhibition has been obtained using this preparation (Ishac et al. 1996; Pertwee and Fernando 1996) and in the myenteric plexus (Coutts and Pertwee 1997; Kulkarni-Narla and Brown 2000). In addition, anandamide was first characterized as an EC based on its actions in the mouse vas deferens (Devane et al. 1992). Subsequently, CB1 receptor-mediated inhibition of release of several neurotransmitters has been documented in various regions of the PNS (see Szabo and Schlicker 2005 for review). Cannabinoids also inhibit neural effects on contraction in the ileum (Croci et al. 1998; Lopez-Redondo et al. 1997), although it is not clear that this is effect involves direct inhibition of neurotransmitter release (Croci et al. 1998). The CB1 receptor has been localized to enteric neurons, and thus the effect on ileum certainly involves actions on these presynaptic neurons. In addition, anandamide produces ileal relaxation via a non-CB1, non-CB2-mediated mechanism (Mang et al. 2001).

In general the effects of synthetic cannabinoids are reversible when rigorous steps are taken to test for reversibility (Auclair et al. 2000; Hajos et al. 2001; Kreitzer and Regehr 2001; Levenes et al. 1998; Maejima et al. 2001; Ronesi et al. 2004; Yin et al. 2006; Zhu and Lovinger 2005, but see Kreitzer and Malenka 2005 and Robbe et al. 2001). However, the cannabinoid drugs and endocannabinoids have molecular properties that make it difficult to demonstrate reversibility in many experimental preparations. The highly lipophilic nature of these compounds yields appreciable affinity for lipid-rich white matter, and for plastics used in experimental settings. This is important to bear in mind when designing and interpreting experiments to examine the effects of CB1 activation or inhibition in neural tissue, particularly in brain slices that contain white matter. Long-lasting effects of the compounds cannot always be attributed to biological actions as they may result from the inability to remove compounds from the experimental preparation. Application of a CB1 antagonist following the cessation of agonist application is the most effective way to determine if the agonist action is reversible in the brain slice setting (Kreitzer and Regehr 2001; Kreitzer and Malenka 2005; Maejima et al. 2001; Ronesi et al. 2004; Yin et al. 2006).

## *3.2 Endocannabinoid Actions*

Application of ECs produces presynaptic depression, but this action has not been demonstrated as consistently as that of synthetic CB1 agonists (see Ameri et al. 1999; Di et al. 2005; Stella et al. 1997; Straiker and Mackie 2005; Szabo et al. 2006, for examples). Two factors likely underlie the relative scarcity of data on EC effects. First, ECs are subject to both cellular uptake and degradation, as mentioned above. Thus, convincing demonstration of effects at some synapses might require blocking either or both of these mechanisms, especially in multicellular preparations such as

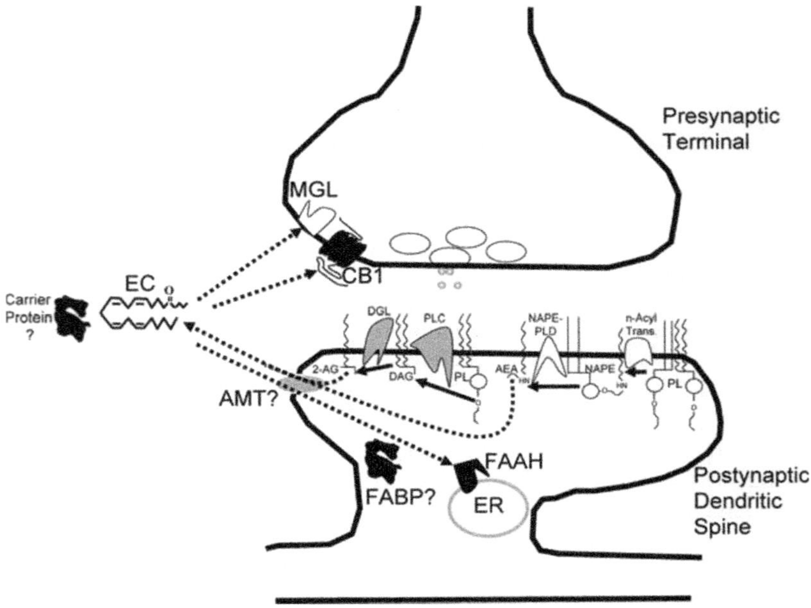


**Fig. 2** Identity and presumed location of endocannabinoid synthetic and degradative enzymes at the synapse. A major synthetic pathway for AEA involves transacylation followed by PLD-catalyzed degradation of the NAPE intermediate. However an alternative pathway involving PLC has been proposed, as discused in the text. Synthesis of 2-AG starts with PLC-catalyzed degradation to DAG, followed by degration to 2-AG involving DGL. EC release from the postsynaptic neuron may involve simple diffusion or an AMT-like transporter. Upon release the EC is thought to traverse the synapse and act on presynaptic CB1 receptors. The lipophilic EC may require some sort of carrier protein to move across the hydrophilic environment of the synaptic cleft. After reuptake that may involve the AMT, ECs are degraded by specific enzymes. FAAH catalyzes AEA breakdown, and there is evidence that this enzyme is located in intracellular membranous organelles such as the ER. If this is the case, then an FABP-like protein may be required to shuttle AEA from the plasma membrane to the ER. The 2-AG catabolic enzyme MGL is thought to reside in the presynaptic plasma membrane, and appears to have a role in termination of 2-AG signaling during DSE and DSI. Abbreviations: 2-AG = 2-arachidonoyl glycerol, AEA = arachidonoylethanolamid (anandamide), AMT = anandamide membrane transporter, CB1 = type 1 cannabinoid receptor, DAG = diacylglycerol, DGL = diacylglycerol lipase, EC = endocannabinoid, ER = endoplasmic reticulum, FAAH = fatty acid amide hydrolase, FABP = fatty acid binding protein, MGL = monoglyceride lipase, NAPE = n-acylphosphatidylethanolamine, NAPE-PLD = NAPE-specific phospholipase D, PL = phospholipid, PLC = phospholipase C.

brain slices. Second, AEA acts as a partial agonist in most experimental settings when CB1 modulation of channels or other effector pathways with little amplification are examined (Breivogel et al. 1998; Burkey et al. 1997; Guo and Ikeda 2004; McAllister et al. 1999; Mackie et al. 1993; but see Glass and Northup 1999). Thus, there is some doubt as to whether AEA-induced short-term synaptic depression can be readily demonstrated. However, synaptic depression induced by application of AEA or 2-AG has been reported several times, and thus it is clear that ECs can initiate presynaptic depression. Interestingly, Straiker and Mackie (2005) examined

EC-induced synaptic depression in an autaptic hippocampal culture preparation, and reported that AEA produces synaptic depression that is poorly reversible, while 2-AG produced larger depression that was rapidly reversible. These findings may indicate additional differences in synaptic depression produced by these two ECs that cannot be explained by a simple difference in relative efficacy at CB1 receptors.

Investigators have made interesting use of AEA in characterizing presynaptic CB1-mediated depression. For example, a compelling recent study used a "caged" form of AEA to demonstrate the action and time course of EC and CB1-mediated presynaptic inhibition at a GABAergic synapse in hippocampal slices (Heinbockel et al. 2005). This tool allowed the investigators to photolytically release the AEA with precise timing using an argon-ion laser for UV chemical uncaging. The AEA uncaging produced presynaptic depression that was CB1 mediated. From these studies, the authors concluded that the mechanisms linking CB1 activation to depression of neurotransmitter release occur on a timescale of 50–175 msec, and these effects were reversed within tens of seconds. Two other investigators have used patch pipettes to fill postsynaptic neurons with AEA and provided evidence that this EC can reach its presynaptic receptor target via a retrograde signaling pathway (Ronesi et al. 2004; Jo et al. 2005). The retrograde signaling function of ECs will now be considered in greater detail. Figure 2 provides a schematic diagram of the likely synaptic arrangement of molecules involved in retrograde endocannabinoid signaling.

# 4 Retrograde Endocannabinoid Signaling in Short-Term Synaptic Depression

The presynaptic location of CB1 receptors, and their implication in agonist and EC-induced presynaptic depression, provided strong clues as to the neuromodulatory roles that might be played by the endocannabinoids. It was also becoming evident that increased intracellular calcium had an important role in endocannabinoid synthesis (Bisogno et al. 1999; Di Marzo et al. 1994). Investigators in a few different laboratories deduced from this information that endocannabinoids could participate in retrograde signaling that was initiated by postsynaptic calcium increases and expressed via endocannabinoid actions on presynaptic CB1 receptors (Kreitzer and Regehr 2001; Maejima et al. 2001; Ohno-Shosaku et al. 2001; Wilson and Nicoll 2001). This provided a crucial breakthrough in our understanding of a major mechanism of endocannabinoid action in the nervous system.

## *4.1 Depolarization-Induced Suppression of Excitatory and Inhibitory Transmission*

As mentioned in the introduction, DSE and DSI involve presynaptic depression of neurotransmitter release brought about by postsynaptic depolarization. The post-synaptic-to-presynaptic signaling involved in DSE and DSI indicated the involvement of a retrograde signal in these forms of synaptic plasticity. Furthermore, it was known that activation of voltage-gated calcium channels and increased postsynaptic calcium levels are necessary and sufficient for DSE/I induction (Llano et al. 1991; Pitler and Alger 1992; Wang and Zucker 2001; Wilson and Nicoll 2001), providing a clue that formation of the retrograde signal depended on a rise in postsynaptic calcium. The idea of retrograde signaling at synapses was not new, as retrograde messengers had been invoked to explain other forms of synaptic plasticity (most notably hippocampal LTP; Madison and Schuman 1995). In the case of DSE/I the retrograde signal hypothesis has held up well under intense scrutiny (see Alger 2002; Chevaleyre et al. 2006 for review). Still, nearly a decade passed between the discovery of the basic phenomena and the discovery of the molecular identity of the messenger.

Several groups nearly simultaneously reported evidence that an EC was the retrograde messenger in DSE and DSI(Kreitzer and Regehr 2001; Maejima et al. 2001; Ohno-Shosaku et al. 2001; Wilson and Nicoll 2001). In at least one laboratory this appeared to be the result of a realization that mechanisms for endocannabinoid production matched those known to be involved in DSI induction (see discussion in Chevaleyre et al. 2006). The main evidence for an EC retrograde signal included (1) the finding that CB1 agonists and/or ECs inhibit transmission at synapses that show DSE/I, and this inhibition involves presynaptic mechanisms (Kreitzer and Regehr 2001; Maejima et al. 2001; Wilson and Nicoll 2001); (2) the demonstration that CB1 antagonists prevent DSE/I (Kreitzer and Regehr 2001; Maejima et al. 2001; Wilson and Nicoll 2001) (see Figure 3 for an example of DSI and blockade by the CB1 antagonist SR141716A); and (3) the demonstration that inhibition of the putative EC transporter mimics and occludes DSI (Wilson and Nicoll 2001). Subsequent studies confirmed and extended these findings (reviewed in Alger 2002; Chevaleyre et al. 2006). It has now been demonstrated that CB1 receptors are present on presynaptic terminals of synapses that show DSE/I, and EC synthetic enzymes are present in postsynaptic elements at these synapses (Freund et al. 2003; Katona et al. 2006; Mackie 2005). Manipulations such as the AEA uncaging and postsynaptic loading of AEA mentioned above have also generated findings consistent with the retrograde signaling idea (Heinbockel et al. 2005; Ronesi et al. 2004).

Many of the mechanisms involved in induction and expression of DSE/I have been elucidated, although some points remain unclear. The depolarization-induced increase in postsynaptic calcium appears to play a role in EC production and/or release (Brenowitz and Regehr 2005; Heibockel et al. 2005; Wang and Zucker 2001; Wilson and Nicoll 2001). However, there is as yet no definitive evidence that loss of any EC synthetic enzyme activity prevents DSE or DSI. Some reports have claimed that DSE/I are unaffected by blockade of PLC or DAG lipase activity, and are not lost following knockout of the PLCβ1 or β4 enzymes (Edwards et al. 2006; Hashimotodani et al. 2005; Maejima et al. 2005; Safo and Regehr 2005). However, a more recent study indicates that blockade of DAG lipase prevents DSI of synapses onto cerebellar Purkinje neurons and neurons in the VTA, but not at inhibitory synapses in the hippocampal CA1 region (Szabo et al. 2006). Inhibition of PLC and DAG lipase does alter other EC-dependent forms of plasticity, as does

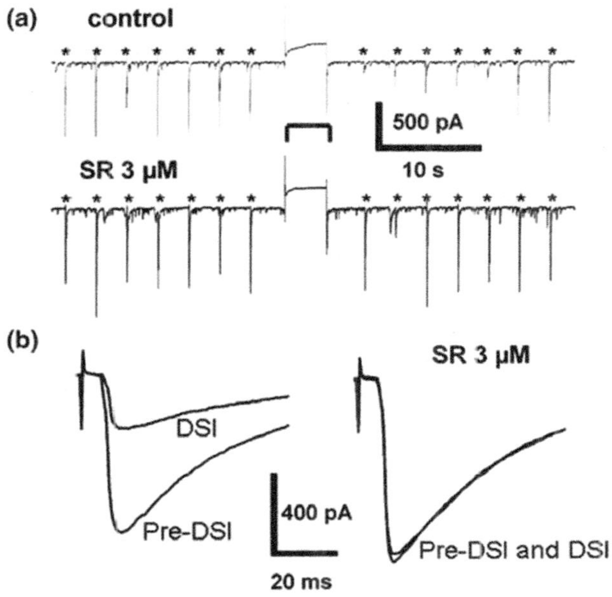


**Fig. 3** Depolarization-induced supression of Inhibition (DSI) is endocannabinoid and CB1 receptor-mediated. (**a**) Current traces showing evoked and spontaneous IPSCs before and after a 2 sec depolarization from −60 to 0 mV in the absence (top) and presence (bottom) of the CB1 antagonist SR141716A (SR). Asterisks denote the evoked IPSCs; bracket indicates the period of depolarization. (**b**) Individual evoked IPSCs evoked before and just after depolarization shown at higher temporal resolution in the absence (left) and presence (right) of SR.

knockout of PLCβ1, and these data will be discussed later in this chapter (see Sections 4.2, 4.3, 6.1). The lack of inhibitor effects at hippocampal synapses has led to speculation that increased postsynaptic calcium plays a role in EC release or "mobilization" during induction of DSE/I (Edwards et al. 2006). However, the mechanisms involved in these processes have yet to be fully elucidated. As mentioned above, Wilson and Nicoll (2001) provided evidence that interference with postsynaptic vesicle fusion steps does not alter hippocampal DSI.

Following their postsynaptic release, ECs are thought to traverse the synaptic cleft and act on presynaptic CB1 receptors (Alger 2002; Chevaleyre et al. 2006). This receptor activation leads to presynaptic depression involving mechanisms described above. Several groups have attempted to determine the spatial specificity of the endocannabinoid signal. The bulk of evidence indicates that ECs released at a single set of synapses have an effective range of 10s of μm, indicating that ECs will mainly influence inputs to the cells and synapses from which they are released during DSE and DSI (Brown et al. 2003; Maejima et al. 2001). However, synaptic inhibition induced by DSE/I has been shown to spread to other neurons (Vincent and Marty 1993; Wilson and Nicoll 2001), and supression of interneuron firing facilitates this spread in cerebellum (Kreitzer et al. 2002).

The temporal characteristics of DSE/I have also been examined in greater detail in recent studies. Estimates from the experiments using AEA uncaging indicate that the time from the postsynaptic depolarization/calcium increase to the onset of synaptic depression is on the order of 400 msec (Heinbockel et al. 2005). The duration of DSE/I ranges from a few sec to 10s of sec (see Jo et al. 2005, Heinbockel et al. 2005 for examples). The duration of these short-lasting forms of plasticity appear to be directly related to the lifetime of the EC messenger in the synaptic cleft. The effects of transport blockers mentioned above support this idea (Wilson and Nicoll 2001). In addition, drugs that inhibit EC catabolic enzymes, MGL inhibitors in particular, prolong DSI lifetime (Kim and Alger 2004; Makara et al. 2005; Szabo et al. 2006). DSE duration decreases as temperature increases (Kreitzer and Regehr 2001), consistent with the fact that EC lifetime in the cleft and related signaling events should proceed faster at higher temperatures. The elegant study by Kreitzer and Regehr (2001) in cerebellum also indicated that the duration of DSE is well correlated with the time course of an associated decrease in the presynaptic calcium transient. This finding supports the idea that CB1 activation inhibits presynaptic VGCC production, and this inhibition is directly linked to DSI expression.

It is still not clear which endocannabinoid mediates DSI/E. As the experiments utilizing PLC and DAG lipase inhibitors have produced ambiguous results, the evidence supporting a role for 2-AG in DSI/E comes mainly from experiments in which Cox-2 or MGL inhibitors prolong DSI (Kim and Alger 2004; Makara et al. 2005; Szabo et al. 2006). Complementary experiments in these same studies indicate that inhibition of FAAH does not alter the time course of DSI, and thus the evidence at present indicates that AEA is not the EC retrograde messenger involved in this type of plasticity, despite the fact that AEA can clearly produce a DSI-like effect (Heinbockel et al. 2005). Full elucidation of the EC responsible for DSE and DSI will require the development of better molecular tools for examination of EC release, reuptake, and degradation.

Interestingly, DSE/I are not observed at all synapses where presynaptic CB1 receptors are found. For example, in the dorsal striatum EC-dependent DSE is not observed at glutamatergic synapses known to express other forms of EC-dependent plasticity (Gerdeman et al. 2002; Kreitzer and Malenka 2005; Yin et al. 2006). Perhaps postsynaptic activation by depolarization alone is not sufficient to induce EC production or release from the postsynaptic elements of these synapses.

The physiological roles of DSE/I are the subject of debate and continued investigation (Alger 2002; Chevaleyre et al. 2006; Hampson and Deadwyler 1998). Regulation of the pattern and timing of neuronal activity in discrete circuits is one attractive possibility. Depression at a subset of synapses onto a given postsynaptic neuron might also boost the relative efficacy or signal/noise ratio of other inputs to those same cells. In this regard, it has been proposed that depression may be one mechanism that helps to equalize synaptic efficacy across extensive dendritic arbors (Ramsey and Abbott 2006). For example, decreasing efficacy at synapses proximal to the soma would yield more equal weighting relative to distal synapses in influencing action potential generation. Evidence has accumulated that the disinhibition produced by DSI in hippocampus enhances neuronal excitability (Wagner and Alger 1996), excitatory transmission and synaptic excitation via an increase in the efficacy of EPSP-spike coupling during disinhibition (Chevaleyre and Castillo 2003). Disinhibition during DSI can also enhance induction of long-term potentiation (LTP), a long-lasting increase in synaptic efficacy at excitatory synapses (Carlson et al. 2002). To date there is little evidence that DSE/I-like mechanisms take place in vivo. Indeed, the depolarization needed to trigger these types of plasticity is strong and prolonged, and perhaps beyond the physiological range of neuronal activity (although see Brown et al. 2004). Nonetheless, endocannabinoid actions in the intact brain are becoming increasingly apparent, and these functions may well involve DSE, DSI or the other EC-dependent forms of synaptic plasticity that will be discussed in the following sections of this chapter (see Sections. 4.2, 6.1).

DSI and DSE have been observed at synapses in several brain regions in addition to the cerebellum and the hippocampal CA1 subregion. These include inhibitory synapses in the basolateral amygdala (Zhu and Lovinger 2005), dentate gyrus of the hippocampus (Isokawa and Alger 2005); hypoglossal interneurons (where DSI alters glycine release) (Mukhtarov et al. 2005), hypothalamus (Di et al. 2005; Hentges et al. 2005; Jo et al. 2005), neocortex (Bodor et al. 2005; Trettel and Levine 2003), substantia nigra (Yanovsky et al. 2003), and the ventral tegmental area (VTA) (Melis et al. 2004a). It is possible that DSE/I and similar EC-dependent transient decreases in synaptic transmission could play roles in brain functions as diverse as addiction, feeding, motor control, and spatial learning. However, much more work is needed to distinguish possible roles of DSE/I from those of the other EC-dependent forms of synaptic plasticity, which will now be discussed.

## *4.2 Synaptically Driven EC-Dependent Short-Term Depression*

Neural production and release of ECs is likely to be stimulated by repetitive firing of neurons or repetitive synaptic input to a neuron rather than the prolonged depolarization used to initiate DSE/I. Investigators have now discovered a number of patterns of neuronal activity that generate short-lasting EC-dependent synaptic depression in different brain regions. The majority of these involve activation of neurotransmitter receptors or brief periods of repetitive synaptic activation (see Chevaleyre et al. 2006 for review).

One motif that has emerged from these investigations is stimulation of EC-dependent plasticity by metabotropic receptors that couple to Gq-like G-proteins. In particular, metabotropic glutamate receptors (mGluRs) of the group I subtypes have been strongly linked to EC production at synapses in several brain regions. Direct activation of mGluRs enhances DSI (Martin and Alger 1999; Morishita et al. 1998; Ohno-Shosaku et al. 2002; Varma et al. 2001) and also directly produces EC-dependent synaptic plasticity (Edwards et al. 2006; Galante and Diana 2004; Kushmerick et al. 2004; Maejima et al. 2001; Martin and Alger 1999; Morishita et al. 1998; Ohno-Shosaku et al. 2002; Varma et al. 2001). In the hippocampal CA1 region, activation of muscarinic acetylcholine receptors produces a similar action

(Edwards et al. 2006; Fukudome et al. 2004; Martin and Alger 1999; Morishita et al. 1998). In many cases, these metabotropic receptor-induced forms of synaptic depression persist only as long as the receptors are activated. However, in some cases depression persists after agonist has been removed (Azad et al. 2004; Chevaleyre and Castillo 2003; Edwards et al. 2006; Kreitzer and Malenka 2005; Robbe et al. 2002; Yin et al. 2006). This long-lasting effect will be discussed in more detail in the section on long-term depression (see Section 6.1).

The idea that Gq activation leads to stimulation of EC formation is logical given that PLC is known to be activated by this class of G-proteins (Sternweiss et al. 1992). Activation of mGluRs has been shown to increase EC levels measured using mass spectrometric analysis of brain tissue (Jung et al. 2005), and activation of PLC and DAG lipase were implicated in this mGluR action.

Short-term EC-dependent synaptic plasticity (eSTD) follows afferent stimulation leading to synaptic activation at synapses in cerebellum (Brown et al. 2003; Galante and Diana 2004; Maejima et al. 2005; Rancillac and Barbara 2005), neocortex (Trettel et al. 2004) and VTA (Melis et al. 2004b). This phenomenon has been observed at excitatory glutamatergic and inhibitory GABAergic synapses. The eSTD at excitatory synapses is homosynaptic. In other words depression is observed at the same synapses whose activation leads to STD induction. In the majority of cases, this eSTD requires activation of a group I mGluR (Brown et al. 2003; Galante and Diana 2004; Maejima et al. 2005; Melis et al. 2004b). At these synapses, the mechanism for eSTD induction involves repetitive activation of glutamate release following afferent activation, glutamate activation of the mGluR, stimulation of PLC or other enzymes implicated in EC formation, retrograde EC signaling to presynaptic CB1 receptors, and ultimately a decrease in glutamate release such as that observed during DSE (Hashimotodani et al. 2005). Activation of mAChR activation can enhance eSTD in neocortex (Trettel et al. 2004).

The eSTD observed at GABAergic inhibitory synapses is heterosynaptic, or at least has a heterosynaptic component (Galante and Diana 2004; Martin and Alger 1999; Morishita et al. 1998; Varma et al. 2001). In this case, activation of glutamatergic synapses appears to be required for STD induction. This type of plasticity usually begins with brief periods of moderate-high frequency activation of glutamatergic synapses or application of a group I mGluR agonist. This leads to mGluR activation and a subsequent increase in postsynaptic EC production and EC release. The EC then acts on presynaptic CB1 receptors on what are presumed to be nearby GABAergic terminals, suppressing GABA release. The reduction in GABA release is thought to persist for the duration of the rise in synaptic EC concentration. However, certain aspects of this scenario are as yet unclear. For example, it is not known if activation of GABAergic synapses plays any role in eSTD initiation. Also, there is little evidence to date concerning the relative locations of the glutamatergic and GABAergic synapses involved in this plasticity. Indeed, it is not yet clear if the EC itself is released from postsynaptic elements associated with the glutamatergic or the GABAergic synapse.

## 4.3 Comparison of DSE/I and eSTD

There are intriguing differences in the mechanisms of induction of DSE/I in comparison to eSTD. Induction of DSE/I does not require activation of metabotropic receptors, such as mGluRs (Kreitzer and Regehr 2001; Maejima et al. 2001; Wilson and Nicoll 2001), while receptor activation does appear to be required for most forms of eSTD (Brown et al. 2003; Edwards et al. 2006; Galante and Diana 2004; Kushmerick et al. 2004; Maejima et al. 2001,2005; Martin and Alger 1999; Melis et al. 2004b; Morishita et al. 1998; Ohno-Shosaku et al. 2002; Varma et al. 2001). In general, blockade of STD with PLC and DAG lipase inhibitors is more consistent than inhibitor effects on DSE or DSI (Edwards et al. 2006; Hashimotodani et al. 2005; Maejima et al. 2005; Safo and Regehr 2005; but see Szabo et al. 2006). This dissociation has been observed even at a single population of inhibitory inputs to hippocampal pyramidal neurons (Edwards et al. 2006). Similar results have also been obtained using gene-targeted mice that lack the PLCβ1 enzyme. In hippocampal neurons from these mice, eSTD induced by receptor activation was lost while DSI persisted (Hashimotodani et al. 2005). In cerebellum, the PLCβ4 enzyme appears to serve a similar role, being necessary for receptor- and synaptically-driven eSTD, but not for DSE (Maejima et al. 2005). The necessary role for increased postsynaptic calcium levels in DSE/I induction is clear. However, the role of this type of signaling in eSTD is less certain. In the cerebellum eSTD can be induced even under conditions of strong postsynaptic calcium chelation (Galante and Diana 2004; Maejima et al. 2001), indicating a lack of necessity for a calcium increase. These findings support a model in which mGluR activation leading to phospholipase-mediated EC formation is necessary for eSTD. In contrast, only activation of voltage-gated calcium channels and a postsynaptic rise in calcium appear to be necessary for induction of DSE/I. Thus, the roles of calcium, EC production, and EC signaling in DSE/I and eSTD differ. The insensitivity of DSI to lipase blockers has led Alger and coworkers to suggest that the ECs involved in DSE/I are part of a pre-existing pool that is "mobilized" by the postsynaptic calcium increase, as mentioned above (Edwards et al. 2006). It is possible that some sort of EC release mechanism, perhaps the transport processes discussed in Section 2.3, plays a role in postsynaptic EC availability for DSE/I, but as yet there is no evidence to support this idea. Evidence from one study suggests that a postsynaptically loaded AMT inhibitor prevents induction of eSTD at synapses onto dopaminergic neurons in the ventral tegmental area (VTA)(Melis et al. 2004b). Thus, some sort of EC mobilization or release mechanism is implicated in this form of synaptic depression.

Synergism between DSE and eSTD has been observed in elegant studies by Brenowitz and Regehr (2005). Parallel fiber synapses onto cerebellar Purkinje neurons exhibited both types of synaptic depression, and stimuli that are subthreshold for induction of either type of depression alone can be co-activated to yield depression. This synergism is dependent on both postsynaptic calcium levels and mGluR activation, but the mGluRs do not appear to participate in the calcium rises crucial for induction of depression.

The evidence available at present implicates 2-AG as the EC involved in agonist and stimulation-induced eSTD. In cerebellar, hippocampal, and VTA neurons eSTD is prevented by blockers of PLC and DAG lipase (Edwards et al. 2006; Galante and Diana et al. 2004; Melis et al. 2004b). The loss of STD in PLC knockout mice is also consistent with a role for 2-AG (Hashimotodani et al. 2005; Maejima et al. 2005). However, the recent evidence for an AEA synthesis mechanism involving PLC (Liu et al. 2006) makes this conclusion less certain. Activation of mGluRs stimluates production of 2-AG in preference to AEA in the hippocampus and striatum (Jung et al. 2005), and both PLC and DAG lipase are involved in this mGluR-stimulated EC production. There may well be a role for AEA in eSTD at some synapses, but the bulk of evidence available at this time favors involvement of 2-AG. Figure 4 summarizes the presynaptic mechanisms thought to participate in transient synaptic depression during DSE/I and eSTD.

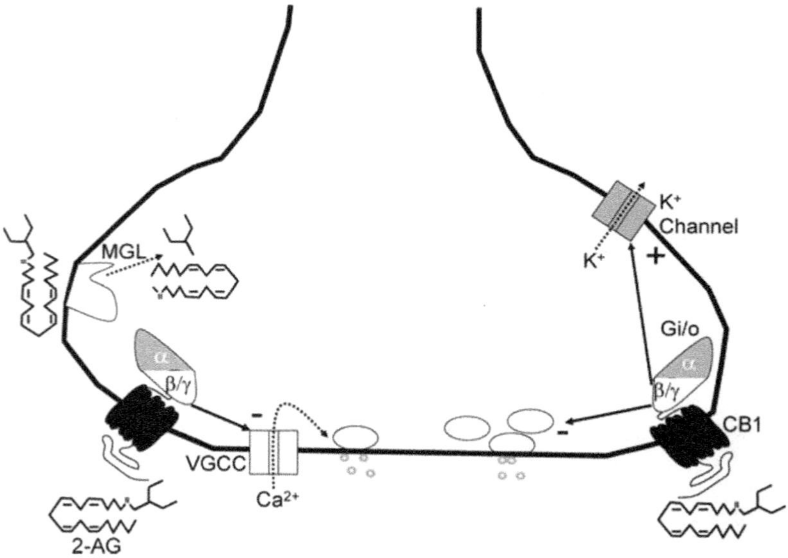


**Fig. 4** Mechanisms underlying presynaptic expression of transient endocannabinoid-mediated synaptic depression. Activation of the CB1 receptor by an EC (most likely 2-AG) stimulates dissociation of Gi/o-type G-proteins. The liberated Gβ/γ subunit inhibits voltage-gated calcium channels (VGCCs), activates potassium channels, and may also directly inhibit the vesicle fusion process. Inhibition of the VGCC reduces the calcium signal involved in excitation-secretion coupling, thereby reducing vesicle fusion and lowering the probability of neurotransmitter release. Activation of potassium channels may hyperpolarize the presynaptic membrane potential, shunt presynaptic current, and/or increase the rate of repolarization of the terminal during excitation. All these effects would reduce calcium entry and decrease the probability of neurotransmitter release. Disruption of vesicle fusion directly reduces the probability of release. EC signaling is terminated, most likely via MGL-catalyzed degradation of 2-AG, and this is a key step in termination of transient EC-mediated synaptic depression such as DSE/I and eSTD.

Endocannabinoid-dependent STD and DSE/I are modulated by other neurotransmitters in a brain region-dependent manner. In the striatum the neuromodulatory transmitter dopamine is abundant (Carlsson 1993). Dopamine activation of D2 receptors produces synaptic depression, especially when paired with activation of glutamatergic synaptic transmission (Calabresi et al. 1992a; Cepeda et al. 1993; Flores-Hernandez et al. 1997; Tang et al. 2001; Bamford et al. 2004). A recent study has implicated retrograde endocannabinoid signaling in this form of STD (Yin and Lovinger 2006). Synaptic activation of mGluRs also appears to play a role in this D2-mediated eSTD. In the lateral hypothalamus DSI is observed at synapses onto neurons that express melanin-concentrating hormone (Jo et al. 2005). As in most forms of DSI, activation of a postsynaptic VGCC is necessary for the induction of plasticity. These lateral hypothalamic neurons express postsynaptic receptors for the neuropeptide leptin. Activation of the leptin receptors suppresses calcium channel function and prevents EC retrograde signaling, presumably by inhibiting EC formation or release. This type of mechanism is likely to be widespread given the large number of GPCRs that can inhibit VGCCs (Dolphin 2003; Elmslie 2003). However, it remains to be determined if there are particular receptors that preferentially inhibit the channels involved in EC production/release during induction of synaptic plasticity.

## 5 Endocannabinoid Tone and Intrinsic CB1 Activity

Given the relationship between neuronal activity, postsynaptic calcium increases, and EC production, it is possible that neurons with especially high levels of activity or periodic or sustained increases in intracellular calcium could tonically produce and/or release EC messengers. This could lead to sustained CB1 activation and a "tone" of decreased transmission. Several reports have now been published that indicate such a tone at inhibitory synapses in neurons from basolateral amygdala (Zhu and Lovinger 2005), hippocampus (Foldy et al. 2006), and hypothalamus (Jo et al. 2005). The main evidence for tonic endocannabinoid suppression of transmission comes from experiments in which CB1 antagonists enhance synaptic transmission. It has been argued that these effects might reflect inverse agonist actions of the drugs, and indeed there is evidence that the most commonly used CB1 antagonists, AM251 and SR131716A, can act as inverse agonists (Bouaboula et al. 1997; Pan et al. 1998). The argument that antagonist actions are due to inhibition of EC effects and not inverse agonist actions is bolstered by experiments involving postsynaptic manipulations designed to prevent EC production/release. Filling the postsynaptic neuron with a calcium chelator, such as BAPTA, prevents the CB1 antagonist action (Foldy et al. 2006; Zhu and Lovinger 2005). It is difficult to see how such a result could reflect interference with an inverse agonist action, as the chelator effects are confined to the interior of the postsynaptic neuron while the antagonist effects appear to be presynaptic. It is known that postsynaptic calcium chelation can block retrograde EC signaling in DSE/I and some forms of EC-STD, as mentioned above. Thus, the finding that chelation blocks antagonist effects is most consistent with prevention of postsynaptic EC generation/release that would normally induce presynaptic CB1 activation. Extracellular application of AMT inhibitors can produce a CB1-mediated decrease in transmission at some synapses (Wilson and Nicoll 2001), as previously discussed in Section 4.1. This finding is also consistent with the idea that ECs can be tonically produced and released, albeit in many cases without accumulation to levels that produce detectable CB1 activation.

At present it is difficult to evaluate the possible contribution of intrinsic CB1 receptor activity to synaptic modulation. It is well known that GPCRs have the intrinsic ability to stimulate their cognate G-proteins in the absence of exogenous or endogenous ligand (see Lefkowitz et al. 1993 for review). Indeed studies of heterologously expressed CB1 indicated that the receptor was tonically active, as measured by inhibition of VGCCs (Pan et al. 1998). However, most, if not all, neurons contain the molecular machinery to synthesize ECs, and thus it is possible that tonic receptor activation might result from actions of an EC produced even in a single cell during a patch-clamp experiment. Pan et al. (1998) examined this possibility using postsynaptic calcium chelation to prevent EC formation and a mutant receptor that is insensitive to AEA activation. The findings in these experiments suggested that the receptor tone is due to intrinsic receptor activity rather than activation by native ECs. More recent studies of tonic GPCR inhibition of calcium channels indicate evidence of tonic receptor activation/channel modulation when receptors are overexpressed at high levels, but not at lower levels of receptor expression (Guo and Ikeda, 2004). Antagonist-induced increase in calcium current was observed in peripheral neurons that express native CB1 receptors (Pan et al. 1998). Nonetheless, the role of intrinsic CB1 activity in altering neuronal function at physiological levels of receptor expression remains unclear, particularly at CNS synapses.

# 6 Endocannabinoid-Dependent Long-Term Synaptic Depression

## *6.1 EC-LTD Induction Mechanisms*

Retrograde signaling involving ECs and the CB1 receptor has now been implicated in long-lasting decreases in synaptic transmission at synapses in many brain regions (see Gerdeman and Lovinger 2003; Chevaleyre et al. 2006 for review). This long-term synaptic plasticity is generally referred to as endocannabinoid-dependent long-term synaptic depression, or EC-LTD. Long-lasting changes in the efficacy of synapses such as long-term potentiation (LTP) and LTD have been postulated to play roles in synaptic development, information storage, homeostatic plasticity, nervous system development, and even pathological phenomena like epilepsy (McEachern and Shaw 1996; Malenka and Bear 2004).

Induction of LTD and other forms of long-lasting synaptic plasticity usually begin with repetitive activation of synapses, or in some cases, deliberately timed coactivation of pre- and postsynaptic neurons (so-called Spike-timing-dependent plasticity, or STDP). Repetitive synaptic activation can initiate several mechanisms that induce changes in synaptic efficacy. At glutamatergic synapses activation of NMDA receptors and mGluRs are widely recognized mechanisms for initiation of plasticity (Malenka and Bear 2004). Most NMDAR-dependent forms of LTP or LTD appear to be both initiated and expressed postsynaptically, with an expression mechanism involving alterations in postsynaptic AMPAR number or function (Malenka and Bear 2004). However, several forms of LTD that involve postsynaptic induction but presynaptic expression have been discovered over the years (Bolshakov and Siegelbaum 1994; Chevaleyre and Castillo 2003; Choi and Lovinger 1997a,b; Marsicano et al. 2002; Robbe et al. 2002; Sjostrom et al. 2003). Activation of mGluRs is a key step in many of these forms of LTD (Boshakov and Siegelbaum 1994; Gubellini et al. 2001; Robbe et al. 2002; Sung et al. 2001). Over the last several years it has become apparent that most forms of LTD that involve this postsynaptic induction-presynaptic expression mechanism are, in fact, EC-LTD.

Initial evidence for EC involvement in LTD came from experiments in which synaptic depression was blocked by CB1 receptor antagonists or lost in CB1-/- mice (Gerdeman et al. 2002, Robbe et al. 2002; Marsicano et al. 2002; Chevaleyre and Castillo 2003; Sjostrom et al. 2003). The LTD expressed at these synapses was associated with changes in transmission indicative of a decrease in probability of neurotransmitter release involved in presynaptic depression (Chevaleyre and Castillo 2003; Choi et al. 1997a,b; Gerdeman et al. 2002; Marsicano et al. 2002; Robbe et al. 2002; Sjostrom et al. 2003). This evidence included increases in PPR and the coefficient of variation of transmission, decreased frequency but not amplitude of miniature synaptic responses, and increases in synaptic failures during paired pre- and postsynaptic recordings. The predominant presynaptic expression of CB1 receptors supported the idea that EC-LTD is expressed presynaptically. Evidence for postsynaptic induction of EC-LTD has come from experiments utilizing postsynaptic manipulations to enhance or block plasticity. Postsynaptic calcium chelation blocks most forms of EC-LTD, presumably by preventing the increase in intracellular calcium needed for EC production/release (Bender et al. 2006; Calabresi et al. 1992b; Choi et al. 1997a; Robbe et al. 2002; Sjostrom et al. 2003; Soler-Llavina and Sabatini 2006). At corticostriatal synapses this blockade can be overcome by extracellular application of an AMT blocker which would enhance the increase in EC levels produced during synaptic activation (Gerdeman et al. 2002). This finding suggests that ECs released from cells other than the postsynaptic neuron can rescue LTD that would normally be blocked by the postsynaptic chelator, lending additional support to the hypothesis that ECs mediate this form of LTD. In layer 3 of visual cortex postsynaptic blockade of NMDA receptors prevents EC-LTD induction (Crozier et al. 2007). Thus, postsynaptic NMDA receptors can participate in some forms of EC-LTD, although this is the exception rather than the rule. Interestingly, EC-LTD in barrel-field cortex is resistant to postsynaptic NMDAR blockade (Bender et al. 2006) indicating that there is heterogeneity in EC-LTD induction mechanisms even within sensory cortex.

The EC-LTD observed at GABAergic synapses in hippocampus and BLA is not blocked by postsynaptic calcium chelation even at levels that prevent EC-mediated DSI (Azad et al. 2004; Chevaleyre and Castillo 2003; Zhu and Lovinger 2005). It is probable that mGluR activation of PLC leading to EC formation bypasses any need for increased intracellular calcium during induction of these forms of plasticity.

Subsequent studies have reinforced the idea that the ECs involved in EC-LTD are released from the postsynaptic cell. Blockers of DAG lipase prevent EC-LTD induction at GABAergic synapses in hippocampus (Chevaleyre and Castillo 2003), and this enzyme is thought to reside postsynaptically at these synapses (Bisogno et al. 2003; Gulyas et al. 2004; Katona et al. 2006). In neocortex, loading the DAG lipase inhibitor directly into the patch pipette prevented LTD induction (Bender et al. 2006). Direct intracellular application of GDPβS, a compound that locks G-proteins in an inactive form, prevents LTD induction (Chevaleyre and Castillo 2003). Postsynaptic G-protein activation is likely necessary to couple mGluRs to EC production during LTD induction. Intracellular postsynaptic application of AMT inhibitors has also been found to prevent LTD induction at synapses in the sensory neocortex and striatum (Bender et al. 2006; Ronesi et al. 2004). These findings support the idea that some sort of postsynaptic transporter or carrier-mediated EC release process is involved in EC-LTD induction. In the striatum blockade of L-type calcium channels prevents EC-LTD induction (Calabresi et al. 1992b, Tang et al. 2001, Wang et al., 2006). A recent study implicates the low-threshold, CaV1.3 form of the L-type channel in this process (Wang et al. 2006). This channel is the likely source of much of the postsynaptic calcium involved in LTD induction given that the channel is expressed postsynaptically in striatal medium spiny neurons (Olson et al. 2005), and L-type channels have little or no direct influence on presynaptic function at these synapses (Lovinger et al. 1994). Compounds that alter intracellular calcium release also prevent EC-LTD in nucleus accumbens (Robbe et al. 2002). This finding is consistent with a scenario in which activation of postsynaptic mGluRs stimulates calcium release from intracellular stores, and this calcium increase contributes to the EC production and release necessary for EC-LTD induction.

Little is known as yet about the molecular mechanisms involved in EC-LTD expression. If CB1 receptor activation is crucial for EC-LTD induction, then it is possible that application of a CB1 agonist might be sufficient to induce LTD. However, the synaptic depression produced by application of CB1 agonists is reversible when transmission is evoked by low-frequency stimulation of afferents or when spontaneous transmission is measured (Auclair et al. 2000; Hajos et al. 2001; Huang et al. 2001; Kreitzer and Regehr 2001; Levenes et al. 1998; Ronesi et al. 2004; Yin et al., 2006; Zhu and Lovinger 2005). Notably, these observations were made at synapses that are known to express EC-LTD. Thus, for the most part experimental evidence suggests that an additional factor acts together with CB1 receptor activation to induce the long-lasting decrease in neurotransmitter release during EC-LTD expression. In the case of EC-LTD within the visual cortex one such factor has been identified, namely presynaptic NMDA receptors (Sjostrom et al. 2003). Receptors that appear to contain the NR2B subunit have presynaptic actions at this synapse.

Antagonists of these receptors prevent EC-LTD, as do CB1 antagonists. The authors suggest that a mechanism involving synergism of the CB1 and NMDAR mechanisms is needed for EC-LTD expression. However, it is not yet clear what molecular mechanisms are involved in this synergism. It must be noted that application of CB1 agonists and ECs appear to induce LTD at some synapses (Bender et al. 2006; Kreitzer and Malenka 2005; Robbe et al. 2001; Sjostrom et al., 2003, Huang et al., 2003). However, Sjostrom et al. (2003) showed that this effect required a certain level of afferent activation. It is possible that during induction of EC-LTD CB1 activation synergizes with molecular mechanisms induced by activation of afferents themselves that are independent of postsynaptic signaling. To date it has been difficult to test this hypothesis because intact glutamatergic transmission is needed to measure EC-LTD expression. However, a prelminary report from Heifets et al. (2006) suggests that both presynaptic activation and CB1 activation are necessary for LTD induction at GABAergic synapses in hippocampus. Application of more direct measures of presynaptic function to the study of EC-LTD would be helpful in this context.

Endocannabinoid-dependent LTD induced by various pharmacological manipulations has been observed in some brain regions. Activation of group I mGluRs with DHPG induces an endocannabinoid-dependent long-lasting decrease in synaptic efficacy that resembles LTD (Azad et al. 2004; Chevaleyre and Castillo 2003; Edwards et al. 2006; Robbe et al. 2002, Kreitzer and Malenka 2005; Yin et al. 2006). This mGluR-induced EC-LTD occludes LTD produced by afferent stimulation, and thus appears to share common mechanisms with EC-LTD (Chevaleyre and Castillo 2003; Robbe et al. 2002; Kreitzer and Malenka 2005). In the dorsal striatum, mGluRs only effectively elicit EC-LTD when the postsynaptic membrane potential is slightly depolarized relative to the resting potential (Kreitzer and Malenka, 2005). This finding suggests the involvement of a voltage-gated channel in mGluR-induced EC-LTD, and indeed a dihydropyridine L-type channel antagonist blocks this form of LTD. It appears that mGluRs can induce LTD by activating endocannabinoid synthesis and that they act in concert with VGCCs , but this has not been explicitly explored at all synapses that show mGluR-induced EC-LTD. Furthermore, the role of presynaptic activation/synaptic transmission in mGluR-EC-LTD has not been fully examined.

In the lateral amygdala, application of amphetamine produces LTD that is dependent on CB1 receptor activation (Huang et al. 2003). The ability of amphetamine to induce LTD is enhanced by extracellular application of an AMT blocker, supporting an EC role. This form of EC-LTD is mimicked and occluded by a CB1 agonist. However, no attempt to reverse the agonist-induced depression was made in this study, and thus it is not entirely clear that the agonist induces true LTD. Postsynaptic calcium chelation prevents the amphetamine-induced EC-LTD, suggesting a postsynaptic locus of EC production/release. Measurement of mEPSCs suggests a presynaptic site of amphetamine-induced LTD expression, consistent with other forms of EC-LTD. This study suggests one possible link between EC-LTD and actions of drugs of abuse.

## 6.2 Mechanisms Involved in Maintained Expression of EC-LTD

Information about the molecular mechanisms involved in the presynaptic expression of EC-LTD is starting to appear. Amphetamine-induced EC-LTD in amygdala is accompanied by a change in the frequency, but not amplitude, of miniature EPSCs (Huang et al. 2003). Thus, expression of this form of LTD is independent of presynaptic action potential firing, and likely involves mechanisms downstream from calcium entry into the terminal. Similar findings were obtained for stimulation-induced EC-LTD in the nucleus accumbens (Robbe et al. 2002). However, in the Robbe et al. (2002) study only spontaneous EPSCs and not true miniature EPSCs were examined, making it difficult to determine if alterations in presynaptic calcium influx have a role in EC-LTD expression at this synapse. Protein kinase inhibitors have been shown to prevent EC-LTD at synapses in striatum (Calabresi et al. 1994), but it is unclear if protein phosphorylation contributes to EC production, as proposed for EC-LTD in amygdala (Azad et al. 2004), or if these signaling mechanisms are involved in presynaptic EC-LTD maintenance. Inhibition of protein translation, but not transcription, prevents EC-LTD at excitatory synapses in striatum (Yin et al. 2006). This finding implicates new protein synthesis in this form of plasticity. Experiments in this study were performed in a preparation lacking the somata of the presynaptic glutamatergic neurons, effectively ruling out contributions of presynaptic nuclei. Inhibition is not effective when applied to the postsynaptic neuron alone, and these inhibitors do not appear to be act on receptors or channels necessary for EC-LTD induction. These findings narrow the potential sites of plasticity-related protein translation to the presynaptic axon or terminal and/or other intact somata in the slice, including glia and interneurons. The possibility of a contribution of presyaptic translation to EC-LTD is intriguing, but further evidence is required to pinpoint the important site of protein expression. Figure 5 presents potential mechanisms involved in the presynaptic expression of EC-LTD.

## 6.3 Endocannabinoids and Cerebellar LTD

The first type of LTD ever described was that which occurs at synapses between parallel fibers and Purkinje neurons in cerebellum (so-called Cerebellar LTD) (Ito and Kano 1982; Ito 2001). Much is known about cerebellar LTD, including the fact that both the mechanisms involved in the initial LTD induction and the ultimate expression of synaptic depression are most likely postsynaptic (see Ito 2001 for review). Thus, it was surprising when Safo and Regehr (2005) recently reported a role for EC retrograde signaling in cerebellar LTD. These investigators demonstrated that cerebellar LTD was blocked by CB1 antagonists and was lost in the CB1 knockout mouse. Application of DAG lipase inhibitors either to the whole preparation, or just to the interior of the postsynaptic neuron, also blocked LTD induction. Activation of CB1 receptors alone produced synaptic depression that was reversed by subsequent antagonist application, indicating that CB1 activation alone is not sufficient

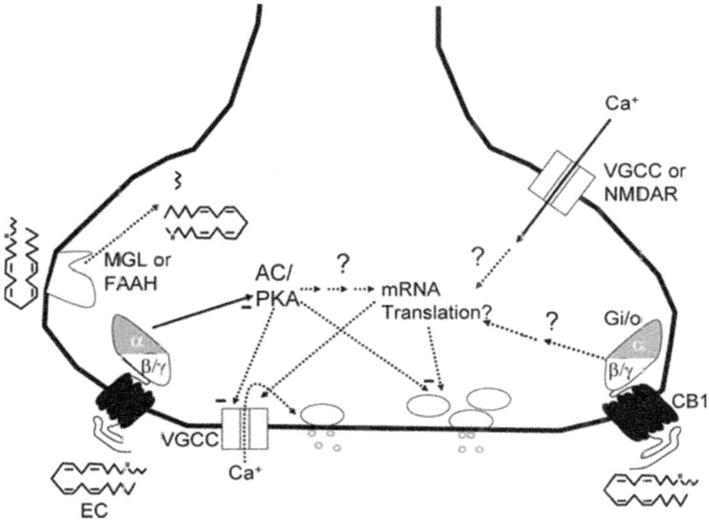


**Fig. 5** Mechanisms that potentially contribute to presynaptic expression of endocannabinoid-mediated LTD. Activation of the presynaptic CB1 receptor may snyergize with other signals in the presynaptic terminal to induce a long-lasting decrease in probability of neurotransmitter release. The CB1-mediated stimulation of Gi/o dissociation likely inhibits adenylate cyclase (AC) which would, in turn, inhibits protein kinase A (PKA). Disruption of PKA-mediated phosphorylation may alter calcium channel function or proteins involved in vesicle fusion. Protein translation has recently been implicated in EC-LTD expression, and it is possible that CB1 signaling may impact this process in the presynaptic terminal. Depolarization of the presynaptic terminal during repetitive afferent activation will stimulate calcium entry via VGCCs, and it is possible that the increase in intraterminal calcium stimulates signaling pathways that add to or synergize with the CB1-stimulated signals. In the neocortex, activation of presynaptic NMDARs due to homosynaptic glutamate release provides a signal that interacts with CB1 activation in LTD induction. Once again, calcium entry is likely to mediate this action, as NMDAR channels are highly calcium permeable. The EC responsible for LTD induction is likely removed from the synapse via uptake and ezymatic breakdown within the first few minutes of EC-LTD expression. However, the decrease in probability of release outlasts the EC signal. The mechanisms directly involved in decreased release probability during EC-LTD expression are not yet known, but long-lasting alterations in calcium channel function and/or vesicle fusion are logical candidate mechanisms.

to induce LTD at this synapse. Evidence supporting the prevailing hypothesis of a postsynaptic locus of cerebellar LTD expression was presented in this paper. Thus, the authors suggested a new model for induction of this most widely studied form of LTD. In this model, the retrograde EC signal stimulates production of another anterograde signal that then helps to produce the final postsynaptic mechanism of LTD. These intriguing findings appear to conflict with earlier studies suggesting that induction and expression of cerebellar LTD do not require any presynaptic participation (Narasimhan and Linden 1996). It is possible that endocannabinoids and CB1 receptors have a modulatory role that can impact LTD under certain experimental conditions (see van Beugen et al. 2006 for discussion of one possibility).

Further studies are clearly needed to sort out the endocannabinoid and CB1 roles in cerebellar LTD, as well as in cerebellar learning and memory.

### *6.4 Physiological Roles of EC-LTD*

There are many potential roles for EC-LTD in the modulation of physiology at the cellular and circuit levels. Long-lasting reduction of inhibitory transmission can increase the relative effectiveness of excitatory drive, as we will consider in discussing endocannabinoids and long-term potentiation (Chevaleyre and Castillo 2004). In this manner, EC-LTD may increase excitatory throughput in circuits like those found in the basolateral amygdala, leading to increased net circuit and behavioral output (e.g., as suggested in Marsicano et al. 2002). Within the visual and barrel field areas of neocortex, EC-LTD may participate in deprivation-induced decreases in neuronal responsiveness, indicating a potential role in sensory plasticity during development (Bender et al. 2006; Crozier et al. 2007; Sjostrom et al. 2003). Decreasing the efficacy of excitatory synapses may also serve as a mechanism for extinction or erasure of memory when LTD takes place at previously potentiated synapses (Marsicano et al. 2002). However, EC-LTD likely has more subtle effects that are important for circuit function and information storage. For example, decreasing efficacy of excitatory inputs may increase the need for spatial or temporal summation in order to reach action potential threshold. This could be important in the maturation of sensory-motor integration in brain regions such as the striatum where such a mechanism might prevent unwanted movements (Choi et al. 1997a; Tang et al. 2001). Long-lasting synaptic depression has also been suggested to be an effective mechanism for equalizing the relative efficacy of synapses on different parts of a large dendritic arbor (Rumsey and Abbott 2006). For example, an EC-LTD-like decrease in synaptic strength at proximal synapses might render them equieffective with distal synapses in terms of their ability to drive the neuron to threshold. Alterations in the efficacy of inhibitory synapses made by local-circuit interneurons may also regulate the timing of circuit input-output relationships. Additional studies of the in vivo impact of EC-dependent synaptic plasticity will be needed to evaluate these possibilities.

## 7 Endocannabinoids and Long-Term Potentiation

Endocannabinoids have also been suggested to participate in long-term synaptic potentiation (LTP), a widely recognized model of synaptic changes underlying learning and memory (Malenka and Bear 2004). Prior to discovery of the CB1 receptor, one study indicated that Δ9-THC alters hippocampal LTP (Nowicky et al. 1987). Activation of CB1 receptors by 2-AG was shown to prevent LTP induction at Schaffer collateral-CA1 pyramidal cell synapses in hippocampus (Stella et al.

1997; Slanina et al. 2005). This inhibition most likely results from suppression of glutamate release at these synapses that decreases the excitatory drive necessary to induce LTP (Misner and Sullivan 1999). Thus, it is presently thought that ECs can modulate LTP in this manner, but are not strictly necessary for this form of plasticity.

Subsequent studies demonstrated another mechanism through which ECs modulate LTP. It has long been known that reducing synaptic inhibition enhances the likelihood of LTP induction (Wigstrom and Gustafsson 1986). The observation that EC retrograde signaling reduces GABAergic inhibition of CA1 pyramidal neurons via DSI and EC-LTD stimulated investigators to explore the possibility that this disinhibition could promote LTP induction. Indeed, Carlson et al. (2002) demonstrated that LTP was more easily induced while DSI was ongoing, and that this effect was blocked by a CB1 antagonist. Subsequently, Chevaleyre and Castillo (2004) showed that EC-LTD of GABAergic synapses could promote LTP induction in a "metaplastic" manner in the hippocampal CA1 region. The term metaplasticity refers to a change in the threshold for plasticity based on the past history of plasticity at that same synapse (Abraham and Bear 1996). In this case, the long-lasting decrease in inhibitory transmission during EC-LTD allowed for stronger excitation during repetitive synaptic activation, thus lowering the stimulus threshold needed for LTP induction. These findings have been somewhat controversial because other investigators report that CB1 antagonists promote LTP (Slanina et al. 2005), rather than having the LTP-blocking effect associated with metaplastic disinhibition (Carlson et al. 2002; Chevaleyre and Castillo 2004). Endocannabinoids and CB1 receptors may participate differently in LTP depending on the relative level of EC release at excitatory versus inhibitory synapses during repetitive synaptic activation. The relative activity of different afferent inputs likely determines which synapses produce ECs. Clearly, there is much work to be done in sorting out the modulatory and metaplastic effects of ECs.

## 8 Endocannabinoids, CB1 Receptors, and Behavior

Endocannabinoids have now been implicated in a variety of brain functions. For example, EC levels increase with exposure to stressful stimuli, and may play a role in neuroadaptations to stress and well as in neural function and behaviors related to anxiety and affective states (Bortolato et al. 2006; Connell et al. 2006; Gobbi et al. 2005; Hohmann et al. 2005).

### *8.1 Endocannabinoid and CB1 Roles in Learning and Memory*

In light of the evidence presented above that endocannabinoids and CB1 play crucial roles in synaptic plasticity, it is interesting to examine their roles in learning and memory. It has long been known that THC and other exogenous CB1 agonists impair learning and memory (Hampson and Deadwyler, 1998; Miller and Branconnier, 1983; Sullivan, 2000), particularly on tasks that involve the hippocampus. However, CB1 agonists will activate all receptors when given peripherally, and all receptors within a given brain region when applied intracerebrally. In contrast, endocannabinoids will likely act on only a subset of synapses within a given brain region during the learning and memory process. Thus, it is important to examine the role of endocannabinoids in learning and memory using experimental tools such as selective CB1 antagonists, CB1 knockout animals, and drugs that prolong endocannabinoid actions at the synapse. This work has been undertaken in several laboratories with mixed results.

CB1 antagonists impair learning in an aversive conditioning paradigm when injected into the medial prefrontal cortex, and CB1 agonists have the opposite effect (Laviolette and Grace 2005). In addition, CB1 receptors have been implicated in extinction of fear conditioning, which is believed to be an active unlearning process (Marsicano et al. 2002). Extinction is blocked by peripheral injection of CB1 antagonist, and is impaired in the CB1-/- mouse. Subsequent studies have also implicated CB1 receptors in extinction in both aversively motivated and appetitively motivated tasks (Suzuki et al. 2004; Varvel and Lichtman 2002; Varvel et al. 2005). In general, these studies suggest the endocannabinoids and CB1 receptors participate in formation of certain types of memories.

Considerable evidence that endocannabinoid actions in the brain impair learning and memory has also been collected. Most of the evidence comes from studies in which CB1 antagonists improve learning or memory (Lichtman 2000; Shiflett et al. 2004; Terranova et al. 2005; Wolf and Leander 2003). These antagonists can also reverse the amnestic effects of agents such as β-amyloid and scopolamine (Mazzola et al. 2003; Takahashi et al. 2005). Thus, it appears that activation of CB1 receptors, presumably following release of brain endocannabinoids, has actions that counteract mechanisms involved in learning and memory.

Little is currently known about the mechanisms that underlie the memory-promoting and memory-impairing functions of endocannabinoids and CB1 receptors. Marsicano and coworkers have attempted to link extinction of fear conditioning to EC-LTD at inhibitory synapses in the basolateral amygdala. However, it is still not clear if activation of CB1 receptors in this brain region is crucial for extinction, and other brain regions could be involved. Facilitation of LTP induction by DSI or EC-LTD in the hippocampal CA1 region has been suggested as a mechanism by which endocannabinoids and CB1 receptors participate in spatial learning and other forms of hippocampal-based memory (Carlson et al. 2003; Chevaleyre and Castillo 2004). It has also been demonstrated that CB1 antagonists block LTD induction in vivo (de Oliveira Alvares et al. 2006). Clearly, behavioral studies are needed to determine if CB1 receptors participate in hippocampal memory formation.

It is tempting to speculate that disinhibitory effects of endocannabinoids would promote learning, while endocannabinoid-mediated reductions in excitatory transmission underly amnestic effects. Thus, the role of the endocannabinoid system in learning and memory would depend on which synaptic inputs within a given brain region contain the receptor, and which inputs are activated during a given phase of learning and memory. However, this disinhibition = memory, inhibition = impaired memory idea is almost certainly too simple. Within the complex circuitry of brain regions such as the cerebellum, hippocampus, neocortex, and striatum both excitatory and inhibitory synapses are likely to participate in learning as well as in supression of learning and memory. Clearly a systematic investigation of the endocannabinoid and CB1 receptor actions within different brain regions during the many varieties of learning and memory is needed to sort out their roles in brain information storage.

## *8.2 The Neural Basis of Cannabinoid Intoxication*

The CB1 agonist Δ9-THC is of course the major psychoactive ingredient in cannabinoid drugs such as marijuana and hashish, and thus discussion of the neural role of CB1 receptors brings to mind questions about acute THC intoxication. The mechanisms underlying the cannabis high and the reinforcing effects of cannabis are not fully understood. Marijuana use produces feelings of stress relief and relaxation in regular users, although the drug can also be anxiogenic, especially in first time users (Parolaro et al. 2005). Blockade of acute marijuana intoxication by SR141716A indicates that these effects are mediated by CB1 receptors (Huestis et al. 2001). Likewise, drug discrimination studies in rodents indicate that the subjective effects of acute THC are mediated by CB1 receptors (Tanda et al. 2003; Rinaldi-Carmona et al. 1994; Mansbach et al. 1996; Perio et al. 1996). From these findings it seems clear that activation of CB1 is crucial to the marijuana high.

There has been some debate as to whether cannabinoid drugs are truly reinforcing, but recent studies indicate that animals will self-administer THC, and will even perform tasks in order to receive self-administration into the VTA and shell of the nucleus accumbens (Tanda et al., 2000; Justinova et al., 2003 refs, Zangen et al. 2006). Adminstration of THC also lowers the threshold for intracranial self-stimulation (Gardner et al., 1988), and can be used as a stimulus for conditioned place preference (Valjent and Maldonado 2000). These findings indicate that cannabinoid drugs are indeed reinforcing. However, negative findings have been reported in both self-administration and conditioned place preference studies (reviewed by Parolaro et al. 2005), and thus it is worth considering what factors may affect THC action in these paradigms.

Most discussion of the neuronal mechanisms underlying acute intoxication and reinforcement has centered on the activation of dopamine release by THC and other CB1 agonists. Indeed, these agonists stimulate the activity of neurons in the VTA and increase extracellular dopamine levels in nucleus accumbens (Ng Cheong Ton et al., 1988; French, 1997; Tanda et al., 1997). It is reasonable to speculate that the increase in dopamine contributes to the reinforcing effects of THC, but the subjective high feeling is more difficult to ascribe to this mechanism. The stress-relieving effects of the drug may in and of themselves enhance relaxation and feelings of well-being, and these mechanisms may involve interactions with GABAergic and glutamatergic transmission in brain regions like the BLA. Interestingly, the opiate receptor antagonist naloxone blocks THC-induced increases in extracellular dopamine (Tanda et al. 1997). Opiate receptor agonists certainly produce intoxication involving strong feelings of well-being, and thus it is possibly that enkephalin or endorphin actions contribute to the cannabinoid high. Ultimately, it will be interesting to determine which brain regions and neurotransmitters participate in the subjective effects of THC. This question could be approached experimentally using the drug discrimination procedure. Indeed, one clue as to the role of GABA in acute intoxication comes from the finding that benzodiazepines produce partial substitution in animals trained to discriminate THC (Wiley and Martin 1999). The tools now seem to be in place to settle this, and other important issues concerning the neural effects of cannabinoids and endocannabinoids in the very near future.

## References

Abraham WC, Bear MF (1996) Metaplasticity: the plasticity of synaptic plasticity. Trends Neurosci 19:126–30

Alexander JP, Cravatt BF (2006) The putative endocannabinoid transport blocker LY2183240 is a potent inhibitor of FAAH and several other brain serine hydrolases. J Am Chem Soc 128(30):9699–9704

Alger BE (2002) Retrograde signaling in the regulation of synaptic transmission: focus on endocannabinoids. Prog Neurobiol 68:247–86

Ameri A, Wilhelm A, Simmet T (1999) Effects of the endogeneous cannabinoid, anandamide, on neuronal activity in rat hippocampal slices. Br J Pharmacol 126(8):1831–9

Auclair N, Otani S, Soubrie P, Crepel F (2000) Cannabinoids modulate synaptic strength and plasticity at glutamatergic synapses of rat prefrontal cortex pyramidal neurons. J Neurophysiol 83:3287–93

Azad SC, Monory K, Marsicano G, Cravatt BF, Lutz B, Zieglgansberger W, Rammes G (2004) Circuitry for associative plasticity in the amygdala involves endocannabinoid signaling. J Neurosci 24(44):9953–61

Bacci A, Huguenard JR, Prince DA (2004) Long-lasting self-inhibition of neocortical interneurons mediated by endocannabinoids. Nature 431(7006):312–16

Bachur NR, Masek K, Melmon KL, Udenfriend S (1965) Fatty acid amides of ethanolamine in mammalian tissues. J Biol Chem 240(3):1019–24

Bamford NS, Zhang H, Schmitz Y, Wu NP, Cepeda C, Levine MS, Schmauss C, Zakharenko SS, Zablow L, Sulzer D (2004) Heterosynaptic dopamine neurotransmission selects sets of corticostriatal terminals. Neuron 42(4):653–63

Bass NM (1988) The cellular fatty acid binding proteins: aspects of structure, regulation, and function. Int Rev Cytol 111:143–84

Bazan NG (2005) Lipid signaling in neural plasticity, brain repair, and neuroprotection. Mol Neurobiol 32(1):89–103

Beckstead MJ, Grandy DK, Wickman K, Williams JT (2004) Vesicular dopamine release elicits an inhibitory postsynaptic current in midbrain dopamine neurons. Neuron 42(6):939–46

Bender VA, Bender KJ, Brasier DJ, Feldman DE (2006) Two Coincidence Detectors for Spike Timing-Dependent Plasticity in Somatosensory Cortex. J Neurosci 26:4166–77

Ben-Shabat S, Fride E, Sheskin T, Tamiri T, Rhee MH, Vogel Z, Bisogno T, De Petrocellis L, Di Marzo V, Mechoulam R (1998) An entourage effect: inactive endogenous fatty acid glycerol esters enhance 2-arachidonoyl-glycerol cannabinoid activity, Eur J Pharm 353(1):23–31

Bisogno T, Sepe N, Melck D, Maurelli S, De Petrocellis L, Di Marzo V (1997) Biosynthesis, release and degradation of the novel endogenous cannabimimetic metabolite 2-arachidonoylglycerol in mouse neuroblastoma cells. Biochem J 322( Pt 2):671–7

Bisogno T, Melck D, De Petrocellis L, Di Marzo V (1999) Phosphatidic acid as the biosynthetic precursor of the endocannabinoid 2-arachidonoylglycerol in intact mouse neuroblastoma cells stimulated with ionomycin. J Neurochem 72(5):2113–19

Bisogno T, Melck D, Bobrov MY, Gretskaya NM, Bezuglov VV, De Petrocellis L, Di Marzo V (2000) N-acyl-dopamines: novel synthetic CB(1) cannabinoid-receptor ligands and inhibitors of anandamide inactivation with cannabimimetic activity in vitro and in vivo. Biochem J 351 Pt 3:817–24

Bisogno T, Howell F, Williams G, Minassi A, Cascio MG, Ligresti A, Matias I, Schiano Moriello A, Paul P, Gangadharan U, Hobbs C, Di Marzo V, Doherty P (2003) Cloning of the first sn 1-DAG lipases points to the spatial and temporal regulation of endocannabinoid signalling in the brain. J Cell Biol 163:463–8

Bodor AL, Katona I, Nyiri G, Mackie K, Ledent C, Hajos N, Freund TF (2005) Endocannabinoid signaling in rat somatosensory cortex: laminar differences and involvement of specific interneuron types. J Neurosci 25:6845–56

Bojesen IN, Hansen HS (2003) Binding of anandamide to bovine serum albumin. J Lipid Res 44(9):1790–4

Bojesen IN, Hansen HS (2005) Membrane transport of anandamide through resealed human red blood cell membranes. J Lipid Res 46(8):1652–9

Bortolato M, Campolongo P, Mangieri RA, Scattoni ML, Frau R, Trezza V, La Rana G, Russo R, Calignano A, Gessa GL, Cuomo V, Piomelli D (2006) Anxiolytic-like properties of the anandamide transport inhibitor AM404. Neuropsychopharm 31(12):2652–9

Bouaboula M, Perrachon S, Milligan L, Canat X, Rinaldi-Carmona M, Portier M, Barth F, Calandra B, Pecceu F, Lupker J, Maffrand JP, Le Fur G, Casellas P (1997) A selective inverse agonist for central cannabinoid receptor inhibits mitogen-activated protein kinase activation stimulated by insulin or insulin-like growth factor 1. Evidence for a new model of receptor/ligand interactions. J Biol Chem 272(35):22330–9

Bolshakov VY, Siegelbaum S (1994) Postsynaptic induction and presynaptic expression of hippocampal long-term depression. Science 264(5162):1148–52

Bramham CR, Messaoudi E (2004) BDNF as a trigger for transsynaptic consolidation in the adult brain. In: Stanton FK, Bramham C, Scharfman H (eds) Synaptic plasticity and transsynaptic signaling. Spinger, New York, pp. 159–84

Breivogel CS, Selley SE, Childers SR (1998) Cannabinoid receptor agonist efficacy for stimulating [35S]GTPgammaS binding to rat cerebellar membranes correlates with agonist-induced decreases in GDP affinity. J Biol Chem 273(27):16865–73

Brenowitz SD, Regehr WG (2005) Associative short-term synaptic plasticity mediated by endocannabinoids. Neuron 45(3):419–31

Brown SP, Brenowitz SD, Regehr WG (2003) Brief presynaptic bursts evoke synapse-specific retrograde inhibition mediated by endogenous cannabinoids. Nat Neurosci 6:1048–57

Brown SP, Safo PK, Regehr WG (2004) Endocannabinoids inhibit transmission at granule cell to Purkinje cell synapses by modulating three types of presynaptic calcium channels. J Neurosci 24(24):5623–31

Brundege JM, Dunwiddie TV (1996) Modulation of excitatory synaptic transmission by adenosine released from single hippocampal pyramidal neurons. J Neurosci 16:5603–12

Burkey TH, Quock RM, Consroe P, Ehlert FJ, Hosohata Y, Roeske WR, Yamamura HI (1997) Relative efficacies of cannabinoid CB1 receptor agonists in the mouse brain. Eur J Pharmacol 336(2–3):295–8

Calabresi P, Maj R, Mercuri NB, Bernardi G (1992a) Coactivation of D1 and D2 dopamine receptors is required for long-term synaptic depression in the striatum. Neurosci Lett 142(1):95–9

Calabresi P, Maj R, Pisani A, Mercuri NB, Bernardi G (1992b) Long-term synaptic depression in the striatum: physiological and pharmacological characterization. J Neurosci 12:4224–33

Calabresi P, Pisani A, Mercuri NB, Bernardi G (1994) Post-receptor mechanisms underlying striatal long-term depression. J Neurosci 14(8):4871–81
Carlson G, Wang Y, Alger BE (2002) Endocannabinoids facilitate the induction of LTP in the hippocampus. Nat Neurosci 5:723–24
Carlsson A (1993) Thirty years of dopamine research. Adv Neurol 60:1–610
Castellucci VF, Carew TJ, Kandel ER (1978) Cellular analysis of long-term habituation of the gill-withdrawal reflex of Aplysia californica. Science 202(4374):1306–8
Cepeda C, Buchwald NA, Levine MS (1993) Neuromodulatory actions of dopamine in the neostriatum are dependent upon the excitatory amino acid receptor subtypes activated. Proc Natl Acad Sci USA 90(20):9576–80
Chevaleyre V, Castillo PE (2003) Heterosynaptic LTD of hippocampal GABAergic synapses: a novel role of endocannabinoids in regulating excitability. Neuron 38:461–72
Chevaleyre V, Castillo PE (2004) Endocannabinoid-mediated metaplasticity in the hippocampus. Neuron 43(6):871–81
Chevaleyre V, Hefeits BD, Kaeser P, Sudhof TC, Castillo PE (2006) Endocannabinoids target the release machinery via protein kinase A to mediate LTD at inhibitory synapses in both hippocampus and amygdala. Soc Neurosci Abstr 41.11
Chevaleyre V, Takahashi KA, Castillo PE (2006) Endocannabinoid-mediated synaptic plasticity in the CNS. Ann Rev Neurosci 29:37–76
Choi S, Lovinger DM (1997a) Decreased probability of neurotransmitter release underlies striatal long-term depression and postnatal development of corticostriatal synapses. Proc Natl Acad Sci USA 94:2665–70
Choi S, Lovinger DM (1997b) Decreased frequency but not amplitude of quantal synaptic responses associated with expression of corticostriatal long-term depression. J Neurosci 17:8613–20
Clapham DE, Neer EJ (1997) G protein beta gamma subunits. Annu Rev Pharmacol Toxicol 37:167–203
Connell K, Bolton N, Olsen D, Piomelli D, Hohmann AG (2006) Role of the basolateral nucleus of the amygdala in endocannabinoid-mediated stress-induced analgesia. Neurosci Lett 397(3):180–4
Coutts AA, Pertwee RG (1997) Inhibition by cannabinoid receptor agonists of acetylcholine release from the guinea-pig myenteric plexus. Br J Pharmacol 121(8):1557–66
Cravatt BF, Giang DK, Mayfield SP, Boger DL, Lerner RA, Gilula NB (1996) Molecular characterization of an enzyme that degrades neuromodulatory fatty-acid amides. Nature (Lond.) 384:83–7
Croci T, Manara L, Aureggi G, Guagnini F, Rinaldi-Carmona M, Maffrand JP, Le Fur G, Mukenge S, Ferla G (1998) In vitro functional evidence of neuronal cannabinoid CB1 receptors in human ileum. Br J Pharmacol 125(7):1393–5
Crozier RA, Wang Y, Liu CH, Bear MF (2007) Deprivation-induced synaptic depression by distinct mechanisms in different layers of mouse visual cortex. Proc Natl Acad Sci USA 2007 104(4):1383–8
Daniel H, Crepel F (2001) Control of Ca(2+) influx by cannabinoid and metabotropic glutamate receptors in rat cerebellar cortex requires K(+) channels. J Physiol 537(Pt 3):793–800
Davis MI, Ronesi J, Lovinger DM (2003) A predominant role for inhibition of the adenylate cyclase/protein kinase A pathway in ERK activation by cannabinoid receptor 1 in N1E-115 neuroblastoma cells. J Biol Chem 278(49):48973–80
Devane WA, Hanus L, Breuer A, Pertwee RG, Stevenson LA, Griffin G, Gibson D, Mandelbaum A, Etinger A, Mechoulam R (1992) Isolation and structure of a brain constituent that binds to the cannabinoid receptor. Science 258:1946–9
Di S, Boudaba C, Popescu IR, Weng FJ, Harris C, Marcheselli VL, Bazan NG, Tasker JG (2005) Activity-dependent release and actions of endocannabinoids in the rat hypothalamic supraoptic nucleus. J Physiol 569:751–60
Diana MA, Marty A (2003) Characterization of depolarization-induced suppression of inhibition using paired interneuron-Purkinje cell recordings. J Neurosci 23:5906–18

Di Marzo V, Fontana A, Cadas H, Schinelli S, Cimino G, Schwartz JC, Piomelli D (1994) Formation and inactivation of endogenous cannabinoid anandamide in central neurons. Nature 372:686–91
Dinh TP, Carpenter D, Leslien FM, Freund TF, Katona I, Sensi SL, Kathuria S, Piomelli D (2002) Brain monoglyceride lipase participating in endocannabinoid inactivation. Proc Natl Acad Sci USA 99(16):10819–24
Dolphin AC (2003) G protein modulation of voltage-gated calcium channels. Pharmacol Rev 55(4):607–27
Domenici MR, Azad SC, Marsicano G, Schierloh A, Wotjak CT, Dodt HU, Zieglgansberger W, Lutz B, Rammes G (2006) Cannabinoid receptor type 1 located on presynaptic terminals of principal neurons in the forebrain controls glutamatergic synaptic transmission. J Neurosci 26(21):5794–9
Dunwiddie TV, Lovinger DM (eds.) (1993) Presynaptic Receptors in the Mammalian Brain. Birkhauser, Boston.
Dunwiddie TV, Masino SA (2001) The role and regulation of adenosine in the central nervous system. Annu Rev Neurosci 24:31–55
Edwards DA, Kim J, Alger BE (2006) Multiple mechanisms of endocannabinoid response initiation in hippocampus. J. Neurophysiol. 95(1):67–75
Elmslie KS (2003) Neurotransmitter modulation of neuronal calcium channels. J Bioenerg Biomembr 35(6):477–89
Fegley D, Kathuria S, Mercier R, Li C, Goutopoulos A, Makriyannis A, Piomelli D (2004) Anandamide transport is independent of fatty-acid amide hydrolase activity and is blocked by the hydrolysis-resistant inhibitor AM1172. Proc Natl Acad Sci USA 101(23):8756–61
Feinmark SJ, Begum R, Tsetkov E, Goussakov I, Funk CD, Siegelbaum SA, Boshakov VY (2003) 12-lipoxygenase metabolites of arachidonic acid mediate metabotropic glutamate receptor-dependent long-term depression at hippocampal CA3-CA1 synapses. J Neurosci 23(36):11427–35
Flores-Hernandez J, Galarraga E, Bargas J (1997) Dopamine selects glutamatergic inputs to neostriatal neurons. Synapse 25(2):185–95
Foldy C, Neu A, Jones MV, Soltesz I (2006) Presynaptic, activity-dependent modulation of cannabinoid type 1 receptor-mediated inhibition of GABA release. J Neurosci 26(5):1465–9
Freund TF, Katona I, Piomelli D (2003) Role of endogenous cannabinoids in synaptic signaling. Physiol Rev 83:1017–66
French ED (1997) 9-Tetrahydrocannabinol excites rat VTA dopamine neurons through activation of cannabinoid CB1 but not opioid receptors. Neurosci Lett 226:159–62
Fukudome Y, Ohno-Shosaku T, Matsui M, Omori Y, Fukaya M, Tsubokawa H, Taketo MM, Watanabe M, Manabe T, Kano M (2004) Two distinct classes of muscarinic action on hippocampal inhibitory synapses: M2-mediated direct suppression and M1/M3-mediated indirect suppression through endocannabinoid signalling. Eur J Neurosci 19:2682–92
Galante M, Diana MA (2004) Group I metabotropic glutamate receptors inhibit GABA release at interneuron-Purkinje cell synapses through endocannabinoid production. J Neurosci 24(20):4865–74
Gardner EL, Paredes W, Smith D, Donner A, Milling C, Cohen D, Morrison D (1988) Facilitation of brain stimulation reward by delta 9-tetrahydrocannabinol. Psychopharmacology (Berl) 96:142–4
Gerdeman G, Lovinger DM (2001) CB1 cannabinoid receptor inhibits synaptic release of glutamate in rat dorsolateral striatum. J Neurophysiol 85:468–71
Gerdeman GL, Ronesi J, Lovinger DM (2002) Postsynaptic endocannabinoid release is necessary for long-term depression in the striatum. Nat Neurosci 5(5):446–51
Gerdeman GL, Lovinger DM (2003) Emerging roles for endocannabinoids in long-term synaptic plasticity. Br J Pharmacol 140(5):781–9
Giuffrida A, Parsons LH, Kerr TM, Rodriguez de Fonseca F, Navarro M, Piomelli D (1999) Dopamine activation of endogenous cannabinoid signaling in dorsal striatum. Nat Neurosci 2:358–63

Glaser ST, Abumrad NA, Fatade F, Kaczocha M, Studholme KM, Deutsch DG (2003) Evidence against the presence of an anandamide transporter. Proc Natl Acad Sci USA 100(7):4269–74
Glass M, Felder CC (1997) Concurrent stimulation of cannabinoid CB1 and dopamine D2 receptors augments cAMP accumulation in striatal neurons: evidence for a Gs linkage to the CB1 receptor. J Neurosci 17(14):5327–33
Glass M, Northup JK (1999) Agonist selective regulation of G proteins by cannabinoid CB(1) and CB(2) receptors. Mol Pharmacol 56(6):1362–9
Gobbi G, Bambico FR, Mangieri R, Bortolato M, Campolongo P, Solinas M, Cassano T, Morgese MG, Debonnel G, Duranti A, Tontini A, Tarzia G, Mor M, Trezza V, Goldberg SR, Cuomo V, Piomelli D (2005) Antidepressant-like activity and modulation of brain monoaminergic transmission by blockade of anandamide hydrolysis. Proc Natl Acad Sci U S A 102(51):18620–5
Gong JP, Onaivi ES, Ishiguro H, Liu QR, Tagliaferro PA, Brusco A, Uhl GR (2006) Cannabinoid CB2 receptors: immunohistochemical localization in rat brain. Br Res 1071(1):10–23
Graham JD, Lewis MJ, Li DM (1974) The effect of delta1-tetrahydrocannabinol on the release of (3H-(-)-noradrenaline from the isolated vas deferens of the rat. Br J Pharmacol 52(2):233–6
Gubellini P, Saulle E, Centonze D, Bonsi P, Pisani A, Bernardi G, Conquet F, Calabresi P (2001) Selective involvement of mGlu1 receptors in corticostriatal LTD. Neuropharm 40:839–46
Gulyas AI, Cravatt BF, Bracey MH, Dinh TP, Piomelli D, Boscia F, Freund TF (2004) Segregation of two endocannabinoid-hydrolyzing enzymes into pre- and postsynaptic compartments in the rat hippocampus, cerebellum and amygdala. Eur J Neurosci 20(2):441–58
Guo J, Ikeda SR (2004) Endocannabinoids modulate N-type calcium channels and G-protein-coupled inwardly rectifying potassium channels via CB1 cannabinoid receptors heterologously expressed in mammalian neurons. Mol Pharmacol 65(3):665–74
Hajos N, Ledent C, Freund TF (2001) Novel cannabinoid-sensitive receptor mediates inhibition of glutamatergic synaptic transmission in the hippocampus. Neurosci 106(1):1–4
Hampson RE, Deadwyler SA (1998) Role of cannabinoid receptors in memory storage. Neurobiol Dis 5(6 Pt B):474–82
Hashimotodani Y, Ohno-Shosaku T, Tsubokawa H, Ogata H, Emoto K, Maejima T, Araishi K, Shin HS, Kano M (2005) Phospholipase Cbeta serves as a coincidence detector through its Ca2+ dependency for triggering retrograde endocannabinoid signal. Neuron 45(2):257–68
Heifets BD, Chevaleyre V, Castillo PE (2006) Interneuron activity dependence of endocannabinoid mediated long term of depression of inhibition (I-LTD) in hippocampus. Soc Neurosci Abstracts 41:12
Heinbockel T, Brager DH, Reich CG, Zhao J, Muralidharan S, Alger BE, Kao J (2005) Endocannabinoid signaling dynamics probed with optical tools. J Neurosci 25(41):9449–59
Hillard CJ, Jarrahian A (2000) The movement of *N*-arachidonoylethanolamine (anandamide) across cellular membranes. Chem Phys Lipids 108:123–34
Henry DJ, Chavkin C (1995) Activation of inwardly rectifying potassium channels (GIRK1) by co-expressed rat brain cannabinoid receptors in Xenopus oocytes. Neurosci Lett 186(2–3):91–4
Hentges ST, Low MJ, Williams JT (2005) Differential regulation of synaptic inputs by constitutively released endocannabinoids and exogenous cannabinoids. J Neurosci 25(42):9746–51
Hoffman AF, Lupica CR (2001) Direct actions of cannabinoids on synaptic transmission in the nucleus accumbens: a comparison with opioids. J Neurophysiol 85:72–83
Hoffman AF, Macgill AM, Smith D, Oz M, Lupica CR (2005) Species and strain differences in the expression of a novel glutamate-modulating cannabinoid receptor in the rodent hippocampus. Eur J Neurosci 22(9):2387–91
Hohmann AG, Suplita RL, Bolton NM, Neely MH, Fegley D, Mangieri R, Krey JF, Walker JM, Holmes PV, Crystal JD, Duranti A, Tontini A, Mor M, Tarzia G, Piomelli D (2005) An endocannabinoid mechanism for stress-induced analgesia. Nature 435(7045):1108–12
Howlett AC, Qualy JM, Khachatrian LL (1986) Involvement of Gi in the inhibition of adenylate cyclase by cannabimimetic drugs. Mol Pharmacol 29(3):307–13
Hsu K-T, Storch J (1996) Fatty acid transfer from liver and intestinal fatty acid-binding proteins to membranes occurs by different mechanisms. J Biol Chem 271(23):13317—23

Huang CC, Lo SW, Hsu KS (2001) Presynaptic mechanisms underlying cannabinoid inhibition of excitatory synaptic transmission in rat striatal neurons. J Physiol 532:731–48

Huang YC, Wang SJ, Chiou LC, Gean PW (2003) Mediation of amphetamine-induced long-term depression of synaptic transmission by CB1 cannabinoid receptors in the rat amygdala. J Neurosci 23(32):10311–20

Huestis MA, Gorelick DA, Heishman SJ, Preston KL, Nelson RA, Moolchan ET, Frank RA, (2001) Blockade of effects of smoked marijuana by the CB1-selective cannabinoid receptor antagonist SR141716. Arch Gen Psychiat 58:322–8

Ishac EJ, Jiang L, Lake KD, Varga K, Abood ME, Kunos G (1996) Inhibition of exocytotic noradrenaline release by presynaptic cannabinoid CB1 receptors on peripheral sympathetic nerves. Br J Pharmacol 118(8):2023–8

Isokawa M, Alger BE (2005) Retrograde endocannabinoid regulation of GABAergic inhibition in the rat dentate gyrus granule cell. J Physiol 567(Pt 3):1001–10

Ito M (2001) Cerebellar long-term depression: characterization, signal transduction, and functional roles. Physiol Rev 81(3):1143–95

Ito M, Kano M (1982) Long-lasting depression of parallel fiber-Purkinje cell transmission induced by conjunctive stimulation of parallel fibers and climbing fibers in the cerebellar cortex. Neurosci Lett 33:253–8

Jo YH, Chen YJ, Chua SC Jr, Talmage DA, Role LW (2005) Integration of endocannabinoid and leptin signaling in an appetite-related neural circuit. Neuron 48(6):1055–66

Jung KM, Mangieri R, Stapleton C, Kim J, Fegley D, Wallace M, Mackie K, Piomelli D (2005) Stimulation of endocannabinoid formation in brain slice cultures through activation of group I metabotropic glutamate receptors. Mol Pharmacol 68(5):1196–1202

Justinova Z, Tanda G, Redhi GH, Goldberg SR (2003) Self-administration of Δ9-tetrahydrocannabinol (THC) by drug naive squirrel monkeys. Psychopharmacology 169: 135–40

Kahn L, Alonso G, Robbe D, Bockaert J, Manzoni OJ (2001) Group 2 metabotropic glutamate receptors induced long term depression in mouse striatal slices. Neurosci Lett 316(3):178–82

Kamprath K, Marsicano G, Tang J, Monory K, Bisogno T, Di Marzo V, Lutz B, Wotjak CT (2006) Cannabinoid CB1 receptor mediates fear extinction via habituation-like processes. J Neurosci 26(25):6677–86

Katona I, Rancz EA, Acsady L, Ledent C, Mackie K, Hajos N, Freund TF (2001) Distribution of CB1 cannabinoid receptors in the amygdala and their role in the control of GABAergic transmission. J Neurosci 21(23):9506–18

Katona I, Urban GM, Wallace M, Ledent C, Jung KM, Piomelli D, Mackie K, Freund TF (2006) Molecular composition of the endocannabinoid system at glutamatergic synapses. J Neurosci 26(21):5628–37

Kawamura Y, Fukaya M, Maejima T, Yoshida T, Miura E, Watanabe M, Ohno-Shosaku T, Kano M (2006) The CB1 cannabinoid receptor is the major cannabinoid receptor at excitatory presynaptic sites in the hippocampus and cerebellum. J Neurosci 26(11):2991–3001

Kim J, Alger BE (2004) Inhibition of cyclooxygenase-2 potentiates retrograde endocannabinoid effects in hippocampus. Nat Neurosci 7(7):697–8

Kobayashi K, Manabe T, Takahashi T (1996) Presynaptic long-term depression at the hippocampal mossy fiber-CA3 synapse. Science 273(5275):648–50

Kozak KR, Crews BC, Morrow JD, Wang LH, Ma YH, Weinander R, Jakobsson PJ, Marnett LJ (2002) Metabolism of the endocannabinoids, 2-arachidonylglycerol and anandamide, into prostaglandin, thromboxane, and prostacyclin glycerol esters and ethanolamides. J Biol Chem 277(47):44877–85

Kreitzer AC, Malenka RC (2005) Dopamine modulation of state-dependent endocannabinoid release and long-term depression in the striatum. J Neurosci 25(45):10537–45

Kreitzer AC, Regehr WG (2001) Retrograde inhibition of presynaptic calcium influx by endogenous cannabinoids at excitatory synapses onto purkinje cells. Neuron 29:717–27

Kreitzer AC, Carter AG, Regehr WG (2002) Inhibition of interneuron firing extends the spread of endocannabinoid signaling in the cerebellum. Neuron 34:787–95

Kulkarni-Narla A, Brown DR (2000) Localization of CB1-cannabinoid receptor immunoreactivity in the porcine enteric nervous system. Cell Tissue Res 302(1):73–80

Kushmerick C, Price GD, Taschenberger H, Puente N, Renden R, Wadiche JI, Duvoisin RM, Grandes P, von Gersdorff H (2004) Retroinhibition of presynaptic Ca2+ currents by endocannabinoids released via postsynaptic mGluR activation at a calyx synapse. J Neurosci 24(26):5955–65

Kyrozis A, Benameur K, Zhang X-L, Winterer J, Muller W, Stanton PK (2004) Retrograde messengers in long-term plasticity of presynaptic glutamate release in hippocampus. In: Stanton PK, Bramham C, Scharfman H (eds) Synaptic plasticity and transsynaptic signaling. Spinger, New York pp. 273–304

Laviolette SR, Grace AA (2006) Cannabinoids Potentiate Emotional Learning Plasticity in Neurons of the Medial Prefrontal Cortex through Basolateral Amygdala Inputs. J Neurosci 26(24):6458–68

Lichtman AH (2000) SR 141716A enhances spatial memory as assessed in a radial-arm maze task in rats. Eur J Pharmacol 404(1–2):175–9

Lichtman AH, Hawkins EG, Griffin G, Cravatt BF (2002) Pharmacological activity of fatty acid amides is regulated, but not mediated, by fatty acid amide hydrolase in vivo. J Pharmacol Exp Ther 302(1):73–9

Liu J, Wang L, Harvey-White J, Osei-Hyiaman D, Razdan R, Gong Q, Chan AC, Zhou Z, Huang B, Kim H-Y, Kunos G (2006) A biosynthetic pathway for anandamide. Proc Natl Acad Sci USA 103(36):13345–50

Llano I, Leresche N, Marty A (1991) Calcium entry increases the sensitivity of cerebellar Purkinje cells to applied GABA and decreases inhibitory synaptic currents. Neuron 6:565–74

Lefkowitz RJ, Cotecchia S, Samama P, Costa T (1993) Constitutive activity of receptors coupled to guanine nucleotide regulatory proteins. Trends Pharmacol Sci 14(8):303–7

Levenes C, Daniel H, Soubrie P, Crepel F (1998) Cannabinoids decrease excitatory synaptic transmission and impair long-term depression in rat cerebellar Purkinje cells. J Physiol 510:867–79

Lopez-Redondo F, Lees GM, Pertwee RG (1997) Effects of cannabinoid receptor ligands on electrophysiological properties of myenteric neurones of the guinea-pig ileum. Br J Pharmacol 122(2):330–4

Lovinger DM, Merritt A, Reyes D (1994) Involvement of N- and non-N-type calcium channels in synaptic transmission at corticostriatal synapses. Neurosci 62(1):31–40

McAllister SD, Griffin G, Satin LS, Abood ME (1999) Cannabinoid receptors can activate and inhibit G protein-coupled inwardly rectifying potassium channels in a xenopus oocyte expression system. J Pharmacol Exp Ther 291(2):618–26

Maccarrone M, Bari M, Battista N, Finazzi-Agrò A (2002) Estrogen stimulates arachidonoylethanolamide release from human endothelial cells and platelet activation. Blood 100:4040–8

McEachern JC, Shaw CA (1996) An alternative to the LTP orthodoxy: a plasticity-pathology continuum model. Brain Res Brain Res Rev 22(1):51–92

Mackie K, Lai Y, Westenbroek R, Mitchell R (1995) Cannabinoids activate an inwardly rectifying potassium conductance and inhibit Q-type calcium currents in AtT20 cells transfected with rat brain cannabinoid receptor. J Neurosci 15(10):6552–61

Mackie K, Devane WA, Hille B (1993) Anandamide, an endogenous cannabinoid, inhibits calcium currents as a partial agonist in N18 neuroblastoma cells. Mol Pharmacol 44(3):498–503

Mackie K, Hille B (1992) Cannabinoids inhibit N-type calcium channels in neuroblastoma-glioma cells. Proc Natl Acad Sci USA 89(9):3825–9

Mackie K (2005) Distribution of cannabinoid receptors in the central and peripheral nervous system. Handbook Exper Pharmacol 168:299–325

Mackie K. (2006) Mechanisms of CB1 receptor signaling: endocannabinoid modulation of synaptic strength. Int. J. Obes. (Lond.) 30 Suppl 1:S19–23

Mckinney MK, Cravatt, BP (2005) Structure and function of fatty acid amide hydrolase. Annu Rev Biochem 74:411–32

Madison DV, Schuman EM (1995) Diffusible messengers and intercellular signaling: locally distributed synaptic potentiation in the hippocampus. Curr Top Microbiol Immunol 196:5–6
Maejima T, Hashimoto K, Yoshida T, Aiba A, Kano M (2001) Presynaptic inhibition caused by retrograde signal from metabotropic glutamate to cannabinoid receptors. Neuron 31:463–75
Maejima T, Oka S, Hashimotodani Y, Ohno-Shosaku T, Aiba A, Wu D, Waku K, Sugiura T, Kano M (2005) Synaptically driven endocannabinoid release requires Ca2+-assisted metabotropic glutamate receptor subtype 1 to phospholipase Cbeta4 signaling cascade in the cerebellum. J Neurosci 25(29):6826–35
Makara JK, Mor M, Fegley D, Szabo SI, Kathuria S, Astarita G, Duranti A, Tontini A, Tarzia G, Rivara S, Freund TF, Piomelli D (2005) Inhibition of cyclooxygenase-2 potentiates retrograde endocannabinoid effects in hippocampus. Nat Neurosci 7(7):697–8
Makriyannis A, Guo J, Tian X (2005a) Albumin enhances the diffusion of lipophilic drugs into the membrane bilayer. Life Sci 77(14):1605–11
Makriyannis A, Tian X, Guo J (2005b) How lipophilic cannabinergic ligands reach their receptor sites. Prostaglandins Other Lipid Mediat 7(1–4):210–18
Malenka RC, Bear MF (2004) LTP and LTD: an embarrassment of riches. Neuron 44(1):5–21
Mallet PE, Beninger RJ (1998) The cannabinoid CB1 receptor antagonist SR141716A attenuates the memory impairment produced by delta9-tetrahydrocannabinol or anandamide. Psychopharm (Berl.) 140(1):11–19
Mang CF, Erbelding D, Kilbinger H (2001) Differential effects of anandamide on acetylcholine release in the guinea-pig ileum mediated via vanilloid and non-CB1 cannabinoid receptors. Br J Pharmacol 134(1):161–7
Mansbach RS, Rovetti CC, Winston EN, Lowe JA 3rd. Effects of the cannabinoid CB1 receptor antagonist SR141716A on the behavior of pigeons and rats.Psychopharmacology (Berl). 1996 Apr; 124(4):315–22
Marsicano G, Wotjak CT, Azad SC, Bisogno T, Rammes G, Cascio MG, Hermann H, Tang J, Hofmann C, Zieglgansberger W, Di Marzo V, Lutz B (2002) The endogenous cannabinoid system controls extinction of aversive memories. Nature 418(6897):530–4
Martin LA, Alger BE (1999) Muscarinic facilitation of the occurrence of depolarization-induced suppression of inhibition in rat hippocampus. Neurosci 92(1):61–71
Mato S, Robbe D, Puente N, Grandes P, Manzoni OJ (2005) Presynaptic homeostatic plasticity rescues long-term depression after chronic Delta 9-tetrahydrocannabinol exposure. J Neurosci 25(50):11619–627
Matsuda LA, Lolait SJ, Brownstein MJ, Young AC, Bonner TI (1990) Structure of a cannabinoid receptor and functional expression of the cloned cDNA. Nature 346:561–4
Mazzola C, Micale V, Drago F (2003) Amnesia induced by beta-amyloid fragments is counteracted by cannabinoid CB1 receptor blockade. Eur J Pharmacol 477(3):219–25
McFarland MJ, Barker EL (2004) Anandamide transport. Pharmacol Ther 104(2):117–35
Mechoulam R, Ben-Shabat S, Hanus L, Ligumsky M, Kaminski NE, Schatz AR, Gopher A, Almog S, Martin BR, Compton DR, Pertwee RG, Griffin G, Bayewitch M, Barg J, Vogel Z (1995) Identification of an endogenous 2-monoglyceride, present in canine gut, that binds to cannabinoid receptors. Biochem Pharmacol 50(1):83–90
Mechoulam R, Deutsch DG (2005) Toward an anandamide transporter. Proc Natl Acad Sci USA 102(49):17541–2
Melis M, Pistis M, Perra S, Muntoni AL, Pillolla G, Gessa GL (2004a) Endocannabinoids mediate presynaptic inhibition of glutamatergic transmission in rat ventral tegmental area dopamine neurons through activation of CB1 receptors. J Neurosci 24(1):53–62
Melis M, Perra S, Muntoni AL, Pillolla G, Lutz B, Marsicano G, Di Marzo V, Gessa GL, Pistis M (2004b) Prefrontal cortex stimulation induces 2-arachidonoyl-glycerol-mediated suppression of excitation in dopamine neurons. J Neurosci 24(47):10707–15
Miller RJ (1998) Presynaptic receptors. Annu Rev Pharmacol Toxicol 38:201–27
Miller LL, Branconnier RJ (1983) Cannabis: effects on memory and the cholinergic limbic system. Psychol. Bull. 93(3) 441–56

Mintz IM, Bean B (1993) GABAB receptor inhibition of P-type Ca2+ channels in central neurons. Neuron 10(5):889–98

Misner DL, Sullivan JM (1999) Mechanism of cannabinoid effects on long-term potentiation and depression in hippocampal CA1 neurons. J Neurosci 19:6795–6805

Moore SA, Nomikos GG, Dickason-Chesterfield AK, Schober DA, Schaus JM, Ying BP, Xu YC, Phebus L, Simmons RM, Li D, Iyengar S, Felder CC (2005) Identification of a high-affinity binding site involved in the transport of endocannabinoids. Proc Natl Acad Sci USA 102(49):17852–7

Morishita W, Kirov SA, Alger BE (1998) Evidence for metabotropic glutamate receptor activation in the induction of depolarization-induced suppression of inhibition in hippocampal CA1. J Neurosci 18(13):4870–82

Mukhtarov M, Ragozzino D, Bregestovski P (2005) Dual Ca2+ modulation of glycinergic synaptic currents in rodent hypoglossal motoneurones. J Physiol 569:817–31

Narasimhan K, Linden DJ (1996) Defining a minimal computational unit for cerebellar long-term depression. Neuron 17(2):333–41

Ng Cheong Ton JM, Gerhardt GA, Friedemann M, Etgen AM, Rose GM, Sharpless NS, Gardner EL (1988) The effects of delta 9-tetrahydrocannabinol on potassium-evoked release of dopamine in the rat caudate nucleus: an in vivo electrochemical and in vivo microdialysis study. Brain Res 451:59–68

Nowicky AV, Teyler TJ, Vardaris RM (1987) The modulation of LTP by delta-9-THC in the rat hippocampus in vitro. Brain Res 19:663–72

Ohno-Shosaku T, Maejima T, Kano M (2001) Endogenous cannabinoids mediate retrograde signals from depolarized postsynaptic neurons to presynaptic terminals. Neuron 29:729–38

Ohno-Shosaku T, Shosaku J, Tsubokawa H, Kano M (2002) Cooperative endocannabinoid production by neuronal depolarization and group I metabotropic glutamate receptor activation. Eur J Neurosci 15:953–61

de Oliveira Alvares L, Genro BP, Vaz Breda R, Pedroso MF, Da Costa JC, Quillfeldt JA (2006) AM251, a selective antagonist of the CB1 receptor, inhibits the induction of long-term potentiation and induces retrograde amnesia in rats. Brain Res 1075(1):60–7

Olson PA, Tkatch T, Hernandez-Lopez S, Ulrich S, Ilijic E, Mugnaini E, Zhang H, Bezprozvanny I, Surmeier DJ (2005) G-protein-coupled receptor modulation of striatal CaV1.3 L-type Ca2+ channels is dependent on a Shank-binding domain. J Neurosci 25(5):1050–62

Orsnes G, Crone C, Krarup C, Petersen N, Nielsen J (2000) The effect of baclofen on the transmission in spinal pathways in spastic multiple sclerosis patients. Clin Neurophysiol 111(8):1372–9

Pan X, Ikeda SR, Lewis DL (1996) Rat brain cannabinoid receptor modulates N-type Ca2+ channels in a neuronal expression system. Mol Pharmacol 49(4):707–14

Pan X, Ikeda SR, Lewis DL (1998) SR 141716A acts as an inverse agonist to increase neuronal voltage-dependent Ca2+ currents by reversal of tonic CB1 cannabinoid receptor activity. Mol Pharmacol 54(6):1064–72

Parolaro D, Vigano D, Rubino T (2005) Endocannabinoids and drug dependence. Curr Drug Targets CNS Neurol Disord 4:643–55

Perio, A, Rinaldi-Carmona, M, Maruani, J, Barth, F, Le Fur, G, Sobrie, P. Central mediation of the cannabinoid cue: activity of a selective CB1 antagonist, SR 141716A. Behav Pharmacol. 1996 Jan; 7(1):65–71

Pertwee RG, Fernando SR (1996) Evidence for the presence of cannabinoid CB1 receptors in mouse urinary bladder. Br J Pharmacol 118(8):2053–8

Pitler TA, Alger BE (1992) Postsynaptic spike firing reduces synaptic $GABA_A$ responses in hippocampal pyramidal cells. J Neurosci 12:4122–32

Porter AC, Sauer JM, Knierman MD, Becker GW, Berna MJ, Bao J, Nomikos GG, Carter P, Bymaster FP, Leese AB, Felder CC (2002) Characterization of a novel endocannabinoid, virodhamine, with antagonist activity at the CB1 receptor. J Pharmacol Exp Ther 301(3):1020–4

Rancillac A, Barbara JG (2005) Frequency-dependent recruitment of inhibition mediated by stellate cells in the rat cerebellar cortex. J Neurosci Res 80(3):414–23

Ralevic V (2003) Cannabinoid modulation of peripheral autonomic and sensory neurotransmission. Eur J Pharmacol 472(1–2):1–21

Reisine T, Pasternak G (1996) Opioid analgesics and antagonists in Goodman & Gilman's The pharmacological basis of therapeutics. McGraw-Hill (New York, USA) pp. 521–56

Rinaldi-Carmona M, Barth F, Heaulme M, Shire D, Calandra B, Congy C, Martinez S, Maruani J, Neliat G, Caput D, Ferrara P, Sobrie P, Breliere JC, Le Fur GL (1994) SR141716A, a potent and selective antagonist of the brain cannabinoid receptor. FEBS Lett 350:240

Robbe D, Alonso G, Duchamp F, Bockaert J, Manzoni OJ (2001) Localization and mechanisms of action of cannabinoid receptors at the glutamatergic synapses of the mouse nucleus accumbens. J Neurosci 21:109–16

Robbe D, Kopf M, Remaury A, Bockaert J, Manzoni OJ (2002) Endogenous cannabinoids mediate long-term synaptic depression in the nucleus accumbens. Proc Natl Acad Sci USA 99:8384–8

Ronesi J, Gerdeman GL, Lovinger DM (2004) Disruption of endocannabinoid release and striatal long-term depression by postsynaptic blockade of endocannabinoid membrane transport. J Neurosci 24(7):1673–9

Rumsey CC, Abbott LF (2006) Synaptic democracy in active dendrites. J Neurophysiol 96(5):2307–18

Saario SM, Savinainen JR, Laitinen JT, Jarvinen T, Niemi R (2004) Monoglyceride lipase-like enzymatic activity is responsible for hydrolysis of 2-arachidonoylglycerol in rat cerebellar membranes. Biochem Pharmacol 67(7):1381–7

Sacchettini JC, Gordon JI (1993) Rat intestinal fatty acid binding protein. A model system for analyzing the forces that can bind fatty acids to proteins. J Biol Chem 268:18399–402

Safo PK, Regehr WG (2005) Endocannabinoids control the induction of cerebellar LTD. Neuron 48:647–59

Sargent Jones L (2005) Recent advances in the role of integrins in developmental and adult synaptic plasticity. In: Stanton PK, Bramham C, Scharfman H (eds) Synaptic plasticity and transsynaptic signaling. Spinger, New York, pp. 457–68

Schmid HHO, Schmid PC, Natarajan V (1990) *N*-Acylated glycerophospholipids and their derivatives. Prog Lipid Res 29:1–43

Shiflett MW, Rankin AZ, Tomaszycki ML, DeVoogd TJ (2004) Cannabinoid inhibition improves memory in food-storing birds, but with a cost. Proc Biol Sci 271(1552):2043–8

Sjostrom PJ, Turrigiano GG, Nelson SB (2003) Neocortical LTD via coincident activation of presynaptic NMDA and cannabinoid receptors. Neuron 39(4):641–54

Slanina KA, Roberto M, Schweitzer P (2005) Endocannabinoids restrict hippocampal long-term potentiation via CB1. Neuropharm 49(5):660–8

Soler-Llavina GJ, Sabatini BL (2006) Synapse-specific plasticity and compartmentalized signaling in cerebellar stellate cells. Nat Neurosci 9(6):798–806

Stanton PK, Bramham C, Scharfman H (eds) (2005) Synaptic plasticity and transsynaptic signaling. Spinger, New York pp. 457–68

Stella N, Schweitzer P, Piomelli D (1997) A second endogenous cannabinoid that modulates long-term potentiation. Nature 388:773–8

Sternweiss PC, Smrcka AV, Gutowski S (1992) Hormone signalling via G-protein: regulation of phosphatidylinositol 4,5-bisphosphate hydrolysis by Gq. Philos Trans R Soc Lond B Biol Sci 336(1276):35–41

Straiker A, Mackie K (2005) Depolarization-induced suppression of excitation in murine autaptic hippocampal neurones. J Physiol 569(Pt 2):501–17

Sullivan JM (2000) Cellular and molecular mechanisms underlying learning and memory impairments produced by cannabinoids. Learn Mem 7(3):132–9

Sugiura T, Kondo S, Sukagawa A, Nakane S, Shinoda A, Itoh K, Yamashita A, Waku K (1995) 2-Arachidonoylglycerol: a possible endogenous cannabinoid receptor ligand in brain. Biochem Biophys Res Commun 215(1):89–97

Sung KW, Choi S, Lovinger DM (2001) Activation of group I mGluRs is necessary for induction of long-term depression at striatal synapses. J Neurophysiol 86:2405–12

Suzuki A, Josselyn SA, Frankland PW, Masushige S, Silva AJ, Kida S (2004) Memory reconsolidation and extinction have distinct temporal and biochemical signatures. J Neurosci 24(20):4787–95

Szabo B, Schlicker E (2005) Effects of cannabinoids on neurotransmission. Handbook Exper Pharmacol (168): 327–65

Szabo B, Urbanski MJ, Bisogno T, Di Marzo V, Mendiguren A, Bar W, Freiman I (2006) Depolarization-induced retrograde synaptic inhibition in the cerebellar cortex is mediated by 2-arachidonoylglycerol. J Physiol in press.

Takahashi KA, Castillo PE (2006) The CB1 cannabinoid receptor mediates glutamatergic synaptic suppression in the hippocampus. Neurosci 139(3):795–802

Takahashi RN, Pamplona FA, Fernandes MS (2005) The cannabinoid antagonist SR141716A facilitates memory acquisition and consolidation in the mouse elevated T-maze. Neurosci Lett 380(3):270–75

Tanda G, Pontieri FE, Di Chiara G (1997) Cannabinoid and heroin activation of mesolimbic dopamine transmission by a common_1 opioid receptor mechanism. Science 27:2048–50

Tanda G, Munzar P, Goldberg SR (2000) Self-administration behavior is maintained by the psychoactive ingredient of marijuana in squirrel monkeys. Nat Neurosci 3:1073–4

Tanda G, Goldberg SR (2003) Cannabinoids: reward, dependence, and underlying neurochemical mechanisms–a review of recent preclinical data. Psychopharmacology 169:115

Tang K, Low MJ, Grandy DK, Lovinger DM (2001) Dopamine-dependent synaptic plasticity in striatum during in vivo development. Proc Natl Acad Sci USA 98(3):1255–60

Terranova JP, Storme JJ, Lafon N, Perio A, Rinaldi-Carmona M, Le Fur G, Soubrie P (1996) Improvement of memory in rodents by the selective CB1 cannabinoid receptor antagonist, SR 141716. Psychopharmacology 126(2):165–72

Trettel J, Levine ES (2003) Endocannabinoids mediate rapid retrograde signaling at interneuron-pyramidal neuron synapses of the neocortex. J Neurophysiol 89:2334–8

Trettel J, Fortin DA, Levine ES (2004) Endocannabinoid signalling selectively targets perisomatic inhibitory inputs to pyramidal neurones in juvenile mouse neocortex. J Physiol 556(Pt 1): 95–107

Tsou K, Brown S, Sanudo-Pena MC, Mackie K, Walker JM (1997) Immunohistochemical distribution of cannabinoid CB1 receptors in the rat central nervous system. Neurosci 83(2):393–411

Van Beugen BJ, Nagaraja RY, Hansel C (2006) Climbing fiber-evoked endocannabinoid signaling heterosynaptically suppresses presynaptic cerebellar long-term potentiation. J Neurosci 26(32):8289–94

Van Sickle MD, Duncan M, Kingsley PJ, Mouihate A, Urbani P, Mackie K, Stella N, Makriyannis A, Piomelli D, Davison JS, Marnett LJ, Di Marzo V, Pittman QJ, Patel KD, Sharkey KA (2005) Identification and functional characterization of brainstem cannabinoid CB2 receptors. Science 310(5746):329–32

Valjent E, Maldonado R (2000) A behavioural model to reveal place preference to delta 9-tetrahydrocannabinol in mice. Psychopharmacology (Berl) 147:436–8

Varma N, Carlson GC, Ledent C, Alger BE (2001) Metabotropic glutamate receptors drive the endocannabinoid system in hippocampus. J Neurosci 21:RC188

Varma N, Brager D, Morishita W, Lenz RA, London B, Alger B (2002) Presynaptic factors in the regulation of DSI expression in hippocampus. Neuropharm 43(4):550–62

Varvel SA, Anum EA, Lichtman AH (2005) Disruption of CB(1) receptor signaling impairs extinction of spatial memory in mice. Psychopharm 179(4):863–72

Varvel SA, Lichtman AH (2002) Evaluation of CB1 receptor knockout mice in the Morris water maze. J Pharmacol Exp Ther 301(3):915–24

Vincent P, Marty A (1993) Neighboring cerebellar Purkinje cells communicate via retrograde inhibition of common presynaptic interneurons. Neuron 11:885–93

Volterra A, Meldolesi J (2005) Astrocytes, from brain glue to communication elements: the revolution continues. Nat Rev Neurosci 6(8):626–40

Wagner JJ, Alger BE (1996) Increased neuronal excitability during depolarization-induced suppression of inhibition in rat hippocampus. J Physiol 495(1):107–12

Walsh JP (1993) Depression of excitatory synaptic input in rat striatal neurons. Brain Res 608(1):123–8

Wang J, Zucker RS (2001) Photolysis-induced suppression of inhibition in rat hippocampal CA1 pyramidal neurons. J Physiol 533:757–63

Wang Z, Kai L, Day M, Ronesi J, Yin HH, Ding J, Tkatch T, Lovinger D, Surmeier DJ (2006) Dopaminergic control of corticostriatal long-term synaptic depression in medium spiny neurons is mediated by cholinergic interneurons. Neuron 50:443–52

Wartmann M, Campbell D, Subramanian A, Burstein SH, Davis RJ (1995) The MAP kinase signal transduction pathway is activated by the endogenous cannabinoid anandamide. FEBS Lett 359(2–3):133–6

Wei BQ, Mikkelsen TS, McKinney MK, Lander ES, Cravatt BF (2006) A second fatty acid amide hydrolase with variable distribution among placental mammals. J Biol Chem 281(48): 36569–78

Weisskopf MG, Zalutsky RA, Nicoll RA (1993) The opioid peptide dynorphin mediates heterosynaptic depression of hippocampal mossy fibre synapses and modulates long-term potentiation. Nature 362(6419):423–7

Wigstrom H, Gustafsson B (1986) Postsynaptic control of hippocampal long-term potentiation. J Physiol (Paris) 81(4):228–36

Wiley JL, Martin BR (1999) Effects of SR141716A on diazepam substitution for delta9-tetrahydrocannabinol in rat drug discrimination. *Pharmacol Biochem Behav* 64:519

Wilson RI, Nicoll RA (2001) Endogenous cannabinoids mediate retrograde signalling at hippocampal synapses. Nature 410:588–92

Wilson RI, Kunos G. Nicoll RA (2001) Presynaptic specificity of endocannabinoid signaling in the hippocampus. Neuron 31(3):453–62

Wolf MC, Leander JD (2003) SR141716A, a cannabinoid CB1 receptor antagonist, improves memory in a delayed radial maze task. Eur J Pharmacol 477(3):213–17

Wu L-G, Saggau P (1997) Presynaptic inhibition of elicited neurotransmitter release. Trends Neurosci 20(5): 204–12

Yanovsky Y, Mades S, Misgeld U (2003) Retrograde signaling changes the venue of postsynaptic inhibition in rat substantia nigra. Neurosci 122:317–238

Yin HH, Ronesi J, Davis M, Lovinger DM (2006) The role of protein synthesis in striatal long-term depression. J Neurosci 26(46):11811–20

Yin HH, Lovinger DM (2006) Frequency-specific and D2 receptor-mediated inhibition of glutamate release by retrograde endocannabinoid signaling. Proc Natl Acad Sci USA 103(21): 8251–6

Yoshida T, Hashimoto K, Zimmer A, Maejima T, Araishi K, Kano M (2002) The cannabinoid CB1 receptor mediates retrograde signals for depolarization-induced suppression of inhibition in cerebellar Purkinje cells. J Neurosci 22(5):1690–7

Zangen A, Solinas M, Ikemoto S, Goldberg SR, Wise RA (2006) Two brain sites for cannabinoid reward. J Neurosci 26(18):4901–7

Zhu PJ, Lovinger DM (2005) Retrograde endocannabinoid signaling in a postsynaptic neuron/synaptic bouton preparation from basolateral amygdala. J Neurosci 25(26):6199–6207

# Presynaptic Ionotropic Receptors

**M.M. Dorostkar and S. Boehm(✉)**



**Abstract** The release of transmitters through vesicle exocytosis from nerve terminals is not constant but is subject to modulation by various mechanisms, including prior activity at the synapse and the presence of neurotransmitters or neuromodulators in the synapse. Instantaneous responses of postsynaptic cells to released transmitters are mediated by ionotropic receptors. In contrast to metabotropic receptors, ionotropic receptors mediate the actions of agonists in a transient manner within milliseconds to seconds. Nevertheless, transmitters can control vesicle exocytosis not only via slowly acting metabotropic, but also via fast acting ionotropic receptors located at the presynaptic nerve terminals. In fact, members of the following subfamilies of ionotropic receptors have been found to control transmitter release: ATP P2X, nicotinic acetylcholine, $GABA_A$, ionotropic glutamate, glycine, 5-$HT_3$, and

S. Boehm
Institute of Experimental and Clinical Pharmacology, Medical University of Graz, Universitätsplatz 4, 8010 Graz, Austria
stefan.boehm@meduni-graz.at

T.C. Südhof, K. Starke (eds.), *Pharmacology of Neurotransmitter Release*.
*Handbook of Experimental Pharmacology 184*.

vanilloid receptors. As these receptors display greatly diverging structural and functional features, a variety of different mechanisms are involved in the regulation of transmitter release via presynaptic ionotropic receptors. This text gives an overview of presynaptic ionotropic receptors and briefly summarizes the events involved in transmitter release to finally delineate the most important signaling mechanisms that mediate the effects of presynaptic ionotropic receptor activation. Finally, a few examples are presented to exemplify the physiological and pharmacological relevance of presynaptic ionotropic receptors.

## 1 Introduction

Information transfer between two neurons or between neurons and effector cells involves the release of chemical substances, which then act on the target cell by binding to appropriate receptors embedded in the plasma membrane. This process, as originally described by Otto Loewi (Loewi 1921), is termed chemical neurotransmission and occurs at contact sites known as synapses. Neurotransmitters exert their effects via members of two major families of receptors: ionotropic and metabotropic neurotransmitter receptors. Activation of ionotropic receptors leads to an increase in the ion conductance of the membrane within a time scale of milliseconds or even less, whereas activation of metabotropic receptors results in slow effects (within seconds or even minutes) which involve more or less complex signaling cascades. Accordingly, information transfer via ionotropic receptors is called fast synaptic transmission, whereas the slow counterpart is called neuromodulation (Kaczmarek and Levitan 1987).

Alterations in the strength of neurotransmission at various synapses underlie complex functions of the nervous system such as learning and memory (Kandel 2001). There are two major types of changes in neurotransmission that occur at single synapses: one depends on the prior activity in the synapse, and the other one is caused by mediators released from neurons or glial cells (Zucker and Regehr 2002). In both cases, alterations in the strength of synaptic transmission may be caused either by changes in the release of the chemical transmitters or by changes in the sensitivity of the according receptors. As these alterations arise at either side of the synapse, they are subsumed under pre- and postsynaptic modulation of synaptic transmission, respectively. Typically, the phenomena of pre- and postsynaptic modulation are experimentally observed on a time scale of seconds or minutes rather than milliseconds and thus represent one prominent aspect of neuromodulation. In support of this categorization, the mechanisms involved are in most instances metabotropic ones.

Presynaptic modulation by neurotransmitters is mediated via receptors, which are accordingly named "presynaptic receptors.". Is, therefore, a presynaptic receptor only a receptor activation of which modulates transmitter release? The answer is no, because activation of a presynaptic receptor may also affect transmitter reuptake via the appropriate carrier proteins (Zahniser and Doolen 2001). This latter effect does not necessarily impinge on transmitter release, unless the carriers are involved in the release process, which is then nonvesicular (Attwell et al. 1993). In this review, however, we will only consider transmitter release mediated by vesicle exocytosis. Hence, by functional means a presynaptic receptor is one, activation of which modulates one or more of the various functions of a presynapse, which also include endocytosis and neurotransmitter synthesis or metabolism. Nevertheless, here we will only deal with receptors that influence the amount of transmitter being released from a nerve terminal.

In morphological terms, presynaptic receptors are those located at the presynapse, i.e., at the neuronal structures from which the neurotransmitters are released. There, one can find a large number of neurotransmitter-containing vesicles which are clustered at specialized regions of the presynapse, called "active zones." At these active zones, vesicle exocytosis takes place, but the active zones occupy only a small part of an entire axon terminal (Südhof 2004). Accordingly, one has to clarify whether a presynaptic receptor is necessarily located at or close to the active zone, or whether it can also be located somewhere else at the nerve terminal. In addition, one has to consider receptors that are found at the axon in close proximity to the nerve terminal. Finally, on a micromorphological level one also has to certify that a receptor within a synapse is indeed embedded at the presynaptic, but not the postsynaptic, side. Thus, it is not easy to define a presynaptic receptor by morphological means. As a consequence, various techniques have been employed to functionally verify whether a receptor is indeed a presynaptic one: (1) the use of synaptosomes (Whittaker 1993); (2) the destruction of nerve terminals by lesioning of the corresponding neuronal cell bodies and the subsequent demonstration of a loss of function or loss of ligand binding; (3) the use of dissociated neurons in cell cultures in which postsynaptic structures are lacking; (4) finally, the blockade of voltage-activated $Na^+$ and/or $Ca^{2+}$ channels. This latter procedure is particularly important when characterizing presynaptic ionotropic receptors. The $Na^+$ channels mediate the propagation of action potentials down the axons into the presynaptic nerve terminals, and the $Ca^{2+}$ channels link the depolarization of the nerve terminal caused by an invading action potential to transmembrane $Ca^{2+}$ entry, which then triggers exocytosis. Vesicle exocytosis typically requires intracellular $Ca^{2+}$ concentrations in the high micromolar range; due to limited diffusion of $Ca^{2+}$ ions in neurons, such concentrations are restricted to the sites of transmembrane $Ca^{2+}$ entry (Augustine 2001). Accordingly, physical interactions between the SNARE proteins syntaxin and SNAP 25, which are involved in vesicle exocytosis and $Ca^{2+}$ channels of the $Ca_V2$ family, provide the proximity between the sites of $Ca^{2+}$ entry and vesicle exocytosis (Zamponi 2003). If this interaction is disrupted, action potential-dependent vesicle exocytosis is greatly reduced (Mochida et al. 1996). Hence, if transmitter release is triggered by the activation of a transmitter-gated ion channel and this effect is insensitive towards a blockade of voltage-gated $Ca^{2+}$ channels (e.g., by $Cd^{2+}$), the ionotropic receptor involved must be $Ca^{2+}$-permeable and in close proximity to the sites of vesicle exocytosis. If this effect is sensitive to a blockade of $Ca^{2+}$ channels but insensitive towards a blockade of voltage-gated $Na^{+}$ channels (e.g., by tetrodotoxin), the receptor must also be near to the sites of exocytosis because the local depolarization due to the activation of that ligand-gated ion channel suffices to activate the $Ca^{2+}$ channels that mediate the depolarization-dependent exocytosis. If, however, the effect is sensitive to a blockade of voltage-gated $Na^{+}$ channels, the ionotropic receptor is too far away from the sites of exocytosis, and the propagation of action potentials is required. In this latter case, it is questionable whether these ionotropic receptors are truly presynaptic ones.

In the light of this latter possibility, one also has to consider the term 'preterminal', which means pre-presynaptic. As specified by this latter expression, this designates a receptor that is located at regions of the axons which are close to the terminals, but not at the terminals themselves. In parallel with the above considerations, the distinction between presynaptic and preterminal receptors cannot be made by morphological criteria. Again, the use of synaptosomes and the blockade of $Na^{+}$-dependent action potentials can help to prevent the contribution of preterminal receptors to presynaptic modulation via ionotropic receptors (Figure 2a).

Although evidence for effects of agonists at ionotropic receptors on nerve terminals was reported more than 80 years ago (Dixon 1924), the first mention of the specific term "presynaptic receptor" was apparently provided by Riker et al (1957), who detected stimulatory effects of quarternary ammonium compounds at the neuromuscular junction. In the same year, Frank and Fuortes (1957) furthered the concept of functional presynaptic ionotropic receptors by implicating γ-aminobutyric acid (GABA) in the reduction of transmitter release from primary afferent nerve endings in the spinal cord. Simultaneously, Trendelenburg (1957) demonstrated that morphine reduced contractions of the nictitating mebrane through a presynaptic site of action, thereby setting the scene for presynaptic metabotropic receptors. Hence, the presynaptic modulation of transmitter release via both ionotropic and metabotropic receptors has been a focus of research for five decades. During the first half of this time period, reports concerning presynaptic G protein-coupled receptors prevailed (see, e.g., Langer 1977; Starke 1981), but more recently the importance of presynaptic transmitter-gated ion channels has been appreciated. Accordingly, several excellent reviews describing presynaptic ionotropic receptors have been published during the last decade (McGehee and Role 1996; MacDermott et al. 1999; Engelman and MacDermott 2004). In particular, the detailed review by MacDermott and colleagues (1999) gives a full account of all the presynaptic ionotropic receptors known at that time, and a shorter update has appeared more recently (Engelman and MacDermott 2004). Therefore, this review does not aim at listing every report that has described some kind of presynaptic ionotropic receptor. We rather try to highlight recent developments in the field of presynaptic transmitter-gated ion channels as a counterbalance to the five preceding chapters of this book which deal with presynaptic metabotropic receptors.

## 2 Types of Presynaptic Ionotropic Receptors

The superfamily of ionotropic receptors can be subclassified according to several principles: by structural means, by functional criteria, and according to the neurotransmitters by which they are activated. Transmitter- and ligand-gated ion channels are the prototypic examples of oligomeric receptors (Figure 1). They are composed of several subunits, and depending on the number of subunits contained within one functional receptor one can discern between three groups: $GABA_A$ receptors, glycine receptors (GlyR), nicotinic acetylcholine receptors (nAChRs), and $5\text{-}HT_3$ serotonin receptors are composed of 5 subunits; the ionotropic glutamate receptors and TRP channels display tetrameric structures; and ATP-gated P2X receptors contain only 3 subunits each. The single subunits contributing to the build-up of members of each of these different receptor classes also show diverging structural characteristics. The pentameric receptors are also called Cys-loop receptors, as each subunit contains a sequence of 13 residues flanked by cysteines which covalently bind to each other and thereby form a closed loop that is situated between the ligand-binding domain and the receptor ionophore. These subunits have a large N-terminal extracellular domain responsible for ligand binding, four transmembrane regions, a large intracellular loop between transmembrane segments 3 and 4, and a short C-terminus. The second transmembrane domain of each of the five subunits contributes to the formation of the ionophore (Lester et al. 2004). The subunits of ionotropic glutamate receptors, in contrast, possess only three transmembrane regions and one re-entrant loop each. The re-entrant loop contributes to the formation of the ionophore and thus determines the functional properties of the receptor (Madden 2002). Subunits for TRPV channels also contain one re-entrant loop contributing to the pore and six transmembrane segments (Liedtke and Kim 2005). Finally, ATP P2X receptors are characterized by two putative transmembrane

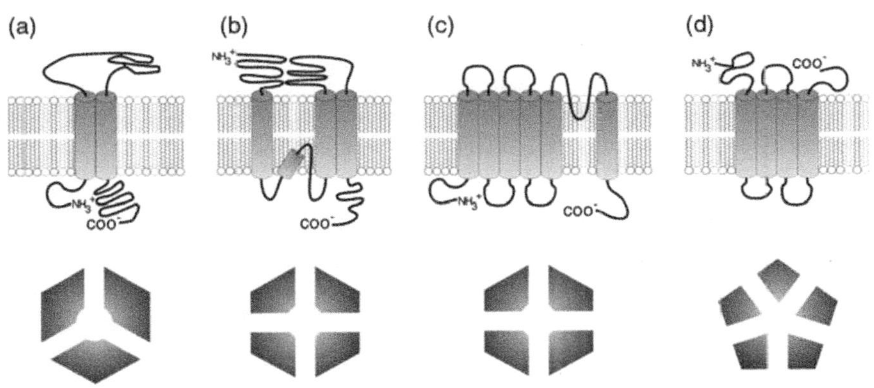


**Fig. 1** Structures of subunits (top) and subunit oligomerization (bottom) in ligand-gated ion channels. (**a**) P2X receptor family. (**b**) Ionotropic glutamate receptor family. (**c**) TRPV channel family. (**d**) Cys-loop superfamily comprising nAChRs, $5\text{-}HT_3$, $GABA_A$, and glycine receptors.

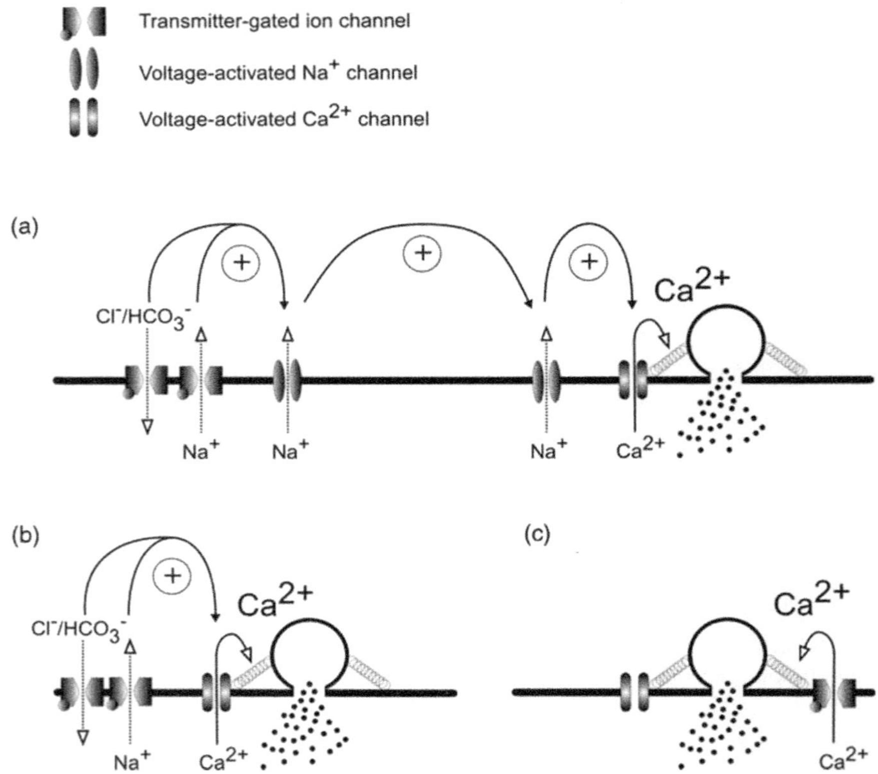


**Fig. 2** Increase in spontaneous transmitter release caused by the activation of ionotropic receptors. (**a**) Activation of a **preterminal** transmitter-gated anion or cation channel depolarizes the membrane either through influx of cations or efflux of anions. The depolarization gates voltage-activated $Na^+$ channels and thereby triggers action potentials which then gate voltage-activated $Ca^{2+}$ channels at the active zone, and the rise in intraterminal $Ca^{2+}$ leads to enhanced vesicle exocytosis. (**b**) Activation of **presynaptic** transmitter-gated anion or cation channels depolarizes the presynaptic membrane as described above and thereby gates voltage-activated $Ca^{2+}$ channels at the active zone. (**c**) Activation of a **presynaptic** transmitter-gated cation channel with a high $Ca^{2+}$ permeability directly raises intraterminal $Ca^{2+}$ at the active zone and thus leads to $Ca^{2+}$-dependent vesicle exocytosis.

segments, which are believed to form the ionophore and which are linked by a large extracellular loop with 10 conserved cysteines that may build disulfide bridges (Khakh and North 2006).

When comparing transmitter-gated ion channels on a functional level, one can discern between anion and cation channels. The former ones comprise the $GABA_A$ and glycine receptors, which display a rank order of anion selectivity of $I^- > Br^- > Cl^-$ and which are also permeable to $HCO_3^-$. All other ionotropic receptors mentioned here are cation channels, which discriminate rather poorly between various monovalent cations, at least when compared with voltage-gated ion channels. Some transmitter-gated cation channels are not only highly permeable to divalent cations (Rogers and Dani 1995), but also to anions (Khakh and North 2006). As the functional characteristics, in particular the ion selectivities of the ionotropic receptors, are of utmost importance for the presynaptic modulation of transmitter release, the receptors will be first categorized by this parameter and subsequently according to their activating neurotransmitter.

## *2.1 Presynaptic Anion Channels*

### 2.1.1 $GABA_A$ Receptors

$\gamma$-Aminobutyric acid is viewed as the major inhibitory transmitter in the brain, and a large part of its action is mediated by ionotropic GABA receptors. As mentioned above, these receptors are composed of five out of a repertoire of at least 19 different subunits which are named by greek letters ($\alpha_1$ to $\alpha_6$, $\beta_1$ to $\beta_3$, $\gamma_1$ to $\gamma_3$, $\delta$, $\varepsilon$, $\theta$, and $\rho_1$ to $\rho_3$). Most $GABA_A$ receptors are made up of $\alpha$, $\beta$, and $\gamma$ subunits. These subunits are differentially targeted to synaptic and extrasynaptic sites, and depending on the composition of the receptors, they display diverging pharmacological properties (Korpi and Sinkkonen 2006). Thus, synaptic and extrasynaptic $GABA_A$ receptors may not only mediate phasic and tonic inhibition, respectively, they can also be affected by specific agents. Receptors containing $\rho$ subunits are designated $GABA_C$ receptors and are pharmacologically different from $GABA_A$ receptors in terms of agonist and antagonist sensitivity (Zhang et al. 2001). Presynaptic inhibition via $GABA_A$ receptors has been demonstrated for the first time more than 40 years ago at the crayfish neuromuscular junction (Dudel and Kuffler 1961), but unequivocal evidence in support of presynaptic modulation via $GABA_C$ receptors has been missing for a long time (Kirischuk et al. 2003). Although a large number of reports have described the various functions of presynaptic $GABA_A$ receptors (MacDermott et al. 1999), the molecular composition and the pharmacological characteristics of such presynaptic receptors remained largely obscure.

In accordance with the general notion that GABA is an inhibitory transmitter, activation of presynaptic $GABA_A$ receptors has been found to reduce depolarization-evoked release of various transmitters (Table 1). This inhibitory action was mainly observed in conjunction with action potential-evoked release (Cerrito et al. 1998; Jang et al. 2002; Belenky et al. 2003; Axmacher and Draguhn 2004; Axmacher et al. 2004). However, in the case of presynaptic $GABA_A$ autoreceptors at hippocampal synapses, their activation did not only reduce action potential-evoked, but also spontaneous, GABA release (Axmacher and Draguhn 2004), and the underlying mechanism was proposed to be a hyperpolarization.

With monoamine (Bonanno and Raitreri 1987; Cerrito et al. 1998) and glycine (Jang et al. 2002) release, however, activation of presynaptic $GABA_A$ receptors was found to enhance spontaneous, but to inhibit stimulated, release. The effect on monoamine release was accompanied by $GABA_A$ receptor-mediated increases

**Table 1** Examples of Presynaptic $GABA_A$ and Glycine Receptors

| Receptor | Spont. | $Na^+$ | $Ca^{2+}$ | Stim. | Transmitter | Location | References |
|---|---|---|---|---|---|---|---|
| $GABA_A$ | ↑ | no | yes | n.d. | glutamate | hippocampus | (Jang et al. 2006) |
| $GABA_A$ | ↑ | yes | yes | ↑ / ↓ | glutamate | hippocampus | (Jang et al. 2005) |
| $GABA_A$ | ↑ | yes | yes | n.d. | glutamate | locus coeruleus | (Koga et al. 2005) |
| $GABA_A$ | n.d. | | | ↑ | glutamate | brain stem | (Turecek and Trussell 2002) |
| $GABA_A$ | ↓ | no | n.d. | ↓ | GABA | hippocampus | (Axmacher and Draguhn 2004; Axmacher et al. 2004) |
| $GABA_A$ | n.d. | | | ↓ | GABA | suprachiasmatic n. | (Belenky et al. 2003) |
| $GABA_A$ | ↑ | no | yes | ↓ | glycine | spinal cord | (Jang et al. 2002) |
| $GABA_A$ | ↑ | n.d. | n.d. | ↓ | NA, DA, 5-HT | cerebellum | (Cerrito et al. 1998) |
| GlyR | ↑ | no | yes | ↑ | glutamate | brain stem | (Turecek and Trussell 2001) |
| GlyR | ↑ | yes | yes | ↑ | GABA | VTA | (Zheng and Johnson 2001; Ye et al. 2004) |
| GlyR | ↓ | | | ↓ | GABA | VTA | (Ye et al. 2004) |
| GlyR | ↑ | yes | yes | ↑ | glycine | spinal cord | (Jeong et al. 2003) |

$Na^+$, are voltage-activated $Na^+$ channels involved?; $Ca^{2+}$, are voltage-activated $Ca^{2+}$ channels involved?; ↑, increase; ↓, decrease; n.d., not determined; NA, noradrenaline; DA, dopamine; VTA, ventral tegmental area.

in $Ca^{2+}$ entry, which were mimicked by depolarizing $K^+$ concentrations (Cerrito et al. 1998). The increase in spontaneous glycine release from spinal cord neurons triggered by $GABA_A$ agonists was abolished by the addition of $Cd^{2+}$ and by the removal of extracellular $Ca^{2+}$, thus indicating that $Ca^{2+}$ entry via voltage-gated $Ca^{2+}$ channels was involved. However, blockade of action potentials by tetrodotoxin had no such effect (Jang et al. 2002). Together, these results indicate that the $GABA_A$ receptor mediated a depolarization of the nerve terminals. $GABA_A$ receptor activation was also found to enhance depolarization-evoked noradrenaline release, but this latter effect may not only involve an activation of $GABA_A$ receptors, but also an action at neurotransmitter transporters (Bonanno et al. 1989; Rosenstein et al. 1990). The functions of $GABA_A$ receptors are modulated by various agents, such as benzodiazepines and barbiturates (Rudolph and Möhler 2004). Accordingly, the facilitation of noradrenaline release via presynaptic $GABA_A$ receptors was reported to be enhanced by diazepam (Martire et al, 2002).

In hippocampal mossy fibers, activation of presynaptic $GABA_A$ receptors augmented basal, but decreased depolarization-evoked, increases in intraaxonal $Ca^{2+}$ concentrations. An antagonist at these receptors left basal $Ca^{2+}$ unchanged, but also reduced depolarization-dependent rises in $Ca^{2+}$ (Ruiz et al. 2003). These results suggested (1) that these presynaptic $GABA_A$ receptors are tonically activated, (2) that they mediate a depolarizing action, and (3) that activation of presynaptic $GABA_A$ receptors may exert opposite effects on transmitter release depending on the concentration of the agonist, as discussed in Section 2.2.3, for presynaptic kainate receptors (see Schmitz et al. 2001).

In the locus coeruleus (Koga et al. 2005) and the hippocampus (Jang et al. 2006), presynaptic $GABA_A$ receptor activation facilitated spontaneous glutamate release via $Ca^{2+}$ influx through voltage-gated $Ca^{2+}$ channels. At Schaffer collaterals, presynaptic $GABA_A$ receptors were also found to mediate a facilitation of spontaneous glutamate release in a tetrodotoxin-sensitive and $Ca^{2+}$-dependent manner, and in that case a $GABA_A$ receptor-mediated increase in action potential-evoked release was also observed (Jang et al. 2005). An enhancement of stimulated glutamate release via $GABA_A$ receptors was also observed at the giant presynaptic nerve terminals of the calyx of Held (Turecek and Trussell 2002). Most recently, evidence has been presented that presynaptic $GABA_C$ receptors control synaptic transmission in the retina and thereby contribute to the fine-tuning of the light response (Sagdullaev et al. 2006).

### 2.1.2 Glycine Receptors

Glycine receptors are ligand-gated anion channels closely related to $GABA_A$ receptors. They are composed of $\alpha$ and $\beta$ subunits. Currently, at least four different $\alpha$ subunits ($\alpha1$ to $\alpha4$) and one $\beta$ subunit are known, and alternative splicing of these may contribute to additional heterogeneity. $\alpha$ And $\beta$ subunits form hetero-oligomers with a stoichiometry of $2\alpha : 3\beta$, but $\alpha$ subunits can also form homomeric receptors. As with $GABA_A$ receptors, glycine receptor subunits are targeted to synaptic and extrasynaptic sites in order to mediate phasic and tonic effects, respectively (Betz and Laube 2006).

Presynaptic glycine receptors were found to operate at several synapses (Table 1). In the spinal cord, for instance, presynaptic glycine autoreceptors mediated a facilitation of spontaneous glycine release in a tetrodotoxin- and $Cd^{2+}$-sensitive manner, which is indicative of presynaptic depolarization. An enhancing effect was also observed with action potential-evoked glycine release (Jeong et al. 2003).

At the calyx of Held in the medial nucleus of the trapezoid body, activation of presynaptic glycine receptors was also shown to facilitate spontaneous glutamate release and to enhance amplitudes of evoked excitatory postsynaptic currents (Turecek and Trussell 2001). The action on spontaneous postsynaptic currents were insensitive towards a blockade of $Na^+$ channels by tetrodotoxin, but abolished by the $Ca^{2+}$ channel blocker $Cd^{2+}$. This indicated that glycine acted by causing a depolarization, which was also corroborated by direct voltage measurements.

In the ventral tegmentum of adult rats, glycine was originally found to reduce the frequency of spontaneous GABAergic inhibitory postynaptic potentials (Zheng and Johnson 2001). In more recent experiments on dopaminergic ventral tegmental neurons of newborn rats, however, activation of presynaptic glycine receptors by either glycine or taurine augmented spontaneous as well as evoked GABA release. These effects were antagonized by strychnine, which on its own caused effects opposite to those of the agonists. The glycine-induced increase in spontaneous GABA release was tetrodotoxin- and $Cd^{2+}$-sensitive, thus indicating a role for glycine receptor-mediated depolarizations. In older animals, glycine inhibited spontaneous as well as evoked GABA release, and strychnine exerted reverse actions (Ye et al. 2004). Hence, the actions of presynaptic glycine receptors may be developmentally regulated.

## *2.2 Presynaptic Cation Channels*

### 2.2.1 Nicotinic Acetycholine Receptors

Nicotinic acetylcholine receptors (nAChRs) are acetylcholine-gated cation channels composed of five subunits, each having four membrane spanning domains. Currently, at least 17 different nicotinic receptor subunits are known which are named by Greek letters ($\alpha_1$ to $\alpha_{10}$, $\beta_1$ to $\beta_4$, $\gamma$, $\delta$, $\varepsilon$). The $\alpha_1$, $\beta_1$, $\gamma$, $\delta$, $\varepsilon$ subunits are found in the musculature, whereas all other subunits are present in the nervous system. The acetylcholine binding sites of the receptors involve the N-termini of $\alpha$ subunits, and they can either assemble to form functional homomers or coassemble with $\beta$ subunits at a 2:3 ratio to build heteromeric receptors (Lukas et al. 1999; Gotti et al. 2006). The most widely expressed nAChR subunits in the brain are $\alpha_4$, $\alpha_7$, and $\beta_2$ and the predominating native nicotinic receptor in the central nervous system is an $\alpha_4/\beta_2$ containing heteropentamer.

In the peripheral (Wessler 1989) as well as central (Wonnacott 1997) nervous system, presynaptic nicotinic autoreceptors were reported to control the release of acetylcholine. In both locations, the consequence of presynaptic nAChR activation most commonly is an increase in both spontaneous and evoked acetylcholine release (MacDermott et al. 1999), whereas presynaptic muscarinic receptors mediate the opposite effect, an autoinhibition. Recent studies have focused on the composition of presynaptic nAChRs (Table 2). In the hippocampus, nicotinic autoreceptors were suggested to be $\alpha_3/\beta_4$ receptors (Tani et al. 1998), but a role of $\beta_2$ subunits has also been implicated (Lloyd et al. 1998). Likewise, in the neocortex, presynaptic nicotinic autoreceptors are likely to be $\alpha_4/\beta_2$ receptors (Marchi et al. 2002). In contrast, in the interpeduncular nucleus the autoreceptors were suggested to mainly contain $\alpha_3$ and $\beta_4$ subunits (Grady et al. 2001).

In addition to the autoreceptor function as described above, presynaptic nAChRs were reported to operate as heteroreceptors, which control the release of various transmitters including glutamate and GABA (Table 2). Early evidence for a presynaptic enhancement of glutamatergic neurotransmission by nicotine acting at nAChRs had been obtained in the medial habenula (McGehee et al. 1995) and in the hippocampus (Gray et al. 1996). There, short applications of nicotine enhanced spontaneous as well as action potential-evoked excitatory postsynaptic currents in a transient $Ca^{2+}$-dependent and tetrodotoxin-insensitive manner, and the receptors involved were suggested to be $\alpha_7$ (Radcliffe and Dani 1998). Subsequently, presynaptic nAChRs containing $\alpha_4/\beta_3$ subunits were found to enhance spontaneous glutamate release in the hippocampal CA1 region (Alkondon and Albuquerque 2002). A role of $\alpha_7$ nAChRs in the presynaptic regulation of glutamate release was

**Table 2** Examples of Presynaptic nAChRs

| Receptor | Spont. | $Na^+$ | $Ca^{2+}$ | Stim. | Transmitter | Location | References |
|---|---|---|---|---|---|---|---|
| α2β4 | ↑ | n.d. | n.d. | n.d. | ACh | hippocampus | (Lloyd et al. 1998) |
| α3β4 | ↑ | n.d. | n.d. | n.d. | ACh | interpeduncular n. | (Grady et al. 2001) |
| α4β2 | ↑ | n.d. | n.d. | n.d. | ACh | hippocampus | (Tani et al. 1998) |
| α4β2 | ↑ | n.d. | n.d. | n.d. | ACh | cortex | (Marchi et al. 2002) |
| nAChR | ↑ | no | n.d. | n.d. | glutamate | area postrema | (Funahashi et al. 2004b) |
| α4β3 | ↑ | no | n.d. | n.d. | glutamate | hippocampus | (Alkondon and Albuquerque 2002) |
| α7 | ↑ | yes | n.d. | n.d. | glutamate | cerebellum | (Reno et al. 2004) |
| α7 | ↑ | no | n.d. | ↑ | glutamate | amygdala | (Alkondon et al. 1996; Barazangi and Role 2001) |
| α7 | ↑ | no | n.d. | ↑ | glutamate | hippocampus | (Radcliffe and Dani 1998) |
| α7 | ↑ | yes | n.d. | n.d. | glutamate | hippocampus | (Barik and Wonnacott 2006) |
| α7 | ↑ | n.d. | n.d. | ↑ | aspartate | cortex | (Marchi et al. 2002; Rousseau et al. 2005) |
| α4β2 | ↑ | no | yes/no | ↑ | GABA | thalamus | (Lena and Changeux 1997) |
| non-α7 | ↑ | no | yes | ↑ | GABA | locus coeruleus | (Seddik et al. 2006) |
| non-α7 | ↑ | yes | yes | n.d. | GABA | interpeduncular n. | (Lena et al. 1993) |
| α4β2 | ↑ | no | n.d. | n.d. | GABA | subst. gelatinosa | (Rashid et al. 2006) |
| α4β2 | ↑ | yes | n.d. | n.d. | GABA | hippocampus | (Alkondon and Albuquerque 2001) |
| α4β2 | ↑ | no | no | n.d. | glycine | subst. gelatinosa | (Kiyosawa et al. 2001) |
| α3β4 | ↑ | yes | n.d. | n.d. | NA | hippocampus | (Sacaan et al. 1995; Luo et al. 1998) |
| α4β2 | ↑ | n.d. | n.d. | n.d. | DA | striatum, olfactory tubercle, n. accumbens, cortex | (Grady et al. 2002) |
| α3β2 | ↑ | no | n.d. | n.d. | DA | striatum | (Sacaan et al. 1995; Luo et al. 1998) |
| α3β4* | ↑ | n.d. | n.d. | n.d. | DA | hippocampus | (Cao et al. 2005b) |
| α4β2* | ↑ | no | n.d. | n.d. | DA | prefrontal cortex | (Cao et al. 2005a) |
| α6β2β3<br>α4α6β2β3<br>α4β2<br>α4α5β2 | ↑ | n.d. | n.d. | n.d. | DA | striatum | (Luetje 2004; Salminen et al. 2004; Cao et al. 2005a) |
| α7 | ↑ | n.d. | n.d. | n.d. | DA | striatum | (Kaiser and Wonnacott 2000) |
| α4β2* | ↑ | no | n.d. | n.d. | 5-HT | striatum | (Reuben and Clarke 2000) |

*additional unidentified subunits are involved; $Na^+$, are voltage-activated $Na^+$ channels involved?; $Ca^{2+}$, are voltage-activated $Ca^{2+}$ channels involved?; ↑, increase; ↓, decrease; n.d., not determined; ACh, acetylcholine; NA, noradrenaline; DA, dopamine.

confirmed by indirect means in experiments originally determining noradrenaline release (Barik and Wonnacott 2006). Presynaptic facilitation of glutamate release via nAChRs has also been observed in the area postrema (Funahashi et al. 2004b), in the amygdala (Barazangi and Role 2001), in the cerebellum (Reno et al. 2004), and in olfactory bulb neurons (Alkondon et al. 1996).

In the hippocampus, activation of presynaptic nAChRs was also reported to enhance spontaneous GABAergic synaptic currents, and depending on the nature of the postsynaptic neuron, the receptors involved were either $\alpha_7$ or $\alpha_4/\beta_2$ (Alkondon and Albuquerque 2001). Before, nAChRs that did not contain $\alpha_7$ subunits were found to facilitate spontaneous GABA release in a tetrodotoxin-sensitive manner in the interpeduncular nucleus (Lena et al. 1993). In the thalamus, nAChRs containing $\beta_2$ subunits enhanced spontaneous and stimulation-evoked GABA release. The former effect was tetrodotoxin-insensitive, and depending on the types of neurons investigated, it was either $Cd^{2+}$-sensitive or not (Lena and Changeux 1997). At synapses between olfactory bulb and amygdala neurons, presynaptic nAChRs also mediated an increase in spontaneous as well as action potential-evoked GABA release (Barazangi and Role 2001). In the spinal cord, spontaneous GABA release onto preganglionic sympathetic neurons was augmented by acetylcholine in a $Cd^{2+}$-sensitive, but not tetrodotoxin-sensitive, manner. The receptor subtype involved was devoid of $\alpha_7$ subunits and also mediated an increase in evoked synaptic GABA responses (Seddik et al. 2006). In substantia gelatinosa neurons of the lumbar spinal cord, activation of $\alpha_4/\beta_2$ receptors facilitated not only spontaneous GABA (Rashid et al. 2006), but also glycine (Kiyosawa et al. 2001) release in a tetrodotoxin-insensitive manner.

A facilitation of release via presynaptic nAChRs was also observed for catecholamines. About 40 years ago, Loffelholz (1970) showed that acetylcholine and nicotinic agonists applied to sympathetically innervated organs caused noradrenaline release (see, e.g., Starke 1977; Fuder and Muscholl 1995). More recently, activation of nAChRs was found to stimulate noradrenaline release in the hippocampus (Sacaan et al. 1995). The noradrenergic input of the hippocampus originates in the locus coeruleus, and locus coeruleus neurons with large somata projecting to the hippocampus were reported to express nAChRs mainly comprised of $\alpha6\beta3\beta2$ subunits. The remaining small-diameter locus coeruleus neurons, in contrast, contain predominantly $\alpha3\beta4$ subunits (Lena et al. 1999). However, evidence has been presented that these latter receptor types were also mediating noradrenaline release in the hippocampus (Luo et al. 1998). Finally, presynaptic nAChRs that regulate noradrenaline release in the hippocampus were found to be composed of differing subunits in different species (Azam and McIntosh 2006).

The control of dopamine release via presynaptic nAChRs has been investigated in great detail. Originally, modulation of dopamine release via nAChRs has been revealed in striatal synaptosomes (Rapier et al. 1990). More recently, the receptors involved have been documented to contain $\alpha_4$, $\alpha_5$, $\alpha_6$, $\beta_2$, and $\beta_3$ subunits (Luetje 2004; Salminen et al. 2004). Although $\alpha_7$ containing receptors were found to contribute to the facilitation of dopamine release, this was due to an indirect effect mediated by increases in glutamate release (Kaiser and Wonnacott 2000). In other brain areas, including the prefrontal cortex (Cao et al. 2005a), the hippocampus (Cao et al. 2005b), the olfactory tubercle, and the thalamus (Grady et al. 2002), presynaptic nAChRs were also reported to control dopamine release. In addition, presynaptic regulation of 5-HT release via nAChRs has also been observed (Reuben and Clarke 2000).

### 2.2.2 5-HT$_3$ Serotonin Receptors

Among the seven families of serotonin receptors, only the 5-HT$_3$ receptors are ligand-gated ion channels, whereas all others are G protein-coupled receptors (Barnes and Sharp 1999). 5-HT$_3$ receptors are rather unselective cation channels with a high permeability for $Ca^{2+}$ (Ronde and Nichols 1998). In the central nervous system, they are expressed abundantly in the neocortex, the anterior olfactory nucleus, the hippocampus, and the amygdala. In addition, 5-HT$_3$ receptors are present in the caudate putamen and the nucleus accumbens, and, at high levels, in the brain stem (Parker et al. 1996; Morales et al. 1998; Chameau and van Hooft 2006).

In rodents, two different isoforms of the 5-HT$_3$ receptor exist, termed 5-HT$_{3A}$ and 5-HT$_{3B}$. While the majority of 5-HT$_3$ receptors, in particular in the central nervous system, seem to be homomers of 5-HT$_{3A}$ subunits, heteromers may exist as well, particularly in the peripheral nervous system. Nevertheless, the expression and functional relevance of 5-HT$_{3B}$ subunits remain to be resolved (Morales and Wang 2002; van Hooft and Yakel 2003). In addition, several novel subunits have been cloned from other species, including humans, but their functional relevance still needs to be established (Karnovsky et al. 2003; Niesler et al. 2003).

In the hippocampus and the amygdala, 5-HT$_3$ receptors are expressed in a subset of GABAergic neurons (Kawa 1994; Morales et al. 1996; Koyama et al. 2000; Koyama et al. 2002; Katsurabayashi et al. 2003; Turner et al. 2004; Somogyi and Klausberger 2005; Dorostkar and Boehm 2007) and were found to control GABA release (Table 3). Receptor activation transiently increased the frequency of miniature inhibitory postsynaptic currents reflecting a presynaptic facilitation of spontaneous GABA release (Kawa 1994; McMahon and Kauer 1997; Koyama et al. 2000; Koyama et al. 2002; Turner et al. 2004; Dorostkar and Boehm 2007). In the hippocampus (Turner et al. 2004; Dorostkar and Boehm 2007), but not in the amygdala (Koyama et al. 2000), this effect was sensitive to a blockade of voltage-gated $Ca^{2+}$ channels by $Cd^{2+}$. In contrast to spontaneous release, presynaptic 5-HT$_3$ receptor activation was reported to inhibit action-potential-dependent GABA release (Dorostkar and Boehm 2007). In cortical neurons, activation presynaptic 5-HT$_3$ receptors also facilitated spontaneous GABA release (Zhou and Hablitz 1999; Ferezou et al. 2002; Puig et al. 2004). A $Cd^{2+}$-insensitive facilitation of glutamate release due to presynaptic 5-HT$_3$ receptor activation has been observed in the area postrema (Funahashi et al. 2004a).

In the cortex, 5-HT$_3$ receptor activation was shown to inhibit stimulation-evoked acetylcholine release (Barnes et al. 1989; Crespi et al. 1997), thus corroborating

**Table 3** Examples of Presynaptic 5-$HT_3$ Feceptors

| Receptor | Spont. | $Na^+$ | $Ca^{2+}$ | Stim. | Transmitter | Location | References |
|---|---|---|---|---|---|---|---|
| 5-$HT_3$ | ↑ | no | no | n.d. | glutamate | area postrema | (Funahashi et al. 2004a) |
| 5-$HT_3$ | ↑ | no | yes | n.d. | GABA | hippocampus | (Kawa 1994; McMahon and Kauer 1997; Koyama et al. 2000) |
| 5-$HT_3$ | ↑ | no | no | n.d. | GABA | hippocampus | (Turner et al. 2004) |
| 5-$HT_3$ | ↑ | no | yes | ↓ | GABA | hippocampus | (Dorostkar and Boehm 2007) |
| 5-$HT_3$ | ↑ | no | no | ↓ | GABA | amygdala | (Koyama et al. 2002) |
| 5-$HT_3$ | ↑ | n.d. | n.d. | n.d. | GABA | cortex | (Zhou and Hablitz 1999) |
| 5-$HT_3$ | n.d. | | | ↓ | ACh | cortex | (Crespi et al. 1997) |
| 5-$HT_3$ | ↑ | no | n.d. | ↑ | NA | hippocampus hypothalamus | (Mongeau et al. 1994) |
| 5-$HT_3$ | ↑ | partly | n.d. | n.d. | DA | striatum | (Blandina et al. 1989) |

$Na^+$, are voltage-activated $Na^+$ channels involved?; $Ca^{2+}$, are voltage-activated $Ca^{2+}$ channels involved?; ↑, increase; ↓, decrease; n.d., not determined; ACh, acetylcholine; NA, noradrenaline; DA, dopamine.

inhibitory effects of presynaptic 5-$HT_3$ receptor activation on evoked release. However, in subsequent experiments an effect of 5-$HT_3$ receptor ligands could not be confirmed (Johnson et al. 1993).

With respect to catecholamines, 5-$HT_3$ receptor activation was found to mediate an enhancement of dopamine release in the striatum (Blandina et al. 1989), but contradictory results have also been presented (Crespi et al. 1997). The question of whether presynaptic 5-$HT_3$ receptors control noradrenaline release also remains controversial: 5-$HT_3$ ligands were reported to either stimulate or inhibit noradrenaline release in perfused hearts (Fozard and Ali 1978; Göthert and Dührsen 1979) and to increase evoked noradrenaline release in brain slices (Mongeau et al. 1994). However, these results must be interpreted with caution as 5-$HT_3$ ligands may interfere with presynaptic $\alpha_2$ autoreceptors (Allgaier et al. 1995).

Similarly, presynaptic 5-$HT_3$ receptors were suggested to control 5-HT release in various brain areas (Blier and Bouchard 1993), but this has also been disputed afterwards (Crespi et al. 1997). Taken together, with the exception of consistent reports concerning the GABAergic system, the possible roles and functions of presynaptic 5-$HT_3$ receptors remains somewhat controversial (van Hooft and Vijverberg 2000).

### 2.2.3 Ionotropic Glutamate Receptors

As mentioned above, members of the superfamily of ionotropic glutamate receptors are tetramers composed out of a repertoire of at least 18 subunits. They can be categorized into three families: N-methyl-D-aspartate (NMDA) receptors, $\alpha$-amino-3-hydroxy-5-methyl-4-isoxazolepropioinic acid (AMPA) receptors, and kainate receptors, the two latter forming the group of non-NMDA receptors. While NMDA receptors are composed of NR1, −2 (A, B, C, and D), and −3 (A and B) subunits, AMPA receptors are made up of GlurA to GlurD (or Glur1 to Glur4) and kainate receptors of Glur5 to Glur7 and KA-1 and −2 subunits (Dingledine et al. 1999; Madden 2002). In addition, there are glutamate receptor δ1 and δ2 subunits, the roles of which in the function of ionotropic glutamate receptors are only beginning to be elucidated (Hirai et al. 2003). Subsequently, we will separately deal with presynaptic ionotropic glutamate receptors: NMDA, AMPA, and kainate. Thereafter, a role of presynaptic glutamate transporters will be discussed in brief.

## NMDA Receptors

Among the ionotropic glutamate receptors, the NMDA receptors are characterized by co-activation by glutamate and glycine, slow gating, weak desensitization, a voltage-dependent block of the ionophore by $Mg^{2+}$, and a high $Ca^{2+}$ permeability (Dingledine et al. 1999). This latter feature renders this receptor subtype particularly interesting with respect to the presynaptic modulation of transmitter release (Table 4). In fact, NMDA receptors were reported to act as presynaptic autoreceptors mediating a tonic facilitation of spontaneous glutamate release in the entorhinal cortex (Berretta and Jones 1996). This effect was tetrodotoxin-insensitive and involved receptors containing the NR2B subunit. Evoked glutamate release is also controlled by presynaptic NMDA receptors. They appear to mediate a frequency-dependent facilitation: evoked excitatory postsynaptic currents are enhanced when the stimulation frequency is raised from 0.5 Hz to 3 Hz, and this enhancement is blocked by NMDA receptor antagonists (Berretta and Jones 1996; Woodhall et al. 2001). In cortical neurons, NR2B containing presynaptic NMDA receptors mediate akin actions on glutamate release (Sjostrom et al. 2003). Interestingly, the function of presynaptic NMDA autoreceptors declines with increasing age (Yang et al. 2006). At Schaffer collaterals, presynaptic NMDA receptors containing NR2A subunits were also found to positively modulate excitatory postsynaptic potentials (Suarez et al. 2005; Suarez and Solis 2006). Moreover, a facilitation of glutamate release via presynaptic NMDA autoreceptors was confirmed in hippocampal synaptosomes (Breukel et al. 1998).

However, presynaptic NMDA autoreceptors may also mediate opposite effects, i.e., a reduction of glutamate release. At parallel fiber Purkinje cell synapses, activation of presynaptic NMDA receptors caused significant reductions in excitatory postsynaptic currents (Casado et al. 2002). Likewise, spontaneous and evoked glutamate release from the nerve endings of primary afferents in the spinal cord was reduced when presynaptic NMDA receptors were activated (Bardoni et al. 2004). In contrast to this, substance P release at such synapses was shown to be enhanced by the activation of presynaptic NMDA receptors (Liu et al. 1997).

Presynaptic NMDA receptors have also been reported to control GABA release. At interneuron synapses onto Purkinje neurons, activation of NMDA receptors raised spontaneous GABA release, and this effect lasted for more than 10 minutes; voltage-gated $Ca^{2+}$ channels were not involved, but a blockade of ryanodine receptors attenuated the enhancement of GABA release and restricted the effect by time

**Table 4** Examples of Presynaptic NMDA and AMPA Receptors

| Receptor | Spont. | $Na^+$ | $Ca^{2+}$ | Stim. | Transmitter | Location | References |
|---|---|---|---|---|---|---|---|
| NR2A | n.d. | | no | ↑ | glutamate | hippocampus | (Suarez et al. 2005; Suarez and Solis 2006) |
| NMDA | | | | ↑ | glutamate | hippocampus | (Breukel et al. 1998) |
| NR1/NR2B | ↑ | no | yes | ↑ | glutamate | entorhinal cortex | (Berretta and Jones 1996; Woodhall et al. 2001) |
| NR2B | ↑ | no | n.d. | ↑ | glutamate | cortex | (Sjostrom et al. 2003) |
| NMDA | n.d. | | | ↓ | glutamate | cerebellum | (Casado et al. 2002) |
| NMDA | ↓ | no | n.d. | ↓ | glutamate | dorsal horn | (Bardoni et al. 2004) |
| NMDA | ↑ | no | no | n.d. | GABA | cerebellum | (Duguid and Smart 2004) |
| NMDA | ↑ | no | | ↓ | GABA | *Xenopus* retinotectum | (Lien et al. 2006) |
| NMDA | ↑ | yes | n.d. | n.d. | NA | cortex | (Fink et al. 1989) |
| NMDA | ↑ | n.d. | n.d. | n.d. | DA | striatum | (Petrov et al. 2001) |
| NR2A KA | ↑ | no | n.d. | n.d. | ACh | zebrafish endplate | (Todd et al. 2004) |
| NMDA | ↑ | n.d. | n.d. | n.d. | substance P | dorsal horn | (Liu et al. 1997) |
| AMPA | n.d. | | | ↓ | glutamate | calyx of Held | (Takago et al. 2005) |
| AMPA | n.d. | | | ↓ | glutamate | dorsal horn | (Lee et al. 2002a) |
| AMPA GluR5 | ↑ | yes | n.d. | n.d. | GABA | hippocampus | (Vignes 2001) |
| AMPA | ↑ | no | n.d. | ↓ | GABA | cerebellum | (Satake et al. 2004; Rusakov et al. 2005) |
| AMPA | ↑ | no | no | ↓ | GABA, glycine | dorsal horn | (Engelman et al. 2006) |
| AMPA, NMDA | ↑ | no | n.d. | ↑ | NA | spinal cord | (Sundström et al. 1998) |

$Na^+$, are voltage-activated $Na^+$ channels involved?; $Ca^{2+}$, are voltage-activated $Ca^{2+}$ channels involved?; ↑, increase; ↓, decrease; n.d., not determined; ACh, acetylcholine; NA, noradrenaline; DA, dopamine.

to last for less than 2 minutes only (Duguid and Smart 2004). Interestingly, a depolarization of the postsynaptic Purkinje neurons also led to the activation of the presynaptic NMDA receptors, indicating that these can be activated by glutamate as a retrograde messenger. This latter phenomenon was called depolarization-induced potentiation of inhibition (Duguid and Smart 2004). In the retinotectal system of *Xenopus* tadpoles, presynaptic NMDA receptor activation enhanced spontaneous GABA release, as reflected by an increase in the frequency of miniature inhibitory postsynaptic currents; simultaneously, amplitudes of action potential-dependent inhibitory postsynaptic currents were reduced (Lien et al. 2006).

NMDA receptor ligands were also observed to control the release of catecholamines via a presynaptic site of action. Noradrenaline release in the cortex and hippocampus, for instance, was enhanced by NMDA agonists, and this effect was augmented by glycine (Fink et al. 1989; Pittaluga and Raiteri 1990). In the spinal cord, NMDA also facilitated noradrenaline release in a partially tetrodotoxin-sensitive manner (Sundström et al. 1998). In the striatum, NMDA agonists were found to increase dopamine release (Petrov et al. 2001).

## AMPA Receptors

In contrast to NMDA receptors, AMPA receptors gate instantaneously, desensitize rapidly, and are much less permeable to $Ca^{2+}$ (Dingledine et al. 1999). Despite these characteristics, which are rather unfavorable for mediating a presynaptic modulation, various actions of presynaptic AMPA receptors have been described (Table 4). Activation of AMPA autoreceptors has been detected to cause a presynaptic inhibition of excitatory postsynaptic currents at the calyx of Held synapse in the brainstem (Takago et al. 2005). In cultured hippocampal neurons, AMPA receptor activation tonically facilitated spontaneous GABA release in the absence of tetrodotoxin (Vignes 2001). In contrast, GABA release in the cerebellum was reduced when presynaptic AMPA heteroreceptors were activated (Satake et al. 2004; Rusakov et al. 2005). At synapses on dorsal horn neurons of the spinal cord, however, AMPA receptor activation increased the spontaneous release of GABA and glycine in a tetrodotoxin- and $Cd^{2+}$-insensitive manner. Evoked inhibitory postsynaptic currents were simultaneously inhibited by AMPA receptor activation (Engelman et al. 2006). At the very same location, presynaptic AMPA receptor activation was reported to also inhibit evoked glutamate release (Lee et al. 2002a).

Release of transmitters other than glutamate or GABA has also been found to be subject to a regulation via presynaptic AMPA receptors. Noradrenaline release, as one example, was reported to be enhanced when presynaptic AMPA receptors were activated (Sundström et al. 1998), and evidence has been presented that the release of other amines is also regulated by AMPA receptors (Ohta et al. 1994).

## Kainate Receptors

In terms of functional characteristics, kainate receptors are similar to AMPA receptors, displaying rapid gating and desensitization. The major difference lies in the recovery from desensitization, which takes about 10 times longer for kainate receptors (Dingledine et al. 1999). For kainate receptors, a large variety of presynaptic functions have been described (Table 5).

Initial evidence for presynaptic kainate autoreceptors was obtained 25 years ago (Ferkany et al. 1982). Although they were originally found to stimulate glutamate release in the hippocampus, later evidence suggested that these presynaptic receptors were inhibitory rather than stimulatory (Chittajallu et al. 1996). It is now clear that hippocampal presynaptic kainate autoreceptors work in a sophisticated manner: at the very same synapse, nanomolar concentrations of kainate receptor agonists enhanced glutamate release, whereas micromolar concentrations elicited an inhibition (Schmitz et al. 2001). Furthermore, within the hippocampus the consequences of presynaptic kainate receptor activation depended on the synapse under investigation: Evoked glutamate release from mossy fibers onto CA3 neurons was inhibited via receptors containing GluR6 subunits, whereas glutamate release from the perforant path onto these neurons was facilitated via receptors with GluR5 and GluR6 subunits. Spontaneous glutamate release onto CA3 neurons was facilitated via receptors

**Table 5** Examples of Presynaptic Kainate Receptors

| Receptor | Spont. | $Na^+$ | $Ca^{2+}$ | Stim. | Transmitte | Location | References |
|---|---|---|---|---|---|---|---|
| GluR5 | n.d. | | | ↓ | glutamate | hippocampus | (Partovi and Frerking 2006) |
| GluR5, | ↓ | n.d. | n.d. | ↓ | glutamate | dorsal horn | (Kerchner et al. 2001b; Kerchner et al. 2002) |
| GluR6/(5) KA2 | n.d. | n.d. | n.d. | ↑ / ↓ | glutamate | hippocampus | (Schmitz et al. 2001; Contractor et al. 2003) |
| GluR6 | ↔ | | | ↓ / ↑ | glutamate | hippocampus | (Contractor et al. 2000) |
| GluR5 | ↑ | no | n.d. | ↑ | | | |
| GluR6 | n.d. | | | ↓ | glutamate | hippocamus | (Partovi and Frerking 2006) |
| GluR5/GluR6 or GluR5/KA2 | ↓ | no | | ↓ | GABA | hippocampus | (Clarke et al. 1997; Rodriguez-Moreno et al. 1997; Rodriguez-Moreno and Lerma 1998; Christensen et al. 2004) |
| non-GluR5 | ↑ | no | no | ↑ | GABA | hippocampus | (Cossart et al. 2001) |
| GluR5, AMPA | ↑ | yes | n.d. | n.d. | GABA | hippocampus | (Vignes 2001) |
| KA | ↑ | no | n.d. | ↑ | GABA | hypothalamus | (Liu et al. 1999) |
| GluR5 | ↑ | no | yes | ↑ / ↓ | GABA, glycine | dorsal horn | (Kerchner et al. 2001a; Xu et al. 2006) |
| KA GluR5 | ↑ / ↓ | no | no | ↑ / ↓ | GABA | basolat. amygdala | (Braga et al. 2003) |

$Na^+$, are voltage-activated $Na^+$ channels involved?; $Ca^{2+}$, are voltage-activated $Ca^{2+}$ channels involved?; ↑, increase; ↓, decrease; n.d., not determined.

containing GluR5 subunits (Contractor et al. 2000). Additional investigations also tried to identify the receptor subunits involved in the presynaptic modulation of glutamate release via kainate receptors, and evidence has been presented for a role of both GluR5 (Lauri et al. 2003; Partovi and Frerking 2006) and GluR6 (Perkinton and Sihra 1999) subunits. In cortical synaptosomes, activation of kainate receptors with GluR6 subunits facilitated evoked glutamate release (Perkinton and Sihra 1999). In the cerebellum, presynaptic kainate receptors located on parallel fibers mediate differing effects depending on the postsynaptic neuron: with Purkinje cells only a facilitation of evoked glutamate release was observed; with stellate cells, moderate receptor activation caused an enhancement and strong receptor activation an inhibition (Delaney and Jahr 2002). In the spinal cord, spontaneous and evoked glutamate release from primary afferent fibers was decreased by presynaptic kainate receptor activation, and the receptors involved were suggested to contain GluR5 subunits (Kerchner et al. 2001b; Kerchner et al. 2002).

Presynaptic kainate receptors have also been reported to control GABA release and to mediate opposing effects. In the hippocampus, spontaneous and evoked release of GABA was originally found to be inhibited by kainate (Clarke et al. 1997; Cunha et al. 1997; Rodriguez-Moreno et al. 1997; Rodriguez-Moreno and Lerma 1998; Clarke and Collingridge 2004). Later on, however, low kainate receptor agonist concentrations were reported to facilitate inhibitory postsynaptic currents, while high concentrations caused the opposite effect (Jiang et al. 2001). The presynaptic receptor subtypes involved were suggested to be heteromers of either GluR6 or KA2 subunits together with GluR5 subunits (Christensen et al. 2004). A facilitation of GABA release via presynaptic kainate receptors was also observed at synapses between different interneurons in the hippocampus, and both spontaneous and evoked GABA release were affected (Cossart et al. 2001). The effect of activated kainate receptors on spontaneous inhibitory synaptic currents in the hippocampus depended on the functions of voltage-gated sodium channels: the frequency of spontaneous inhibitory postsynaptic currents in the absence of tetrodotoxin was increased by kainate receptor activation, while the frequency of miniature inhibitory postsynaptic currents was decreased (Maingret et al. 2005). Apparently contradictory effects of kainate receptor agonists on GABA release can be explained by two distinct receptor subtypes being involved: an inhibitory receptor with high affinity for glutamate and a stimulatory receptor with lower affinity (Rodriguez-Moreno et al. 2000).

In dorsal horn neurons, spontaneous release of GABA and glycine was facilitated by kainate in a tetrodotoxin-insensitive but $Cd^{2+}$-sensitive manner. Evoked inhibitory postsynaptic currents were also modulated biphasically: low concentrations of kainate caused a facilitation, whereas higher concentrations caused an inhibition. This inhibition, however, was suggested to involve $GABA_B$ receptors (Kerchner et al. 2001a). The facilitatory mechanism has been shown to be mediated by GluR5 containing receptors (Xu et al. 2006).

In the basolateral amygdala, the effect of presynaptic GluR5 receptor activation also depended on the agonist concentration: at low concentrations, spontaneous and evoked GABA release was facilitated, the first in a tetrodotoxin- and $Cd^{2+}$-insensitive manner; at high agonist concentrations, both effects were reversed to inhibition. The receptors involved were suggested to be various GluR5-containing heteromers (Braga et al. 2003). In hypothalamic neurons, presynaptic kainate receptor activation facilitated spontaneous and evoked GABA release (Liu et al. 1999).

Presynaptic kainate receptors control not only the release of glutamate and GABA.. In the striatum, for instance, GluR6 activation facilitated adenosine release, which then acted onto $A_{2A}$ receptors to inhibit evoked GABA release, whereas spontaneous GABA release was directly facilitated via kainate receptors (Chergui et al. 2000). At neuromuscular junctions of zebrafish larvae, activation of NR2A and kainate receptors increased spontaneous acetylcholine release (Todd et al. 2004).

## Excitatory Amino Acid Transporters

The postsynaptic actions of transmitters being released from presynaptic nerve terminals depend not only on the kinetics and the extent of vesicle exocytosis, but also on the functions of reuptake mechanisms, which are provided inter alia by transporter proteins embedded in the presynaptic plasma membrane. At glutamatergic synapses, reuptake is mediated by members of the family of excitatory amino acid transporters (Amara and Fontana 2002). Interestingly enough, all neurotransmitter transporters mediate not only the reuptake of previously released transmitters, but simultaneously provide a significant ion conductance. Accordingly, currents through these proteins can be measured, and therefore these transporters can also be viewed as transmitter-gated ion channels (Gerstbrein and Sitte 2006). Excitatory amino acid transporters 4 and 5 provide particularly large $Cl^-$ conductances (Amara and Fontana 2002), and the latter protein is found in the retina. There, the excitatory amino acid transporter 5 was found to be activated by endogenously released glutamate and to mediate $Cl^-$-dependent changes in the electrical properties of rod bipolar cells that led to a reduction in transmitter release (Veruki et al. 2006; Wersinger et al. 2006). Hence, the presynaptic excitatory amino acid transporter can be included in the list of ionotropic "receptors."

### 2.2.4 ATP P2X Receptors

P2X receptors are transmitter-gated ion channels activated by micromolar concentrations of ATP that consist of three out of a repertoire of seven subunits. The P2X subunits may form homomeric and/or heteromeric receptors, which are characterized by their ligand specificity and current kinetics (North 2002) and which all display a high $Ca^{2+}$-permeability (Khakh and North 2006), thus providing these receptors with a function important for presynaptic modulation (Table 6). First evidence for a role of presynaptic P2X receptors in the regulation of transmitter release was obtained in the spinal cord (Gu and MacDermott 1997) and brain stem (Khakh and Henderson 1998), respectively. In both locations, application of ATP raised the frequency of spontaneous excitatory postsynaptic currents, and this effect was not prevented when $Na^+$-carried action potentials were blocked. In the brain stem, the activation of presynaptic P2X receptors enhanced spontaneous glutamate release even when voltage-activated $Ca^{2+}$ channels were blocked (Khakh and Henderson 1998), and these results were later corroborated in the spinal cord (Nakatsuka and Gu 2001). There, evoked glutamate release onto dorsal horn neurons was also facilitated by presynaptic P2X receptor activation, and the blockade of ATPases which degrade endogenously released ATP was sufficient achieve receptor activation (Nakatsuka and Gu 2001).

In the brain stem, the ATP facilitation of spontaneous glutamate release in a tetrodotoxin- and $Cd^{2+}$-insensitive, but $Ca^{2+}$-dependent, manner was mediated by presynaptic $P2X_3$ homomers (Jin et al. 2004; Shigetomi and Kato 2004). In addition, ATP reduced evoked glutamate release, but this was due to its degradation towards adenosine and subsequent activation of inhibitory $A_1$ receptors (Kato and Shigetomi 2001). In hypoglossal motoneurons (Ireland et al. 2004) and cortical synaptosomes (Patti et al. 2006), activation of P2X7 receptors was reported to facilitate glutamate release.

**Table 6** Examples of Presynaptic P2X Receptors

| Receptor | Spont. | $Na^+$ | $Ca^{2+}$ | Stim. | Transmitter | Location | References |
|---|---|---|---|---|---|---|---|
| P2X | ↑ | no | yes | n.d. | glutamate | trigeminal mesencephalic n. | (Khakh and Henderson 1998) |
| P2X | ↑ | no | no | ↑ | glutamate | dorsal horn | (Nakatsuka and Gu 2001) |
| P2X | ↑ | yes | n.d. | n.d. | glutamate | spinal cord | (Gu and MacDermott 1997) |
| P2X2 | ↑ | no | n.d. | n.d. | glutamate | hippocampus | (Khakh et al. 2003) |
| P2X3 | ↑ | no | no | ↔ | glutamate | n. tractus solitarius | (Kato and Shigetomi 2001; Jin et al. 2004) |
| P2X7 | n.d. | | | ↓ | glutamate | hippocampus | (Armstrong et al. 2002) |
| P2X7 | ↑ | n.d. | n.d. | n.d. | glutamate, aspartate | neocortex | (Patti et al. 2006) |
| P2X7 | ↔ | | n.d. | ↑ | glutamate | hypoglossus motoneurons | (Ireland et al. 2004) |
| P2X2 | ↑ | no | no | n.d. | glycine | subst. gelatinosa | (Rhee et al. 2000) |
| P2X2 | ↑ | partially | partially | n.d. | NA | SCG | (Boehm 1999) |
| P2X7 | ↑ | n.d. | yes | n.d. | ACh | neuromuscular junction | (Fu and Huang 1994; Moores et al. 2005) |

$Na^-$, are voltage-activated $Na^+$ channels involved?; $Ca^{2+}$, are voltage-activated $Ca^{2+}$ channels involved?; ↑, increase; ↓, decrease; n.d., not determined; NA, noradrenaline; ACh, acetylcholine; SCG, superior cervical ganglion.

In the hippocampus, presynaptic $P2X_2$ receptors mediated increases in the frequencies of miniature excitatory postsynaptic currents in stratum radiatum interneurons, but not in pyramidal neurons (Khakh et al. 2003). Glutamate release from hippocampal mossy fibers, in contrast, was found to be inhibited by $P2X_7$ receptor activation (Armstrong et al. 2002).

Evidence for the regulation of the release of inhibitory transmitters via presynaptic P2X receptors has also been obtained. In the spinal cord, $P2X_2$ receptor activation facilitated spontaneous glycine release in a tetrodotoxin- and $Cd^{2+}$-insensitive, but $Ca^{2+}$-dependent, manner (Rhee et al. 2000).

ATP is a well-established co-transmitter to noradrenaline in the central (Poelchen et al. 2001) as well as peripheral (von Kugelgen and Starke 1991) nervous system. Therefore, the functions of presynaptic P2X receptors have been investigated in various preparations containing noradrenergic nerve terminals (Cunha and Ribeiro 2000). Early evidence for a stimulation and positive feedback modulation of noradrenaline via presynaptic P2X receptors has been obtained in sympathetic neurons (Boehm 1999). At the neuromuscular junction, ATP stimulated the release of acetylcholine (1994), and this effect was suggested to be mediated by presynaptic $P2X_7$ receptors (Moores et al. 2005).

### 2.2.5 Vanilloid Receptors (TRPV)

The family of mammalian vanilloid receptors (TRPV) has at least six members and is part of the superfamily of transient receptor potential (TRP) ion channels. TRPV1 is a $Ca^{2+}$-permeable ion channel that is gated by capsaicin and other vanilloids, by protons, and by heat (Liedtke and Kim 2005). It is thus involved in transmitting sensory information. Accordingly, initial evidence for presynaptic vanilloid receptors has been obtained in sensory neurons (Table 7). The spontaneous release of glutamate from axon terminals of these neurons in the spinal cord was enhanced by capsaicin in a tetrodotoxin-insensitive way (Yang et al. 1998), whereas evoked release was inhibited (Yang et al. 1999). The increase in spontaneous glutamate release was also observed when other activators of TRPV1 were used instead of capsaicin (Yue et al. 2004). After peripheral inflammation, but not in naïve tissue, a TRPV1 antagonist exerted effects opposite to those mentioned above, thus indicating that the presynaptic vanilloid receptors were activated by some endogenous mechanism (Lappin et al. 2006).

In the brain stem, TRPV1 receptor activation, similarly to $P2X_3$ receptor activation (see above), facilitated spontaneous glutamate release in a tetrodotoxin- and $Cd^{2+}$-insensitive manner, even though the two presynaptic receptors were located on different primary afferent fibres (Jin et al. 2004). Action potential-dependent glutamate release, in contrast, was reduced by TRPV1 activation (Doyle et al. 2002). In paraventricular neurons, TRPV1 receptor activation facilitated spontaneous glutamate release in a tetrodotoxin- and $Cd^{2+}$-insensitive way, but in that case evoked glutamate release was enhanced too (Li et al. 2004).

According to the above reports, only glutamate release is subject to a modulation by presynaptic vanilloid receptors, as GABA release, for instance, was not affected (Yang et al. 1998). In apparent contradiction, activation of vanilloid receptors was

**Table 7** Examples of Presynaptic Vannilloid Receptors

| Receptor | Spont. | $Na^+$ | $Ca^{2+}$ | Stim. | Transmitter | Location | References |
|---|---|---|---|---|---|---|---|
| TRPV1 | ↑ | no | n.d. | ↑ | glutamate | subst. gelatinosa | (Lappin et al. 2006) |
| TRPV1 | ↑ | no | n.d. | ↓ | glutamate | subst. gelatinosa | (Yang et al. 1998; Yang et al. 1999; Yue et al. 2004) |
| TRPV1 | ↑ | no | no | ↑ | glutamate | paraventricular nucleus | (Li et al. 2004) |
| TRPV1 | ↑ | no | no | ↓ | glutamate | nucleus tractus solitarius | (Doyle et al. 2002; Jin et al. 2004) |
| TRPV1 | ↑ | no | n.d. | ↓ | GABA | dorsal motor nucleus of vagus | (Derbenev et al. 2006) |
| TRPV1 | ↑ | n.d. | n.d. | n.d. | CGRP | trigeminal sensory neurons | (Price et al. 2004) |
| TRPV1 | ↑ | n.d. | n.d. | n.d. | CGRP | DRG | (Ahluwalia et al. 2003) |

$Na^+$, are voltage-activated $Na^+$ channels involved?; $Ca^{2+}$, are voltage-activated $Ca^{2+}$ channels involved?; ↑, increase; ↓, decrease; n.d., not determined; CGRP, calcitonin gene related peptide; DRG, dorsal root ganglion.

reported to control the spontaneous release of GABA in the dorsal motor nucleus of the vagus. However, these effects were blocked by glutamate receptor antagonists and were thus mediated by an enhancement of glutamate release (Derbenev et al. 2006). In cultures of sensory neurons, however, TRPV1 receptor activation was shown to facilitate the release of a transmitter other than glutamate, namely calcitonin gene-related peptide (Ahluwalia et al. 2003).

# 3 Mechanisms of Action

## *3.1 Spontaneous and Stimulated Release*

In principle, the activation of a presynaptic ionotropic receptor may either increase or decrease the amount of transmitter being released. However, the effect may depend on the situation prior to the activation of the receptor. In this context one has to mention that transmitter release occurs either due to spontaneous fusion of vesicles with the plasma membrane or as a consequence of an event that triggers increases in the intracellular $Ca^{2+}$ concentration at the active zone where vesicle exocytosis takes place. The $Ca^{2+}$ concentrations required to promote exocytosis towards maximal rates lie between approximately 10 and 100 μM (Augustine 2001; Schneggenburger and Neher 2005), depending on the synapse being investigated. These two types of transmitter release are usually named spontaneous and stimulated (or stimulation-evoked) release, respectively.

Several stimulation paradigms are used to experimentally promote transmitter release: (1) the triggering of action potentials by short (micro- to millisecond) depolarizations evoked via intra- or extracellular electrodes; (2) the application of depolarizing $K^+$ concentrations: this usually takes several seconds, but can be optimized to also determine $K^+$-evoked release on a subsecond time scale (Turner and Dunlap 1995); (3) the application of drugs that cause depolarization through an opening of $Na^+$ channels (Bouron and Reuter 1996) or via a blockade of $K^+$ channels (Raffaelli et al. 2004); (4) flash photolysis of intracellular caged $Ca^{2+}$ (Heidelberger et al. 1994). With the former three types of manipulations, the stimulation will promote exocytotic transmitter release only in the presence of extracellular $Ca^{2+}$ and thus depends on transmembrane $Ca^{2+}$ entry via voltage-gated $Ca^{2+}$ channels (Heidelberger et al. 1994). In the latter case, however, stimulated exocytosis is independent of transmembrane $Ca^{2+}$ entry pathways and thus of the local arrangement of ion channels. Likewise, spontaneous vesicle exocytosis is independent of extracellular $Ca^{2+}$ and transmembrane $Ca^{2+}$ entry (Scanziani et al. 1992; Scholz and Miller 1992). Hence, the difference between spontaneous and stimulated release is the $Ca^{2+}$ dependence of the latter.

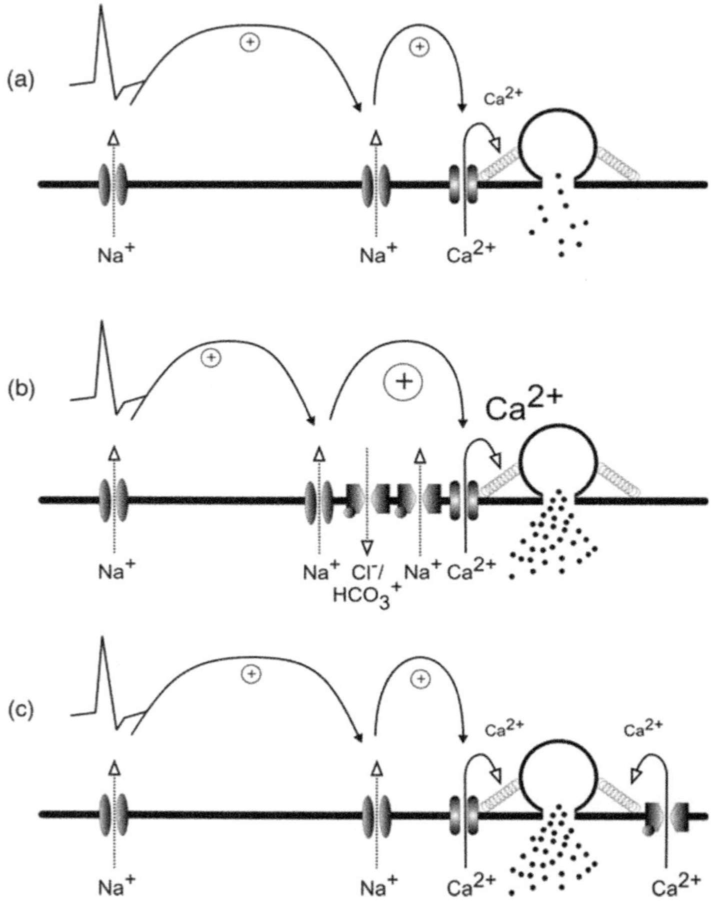


**Fig. 3** Facilitation of stimulated transmitter release caused by the activation of ionotropic receptors. (**a**) Action potentials are propagated via voltage-activated $Na^+$ channels to the presynaptic active zone. There, voltage-activated $Ca^{2+}$ channels are gated, and the rise in intraterminal $Ca^{2+}$ leads to $Ca^{2+}$-dependent vesicle exocytosis. (**b**) Concurrent activation of presynaptic transmitter-gated anion or cation channels, together with an invading action potential, enhances the depolarization of the presynaptic membrane, which leads to the increased gating of voltage-activated $Ca^{2+}$ channels. The additional rise in intraterminal $Ca^{2+}$ at the active zone may cause further increases in $Ca^{2+}$-dependent vesicle exocytosis. (**c**) Concurrent activation of presynaptic transmitter-gated cation channels with high $Ca^{2+}$ permeabilities together with invading action potentials leads to an additional rise in intraterminal $Ca^{2+}$ at the active zone which may cause further increases in $Ca^{2+}$-dependent vesicle exocytosis.

## *3.2 Excitation-Secretion Coupling*

Under physiological conditions, release of neurotransmitters is initiated by a propagating action potential that invades the nerve terminal and triggers vesicle exocytosis

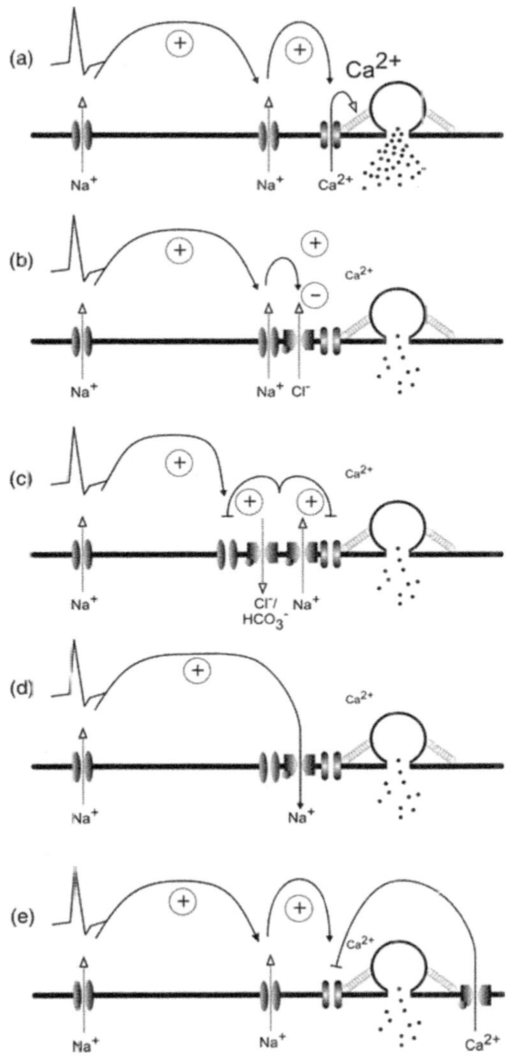


**Fig. 4** Inhibition of stimulated transmitter release caused by the activation of ionotropic receptors. (**a**) Action potentials are propagated via voltage-activated $Na^+$ channels to the presynaptic active zone. There, voltage-activated $Ca^{2+}$ channels are gated, and the rise in intraterminal $Ca^{2+}$ leads to $Ca^{2+}$-dependent vesicle exocytosis. (**b**) Activation of presynaptic transmitter-gated anion channels hyperpolarizes the presynaptic membrane through anion influx thereby counteracting the depolarization of the invading action potential. This may reduce or even prevent the gating of voltage-activated $Ca^{2+}$ channels at the active zone. (**c**) Activation of presynaptic transmitter-gated anion or cation channels depolarizes the presynaptic membrane and thereby inactivates voltage-activated $Na^+$ and/or $Ca^{2+}$ channels. Consequently, action potential propagation and/or $Ca^{2+}$ influx at the active zone is prevented. (**d**) Activation of presynaptic transmitter-gated anion or cation channels shunt an invading action potential as the membrane resistance is too low. (**e**) Activation of presynaptic transmitter-gated cation channels with high $Ca^{2+}$ permeabilities raise the intraterminal $Ca^{2+}$ levels and thereby inactivate voltage-activated $Ca^{2+}$ channels, which reduces or prevents action potential-evoked $Ca^{2+}$ entry at the active zone.

within less than 1 millisecond. There are three crucial steps involved in this excitation-secretion coupling (Figures 3a and 4A): (1) the transient depolarization of the nerve terminal by a $Na^+$-carried action potential; (2) the concomitant transmembrane $Ca^{2+}$ entry due to the opening of voltage-gated $Ca^{2+}$ channels; (3) the rise of free intracellular $Ca^{2+}$ up to high micromolar concentrations which trigger vesicle exocytosis (Augustine 2001; Schneggenburger and Neher 2005). These $Ca^{2+}$ channels are clustered at active zones, and after opening, high $Ca^{2+}$ concentrations accumulate in the proximity of the vesicle fusion machinery (Zhai and Bellen 2004; Schneggenburger and Neher 2005). However, $Ca^{2+}$ elevations due to individual action potentials do not saturate the $Ca^{2+}$ sensor of the vesicle fusion machinery, as sustained micromolar $Ca^{2+}$ concentrations may trigger transmitter release more effectively than $Ca^{2+}$ rises produced by presynaptic action potentials. This and the high cooperativity of $Ca^{2+}$ ions in eliciting vesicle exocytosis suggest that a minor inhibition of voltage-gated $Ca^{2+}$ channels may have pronounced effects on transmitter release (Schneggenburger and Neher 2005). In addition, during a millisecond action potential the presynaptic membrane is just briefly depolarized. Therefore, an only slight inhibition of the gating of $Ca^{2+}$ channels may profoundly affect the presynaptic $Ca^{2+}$ elevation (Garcia et al. 1998). Accordingly, appropriate $Ca^{2+}$ channel blockers reduce or abolish stimulation-evoked release (e.g., Hirning et al. 1988; Scanziani et al. 1992), and so do all other mechanisms that hamper the gating of voltage-gated $Ca^{2+}$ channels involved in excitation-secretion coupling.

Various subtype-specific $Ca^{2+}$ channel blockers were found to be efficient inhibitors of stimulation-evoked transmitter release. For example, the release of monoamines in the hippocampus is greatly reduced by ω-conotoxin GVIA (Dooley et al. 1987), which selectively blocks $Ca_V$ 2.2 $Ca^{2+}$ channels, whereas the release of glutamate involves both, $Ca_V$ 2.2 and $Ca_V$ 2.1 channels (Wheeler et al. 1994). In contrast, the release of GABA at hippocampal synapses is either abolished by a $Ca_V$ 2.2 channel blocking conotoxin, or by a $Ca_V$ 2.1 channel blocking agatoxin, depending on the synapse (Poncer et al. 1997). In addition, there is increasing evidence to suggest that transmitter release may also involve $Ca_V$ 2.3 channels (Kamp et al. 2005). At any rate, $Ca_V$ 2, but not $Ca_V$ 1 or $Ca_V$ 3 $Ca^{2+}$ channels are involved in stimulated transmitter release, and all the members of the $Ca_V$ 2 family interact directly with proteins involved in vesicle exocytosis (Spafford and Zamponi 2003; Kamp et al. 2005). Disruption of the interaction between $Ca_V$ 2 channels and SNARE proteins reduces depolarization-evoked transmitter release (Mochida et al. 1996; Mochida et al. 2003). This suggests that a physical link between the route of transmembrane $Ca^{2+}$ entry and the vesicle exocytosis machinery is important for action potential-dependent transmitter release.

$K^+$ channels are responsible for setting the resting membrane potential and for the repolarizing phase of action potentials. In addition, $K^+$ channels mediate afterhyperpolarizations to terminate periods of high neuronal activity and to modulate firing rates. When located in presynaptic nerve terminals, these actions of $K^+$ channels will contribute to the regulation of transmitter release. In fact, most types of the huge superfamily of $K^+$ channels, in particular delayed rectifier, fast transient, and $Ca^{2+}$-sensitive $K^+$ channels, have been found in a variety of nerve terminals and there contribute to transmitter release (Meir et al. 1999). In contrast, G protein-gated inward rectifier $K^+$ channels (GIRK) were reported to mediate hyperpolarizations at somatodendritic regions, but not in presynaptic axon terminals. More recently, transmitter release was suggested to be under the control of presynaptic KCNQ $K^+$ channels (Martire et al. 2004), although contradictory evidence has also been presented (Kristufek et al. 1999a; Lechner et al. 2003). Interestingly, certain $K^+$ channel subtypes have also been found to interact with proteins involved in vesicle exocytosis (Michaelevski et al. 2002).

The actions of voltage-activated ion channels in excitation-secretion coupling as described in Section 3.2, will help to reason about signaling mechanisms employed by presynaptic ionotropic receptors. Receptors, the activation of which mimics one of the functions of voltage-activated channels, will mediate the same effect as that channel. In contrast, if receptor activation interferes with one of the actions of the voltage-sensitive channels, the result will be the opposite.

## 3.3 *Ionotropic Mechanisms*

Generally, opening of an ion channel located at axon terminals, i.e., directly at or near the sites of vesicle exocytosis, provides an increase in the membrane conductance and may thus cause changes in the membrane potential, i.e., hyper- or depolarization. Depolarization may exert multiple actions on spontaneous as well as stimulated release. With respect to spontaneous release, one has to consider that increasing depolarization will bring the membrane potential closer to and then beyond the activation threshold of voltage-activated $Na^+$ and $Ca^{2+}$ channels (Figure 2a). Accordingly, increases in depolarizing $K^+$ concentrations lead to increases in the rate of vesicle exocytosis and transmitter release, whether $Na^+$ channels are blocked or not (Kristufek et al. 1999a; Dorostkar and Boehm 2007). Only at moderate (25 mM), but not at high (40 mM), $K^+$ concentrations, the blockade of voltage-gated $Na^+$ channels by, e.g., tetrodotoxin may cause a reduction in $K^+$-evoked transmitter release (Kristufek et al. 1999a). This indicates that larger depolarizations per se block action potential propagation ("depolarization block"), but nevertheless permit ongoing $Ca^{2+}$-triggered vesicle exocytosis which can be abolished by $Ca^{2+}$ channel inhibitors or by the removal of extracellular $Ca^{2+}$. Hence, if activation of presynaptic ionotropic receptors depolarizes nerve terminals to an extent that is sufficient to open voltage-dependent $Na^+$ and/or $Ca^{2+}$ channels, this will finally cause transmembrane $Ca^{2+}$ entry and thereby elicit $Ca^{2+}$-dependent exocytosis (Figure 2b). For presynaptic ligand-gated cation channels, there is an additional possibility of how to raise spontaneous release. If the ionophores of the receptors are sufficiently $Ca^{2+}$-permeable (Rogers and Dani 1995), their activation will lead to $Ca^{2+}$ influx independently of voltage-activated $Ca^{2+}$ channels (Figure 2c), and the rise in intraterminal $Ca^{2+}$ promotes $Ca^{2+}$-dependent exocytosis.

The mechanisms by which activation of presynaptic ligand-gated cation channels elicits depolarization are obvious: at negative membrane potentials, a more or less unselective cation conductance leads to influx of positive charges, since permeant cations are equally distributed at both sides of the membrane and thus have an equilibrium potential of about 0 mV. The situation is more complex with regard to ligand-gated anion channels. Whether the activation of an anion, more precisely a $Cl^-$, conductance results in a hyper- or depolarization depends on the intracellular $Cl^-$ concentration and the resulting $Cl^-$ equilibrium potential. With low intracellular $Cl^-$, the equilibrium potential is negative to the resting membrane potential. Accordingly, a $Cl^-$ conductance hyperpolarizes the membrane. With high intracellular $Cl^-$, however, the equilibrium potential is positive to the resting membrane potential. Then, the Cl-conductance depolarizes the membrane (e.g., Staley 1992; Owens et al. 1996). If the $Cl^-$ equilibrium potential is more positive than the threshold for action potential firing, activation of a $Cl^-$ conductance leads to the generation of action potentials (Staley 1992). In spinal cord neurons, for example, inhibition of intracellular $Cl^-$ accumulation (by the $Na^+/K^+/2Cl^-$ cotransporter blocker bumetanide) reduced the enhancement of spontaneous glycine release via $GABA_A$ receptors (Jang et al. 2002). This corroborates the importance of high intracellular $Cl^-$ for stimulatory actions of presynaptic ligand-gated anion channels. In this study, however, evoked glycine release was reduced by presynaptic $GABA_A$ receptor activation. Hence, activation of a $Cl^-$ conductance may have diverging effects on the resting membrane potential and on action potentials, respectively. This might be explained by the fact that an active $Cl^-$ conductance, even with high intracellular $Cl^-$, provides a $Cl^-$ influx as soon as the membrane voltage during the action potential exceeds the reversal potential for $Cl^-$. Such a transient $Cl^-$ influx counteracts the full discharge of action potentials and may thereby provide an inhibitory effect. In addition, a depolarization may per se cause opposing effects on spontaneous and action potential-evoked transmitter release, as described further below.

Considering that high intracellular $Cl^-$ concentrations are a prerequisite for the depolarizing action mediated by ligand-gated anion channels, one needs to clarify how intracellular $Cl^-$ is controlled. During the postnatal development of the nervous system, various growth factors drive the expression of the $Cl^-$-extruding $Cl^-/K^+$ co-transporter KCC2. About two weeks after birth, this switches the depolarizing action of $GABA_A$ receptor activation to a hyperpolarizing one (Rivera et al. 1999; Rivera et al. 2005). Nevertheless, even at later developmental stages, depolarizing $GABA_A$ receptor-mediated effects have been observed. However, in these cases the permeant anion determining the nature of the $GABA_A$ response is $HCO_3^-$ rather than $Cl^-$, and the intracellular supply of $HCO_3^-$ is mediated by cytosolic carbonic anhydrases (Rivera et al. 2005).

Alternatively, opening of ligand-gated anion channels can elicit membrane hyperpolarizations, and these may lead to an inhibition of transmitter release (Figure 4b). In theory, such an action increases the difference between the resting membrane voltage and the activation threshold of voltage-gated $Na^+$ or $Ca^{2+}$ channels. In accordance, 20 mV hyperpolarizations rendered single rat sympathetic neurons less excitable, and they fired less action potentials in response to scaled current injections. Nevertheless, despite such hyperpolarizations the overall amount of action potential-mediated transmitter release from these neurons remained unchanged (Lechner et al. 2003).

With respect to spontaneous transmitter release, neither increases in the membrane conductance nor hyerpolarizations can be expected to cause significant changes, as the rate of vesicle exocytosis is only determined by the intracellular $Ca^{2+}$ concentration but not by the membrane potential (Felmy et al. 2003).

Opening of an ion channel located at axon terminals provides an increase in the membrane conductance independently of the charge of the permeating ions. If the increase in the membrane conductance is sufficiently large, action potentials might become shunted and the propagation of an action potential could be hampered or even prevented (Figure 4d). This mechanism could provide an inhibitory effect on action potential-dependent transmitter release, whether the transmitter-gated ion channel is anionic or cationic. However, the conductance changes caused by the activation of $GABA_A$ receptors were found to exert only a minor impact on action potential propagation (Zhang and Jackson 1995). Likewise, simulations of action potentials in synaptic boutons suggested that mere conductance increases are unlikely to shunt action potentials but likely to mediate changes in transmitter release by other mechanisms (Graham and Redman 1994). Nevertheless, there is also evidence in support of the notion that shunting of action potentials may represent the predominant mechanism in the inhibition of action potential-dependent transmitter release (Cattaert and El Manira 1999).

One issue remains to be discussed: how does depolarization affect action potential-evoked transmitter release? As the membrane is increasingly depolarized from −80 to −70, −60, and −50 mV, $Ca_V$ 2 channels become slightly activated and permit tiny $Ca^{2+}$ currents and small increases in intraterminal $Ca^{2+}$ concentrations (Figure 3b). This phenomenon does not require activation of $Na^+$ channels. As a consequence of the intraterminal $Ca^{2+}$ increase, action potential-dependent transmitter release is enhanced with increasing depolarizations (Awatramani et al. 2005). However, if the depolarization exceeds a certain threshold, this facilitatory effect is converted into an inhibition. This becomes evident from the fact that the addition of up to 5 mM $K^+$ caused an increase in action potential-evoked glutamate release at hippocampal synapses, whereas higher $K^+$ concentrations (8 to 12 mM) caused an inhibition (Schmitz et al. 2001). Likewise, action potential-evoked GABA release from hippocampal neurons remained largely unaltered when 10 mM $K^+$ were added, but was almost abolished by the addition of 20 mM $K^+$ (Dorostkar and Boehm 2007). Thus, depolarizations by up to about 30 mV do not prevent action potential propagation and may lead to an increase in elicited transmitter release due to small increases in intraterminal $Ca^{2+}$ (Figure 3b). Larger depolarizations, however, lead to a voltage-dependent inactivation of $Na^+$ channels and/or to a voltage- and $Ca^{2+}$-dependent inactivation of $Ca^{2+}$ channels, and both mechanisms result in an inhibition of action potential-evoked transmitter release (Figure 4c).

In addition to causing presynaptic depolarization, activation of presynaptic ionotropic receptors may provide another way of transmembrane $Ca^{2+}$ entry in order to promote $Ca^{2+}$-dependent vesicle exocytosis: the $Ca^{2+}$-permeable ionophores of transmitter-gated cation channels (Rogers and Dani 1995). In this latter case, $Ca^{2+}$ entering the nerve terminal via an ionotropic receptor may bypass voltage-gated $Ca^{2+}$ channels and directly stimulate $Ca^{2+}$-dependent vesicle exocytosis. If action potentials invade while such $Ca^{2+}$ permeable receptors are activated, at least two different scenarios can be envisaged: (1) the $Ca^{2+}$ entering the axon terminal via an ionotropic receptor can sum up with the $Ca^{2+}$ entering via voltage-gated $Ca^{2+}$ channels; if the $Ca^{2+}$ sensor for vesicle exocytosis is not saturated (see Section 3.3), this results in an enhancement of action potential-evoked transmitter release (Figure 3c). (2) the $Ca^{2+}$ entering via the ionotropic receptor can lead to $Ca^{2+}$-dependent inactivation of the voltage-gated $Ca^{2+}$ channels, and the combined $Ca^{2+}$ influx may then be less than the $Ca^{2+}$ entry provided by only the voltage-gated channels; such a mechanism results in an inhibition of action potential-evoked release (Figure 4e). (3) A similar effect can be achieved if the depolarization caused by the activation of the ligand-gated ion channel causes voltage-dependent inactivation of either $Na^+$ or $Ca^{2+}$ channels (Figure 4c).

## 3.4 Metabotropic Mechanisms

Although this review deals with ionotropic receptors, we need to consider some signaling mechanisms associated with G protein-coupled receptors (GPCRs), as these may also be employed by some so-called ionotropic receptors, such as AMPA and kainate receptors. Presynaptic inhibition via GPCRs frequently operates via changes in excitation-secretion coupling. It has been recognized quite some time ago that voltage-gated $Ca^{2+}$ and $K^+$ channels are regulated by a variety of G proteins (Brown and Birnbaumer 1990). A large number of neurotransmitters and agonists at GPCRs have been found to inhibit voltage-gated $Ca^{2+}$ channels in various types of neurons (Hille 1994; Elmslie 2003), and many GPCRs that were found to mediate an inhibition of voltage-dependent $Ca^{2+}$ currents were also reported to cause presynaptic inhibition (Lipscombe et al. 1989; Boehm et al. 1991, 1992; Koh and Hille 1997; Wu and Saggau 1997). The G protein-mediated modulation relevant for presynaptic inhibition targets primarily the channels involved in excitation-secretion coupling, i.e., $Ca_V$ 2 family channels; members of the $Ca_V$ 1 channel family are affected to only a minor extent (Hille 1994; Elmslie 2003). There are two major G protein-linked signaling pathways that control $Ca^{2+}$ channels: one is membrane-delimited, the other is not and rather involves diffusible second messengers. The membrane-delimited pathway involves pertussis toxin-sensitive inhibitory G proteins, achieves full inhibition within one second, slows the activation kinetics of the currents, and is characterized by voltage-dependence (the more the membrane is depolarized, the less the G proteins can inhibit $Ca^{2+}$ channel gating). The underlying mechanism is a direct binding of G protein $\beta\gamma$ subunits to the $\alpha_1$ subunits of voltage-gated $Ca^{2+}$ channels at a 1:1 stoichiometry. Through this interaction, $Ca^{2+}$ channels become unwilling to open, and the current-voltage dependence of the channels is shifted to more positive values (Elmslie 2003; De Waard et al. 2005). The diffusible second messenger pathway involves pertussis toxin-insensitive G proteins of the Gq family, is much slower, does not affect current kinetics, and controls, in addition to $Ca_V$ 2 family channels, the opening of $Ca_V$ 1 channels (Hille 1994). This pathway involves activation of phospholipase C and depletion of membrane phosphatidylinositol 4,5-bisphosphate (Gamper et al. 2004; Lechner et al. 2005; Liu et al. 2006).

The first evidence for metabotropic signaling of a priori ionotropic receptors has been obtained with kainate receptors involved in the inhibition of GABA release in the hippocampus. This action was pertussis toxin-sensitive and appeared to involve phospholipase C and proteinkinase C (Rodriguez-Moreno and Lerma 1998). A role for G proteins in the presynaptic inhibition via kainate receptors has also been described for glutamate release from hippocampal neurons (Frerking et al. 2001), and the signaling cascade was found to be abrogated by pertussis toxin and proteinkinase C inhibitors (Lauri et al. 2005). In independent experiments, GluR5 containing kainate receptors were found to mediate a pertussis toxin-sensitive rise in intracellular $Ca^{2+}$ and an inhibition of voltage-gated $Ca^{2+}$ channels (Rozas et al. 2003). This latter effect involved the diffusible second messenger rather than the membrane-delimited pathway. The kainate receptor-mediated inhibition of evoked glutamate release from hippocampal neurons was shown to be occluded when presynaptic $GABA_B$ and adenosine $A_1$ receptors were activated (Partovi and Frerking 2006). This is in line with the frequently observed phenomenon that the effects of two inhibitory presynaptic receptors are not additive if they share common signaling mechanisms, such as the inhibition of voltage-gated $Ca^{2+}$ channels (Schlicker and Göthert 1998; Kubista and Boehm 2006).

In cortical neurons, activation of AMPA receptors was found to activate inhibitory G proteins, which led to the reduction of forskolin-stimulated cyclic AMP formation (Wang et al. 1997). Later on, AMPA receptor activation in the cerebellum was reported to limit evoked GABA release through an inhibition of voltage-activated $Ca^{2+}$ channels of the $Ca_V$ 2.1 subfamily via N-ethylmaleimide sensitive G proteins (Satake et al. 2004; Rusakov et al. 2005), thus suggesting that the membrane-delimited pathway was involved. Likewise, the inhibition of evoked glutamate release via presynaptic AMPA receptors at the calyx of Held has been shown to involve a G protein-dependent and voltage-sensitive inhibition of voltage-gated calcium channels (Takago et al. 2005), which again favors a role of the fast membrane pathway.

Another example of metabotropic mechanisms involved in the presynaptic inhibition via ionotropic receptors was revealed when the reduction of glutamate release via $P2X_7$ receptors was investigated (Armstrong et al. 2002). The inhibition via this ATP-gated ion channel required an activation of p38 MAP kinase and was suggested to involve an inhibition of voltage-gated $Ca^{2+}$ channels.

Not only the inhibition, but also the facilitation, of transmitter release via presynaptic ionotropic receptors occasionally involves metabotropic mechanisms. The enhancement of evoked glutamate release at mossy fiber-CA3 synapses via presynaptic kainate receptors was shown to involve adenylyl cyclase and proteinkinase A (Rodriguez-Moreno and Sihra 2004). This corroborates that increases in cyclic AMP and concomitant activation of proteinkinase A leads to an enhancement of transmitter release A and raises the question concerning the underlying mechanisms. Since transmembrane $Ca^{2+}$ influx is decisive in excitation-secretion coupling, activation of PKA might operate through an enhancement of voltage-dependent $Ca^{2+}$

currents (e.g., Artalejo et al. 1990). In addition, proteinkinase A may act downstream of $Ca^{2+}$ entry at the level of exocytosis. This enzyme is known to mediate the phosphorylation of some of the SNARE proteins and interaction partners involved in vesicle exocytosis. α-SNAP, for instance, was identified as a good substrate for phosphorylation by protein kinase A, and phosphorylated α-SNAP displayed reduced binding to the SNARE core complex (Hirling and Scheller 1996). Snapin, in contrast, shows increased binding to SNAP-25 when phosphorylated by protein kinase A (Chheda et al. 2001). And the interaction between syntaphilin and syntaxin-1A is inhibited by protein kinase A-dependent phosphorylation (Boczan et al. 2004). All these effects may contribute to the direct modulation of vesicle exocytosis by the cyclic AMP – protein kinase A pathway, which was found to enhance the $Ca^{2+}$ sensitivity of the release machinery and to increase the readily releasable pool of vesicles (Saitow et al. 2005).

Taken together, there is ample evidence that presynaptic ionotropic receptors, in particular kainate receptors, may control transmitter release via signaling mechanisms that are usually employed by GPCRs and involve heterotrimeric GTP binding proteins. The big question as to how ionotropic receptors are coupled to G proteins remained unanswered (Lerma, 2006).

## 4 Interactions Between Ionotropic Receptors

It has been demonstrated for a variety of transmitter-gated ion channels that their functions are regulated by a plethora of other membrane proteins, in most instances GPCRs. Most frequently, this modulation via G proteins involves one or more phosphorylation reactions (Swope et al. 1999). In addition, transmitter-gated ion channels can be controlled by G protein subunits that directly bind to the channel protein (Yevenes et al. 2003). However, the functions of ligand-gated ion channels can also be influenced by membrane receptors independently of cytosolic or membrane-associated signaling cascades: GPCRs have been found to interfere with ionotropic receptors by direct protein-protein interactions (Liu et al. 2000; Lee et al. 2002b). In addition, it has been shown for various ligand-gated ion channels that their gating is directly influenced by other ionotropic receptors.

The latter type of interaction has been demonstrated for the following pairs of ionotropic receptors: P2X and nAChRs (Nakazawa 1994), P2X and $GABA_A$ (Boue-Grabot et al. 2004), ATP and $5HT_3$ (Barajas-Lopez et al. 2002), and $GABA_A$ and glycine receptors (Li and Xu 2002). In general, the opening of one ionotropic receptor was found to impede the gating of the other and vice versa. Such a cross-inhibition was not only found with recombinant receptors (Khakh et al. 2000; Boue-Grabot et al. 2004), but also with receptors natively expressed in neurons (Nakazawa 1994), and even with postsynaptic receptors activated by endogenously released transmitters (Khakh et al. 2000). These interactions can be revealed on a functional level by measuring currents through the receptor ionophores (Nakazawa 1994), but also by biochemical methods (Boue-Grabot et al. 2004) and fluorescence resonance energy transfer (Khakh et al. 2005), which both indicate direct physical interactions. Taken together, it is well established that several ionotropic receptors co-localize in neuronal membranes and at postsynaptic sites and cause mutual inhibition.

It is well known that nerve terminals harbor more than one type of presynaptic receptor. As a consequence, there are multiple interactions between various presynaptic receptors, and the activation of one can positively or negatively interfere with the function of another one. Such interactions were not only described for two different types of presynaptic GPCRs, but also for GPCRs and ionotropic receptors (Kubista and Boehm 2006). Nerve terminals may also possess more than one type of ionotropic receptor. The axon endings of primary sensory afferents in the dorsal horn of the spinal cord, for instance, are known to be equipped with $GABA_A$ receptors and with members of each of the subfamilies of ionotropic glutamate receptors (Rustioni 2005). Nevertheless, reports on functional interactions between these receptors are scarce. The interaction between two different ionotropic receptors that was detected first, between P2X receptors and nAChRs, has also been investigated at the presynaptic level. In contrast to what has been observed with postsynaptic receptors (Khakh and Henderson 2000), activation of presynaptic P2X receptors on neocortical glutamatergic nerve terminals potentiated glutamate release due to the activation of nAChRs and vice versa (Patti et al. 2006). The mechanisms underlying both the inhibitory postsynaptic and the facilitatory presynaptic, interactions remained unknown, and it will be interesting to learn more about the interactions of ionotropic receptors in presynaptic nerve terminals in the future.

## 5 Physiological and Pharmacological Relevance of Presynaptic Ionotropic Receptors

As evident from the large variety of presynaptic ionotropic receptors, one can expect that these impinge on various aspects of higher-order functions in the nervous system. However, it is beyond the scope of this review to give a full account of all the possible implications. We will rather select a few examples to highlight how presynaptic transmitter-gated ion channels may contribute to the integrated organization of the nervous system.

As monosynaptic reflexes were among the first experimental systems in which presynaptic inhibition has been observed (Frank and Fuortes 1957), it is not astonishing that considerable progress has been made in that field. In the spinal cord, the presynaptic nerve terminals of primary afferent fibers that are in direct contact with motoneurons receive themselves a synaptic input from GABAergic interneurons: an axo-axonic synapse (Conradi 1969). There, the opening of $GABA_A$ receptors causes primary afferent depolarization (Eccles et al. 1962) due to $Cl^-$ efflux and concomitant presynaptic inhibition. Tonic presynaptic inhibition via such axo-axonic synapses is reduced at the onset of voluntary movements in humans at the afferents from the muscle-spindles of the muscle that is to be contracted. In contrast, there is a simultaneous increase in presynaptic inhibition at the afferents innervating the muscles that are to remain relaxed (Hultborn et al. 1987). Evidence has been presented that similar presynaptic mechanisms are operating when we walk (Faist et al. 1996). Thus, presynaptic $GABA_A$ receptors are well established as most important entities in daily human life. After spinal cord injuries, a disturbance in the presynaptic inhibition via these receptors is believed to contribute to the clinical symptoms of spasticity (Morita et al. 2001). Accordingly, an increase in the function of presynaptic $GABA_A$ receptors due to the application of benzodiazepines contributes to the well-known muscle relaxant properties of these anxiolytics (Rudolph and Möhler 2004).

Similarly to $GABA_A$ receptors, activation of presynaptic kainate receptors was found to reduce glutamate release from primary afferents in the dorsal horn of the spinal cord (Youn and Randic 2004), where nociceptive stimuli are processed. Although this effect involves GluR5 containing receptors, the nociceptive thresholds in GluR5-deficient mice were not significantly different from those in wild-type animals (Sailer et al. 1999). Nevertheless, responses to inflammatory pain were significantly reduced in these transgenic mice (Ko et al. 2005). In agreement with these latter results, a GluR5-selective antagonist was found to attenuate experimental hyperalgesia in humans (Sang et al. 1998) and proved to be efficacious in acute migraine (Sang et al. 2004).

As indicated above, presynaptic kainate receptors operate not only in the spinal cord but also in the brain, for instance, the hippocampus. The hippocampal receptors are decisive in two different aspects, namely development and synaptic plasticity. In immature hippocampal neurons, the filopodia of axons move rapidly to "search" for their targets, and this movement is terminated by synapse formation. Activation of presynaptic kainate receptors by low glutamate concentrations enhances the filopodia mobility, whereas high concentrations have the opposite effect. As the perisynaptic space becomes tighter and tighter when synapses are formed, the concentration of released glutamate will increase as synapse formation proceeds. Thus, the bidirectional modulation of filopodia mobility via presynaptic kainate receptor was suggested to be essential for synapse formation in the hippocampus (Tashiro et al. 2003). In addition, the opposite effects of presynaptic kainate receptor activation on glutamate release at various hippocampal synapses, together with the modulation of GABA release, determine the network activity in the neonatal hippocampus (Lauri et al. 2005). As such network oscillations are important for the development of proper network circuitry (Zhang and Poo 2001), these actions of presynaptic kainate receptors also determine the coordinated hippocampal development.

In the hippocampus, the mossy fiber synapses onto CA3 pyramidal neurons are characterized by a particularly high degree of synaptic plasticity. On a short-term time scale in the range of seconds, one can observe pronounced paired pulse as well as frequency facilitation. The latter phenomenon reflects a multifold increase in synaptic strength when the stimulation frequency is enhanced and involves activation of presynaptic kainate autoreceptors (Schmitz et al. 2001). On a longer time scale in the range of minutes, one can observe long-term potentiation (LTP). At the mossy fiber synapses, LTP is independent of NMDA receptors, but there is sufficient experimental evidence to suggest that presynaptic kainate recptors do participate (Nicoll and Schmitz 2005).

It is well established that presynaptic nAChRs can function as auto- as well as heteroreceptors. As presynaptic autoreceptors controlling the release of acetylcholine at the neuromuscular junction, nAChRs are believed to be presynaptic amplifiers of neuromuscular transmission (Wessler 1989): by activating these autoreceptors, endogenously released acetylcholine mediates a positive feedback, in particular during pronounced muscle activity. Blockade of the presynaptic nicotinic autoreceptors was also inferred in the phenomenon of titanic fade or train-of-four-fade of neuromuscular transmission which is observed when nAChR blocking agents, such as tubocurarin, are applied (Bowman et al. 1986): when tetanic stimuli or trains of four stimuli are applied to motor nerves, muscle contractions remain rather constant, but decrease in the presence of nAChR antagonists. This is explained by the interruption of the positive feedback which then results in an increasing decline of acetylcholine release (Bowman et al. 1986), but there also are experimental data that are incompatible with this explanation (Wessler 1989).

A major source for nicotine in humans is smoking, and first- as well as second-hand smoke are major causes for a multitude of cardiovascular diseases (Barnoya and Glantz 2005). However, only a few of the various aspects of cardiovasular pathology caused by smoking are believed to be mediated by nicotine (Benowitz 1997). Accordingly, various types of nicotine replacement therapies have been devised which aim at keeping smokers away from cigarettes by delivering nicotine via other routes of administration instead (Henningfield 1995). Nevertheless, nasal and transdermal nicotine still exert powerful effects on the cardiovascular system, and raise, for instance, systolic as well as diastolic blood pressure and the heart rate (Benowitz et al. 2002). As nicotine itself mediates detrimental cardiovascular effects, one has to pose the question as to how they arise: the effects of nicotine are known to be mediated by the sympathetic nervous system (Haass and Kubler 1997), but nAChRs are located in the sympathetic ganglia and at the axon terminals within the innervated tissues (Löffelholz 1970), such as the heart. Hence, nicotine might either enhance the firing rate of the postganglionic sympathetic neurons (by a central effect or by ganglionic stimulation) or directly cause noradrenaline release from the nerve endings. In early investigations related to this topic, smoking was found to raise the blood pressure and the heart rate, but to simultaneously reduce sympathetic nerve activity (Grassi et al. 1994). This appeared to indicate that the effects of nicotine were rather mediated by direct effects on the nerve terminals, i.e., by presynaptic nAChRs. However, subsequent studies revealed that smoking does increase sympathetic nerve activity as long as this phenomenon is not counteracted by the baroreflex (Narkiewicz et al. 1998). Moreover, nicotine uptake by routes other than smoking was also found to stimulate the sympathetic outflow (Wolk et al. 2005). Thus, the effects of nicotine involve central or ganglionic stimulation rather than activation of presynaptic nAChRs on sympathetic axons. This is also indicated by experiments on postganglionic sympathetic neurons, where nicotine was found to first stimulate noradrenaline release via action potential propagation, and to cause a direct presynaptic effect at higher concentrations only (Kristufek et al. 1999b).

## 6 Conclusion

It is now 50 years since first evidence for presynaptic modulation via ligand-gated ion channels was obtained. During the last decade, the list of ionotropic receptors has been growing rapidly, and it does not appear too speculative to predict that many more examples will be added in the future. Members of each of the known families of the transmitter-agetd ion channels have been detected at various types of presynaptic nerve terminals. Their activation may cause an increase in a priori spontaneous vesicle exocytosis and/or a facilitation or inhibition of activity-dependent transmitter release. Thus, the superfamily of ionotropic receptors provides presynaptic regulators, which are as multifunctional and versatile as members of the superfamily of presynaptic GPCRs. Considering that each ionotropic receptor is composed of at least three subunits, and that there is a selection of more than 70 diverging subunits, one can easily imagine that an unpredictably huge number of different presynaptic ionotropic receptors exist which by far outnumber the various presynaptic GPCRs. In light of the numerous GPCRs involved in the presynaptic modulation of transmitter release as summarized in the previous five chapters of this handbook, we can expect that the number of reports concerning presynaptic ionotropic receptors will continue to grow in the future.

**Acknowledgements** Work in the authors' laboratory is supported by the Austrian Science Fund (FWF; P15797, P17611, W1205) and by the Virologiefonds of the Medical University of Vienna.

## References

Ahluwalia J, Urban L, Bevan S, Nagy I (2003) Anandamide regulates neuropeptide release from capsaicin-sensitive primary sensory neurons by activating both the cannabinoid1 receptor and the vanilloid receptor1 in vitro. Eur J Neurosci 17:2611–18

Alkondon M, Albuquerque EX (2001) Nicotinic acetylcholine receptor alpha 7 and alpha 4beta 2 subtypes differentially control GABAergic input to CA1 neurons in rat hippocampus. J Neurophysiol 86:3043–55

Alkondon M, Albuquerque EX (2002) A non-alpha 7 nicotinic acetylcholine receptor modulates excitatory input to hippocampal CA1 interneurons. J Neurophysiol 87: 651–4

Alkondon M, Rocha ES, Maelicke A, Albuquerque EX (1996) Diversity of nicotinic acetylcholine receptors in rat brain. V. alpha-Bungarotoxin-sensitive nicotinic receptors in olfactory bulb neurons and presynaptic modulation of glutamate release. J Pharmacol Exp Ther 278:1460–71

Allgaier C, Warnke P, Stangl AP, Feuerstein TJ (1995) Effects of 5-HT receptor agonists on depolarization-induced [3H]-noradrenaline release in rabbit hippocampus and human neocortex. Br J Pharmacol 116:1769–74

Amara SG, Fontana ACK (2002) Excitatory amino acid transporters: keeping up with glutamate. Neurochem Internat 41:313

Armstrong JN, Brust TB, Lewis RG, MacVicar BA (2002) Activation of presynaptic P2X7-like receptors depresses mossy fiber-CA3 synaptic transmission through p38 mitogen-activated proteink. J Neurosci 22:5938–45

Artalejo CR, Ariano MA, Perlman RL, Fox AP (1990) Activation of facilitation calcium channels in chromaffin cells by D1 dopamine receptors through a cAMP/protein kinase A-dependent mechanism. Nature 348:239–42

Attwell D, Barbour B, Szatkowski M (1993) Nonvesicular release of neurotransmitter. Neuron 11:401–7

Augustine GJ (2001) How does calcium trigger neurotransmitter release? Curr Opin Neurobiol 11:320–6

Awatramani GB, Price GD, Trussell LO (2005) Modulation of transmitter release by presynaptic resting potential and background calcium levels. Neuron 48:109–21

Axmacher N, Draguhn A (2004) Inhibition of GABA release by presynaptic ionotropic GABA receptors in hippocampal CA3. Neuroreport 15:329–34

Axmacher N, Winterer J, Stanton PK, Draguhn A, Muller W (2004) Two-photon imaging of spontaneous vesicular release in acute brain slices and its modulation by presynaptic GABAA receptors. NeuroImage 22:1014

Azam L, McIntosh JM (2006) Characterization of nicotinic acetylcholine receptors that modulate nicotine-evoked [3H]norepinephrine release from mouse hippocampal synaptosomes. Mol Pharmacol 70:967–76

Barajas-Lopez C, Montano LM, Espinosa-Luna R (2002) Inhibitory interactions between 5-HT3 and P2X channels in submucosal neurons. Am J Physiol Gastrointest Liver Physiol 283:G1238–48

Barazangi N, Role LW (2001) Nicotine-induced enhancement of glutamatergic and GABAergic synaptic transmission in the mouse amygdala. J Neurophysiol 86:463–74

Bardoni R, Torsney C, Tong C-K, Prandini M, MacDermott AB (2004) Presynaptic NMDA receptors modulate glutamate release from primary sensory neurons in rat spinal cord dorsal horn. J Neurosci 24: 2774–81

Barik J, Wonnacott S (2006) Indirect modulation by {alpha}7 nicotinic acetylcholine receptors of noradrenaline release in rat hippocampal slices: interaction with glutamate and GABA systems and effect of nicotine withdrawal. Mol Pharmacol 69: 618–28

Barnes JM, Barnes NM, Costall B, Naylor RJ, Tyers MB (1989) 5-HT3 receptors mediate inhibition of acetylcholine release in cortical tissue. Nature 338:762–3

Barnes NM, Sharp T (1999) A review of central 5-HT receptors and their function. Neuropharmacology 38:1083–1152

Barnoya J, Glantz SA (2005) Cardiovascular effects of secondhand smoke: nearly as large as smoking. Circulation 111:2684–98

Belenky MA, Sagiv N, Fritschy JM, Yarom Y (2003) Presynaptic and postsynaptic GABAA receptors in rat suprachiasmatic nucleus. Neuroscience 118:909

Benowitz NL (1997) The role of nicotine in smoking-related cardiovascular disease. Prev Med 26:412–17

Benowitz NL, Hansson A, Jacob P, 3rd (2002) Cardiovascular effects of nasal and transdermal nicotine and cigarette smoking. Hypertension 39:1107–12

Berretta N, Jones RSG (1996) Tonic facilitation of glutamate release by presynaptic N-methyl-D-aspartate autoreceptors in the entorhinal cortex. Neuroscience 75:339

Betz H, Laube B (2006) Glycine receptors: recent insights into their structural organization and functional diversity. J Neurochem 97:1600–10

Blandina P, Goldfarb J, Craddock-Royal B, Green JP (1989) Release of endogenous dopamine by stimulation of 5-hydroxytryptamine3 receptors in rat striatum. J Pharmacol Exp Ther 251:803–9

Blier P, Bouchard C (1993) Functional characterization of a 5-HT3 receptor which modulates the release of 5-HT in the guinea-pig brain. Br J Pharmacol 108:13–22

Boczan J, Leenders AG, Sheng ZH (2004) Phosphorylation of syntaphilin by cAMP-dependent protein kinase modulates its interaction with syntaxin-1 and annuls its inhibitory effect on vesicle exocytosis. J Biol Chem 279:18911–19

Boehm S (1999) ATP stimulates sympathetic transmitter release via presynaptic P2X purinoceptors. J Neurosci 19: 737–46

Boehm S, Drobny H, Singer E (1991) Electrically evoked noradrenaline release from cultured chick sympathetic neurons: modulation via presynaptic α2-adrenoceptors and lack of autoinhibition. Naunyn Schmiedebergs Arch Pharmacol 344:130–2

Boehm S, Huck S, Drobny H, Singer EA (1992) Pertussis toxin abolishes the inhibition of Ca2+ currents and of noradrenaline release via alpha 2-adrenoceptors in chick sympathetic neurons. Naunyn Schmiedebergs Arch Pharmacol 345:606–9

Bonanno G, Raiteri M (1987) Release-regulating GABAA receptors are present on noradrenergic nerve terminals in selective areas of the rat brain. Synapse 1: 254–7

Bonanno G, Fontana G, Fedele E, Robino G, Raiteri M (1989) Presynaptic mechanisms underlying the gamma-aminobutyric acid-evoked receptor-independent release of [3H]norepinephrine in rat hippocampus. J Neurochem 52:1854–8

Boue-Grabot E, Emerit MB, Toulme E, Seguela P, Garret M (2004) Cross-talk and co-trafficking between rho1/GABA receptors and ATP-gated channels. J Biol Chem 279:6967–75

Bouron A, Reuter H (1996) A role of intracellular Na+ in the regulation of synaptic transmission and turnover of the vesicular pool in cultured hippocampal cells. Neuron 17:969–78

Bowman W, Gibb A, Harvey A, Marshall I (1986) Prejunctional actions of cholinoceptor agonists and antagonists, and of anticholinesterase drugs. Handb Exp Pharmacol 79:141–70

Braga MF, Aroniadou-Anderjaska V, Xie J, Li H (2003) Bidirectional modulation of GABA release by presynaptic glutamate receptor 5 kainate receptors in the basolateral amygdala. J Neurosci 23:442–52

Breukel AIM, Besselsen E, da Silva FHL, Ghijsen WEJM (1998) A presynaptic N methyl d aspartate autoreceptor in rat hippocampus modulating amino acid release from a cytoplasmic pool. Eur J Neurosci 10:106–14

Brown AM, Birnbaumer L (1990) Ionic channels and their regulation by G protein subunits. Annu Rev Physiol 52:197–213

Cao YJ, Surowy CS, Puttfarcken PS (2005a) Different nicotinic acetylcholine receptor subtypes mediating striatal and prefrontal cortical [3H]dopamine release. Neuropharmacology 48:72–9

Cao YJ, Surowy CS, Puttfarcken PS (2005b) Nicotinic acetylcholine receptor-mediated [3H]dopamine release from hippocampus. J Pharmacol Exp Ther 312: 1298–1304

Casado M, Isope P, Ascher P (2002) Involvement of presynaptic N-methyl-D-aspartate receptors in cerebellar long-term depression. Neuron 33:123

Cattaert D, El Manira A (1999) Shunting versus inactivation: analysis of presynaptic inhibitory mechanisms in primary afferents of the crayfish. J Neurosci 19:6079–89

Cerrito F, Aloisi G, Arminio P, Fanini D (1998) A new GABA-A receptor subtype coupled with Ca++/Cl-synporter modulates aminergic release from rat brain neuron terminals. J Neurosci Res 51:15–22

Chameau P, van Hooft JA (2006) Serotonin 5-HT(3) receptors in the central nervous system. Cell Tissue Res 326:573–81

Chergui K, Bouron A, Normand E, Mulle C (2000) Functional GluR6 kainate receptors in the striatum: indirect downregulation of synaptic transmission. J Neurosci 20:2175–82

Chheda MG, Ashery U, Thakur P, Rettig J, Sheng ZH (2001) Phosphorylation of Snapin by PKA modulates its interaction with the SNARE complex. Nat Cell Biol 3: 331–8

Chittajallu R, Vignes M, Dev KK, Barnes JM, Collingridge GL, Henley JM (1996) Regulation of glutamate release by presynaptic kainate receptors in the hippocampus. Nature 379:78

Christensen JK, Paternain AV, Selak S, Ahring PK, Lerma J (2004) A mosaic of functional kainate receptors in hippocampal interneurons. J Neurosci 24:8986–93

Clarke VRJ, Collingridge GL (2004) Characterisation of the effects of ATPA, a GLUK5 kainate receptor agonist, on GABAergic synaptic transmission in the CA1 region of rat hippocampal slices. Neuropharmacology 47:363

Clarke VRJ, Ballyk BA, Hoo KH , Mandelzys A, Pellizzari A, Bath CP, Thomas J, Sharpe EF, Davies CH, Ornstein PL, Schoepp DD, Kamboj RK, Collingridge GL, Lodge D, Bleakman D (1997) A hippocampal GluR5 kainate receptor regulating inhibitory synaptic transmission. Nature 389:599

Conradi S (1969) Ultrastructure of dorsal root boutons on lumbosacral motoneurons of the adult cat, as revealed by dorsal root section. Acta Physiol Scand Suppl 332: 85–115

Contractor A, Sailer AW, Darstein M, Maron C, Xu J, Swanson GT, Heinemann SF (2003) Loss of kainate receptor-mediated heterosynaptic facilitation of mossy-fiber synapses in KA2-/-mice. J Neurosci 23:422–9
Contractor A, Swanson GT, Sailer A, O'Gorman S, Heinemann SF (2000) Identification of the kainate receptor subunits underlying modulation of excitatory synaptic transmission in the CA3 region of the hippocampus. J Neurosci 20:8269–78
Cossart R, Tyzio R, Dinocourt C, Esclapez M, Hirsch JC, Ben-Ari Y, Bernard C (2001) Presynaptic kainate receptors that enhance the release of GABA on CA1 hippocampal interneurons. Neuron 29:497
Crespi D, Gobbi M, Mennini T (1997) 5-HT3 serotonin hetero-receptors inhibit [3H]acetylcholine release in rat cortical synaptosomes. Pharmacological Res 35:351
Cunha RA, Ribeiro JA (2000) ATP as a presynaptic modulator. Life Sci 68:119–37
De Waard M, Hering J, Weiss N, Feltz A (2005) How do G proteins directly control neuronal Ca2+ channel function? Trends Pharmacol Sci 26:427–36
Cunha RA, Constantino MD, Ribeiro JA (1997) Inhibition of [3H][gamma]-aminobutyric acid release by kainate receptor activation in rat hippocampal synaptosomes. Eur J Pharmacol 323:167
Delaney AJ, Jahr CE (2002) Kainate receptors differentially regulate release at two parallel fiber synapses. Neuron 36:475–82
Derbenev AV, Monroe MJ, Glatzer NR, Smith BN (2006) Vanilloid-mediated heterosynaptic facilitation of inhibitory synaptic input to neurons of the rat dorsal motor nucleus of the vagus. J Neurosci 26:9666–72
Dingledine R, Borges K, Bowie D, Traynelis SF (1999) The glutamate receptor ion channels. Pharmacol Rev 51:7–61
Dixon W (1924) Nicotin, Coniin, Piperidin, Lupetidin, Cytisin, Lobelin, Spartein, Gelsemin. Mittel, welche auf bestimmte Nervenzellen wirken. Handb Exp Pharmacol 2:656–36
Dooley DJ, Lupp A, Hertting G (1987) Inhibition of central neurotransmitter release by omega-conotoxin GVIA, a peptide modulator of the N-type voltage-sensitive calcium channel. Naunyn Schmiedebergs Arch Pharmacol 336:467–70
Dorostkar MM, Boehm S (2007) Opposite effects of presynaptic 5-HT3 receptor activation on spontaneous and action potential-evoked GABA release at hippocampal synapses. J Neurochem, in press
Doyle MW, Bailey TW, Jin Y-H, Andresen MC (2002) Vanilloid receptors presynaptically modulate cranial visceral afferent synaptic transmission in nucleus tractus solitarius. J Neurosci 22:8222–9
Dudel J, Kuffler SW (1961) Presynaptic inhibition at the crayfish neuromuscular junction. J Physiol 155:543–62
Duguid IC, Smart TG (2004) Retrograde activation of presynaptic NMDA receptors enhances GABA release at cerebellar interneuron-Purkinje cell synapses. Nat Neurosci 7:525
Eccles JC, Kostyuk PG, Schmidt RF (1962) Central pathways responsible for depolarization of primary afferent fibres. J Physiol 161:237–57
Elmslie KS (2003) Neurotransmitter modulation of neuronal calcium channels. J Bioenerg Biomembr 35:477–89
Engelman HS, Anderson RL, Daniele C, MacDermott AB (2006) Presynaptic alpha-amino-3-hydroxy-5-methyl-4-isoxazolepropionic acid (AMPA) receptors modulate release of inhibitory amino acids in rat spinal cord dorsal horn. Neuroscience 139: 539
Engelman HS, MacDermott AB (2004) Presynaptic ionotropic receptors and control of transmitter release. Nat Rev Neurosci 5:135–45
Faist M, Dietz V, Pierrot-Deseilligny E (1996) Modulation, probably presynaptic in origin, of monosynaptic Ia excitation during human gait. Exp Brain Res 109:441–9
Felmy F, Neher E, Schneggenburger R (2003) The timing of phasic transmitter release is Ca2+-dependent and lacks a direct influence of presynaptic membrane potential. Proc Natl Acad Sci USA 100:15200–5

Ferezou I, Cauli B, Hill EL, Rossier J, Hamel E, Lambolez B (2002) 5-HT3 receptors mediate serotonergic fast synaptic excitation of neocortical vasoactive intestinal peptide/cholecystokinin interneurons. J Neurosci 22:7389–97

Ferkany JW, Zaczek R, Coyle JT (1982) Kainic acid stimulates excitatory amino acid neurotransmitter release at presynaptic receptors. Nature 298:757–9

Fink K, Gothert M, Molderings G, Schlicker E (1989) N-methyl-D-aspartate (NMDA) receptor-mediated stimulation of noradrenaline release, but not release of other neurotransmitters, in the rat brain cortex: receptor location, characterization and desensitization. Naunyn Schmiedeberg's Arch Pharmacol 339:514–21

Fozard JR, Ali AT (1978) Receptors for 5-hydroxytryptamine on the sympathetic nerves of the rabbit heart. Naunyn Schmiedeberg's Arch Pharmacol 301:223–35

Frank K, Fuortes MGF (1957) Presynaptic and postsynaptic inhibition of neurosynaptic reflex. Fed Proc 16:39–40

Frerking M, Schmitz D, Zhou Q, Johansen J, Nicoll RA (2001) Kainate receptors depress excitatory synaptic transmission at CA3{right-arrow}CA1 synapses in the hippocampus via a direct presynaptic action. J Neurosci 21:2958–66

Fu WM, Huang FL (1994) Potentiation by endogenously released ATP of spontaneous transmitter secretion at developing neuromuscular synapses in Xenopus cell cultures. Br J Pharmacol 111:880–6

Fuder H, Muscholl E (1995) Heteroreceptor-mediated modulation of noradrenaline and acetylcholine release from peripheral nerves. Rev Physiol Biochem Pharmacol 126:265–412

Funahashi M, Mitoh Y, Matsuo R (2004a) Activation of presynaptic 5-HT3 receptors facilitates glutamatergic synaptic inputs to area postrema neurons in rat brain slices. Methods Find Exp Clin Pharmacol 26:615–22

Funahashi M, Mitoh Y, Matsuo R (2004b) Nicotinic modulation of area postrema neuronal excitability in rat brain slices. Brain Res 1017:227

Gamper N, Reznikov V, Yamada Y, Yang J, Shapiro MS (2004) Phosphatidylinositol 4,5-bisphosphate signals underlie receptor-specific Gq/11-mediated modulation of N-type Ca2+ channels. J Neurosci 24:10980–92

Garcia AG, Albillos A, Cano-Abad MF, Garcia-Palomero E, Hernandez-Guijo M, Herrero CJ, Lomax RB, Gandia L (1998) Calcium channels for exocytosis in chromaffin cells. Adv Pharmacol 42:91–4

Gerstbrein K, Sitte HH (2006) Currents in neurotransmitter transporters. Handb Exp Pharmacol:95–111

Göthert M, Duhrsen U (1979) Effects of 5-hydroxytryptamine and related compounds on the sympathetic nerves of the rabbit heart. Naunyn Schmiedebergs Arch Pharmacol 308:9–18

Gotti C, Zoli M, Clementi F (2006) Brain nicotinic acetylcholine receptors: native subtypes and their relevance. Trends Pharmacol Sci 27:482–91

Grady SR, Meinerz NM, Cao J, Reynolds AM, Picciotto MR, Changeux JP, McIntosh JM, Marks MJ, Collins AC (2001) Nicotinic agonists stimulate acetylcholine release from mouse interpeduncular nucleus: a function mediated by a different nAChR than dopamine release from striatum. J Neurochem 76:258–68

Grady SR, Murphy KL, Cao J, Marks MJ, McIntosh JM, Collins AC (2002) Characterization of nicotinic agonist-induced [(3)H]dopamine release from synaptosomes prepared from four mouse brain regions. J Pharmacol Exp Ther 301: 651–60

Graham B, Redman S (1994) A simulation of action potentials in synaptic boutons during presynaptic inhibition. J Neurophysiol 71:538–49

Grassi G, Seravalle G, Calhoun DA, Bolla GB, Giannattasio C, Marabini M, Del Bo A, Mancia G (1994) Mechanisms responsible for sympathetic activation by cigarette smoking in humans. Circulation 90:248–53

Gray R, Rajan AS, Radcliffe KA, Yakehiro M, Dani JA (1996) Hippocampal synaptic transmission enhanced by low concentrations of nicotine. Nature 383:713–16

Gu JG, MacDermott AB (1997) Activation of ATP P2X receptors elicits glutamate release from sensory neuron synapses. Nature 389:749–53

Haass M, Kubler W (1997) Nicotine and sympathetic neurotransmission. Cardiovasc Drugs Ther 10:657–65

Heidelberger R, Heinemann C, Neher E, Matthews G (1994) Calcium dependence of the rate of exocytosis in a synaptic terminal. Nature 371:513–15

Henningfield JE (1995) Nicotine medications for smoking cessation. N Engl J Med 333:1196–1203

Hille B (1994) Modulation of ion-channel function by G-protein-coupled receptors. Trends Neurosci 17:531–6

Hirai H, Launey T, Mikawa S, Torashima T, Yanagihara D, Kasaura T, Miyamoto A, Yuzaki M (2003) New role of delta2-glutamate receptors in AMPA receptor trafficking and cerebellar function. Nat Neurosci 6:869–76

Hirling H, Scheller RH (1996) Phosphorylation of synaptic vesicle proteins: modulation of the alpha SNAP interaction with the core complex. Proc Natl Acad Sci USA 93:11945–9

Hirning LD, Fox AP, McCleskey EW, Olivera BM, Thayer SA, Miller RJ, Tsien RW (1988) Dominant role of N-type Ca2+ channels in evoked release of norepinephrine from sympathetic neurons. Science 239:57–61

Hultborn H, Meunier S, Pierrot-Deseilligny E, Shindo M (1987) Changes in presynaptic inhibition of Ia fibres at the onset of voluntary contraction in man. J Physiol 389:757–72

Ireland MF, Noakes PG, Bellingham MC (2004) P2X7-like receptor subunits enhance excitatory synaptic transmission at central synapses by presynaptic mechanisms. Neuroscience 128:269

Jang I-S, Jeong H-J, Katsurabayashi S, Akaike N (2002) Functional roles of presynaptic GABAA receptors on glycinergic nerve terminals in the rat spinal cord. J Physiol (Lond) 541:423–34

Jang IS, Ito Y, Akaike N (2005) Feed-forward facilitation of glutamate release by presynaptic GABAA receptors. Neuroscience 135:737

Jang IS, Nakamura M, Ito Y, Akaike N (2006) Presynaptic GABAA receptors facilitate spontaneous glutamate release from presynaptic terminals on mechanically dissociated rat CA3 pyramidal neurons. Neuroscience 138:25

Jeong H-J, Jang I-S, Moorhouse AJ, Akaike N (2003) Activation of presynaptic glycine receptors facilitates glycine release from presynaptic terminals synapsing onto rat spinal sacral dorsal commissural nucleus neurons. J Physiol (Lond) 550: 373–83

Jiang L, Xu J, Nedergaard M, Kang J (2001) A kainate receptor increases the efficacy of GABAergic synapses. Neuron 30:503

Jin Y-H, Bailey TW, Li B-y, Schild JH, Andresen MC (2004) Purinergic and vanilloid receptor activation releases glutamate from separate cranial afferent terminals in nucleus tractus solitarius. J Neurosci 24:4709–17

Johnson RM, Inouye GT, Eglen RM, Wong EH (1993) 5-HT3 receptor ligands lack modulatory influence on acetylcholine release in rat entorhinal cortex. Naunyn Schmiedebergs Arch Pharmacol 347:241–7

Kaczmarek L, Levitan I (1987) Neuromodulation: the biochemical control of neuronal excitability. Oxford University Press, New York

Kaiser S, Wonnacott S (2000) alpha -Bungarotoxin-sensitive nicotinic receptors indirectly modulate [3H]dopamine release in rat striatal slices via glutamate release. Mol Pharmacol 58:312–318

Kamp MA, Krieger A, Henry M, Hescheler J, Weiergraber M, Schneider T (2005) Presynaptic 'Ca2.3-containing' E-type Ca channels share dual roles during neurotransmitter release. Eur J Neurosci 21:1617–25

Kandel ER (2001) The molecular biology of memory storage: a dialogue between genes and synapses. Science 294:1030–8

Karnovsky AM, Gotow LF, McKinley DD, Piechan JL, Ruble CL, Mills CJ, Schellin KAB, Slightom JL, Fitzgerald LR, Benjamin CW, Roberds SL (2003) A cluster of novel serotonin receptor 3-like genes on human chromosome 3. Gene 319:137

Kato F, Shigetomi E (2001) Distinct modulation of evoked and spontaneous EPSCs by purinoceptors in the nucleus tractus solitarii of the rat. J Physiol (Lond) 530:469–86

Katsurabayashi S, Kubota H, Tokutomi N, Akaike N (2003) A distinct distribution of functional presynaptic 5-HT receptor subtypes on GABAergic nerve terminals projecting to single hippocampal CA1 pyramidal neurons. Neuropharmacology 44: 1022–30
Kawa K (1994) Distribution and functional properties of 5-HT3 receptors in the rat hippocampal dentate gyrus: a patch-clamp study. J Neurophysiol. 71:1935–47
Kerchner GA, Wang GD, Qiu CS, Huettner JE, Zhuo M (2001a) Direct presynaptic regulation of GABA/glycine release by kainate receptors in the dorsal horn: an ionotropic mechanism. Neuron 32:477–88
Kerchner GA, Wilding TJ, Huettner JE, Zhuo M (2002) Kainate receptor subunits underlying presynaptic regulation of transmitter release in the dorsal horn. J Neurosci 22:8010–17
Kerchner GA, Wilding TJ, Li P, Zhuo M, Huettner JE (2001b) Presynaptic kainate receptors regulate spinal sensory transmission. J Neurosci 21:59–66
Khakh BS, Henderson G (1998) ATP receptor-mediated enhancement of fast excitatory neurotransmitter release in the brain. Mol Pharmacol 54:372–8
Khakh BS, Henderson G (2000) Modulation of fast synaptic transmission by presynaptic ligand-gated cation channels. J Autonom Nerv Sys 81:110
Khakh BS, North RA (2006) P2X receptors as cell-surface ATP sensors in health and disease. Nature 442:527
Khakh BS, Zhou X, Sydes J, Galligan JJ, Lester HA (2000) State-dependent cross-inhibition between transmitter-gated cation channels. Nature 406:405–10
Khakh BS, Gittermann D, Cockayne DA, Jones A (2003) ATP Modulation of excitatory synapses onto interneurons. J Neurosci 23:7426–37
Khakh BS, Fisher JA, Nashmi R, Bowser DN, Lester HA (2005) An angstrom scale interaction between plasma membrane ATP-gated P2X2 and alpha4beta2 nicotinic channels measured with fluorescence resonance energy transfer and total internal reflection fluorescence microscopy. J Neurosci 25:6911–20
Kirischuk S, Akyeli J, Iosub R, Grantyn R (2003) Pre- and postsynaptic contribution of GABAC receptors to GABAergic synaptic transmission in rat collicular slices and cultures. Eur J Neurosci 18:752–8
Kiyosawa A, Katsurabayashi S, Akaike N, Pang ZP, Akaike N (2001) Nicotine facilitates glycine release in the rat spinal dorsal horn. J Physiol (Lond) 536:101–10
Ko S, Zhao MG, Toyoda H, Qiu CS, Zhuo M (2005) Altered behavioral responses to noxious stimuli and fear in glutamate receptor 5 (GluR5)- or GluR6-deficient mice. J Neurosci 25:977–84
Koga H, Ishibashi H, Shimada H, Jang I-S, Nakamura TY, Nabekura J (2005) Activation of presynaptic GABAA receptors increases spontaneous glutamate release onto noradrenergic neurons of the rat locus coeruleus. Brain Res 1046:24
Koh DS, Hille B (1997) Modulation by neurotransmitters of catecholamine secretion from sympathetic ganglion neurons detected by amperometry. Proc Natl Acad Sci USA 94:1506–11
Korpi ER, Sinkkonen ST (2006) GABA(A) receptor subtypes as targets for neuropsychiatric drug development. Pharmacol Ther 109:12–32
Koyama S, Matsumoto N, Kubo C, Akaike N (2000) Presynaptic 5-HT3 receptor-mediated modulation of synaptic GABA release in the mechanically dissociated rat amygdala neurons. J Physiol (Lond) 529:373–83
Koyama S, Matsumoto N, Murakami N, Kubo C, Nabekura J, Akaike N (2002) Role of presynaptic 5-HT1A and 5-HT3 receptors in modulation of synaptic GABA transmission in dissociated rat basolateral amygdala neurons. Life Sciences 72:375–87
Kristufek D, Koth G, Motejlek A, Schwarz K, Huck S, Boehm S (1999a) Modulation of spontaneous and stimulation-evoked transmitter release from rat sympathetic neurons by the cognition enhancer linopirdine: insights into its mechanisms of action. J Neurochem 72:2083–91
Kristufek D, Stocker E, Boehm S, Huck S (1999b) Somatic and prejunctional nicotinic receptors in cultured rat sympathetic neurones show different agonist profiles. J Physiol 516 (Pt 3):739–56
Kubista H, Boehm S (2006) Molecular mechanisms underlying the modulation of exocytotic noradrenaline release via presynaptic receptors. Pharmacol Ther 112: 213–42

Langer S (1977) Presynaptic receptors and their role in the regulation of transmitter release. Br J Pharmacol 60:481–97

Lappin SC, Randall AD, Gunthorpe MJ, Morisset V (2006) TRPV1 antagonist, SB-366791, inhibits glutamatergic synaptic transmission in rat spinal dorsal horn following peripheral inflammation. Eur J Pharmacol 540:73

Lauri SE, Bortolotto ZA, Nistico R, Bleakman D, Ornstein PL, Lodge D, Isaac JT, Collingridge GL (2003) A role for Ca2+ stores in kainate receptor-dependent synaptic facilitation and LTP at mossy fiber synapses in the hippocampus. Neuron 39:327–41

Lauri SE, Segerstrale M, Vesikansa A, Maingret F, Mulle C, Collingridge GL, Isaac JT, Taira T (2005) Endogenous activation of kainate receptors regulates glutamate release and network activity in the developing hippocampus. J Neurosci 25:4473–84

Lechner SG, Hussl S, Schicker KW, Drobny H, Boehm S (2005) Presynaptic inhibition via a phospholipase C- and phosphatidylinositol bisphosphate-dependent regulation of neuronal Ca2+ channels. Mol Pharmacol 68:1387–96

Lechner SG, Mayer M, Boehm S (2003) Activation of M1 muscarinic receptors triggers transmitter release from rat sympathetic neurons through an inhibition of M-type K+ channels. J Physiol 553:789–802

Lee CJ, Bardoni R, Tong C-K, Engelman HS, Joseph DJ, Magherini PC, MacDermott AB (2002a) Functional expression of AMPA receptors on central terminals of rat dorsal root ganglion neurons and presynaptic inhibition of glutamate release. Neuron 35:135

Lee FJ, Xue S, Pei L, Vukusic B, Chery N, Wang Y, Wang YT, Niznik HB, Yu XM, Liu F (2002b) Dual regulation of NMDA receptor functions by direct protein-protein interactions with the dopamine D1 receptor. Cell 111:219–30

Lena C, Changeux JP (1997) Role of Ca2+ ions in nicotinic facilitation of GABA release in mouse thalamus. J Neurosci 17:576–85

Lena C, Changeux JP, Mulle C (1993) Evidence for "preterminal" nicotinic receptors on GABAergic axons in the rat interpeduncular nucleus. J Neurosci 13:2680–8

Lena C, de Kerchove D'Exaerde A, Cordero-Erausquin M, Le Novere N, del Mar Arroyo-Jimenez M, Changeux JP (1999) Diversity and distribution of nicotinic acetylcholine receptors in the locus ceruleus neurons. Proc Natl Acad Sci USA 96: 12126–31

Lerma J (2006) Kainate receptor physiology. Cur Op Pharmacol 6:89–97

Lester HA, Dibas MI, Dahan DS, Leite JF, Dougherty DA (2004) Cys-loop receptors: new twists and turns. Trends Neurosci 27:329–36

Li D-P, Chen S-R, Pan H-L (2004) VR1 Receptor activation induces glutamate release and postsynaptic firing in the paraventricular nucleus. J Neurophysiol 92: 1807–16

Li Y, Xu TL (2002) State-dependent cross-inhibition between anionic GABA(A) and glycine ionotropic receptors in rat hippocampal CA1 neurons. Neuroreport 13: 223–6

Liedtke W, Kim C (2005) Functionality of the TRPV subfamily of TRP ion channels: add mechano-TRP and osmo-TRP to the lexicon! Cell Mol Life Sci 62:2985–3001

Lien C-C, Mu Y, Vargas-Caballero M, Poo M-m (2006) Visual stimuli-induced LTD of GABAergic synapses mediated by presynaptic NMDA receptors. Nat Neurosci 9:372

Lipscombe D, Kongsamut S, Tsien RW (1989) Alpha-adrenergic inhibition of sympathetic neurotransmitter release mediated by modulation of N-type calcium-channel gating. Nature 340:639–42

Liu F, Wan Q, Pristupa ZB, Yu XM, Wang YT, Niznik HB (2000) Direct protein-protein coupling enables cross-talk between dopamine D5 and gamma-aminobutyric acid A receptors. Nature 403:274–80

Liu H, Mantyh PW, Basbaum AI (1997) NMDA-receptor regulation of substance P release from primary afferent nociceptors. Nature 386:721

Liu L, Zhao R, Bai Y, Stanish LF, Evans JE, Sanderson MJ, Bonventre JV, Rittenhouse AR (2006) M1 muscarinic receptors inhibit L-type Ca2+ current and M-current by divergent signal transduction cascades. J Neurosci 26:11588–98

Liu Q-S, Patrylo PR, Gao X-B, van den Pol AN (1999) Kainate acts at presynaptic receptors to increase GABA release from hypothalamic neurons. J Neurophysiol 82: 1059–62

Lloyd GK, Menzaghi F, Bontempi B, Suto C, Siegel R, Akong M, Stauderman K, Velicelebi G, Johnson E, Harpold MM, Rao TS, Sacaan AI, Chavez-Noriega LE, Washburn MS, Vernier JM, Cosford ND, McDonald LA (1998) The potential of subtype-selective neuronal nicotinic acetylcholine receptor agonists as therapeutic agents. Life Sci 62:1601–6

Loewi O (1921) Über humorale Übertragbarkeit der Herznervenwirkung. Pflugers Arch 189:239–42

Löffelholz K (1970) Autoinhibition of nicotinic release of noradrenaline from postganglionic sympathetic nerves. Naunyn Schmiedebergs Arch Pharmakol 267: 49–63

Luetje CW (2004) Getting past the asterisk: the subunit composition of presynaptic nicotinic receptors that modulate striatal dopamine release. Mol Pharmacol 65:1333–5

Lukas RJ, Changeux J-P, le Novere N, Albuquerque EX, Balfour DJK, Berg DK, Bertrand D, Chiappinelli VA, Clarke PBS, Collins AC, Dani JA, Grady SR, Kellar KJ, Lindstrom JM, Marks MJ, Quik M, Taylor PW, Wonnacott S (1999) International Union of Pharmacology. XX. Current Status of the Nomenclature for Nicotinic Acetylcholine Receptors and Their Subunits. Pharmacol Rev 51:397–401

Luo S, Kulak JM, Cartier GE, Jacobsen RB, Yoshikami D, Olivera BM, McIntosh JM (1998) alpha-conotoxin AuIB selectively blocks alpha3 beta4 nicotinic acetylcholine receptors and nicotine-evoked norepinephrine release. J Neurosci 18:8571–9

MacDermott AB, Role LW, Siegelbaum SA (1999) Presynaptic ionotropic receptors and the control of transmitter release. Annu Rev Neurosci 22:443–85

Madden DR (2002) The structure and function of glutamate receptor ion channels. Nat Rev Neurosci 3:91–101

Maingret F, Lauri SE, Taira T, Isaac JT (2005) Profound regulation of neonatal CA1 rat hippocampal GABAergic transmission by functionally distinct kainate receptor populations. J Physiol 567:131–42

Marchi M, Risso F, Viola C, Cavazzani P, Raiteri M (2002) Direct evidence that release-stimulating alpha7* nicotinic cholinergic receptors are localized on human and rat brain glutamatergic axon terminals. J Neurochem 80:1071–8

Martire M, Altobelli D, Cannizzaro C, Maurizi S, Preziosi P (2002) Prenatal diazepam exposure functionally alters the GABA(A) receptor that modulates [3H]noradrenaline release from rat hippocampal synaptosomes. Dev Neurosci 24:71–8

Martire M, Castaldo P, D'Amico M, Preziosi P, Annunziato L, Taglialatela M (2004) M channels containing KCNQ2 subunits modulate norepinephrine, aspartate, and GABA release from hippocampal nerve terminals. J Neurosci 24:592–7

McGehee DS, Heath MJ, Gelber S, Devay P, Role LW (1995) Nicotine enhancement of fast excitatory synaptic transmission in CNS by presynaptic receptors. Science 269:1692–6

McGehee DS, Role LW (1996) Presynaptic ionotropic receptors. Curr Opin Neurobiol 6:342

McMahon LL, Kauer JA (1997) Hippocampal interneurons are excited via serotonin-gated ion channels. J Neurophysiol 78:2493–2502

Meir A, Ginsburg S, Butkevich A, Kachalsky SG, Kaiserman I, Ahdut R, Demirgoren S, Rahamimoff R (1999) Ion channels in presynaptic nerve terminals and control of transmitter release. Physiol Rev 79:1019–88

Michaelevski I, Chikvashvili D, Tsuk S, Fili O, Lohse MJ, Singer-Lahat D, Lotan I (2002) Modulation of a brain voltage-gated K+ channel by syntaxin 1A requires the physical interaction of Gbetagamma with the channel. J Biol Chem 277:34909–17

Mochida S, Sheng ZH, Baker C, Kobayashi H, Catterall WA (1996) Inhibition of neurotransmission by peptides containing the synaptic protein interaction site of N-type Ca2+ channels. Neuron 17:781–8

Mochida S, Westenbroek RE, Yokoyama CT, Zhong H, Myers SJ, Scheuer T, Itoh K, Catterall WA (2003) Requirement for the synaptic protein interaction site for reconstitution of synaptic transmission by P/Q-type calcium channels. Proc Natl Acad Sci USA 100:2819–24

Mongeau R, De Montigny C, Blier P (1994) Activation of 5-HT3 receptors enhances the electrically evoked release of [3H]noradrenaline in rat brain limbic structures. Eur J Pharmacol 256:269–79

Moores TS, Hasdemir B, Vega-Riveroll L, Deuchars J, Parson SH (2005) Properties of presynaptic P2X7-like receptors at the neuromuscular junction. Brain Res 1034:40
Morales M, Wang SD (2002) Differential composition of 5-hydroxytryptamine3 receptors synthesized in the rat CNS and peripheral nervous system. J Neurosci 22: 6732–41
Morales M, Battenberg E, de Lecea L, Bloom FE (1996) The type 3 serotonin receptor is expressed in a subpopulation of GABAergic neurons in the rat neocortex and hippocampus. Brain Res 731:199–202
Morales M, Battenberg E, Bloom FE (1998) Distribution of neurons expressing immunoreactivity for the 5HT3 receptor subtype in the rat brain and spinal cord. J Comp Neurol 402:385–401
Morita H, Crone C, Christenhuis D, Petersen NT, Nielsen JB (2001) Modulation of presynaptic inhibition and disynaptic reciprocal Ia inhibition during voluntary movement in spasticity. Brain 124:826–37
Nakatsuka T, Gu JG (2001) ATP P2X Receptor-mediated enhancement of glutamate release and evoked EPSCs in dorsal horn neurons of the rat spinal cord. J Neurosci 21:6522–31
Nakazawa K (1994) ATP-activated current and its interaction with acetylcholine-activated current in rat sympathetic neurons. J Neurosci 14:740–50
Narkiewicz K, van de Borne PJ, Hausberg M, Cooley RL, Winniford MD, Davison DE, Somers VK (1998) Cigarette smoking increases sympathetic outflow in humans. Circulation 98:528–34
Nicoll RA, Schmitz D (2005) Synaptic plasticity at hippocampal mossy fibre synapses. Nat Rev Neurosci 6:863–76
Niesler B, Frank B, Kapeller J, Rappold GA (2003) Cloning, physical mapping and expression analysis of the human 5-HT3 serotonin receptor-like genes HTR3C, HTR3D and HTR3E. Gene 310:101
North RA (2002) Molecular Physiology of P2X Receptors. Physiol Rev 82:1013–67
Ohta K, Araki N, Shibata M, Komatsumoto S, Shimazu K, Fukuuchi Y (1994) Presynaptic ionotropic glutamate receptors modulate in vivo release and metabolism of striatal dopamine, noradrenaline, and 5-hydroxytryptamine: involvement of both NMDA and AMPA/kainate subtypes. Neurosci Res 21:83–9
Owens DF, Boyce LH, Davis MB, Kriegstein AR (1996) Excitatory GABA responses in embryonic and neonatal cortical slices demonstrated by gramicidin perforated-patch recordings and calcium imaging. J Neurosci 16:6414–23
Parker RM, Barnes JM, Ge J, Barber PC, Barnes NM (1996) Autoradiographic distribution of [3H]-(S)-zacopride-labelled 5-HT3 receptors in human brain. J Neurol Sci 144:119–27
Partovi D, Frerking M (2006) Presynaptic inhibition by kainate receptors converges mechanistically with presynaptic inhibition by adenosine and GABAB receptors. Neuropharmacology 51:1030
Patti L, Raiteri L, Grilli M, Parodi M, Raiteri M, Marchi M (2006) P2X7 receptors exert a permissive role on the activation of release-enhancing presynaptic [alpha]7 nicotinic receptors co-existing on rat neocortex glutamatergic terminals. Neuropharmacology 50:705
Perkinton MS, Sihra TS (1999) A high-affinity presynaptic kainate-type glutamate receptor facilitates glutamate exocytosis from cerebral cortex nerve terminals (synaptosomes). Neuroscience 90:1281–92
Petrov VI, Sergeyev VS, Onishchenko NV (2001) The effect of N-acetyl–aspartic acid dilithium salt on dopamine release and synthesis in the rat striatum in vivo. Eur J Pharmacol 416:69
Pittaluga A, Raiteri M (1990) Release-enhancing glycine-dependent presynaptic NMDA receptors exist on noradrenergic terminals of hippocampus. Eur J Pharmacol 191:231–4
Poelchen W, Sieler D, Wirkner K, Illes P (2001) Co-transmitter function of ATP in central catecholaminergic neurons of the rat. Neuroscience 102:593–602
Poncer JC, McKinney RA, Gahwiler BH, Thompson SM (1997) Either N- or P-type calcium channels mediate GABA release at distinct hippocampal inhibitory synapses. Neuron 18:463–72
Price TJ, Patwardhan A, Akopian AN, Hargreaves KM, Flores CM (2004) Modulation of trigeminal sensory neuron activity by the dual cannabinoid-vanilloid agonists anandamide, N-arachidonoyl-dopamine and arachidonyl-2-chloroethylamide. 141: 1118

Puig MV, Santana N, Celada P, Mengod G, Artigas F (2004) In vivo excitation of GABA interneurons in the medial prefrontal cortex through 5-HT3 receptors. Cereb Cortex 14:1365–75
Radcliffe KA, Dani JA (1998) Nicotinic stimulation produces multiple forms of increased glutamatergic synaptic transmission. J Neurosci 18:7075–83
Raffaelli G, Saviane C, Mohajerani MH, Pedarzani P, Cherubini E (2004) BK potassium channels control transmitter release at CA3-CA3 synapses in the rat hippocampus. J Physiol 557:147–57
Rapier C, Lunt GG, Wonnacott S (1990) Nicotinic modulation of [3H]dopamine release from striatal synaptosomes: pharmacological characterisation. J Neurochem 54:937–45
Rashid MH, Furue H, Yoshimura M, Ueda H (2006) Tonic inhibitory role of [alpha]4[beta]2 subtype of nicotinic acetylcholine receptors on nociceptive transmission in the spinal cord in mice. Pain 125:125
Reno LAC, Zago W, Markus RP (2004) Release of [3H]glutamate by stimulation of nicotinic acetylcholine receptors in rat cerebellar slices. Neuroscience 124:647
Reuben M, Clarke PBS (2000) Nicotine-evoked [3H]5-hydroxytryptamine release from rat striatal synaptosomes. Neuropharmacology 39:290
Rhee JS, Wang ZM, Nabekura J, Inoue K, Akaike N (2000) ATP facilitates spontaneous glycinergic IPSC frequency at dissociated rat dorsal horn interneuron synapses. J Physiol (Lond) 524:471–83
Riker WF, Jr., Roberts J, Standaert FG, Fujimori H (1957) The motor nerve terminal as the primary focus for drug-in-duced facilitation of neuromuscular transmission. J Pharmacol Exp Ther 121:286–312
Rivera C, Voipio J, Kaila K (2005) Two developmental switches in GABAergic signalling: the K+-Cl− cotransporter KCC2 and carbonic anhydrase CAVII. J Physiol (Lond) 562:27–36
Rivera C, Voipio J, Payne JA, Ruusuvuori E, Lahtinen H, Lamsa K, Pirvola U, Saarma M, Kaila K (1999) The K+/Cl− co-transporter KCC2 renders GABA hyperpolarizing during neuronal maturation. Nature 397:251
Rodriguez-Moreno A, Herreras O, Lerma J (1997) Kainate receptors presynaptically downregulate GABAergic inhibition in the rat hippocampus. Neuron 19:893
Rodriguez-Moreno A, Lerma J (1998) Kainate receptor modulation of GABA release involves a metabotropic function. Neuron 20:1211
Rodriguez-Moreno A, Lopez-Garcia JC, Lerma J (2000) Two populations of kainate receptors with separate signaling mechanisms in hippocampal interneurons. PNAS 97:1293–8
Rodriguez-Moreno A, Sihra TS (2004) Presynaptic kainate receptor facilitation of glutamate release involves protein kinase A in the rat hippocampus. J Physiol (Lond) 557:733–45
Rogers M, Dani JA (1995) Comparison of quantitative calcium flux through NMDA, ATP, and ACh receptor channels. Biophys J 68:501–6
Ronde P, Nichols RA (1998) High calcium permeability of serotonin 5-HT3 receptors on presynaptic nerve terminals from rat striatum. J Neurochem 70:1094–1103
Rosenstein RE, Chuluyan HE, Cardinali DP (1990) Presynaptic effects of gamma-aminobutyric acid on norepinephrine release and uptake in rat pineal gland. J Neural Transm 82:131–40
Rousseau SJ, Jones IW, Pullar IA, Wonnacott S (2005) Presynaptic [alpha]7 and non-[alpha]7 nicotinic acetylcholine receptors modulate [3H]d-aspartate release from rat frontal cortex in vitro. Neuropharmacology 49:59
Rozas JL, Paternain AV, Lerma J (2003) Noncanonical signaling by ionotropic kainate receptors. Neuron 39:543–53
Rudolph U, Möhler H (2004) Analysis of GABAA receptor function and dissection of the pharmacology of benzodiazepines and general anesthetics through mouse genetics. Annu Rev Pharmacol Toxicol 44:475–98
Ruiz A, Fabian-Fine R, Scott R, Walker MC, Rusakov DA, Kullmann DM (2003) GABAA Receptors at hippocampal mossy fibers. Neuron 39:961
Rusakov DA, Saitow F, Lehre KP, Konishi S (2005) Modulation of presynaptic Ca2+ entry by AMPA receptors at individual GABAergic synapses in the cerebellum. J Neurosci 25:4930–40
Rustioni A (2005) Modulation of sensory input to the spinal cord by presynaptic ionotropic glutamate receptors. Arch Ital Biol 143:103–12

Sacaan AI, Dunlop JL, Lloyd GK (1995) Pharmacological characterization of neuronal acetylcholine gated ion channel receptor-mediated hippocampal norepinephrine and striatal dopamine release from rat brain slices. J Pharmacol Exp Ther 274:224–30

Sagdullaev BT, McCall MA, Lukasiewicz PD (2006) Presynaptic inhibition modulates spillover, creating distinct dynamic response ranges of sensory output. Neuron 50: 923–35

Sailer A, Swanson GT, Perez-Otano I, O'Leary L, Malkmus SA, Dyck RH, Dickinson-Anson H, Schiffer HH, Maron C, Yaksh TL, Gage FH, O'Gorman S, Heinemann SF (1999) Generation and analysis of GluR5(Q636R) kainate receptor mutant mice. J Neurosci 19:8757–64

Saitow F, Suzuki H, Konishi S (2005) beta-Adrenoceptor-mediated long-term up-regulation of the release machinery at rat cerebellar GABAergic synapses. J Physiol 565:487–502

Salminen O, Murphy KL, McIntosh JM, Drago J, Marks MJ, Collins AC, Grady SR (2004) Subunit composition and pharmacology of two classes of striatal presynaptic nicotinic acetylcholine receptors mediating dopamine release in mice. Mol Pharmacol 65: 526–35

Sang CN, Hostetter MP, Gracely RH, Chappell AS, Schoepp DD, Lee G, Whitcup S, Caruso R, Max MB (1998) AMPA/kainate antagonist LY293558 reduces capsaicin-evoked hyperalgesia but not pain in normal skin in humans. Anesthesiology 89: 1060–7

Sang CN, Ramadan NM, Wallihan RG, Chappell AS, Freitag FG, Smith TR, Silberstein SD, Johnson KW, Phebus LA, Bleakman D, Ornstein PL, Arnold B, Tepper SJ, Vandenhende F (2004) LY293558, a novel AMPA/GluR5 antagonist, is efficacious and well-tolerated in acute migraine. Cephalalgia 24:596–602

Satake SI, Saitow F, Rusakov D, Konishi S (2004) AMPA receptor-mediated presynaptic inhibition at cerebellar GABAergic synapses: a characterization of molecular mechanisms. Eur J Neurosci 19:2464–74

Scanziani M, Capogna M, Gahwiler BH, Thompson SM (1992) Presynaptic inhibition of miniature excitatory synaptic currents by baclofen and adenosine in the hippocampus. Neuron 9:919–27

Schlicker E, Göthert M (1998) Interactions between the presynaptic alpha2-autoreceptor and presynaptic inhibitory heteroreceptors on noradrenergic neurones. Brain Res Bull 47:129–32.

Schmitz D, Mellor J, Nicoll RA (2001) Presynaptic kainate receptor mediation of frequency facilitation at hippocampal mossy fiber synapses. Science 291:1972–6

Schneggenburger R, Neher E (2005) Presynaptic calcium and control of vesicle fusion. Curr Opin Neurobiol 15:266–74

Scholz KP, Miller RJ (1992) Inhibition of quantal transmitter release in the absence of calcium influx by a G protein-linked adenosine receptor at hippocampal synapses. Neuron 8:1139–50

Seddik R, Schlichter R, Trouslard J (2006) Modulation of GABAergic synaptic transmission by terminal nicotinic acetylcholine receptors in the central autonomic nucleus of the neonatal rat spinal cord. Neuropharmacology 51:77

Shigetomi E, Kato F (2004) Action potential-independent release of glutamate by Ca2+ entry through presynaptic P2X receptors elicits postsynaptic firing in the brainstem autonomic network. J Neurosci 24:3125–35

Sjostrom PJ, Turrigiano GG, Nelson SB (2003) Neocortical LTD via coincident activation of presynaptic NMDA and cannabinoid receptors. Neuron 39: 41

Somogyi P, Klausberger T (2005) Defined types of cortical interneurone structure space and spike timing in the hippocampus. J Physiol (Lond) 562:9–26

Spafford JD, Zamponi GW (2003) Functional interactions between presynaptic calcium channels and the neurotransmitter release machinery. Curr Opin Neurobiol 13:308–14

Staley K (1992) Enhancement of the excitatory actions of GABA by barbiturates and benzodiazepines. Neurosci Lett 146:105–7

Starke K (1977) Regulation of noradrenaline release by presynaptic receptor systems. Rev Physiol Biochem Pharmacol 77:1–124

Starke K (1981) Presynaptic receptors. Annu Rev Pharmacol Toxicol 21:7–30

Suarez LM, Solis JM (2006) Taurine potentiates presynaptic NMDA receptors in hippocampal Schaffer collateral axons. European Journal of Neuroscience 24:405–18

Suarez LM, Suarez F, del Olmo N, Ruiz M, Gonzalez-Escalada JR, Solis JM (2005) Presynaptic NMDA autoreceptors facilitate axon excitability: a new molecular target for the anticonvulsant gabapentin. European Journal of Neuroscience 21:197–209

Südhof TC (2004) The synaptic vesicle cycle. Annu Rev Neurosci 27:509–47

Sundström E, Holmberg L, Souverbie F (1998) NMDA and AMPA receptors evoke transmitter release from noradrenergic axon terminals in the rat spinal cord. Neurochem Res V23:1501

Swope SL, Moss SJ, Raymond LA, Huganir RL (1999) Regulation of ligand-gated ion channels by protein phosphorylation. Adv Sec Mess Phosph Res 33:49–78

Takago H, Nakamura Y, Takahashi T (2005) G protein-dependent presynaptic inhibition mediated by AMPA receptors at the calyx of Held. PNAS 102:7368–73

Tani Y, Saito K, Imoto M, Ohno T (1998) Pharmacological characterization of nicotinic receptor-mediated acetylcholine release in rat brain–an in vivo microdialysis study. Eur J Pharmacol 351:181–8

Tashiro A, Dunaevsky A, Blazeski R, Mason CA, Yuste R (2003) Bidirectional regulation of hippocampal mossy fiber filopodial motility by kainate receptors: a two-step model of synaptogenesis. Neuron 38:773–84

Todd KJ, Slatter CAB, Ali DW (2004) Activation of ionotropic glutamate receptors on peripheral axons of [primary motoneurons mediates transmitter release at the zebrafish NMJ. J Neurophysiol 91:828–40

Trendelenburg U (1957) The action of morphine on the superior cervical ganglion and on the nictitating membrane of the cat. Br J Pharmacol Chemother 12:79–85

Turecek R, Trussell LO (2001) Presynaptic glycine receptors enhance transmitter release at a mammalian central synapse. Nature 411:587

Turecek R, Trussell LO (2002) Reciprocal developmental regulation of presynaptic ionotropic receptors. Proc Natl Acad Sci USA 99:13884–9

Turner TJ, Dunlap K (1995) Pharmacological characterization of presynaptic calcium channels using subsecond biochemical measurements of synaptosomal neurosecretion. Neuropharmacology 34:1469–78

Turner TJ, Mokler DJ, Luebke JI (2004) Calcium influx through presynaptic 5-HT3 receptors facilitates GABA release in the hippocampus: in vitro slice and synaptosome studies. Neuroscience 129:703–18

van Hooft JA, Vijverberg HPM (2000) 5-HT3 receptors and neurotransmitter release in the CNS: a nerve ending story? Trends Neurosci 23:605–10

van Hooft JA, Yakel JL (2003) 5-HT3 receptors in the CNS: 3B or not 3B? Trends iPharmacolog Sci 24:157–60

Veruki ML, Morkve SH, Hartveit E (2006) Activation of a presynaptic glutamate transporter regulates synaptic transmission through electrical signaling. Nat Neurosci 9:1388–96

Vignes M (2001) Regulation of spontaneous inhibitory synaptic transmission by endogenous glutamate via non-NMDA receptors in cultured rat hippocampal neurons. Neuropharmacology 40:737

von Kugelgen I, Starke K (1991) Noradrenaline-ATP co-transmission in the sympathetic nervous system. Trends Pharmacol Sci 12:319–24

Wang Y, Small DL, Stanimirovic DB, Morley P, Durkin JP (1997) AMPA receptor-mediated regulation of a Gi-protein in cortical neurons. Nature 389:502

Wersinger E, Schwab Y, Sahel J-A, Rendon A, Pow DV, Picaud S, Roux MJ (2006) The glutamate transporter EAAT5 works as a presynaptic receptor in mouse rod bipolar cells. J Physiol (Lond) 577:221–34

Wessler I (1989) Control of transmitter release from the motor nerve by presynaptic nicotinic and muscarinic autoreceptors. Trends Pharmacol Sci 10:110–14

Wheeler DB, Randall A, Tsien RW (1994) Roles of N-type and Q-type Ca2+ channels in supporting hippocampal synaptic transmission. Science 264:107–11

Whittaker VP (1993) Thirty years of synaptosome research. J Neurocytol 22:735–42

Wolk R, Shamsuzzaman AS, Svatikova A, Huyber CM, Huck C, Narkiewicz K, Somers VK (2005) Hemodynamic and autonomic effects of smokeless tobacco in healthy young men. J Am Coll Cardiol 45:910–14
Wonnacott S (1997) Presynaptic nicotinic ACh receptors. Trends Neurosci 20:92
Woodhall G, Evans DI Cunningham MO, Jones RSG (2001) NR2B-containing NMDA autoreceptors at synapses on entorhinal cortical neurons. J Neurophysiol 86: 1644–51
Wu LG, Saggau P (1997) Presynaptic inhibition of elicited neurotransmitter release. Trends Neurosci 20:204–12
Xu H, Wu L-J, Zhao M-G, Toyoda H, Vadakkan K, Jia Y, Pinaud R, Zhuo M (2006) Presynaptic regulation of the inhibitory transmission by GluR5-containing kainate receptors in spinal substantia gelatinosa. Molec Pain 2:29
Yang J, Woodhall GL, Jones RSG (2006) Tonic facilitation of glutamate release by presynaptic NR2B-containing NMDA receptors Is increased in the entorhinal cortex of chronically epileptic rats. J Neurosci 26:406–10
Yang K, Kumamoto E, Furue H, Li Y-Q, Yoshimura M (1999) Action of capsaicin on dorsal root-evoked synaptic transmission to substantia gelatinosa neurons in adult rat spinal cord slices. Brain Res 830:268
Yang K, Kumamoto E, Furue H, Yoshimura M (1998) Capsaicin facilitates excitatory but not inhibitory synaptic transmission in substantia gelatinosa of the rat spinal cord. Neurosci Lett 255:135
Ye J-H, Wang F, Krnjevic K, Wang W, Xiong Z-G, Zhang J (2004) Presynaptic glycine receptors on GABAergic terminals facilitate discharge of dopaminergic neurons in ventral tegmental area. J Neurosci 24:8961–74
Yevenes GE, Peoples RW, Tapia JC, Parodi J, Soto X, Olate J, Aguayo LG (2003) Modulation of glycine-activated ion channel function by G-protein betagamma subunits. Nat Neurosci 6:819–24
Youn DH, Randic M (2004) Modulation of excitatory synaptic transmission in the spinal substantia gelatinosa of mice deficient in the kainate receptor GluR5 and/or GluR6 subunit. J Physiol 555:683–98
Yue H-Y, Fujita T, Kawasaki Y, Kumamoto E (2004) AM404 enhances the spontaneous release of -glutamate in a manner sensitive to capsazepine in adult rat substantia gelatinosa neurones. Brain Res 1018:283
Zahniser NR, Doolen S (2001) Chronic and acute regulation of Na+/Cl− -dependent neurotransmitter transporters: drugs, substrates, presynaptic receptors, and signaling systems. Pharmacol Ther 92:21–55
Zamponi GW (2003) Regulation of presynaptic calcium channels by synaptic proteins. J Pharmacol Sci 92:79–83
Zhai RG, Bellen HJ (2004) The architecture of the active zone in the presynaptic nerve terminal. Physiology (Bethesda) 19:262–70
Zhang D, Pan ZH, Awobuluyi M, Lipton SA (2001) Structure and function of GABA(C) receptors: a comparison of native versus recombinant receptors. Trends Pharmacol Sci 22:121–32
Zhang LI, Poo MM (2001) Electrical activity and development of neural circuits. Nat Neurosci 4 Suppl:1207–14
Zhang SJ, Jackson MB (1995) GABAA receptor activation and the excitability of nerve terminals in the rat posterior pituitary. J Physiol 483 (Pt 3):583–95
Zheng F, Johnson SW (2001) Glycine receptor-mediated inhibition of dopamine and non-dopamine neurons of the rat ventral tegmental area in vitro. Brain Res 919: 13
Zhou F-M, Hablitz JJ (1999) Activation of serotonin receptors modulates synaptic transmission in rat cerebral cortex. J Neurophysiol 82:2989–99
Zucker RS, Regehr WG (2002) Short-term synaptic plasticity. Annu Rev Physiol 64: 355–405

G protein signalling. NO has further been suggested to increase transmitter release through presynaptic clustering of α-synuclein. Alternative modes of NO/cGMP signalling resulting in inhibition of transmitter release and long-term depression of synaptic activity will also be addressed, as well as anterograde NO signalling in the cerebellum. Finally, emerging evidence for cGMP signalling through CNG channels and hyperpolarization-activated cyclic nucleotide-gated (HCN) channels will be discussed.

# 1 Introduction

The gas nitric oxide (NO) is well suited to serve as a messenger molecule because of its capacity for rapid diffusion in both aqueous and lipid environments. A myriad of reports document a broad range of NO effects in the CNS, such as the modulation of neuronal development, nociception, apoptosis, synaptic plasticity, and complex behavioral responses (for review see Prast and Philippu 2001; Hofmann et al. 2003). Nearly 20 years ago, Garthwaite and co-workers (1988) recognized for the first time the potential of NO as a neuromodulator. They observed that activation of $Ca^{2+}$-permeable ionotropic glutamate receptors in cultured cerebellar granule cells triggers the release of a messenger similar to endothelium-derived relaxing factor, which is identical to NO (Ignarro et al. 1987; Furchgott 1996). Based on these results, they put forward the truly inspiring hypothesis that this factor may provide a universal link from postsynaptic activity to functional modifications of neighbouring presynaptic terminals and glial cells. Ironically, the concept of retrograde signalling via NO has been proven in many regions of the brain, but does not hold true for glutamatergic parallel fiber synapses in the cerebellum (see Section 3.2.2). This chapter aims to summarize the current knowledge about NO-dependent mechanisms implicated in the modulation of synaptic transmission. First, the cellular and molecular components involved in NO signalling in the brain will be described. Then, we will provide a synopsis of NO/cGMP-dependent effects on synaptic transmission and corresponding signal transduction based on observations from electrophysiological studies and studies with radioactive or fluorescent markers for presynaptic function. With regard to the general topic of this volume, which is the control of transmitter release, the focus will be set on presynaptic signalling mechanisms of NO and its intracellular second messenger cyclic guanosine-3′,5′-monophosphate (cGMP).

# 2 Overview of the NO/cGMP Signalling System and Its Expression in the CNS

## *2.1 Sources of NO and cGMP*

NO is produced by a complex reaction via oxidative release from L-arginine giving rise to L-citrulline. Three mammalian isozymes catalyzing this reaction have been identified in various cell types, the constitutively expressed neuronal and endothelial NO synthase (nNOS/NOS1 and eNOS/NOS3), and the inducible NO synthase (iNOS /NOS2). The nNOS is ubiquitously and abundantly expressed throughout the CNS and represents the principal source of NO in many neuronal populations (Bredt and Snyder 1990; Dawson and Dawson 1996; Prast and Philippu 2001). It is a $Ca^{2+}$/calmodulin-regulated enzyme, which can be activated by $Ca^{2+}$ influx via N-methyl-D-aspartate (NMDA) receptors (Garthwaite et al. 1988). This functional relationship is thought to be especially effective, because the scaffolding molecule PSD-95 keeps the nNOS protein in close proximity to NMDA receptors (Christopherson et al. 1999; Valtschanoff and Weinberg 2001). The eNOS, another $Ca^{2+}$/calmodulin-dependent NOS initially detected in endothelial cells, has also been reported to be expressed in hippocampal pyramidal cells and neurons of other brain regions (Dinerman et al. 1994; O'Dell et al. 1994). However, this observation has been challenged by others (Stanarius et al. 1997; Demas et al. 1999; Blackshaw et al. 2003). It is now suggested that eNOS expression is confined to endothelial cells, and that NO released within the cerebral vasculature is able to signal to axons located in close proximity (Garthwaite et al. 2006). In contrast to the two constitutive forms of NOS, the iNOS is normally not detectable in the CNS, but upregulated following toxic or inflammatory stimuli.

In many cases, NO acts in the brain by increasing the concentration of cGMP, which is achieved through activation of the canonical NO receptor, the soluble guanylyl cyclase (sGC) (Figure 1) (Friebe and Koesling 2003). The NO-sensitive sGC is a heme-containing enzyme consisting of two different subunits, $\alpha$ and $\beta$. Heterodimers of the composition $\alpha_1\beta_1$ and $\alpha_2\beta_1$ have been shown to be functional enzymes, while the role of a putative $\beta_2$ subunit remains unclear (for review see Koesling et al. 2005). The $\alpha_1\beta_1$ isoform is ubiquitously expressed in the CNS, whereas expression of $\alpha_2\beta_1$ may be restricted to specific areas in the brain (Mergia et al. 2003; Ding et al. 2004; Russwurm and Koesling 2004). Taken together, it appears that principally every neuron in the CNS can generate NO and/or cGMP. Note that, due to the diffusible nature of NO, the sites of NO and cGMP synthesis must not necessarily overlap. A useful tool to analyze and compare the expression patterns of components of the NO/cGMP signalling pathway in the brain is the recently released Allen Brain Atlas (http://www.brainatlas.org/aba/), which contains the mRNA expression profiles of virtually all mouse genes in the brain (Lein et al. 2006)

It is important to note that many cell types possess alternative NO-independent mechanisms to generate cGMP (Figure 1), namely through activation of membrane-

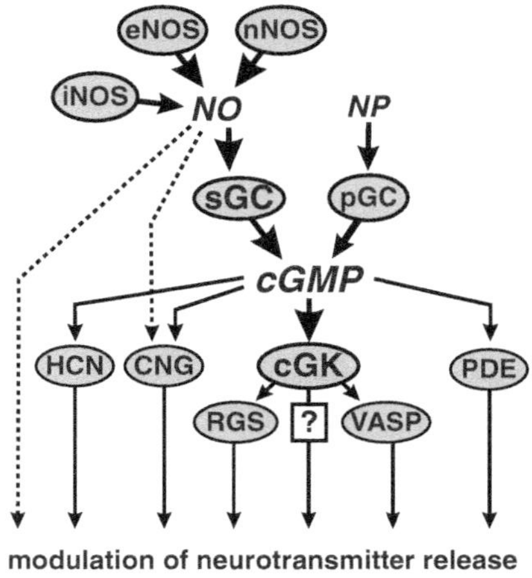


**Fig. 1** NO signalling pathways. Nitric oxide (NO) can be generated by endothelial (eNOS), neuronal (nNOS), and inducible (iNOS). NO synthase and activates the cytosolic soluble guanylyl cyclase (sGC), leading to a rise of intracellular cGMP, which in turn activates cGMP-dependent protein kinase (cGK). An increase of cGMP can also be induced by natriuretic peptides (NP) through activation of membrane-bound particulate guanylyl cyclases (pGC). cGK substrates that may be functionally important in the brain include the vasodilator-stimulated phosphoprotein (VASP protein) and regulators of G protein signalling (RGS proteins). The question mark indicates unknown cGK substrates. cGMP can also signal via cGK-independent mechanisms, e.g., by activating cyclic nucleotide-gated (CNG) channels, hyperpolarization-activated cyclic nucleotide-gated (HCN) channels, or by modulating various phosphodiesterases (PDE). Note that the mechanisms that modulate transmitter release downstream of cGMP receptors are not well understood. There have been proposed signalling mechanisms of NO not related to the cGMP pathway (dotted lines). For example, NO may trigger $Ca^{2+}$ entry by S-nitrosylation of a CNG channel subunit.

bound particulate guanylyl cyclases (pGCs), e.g., GC-A (NPR-A) and GC-B (NPR-B), by natriuretic peptides (NPs) (for review see Kuhn 2004). NPs comprise a family of three homologous members, atrial (ANP), brain (BNP), and C-type (CNP) natriuretic peptide. Despite the fact that BNP and CNP as well as their receptors (GC-A and GC-B) are found in the brain (Sudoh et al. 1988, 1990; Komatsu et al. 1991; Herman et al. 1993; DiCicco-Bloom et al. 2004), their function in the CNS remains largely elusive and, therefore, NPs and pGCs will not be discussed further in the following sections. Recent findings suggest a modulation of α-amino-3-hydroxy-5-methylisoxazole-4-propionic acid (AMPA) receptors and ionotropic γ-amino butyric acid ($GABA_A$) receptors by NPs in retinal neurons (Tian and Yang 2006; Yu et al. 2006).

## *2.2 cGMP Receptors*

Extensive analysis of the cGMP system has led to the identification of several effectors for this intracellular second messenger. The classical cGMP receptors include cGMP-dependent protein kinases (cGKs), cyclic nucleotide-gated (CNG) cation channels, members of the family of hyperpolarization-activated cyclic nucleotide-gated (HCN) channels, and of the family of cyclic nucleotide hydrolysing phosphodiesterases (PDEs). We will not discuss mechanisms related to cGMP-mediated cross-activation of cAMP-dependent protein kinase.

### 2.2.1 cGMP-Dependent Protein Kinases (cGKs)

A major role in the signal transduction of NO and cGMP in the nervous system has been assigned to cGKs (for review see Feil R et al. 2005b; Hofmann et al. 2006). cGKs belong to a family of serine/threonine kinases found in diverse eukaryotic organisms and various tissues, including the nervous system (Muller 1997; Wang and Robinson 1997; Francis and Corbin 1999; Pfeifer et al. 1999). Mammals possess two cGK genes, *prkg1* and *prkg2*, that encode cGKI and cGKII, respectively. The cGKI gene encodes two isoforms, cGKIα and cGKIβ. These isoforms are identical except for their N-termini (≈90–100 amino acid residues), which are encoded by two alternatively used 5′-exons of the cGKI gene. In vitro, cGKs are activated at submicromolar to micromolar concentrations of cGMP, the cGKIα isoform being ≈10-fold more sensitive to activation by cGMP than cGKIβ (Ruth et al. 1991). Both the cGKI and cGKII are homodimers of subunits with a molecular weight of ≈75 kDa and ≈85 kDa, respectively. Each subunit is composed of three functional domains. The *N-terminal domain* comprises regions that control dimerization, targeting, and the activity of the catalytic center. In contrast to the cytosolic cGKI, the N-terminus of the cGKII is myristoylated, thereby anchoring the enzyme to the plasma membrane. The *regulatory domain* contains two tandem cGMP-binding sites that bind cGMP with high and low affinity and interact allosterically. Occupation of both binding sites induces a large change in secondary structure (Landgraf et al. 1990) to yield a more elongated protein molecule (Wall et al. 2003). The *catalytic domain* bears the kinase activity and contains the MgATP- and peptide-binding pockets. Binding of cGMP to both sites in the regulatory domain releases the inhibition of the catalytic center by an N-terminal autoinhibitory/pseudosubstrate site and allows the phosphorylation of serine/threonine residues in target proteins and in the N-terminal autophosphorylation site.

Both cGKs are expressed in the nervous system, while the cGKI appears to be the dominating isoform (http://www.brainatlas.org/aba/). Correlating with the maximum of enzymatic cGK activity in the brain, the cGKIα isoform is highly expressed in cerebellar Purkinje cells (Hofmann and Sold 1972; Lohmann et al. 1981; Geiselhoringer et al. 2004; Feil S et al. 2005). Significant levels of cGKIα have also been observed in nociceptive neurons of the dorsal root ganglia (Qian et al. 1996; Schmidt et al. 2002; Tegeder et al. 2004). Other brain regions with cGKI expression include the dorsomedial nucleus of the hypothalamus, the suprachiasmatic nucleus, the hippocampus, the amygdala, the olfactory bulb, and the cerebral cortex (El-Husseini et al. 1999; Kleppisch et al. 1999; Revermann et al. 2002; Feil S et al. 2005). Noteworthy, cGKIβ appears to be the prevailing isozyme in these latter regions (Geiselhoringer et al. 2004; Feil S et al. 2005). For the cGKII, low to moderate levels of expression have been reported in the hippocampus, the olfactory tubercle, the suprachiasmatic nucleus, and the thalamus (El-Husseini et al. 1995, 1999; Kleppisch et al. 1999; De Vente et al. 2001; Oster et al. 2003).

Several substrates of cGKs have been identified unequivocally in non-neuronal tissues (Hofmann et al. 2006), while evidence for functional cGK substrates in the nervous system is lagging behind. Table 1 summarizes proteins phosphorylated by

**Table 1** cGK Substrates with Possible **Funtion** in the Nervous System

| Substrate | Possible Function of Phosphorylation in the CNS | Brain Region Experimental System | References |
|---|---|---|---|
| G-substrate | Protein phosphatase inhibition<br>Initiation of cerebellar LTD | Cerebellum | Aswad and Greengard 1981; Ajima and Ito 1995; Hall et al. 1999; Endo et al. 2003 |
| DARPP-32 | Protein phosphatase inhibition<br>Modulation of signallingr pathways | Substantia nigra | Tsou et al. 1993; Nishi et al. 2005 |
| $IP_3$ receptor type I | Stimulation of $Ca^{2+}$ release from $IP_3$ sensitive stores | Cerebellum | Haug et al. 1999; Wagner et al. 2003 |
| VASP | Regulation of actin dynamics<br>Vesicle trafficking<br>Aggregation of vesicle proteins | Cerebellum, Hippocampus | Hauser et al. 1999; Arancio et al. 2001; Wang et al. 2005 |
| RhoA | Regulation of actin dynamics | Hippocampus | Wang et al. 2005[#] |
| | Vesicle trafficking | Rat C6 glioma cells | Zhuang et al. 2004 |
| | Aggregation of vesicle proteins | Vascular smooth muscle | Sauzeau et al. 2000 |
| | | NIH 3T3 cells | Ellerbroek et al. 2003 |
| Septin-3 | Vesicle trafficking | Hippocampus<br>COS7 cells | Xue et al. 2000, 2004 |
| PDE5 | Enhanced cGMP degradation | Cerebellar Purkinje cells | Shimizu-Albergine et al. 2003 |
| | Accelerated termination of cGMP signal | Smooth muscle | Rybalkin et al. 2002 |

(*Continued*)

**Table 1** Continued

| Substrate | Possible Function of Phosphorylation in the CNS | Brain Region Experimental System | References |
|---|---|---|---|
| RGS3 and RGS4 | Termination of GPCR activation | Astrocytes | Pedram et al. 2000 |
| RGS2 | Termination of GPCR activation | Hippocampus | Oliveira-Dos-Santos et al. 2000, Han et al. 2006[§] |
| | Downregulation of G protein-mediated presynaptic inhibition | Vascular smooth muscle | Tang et al. 2003 |
| | Increase in transmitter release | | |
| ADP ribosyl cyclase | Decreased transmitter release<br>Initiation of hippocampal LTD | Hippocampus | Fossier et al. 1999; Reyes-Harde et al. 1999a,b |
| $BK_{Ca}$ | Increased open probability | Pituitary nerve terminals | Klyachko et al. 2001 |
| | Enhances afterhyperpolarisation | Smooth muscle | Sausbier et al. 2000 |

Substrates were suggested according to *in vivo* or *in vitro* phosphorylation studies or based on functional studies. Note that most of the proteins were described as substrates of cGKI; cGKII substrates in the brain are poorly characterized.

[#] Wang et al. (2005) suggest that RhoA, though involved in synaptic plasticity, is not a downstream target of cGK. RhoA has been identified as a cGK substrate in other tissues.

[§] Studies in the hippocampus solely support a possible function of RGS2 in synaptic transmission, but do not prove its phosphorylation by cGK. RGS2 has been identified as a cGK substrate in other tissues.

cGKs in neurons and known substrates of cGKs that are expressed in the brain and known to modulate transmitter release. In general, cGK substrates identified so far include ion channels, G proteins and associated regulators, and cytoskeleton-associated proteins.

### 2.2.2 Cyclic Nucleotide-Gated (CNG) Channels

Cytosolic cGMP can activate CNG channels. These ion channels are not only an important part of the signal transduction pathway in the visual and olfactory system (Biel et al. 1999; Hofmann et al. 2003), but are also expressed in other neurons where they may be involved in the modulation of various functions by NO. The CNG channel family comprises six homologous subunits. Based on phylogenetic relationship, these proteins are classified as A subunits (CNGA1-A4) and B subunits (CNGB1 and CNGB3) (Biel et al. 1999; Bradley et al. 2001). Native CNG channels are heterotetramers composed of A and B subunits, and their sensitivity for cGMP largely depends on the individual subunit composition. For instance, the cGMP-gated rod channel consists of CNGA1 (Kaupp et al. 1989) and a long splice variant of CNGB1 (CNGB1a) (Korschen et al. 1995). Activation of CNG channels causes depolarization and, thus, facilitates excitation of neurons. In addition, CNG channels have been suggested to permit $Ca^{2+}$ influx relevant for synaptic function (Broillet and Firestein 1997; Parent et al. 1998).

### 2.2.3 Phosphodiesterases (PDEs)

Cyclic nucleotide signalling is controlled not only at the level of cAMP and cGMP synthesis, but also by the rate of cyclic nucleotide degradation via PDEs (Sonnenburg and Beavo 1994). The PDE superfamily comprises 11 subfamilies with 21 genes that are transcribed into perhaps more than 50 enzyme species by alternative splicing, and virtually all PDEs are expressed in the CNS (Menniti et al. 2006). The families are classified according to substrate specificity, mechanism of regulation, sensitivity to inhibitors, and amino-acid sequence. Cyclic GMP is degraded by a number of cGMP-specific PDEs as well as by dual-substrate PDEs that hydrolyse both cAMP and cGMP. Importantly, cGMP can also inhibit or activate specific PDE subtypes via binding to their regulatory domains. Thereby, cGMP can modulate its own level and/or that of cAMP. For instance, cGMP can increase cAMP levels via binding to the cGMP-inhibited cAMP-PDE, PDE3A; it can lower cyclic nucleotide levels by binding to the cGMP-stimulated cAMP/cGMP-PDE, PDE2A, or by binding to the cGMP-specific PDE5A, the target of sildenafil. Thus, the cGMP-regulated PDEs are important cGMP receptors helping to shape the spatiotemporal profile of cyclic nucleotide signals. Particularly, they might represent "switches" that transform cGMP signals into cAMP signals, allowing cyclic nucleotide cross-talk in the brain. Another interesting mode of PDE regulation is represented by the dual-substrate PDE1 family. This family has the unique feature of being activated by the binding of $Ca^{2+}$/calmodulin, providing a mechanism for cross-talk between the $Ca^{2+}$ and cyclic nucleotide signalling pathways.

## *2.3 cGMP-Independent NO Signalling*

It is important to note that not all effects of NO on synaptic transmission can be accounted for by cGMP-dependent signalling, i.e., NO can also signal through alternative mechanisms independent of cGMP. Owing to the limited space and the general scope of this chapter, we will not cover these mechanisms in detail but just discuss them briefly. First, NO can give rise to free radical species that, in turn, may affect cellular functions through various mechanisms (for review see Stamler et al. 1997). Second, NO can modulate proteins controlling neuronal functions by stimulating their ADP ribosylation (Schuman et al. 1994; Zhang et al. 1994; Sullivan et al.

1997). This mechanism may account partially for the stimulatory action of NO on transmitter release in hippocampal synapses (Schuman et al. 1994). Last but not least, NO may modulate cellular functions by S-nitrosylation of various proteins. For example, S-nitrosylation of nuclear proteins associated with cAMP response element (CRE) binding protein is involved in regulation of its DNA binding and, hence, CRE-mediated gene expression (Riccio et al. 2005). S-nitrosylation has also been reported to modify the function of various ion channels (Broillet and Firestein 1996, 1997; Xu et al. 1998; Jaffrey et al. 2001).

# 3 NO and cGMP as Modulators of Synaptic Transmission

## 3.1 *Effects of NO/cGMP on Transmitter Release*

The proposal that NO might act presynaptically to modify transmitter release (Garthwaite et al. 1988) prompted a plethora of investigations into the effects of NO donors and inhibitors of NOS and sGC on transmitter release. Studies were performed in vitro, e.g., on brain slices, synaptosomes, or cultured neurons, or in vivo using microdialysis or the push-pull superfusion technique. The effects of NO donors suggest that the release of several transmitters, including acetylcholine, catecholamines, and excitatory and inhibitory amino acids, may be influenced by endogenous NO. NO donors also inhibit the uptake of various radiolabeled transmitters (Pogun et al. 1994). However, these NO donor studies usually lacked important controls for potential unspecific effects of NO carrier molecules (Guevara-Guzman et al. 1994). Other experiments more convincingly show that authentic NO (Guevara-Guzman et al. 1994) as well as endogenous NO that is "tonically" present in unstimulated tissue (Prast and Philippu 1992; Bugnon et al. 1994) or "phasically" generated by NMDA receptor activation (Dickie et al. 1992; Hanbauer et al. 1992; Sorkin 1993; Montague et al. 1994) modulates transmitter release. Recent in vitro studies suggest that tonic NO might come from eNOS in blood vessels, while the phasic NO signals are produced by nNOS in neurons (Garthwaite et al. 2006; Hopper and Garthwaite 2006). The concept of vessel-to-neuron signalling by NO is supported by the finding that mice lacking eNOS in the cerebral vasculature exhibit various neuronal phenotypes, including altered transmitter release or turnover (Kano et al. 1998; Frisch et al. 2000).

It appears that NO, predominantly via cGMP-dependent mechanisms, regulates primarily the release of glutamate, which then modulates the release of various other transmitters in several brain regions, such as the hippocampus, the striatum, the hypothalamus, and the locus coeruleus (Prast and Philippu 2001). Depending on the brain region and NO concentration, NO can both stimulate and inhibit the release of a particular transmitter. Based on the observations of many groups, Prast and Philippu (2001) have proposed the following model for the modulation of acetylcholine release by endogenous NO: NO stimulates cGMP synthesis within glutamatergic neurons. Depending on the NO concentration, glutamate release is either enhanced (high NO) or decreased (low NO), resulting in either enhanced or decreased activation of postsynaptic NMDA or AMPA/kainate receptors on cholinergic neurons followed by enhanced or decreased acetylcholine release. NO may influence cholinergic transmission not only via excitatory glutamatergic neurons, but also via effects on inhibitory GABAergic neurons. The release of catecholamines, histamine, and serotonin may be modulated in a similar manner.

Basically, then, NO seems to be able to modulate vesicular release of transmitter in either direction, or not at all, depending on the coincident level of presynaptic activity and NO concentration. The concept of an activity-dependent retrograde NO signal that is generated in the postsynapse and then diffuses into the presynapse to regulate transmitter release has been investigated extensively, due to its possible involvement in neuronal excitability and memory processes (see Section 3.2.1).

Although many studies have indicated that NO/cGMP can modulate transmitter release (Arancio et al. 1995; Meffert et al. 1996; Sporns and Jenkinson 1997; Klyachko et al. 2001), the molecular mechanisms and functional relevance of presynaptic NO/cGMP signalling are not well understood. It is generally accepted that activity-dependent changes in synaptic strength, such as long-term potentiation (LTP) and long-term depression (LTD), underlie learning and memory (Pastalkova et al. 2006; Whitlock et al. 2006). However, there are ongoing debates whether a given form of synaptic plasticity in a given brain region results from presynaptic and/or postsynaptic alterations, as will be exemplified below for the hippocampus. Conflicting evidence for presynaptic changes has been obtained largely from the classical electrophysiological approach of quantal analysis of pairs of monosynaptically connected neurons, vesicular antibody uptake, and postsynaptic drug-infusion studies (Stanton et al. 2003 and refs. therein). A more direct visualization of presynaptic vesicle release has been achieved with the fluorescent styryl dye FM1-43 (Betz and Bewick 1992; Ryan et al. 1993). FM1-43 is taken up into synaptic vesicles in an activity-dependent manner as a result of endocytosis after transmitter release. Subsequent synaptic stimulation evokes the release of the dye by exocytosis, which is visualized as a destaining whose rate is a direct measure of presynaptic release efficacy. FM1-43 imaging of cultured neurons and acute brain slices demonstrated that certain forms of LTP or LTD in the hippocampus depend, at least in part, on increased or decreased transmitter release from the presynaptic neuron. Interestingly, retrograde NO signalling appears to contribute to both presynaptic LTP and LTD of transmitter release.

A model of retrograde NO signalling in a glutamatergic synapse is presented in Figure 2. The pathway comprises (1) an action potential-driven release of glutamate from presynaptic axon terminals, (2) activation of postsynaptic NMDA receptors, which trigger the synthesis of NO, (3) diffusion of NO to a presynaptic site, where (4) it stimulates cGMP production and activation of cGK leading to (5) increase (as illustrated in Figure 2) or decrease (not shown in Figure 2) of transmitter release (Sporns and Jenkinson 1997; Stanton et al. 2001, 2003, 2005). In line with such a model, increases in cGMP have been implicated in the induction of both hippocampal LTP (Arancio et al. 2001) and LTD (Gage et al. 1997; Reyes-Harde et al. 1999a).

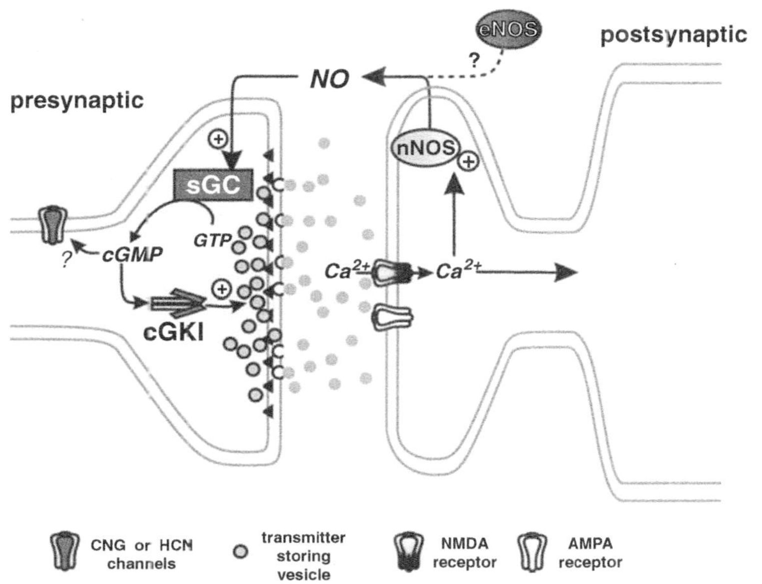


**Fig. 2** Retrograde NO signalling in a glutamatergic synapse. Nitric oxide (NO) generated postsynaptically by $Ca^{2+}$/calmodulin-activated neuronal NO synthase (nNOS) and/or derived from endothelial NO synthase (eNOS) in nearby vessels diffuses to the presynaptic terminal, where it activates the soluble guanylyl cyclase (sGC). The resulting increase of the intracellular second messenger cGMP activates various receptors, including the cGMP-dependent protein kinase I (cGKI). Through phosphorylation of its substrates, cGKI leads to an increase in presynaptic transmitter release. This action is thought to involve clustering of vesicular proteins and other proteins of the docking/fusion machinery (black triangles) at the release sites. In addition, cGMP may modulate transmitter release by activating presynaptic ion channels regulated by cyclic nucleotides (CNG and HCN channels).

However, the mechanisms or mediators controlling the switch from one to the other effect dominating the net change remain to be identified (see also Section 3.2.1). The dual role of the NO/cGMP/cGK pathway in both LTP and LTD may also explain some contradictory reports on the role of this pathway in synaptic plasticity.

What are the molecular mechanisms of presynaptic NO/cGMP signalling that alter transmitter release? In some systems, NO-induced transmitter release occurs independently of an increase in presynaptic $Ca^{2+}$ (Schuman et al. 1994; Stewart et al. 1996), perhaps by modulating the interaction of components of the vesicle docking/fusion machinery (Meffert et al. 1996). On the other hand, it has been reported that one action of presynaptic cGMP required for LTD induction is the stimulation of presynaptic ryanodine receptor-mediated $Ca^{2+}$ release (Reyes and Stanton 1996; Reyes-Harde et al. 1999a), but how an *increase* in presynaptic $Ca^{2+}$ results in long-term *reduction* of release probability is not clear. An alternative mechanism to increase $Ca^{2+}$ is the stimulation of presynaptic $Ca^{2+}$ channels. In the brain stem, for instance, the NO/cGMP/cGK pathway might facilitate glutamate release through an enhancement of presynaptic N-type $Ca^{2+}$ channel activity (Huang et al. 2003).

In posterior pituitary nerve terminals, NO stimulates $Ca^{2+}$-activated $K^+$ (BK) channel activity via a cGMP/cGK-dependent mechanism (Klyachko et al. 2001). Opening of BK channels results in $K^+$ efflux and membrane hyperpolarization. The enhancement of BK channel activity by cGMP was greatest at depolarized potentials, so it will manifest during and immediately after action potentials and, thus, increase the magnitude of the spike afterhyperpolarization, which in turn promotes $Na^+$ channel recovery from inactivation. This mechanism would reduce action potential failures and allow more $Ca^{2+}$ to enter followed by increased transmitter release. The stimulation of presynaptic $K^+$ channels by the NO/cGMP/cGK cascade could generate a short-term, use-dependent enhancement of release at nerve terminals and may also operate at a certain stage of LTP induction in the hippocampus (Klyachko et al. 2001).

After exocytosis, synaptic vesicles must be retrieved and refilled with transmitters to supply the needs of an active neuron. Work by Micheva and colleagues (2001, 2003) showed that retrograde signalling via NO promotes synaptic vesicle recycling in CNS neurons, especially at synapses with high levels of recycling, i.e., in an activity-dependent manner. The authors monitored exo- and endocytosis in cultured hippocampal neurons using a pH-sensitive green fluorescent protein (GFP) reporter protein and the fluorescent dye FM4-64. Based on fluorescence imaging and pharmacological analysis they proposed that NMDA receptor-dependent postsynaptic production of NO accelerates vesicle endocytosis by increasing the level of presynaptic phosphatidylinositol 4,5-bisphosphate ($PIP_2$) via a cGMP-dependent mechanism. $PIP_2$ has been extensively implicated in vesicle cycling (Martin 2001; Osborne et al. 2001), but how NO/cGMP leads to an increase in $PIP_2$ at the synapse is not known. Studies in non-neuronal cells suggest that the NO/cGMP/cGK pathway can inhibit the $PIP_2$-hydrolysing phospholipase C (Clementi et al. 1995; Xia et al. 2001).

Recent results support the view that microstructural changes occur already at early stages of long-lasting synaptic plasticity (Engert and Bonhoeffer 1999; Maletic-Savatic et al. 1999). For instance, glutamate-induced potentiation of synaptic transmission in cultured hippocampal neurons is accompanied by a rapid increase in clusters of presynaptic synaptophysin and postsynaptic GluR1 protein (Antonova et al. 2001). A recent study by Wang and colleagues (2005) suggests that NO/cGMP/cGKI signalling, actin, and the small GTPase RhoA play important roles in potentiation in cultured hippocampal neurons. Surprisingly, cGMP and cGKI appear to act directly in both the presynaptic and postsynaptic neurons, where they contribute to an increased aggregation of synaptic proteins, perhaps via phosphorylation of the cGKI substrate vasodilator-stimulated phosphoprotein (VASP) and regulation of the actin cytoskeleton. The structural changes promoted by the NO/cGMP/cGKI pathway may not only affect transmitter release and strengthen existing synapses, but may also trigger the activity-dependent formation of new synapses (cf. Wang et al. 2005). These mechanisms will be discussed in more detail later in the context of hippocampal LTP (see Section 3.2.1).

### 3.2 *Long-term Potentiation (LTP) and Long-Term Depression (LTD) as Model Systems to Study NO/cGMP Signalling in Central Synapses*

Long-lasting activity-dependent changes of synaptic transmission in a neural network are thought to serve as cellular mechanisms for learning and memory (Hebb 1949). Two prominent forms of synaptic plasticity, LTP and LTD, are indeed associated with different types of learning (Rogan et al. 1997; Manahan-Vaughan and Braunewell 1999; Ito 2002; Whitlock et al. 2006) and have become well-established model systems for studying the modulation of synaptic transmission in various brain regions (Malenka and Bear 2004). It is now generally acknowledged that both pre- and postsynaptic processes can contribute to the changes in glutamatergic synaptic transmission seen after induction of LTP and LTD (for review see Bear and Malenka 1994; Larkman and Jack 1995; Lisman and Raghavachari 2006). Postsynaptic mechanisms ultimately lead to changes in the density and/or function of AMPA type glutamate receptors. For example, LTP is associated with the insertion of AMPA receptors directly into the synaptic region and/or into the extrasynaptic membrane with subsequent diffusional redistribution to the synaptic region. Conversely, internalization of AMPA receptors from these sites plays an important role in LTD. Presynaptic mechanisms lead to a modification of glutamate release, e.g., by changing the mode of vesicle fusion and/or the number of vesicles released in response to afferent stimulation. In addition, aligning presynaptic release sites with postsynaptic receptors may enhance synaptic transmission (Antonova et al. 2001; Wang et al. 2005). NO and cGMP have been reported to signal through many of these mechanisms (see also Section 3.1). The subsequent section is dedicated to the function of NO as a retrograde messenger, i.e., its presynaptic signalling.

#### 3.2.1 NO as a Retrograde Messenger in Hippocampal LTP

Considerable efforts have been made to elucidate the role of NO for synaptic plasticity in various brain regions (for review see Larkman and Jack 1995; Huang 1997; Daniel et al. 1998; Hawkins et al. 1998; Feil R et al. 2005b). The following discussion focuses on findings about the role of NO for LTP in Schaffer collateral/CA1 synapses of the hippocampus, a NMDA receptor-dependent form of synaptic plasticity characterized in great detail (for review see Malenka and Bear 2004; Lisman and Raghavachari 2006). In spite of some controversial findings, it is accepted that NO (1) is involved in LTP in the Schaffer collateral pathway and (2) acts as a retrograde messenger (Garthwaite and Boulton 1995; Arancio et al. 1996; Huang 1997; Hawkins et al. 1998). Both nNOS and eNOS have been detected in postsynaptic pyramidal neurons (Dinerman et al. 1994; O'Dell et al. 1994; Brenman et al. 1996) and various NOS inhibitors suppress LTP in the CA1 region (Schuman and Madison 1991; Arancio et al. 1996). Analysis of mice lacking nNOS and/or eNOS revealed the functional contribution of either NOS isoform to hippocampal LTP (Table 2). Mice deficient in either nNOS or eNOS appear to be capable of normal LTP in the CA1 region (O'Dell et al. 1994; Son et al. 1996), while a strong impairment was observed in compound nNOS/eNOS knockout mice (Son et al. 1996). Other groups have shown that the lack of eNOS alone can result in reduced LTP in the Schaffer collateral pathway (Wilson et al. 1999) and other regions (Haul et al. 1999; Doreulee et al. 2001). A major role for this isoform has also been suggested by Kantor et al. (1996), who reported that disrupting the localization of eNOS to the membrane causes impairment of LTP, and that this effect could be overcome by expressing a chimeric form of eNOS constitutively targeted to the membrane. On the other hand, the function of nNOS has been underscored by a recent report that selective inhibitors of this isoform reduce hippocampal LTP (Hopper and Garthwaite 2006). It was further suggested that phasic and tonic NO signals are needed for LTP, and that these signals are derived from nNOS in hippocampal pyramidal cells in response to neural activity and eNOS expressed in nearby vascular endothelial cells, respectively (Garthwaite et al. 2006; Hopper and Garthwaite 2006). The findings discussed so far demonstrate the ability of NO to support LTP, but they do not discriminate between possible sites of its action. The observation that bath application of hemoglobin, a membrane-impermeable scavenger of NO, blocks LTP (O'Dell et al. 1991) suggests that NO travels extracellularly through the synaptic cleft. Additional evidence for a presynaptic action of NO comes from studies of miniature excitatory postsynaptic currents (EPSC) (O'Dell et al. 1991) and NMDA receptor-dependent LTP in cultured hippocampal neurons (Arancio et al. 1995, 1996, 2001; Wang et al. 2005). NO increases the frequency of spontaneous miniature EPSCs. Oxymyoglobin, another membrane-impermeable NO scavenger, is able to suppress LTP between pairs of cultured neurons following a tetanus, but its extracellular application has no effect on LTP induced by a weak tetanus in conjunction with photolytic uncaging of NO in the presynaptic cell. These data can be compiled into a scheme with NO as a retrograde messenger in hippocampal LTP (Figure 2). The following section will review efforts made to unravel presynaptic signalling mechanisms of NO involved in the strengthening of synaptic transmission.

## Role of the Canonical NO/cGMP/cGK Pathway in Retrograde Signalling

There are some reports that NO can support LTP through cGMP-independent mechanisms, e.g., by stimulating the ADP ribosylation of proteins regulating synaptic transmission (Schuman et al. 1994; Zhang et al. 1994; Sullivan et al. 1997). Also, direct modulation of a CNG channel subunit by NO may cause a $Ca^{2+}$ influx with functional impact on synaptic function (Broillet and Firestein 1997; Parent et al. 1998). But it is no big surprise that the bulk of pharmacological studies as well as the phenotypes of transgenic mouse models (Table 2) support the conclusion that NO signals through its canonical pathway via the sGC, cGMP, and cGKs. LTP-inducing stimuli elicit an increase of cGMP in the hippocampus that is sensitive to NOS inhibitors and NO scavengers (Chetkovich et al. 1993). Moreover,

**Table 2** Phenotypes Related to Synaptic Plasticity in Transgenic Mice with Genetic Alterations of NO/cGMP Signalling

| Gene | Mouse Model | Phenotypes | References |
|---|---|---|---|
| nNOSα | Null mutation* | Reduced LTD in cerebellar parallel fiber synapses | Lev-Ram et al. 1997b |
| | | Normal LTP in Schaffer collateral pathway | O'Dell et al. 1994 |
| eNOS | Null mutation | Defective LTP in Schaffer collateral pathway | Wilson et al. 1999 |
| | | Defective LTP mossy fiber pathway | Doreulee et al. 2001 |
| | | Defective LTP in cerebral cortex | Haul et al. 1999 |
| eNOS + nNOSα | Null mutation# | Defective LTP in Schaffer collateral pathway<br>Normal LTP in stratum oriens | Son et al. 1996 |
| sGC α1 | Null mutation | Reduced LTP in visual cortex | Mergia et al. 2006 |
| sGC α2 | Null mutation | Reduced LTP in visual cortex | Mergia et al. 2006 |
| cGKI | Null mutation | Normal LTP in Schaffer collateral pathway of young mice (4 weeks old) | Kleppisch et al. 1999 |
| cGKII | Null mutation | Normal LTP in Schaffer collateral pathway of adult mice (12 weeks old) | Kleppisch et al. 1999 |
| cGKI + cGKII | Null mutation | Normal LTP in Schaffer collateral pathway of young mice (4 weeks old) | Kleppisch et al. 1999 |
| cGKI | Hippocampus-specific mutation | Reduced L-LTP in Schaffer collateral pathway of adult mice (12 weeks old) | Kleppisch et al. 2003 |
| cGKI | Purkinje cell-specific mutation | Reduced LTD at the parallel fiber-Purkinje cell synapse | Feil R et al. 2003 |

*The nNOS knockout mice reported were created by a deletion of exon 2 eliminating the splice variant nNOSα, which accounts for about 95% of the catalytic activity in the whole brain. nNOS mutants retain significant NOS activity in the brain, perhaps reflecting the function of the remaining isoform, eNOS. However, this interpretation is complicated by the expression of two alternative splice variants, nNOSβ and nNOSγ, lacking exon 2 in these mice (Eliasson et al. 1997).

the membrane-permeable cGMP analogue dibutyryl cGMP partially restores LTP blocked by a NOS inhibitor (Haley et al. 1992), and various sGC inhibitors suppress LTP (Zhuo et al. 1994; Arancio et al. 1995; Boulton et al. 1995). Evidence for the presynaptic localization of sGC came from studies of LTP between pairs of cultured hippocampal neurons (Arancio et al. 1995). Importantly, cGMP produces activity-dependent LTP when it is injected into the presynaptic neuron, but not when it is injected into the postsynaptic neuron. Results from quantal analysis together with the observation that 8-Br-cGMP mimicks the NO effect on the frequency of spontaneous miniature EPSCs support the conclusion that cGMP-dependent potentiation of synaptic transmission is due to an increase in presynaptic transmitter release.

As described above, the cGKs are major receptors of cGMP in the CNS. The first pharmacological evidence for a role of cGK in hippocampal LTP came from a study of Zhuo and coworkers (1994). These authors observed an enhancement of LTP following a weak tetanic stimulation in the presence of cGK activators, and, conversely, a suppression of LTP in the presence of cGK inhibitors. However, due to the inadequate selectivity of the pharmacological tools, the functional relevance of individual cGK isoforms was not clear. Two pieces of evidence hint at a predominant function of cGKI. First, the expression level of cGKII in the hippocampus appears substantially lower than that of cGKI (El-Husseini et al. 1995; Kleppisch et al. 1999; De Vente et al. 2001). Second, in line with the idea of NO as a retrograde messenger, cGKI has been detected in presynaptic terminals colocalized with the vesicle protein synaptophysin (Arancio et al. 2001). The concept of presynaptic NO/cGMP signalling via cGKI was further supported by studies of NMDA receptor-dependent LTP between pairs of cultured hippocampal neurons (Arancio et al. 2001): infusion of purified cGKI protein into the presynaptic neuron facilitated LTP in response to a weak tetanus, while a cGK inhibitor injected into the presynaptic neuron blocked LTP. Both substances failed to affect LTP when applied postsynaptically, underlining that cGK functions as a presynaptic target of NO/cGMP (Arancio et al. 2001). Apparently at odds with these findings, conventional knockout mice lacking cGKI, cGKII, or both (double knockout mice) are capable of normal LTP in the Schaffer collateral pathway (Kleppisch et al. 1999; Table 2). This may be due to the fact that null mutants lack the cGKs during their entire ontogenesis and in all tissues, so that other phenotypes or functional compensation could occlude potential defects of LTP. Notably, the analysis of synaptic plasticity in the conventional cGKI knockout was limited to juvenile mice (up to ~4 weeks of age) due to their premature death caused by cardiovascular and gastrointestinal defects (Pfeifer et al. 1998). Hippocampus-specific cGKI knockout ($cGKI^{hko}$) mice, which exhibit a normal life expectancy, also show normal hippocampal LTP in response to a single theta burst stimulation (Kleppisch et al. 2003). NO and cGMP have been shown to support late phase LTP (L-LTP) in the CA1 region (Lu et al. 1999; Lu and Hawkins 2002), a protein synthesis-dependent form of LTP that requires multiple strong tetanic stimulation of Schaffer collaterals (Frey et al. 1988; Barco et al. 2002). The following findings in $cGKI^{hko}$ mice support the view that cGKI may act as a downstream target of NO/cGMP in L-LTP (Kleppisch et al. 2003). First, LTP following multiple episodes of strong theta burst stimulation is decreased in adult $cGKI^{hko}$ mice (~12 weeks of age). Second, the protein synthesis inhibitor anisomycin has no effect on LTP in $cGKI^{hko}$ mice and decreases LTP in control mice to the level observed in the mutants. It remains unresolved whether this cGKI-dependent form of L-LTP is related to presynpatic and/or postsynaptic changes in protein synthesis.

## Signal Transduction Mechanisms Downstream of cGK

What are the functionally relevant substrates and downstream effectors of cGK involved in LTP? Studies in cultured hippocampal neurons have shed light on potential presynaptic effector molecules of the NO/cGMP/cGK signalling cascade, such as the $Ca^{2+}$/calmodulin-dependent protein kinase II (CaMKII) (Ninan and Arancio 2004), the VASP protein (Wang et al. 2005), and α-synuclein (Liu et al. 2004). Some regulators of G protein signalling (RGS proteins) and a few other synaptic proteins thought to modulate transmitter release are candidate substrates for cGKs in the brain (see Table 1). Differently from the VASP protein and RGS proteins, CaMKII and α-synuclein have not been shown to be phosphorylated directly by cGK.

The CaMKII is expressed in hippocampal neurons and plays an important role in synaptic plasticity. Primary interest was focussed on the postsynaptic function of the enzyme (for review see Lisman et al. 2002). A recent study with cultured hippocampal neurons has elucidated the presynaptic function of CaMKII in synaptic plasticity in the context of NO/cGMP/cGK signalling (Ninan and Arancio 2004). Presynaptic injection of the membrane-impermeable CaMKII inhibitor peptide 281-309 blocks long-lasting potentiation of synaptic transmission induced by combining a weak tetanus with brief superfusion of either the NO donor DEA/NO or the cGK activator 8-pCPT-cGMP. The CaMKII inhibitor peptide also suppresses the increase in the number of active presynaptic boutons occurring normally under these conditions. Thus, CaMKII appears to be essential for potentiation of synaptic transmission induced by retrograde NO/cGMP/cGK signalling. Additional findings show that the enzyme can also be activated independent of NO/cGMP/cGK signalling, e.g., by increased presynaptic $Ca^{2+}$ entry, and that this is sufficient to induce long-lasting potentiation. However, few data support a function of CaMKII in transmitter release. A study in conditional knockout mice lacking the CaMKII α subunit selectively in presynaptic CA3 neurons of the Schaffer collateral pathway even suggests that the enzyme *inhibits* the activity-dependent increase of transmitter release (Hinds et al. 2003). Future studies are needed to improve our understanding of the functional relationship between NO/cGMP/cGK signalling, CaMKII activation, and presynaptic transmitter release.

Two additional putative downstream components of presynaptic NO/cGMP/cGK signalling in the course of synaptic plasticity are VASP and the small GTP binding protein RhoA (Wang et al. 2005). VASP has been detected in the brain and shown to serve as an endogenous substrate of cGK in hippocampal neurons (Hauser et al. 1999; Arancio et al. 2001). It is also well known that VASP and related proteins are crucial in the regulation of actin dynamics, which have been linked to synaptic plasticity (for review see Lisman 2003; Lisman and Raghavachari 2006).

Activity patterns leading to LTP are associated with actin filament reorganization in dendritic spines (Fukazawa et al. 2003), and actin filament reorganization is essential to enhance AMPA receptor-mediated transmission during LTP (for review see Lisman and Raghavachari 2006). How is this related to presynaptic NO/cGMP/cGK signalling? It has emerged that similar cytoskeletal changes occur in an activity-dependent manner at the presynaptic site and support a rapid increase in clusters of the vesicular proteins synaptophysin and synapsin I (Antonova et al. 2001; Wang et al. 2005). Evidence from studies in cultured hippocampal neurons further suggests that this presynaptic process can be induced through retrograde NO/cGMP/cGK signalling with cGK-dependent VASP phosphorylation as a critical step (Wang et al. 2005). Elements of the NO signalling cascade, including sGC, cGKI, and VASP, are expressed in presynaptic terminals, where cGKI colocalizes with synaptophysin. Finally, VASP is phosphorylated at the site preferentially phosphorylated by cGK (Ser-239) during synaptic potentiation (Arancio et al. 2001). Hence, cGK, via phosphorylation of VASP, appears to modulate presynaptic cytoskeletal structures linked to vesicle function and transmitter release. Unexpectedly, mice lacking VASP or related proteins do not show severe phenotypes related to synaptic plasticity (Hauser et al. 1999; Reinhard et al. 2001).

Another interesting substrate of cGK is the small GTP binding protein RhoA (Sauzeau et al. 2000; Ellerbroek et al. 2003). Its cGK-dependent phosphorylation, similar to that of VASP, may contribute to the regulation of actin filament dynamics in presynaptic terminals. The finding that the general inhibitor of Rho GTPases, *Clostridium difficile* toxin B, and the inhibitor of RhoA-dependent kinase Y27632 reduce glutamate-induced synaptic potentiation in cultured hippocampal neurons (Wang et al. 2005) would indeed support this idea. However, toxin B has no effect in the presence of the cGK activator 8p-CPT-cGMP, which argues against a role of RhoA as a downstream target of cGK.

It has been reported recently that α-synuclein, a protein linked to neurodegeneration, contributes to synaptic potentiation by increasing transmitter release and that activation of the NO signalling cascade may promote this effect (Liu et al. 2004). Long-lasting-potentiation of synaptic transmission between cultured hippocampal neurons is accompanied by an increase in the number of presynaptic α-synuclein clusters, while turning down the expression of α-synuclein expression (e.g., by antisense nucleotides or knockout techniques) blocks potentiation and the increase in the presynaptic number of functional boutons normally associated with it. Moreover, presynaptic injection of α-synuclein causes rapid and long-lasting enhancement of synaptic transmission and rescues potentiation in cultures from mice carrying a α-synuclein null mutation (Liu et al. 2004). A link between NO signalling and α-synuclein has been suggested based on findings that exogenous NO can also increase the number of α-synuclein clusters and that NOS inhibitors can block glutamate-induced increase in their density. In line with these data, α-synuclein may function as a downstream effector linking NO/cGMP/cGK signalling to increased transmitter release from presynaptic terminals. However, without doubt more experiments are needed to substantiate this view. The discussion above underlines that mechanisms of synaptic potentiation involve presynaptic changes. Retrograde

NO-dependent signalling mechanisms promote coordinate changes in the distribution of proteins in the presynaptic terminal which, ultimately, increase presynaptic transmitter release, e.g., by altering vesicle function. In addition, these mechanisms may support the alignment of presynaptic release sites with postsynaptic receptors to increase the efficacy of synaptic transmission.

G protein-mediated signalling pathways are also critically involved in the regulation of transmitter release, e.g., by a negative feedback via presynaptic autoreceptors (cf. Starke 1981). Besides the signal-receiving G protein-coupled receptor (GPCR), a linked heterotrimeric G protein and its downstream effectors, this signal transduction machinery includes modulatory RGS proteins. The more than 30 currently known RGS proteins, which have been classified into seven subfamilies (Hollinger and Hepler 2002), promote the termination of GPCR-mediated signalling by interacting with Gα–GTP and accelerating GTP hydrolysis. Studies in astrocytes and vascular myocytes indicate that members of subfamilies 2, 3, and 4 are cGK substrates (Pedram et al. 2000; Tang et al. 2003). Moreover, cGKI-dependent RGS2 phosphorylation might play a functional role for NO-dependent effects in the vasculature (Tang et al. 2003; Sun et al. 2005). Morphological and electrophysiological analyses support the idea that RGS2 is an important regulator of synaptic plasticity in hippocampal CA1 neurons (Ingi et al. 1998; Oliveira-Dos-Santos et al. 2000; Han et al. 2006). Given these findings, it is tempting to speculate that the NO/cGMP/cGK signalling cascade can modulate presynaptic transmitter release by phosphorylating the RGS2 protein.

Autoinhibition of transmitter release has been largely attributed to GPCR-dependent suppression of $Ca^{2+}$ influx via presynaptic voltage-dependent $Ca^{2+}$ channels (for review see Reid et al. 2003). The intracellular $Ca^{2+}$ concentration and, hence, transmitter release are also controlled by $Ca^{2+}$ discharged from intracellular stores (Fossier et al. 1999; Collin et al. 2005). Ryanodine-sensitive stores in presynaptic terminals have been reported to mobilize $Ca^{2+}$ needed to induce LTD at Schaffer collateral/CA1 synapses (Reyes and Stanton 1996; Reyes-Harde et al. 1999a,b). $Ca^{2+}$ release causing LTD can be induced by an endogenous activator of ryanodine receptors, cyclic ADP ribose (cADPR). This messenger is synthesized by ADP-ribosyl cyclase, which is expressed in various types of neurons (for review see Higashida et al. 2001). Importantly, the activity of the enzyme is regulated through the NO/cGMP/cGK signalling cascade (Galione et al. 1993; Galione 1994; Reyes-Harde et al. 1999a,b; Higashida et al. 2001). A rise in cGMP leads to an increase of the cADPR concentration in hippocampal slices. It is assumed that ADP-ribosyl cyclase is a presynaptic target of cGK. Its phosphorylation boosts cADPR synthesis, which leads to enhanced $Ca^{2+}$ release from ryanodine-sensitive stores. This store-dependent presynaptic $Ca^{2+}$ pool, ultimately, causes a decrease in transmitter release through unknown mechanisms. Collectively, the observations discussed above indicate that presynaptic target molecules with opposite functional impact on transmitter release, e.g., VASP (increase in release) and ADP ribosyl cyclase (decrease in release), may become activated through the NO/cGMP/cGK signalling cascade. Differential expression of these signal transduction elements may provide a mechanism enabling discrete recruitment of effectors that either facilitate or inhibit transmitter release.

### cGK-Independent Presynaptic cGMP Signalling

In most of the aforementioned studies the involvement of cGKs as mediators of NO/cGMP effects has been analyzed by using pharmacological "cGK inhibitors." However, it is emerging that the presently available "cGK inhibitors" might be less specific than previously thought and, therefore, produce effects not related to cGK inhibition (Burkhardt et al. 2000; Gambaryan et al. 2004; Marshall et al. 2004; Feil R et al. 2005a). In the future, the analysis of cGK-deficient mouse models should help to dissect presynaptic NO/cGMP signalling without the use of uncertain pharmacological inhibitors. It is likely that not all effects of NO/cGMP are mediated by cGKs. Cytosolic cGMP can activate members of two ion channel families, CNG and HCN channels (Hofmann et al. 2005) leading to membrane depolarization and excitation. Moreover, both CNG and HCN channels have been reported to permit $Ca^{2+}$ influx that may contribute to modulation of neuronal excitability and synaptic function (Koutalos and Yau 1996; Broillet and Firestein 1997; Parent et al. 1998; Yu et al. 2004). This section shortly summarizes evidence supporting the view that cGMP can indeed modulate transmitter release through these ion channels clustered in presynaptic terminals.

Recordings of membrane currents and capacitance changes from cones in the salamander retina have demonstrated that exocytosis can be triggered by a $Ca^{2+}$ influx mediated through a cGMP-gated channel in the inner segment and in synaptic processes of cones (Rieke and Schwartz 1994). These CNG channels are clustered at high-density in cone terminals and can be activated by endogenous NO through stimulation of sGC-dependent cGMP synthesis (Savchenko et al. 1997). Furthermore, the sGC inhibitors, ODQ and LY-83583, suppress NO-induced glutamate release from cones. CNG channels are expressed in a variety of brain regions, including the hippocampus and, accordingly, have been suggested to serve as universal presynaptic transducers linking activity-dependent generation of NO and cGMP to enhanced transmitter release (Kingston et al. 1996; Bradley et al. 1997).

Garthwaite and co-workers (2006) have recently shown that rat optic nerve axons express various members of the HCN family and that HCN channels can serve as presynaptic transducers for the NO/cGMP signalling cascade. Actually, NO can cause a massive rise in cGMP levels sufficient to saturate the cyclic nucleotide binding site of HCN channels (Garthwaite 2005). The HCN channel blocker ZD7288 diminishes depolarization induced by the NO donor PAP/NO or by 8-Br-cGMP in optic nerve axons. NO-induced effects in this preparation are associated with an increase in cytosolic cGMP but not cAMP. There are a few reports suggesting that NO/cGMP can activate HCN channels also in other neurons (Pape and Mager 1992; Ingram and Williams 1996; Abudara et al. 2002; Pose et al. 2003; Kim et al. 2005). Within the HCN family, the HCN2 and HCN4 isoforms exhibit the most prominent regulation by cyclic nucleotides. The HCN2 subtype is expressed at considerable levels throughout the brain, while substantial expression of HCN4 channels is limited to a few areas, including the thalamus (Moosmang et al. 1999; Bender et al. 2001; Notomi and Shigemoto 2004). This hints at the HCN2 isoform as a possible general effector of NO signalling in the brain. As for CNG channels,

NO/cGMP-dependent activation of HCN channels leads to depolarization, increased excitability, and increased ability to conduct action potential bursts (for review see Robinson and Siegelbaum 2003).

### 3.2.2 NO as an Anterograde Messenger in Cerebellar LTD

LTD of the parallel fiber (PF)-Purkinje cell (PC) synapse in the cerebellum is a cellular model system, which has been suggested to underlie certain forms of motor learning. Induction of cerebellar LTD requires a postsynaptic kinase limb and a complementary phosphatase limb. The latter involves NMDA receptor-dependent generation of NO and postsynaptic cGMP synthesis followed by inhibition of protein phosphatases. Activation of both limbs promotes AMPA receptor phosphorylation and internalization resulting in LTD of synaptic transmission (Ito 2002). A postsynaptic source of NO is highly unlikely, since neither NMDA receptors nor nNOS were detected in postsynaptic PCs. Instead, and in contrast to the hippocampus, it is thought that NO acts in the cerebellum as an anterograde messenger that is produced in PF terminals or interneurons (Shin and Linden 2005) and then diffuses to the postsynaptic PC to induce LTD via a cGMP/cGKI-dependent mechanism. These postsynaptic actions will be discussed only briefly, because the focus of this review is presynaptic NO/cGMP signalling. A comprehensive review of the role of the NO/cGMP/cGKI cascade in cerebellar LTD and learning can be found elsewhere (Feil R et al. 2005b).

NO or cGMP present coincidently with a postsynaptic $Ca^{2+}$ signal are sufficient to induce LTD (Lev-Ram et al. 1995, 1997a). LTD is blocked by inhibitors of NO synthase or guanylyl cyclase, can be restored by exogenous NO and cGMP, and is abolished in nNOS mutant mice (Hartell 1994; Boxall and Garthwaite 1996; Lev-Ram et al. 1997a,b). Among the multiple receptors for cGMP, the cGKI might serve a key function in cerebellar LTD. It is highly expressed in PCs (Hofmann and Sold 1972; Lohmann et al. 1981; Feil S et al. 2005), while cGKII has not been detected (El-Husseini et al. 1995; Geiselhoringer et al. 2004). A role for cGKI was shown by inhibitor studies (Hartell 1994; Lev-Ram et al. 1997a) and, more recently, by the analysis of PC-specific cGKI knockout mice (Feil R et al. 2003). These conditional mouse mutants are fully viable and show no generalized structural or physiological abnormalities in the cerebellum. However, they exhibit a near complete loss of cerebellar LTD. Interestingly, this LTD defect does not affect general motor performance of the mutants, but results in a rather specific defect in the adaptation of the vestibulo-ocular reflex (VOR), a simple form of motor learning (Feil R et al. 2003). Similar phenotypes, defective cerebellar LTD but relatively mild and specific defects in motor learning, have also been reported for other transgenic mouse models, including nNOS-deficient mice (Lev-Ram et al. 1997b) and mice overexpressing a protein kinase C inhibitor peptide selectively in PCs (De Zeeuw et al. 1998). On the other hand, it appears that the impaired general motor performance observed in many conventional knockout mouse models is usually associated with structural alterations of the cerebellum caused by the gene knockout (Ito 2001). Together, these results unequivocally identify cGKI as a critical component in cerebellar LTD and, moreover, suggest that this type of synaptic plasticity is involved in specific forms of motor learning, such as VOR adaptation, rather than in general motor performance. This concept has been recently supported by the observation that pharmacological prevention of cerebellar LTD in rodents does not affect motor learning in several standard tests (Welsh et al. 2005). Indeed, there is little doubt that synaptic plasticity in other brain regions, such as the striatum, plays also an important role in motor learning (Dang et al. 2006).

How could activation of cGKI in PCs contribute to LTD and cerebellum-dependent learning? As described above, the induction of LTD is thought to depend on the balance between protein kinases and phosphatases that regulates the level of postsynaptic AMPA receptor phosphorylation and internalization. The NO/cGMP/cGKI cascade has been linked to the phosphatase limb (Ito 2002). LTD can be facilitated by inhibiting the protein phosphatase 1/2A (Ajima and Ito 1995). Intriguingly, the phosphorylated form of G-substrate, a well-characterized cGK substrate expressed in PCs, reduces the activity of protein phosphatase 1/2A (Hall et al. 1999). Decreased phosphatase activity should result in increased levels of AMPA receptor phosphorylation, which is assumed to facilitate its clathrin-mediated endocytotic removal from the postsynaptic membrane (Wang and Linden 2000; Chung et al. 2003). Thus, a likely signalling pathway is as follows: NO/cGMP-dependent activation of cGKI in PCs results in phosphorylation of G-substrate followed by inhibition of protein phosphatases 1/2A and increased phosphorylation and endocytosis of AMPA receptors. The removal of AMPA receptors from the postsynaptic site induces LTD associated with specific forms of motor learning.

# 4 Conclusion

Presynaptic NO/cGMP signalling modulates transmitter release and thereby supports synaptic plasticity important for learning and memory. Dysfunction of this signal transduction cascade may also contribute to various neurological disorders that involve an altered release of transmitters. Thus, components of the presynaptic NO/cGMP signalling pathway represent interesting targets for the development of new drugs improving cognition and drugs effective in neurodegenerative and psychiatric diseases, e.g., Morbus Alzheimer and schizophrenia. Indeed, initial promising findings in this direction have been reported for PDE inhibitors as well as for sGC stimulators (Chien et al. 2005; Feil R and Kemp-Harper 2006; Hebb and Robertson 2006; Menniti et al. 2006). Certainly, future therapeutic efforts will include further elements of NO/cGMP signalling involved in the modulation of transmitter release.

**Acknowledgements** Our apologies go to all colleagues who may not find their results represented in this chapter. We concentrated on the canonical NO/cGMP/cGK signalling cascade and its role for synaptic plasticity. The work of *RF* and *TK* was supported by the DFG and the VolkswagenStiftung.

## References


Abudara V, Alvarez AF, Chase MH, Morales FR (2002) Nitric oxide as an anterograde neurotransmitter in the trigeminal motor pool. J Neurophysiol 88:497–506

Ajima A, Ito M (1995) A unique role of protein phosphatases in cerebellar long-term depression. Neuroreport 6:297–300

Antonova I, Arancio O, Trillat AC, Wang HG, Zablow L, Udo H, Kandel ER, et al. (2001) Rapid increase in clusters of presynaptic proteins at onset of long-lasting potentiation. Science 294:1547–50

Arancio O, Kandel ER, Hawkins RD (1995) Activity-dependent long-term enhancement of transmitter release by presynaptic 3′,5′-cyclic GMP in cultured hippocampal neurons. Nature 376:74–80

Arancio O, Kiebler M, Lee CJ, Lev-Ram V, Tsien RY, Kandel ER, Hawkins RD (1996) Nitric oxide acts directly in the presynaptic neuron to produce long-term potentiation in cultured hippocampal neurons. Cell 87:1025–35

Arancio O, Antonova I, Gambaryan S, Lohmann SM, Wood JS, Lawrence DS, Hawkins RD (2001) Presynaptic role of cGMP-dependent protein kinase during long-lasting potentiation. J Neurosci 21:143–9

Aswad DW, Greengard P (1981) A specific substrate from rabbit cerebellum for guanosine 3′,5′-monophosphate-dependent protein kinase. I. Purification and characterization. J Biol Chem 256:3487–93

Barco A, Alarcon JM, Kandel ER (2002) Expression of constitutively active CREB protein facilitates the late phase of long-term potentiation by enhancing synaptic capture. Cell 108:689–703

Bear MF, Malenka RC (1994) Synaptic plasticity:LTP and LTD. Curr Opin Neurobiol 4:389–99

Bender RA, Brewster A, Santoro B, Ludwig A, Hofmann F, Biel M, Baram TZ (2001) Differential and age-dependent expression of hyperpolarization-activated, cyclic nucleotide-gated cation channel isoforms 1–4 suggests evolving roles in the developing rat hippocampus. Neuroscience 106:689–98

Betz WJ, Bewick GS (1992) Optical analysis of synaptic vesicle recycling at the frog neuromuscular junction. Science 255:200–3

Biel M, Zong X, Ludwig A, Sautter A, Hofmann F (1999) Structure and function of cyclic nucleotide-gated channels. Rev Physiol Biochem Pharmacol 135:151–71

Blackshaw S, Eliasson MJ, Sawa A, Watkins CC, Krug D, Gupta A, Arai T, et al. (2003) Species, strain and developmental variations in hippocampal neuronal and endothelial nitric oxide synthase clarify discrepancies in nitric oxide-dependent synaptic plasticity. Neuroscience 119:979–90

Boulton CL, Southam E, Garthwaite J (1995) Nitric oxide-dependent long-term potentiation is blocked by a specific inhibitor of soluble guanylyl cyclase. Neuroscience 69:699–703

Boxall AR, Garthwaite J (1996) Long-term depression in rat cerebellum requires both NO synthase and NO-sensitive guanylyl cyclase. Eur J Neurosci 8:2209–12

Bradley J, Zhang Y, Bakin R, Lester HA, Ronnett GV, Zinn K (1997) Functional expression of the heteromeric "olfactory" cyclic nucleotide-gated channel in the hippocampus: a potential effector of synaptic plasticity in brain neurons. J Neurosci 17:1993–2005

Bradley J, Frings S, Yau KW, Reed R (2001) Nomenclature for ion channel subunits. Science 294:2095–6

Bredt DS, Snyder SH (1990) Isolation of nitric oxide synthetase, a calmodulin-requiring enzyme. Proc Natl Acad Sci USA 87:682–5

Brenman JE, Chao DS, Gee SH, McGee AW, Craven SE, Santillano DR, Wu Z, et al. (1996) Interaction of nitric oxide synthase with the postsynaptic density protein PSD-95 and alpha 1-syntrophin mediated by PDZ domains. Cell 84:757–67

Broillet MC, Firestein S (1996) Direct activation of the olfactory cyclic nucleotide-gated channel through modification of sulfhydryl groups by NO compounds. Neuron 16:377–85

Broillet MC, Firestein S (1997) Beta subunits of the olfactory cyclic nucleotide-gated channel form a nitric oxide activated Ca2+ channel. Neuron 18:951–8

Bugnon O, Schaad NC, Schorderet M (1994) Nitric oxide modulates endogenous dopamine release in bovine retina. Neuroreport 5:401–4

Burkhardt M, Glazova M, Gambaryan S, Vollkommer T, Butt E, Bader B, Heermeier K, et al. (2000) KT5823 inhibits cGMP-dependent protein kinase activity in vitro but not in intact human platelets and rat mesangial cells. J Biol Chem 275:33536–41

Chetkovich DM, Klann E, Sweatt JD (1993) Nitric oxide synthase-independent long-term potentiation in area CA1 of hippocampus. Neuroreport 4:919–22

Chien WL, Liang KC, Teng CM, Kuo SC, Lee FY, Fu WM (2005) Enhancement of learning behaviour by a potent nitric oxide-guanylate cyclase activator YC-1. Eur J Neurosci 21:1679–88

Christopherson KS, Hillier BJ, Lim WA, Bredt DS (1999) PSD-95 assembles a ternary complex with the N-methyl-D-aspartic acid receptor and a bivalent neuronal NO synthase PDZ domain. J Biol Chem 274:27467–73

Chung HJ, Steinberg JP, Huganir RL, Linden DJ (2003) Requirement of AMPA receptor GluR2 phosphorylation for cerebellar long-term depression. Science 300:1751–5

Clementi E, Sciorati C, Riccio M, Miloso M, Meldolesi J, Nistico G (1995) Nitric oxide action on growth factor-elicited signals. Phosphoinositide hydrolysis and [Ca2+]i responses are negatively modulated via a cGMP-dependent protein kinase I pathway. J Biol Chem 270:22277–82

Collin T, Marty A, Llano I (2005) Presynaptic calcium stores and synaptic transmission. Curr Opin Neurobiol 15:275–81

Dang MT, Yokoi F, Yin HH, Lovinger DM, Wang Y, Li Y (2006) Disrupted motor learning and long-term synaptic plasticity in mice lacking NMDAR1 in the striatum. Proc Natl Acad Sci USA 103:15254–9

Daniel H, Levenes C, Crepel F (1998) Cellular mechanisms of cerebellar LTD. Trends Neurosci 21:401–7

Dawson TM, Dawson VL (1996) Nitric oxide synthase:role as a transmitter/mediator in the brain and endocrine system. Annu Rev Med 47:219–27

De Vente J, Asan E, Gambaryan S, Markerink-van Ittersum M, Axer H, Gallatz K, Lohmann SM, et al. (2001) Localization of cGMP-dependent protein kinase type II in rat brain. Neuroscience 108:27–49

De Zeeuw CI, Hansel C, Bian F, Koekkoek SK, van Alphen AM, Linden DJ, Oberdick J (1998) Expression of a protein kinase C inhibitor in Purkinje cells blocks cerebellar LTD and adaptation of the vestibulo-ocular reflex. Neuron 20:495–508

Demas GE, Kriegsfeld LJ, Blackshaw S, Huang P, Gammie SC, Nelson RJ, Snyder SH (1999) Elimination of aggressive behavior in male mice lacking endothelial nitric oxide synthase. J Neurosci 19:RC30

DiCicco-Bloom E, Lelievre V, Zhou X, Rodriguez W, Tam J, Waschek JA (2004) Embryonic expression and multifunctional actions of the natriuretic peptides and receptors in the developing nervous system. Dev Biol 271:161–75

Dickie BG, Lewis MJ, Davies JA (1992) NMDA-induced release of nitric oxide potentiates aspartate overflow from cerebellar slices. Neurosci Lett 138:145–8

Dinerman JL, Dawson TM, Schell MJ, Snowman A, Snyder SH (1994) Endothelial nitric oxide synthase localized to hippocampal pyramidal cells:implications for synaptic plasticity. Proc Natl Acad Sci USA 91:4214–8

Ding JD, Burette A, Nedvetsky PI, Schmidt HH, Weinberg RJ (2004) Distribution of soluble guanylyl cyclase in the rat brain. J Comp Neurol 472:437–48

Doreulee N, Brown RE, Yanovsky Y, Godecke A, Schrader J, Haas HL (2001) Defective hippocampal mossy fiber long-term potentiation in endothelial nitric oxide synthase knockout mice. Synapse 41:191–4

El-Husseini AE, Bladen C, Vincent SR (1995) Molecular characterization of a type II cyclic GMP-dependent protein kinase expressed in the rat brain. J Neurochem 64:2814–7

El-Husseini AE, Williams J, Reiner PB, Pelech S, Vincent SR (1999) Localization of the cGMP-dependent protein kinases in relation to nitric oxide synthase in the brain. J Chem Neuroanat 17:45–55

Eliasson MJ, Blackshaw S, Schell MJ, Snyder SH (1997) Neuronal nitric oxide synthase alternatively spliced forms:prominent functional localizations in the brain. Proc Natl Acad Sci USA 94:3396–401

Ellerbroek SM, Wennerberg K, Burridge K (2003) Serine phosphorylation negatively regulates RhoA in vivo. J Biol Chem 278:19023–31

Endo S, Nairn AC, Greengard P, Ito M (2003) Thr123 of rat G-substrate contributes to its action as a protein phosphatase inhibitor. Neurosci Res 45:79–89

Engert F, Bonhoeffer T (1999) Dendritic spine changes associated with hippocampal long-term synaptic plasticity. Nature 399:66–70

Feil R, Kemp-Harper B (2006) cGMP signalling:from bench to bedside. Conference on cGMP generators, effectors and therapeutic implications. EMBO Rep 7:149–53

Feil R, Hartmann J, Luo C, Wolfsgruber W, Schilling K, Feil S, Barski JJ, et al. (2003) Impairment of LTD and cerebellar learning by Purkinje cell-specific ablation of cGMP-dependent protein kinase I. J Cell Biol 163:295–302

Feil R, Feil S, Hofmann F (2005a) A heretical view on the role of NO and cGMP in vascular proliferative diseases. Trends Mol Med 11:71–5

Feil R, Hofmann F, Kleppisch T (2005b) Function of cGMP-dependent protein kinases in the nervous system. Rev Neurosci 16:23–41

Feil S, Zimmermann P, Knorn A, Brummer S, Schlossmann J, Hofmann F, Feil R (2005) Distribution of cGMP-dependent protein kinase type I and its isoforms in the mouse brain and retina. Neuroscience 135:863–8

Fossier P, Tauc L, Baux G (1999) Calcium transients and neurotransmitter release at an identified synapse. Trends Neurosci 22:161–6

Francis SH, Corbin JD (1999) Cyclic nucleotide-dependent protein kinases:intracellular receptors for cAMP and cGMP action. Crit Rev Clin Lab Sci 36:275–328

Frey U, Krug M, Reymann KG, Matthies H (1988) Anisomycin, an inhibitor of protein synthesis, blocks late phases of LTP phenomena in the hippocampal CA1 region in vitro. Brain Res 452:57–65

Friebe A, Koesling D (2003) Regulation of nitric oxide-sensitive guanylyl cyclase. Circ Res 93:96–105

Frisch C, Dere E, Silva MA, Godecke A, Schrader J, Huston JP (2000) Superior water maze performance and increase in fear-related behavior in the endothelial nitric oxide synthase-deficient mouse together with monoamine changes in cerebellum and ventral striatum. J Neurosci 20:6694–700

Fukazawa Y, Saitoh Y, Ozawa F, Ohta Y, Mizuno K, Inokuchi K (2003) Hippocampal LTP is accompanied by enhanced F-actin content within the dendritic spine that is essential for late LTP maintenance in vivo. Neuron 38:447–60

Furchgott RF (1996) The 1996 Albert Lasker Medical Research Awards. The discovery of endothelium-derived relaxing factor and its importance in the identification of nitric oxide. JAMA 276:1186–8

Gage AT, Reyes M, Stanton PK (1997) Nitric-oxide-guanylyl-cyclase-dependent and -independent components of multiple forms of long-term synaptic depression. Hippocampus 7:286–95

Galione A (1994) Cyclic ADP-ribose, the ADP-ribosyl cyclase pathway and calcium signalling. Mol Cell Endocrinol 98:125–31

Galione A, White A, Willmott N, Turner M, Potter BV, Watson SP (1993) cGMP mobilizes intracellular Ca2+ in sea urchin eggs by stimulating cyclic ADP-ribose synthesis. Nature 365:456–9

Gambaryan S, Geiger J, Schwarz UR, Butt E, Begonja A, Obergfell A, Walter U (2004) Potent inhibition of human platelets by cGMP analogs independent of cGMP-dependent protein kinase. Blood 103:2593–600

Garthwaite J (2005) Dynamics of cellular NO-cGMP signaling. Front Biosci 10 1868–80

Garthwaite J, Boulton CL (1995) Nitric oxide signaling in the central nervous system. Annu Rev Physiol 57:683–706
Garthwaite J, Charles SL, Chess-Williams R (1988) Endothelium-derived relaxing factor release on activation of NMDA receptors suggests role as intercellular messenger in the brain. Nature 336:385–8
Garthwaite G, Bartus K, Malcolm D, Goodwin D, Kollb-Sielecka M, Dooldeniya C, Garthwaite J (2006) Signaling from blood vessels to CNS axons through nitric oxide. J Neurosci 26:7730–40
Geiselhoringer A, Gaisa M, Hofmann F, Schlossmann J (2004) Distribution of IRAG and cGKI-isoforms in murine tissues. FEBS Lett 575:19–22
Guevara-Guzman R, Emson PC, Kendrick KM (1994) Modulation of in vivo striatal transmitter release by nitric oxide and cyclic GMP. J Neurochem 62:807–10
Haley JE, Wilcox GL, Chapman PF (1992) The role of nitric oxide in hippocampal long-term potentiation. Neuron 8:211–6
Hall KU, Collins SP, Gamm DM, Massa E, DePaoli-Roach AA, Uhler MD (1999) Phosphorylation-dependent inhibition of protein phosphatase-1 by G-substrate. A Purkinje cell substrate of the cyclic GMP-dependent protein kinase. J Biol Chem 274:3485–95
Han J, Mark MD, Li X, Xie M, Waka S, Rettig J, Herlitze S (2006) RGS2 determines short-term synaptic plasticity in hippocampal neurons by regulating Gi/o-mediated inhibition of presynaptic Ca2+ channels. Neuron 51:575–86
Hanbauer I, Wink D, Osawa Y, Edelman GM, Gally JA (1992) Role of nitric oxide in NMDA-evoked release of [3H]-dopamine from striatal slices. Neuroreport 3:409–12
Hartell NA (1994) cGMP acts within cerebellar Purkinje cells to produce long term depression via mechanisms involving PKC and PKG. Neuroreport 5:833–6
Haug LS, Jensen V, Hvalby O, Walaas SI, Ostvold AC (1999) Phosphorylation of the inositol 1,4,5-trisphosphate receptor by cyclic nucleotide-dependent kinases in vitro and in rat cerebellar slices in situ. J Biol Chem 274:7467–73
Haul S, Godecke A, Schrader J, Haas HL, Luhmann HJ (1999) Impairment of neocortical long-term potentiation in mice deficient of endothelial nitric oxide synthase. J Neurophysiol 81:494–7
Hauser W, Knobeloch KP, Eigenthaler M, Gambaryan S, Krenn V, Geiger J, Glazova M, et al. (1999) Megakaryocyte hyperplasia and enhanced agonist-induced platelet activation in vasodilator-stimulated phosphoprotein knockout mice. Proc Natl Acad Sci USA 96:8120–5
Hawkins RD, Son H, Arancio O (1998) Nitric oxide as a retrograde messenger during long-term potentiation in hippocampus. Prog Brain Res 118:155–72
Hebb AL, Robertson HA (2006) Role of phosphodiesterases in neurological and psychiatric disease. Curr Opin Pharmacol (doi:10.1016/j.coph.2006.08.014)
Hebb DO (1949) The organisation of behavior. A neurophsychological theory. Wiley, New York
Herman JP, Langub MC, Jr., Watson RE, Jr. (1993) Localization of C-type natriuretic peptide mRNA in rat hypothalamus. Endocrinology 133:1903–6
Higashida H, Hashii M, Yokoyama S, Hoshi N, Chen XL, Egorova A, Noda M, et al. (2001) Cyclic ADP-ribose as a second messenger revisited from a new aspect of signal transduction from receptors to ADP-ribosyl cyclase. Pharmacol Ther 90:283–96
Hinds HL, Goussakov I, Nakazawa K, Tonegawa S, Bolshakov VY (2003) Essential function of alpha-calcium/calmodulin-dependent protein kinase II in neurotransmitter release at a glutamatergic central synapse. Proc Natl Acad Sci USA 100:4275–80
Hofmann F, Sold G (1972) A protein kinase activity from rat cerebellum stimulated by guanosine-3′5′-monophosphate. Biochem Biophys Res Commun 49:1100–7
Hofmann F, Biel M, Feil R, Kleppisch T (2003) Mouse models of NO/natriuretic peptide/cGMP kinase signaling. In:Hein L, Offermanns S (eds) Handbook of Experimental Pharmacology, pp 95–130
Hofmann F, Biel M, Kaupp UB (2005) International Union of Pharmacology. LI. Nomenclature and structure-function relationships of cyclic nucleotide-regulated channels. Pharmacol Rev 57:455–62

Hofmann F, Feil R, Kleppisch T, Schlossmann J (2006) Function of cGMP-dependent protein kinases as revealed by gene deletion. Physiol Rev 86:1–23

Hollinger S, Hepler JR (2002) Cellular regulation of RGS proteins:modulators and integrators of G protein signaling. Pharmacol Rev 54:527–59

Hopper RA, Garthwaite J (2006) Tonic and phasic nitric oxide signals in hippocampal long-term potentiation. J Neurosci 26:11513–21

Huang CC, Chan SH, Hsu KS (2003) cGMP/protein kinase G-dependent potentiation of glutamatergic transmission induced by nitric oxide in immature rat rostral ventrolateral medulla neurons in vitro. Mol Pharmacol 64:521–32

Huang EP (1997) Synaptic plasticity:a role for nitric oxide in LTP. Curr Biol 7:R141–3

Ignarro LJ, Buga GM, Wood KS, Byrns RE, Chaudhuri G (1987) Endothelium-derived relaxing factor produced and released from artery and vein is nitric oxide. Proc Natl Acad Sci USA 84:9265–9

Ingi T, Krumins AM, Chidiac P, Brothers GM, Chung S, Snow BE, Barnes CA, et al. (1998) Dynamic regulation of RGS2 suggests a novel mechanism in G-protein signaling and neuronal plasticity. J Neurosci 18:7178–88

Ingram SL, Williams JT (1996) Modulation of the hyperpolarization-activated current (Ih) by cyclic nucleotides in guinea-pig primary afferent neurons. J Physiol 492 ( Pt 1):97–106

Ito M (2001) Cerebellar long-term depression:characterization, signal transduction, and functional roles. Physiol Rev 81:1143–95

Ito M (2002) Historical review of the significance of the cerebellum and the role of Purkinje cells in motor learning. Ann N Y Acad Sci 978:273–88

Jaffrey SR, Erdjument-Bromage H, Ferris CD, Tempst P, Snyder SH (2001) Protein S-nitrosylation:a physiological signal for neuronal nitric oxide. Nat Cell Biol 3:193–7

Kano T, Shimizu-Sasamata M, Huang PL, Moskowitz MA, Lo EH (1998) Effects of nitric oxide synthase gene knockout on neurotransmitter release in vivo. Neuroscience 86:695–9

Kantor DB, Lanzrein M, Stary SJ, Sandoval GM, Smith WB, Sullivan BM, Davidson N, et al. (1996) A role for endothelial NO synthase in LTP revealed by adenovirus-mediated inhibition and rescue. Science 274:1744–8

Kaupp UB, Niidome T, Tanabe T, Terada S, Bönigk W, Stühmer W, Cook NJ, et al. (1989) Primary structure and functional expression from complementary DNA of the rod photoreceptor cyclic GMP-gated channel. Nature 342:762–6

Kim HY, Kim SJ, Kim J, Oh SB, Cho H, Jung SJ (2005) Effect of nitric oxide on hyperpolarization-activated current in substantia gelatinosa neurons of rats. Biochem Biophys Res Commun 338:1648–53

Kingston PA, Zufall F, Barnstable CJ (1996) Rat hippocampal neurons express genes for both rod retinal and olfactory cyclic nucleotide-gated channels:novel targets for cAMP/cGMP function. Proc Natl Acad Sci USA 93:10440–5

Kleppisch T, Pfeifer A, Klatt P, Ruth P, Montkowski A, Fassler R, Hofmann F (1999) Long-term potentiation in the hippocampal CA1 region of mice lacking cGMP-dependent kinases is normal and susceptible to inhibition of nitric oxide synthase. J Neurosci 19:48–55

Kleppisch T, Wolfsgruber W, Feil S, Allmann R, Wotjak CT, Goebbels S, Nave K-A, et al. (2003) Hippocampal cyclic GMP-dependent protein kinase I supports an age- and protein synthesis-dependent component of long-term potentiation but is not essential for spatial reference and contextual memory. J Neurosci 23:6005–12

Klyachko VA, Ahern GP, Jackson MB (2001) cGMP-mediated facilitation in nerve terminals by enhancement of the spike afterhyperpolarization. Neuron 31:1015–25

Koesling D, Mullershausen F, Lange A, Friebe A, Mergia E, Wagner C, Russwurm M (2005) Negative feedback in NO/cGMP signalling. Biochem Soc Trans 33:1119–22

Komatsu Y, Nakao K, Suga S, Ogawa Y, Mukoyama M, Arai H, Shirakami G, et al. (1991) C-type natriuretic peptide (CNP) in rats and humans. Endocrinology 129:1104–6

Korschen HG, Illing M, Seifert R, Sesti F, Williams A, Gotzes S, Colville C, et al. (1995) A 240 kDa protein represents the complete beta subunit of the cyclic nucleotide-gated channel from rod photoreceptor. Neuron 15:627–36

Koutalos Y, Yau KW (1996) Regulation of sensitivity in vertebrate rod photoreceptors by calcium. Trends Neurosci 19:73–81
Kuhn M (2004) Molecular physiology of natriuretic peptide signalling. Basic Res Cardiol 99:76–82
Landgraf W, Hofmann F, Pelton JT, Huggins JP (1990) Effects of cyclic GMP on the secondary structure of cyclic GMP dependent protein kinase and analysis of the enzyme's amino-terminal domain by far-ultraviolet circular dichroism. Biochemistry 29:9921–8
Larkman AU, Jack JJ (1995) Synaptic plasticity:hippocampal LTP. Curr Opin Neurobiol 5:324–34
Lein ES, Hawrylycz MJ, Ao N, Ayres M, Bensinger A, Bernard A, Boe AF, et al. (2006) Genome-wide atlas of gene expression in the adult mouse brain. Nature 445:168–76
Lev-Ram V, Makings LR, Keitz PF, Kao JP, Tsien RY (1995) Long-term depression in cerebellar Purkinje neurons results from coincidence of nitric oxide and depolarization-induced Ca2+ transients. Neuron 15:407–15
Lev-Ram V, Jiang T, Wood J, Lawrence DS, Tsien RY (1997a) Synergies and coincidence requirements between NO, cGMP, and Ca2+ in the induction of cerebellar long-term depression. Neuron 18:1025–38
Lev-Ram V, Nebyelul Z, Ellisman MH, Huang PL, Tsien RY (1997b) Absence of cerebellar long-term depression in mice lacking neuronal nitric oxide synthase. Learn Mem 4:169–77
Lisman J (2003) Actin's actions in LTP-induced synapse growth. Neuron 38:361–2
Lisman J, Raghavachari S (2006) A unified model of the presynaptic and postsynaptic changes during LTP at CA1 synapses. Sci STKE 2006:re11
Lisman J, Schulman H, Cline H (2002) The molecular basis of CaMKII function in synaptic and behavioural memory. Nat Rev Neurosci 3:175–90
Liu S, Ninan I, Antonova I, Battaglia F, Trinchese F, Narasanna A, Kolodilov N, et al. (2004) alpha-Synuclein produces a long-lasting increase in neurotransmitter release. Embo J 23:4506–16
Lohmann SM, Walter U, Miller PE, Greengard P, De Camilli P (1981) Immunohistochemical localization of cyclic GMP-dependent protein kinase in mammalian brain. Proc Natl Acad Sci USA 78:653–7
Lu YF, Hawkins RD (2002) Ryanodine receptors contribute to cGMP-induced late-phase LTP and CREB phosphorylation in the hippocampus. J Neurophysiol 88:1270–8
Lu YF, Kandel ER, Hawkins RD (1999) Nitric oxide signaling contributes to late-phase LTP and CREB phosphorylation in the hippocampus. J Neurosci 19:10250–61
Malenka RC, Bear MF (2004) LTP and LTD:an embarrassment of riches. Neuron 44:5–21
Maletic-Savatic M, Malinow R, Svoboda K (1999) Rapid dendritic morphogenesis in CA1 hippocampal dendrites induced by synaptic activity. Science 283:1923–7
Manahan-Vaughan D, Braunewell KH (1999) Novelty acquisition is associated with induction of hippocampal long-term depression. Proc Natl Acad Sci USA 96:8739–44
Marshall SJ, Senis YA, Auger JM, Feil R, Hofmann F, Salmon G, Peterson JT, et al. (2004) GPIb-dependent platelet activation is dependent on Src kinases but not MAP kinase or cGMP-dependent kinase. Blood 103:2601–9
Martin TF (2001) PI(4,5)P(2) regulation of surface membrane traffic. Curr Opin Cell Biol 13:493–9
Meffert MK, Calakos NC, Scheller RH, Schulman H (1996) Nitric oxide modulates synaptic vesicle docking fusion reactions. Neuron 16:1229–36
Menniti FS, Faraci WS, Schmidt CJ (2006) Phosphodiesterases in the CNS:targets for drug development. Nat Rev Drug Discov 5:660–70
Mergia E, Russwurm M, Zoidl G, Koesling D (2003) Major occurrence of the new alpha2beta1 isoform of NO-sensitive guanylyl cyclase in brain. Cell Signal 15:189–95
Mergia E, Haghikia A, Mittmann T, Friebe A, Koesling D (2006) Vascular and neuronal functions of the NO-sensitive guanylyl cyclase isoforms (alpha(1)-GC, alpha(2)-GC). Naunyn Schmiedebergs Arch Pharmacol 372 (Suppl. 1):45–46
Micheva KD, Holz RW, Smith SJ (2001) Regulation of presynaptic phosphatidylinositol 4,5-biphosphate by neuronal activity. J Cell Biol 154:355–68
Micheva KD, Buchanan J, Holz RW, Smith SJ (2003) Retrograde regulation of synaptic vesicle endocytosis and recycling. Nat Neurosci 6:925–32

Montague PR, Gancayco CD, Winn MJ, Marchase RB, Friedlander MJ (1994) Role of NO production in NMDA receptor-mediated neurotransmitter release in cerebral cortex. Science 263:973–7
Moosmang S, Biel M, Hofmann F, Ludwig A (1999) Differential distribution of four hyperpolarization-activated cation channels in mouse brain. Biol Chem 380:975–80
Muller U (1997) The nitric oxide system in insects. Prog Neurobiol 51:363–81
Ninan I, Arancio O (2004) Presynaptic CaMKII is necessary for synaptic plasticity in cultured hippocampal neurons. Neuron 42:129–41
Nishi A, Watanabe Y, Higashi H, Tanaka M, Nairn AC, Greengard P (2005) Glutamate regulation of DARPP-32 phosphorylation in neostriatal neurons involves activation of multiple signaling cascades. Proc Natl Acad Sci USA 102:1199–204
Notomi T, Shigemoto R (2004) Immunohistochemical localization of Ih channel subunits, HCN1-4, in the rat brain. J Comp Neurol 471:241–76
O'Dell TJ, Hawkins RD, Kandel ER, Arancio O (1991) Tests of the roles of two diffusible substances in long-term potentiation:evidence for nitric oxide as a possible early retrograde messenger. Proc Natl Acad Sci USA 88:11285–9
O'Dell TJ, Huang PL, Dawson TM, Dinerman JL, Snyder SH, Kandel ER, Fishman MC (1994) Endothelial NOS and the blockade of LTP by NOS inhibitors in mice lacking neuronal NOS. Science 265:542–6
Oliveira-Dos-Santos AJ, Matsumoto G, Snow BE, Bai D, Houston FP, Whishaw IQ, Mariathasan S, et al. (2000) Regulation of T cell activation, anxiety, and male aggression by RGS2. Proc Natl Acad Sci USA 97:12272–7
Osborne SL, Meunier FA, Schiavo G (2001) Phosphoinositides as key regulators of synaptic function. Neuron 32:9–12
Oster H, Werner C, Magnone MC, Mayser H, Feil R, Seeliger MW, Hofmann F, et al. (2003) cGMP-dependent protein kinase II modulates mPer1 and mPer2 gene induction and influences phase shifts of the circadian clock. Curr Biol 13:725–33
Pape HC, Mager R (1992) Nitric oxide controls oscillatory activity in thalamocortical neurons. Neuron 9:441–8
Parent A, Schrader K, Munger SD, Reed RR, Linden DJ, Ronnett GV (1998) Synaptic transmission and hippocampal long-term potentiation in olfactory cyclic nucleotide-gated channel type 1 null mouse. J Neurophysiol 79:3295–301
Pastalkova E, Serrano P, Pinkhasova D, Wallace E, Fenton AA, Sacktor TC (2006) Storage of spatial information by the maintenance mechanism of LTP. Science 313:1141–4
Pedram A, Razandi M, Kehrl J, Levin ER (2000) Natriuretic peptides inhibit G protein activation. Mediation through cross-talk between cyclic GMP-dependent protein kinase and regulators of G protein-signaling proteins. J Biol Chem 275:7365–72
Pfeifer A, Klatt P, Massberg S, Ny L, Sausbier M, Hirneiss C, Wang GX, et al. (1998) Defective smooth muscle regulation in cGMP kinase I-deficient mice. Embo J 17:3045–51
Pfeifer A, Ruth P, Dostmann W, Sausbier M, Klatt P, Hofmann F (1999) Structure and function of cGMP-dependent protein kinases. Rev Physiol Biochem Pharmacol 135:105–49
Pogun S, Baumann MH, Kuhar MJ (1994) Nitric oxide inhibits [3H]dopamine uptake. Brain Res 641:83–91
Pose I, Sampogna S, Chase MH, Morales FR (2003) Mesencephalic trigeminal neurons are innervated by nitric oxide synthase-containing fibers and respond to nitric oxide. Brain Res 960:81–9
Prast H, Philippu A (1992) Nitric oxide releases acetylcholine in the basal forebrain. Eur J Pharmacol 216:139–40
Prast H, Philippu A (2001) Nitric oxide as modulator of neuronal function. Prog Neurobiol 64:51–68
Qian Y, Chao DS, Santillano DR, Cornwell TL, Nairn AC, Greengard P, Lincoln TM, et al. (1996) cGMP-dependent protein kinase in dorsal root ganglion:relationship with nitric oxide synthase and nociceptive neurons. J Neurosci 16:3130–8
Reid CA, Bekkers JM, Clements JD (2003) Presynaptic Ca2+ channels:a functional patchwork. Trends Neurosci 26:683–7

Reinhard M, Jarchau T, Walter U (2001) Actin-based motility:stop and go with Ena/VASP proteins. Trends Biochem Sci 26:243–9

Revermann M, Maronde E, Ruth P, Korf HW (2002) Protein kinase G I immunoreaction is colocalized with arginine-vasopressin immunoreaction in the rat suprachiasmatic nucleus. Neurosci Lett 334:119–22

Reyes M, Stanton PK (1996) Induction of hippocampal long-term depression requires release of Ca2+ from separate presynaptic and postsynaptic intracellular stores. J Neurosci 16:5951–60

Reyes-Harde M, Empson R, Potter BV, Galione A, Stanton PK (1999a) Evidence of a role for cyclic ADP-ribose in long-term synaptic depression in hippocampus. Proc Natl Acad Sci USA 96:4061–6

Reyes-Harde M, Potter BV, Galione A, Stanton PK (1999b) Induction of hippocampal LTD requires nitric-oxide-stimulated PKG activity and Ca2+ release from cyclic ADP-ribose-sensitive stores. J Neurophysiol 82:1569–76

Riccio A, Alvania RS, Lonze BE, Ramanan N, Kim T, Huang Y, Dawson TM, et al. (2006) A nitric oxide signaling pathway controls CREB-mediated gene expression in neurons. Mol Cell 21:283–94

Rieke F, Schwartz EA (1994) A cGMP-gated current can control exocytosis at cone synapses. Neuron 13:863–73

Robinson RB, Siegelbaum SA (2003) Hyperpolarization-activated cation currents:from molecules to physiological function. Annu Rev Physiol 65:453–80

Rogan MT, Staubli UV, LeDoux JE (1997) Fear conditioning induces associative long-term potentiation in the amygdala. Nature 390:604–7

Russwurm M, Koesling D (2004) Guanylyl cyclase:NO hits its target. Biochem Soc Symp:51–63

Ruth P, Landgraf W, Keilbach A, May B, Egleme C, Hofmann F (1991) The activation of expressed cGMP-dependent protein kinase isozymes I alpha and I beta is determined by the different amino-termini. Eur J Biochem 202:1339–44

Ryan TA, Reuter H, Wendland B, Schweizer FE, Tsien RW, Smith SJ (1993) The kinetics of synaptic vesicle recycling measured at single presynaptic boutons. Neuron 11:713–24

Rybalkin SD, Rybalkina IG, Feil R, Hofmann F, Beavo JA (2002) Regulation of cGMP-specific phosphodiesterase (PDE5) phosphorylation in smooth muscle cells. J Biol Chem 277:3310–7

Sausbier M, Schubert R, Voigt V, Hirneiss C, Pfeifer A, Korth M, Kleppisch T, et al. (2000) Mechanisms of NO/cGMP-dependent vasorelaxation. Circ Res 87:825–30

Sauzeau V, Le Jeune H, Cario-Toumaniantz C, Smolenski A, Lohmann SM, Bertoglio J, Chardin P, et al. (2000) Cyclic GMP-dependent protein kinase signaling pathway inhibits RhoA-induced Ca2+ sensitization of contraction in vascular smooth muscle. J Biol Chem 275:21722–9

Savchenko A, Barnes S, Kramer RH (1997) Cyclic-nucleotide-gated channels mediate synaptic feedback by nitric oxide. Nature 390:694–8

Schmidt H, Werner M, Heppenstall PA, Henning M, Moré MI, Kühbandner S, Lewin GR, et al. (2002) cGMP-mediated signalling via cGKIalpha is required for the guidance and connectivity of sensory axons. J Cell Biol 159:489–98

Schuman EM, Madison DV (1991) A requirement for the intercellular messenger nitric oxide in long-term potentiation. Science 254:1503–6

Schuman EM, Meffert MK, Schulman H, Madison DV (1994) An ADP-ribosyltransferase as a potential target for nitric oxide action in hippocampal long-term potentiation. Proc Natl Acad Sci USA 91:11958–62

Shimizu-Albergine M, Rybalkin SD, Rybalkina IG, Feil R, Wolfsgruber W, Hofmann F, Beavo JA (2003) Individual cerebellar Purkinje cells express different cGMP phosphodiesterases (PDEs):in vivo phosphorylation of cGMP-specific PDE (PDE5) as an indicator of cGMP-dependent protein kinase (PKG) activation. J Neurosci 23:6452–9

Shin JH, Linden DJ (2005) An NMDA receptor/nitric oxide cascade is involved in cerebellar LTD but is not localized to the parallel fiber terminal. J Neurophysiol 94:4281–9

Son H, Hawkins RD, Martin K, Kiebler M, Huang PL, Fishman MC, Kandel ER (1996) Long-term potentiation is reduced in mice that are doubly mutant in endothelial and neuronal nitric oxide synthase. Cell 87:1015–23

Sonnenburg WK, Beavo JA (1994) Cyclic GMP and regulation of cyclic nucleotide hydrolysis. Adv Pharmacol 26:87–114
Sorkin LS (1993) NMDA evokes an L-NAME sensitive spinal release of glutamate and citrulline. Neuroreport 4:479–82
Sporns O, Jenkinson S (1997) Potassium ion- and nitric oxide-induced exocytosis from populations of hippocampal synapses during synaptic maturation in vitro. Neuroscience 80:1057–73
Stamler JS, Toone EJ, Lipton SA, Sucher NJ (1997) (S)NO signals translocation, regulation, and a consensus motif. Neuron 18:691–6
Stanarius A, Topel I, Schulz S, Noack H, Wolf G (1997) Immunocytochemistry of endothelial nitric oxide synthase in the rat brain:a light and electron microscopical study using the tyramide signal amplification technique. Acta Histochem 99:411–29
Stanton PK, Heinemann U, Muller W (2001) FM1-43 imaging reveals cGMP-dependent long-term depression of presynaptic transmitter release. J Neurosci 21:RC167
Stanton PK, Winterer J, Bailey CP, Kyrozis A, Raginov I, Laube G, Veh RW, et al. (2003) Long-term depression of presynaptic release from the readily releasable vesicle pool induced by NMDA receptor-dependent retrograde nitric oxide. J Neurosci 23:5936–44
Stanton PK, Winterer J, Zhang XL, Muller W (2005) Imaging LTP of presynaptic release of FM1-43 from the rapidly recycling vesicle pool of Schaffer collateral-CA1 synapses in rat hippocampal slices. Eur J Neurosci 22:2451–61
Starke K (1981) Presynaptic receptors. Annu Rev Pharmacol Toxicol 21:7–30
Stewart TL, Michel AD, Black MD, Humphrey PP (1996) Evidence that nitric oxide causes calcium-independent release of [3H] dopamine from rat striatum in vitro. J Neurochem 66:131–7
Sudoh T, Minamino N, Kangawa K, Matsuo H (1988) Brain natriuretic peptide-32:N-terminal six amino acid extended form of brain natriuretic peptide identified in porcine brain. Biochem Biophys Res Commun 155:726–32
Sudoh T, Minamino N, Kangawa K, Matsuo H (1990) C-type natriuretic peptide (CNP):a new member of natriuretic peptide family identified in porcine brain. Biochem Biophys Res Commun 168:863–70
Sullivan BM, Wong S, Schuman EM (1997) Modification of hippocampal synaptic proteins by nitric oxide-stimulated ADP ribosylation. Learn Mem 3:414–24
Sun X, Kaltenbronn KM, Steinberg TH, Blumer KJ (2005) RGS2 is a mediator of nitric oxide action on blood pressure and vasoconstrictor signaling. Mol Pharmacol 67:631–9
Tang KM, Wang GR, Lu P, Karas RH, Aronovitz M, Heximer SP, Kaltenbronn KM, et al. (2003) Regulator of G-protein signaling-2 mediates vascular smooth muscle relaxation and blood pressure. Nat Med 9:1506–12
Tegeder I, Del Turco D, Schmidtko A, Sausbier M, Feil R, Hofmann F, Deller T, et al. (2004) Reduced inflammatory hyperalgesia with preservation of acute thermal nociception in mice lacking cGMP-dependent protein kinase I. Proc Natl Acad Sci USA 101:3253–7
Tian M, Yang XL (2006) C-type natriuretic peptide modulates glutamate receptors on cultured rat retinal amacrine cells. Neuroscience 139:1211–20
Tsou K, Snyder GL, Greengard P (1993) Nitric oxide/cGMP pathway stimulates phosphorylation of DARPP-32, a dopamine- and cAMP-regulated phosphoprotein, in the substantia nigra. Proc Natl Acad Sci USA 90:3462–5
Valtschanoff JG, Weinberg RJ (2001) Laminar organization of the NMDA receptor complex within the postsynaptic density. J Neurosci 21:1211–7
Wagner LE, 2nd, Li WH, Yule DI (2003) Phosphorylation of type-1 inositol 1,4,5-trisphosphate receptors by cyclic nucleotide-dependent protein kinases:a mutational analysis of the functionally important sites in the S2+ and S2− splice variants. J Biol Chem 278:45811–7
Wall ME, Francis SH, Corbin JD, Grimes K, Richie-Jannetta R, Kotera J, Macdonald BA, et al. (2003) Mechanisms associated with cGMP binding and activation of cGMP-dependent protein kinase. Proc Natl Acad Sci USA 100:2380–5

Wang HG, Lu FM, Jin I, Udo H, Kandel ER, de Vente J, Walter U, et al. (2005) Presynaptic and postsynaptic roles of NO, cGK, and RhoA in long-lasting potentiation and aggregation of synaptic proteins. Neuron 45:389–403

Wang X, Robinson PJ (1997) Cyclic GMP-dependent protein kinase and cellular signaling in the nervous system. J Neurochem 68:443–56

Wang YT, Linden DJ (2000) Expression of cerebellar long-term depression requires postsynaptic clathrin-mediated endocytosis. Neuron 25:635–47

Welsh JP, Yamaguchi H, Zeng XH, Kojo M, Nakada Y, Takagi A, Sugimori M, et al. (2005) Normal motor learning during pharmacological prevention of Purkinje cell long-term depression. Proc Natl Acad Sci USA 102:17166–71

Whitlock JR, Heynen AJ, Shuler MG, Bear MF (2006) Learning induces long-term potentiation in the hippocampus. Science 313:1093–7

Wilson RI, Godecke A, Brown RE, Schrader J, Haas HL (1999) Mice deficient in endothelial nitric oxide synthase exhibit a selective deficit in hippocampal long-term potentiation. Neuroscience 90:1157–65

Xia C, Bao Z, Yue C, Sanborn BM, Liu M (2001) Phosphorylation and regulation of G-protein-activated phospholipase C-beta 3 by cGMP-dependent protein kinases. J Biol Chem 276:19770–7

Xu L, Eu JP, Meissner G, Stamler JS (1998) Activation of the cardiac calcium release channel (ryanodine receptor) by poly-S-nitrosylation. Science 279:234–7

Xue J, Wang X, Malladi CS, Kinoshita M, Milburn PJ, Lengyel I, Rostas JA, et al. (2000) Phosphorylation of a new brain-specific septin, G-septin, by cGMP-dependent protein kinase. J Biol Chem 275:10047–56

Xue J, Milburn PJ, Hanna BT, Graham ME, Rostas JA, Robinson PJ (2004) Phosphorylation of septin 3 on Ser-91 by cGMP-dependent protein kinase-I in nerve terminals. Biochem J 381:753–60

Yu X, Duan KL, Shang CF, Yu HG, Zhou Z (2004) Calcium influx through hyperpolarization-activated cation channels (I(h) channels) contributes to activity-evoked neuronal secretion. Proc Natl Acad Sci USA 101:1051–6

Yu YC, Cao LH, Yang XL (2006) Modulation by brain natriuretic peptide of GABA receptors on rat retinal ON-type bipolar cells. J Neurosci 26:696–707

Zhang J, Dawson VL, Dawson TM, Snyder SH (1994) Nitric oxide activation of poly(ADP-ribose) synthetase in neurotoxicity. Science 263:687–9

Zhuang S, Nguyen GT, Chen Y, Gudi T, Eigenthaler M, Jarchau T, Walter U, et al. (2004) Vasodilator-stimulated phosphoprotein activation of serum-response element-dependent transcription occurs downstream of RhoA and is inhibited by cGMP-dependent protein kinase phosphorylation. J Biol Chem 279:10397–407

Zhuo M, Hu Y, Schultz C, Kandel ER, Hawkins RD (1994) Role of guanylyl cyclase and cGMP-dependent protein kinase in long-term potentiation. Nature 368:635–9

# Therapeutic Use of Release-Modifying Drugs

**S.Z. Langer**



**Abstract** Presynaptic inhibitory or facilitatory autoreceptors are targets for the endogenous neurotransmitter of the respective neuron, and also for exogenous agonists, partial agonists and antagonists which can produce pharmacological actions through changes in transmitter release. In addition, presynaptic inhibitory or facilitatory heteroreceptors can also be acted upon by exogenous agonists, partial agonists or antagonists to induce changes in transmitter release with useful therapeutic effects. This article summarizes drugs that are known or likely to produce their therapeutic effects through presynaptic modulation of neurotransmitter release. Included are drugs acting on α and β adrenoceptors, dopamine receptors,

S.Z. Langer
Alpha2 Pharmaceutica AB, 8 H. Rosenblum Street, Apt. 4650, Tel-Aviv 69379, Israel
langer.salomon@gmail.com

T.C. Südhof, K. Starke (eds.), *Pharmacology of Neurotransmitter Release.*
*Handbook of Experimental Pharmacology 184.*

angiotensin, opioid, cannabinoid, and nicotinic acetylcholine receptors. Also discussed are changes in presynaptic receptor mechanisms produced by drugs that inhibit transmitter re-uptake.

## 1 Introduction

A considerable number of inhibitory and facilitatory presynaptic autoreceptors were characterized in the peripheral and central nervous systems. The available evidence indicates that these autoreceptors correspond to a specific subtype of transmitter receptor (Table 1). Since presynaptic autoreceptors play a physiological role in the regulation of transmitter release, it follows that their selective activation by agonists will lead to a decrease (inhibitory autoreceptors) or increase (facilitatory autoreceptors) in transmitter output, while blockade by antagonists will produce the opposite change. These presynaptic effects on neurotransmitter release can result in mild pharmacological effects possessing useful therapeutic properties. It can be argued that the side effect profile of drugs acting presynaptically on transmitter release may be different and probably milder than that obtained with the classical approach of agonists or antagonists acting at postsynaptic receptors and therefore modifying effector organ responses in every tissue where such postsynaptic receptor subtype is located.

The number of presynaptic inhibitory or facilitatory heteroreceptors is larger than that of autoreceptors (Table 2), but these presynaptic receptors are mainly the target of action for exogenous agonists or partial agonists, unless the presynaptic heteroreceptor is activated under physiological or pathological conditions by a locally formed autacoid or by the transmitter released from neighboring nerve terminals.

An additional mechanism whereby presynaptic autoreceptors can be acted upon corresponds to drug–induced increases in the synaptic concentration of the neurotransmitter, as observed with antidepressants which inhibit the neuronal uptake of

**Table 1** Presynaptic autoreceptors. Shown are the transmitters for which a negative or positive feedback mechanism mediated by presynaptic autoreceptors has been established. Note that in each case a specific receptor subtype is involved in the presynaptic modulation of transmitter release

| Transmitter | Inhibitory Receptor Subtype | Facilitatory Receptor Subtype |
|---|---|---|
| NA | $\alpha_{2A}$-Adrenoceptors (mainly) | $\beta_2$-Adrenoceptors (periphery) |
| DA | $D_2$ / $D_3$ | |
| ACh | $M_2$ | Nicotinic |
| 5-HT | 5-$HT_{1D}$ | |
| Histamine | $H_3$ | |
| GABA | $GABA_B$ | |
| Glutamate | mGluR3 – mGluR5 | |

**Table 2** Presynaptic heteroreceptors. The list is not exhaustive. Agonists acting on the presynaptic heteroreceptor subtypes inhibit or facilitate transmitter release. Selective antagonists block the effect of the agonist. Given alone, antagonists modify transmitter release only if there is activation of the heteroreceptors by an endogenous agonist from a neighboring synapse or a locally produced or circulating endocoid

| Transmitter | Inhibitory Receptor Subtype | Facilitatory Receptor Subtype |
|---|---|---|
| NA | Opioid, $M_2$, $H_3$, $CB_1$, PGE, $D_2$ | ANG AT-1, Nicotinic, NMDA |
| DA | $M_2$, $CB_1$ | Nicotinic, NMDA |
| ACh | $\alpha_2$, $D_2$, Opioid, $CB_1$ | |
| 5-HT | $\alpha_2$ | |
| Glutamate | 5-$HT_{1D}$, $GABA_B$, $CB_1$ | Nicotinic, NMDA |
| CCK | $GABA_B$ | |
| GABA | $CB_1$ | |
| CGRP | 5-$HT_{1D}$ | |
| Somatostatin | $GABA_B$ | |

ANG: angiotensin; CCK: cholecystokinin; CGRP: calcitonin-gene-related peptide; NMDA: N-methyl–D-aspartate; PGE: prostaglandin E; CB: cannabinoid.

noradrenaline or serotonin. Such antidepressant drugs initially induce a decrease in transmitter release by overactivation of presynaptic inhibitory autoreceptors. Chronic activation of inhibitory autoreceptors by the endogenous agonist, however, leads to downregulation and desensitization, and the net result then is a facilitation of noradrenergic or serotonergic neurotransmission.

In some cases, the discovery and characterization of presynaptic receptors established the existence of a new receptor subtype, like the $\alpha_2$ adrenoceptor (Dubocovich and Langer 1974; Langer 1974) and histamine $H_3$ receptor (Arrang et al. 1987). It should be noted, however, that the receptor subtypes located presynaptically are also located postsynaptically. The latter should be taken into account in the strategy of discovery and development of new drugs with preferential presynaptic effects.

## 2 Presynaptic $\alpha_2$ Autoreceptor Agonists and Antagonists

The $\alpha_2$ subtype of adrenoceptors was first discovered on peripheral noradrenergic nerve endings (Dubocovich and Langer 1974; Langer 1974; Cubeddu et al. 1974) and on cell bodies and dendrites of central noradrenergic neurons (Svensson et al. 1975). Following the classification of adrenoceptors into $\alpha$ and $\beta$ subtypes by Alquist (1948) it was generally accepted for many years that the $\alpha$ adrenoceptors represented a homogeneous population. In 1974, however, pharmacological differences were established between $\alpha_1$ and $\alpha_2$ adrenoceptors, with the potency of the antagonist phenoxybenzamine differing by 100-fold in blocking postsynaptic $\alpha_1$ adrenoceptors when compared to the presynaptic $\alpha_2$ adrenoceptors in the perfused cat spleen (Dubocovich and Langer 1974 ; Cubbedu et al. 1974). Subsequently, the pharmacological differences between $\alpha_1$ and $\alpha_2$ adrenoceptors were further confirmed by establishing differences in the relative order of potencies of many agonists as well of antagonists. More than 10 years later three $\alpha_1$ and three $\alpha_2$ adrenoceptor subtypes were cloned and expressed: $\alpha_{1A}$, $\alpha_{1B}$ and $\alpha_{1D}$ on the one hand, $\alpha_{2A}$, $\alpha_{2B}$ and $\alpha_{2C}$ on the other hand (for review see Bylund et al. 1994). It is of interest to note that the majority of the presynaptic inhibitory $\alpha_2$ autoreceptors belong to the $\alpha_{2A}$ subtype (Table 1; Starke 2001), although $\alpha_{2B}$ and $\alpha_{2C}$ autoreceptors also exist (Trendelenburg et al. 2003; Gilsbach and Hein, this volume).

## *2.1 Agonists: Clonidine and Analogs*

Clonidine is a classical reference compound, discovered by serendipity (Hoefke and Kobinger, 1966) and developed for its antihypertensive and bradycardic properties. Subsequently, clonidine was shown to possess $\alpha_2$ adrenoceptor agonist properties, both in the periphery and in the central nervous system (Kobinger 1978). The therapeutic effects of clonidine in hypertension are accompanied by sedation and dry mouth as side effects, which limit its wide clinical use. The sympathoinhibitory effects of clonidine and clonidine–like drugs are due to activation of postsynaptic $\alpha_2$ adrenoceptors in the cardiovascular centers of the medulla oblongata. Peripheral inhibition of neurotransmitter release from postganglionic sympathetic neurons by clonidine contributes to the bradychardic and hypotensive effects, through activation of presynaptic $\alpha_2$ autoreceptors (Langer et al. 1980; Urban et al. 1995). It is of interest that dry mouth is mainly due to the activation by clonidine of presynaptic $\alpha_2$ heteroreceptors on the cholinergic nerves innervating the salivary gland (Izumi et al. 1995).

A second hypothesis to explain the cardiovascular effects of clonidine involves an interaction with imidazoline-1 receptors (Bousquet and Feldman 1999). However, the available evidence at present does not prove conclusively the imidazoline hypothesis for the mechanism of action of clonidine and analog drugs, like rilmenidine and moxonidine (Szabo 2002).

## *2.2 Antagonists: Mianserin, Mirtazapine, Idazoxan, Efaroxan*

Both mianserin and mirtazapine are antidepressant drugs which possess central $\alpha_2$ adrenoceptor blocking properties ($pA_2$ 7.3). However, mirtazapine is much more potent at histamine $H_1$ receptors ($pA_2$ 9.1) and at 5-$HT_2$ and 5-$HT_3$ receptors ($pA_2$ 8.2). Blocking of $H_1$ receptors explains the main side effects of mirtazapine, which produces marked sedation and weight gain. Blockade of presynaptic inhibitory $\alpha_2$ autoreceptors increases the release of NA, while blockade of presynaptic $\alpha_2$ inhibitory heteroreceptors on serotonin nerve terminals (Table 2) is likely to increase the release of serotonin.

These facilitatory effects on NA and serotonin are likely to contribute to the antidepressant actions of these two drugs, since a decrease in monoaminergic neurotransmission involving both NA and serotonin in the brain is associated with the physiopathology of depression (Svensson 2000; Mongeau et al. 1997).

Based on the enhancement of noradrenergic and serotonergic transmission induced by central $\alpha_2$ adrenoceptor antagonists, it was proposed that this class of compounds may possess antidepressant properties (Langer 1978). Idazoxan is a potent and selective $\alpha_2$ adrenoceptor antagonist that penetrates readily the blood-brain barrier (De Boer et al. 1996), yet only a few studies exist in the literature where idazoxan was administered in monotherapy in monopolar depressed patients, and the antidepressant effects were present only in some of the patients. There are no published double-blind studies of idazoxan against placebo in monopolar depression. There is one double-blind study comparing idaxozan and bupropion during six weeks in bipolar depressed patients but without a placebo group (Grossman et al. 1999). In this study, both idaxozan and bupropion demonstrated significant improvement over time with reductions in the Hamilton Depression Rating Scale of 50% (Grossman et al. 1999).

Use of idaxozan as add-on therapy to antipsychotics in the treatment of schizophrenia is supported by preclinical data (Hertel et al. 1999; Wadenberg et al. 2007) and clinical results (Litman et al. 1996).

It should be noted that idazoxan has a short duration of action in humans and has to be administered three times a day. In addition, only racemic idazoxan has been administered to humans. The two enantiomers of idazoxan were studied on both $\alpha_1$ and $\alpha_2$ adrenoceptors (Dabire et al. 1986). It was shown that (+) idazoxan was more potent than (−) idazoxan. Furthermore, (+) idazoxan antagonizes both $\alpha_1$ and $\alpha_2$ adrenoceptors, while (−) idazoxan possesses both $\alpha_1$ and $\alpha_2$ adrenoceptor agonist effects (Dabire et al. 1986).

It is of interest that idazoxan was tested in Parkinson's disease, although this indication was abandoned for lack of efficacy (F. Colpaert, personal communication). Efaroxan and its active dextro enantiomer dexefaroxan are also potent and selective centrally acting $\alpha_2$ adrenoceptor antagonists (Bauer et al. 2003; Rizk et al. 2006). Efaroxan was developed for the treatment of chronic neurodegenerative diseases but abandoned because of side effects (P. Sokoloff, personal communication).

Activation of $\alpha_2$-adrenoceptors in pancreatic β-cells inhibits insulin release, and this effect is blocked by $\alpha_2$ adrenoceptor antagonists (Angel and Langer. 1988; Lorrain et al. 1992). A peripherally active $\alpha_2$ adrenoceptor antagonist, SL 84.0148, was developed for the treatment of type II diabetes (Angel et al. 1992;1996) but was abandoned in Phase II (S.Z. Langer, unpublished).

$\alpha_2$ Adrenoceptor antagonists may increase heart rate as well as blood pressure. These side effects are due to increased release of NA resulting from the blockade of presynaptic terminal $\alpha_2$ autoreceptors on peripheral sympathetic neurons (Langer 1981). These cardiovascular side effects are minimal or absent with idazoxan (Schmidt et al. 1997).

## 3 Presynaptic Central Dopamine Autoreceptor Partial Agonist: Aripiprazole

Aripiprazole is a new drug for the treatment of schizophrenia and was approved by the Food and Drug Administration in November, 2002. Aripiprazole possesses presynaptic $D_2$ autoreceptor partial agonist activity and postsynaptic $D_2$ receptor antagonist properties (Kane et al. 2002; Kasper et al. 2003; Potkin et al. 2003). In contrast to most antipsychotic drugs, aripiprazole does not elevate the circulating prolactin levels, one of the undesirable side effects of this class of drugs. Like many of the presynaptically acting compounds, aripiprazole acts also on other receptors, possessing 5-$HT_{1A}$ partial agonist properties and 5-$HT_{2A}$ antagonist actions.

## 4 Presynaptic Peripheral Dopamine Heteroreceptor Agonist: Lergotrile

Presynaptic inhibitory dopamine heteroreceptors on noradrenergic nerve endings were first reported in in vitro studies and shown to be of the $D_2$ receptor subtype (Table 2; Enero and Langer 1975; Dubocovich and Langer 1980a). In vivo studies in the dog showed that the hypotensive and bradycardic effects of lergotrile, a $D_2$ receptor agonist, were due to the stimulation of presynaptic DA receptors on noradrenergic nerve terminals (Barret and Lockandwala 1981). In support of this view, lergotrile inhibited the positive chronotropic and the renal vasoconstrictor effects elicited by postganglionic sympathetic stimulation. These effects were prevented by prior treatment with sulpiride, a selective $D_2$ receptor antagonist (Barret and Lockandwala 1981). Additional in vivo evidence confirmed that compounds with $D_2$ receptor agonist effects like pergolide and N-n-propyl-N-n-butyl dopamine possess bradycardic and antihypertensive actions due to a reduction in NA output from peripheral sympathetic nerve fibers (Hamed et al. 1981 ; Cavero et al. 1985). However, the main side effect of $D_2$ agonists, nausea and vomiting, represents an important obstacle in the development of compounds in this category for the treatment of hypertension.

## 5 Presynaptic 5-$HT_{1D}$ Heteroreceptor Agonists: Sumatriptan and Analogs

Presynaptic 5-$HT_{1D}$ inhitory heteroreceptors exist on substance P- and CGRP-containing terminals (Smith et al. 2002; Hou et al. 2001; Table 2). Activation of these receptors by agonists like sumatriptan and related analogs leads to inhibition of the release of these two neurotransmitters (Williamson et al. 1997, 2001). The therapeutic efficacy of sumatriptan and analogous drugs in the treatment of migraine is associated at least partially with these inhibitory presynaptic effects on substance P and CGRP release and reflects a significant therapeutic improvement when compared with other available medications – an important recent example of the suitability of pharmacological presynaptic modulation for therapeutic purposes.

## 6 Presynaptic Angiotensin Heteroreceptors: Angiotensin Receptor Antagonists

Blockade of presynaptic facilitatory angiotensin $AT_1$ receptors on noradrenergic nerve endings in the cardiovascular system (Table 2) may contribute to the antihypertensive effects of angiotensin receptor antagonists. By reducing the enhanced release of NA induced by locally produced or circulating angiotensin, the antagonists reduce sympathetic tone in the cardiovascular system.

## 7 Presynaptic β Autoreceptors: β Adrenoceptor Antagonists

Peripheral but not central noradrenergic nerve endings possess facilitatory $\beta_2$ in addition to the inhibitory $\alpha_2$ autoreceptors (Adler-Graschinsky and Langer 1975; Table 1). The presynaptic facilitatory β adrenoceptors in noradrenergic nerve endings of peripheral blood vessels play a physiopathological role in Raynaud's disease. It was reported that infusion of a very low dose of isoprenaline (1 µg/minute) in patients with Raynaud's disease produced a highly significant reduction in blood flow when compared to healthy controls (Giovanni et al. 1984). These low doses of isoprenaline enhance NA release but are not high enough to produce $\beta_2$ adrenoceptor-mediated vasodilation postsynaptically. Similar conclusions were reached by Brotzu et al. (1989), who administered atenolol and/or flunnarizine to 40 patients with Raynaud's disease.

Increased responsiveness to presynaptic facilitatory β adrenoceptors on peripheral noradrenergic nerve endings has been suggested to contribute to the development, rather than the maintenance, of hypertension (Borkowski, 1990). Accordingly, enhanced vasoconstriction would result from the increased release of NA through activation of presynaptic facilitatory β-receptors by the transmitter NA and by circulating adrenaline. The antihypertensive effect of β adrenoceptor antagonists is well established. It was suggested that upon chronic administration of β adrenoceptor antagonist drugs there is a reduction in NA release due to blockade of the presynaptic facilitatory β adrenoceptors.

# 8 Presynaptic Opioid Heteroreceptor Agonists: Morphine and Related Drugs

The inhibition by opioid receptor agonists of peripheral noradrenergic neurotransmission was reported as early as 1957 (Trendelenburg 1957). Subsequently, it was established that presynaptic inhibitory opioid heteroreceptors exist in many peripheral postganglionic sympathetic neurons (Dubocovich and Langer 1980b). It is of interest that the presynaptic inhibitory opioid receptors in the vas deferens were used successfully as a bioassay in the discovery of the endogenous pentapeptides leu- and met-enkephalin. Presynaptic opioid inhibitory receptors also exist in the central nervous system (Arbilla and Langer 1978). One therapeutically important effect is inhibition of the release of substance P from primary afferent pain fibers (Jessel and Iversen 1977). The possible role of presynaptic opioid receptors in the development of addiction or in the symptomatology of withdrawal remains an area of interest. Such a role is supported by the finding that in opioid withdrawal, contrary to the acute effect of opioids, release of noradrenaline from locus coeruleus neurons is increased, both in situ (Montel et al. 1975) and when the neurons are kept in culture (Ronken et al. 1993).

# 9 Presynaptic Cannabinoid Heteroreceptor Agonists: Cannabis and Related Drugs

Presynaptic inhibition through cannabinoid subtype-1 ($CB_1$) receptors is the most prominent neuronal effect of cannabinoids. Inhibitory $CB_1$ occur on terminals of noradrenergic, dopaminergic, cholinergic, GABAergic, and glutamatergic neurons (Katona et al. 1999; Gobel et al. 2000; Schlicker and Kathmann 2001; Szabo and Schlicker 2005; Domenici et al. 2006; Table 2). Endocannabinoids regulate the function of many synapses as retrograde messengers (Harkany et al. 2007; Lovinger, this book). It is very likely, therefore, that presynaptic inhibition also underlies the proposed therapeutic effects of cannabinoids – as well as the effects sought by cannabis abusers. For example, cannabinoids, like opioids, inhibit transmitter release from primary nociceptive afferents (Liang et al. 2003) – a basis for their analgesic properties. Sativex®, a whole cannabis plant extract, has been approved in Canada for neuropathic pain associated with multiple sclerosis.

# 10 Presynaptic Nicotinic Heteroreceptor Agonists: Nicotine as a Drug of Addiction

Nicotine enhances dopamine release by acting on presynaptic facilitatory heteroreceptors located on the terminal regions of dopaminergic neurons (Marshall et al. 1997). It is tempting to associate these effects of nicotine and the fact that tobacco dependence, the most common substance abuse disorder, is due to nicotine. Nicotine is thought to interact with the mesocorticolimbic DA system in a manner similar to that of other drugs of abuse, such as cocaine (Corrigal et al. 1992). This may explain the tolerance, dependence, and withdrawal syndrome in both animals and humans (Stein et al. 1998).

## 11 Changes in Sensitivity of Presynaptic Autoreceptors Induced by Antidepressant Drugs

It is generally accepted that a decrease in monoaminergic neurotransmission involving NA and serotonin is central to the pathophysiology of depression. This view is the rationale for the clinical use of inhibitors of neuronal uptake of NA and/or serotonin, which augment and prolong the synaptic actions of these transmitters in the central nervous system. However, while the inhibition of neuronal uptake of NA or serotonin is established within 24 hours of starting the treatment, the clinical improvement takes three to four weeks to develop – a paradoxical phenomenon, which can be explained through effects on somadendritic and presynaptic autoreceptors. The acute administration of drugs that inhibit neuronal uptake (NA or serotonin) increases the synaptic concentration of the corresponding transmitter, which produces overactivation of somatodendritic and terminal autoreceptors. Such effects result in suppression of the firing rate of NA or serotonin neurons or both, and in a decrease in transmitter release from the nerve terminals. Chronic activation of inhibitory autoreceptors by endogenous transmitter in turn leads to their downregulation; after a few weeks, the firing rate and nerve terminal release of the transmitter recover, and the result is a facilitation of noradrenergic or serotonergic neurotransmission (Svensson and Usdin 1978; Langer 1997).

On this basis, the potential antidepressant action of central $\alpha_2$ adrenoceptor antagonists was predicted in 1978 (Langer 1978). However, as mentioned above, idazoxan, a selective antagonist on central $\alpha_2$ adrenoceptors, does not possess antidepressant effects in monotherapy in unipolar depression. The antidepressant effects of idazoxan alone in bipolar depression remain to be confirmed by additional double blind studies against placebo.

It is possible that the therapeutic indication for idazoxan in depression may be as add-on therapy in patients who are non-responders to inhibitors of neuronal uptake of serotonin or of NA. In addition, drugs which block presynaptic autoreceptors on serotonergic neurons may possess antidepressant properties in monotherapy or as add-on therapy to serotonin uptake inhibitors.

## 12 Conclusions and Future Perspectives

The original observations made in the early 1970s led to the hypothesis that neurotransmitter release was regulated at the level of the nerve terminals by presynaptic auto- and heteroreceptors. These discoveries were confirmed and extended during the following 30 years. Presynaptic release-modulating receptors represent suitable targets for pharmacological intervention by exogenous compounds acting as agonists, partial agonists, or antagonists, with therapeutic value by influencing transmitter release. Such drugs may have useful therapeutic properties and, in some cases, like depression, address unmet medical needs. In addition, by influencing transmitter release presynaptically, this novel family of drugs may possess fewer side effects than drugs which stimulate or block the classical postsynaptic receptors. In addition to the development of increasingly selective drugs acting on presynaptic release-modulating receptors, it is important to consider the relevance of changes in sensitivity of presynaptic autoreceptors induced by antidepressant drugs which inhibit the neuronal transporters in noradrenergic and serotonergic neurons.

Several families of drugs of addiction (opioid receptor agonists, cannabinoid receptor agonists, and nicotine) possess important presynaptic effects which result in inhibition (opioids and cannabinoids) or facilitation (nicotine) of the release of several neurotransmitters in the periphery and the central nervous system. These effects are mediated through specific receptor subtypes and may be relevant to tolerance, dependence, or withdrawal syndromes. It is tempting to speculate that presynaptic effects of these three families of drugs of addiction may provide useful insights in understanding neurochemical mechanisms in addiction and developing novel drug treatments for these patients.

## References


Adler-Graschinsky E, Langer SZ (1975) Possible role of a β-adrenoceptor in the regulation of noradrenaline release by nerve stimulation through a positive feedback mechanism. Br J Pharmacol 53:43–50

Ahlquist RP (1948) A study of the adrenotropic receptors. Am J Physiol 153:586–600

Angel I, Langer SZ (1988 Adrenergic induced hyperglycemia in anaesthetized rats: involvement of peripheral α-2 adrenoceptors. Eur J Pharmacol 154 :191–6

Angel I, Schoemaker H, Arbilla S, Galzin AM, Berry C, Niddam R, Pimoule C, Sevrin M, Wick A, Langer SZ (1992) SL 84.0418: A novel potent and selective α-2 adrenoceptor antagonist: I. In vitro pharmacological profile. J Pharmacol Exp Ther 263:1327–33

Angel I, Burcelin R, Prouteau M, Girard J, Langer SZ (1996) Normalization of insulin secretion by a selective α-2 adrenoceptor antagonist restores GLUT-4 glucose transporter expression in adipose tissue of type II diabetic rats. Endocrinology 137:1–6

Arbilla S, Langer SZ (1978) Morphine and β-endorphin inhibit release of noradrenaline from cerebral cortex but not of dopamine from rat striatum. Nature 271:559–61

Arrang JM, Garbarg M, Lancelot JC, Lecomte J, Pollard H, Robba M, Shunack W, Schwartz JC (1987) Highly potent and selective ligands for histamine H3-receptors. Nature 327:117–23

Barret RJ, Lockhandwala MF (1981) Presynaptic dopamine receptor stimulation in the cardiovascular system actions of lergotrile. J Pharmacol Exp Ther 217:660–5

Bauer S, Moyse E, Jourdan F, Colpaert F, Martel JC, Marien M (2003) Effects of the α-2 adrenoceptor antagonist dexefaroxan on neurogenesis in the olfactory bulb of the adult rat in vivo: selective protection agains neuronal death. Neuroscience 117:281–91

Bousquet P, Feldman J (1999) Drugs acting on imidazoline receptors. A review of their pharmacology, their use in blood pressure control and their potential interest in cardioprotection. Drugs 58:799–812

Berkowski KA (1990) Presynaptic receptors in hypertension. Ann NY Acad Sci 604:389–97
Brotzu G, Falchi S, Mannu B, Montisci R, Petruzzo P, Staico R (1989) The importance of presynaptic $\beta$ receptors in Raynaud's disease. J Vasc Surg 9:767–71
Bylund DB, Eikenberg C, Hieble JP, Langer SZ, Lefkowitz RJ, Minneman KP, Molinoff PB, Ruffolo RR, Trendelenburg U (1994) Adrenoceptor subtypes. Pharmacol Rev 46:121–36
Cavero I, Lorrain J, Di Paola ED, Lhoste F, Payen B, Dennis T, Scatton B (1985) Pharmacological, hemodynamic and biochemical mechanisms involved in the blood pressure lowering effects of pergolide, in normotensive and hypertensive dogs. J Pharmacol Exp Ther 235:798–809
Corrigall WA, Franklin KBJ, Coen KM, Clarke PBS (1992) The mesolimbic dopaminergic system is implicated in the reinforcing effects of nicotine. Psychopharmacology 107:285–9
Cubeddu LX, Barnes EM, Langer SZ, Weiner N (1974) Release of norepinephrine and dopamine-$\beta$-hydroxylase by nerve stimulation. I. Role of neuronal and extraneuronal uptake and of $\alpha$-presynaptic receptors. J Pharmacol Exp Ther 190:431–50
Dabire H, Dausse JP, Mouille P, Fournier B, Cardot A, Meyer P, Schmitt H (1986) Pharmacological properties of the enantiomers of idazoxan: possible separation between their $\alpha$-adrenoceptor blocking effects. Clin Exp Hypertens A 8:387–409
De Boer T, Nefkens F, Van Helvoirt A, Van Delft AML (1996) Differences in modulation of noradrenergic and serotonergic transmission by the $\alpha$-2 adrenoceptor antagonists, mirtazapine, mianserine and idazoxan. J Pharmacol Exp Ther 277:852–60
Domenici MR, Azad SC, Marsicano G, Schierloh A, Wotjak CT, Dodt HU, Zieglgansberger W, Lutz B, Rammes G (2006) Cannabinoid receptor type 1 located on presynaptic terminals of principal neurons in the forebrain controls glutamatergic synaptic transmission. J Neurosci 26:5794–9
Dubocovich ML, Langer SZ (1974) Negative feedback regulation of noradrenaline release by nerve stimulation in the perfused cat's spleen: differences in potency of phenoxybenzamine in blocking the pre and post-synaptic adrenergic receptors. J Physiol 237:505–19
Dubocovich ML, Langer SZ (1980a) Dopamine and $\alpha$-adrenoceptor agonists inhibit neurotransmission in the cat spleen through different presynaptic receptors. J Pharmacol Exp Ther 212:144–52
Dubocovich ML, Langer SZ (1980b) Pharmacological differentiation of presynaptic inhibitory $\alpha$-adrenoceptors and opiate receptors in the cat nictitating membrane. Br J Pharmacol 70:383–93
Enero MA, Langer SZ (1975) Inhibition by dopamine of 3H-noradrenaline release elicited by nerve stimulation in the isolated cat's nictitating membrane. Naunyn-Schmiedeberg's Arch Pharmacol 289:179–203
Giovanni B, Giuseppina CM, Susanna F, Roberto M (1984) Altered regulator mechanisms of presynaptic adrenergic nerve: a new physiopathological hypothesis in Raynaud's disease. Microvasc Res 27:110–13
Göbel I, Trendelenburg AU, Cox SL, Meyer A, Starke K (2000) Electrically evoked release of 3H-noradrenaline from mouse cultured sympathetic neurons: release-modulating heteroreceptors. J Neurochem 75:2087–94
Grossman F, Potter WZ, Brown EA, Maislin G (1999) A double-blind study comparing idazoxan and bupropion in bipolar depressed patients. J Affect Disord 56:237–43
Hamed AT, Jandhyala BS, Ginos JZ, Lockhandwala MF (1981) Presynaptic dopamine receptors and $\alpha$-adrenoceptors as mediators of the bradychardic action of N-n-propyl-N-n-butyl dopamine. Eur J Pharmacol 74:83–90
Harkany T, Guzman M, Galve-Ropert I, Berghuis P, Devi LA, Mackie K (2007) The emerging functions of endocannabinoid signaling during CNS development Trends Pharmacol Sci 28:83–92
Hertel P, Nomikos GG, Svensson TH (1999) Idazoxan preferentially increases dopamine output in the rat medial prefrontal cortex at nerve terminal level. Eur J Pharmacol 371:153–8
Hoefke W, Kobinger W (1966) Pharmakologische Wirkungen des 2-(2,6 dichlorphenylamino)-2-imidazolin-hydrochlorids, einer neuen, antihypertensiven Substance. Arzneimittelforschung 16:1038–50

Hou M, Kanje M, Longmore J, Tajti J, Uddman R, Edvinsson L (2001) 5-HT(1B) and (1D) receptors in the human trigeminal ganglion:co-localization with calcitonin gene-related peptide, substance P and nitric oxide synthase. Brain Res 909:112–20

Izumi H, Nakamura I, Karita K (1995) Effects of clonidine and yohimbine on parasympathetic reflex salivation and vasodilatation in cat SMG. Am J Physiol 268:R1197–R1202

Jessel TM, Iversen LL (1977) Opiate analgesics inhibit substance P release from rat trigeminal nucleus. Nature 268:549–51

Kane JM, Carson WH, Saha AR, Saha AR, McQuade RD, Ingenito GG, Zumbroff DL, Ali MW (2002) Efficacy and safety of aripiprazole and haloperidol versus placebo in patients with schizophrenia and schizoaffective disorder. J Clin Psychiat 63:763–71

Kasper S, Lerman MN, McQuade RD, Saha A, Carson WH, Ali M, Archibald D, Ingenito G, Marcus R, Pigott T (2003). Efficacy and safety of aripiprazole vs. haloperidol for long-term maintenance treatment following acute relapse of schizophrenia. Int J Neuropsychopharmacol 6:325–37

Katona I, Sperlagh B, Sik A, Kafalvi A, Vizi ES, Mackie K, Frend TF (1999) Presynaptically located CB1 Cannabinoid receptors regulate GABA release from axon terminals of specific hippocampal neurons. J Neurosci 19:4544–58

Kobinger W (1978) Central α adrenergic systems as targets of action for hypotensive drugs. Rev Physiol Biochem Pharmacol 8:40–100

Langer SZ (1974) Presynaptic regulation of catecholamine release. Biochem Pharmacol 23: 1793–1800

Langer SZ (1978) Presynaptic receptors. Nature 275:479–80

Langer SZ (1981) Presynaptic regulation of the release of catecholamines. Pharmacol Rev 32:337–62

Langer SZ (1997) 25 Years since the discovery of presynaptic receptors:present knowledge and future perspectives. Trends Pharmacol Scie 18:95–9

Langer SZ, Cavero I, Massingham R (1980) Recent developments in noradrenergic neurotransmission and its relevance to the mechanism of action of certain antihypertensive agents. Hypertension 2:372–82

Liang YC, Huang CC, Hsu KS, Takahashi T (2003) Cannabinoid-induced presynaptic inhibition at the primary afferent trigeminal synapse of juvenile rat brainstem slices. J Physiol 555:85–96

Litman RE, Su TP, Potter WZ, Hong WW, Pickar D (1996) Idazoxan and response to typical neuroleptics in treatment-resistant schizophrenia. Comparison with the atypical neuroleptic, clozapine. Br J Psychiatr 168:571–9

Lorrain J, Angel A, Duval N, Eon MT, Oblin A, Langer SZ (1992) Adrenergic and noradrenergic cotransmitters inhibit insulin secretion during sympathetic stimulation in dogs. Am J Physiol 263:E72–E76

Marshall DL, Redfern PH, Wonnacott S (1997) Presynaptic nicotinic modulation of dopamine release in the three ascending pathways studied by in vivo microdialysis: comparison of naïve and chronic nicotine-treated rats. J Neurochem 68:1511–19

Mongeau R, Blier P, De Montigny C (1997) The serotonergic and noradrenergic systems of the hippocampus: their interactions and the effects of antidepressant treatments. Brain Res Rev 23:145–95

Montel H, Starke K, Taube HD (1975) Morphine tolerance and dependence in noradrenaline neurons of the rat cerebral cortex. Naunyn-Schmiedeberg's Arch Pharmacol 258:415–26

Potkin SG, Saha AR, Kujawa MJ, Carson WH, Ali M, Stock E, Strinfellow J, Ingenito G, Marder SR (2003) Aripiprazole, an antipsychotic with a novel mechanism of action and risperidone vs. placebo in patients with schizophrenia and psychoaffective disorder. Arch Gen Psychiat 60:681–90

Raiteri M (2006) Functional pharmacology in human brain. Pharmacol Rev 58:162–93

Rizk P, Salazar J, Raisman-Vozari R, Marien M, Ruberg M, Colpaert F, Debeir T (2006) The α-2 adrenoceptor antagonist dexefaroxan enhances hippocampal neurogenesis by increasing the survival and differentiation of new granule cells. Neuropharmacology 31:1146–57

Ronken E, Mulder AH, Schoffelmeer ANM (1993) Chronic activation of mu and kappa-opioid receptors in cultured catecholaminergic neurons from rat brain causes neuronal supersensitivity without receptor desensitization. J Pharmacol Exp Ther 268:595–9

Schlicker E, Kathmann M (2001) Modulation of transmitter release via presynaptic cannabinoid receptors. Trends Pharmacol Sci 22:565–572

Schmidt ME, Risinger RC, Hauger RL, Potter WZ (1997) Responses to $\alpha_2$ adrenoceptor blockade by idazoxan in healthy male and female volunteers. Psychoneuroendocrinology 22:177–88

Smith D, Hill RG, Edvinsson L, Longmore J (2002) An immunohistochemical investigation of human trigeminal nucleus caudalis: CGRP, substance P and 5-HT1D-receptor immunoreactivities are expressed by trigeminal sensory fibers. Cephalalgia 22 :424–31

Starke K (2001) Presynaptic autoreceptors in the third decade:focus on $\alpha$-2 adrenoceptors. J Neurochem 78:685–93

Stein EA, Pankiewicz J, Harsch HH, Cho JK, Fuller SA, Hoffmann RG, Hawkins M, Rao SM, Bandettini PA, Bloom AS (1998) Nicotine-induced limbic cortical activation in the human brain: a functional MRI study. Am J Psychiatr 155:1009–15

Svensson TH (2000) Brain noradrenaline and the mechanisms of action of antidepressant drugs. Acta Psychiatrica Scandinavica Supplementum 402:18–27.

Svensson TH, Usdin T (1978) Feedback inhibition of brain noradrenaline neurons by tricyclic antidepressants: $\alpha$-receptor mediation. Science 202:1089–91

Svensson TH, Bunney BS, Aghajanian G (1975) Inhibition of both noradrenergic and serotonergic neurons in the brain by the $\alpha$-adrenergic agonist clonidine. Brain Res 92:291–306

Szabo B (2002) Imidazoline antihypertensive drugs: a critical review on their mechanism of action. Pharmacol Therapeut, 93:1–35

Szabo B, Schlicker E (2005) Effects of cannabinoids on neurotransmission. Handb Exp Pharmacol 168:327–65

Trendelenburg U (1957) The action of morphine on the superior cervical ganglion and on the nictitating membrane of the cat. Br J Pharmacol 12:79–85

Trendelenburg AU, Philipp M, Meyer A, Klebroff W, Hein L, Starke K (2003) All three $\alpha_2$-adrenoceptor types serve as autoreceptors in postganglionic sympathetic neurons. Naunyn-Schmiedeberg's Arch Pharmacol 368:504–12

Urban R, Szabo B, Starke K (1995) Involvement of peripheral presynaptic inhibition in the reduction of sympathetic tone by moxonidine, rilmenidine and UK 14304. Eur J Pharmacol 282:29–37

Wadenberg ML, Wiker C, Svensson TH (2007) Enhanced efficacy of both typical and atypical antipsychotic drugs by adjunctive $\alpha$-2 adrenoceptor blockade: experimental evidence. Int J Neuropsychopharmacol 10:191–202

Williamson DJ, Hargreaves RJ, Hill RJ, Shepheard SL (1997) Sumatriptan inhibits neurogenic vasodilation of dural blood vassels in the anaesthetized rat—intravital microscope studies. Cephalalgia 17:525–31

Williamson DJ, Hill RG, Shepheard SL, Hargreaves RJ (2001) The anti-migrane 5-HT (1B/1D) agonist rizatriptan inhibits neurogenic dural vasodilation in anaesthetized guinea-pigs. Br J Pharmacol 133:1029–34

# Index

Printed by Printforce, the Netherlands